System Reliability Modeling

IEEE Press
445 Hoes Lane
Piscataway, NJ 08854

System Reliability Modeling

A Matrix-Algorithmic Approach in MATLAB

Abdullah Alkaff
Sepuluh Nopember Institute of Technology (ITS)
Surabaya, Indonesia

IEEE Press
WILEY

Published by John Wiley & Sons, Inc., Hoboken, New Jersey.

Library of Congress Cataloging-in-Publication Data applied for:

Hardback ISBN: 9781394371129

Cover Design: Wiley
Cover Image: Yuichiro Chino/Getty Images

Set in 9.5/12.5pt STIXTwoText by Straive, Chennai, India

To Kak, Bang, Dik
I raised the bar for you to have a more reliable life, here and after.

Contents

About the Author

Abdullah Alkaff is a retiring Professor at the Laboratory of Systems and Cybernetics, the Electrical Engineering Department, Sepuluh Nopember Institute of Technology (ITS), Surabaya, Indonesia, where he teaches System Reliability Engineering, Network Analysis, Control System Signal Processing, Systems Identification, Perception and Navigation for Autonomous Vehicles, and Introduction to Research in Control and System Engineering. Prior to that, he was a Lecturer at the University of Florida, USA; an Assistant Professor at Syracuse University, USA; Project Director for the design and implementation of country-wide radar integration, an integrated maritime surveillance system for the Strait of Malacca, and a national education data center; and advisor to the ministers of ICT and Educations and Culture, Indonesia. Professor Alkaff holds a Doctor of Philosophy degree in Systems Engineering from the University of Florida, USA.

Preface

Charlie made the right decision during his undergraduate program by taking a reliability engineering elective. After completing his master's program, Charlie landed a job at one of the two dream companies for engineering graduates at the time. Perhaps because of the reliability engineering course he took, Charlie was entrusted with developing the company's large-scale system reliability analysis software. The system reliability analysis software used the Monte Carlo simulation method. Charlie found the program very slow, which made him interested in digging it deeper. The results of his investigation found that the software used the failure rate of each component, which was a constant value, as input. Based on this input, the Monte Carlo simulation program generated a series of random numbers to estimate the reliability value of each component. This process was repeated for all system components, and the system reliability was calculated based on the reliability values of the components and the system structure. Armed with his reliability engineering course, Charlie immediately had the idea to apply his knowledge to make the large-scale system reliability analysis software faster, even much faster. After the appropriate system reliability concepts were applied, the modified software was able to complete simulations to estimate system reliability approximately 10 times faster than before. This achievement was enough to earn Charlie a special bonus and a business trip to attend a conference.

I met Charlie at the conference where he presented his simulation model. As former graduate school classmates who shared academic interests, as evidenced by our shared electives, we enjoyed eating and sightseeing around the city. Charlie shared the story behind his experiences developing the software he was assigned. He didn't want to take too much credit for the success of his design program because, he said, what he was doing was actually trivial for someone who had taken a basic reliability engineering course. Large-scale systems have many components connected in series, he said. The course had explained that components connected in series can be represented as a subsystem with a failure rate equal to the sum of the failure rates of its constituent components. Large-scale system simulation software doesn't need to generate random numbers to estimate the reliability of each component based on its failure rate; it should simply generate random numbers to estimate the reliability of the subsystem using the total failure rate of its constituent components. The addition of a simple subprogram to identify the presence of series structures (already available in the original program), representing each as a subsystem with a failure rate equal to the sum of the failure rates of its constituent components, combined with an existing Monte Carlo simulation program applied to these subsystems, yielded a very efficient simulation program. Charlie further developed his simulation software by making the failure rate of each component a function of time, not a constant one, and the result was still faster than the original program, which only accepted a constant failure rate for each component. Accepting only a constant failure rate would have been the same as assuming all components had exponentially distributed lifetimes, which was unrealistic. Expanding it to accept failure rates as a function of time, meaning the components' lifetimes follow Weibull distribution, made the system model more realistic. Charlie had generalized the software, not only made it more efficient.

Charlie's story conveys a moral message: understanding system reliability concepts can significantly help determine system reliability more efficiently, even with the support of simulation software. When I presented my research papers at seminars, some participants commented on why I bothered to derive mathematical equations for modeling system reliability when system reliability simulation software is readily available. My resolve remained firm. While my colleagues were puzzled as to why I would write a book containing complex mathematical models for calculating system reliability, I remained confident that this book would be beneficial for more comprehensive and efficient system reliability analysis. To ensure its usefulness, this "imaginary" book should discuss new methods for modeling and analyzing system reliability, some of which I believe are not currently available in system reliability simulation software packages. The book should explain that not only series structures can be replaced by a single subsystem with reliability parameters calculated from the reliability parameters of its constituent components, as Charlie did. All structures can be done in the same way. Furthermore, the methods discussed in the book must be able to offer more in-depth, comprehensive, and accurate analysis results than those produced by simulation software packages. The most appropriate choice to achieve this goal is to develop dynamic models of system reliability and the methods that can be used to produce dynamic reliability measures of a system as functions of time. Simulations will have difficulty displaying analysis results as functions of time because they calculate system reliability values at several times and arrange them into a function graph through a smoothing process; however, the results are still not the exact mathematical equations that underline the graph and, therefore, cannot be used to analyze its characteristics more accurately and further.

Of course, a system reliability book must do more than simply offer a new method for modeling and analyzing system reliability; such material would be more suitable as a research article. Furthermore, many system reliability books have already been published, each with its own unique characteristics. A strong distinction is needed so that a new book on a classic topic can complement them by filling a gap untouched by existing books. This gap is the lack of a system reliability book with the following description:

- A system reliability book that discusses methods of dynamic analysis of system reliability in the form of algorithms that can be easily implemented in computer software. The software must be ready to use and easy to use by the readers of the book. Throughout the book, the software shows its use in solving system reliability problems. As a method of dynamic analysis of system reliability, the output of the software is dynamic measures of system reliability in the form of time functions that express changes in reliability, failure, residual reliability, and the expected remaining life of the system against the duration of its use.
- A system reliability book that discusses techniques for modeling complex systems by expressing them in a series of computer program functions according to the structures contained within the complex system. These functions can be called by the main program to analyze the reliability of the complex system as a whole. In fact, a complex system can simply be expressed in a series of the same functions that apply generally to all system structures. The results of these functions are the parameters of the system reliability functions, not the system reliability values.
- A system reliability book that discusses methods for calculating system reliability parameters from the reliability parameters of its constituent components, use them for calculating system reliability measures, all of which are carried out without using calculus (differential and integral),

but rather the calculations are simply done using algebra so that they are easier to program and provide real values, not estimations or approximations.

- A system reliability book that discusses multistate systems whose conditions cannot be said to be either good or failed, but rather there are many gradations of conditions between these two extreme conditions.

Fortunately, my research aligns with the aforementioned concept of the book. The research articles I have published can be used as raw material for compiling the book, in addition to a concise material for the System Reliability Engineering course I teach that I self-published in my native language to fulfill a requirement to maintain my professorship. The presentation is tailored to suit lecture purposes, complemented by examples performed manually and with computer assistance. Each example requiring computer assistance is equipped with a main program to call the functions required to model the exemplary system, to model the lifetime distribution of each component, and to model the desired output. Analysis of the modeled system is sufficiently achieved by running the main program. To do this, the computer programs for my published research articles must be rewritten, organized, and comprehended. The functions prepared for modeling and analyzing system reliability are provided in the program library, including models of the lifetime distribution of system components commonly used in reliability analysis and models of the output generally desired from a system reliability analysis. Users simply create the main program according to the instructions provided in the numerous example problems discussed in this book. With the help of these functions, modeling and analyzing system reliability using analytical dynamic models is as easy, and can even be easier, than using simulation models, but with more accurate results because it does not use statistical approaches that depend on the number of samples. In fact, those strong supporters of simulation models can use the methods in this book to do the same idea as Charlie did but not just limited to subsystems with a series structure. Thus, system reliability simulation models can be made faster, if not much faster, and more accurate than the existing simulation software.

This book and its supporting software package are a single unit. The book provides the derivation of the equations needed to model and analyze system reliability, along with explanations of their use. These equations are then programmed using MATLAB into functions in a ready-to-use software package that can be used to model system reliability and calculate reliability measures. Those who wish to understand the background of the algorithms used in the software, including their mathematical derivation and proof, can find them in the book. With this understanding, they can create additional programs tailored to their specific needs. With the availability of source code, these additional programs can be easily integrated with the pre-prepared software package. Practitioners who want to take a shortcut simply by using the software package can learn from the software user manual and main program in each example. This book can be used by lecturers and students to teach and learn mathematical modeling and analysis of system reliability through systems reliability engineering courses. For course learning purposes, this book is supplemented with lecture notes in the form of presentations, easing the burden on lecturers in preparing course materials. The availability of both the book and the software package also allows practitioners and professionals to use them for self-study. They are also possible to be utilized for practical training as part of credit-earning activities or just skill improvement. The embedded exercises and problems at the end of each chapter can be used to test users' skills. Efforts have been made to ensure the book and software package are self-contained; all the material necessary for understanding them is available within the book and software package. Thus, users do not need the support of other books or programs to understand them, although they must still have a MATLAB license.

MATLAB was chosen because the calculations used in this book rely heavily on matrix operations, and MATLAB, according to my son Hilfi who worked in Silicon Valley, is more consistent in using matrices than other software, such as Python, which interprets matrices as multidimensional lists, despite he is using Python all the time for his works. However, the software package supporting this book was created by academia, not professional programmers. This software package is more accurately described as a student version, not a professional version. With the source code available, users with professional programming skills can develop it further to be equivalent to the professional version. This development could include adding a reliability block diagram-based user interface, improving computer memory management, which does require specialized programming skills beyond the capabilities of reliability engineers, and perhaps streamlining the algorithms used to make them more efficient (although we have tried to make them as efficient as possible). I have deployed students to run all the example programs in the book. Users' feedback on their experience running the program is very much needed for improvement in case there are still any that do not meet their expectations.

To serve as a self-contained book, the presentation of the book begins with an explanation of matrix-based distributions and their use in modeling system reliability (Chapter 2). Systems are discussed starting with single-component systems (Chapter 3), then moving on to systems with a single basic structure and combinations of several basic structures (Chapter 4). The last half of Chapter 3 is devoted to explaining the use of matrix-based distributions to design maintenance and replacement programs. The discussion continues with a general standby structure with varying degrees of warmness from cold to hot (Chapter 5). This general standby structure is shown to be useful to model system reliability with a general structure applicable to all basic structures. Complex systems in the form of networks, both without flow constraints and with flow constraints, are discussed in Chapter 6. With the increasing importance of computers that calculate system reliability in discrete-time, all continuous-time models discussions in Chapters 2–6 are translated to discrete-time systems (Chapter 7). Simple optimization problems relevant to that chapter are presented in each chapter as a complement. Given the increasing popularity of AI in all fields since the introduction of ChatGPT, it is certainly interesting to understand its capabilities in solving the problems presented in this book. The final chapter of the book discusses the use of AI for modeling and analyzing system reliability as a comparison to the methods developed in this book (Chapter 8). The AI used is the free version of ChatGPT from OpenAI, which is easy to use for those just wanting to try AI. Perhaps because it is free, the solutions proposed by the AI generally use classic methods, such as scalar reliability functions, approximation, and simulation, so they are different than provided in the book. In addition, as a convenience for readers, the list of MATLAB functions created using the methods developed in the book is given in the Appendix.

I have a moral obligation to acknowledge individuals who made the book into being in closing this preface. They provided me with sincere assistance in producing this book by compiling the materials from my system reliability engineering research publications and lecture notes. I would like to take this opportunity to express my sincere gratitude to them. The greatest support came from NurQ, a programmer and researcher in the Systems and Cybernetics Laboratory in the Department of Electrical Engineering, ITS, where I work. He translated the algorithms described in the book into ready-to-use software. If needed, NurQ can be contacted via email listed in the source code and operating instructions for the supporting software. Early drafts of several chapters of this book were edited and laid out by Eko Stefi, a lecturer in the Department of Industrial and Systems Engineering. He still found time to help prepare the book template and initial drafts despite his busy schedule as one of the ADB project managers at the university. Yus, a colleague in the Laboratory of Systems and Cybernetics, helped prepare the lecture notes. All the students taking the System

Reliability Engineering course in the Electrical Engineering Department were very helpful with their assignments and with running the main programs for each example in the book, especially Izza and Fardli who also read the final draft of this book and filled in the forgotten hyperlinks, and Lukman who meticulously checked the hyperlinks, tried the programs, and solved the problems. Anonymous reviewers have been very helpful in suggesting improvements. The best parts of this book are thanks to their efforts, while the worst parts are entirely due to my negligence. Although this book is a personal side project that I worked on at home, I feel the need to thank the Rector and the Head of the Electrical Engineering Department of ITS who still trusted me to work again in the Systems and Cybernetics Laboratory after a long absence from campus. My thanks should not be forgotten, although mentioned last, to editorial team of Wiley (Kavipriya Ramachandran, Kavin Shanmughasundaram, and others who work behind the scene), and especially to Sandra Grayson of Wiley, whose confidence in this book let me have confidence in writing it ("you don't need luck, it is a good book" was her reply when I asked her to wish me luck at the time the proposal was sent to reviewers!)

At last, I have something to further motivate *dzurriyati*: never too late to start!

January 2026

Abdullah Alkaff
Surabaya

About the Companion Website

This book is accompanied by a companion website.

www.wiley.com/go/systemreliabilitymodeling1e

To access the website materials, click on the link under Related Resources on the book's page on wiley.com. You can also find this by searching www.wiley.com for the ISBN 9781394371129.

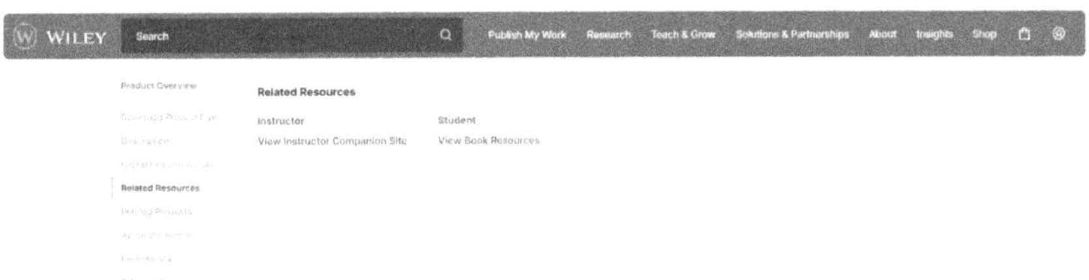

This content includes the MATLAB scripts and the presentation material for Lecture Notes.

1
Introduction

Learning objectives:

1) Understand the existence of several system reliability models.
2) Understand the importance of dynamic system reliability models.
3) Understand the advantages of using a matrix-form reliability function.

Abbreviations

MTTF mean time to failure

Notations

λ	rate of exponential reliability function
f	structure function
$R_i(t)$	reliability function of component i
$R_S(t)$	system reliability function
$h(t)$	hazard function
$\boldsymbol{a}, \boldsymbol{A}$	parameters of a matrix-form reliability function

The book reforms the teaching and learning of system reliability from calculus based to matrix algebra based. This reformation is necessary for dynamic system reliability modeling and analysis with computer assistance. Previously, computer assistance was possible only to support analytical modeling for static system reliability analysis, which is to derive the numerical value of system reliability from the numerical values of reliability of its components. However, system reliability needs to be presented as a function of time to understand its behaviors. In case of the reliability of components given as functions, computer assistance to produce system reliability using analytical models is too complicated since they involve calculus and functional operations. The way out is using simulation models that rely on random numbers and statistical estimation. The results are estimates of the real values, which need to be represented in confidence intervals to make them consistent across several experiments. With the reformation proposed in this book, all popular equations in analytical system reliability models involving calculus operations and functional operations are translated into equations involving matrix algebra operations. Algorithms for system reliability analysis are easier to program if they are derived from matrix algebra formulation than from calculus or functional formulations. With that simplicity, dynamic system reliability modeling and analysis are made possible and easier to support with computer assistance.

1.1 Evolution of System Reliability Modeling

System reliability is defined as the probability that a system will successfully perform its mission or will not experience failures during its mission. The uncertainty surrounding the mission execution is due to the system lifetime being a random variable whose value cannot be precisely determined. The randomness of a systems lifetime makes it difficult to guarantee that the system will be able to perform its mission. The best that can be provided is a probability value for the system to be able to perform its mission; this probability value is then termed system reliability. The reliability of a system depends on the reliability of its components; the higher the component reliability, the higher the system reliability. Other than that, system reliability also depends on the arrangement of its components, which is referred to as the system structure. The system structure and components are determined at the design stage of the system; hence, one consideration of system design is the design for reliability. The system structure determines the function that relates the system reliability to the reliability of its components.

System reliability analysis is the process of determining component reliability and the system structure and using them to evaluate the system reliability. System reliability analysis uses system reliability models to represent the relationship between component reliability, system structure, and system reliability. The static system reliability model is formulated for the purpose of performing static system reliability analysis, which is an analysis to determine the system reliability value from the component reliability values based on the system structure. The static system reliability model is illustrated in Figure 1.1, where R_i, $i = 1, 2, \ldots, N$, is the reliability of component i, N is the number of components in the system, R_S is the system reliability, and f is a function that represents the system structure, which is commonly referred to as the system structure function.

Complex systems contain subsystems in different structures, and each subsystem is built using several components. Finding the system reliability requires the calculation of each subsystems reliability from the reliability of its components. If the subsystems have combinations of basic structures, such as series and parallel, the calculation is straightforward using arithmetic operators on component reliability. Sometimes, the system needs a longer hierarchy from the component level to the system level. In such systems, computer assistance is necessary to make system reliability calculation not a burden. Despite its simplicity, the static reliability model cannot be used to determine statistics of time to system failure since the model does not involve time; it is only applicable to determine the probability of system success for a given mission duration. In fact, the static reliability model cannot be applied to a system that contains a standby structure.

However, system reliability depends on the mission duration; the longer the mission, the higher the probability of the system experiencing failure, which makes the system reliability decrease over time. There are a few cases where the system gets better while being used. However, whatever the

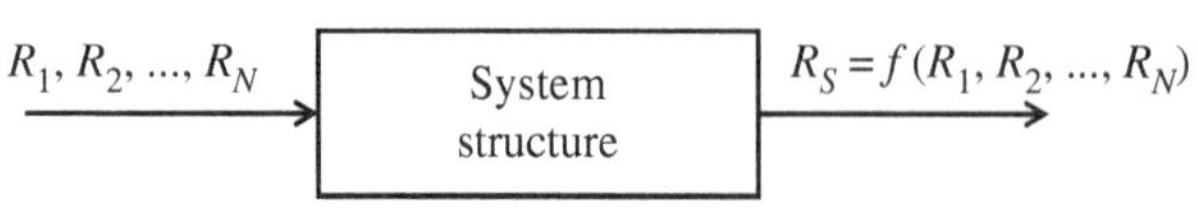

Figure 1.1 Static system reliability model.

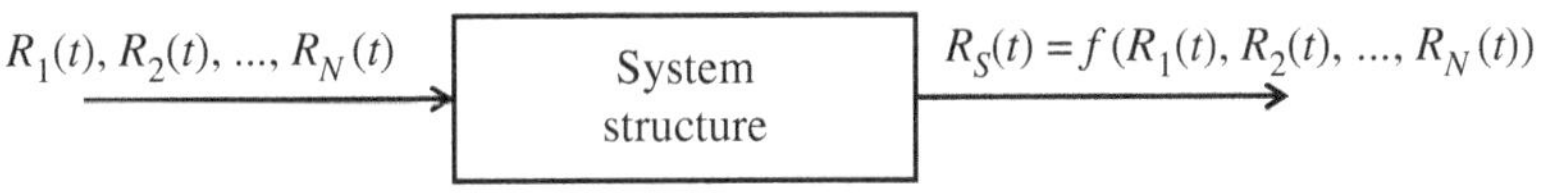

Figure 1.2 Dynamic system reliability model.

cases are, system reliability is a dynamic quantity whose value is a function of time. A function of time that expresses the system reliability at any time instant is called a reliability function. The reliability function, in general, is a monotonic nonincreasing function that takes the value 0 as time approaches infinity, because any system will eventually fail in the future. It is possible that at the beginning of operation, the system will get better, but after a while, the system gets worse and eventually fails. The system reliability function also depends on the reliability of its components and the system structure function. Their relationship is represented in a dynamic system reliability model as exhibited in Figure 1.2, where $R_i(t)$, $i = 1, 2, \ldots, N$, is the reliability function of component i and $R_S(t)$ is the system reliability function.

System reliability function is versatile; it can provide a wide range of information on any aspects related not only to system reliability but also to system failure. The value of a system reliability at any time, generated from its reliability function, can be used to determine the realistic duration of its use with respect to an acceptable risk of failure. The system reliability function contains information on statistics of time to failure, such as mean time to failure ($MTTF$) and standard deviation of time to failure, which are easier to understand and useful in practice. The information on the change in reliability values over time is available in the system reliability function, which can be utilized to determine the hazard, or failure rate, function of the system and the residual reliability function including its residual life expectancy as a function of the age of the system. The system reliability function can also be developed further to determine the effect of maintenance or replacement on system reliability, including the number of failures in a duration that can be utilized to determine the number of spare parts that satisfy a certain requirement of stockout risk. By including costs, the optimum maintenance or replacement interval can be determined. Such a versatile analysis is impossible to achieve using static system reliability analysis. In short, once the system reliability function is obtained, then other system reliability measures can be derived, even finding the most suitable maintenance and replacement programs. Using reliability functions is highly desirable because it can provide more detailed information that cannot be obtained from static analysis [1, 2].

Unfortunately, applying dynamic system reliability analysis is a burden. Even though the system has subsystems in a combination of series and parallel structures, which require only algebraic operations, generating the system reliability function leads to more complicated procedures. The complication increases if it is implemented in computer programs, considering the procedure is impossible to be performed manually for realistic complexities of component reliability functions, number of components, and system structures, especially if the objective is to generate the system reliability function. When a standby structure exists among complex combinations of system structures, the operation is more complicated since the reliability of a system in standby structure must be generated using the convolution integrals of reliability functions of its components, folded as many times as the number of components.

Figure 1.2 points out that static system reliability analysis is a special case of dynamic system reliability analysis. At a time-instance a, reliability function of component i, $R_i(t)$, has a constant value $R_i(a)$, which can be written simply as R_i. These values are the input to static system reliability analysis to determine the system reliability value R_S. Repeating this process at other time-instances produces system reliability values at those time-instances, which then can be combined to obtain the graph of system reliability function. Having the graph of system reliability function does not immediately lead to the system reliability function; it needs a curve-fitting process to come up with a function with some degree of inaccuracy. This inaccuracy will be carried over and amplified in the calculation of system reliability measures. This weakness, however, does not prevent system reliability analysts from using static reliability analysis. Its simplicity weighed heavily than its weakness, and thus, static reliability analysis prevails. Even so, it is still difficult to apply if the component reliability functions in the standby subsystem are scalar reliability functions.

As a way out, simulation models based on Monte Carlo methods are proposed. Random numbers are generated using component reliability functions to determine each component status and use the status of all components to determine the system status. By sufficiently repeating this process, the probability that the system is in good status can be produced based on the relative frequency that the system is in good status. Again, repeating this process for other time-instances can produce the probability that the system is in good status. The system reliability values obtained from simulation are calculated using statistical methods; therefore, the values are not exact. They are estimates of the real values, and as estimates, they are equipped with confidence intervals. If each point is an interval, a function that represents those values most likely is not smooth and using it to calculate system reliability measures will be inaccurate. Accuracy can be increased by generating more random numbers at the price of computing time, and yet the inaccuracy cannot be eliminated. Such weakness is amplified by the time required for executing simulation models to produce high-accuracy results.

To overcome the weakness of the conventional approach, it is necessary to design an algorithm to generate a system reliability function from the input in the form of a reliability function for each component and the system structure. Designing such an algorithm is not easy because it must process the input in the form of functions to produce an output in the form of a function. Currently available computer programs support calculations from the input of component reliability values to producing the system reliability value or static reliability analysis. To perform dynamic system reliability analysis, symbolic processing is required, which is certainly not easy and not widely understood. Let alone the resulting function is long and unstructured that is difficult to digest. A different method that no longer requires symbolic processing to generate the system reliability function from the component reliability functions and system structure as shown in Figure 1.2 is a necessity. An approach that can be used for this purpose is a matrix-form distribution. The matrix-form distribution function approach in system reliability modeling and analysis is new. Recently published reliability textbooks, such as Elsayed [3], Navarro [4], Rausand et al. [5], Bernstein, Bensoussan, and Bender [6], and Saket and Sanjeevikumar [7], still use scalar distributions. Yan-Fu Li, Enrico Zio, Andre V. Kleyner [8] mentioned multistate systems in one of the chapters but did not use the matrix-form reliability function. Trivedi and Bobbio [9] provided only one chapter among the 17 chapters in the book that utilized a matrix-form distribution, and it is for small-scale systems and is not presented in an algorithmic manner. Eryilmaz [10] used the matrix-form distribution for reliability and quality engineering, but it is in discrete time, suitable for research, and not at the system level.

Figure 1.3 provides an overview of the various models for system reliability analysis and the scope of this book.

The ability to design simple algorithms to accommodate the procedures of generating system reliability functions from the component reliability functions for various system structures is one of the objectives of this book. As a prerequisite to achieve that objective, having the ability to represent the system reliability in a model that guides the analyst to the right algorithms is necessary.

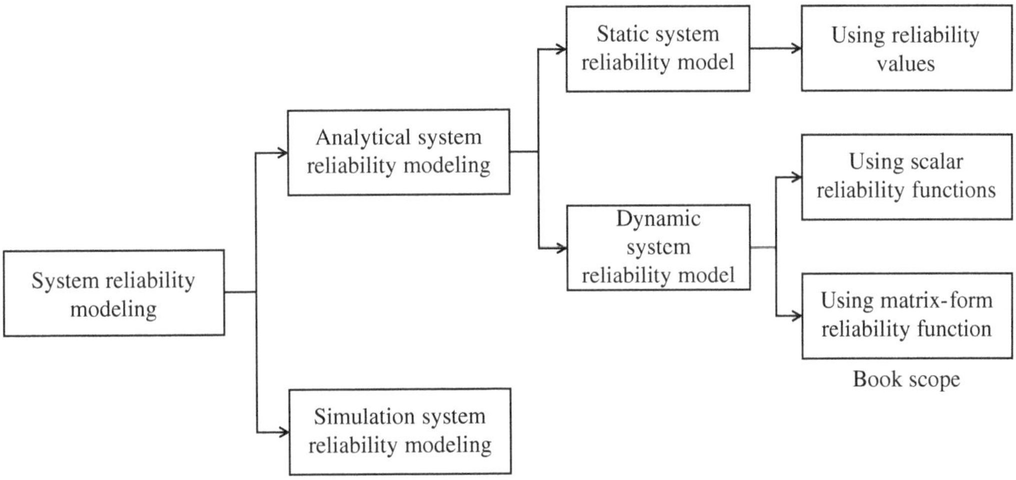

Figure 1.3 Various system reliability modeling.

1.2 Difficulty in Using Scalar Distribution for Dynamic System Reliability Modeling

The system reliability function is obtained from the reliability functions of its components using mathematical operations appropriate to the system structure. For certain structures, mathematical operations are simple enough to be given in closed-form expressions involving algebraic operations. However, the expression for the reliability function of a system in standby structure requires the convolution integral operations of the reliability functions of its components. The convolution integral operations are more complicated for imperfect switch standby structure. Table 1.1 provides examples of expressions for the reliability function of systems with basic structures in terms of the reliability functions of their components.

The existence of these mathematical expressions does not make the process of generating system reliability functions easy, especially if the component reliability function is not simple, the number of components is large, and the system structure is a combination of several structures. Generating the system reliability function from the reliability functions of its components must be done manually or using symbolic processing. The resulting system reliability function is also not easily discernible. Numerous simplification steps are still required.

As an illustration, systems in several structures are constructed using three components, all of which have lifetimes following a nonidentical 2-stage Erlang (2-Erlang) distribution. Following the 2-Erlang distribution function, the component reliability functions are given by

$$R_i(t) = e^{-\lambda_i t} + \lambda_i t e^{-\lambda_i t}, \quad i = 1, 2, 3, \quad t \geq 0. \tag{1.1}$$

The system reliability functions for various structures can be derived using the equations in Table 1.1. The results are given in Table 1.2.

Table 1.2 shows that the resulting system reliability function expressions are not easily understood even for relatively small-sized systems with only three components and simple component reliability functions such as a 2-Erlang distribution.

Table 1.1 Expression of reliability functions of systems in basic structures.

Structure	System reliability function
Series	$R_s(t) = \prod_{i=1}^{N} R_i(t)$
Parallel	$R_s(t) = 1 - \prod_{i=1}^{N}(1 - R_i(t))$
K-out-of-N	$R_s(t) = 1 - \sum_{\substack{r_1,r_2,\ldots,r_n=0 \\ r_1+r_2+\cdots+r_n \leq K-1}} \prod_{i=1}^{N}[R_i(t)]^{r_i}[1 - R_i(t)]^{1-r_i}$
Perfect switch standby	$R_s(t) = R_1(t) * R_2(t) * \cdots * R_N(t)$ $*$ Convolution $\rightarrow R_1(t) * R_2(t) = 1 + \int_0^t (1 - R_1(t-u))dR_2(u)$
Imperfect switch standby	$R_s(t) = (1-p_1)R_1(t) + p_1(1-p_2)R_1(t) * R_2(t) + \cdots$ $+\prod_{j=1}^{N-2} p_j(1 - p_{N-1})R_1(t) * R_2(t) * \cdots * R_{N-1}(t)$ $+\prod_{j=1}^{N-1} p_j R_1(t) * R_2(t) * \cdots * R_N(t)$

Table 1.2 Reliability functions of systems with three components having 2-Erlang distributed lifetimes.

Component reliability functions: $R_i(t) = e^{-\lambda_i t} + \lambda_i t e^{-\lambda_i t}, \quad i = 1, 2, 3, \quad t \geq 0$	
Structure	**System reliability function**
3-Series	$R_s(t) = \left(e^{-\lambda_1 t} + \lambda_1 t e^{-\lambda_1 t}\right)\left(e^{-\lambda_2 t} + \lambda_2 t e^{-\lambda_2 t}\right)\left(e^{-\lambda_3 t} + \lambda_3 t e^{-\lambda_3 t}\right)$
3-Parallel	$R_s(t) = 1 - \left(1 - \left(e^{-\lambda_1 t} + \lambda_1 t e^{-\lambda_1 t}\right)\right)\left(1 - \left(e^{-\lambda_2 t} + \lambda_2 t e^{-\lambda_2 t}\right)\right)\left(1 - \left(e^{-\lambda_3 t} + \lambda_3 t e^{-\lambda_3 t}\right)\right)$
2-out-of-3	$R_s(t) = \left(e^{-\lambda_1 t} + \lambda_1 t e^{-\lambda_1 t}\right)\left(e^{-\lambda_2 t} + \lambda_2 t e^{-\lambda_2 t}\right)$ $+\left(e^{-\lambda_1 t} + \lambda_1 t e^{-\lambda_1 t}\right)\left(1 - e^{-\lambda_2 t} - \lambda_2 t e^{-\lambda_2 t}\right)\left(e^{-\lambda_3 t} + \lambda_3 t e^{-\lambda_3 t}\right)$ $+\left(1 - e^{-\lambda_1 t} - \lambda_1 t e^{-\lambda_1 t}\right)\left(e^{-\lambda_2 t} + \lambda_2 t e^{-\lambda_2 t}\right)\left(e^{-\lambda_3 t} + \lambda_3 t e^{-\lambda_3 t}\right)$
3-standby with perfect switch	$R_s(t) = \left(e^{-\lambda_1 t} + \lambda_1 t e^{-\lambda_1 t}\right) * \left(e^{-\lambda_2 t} + \lambda_2 t e^{-\lambda_2 t}\right) * \left(e^{-\lambda_3 t} + \lambda_3 t e^{-\lambda_3 t}\right)$
3-standby with imperfect switch	$R_s(t) = (1-p_1)R_1(t) + p_1(1-p_2)R_1(t) * R_2(t) + p_1 p_2 R_1(t) * R_2(t) * R_3(t)$

Example 1.1 A system in standby structure with perfect switch is constructed using two components. Component 1 has lifetime following a 2-hypoexponential distribution whose reliability function is given by

$$R_1(t) = 2e^{-t} - e^{-2t}, \quad t \geq 0.$$

Component 2 has lifetime following a 2-hyperexponential distribution whose reliability function is given by

$$R_2(t) = 0.8e^{-t} + 0.2e^{-4t}, \quad t \geq 0.$$

The system reliability function can be derived using the convolution of two reliability functions [11] as follows:

$$\begin{aligned} R_S(t) &= R_1(t) * R_2(t) = 1 + \int_{u=0}^{t} (1 - R_1(t-u))dR_2(u) \\ &= 1 + \int_{u=0}^{t} (1 - 2e^{-(t-u)} + e^{-2(t-u)})d(0.8e^{-u} + 0.2e^{-4u}) \\ &= 1 + \int_{u=0}^{t} d(0.8e^{-u} + 0.2e^{-4u}) - 2\int_{u=0}^{t} e^{-(t-u)}(-0.8e^{-u} - 0.8e^{-4u})du \\ &\quad + \int_{u=0}^{t} e^{-2(t-u)}(-0.8e^{-u} - 0.8e^{-4u})du \\ &= 0.8e^{-t} + 0.2e^{-4t} + 1.6e^{-t}\left(t - \frac{1}{3}e^{-3t} + \frac{1}{3}\right) - 0.8e^{-2t}\left(e^{t} - 1 - \frac{1}{2}e^{-2t} + \frac{1}{2}\right) \\ &= \frac{1.6}{3}e^{-t} + 1.6te^{-t} + 0.4e^{-2t} + \left(0.2 - \frac{1.6}{3} + 0.4\right)e^{-4t}. \end{aligned}$$

The resulting function can be verified to be a valid reliability function; that is, it is monotonic nonincreasing function with a value 1 at $t = 0$ and 0 as $t \to \infty$. However, it is an unstructured combination of exponential functions with difficult-to-recognize patterns. The system reliability function will be more complicated for systems with more than two components or using components with more complex reliability functions. ■

The difficulty encountered in using scalar reliability functions is not only in generating the system reliability function but also in other reliability measures that are even more complicated.

Besides the reliability function, another important functional reliability measure is the hazard function, which is a function that represents the rate of failure at any time. It can be derived from the reliability function as

$$h(t) = -\frac{R'(t)}{R(t)}, \tag{1.2}$$

where $R'(t) = dR(t)/dt$.

Equation (1.2) shows that finding the hazard function requires the use of differential operators on the reliability function. The differential operator will produce a more complex mathematical expression for the hazard function than for the reliability function. The derivation must also be done manually or using computerized symbolic processing, which is also cumbersome and produces uninformative results.

Example 1.2 The hazard function of the 2-standby system in Example 1.1 can be derived as follows:

$$h_S(t) = -\frac{\frac{d}{dt}R_S(t)}{R_S(t)} = -\frac{\frac{d}{dt}\left(\frac{1.6}{3}e^{-t} + 1.6te^{-t} + 0.4e^{-2t} + \left(0.2 - \frac{1.6}{3} + 0.4\right)e^{-4t}\right)}{\frac{1.6}{3}e^{-t} + 1.6te^{-t} + 0.4e^{-2t} + \left(0.2 - \frac{1.6}{3} + 0.4\right)e^{-4t}},$$

which is complicated, and the resulting expression cannot be easily simplified. ■

One of the statistical reliability measures is the expected lifetime, commonly known as mean time to failure ($MTTF$). This statistic can be calculated using the integration of the reliability function,

$$MTTF = \int_{0}^{\infty} R(t)dt. \tag{1.3}$$

Example 1.3 The expected lifetime of the 2-standby system with perfect switch in Example 1.1 can be calculated using Equation (1.3) as follows:

$$MTTF_S = \int_0^\infty R_S(t)dt = \int_0^\infty \left(\frac{1.6}{3}e^{-t} + 1.6te^{-t} + 0.4e^{-2t} + \left(0.2 - \frac{1.6}{3} + 0.4\right)e^{-4t}\right) dt.$$

The integral can be calculated using numerical integration to obtain an estimate of the *MTTF*. The accuracy of the estimated value can be made higher by making the step size in numerical integration smaller. Other statistics, such as variance time to failure, require different and more complicated integration than for calculating *MTTF*. ■

1.3 Advantages of Using Matrix-Form Distribution in System Reliability Modeling

The reliability function in matrix form is a way to express the reliability function in a simple form by involving vector and matrix parameters instead of scalar parameters as in the scalar reliability function. The matrix-form reliability function is very rich in variations so that it can represent many kinds of scalar reliability functions. The matrix-form reliability function uses the exponential function of a square matrix premultiplied by a row vector and postmultiplied by a column vector so that the final result is a scalar-valued function that expresses the reliability value at any time. The most popular matrix-form reliability function is phase-type (PH) distribution [12, 13]:

$$R(t) = \boldsymbol{a}e^{\boldsymbol{A}t}\boldsymbol{u} \quad \text{for} \quad t \geq 0, \tag{1.4}$$

where $\boldsymbol{A}$ is a square matrix, $\boldsymbol{a}$ is a row vector with the number of rows equal to the number of columns of the matrix $\boldsymbol{A}$, and $\boldsymbol{u}$ is a unitary column vector (a vector with all elements equal to 1) with the number of rows equal to the number of columns of the matrix $\boldsymbol{A}$. The reliability function in matrix form must also satisfy the properties of reliability function, namely, monotonic non-increasing function, which has a value of 0 when $t \to \infty$, or $e^{\boldsymbol{A}t} \to 0$ when $t \to \infty$, and has finite moments.

The matrix-form reliability function given by Equation (1.4) is versatile. It can be used to represent many well-known scalar reliability functions.

Example 1.4 The simplest form of the matrix-form reliability function is an exponential reliability function, assuming the parameters of the function as matrix and vectors with dimension 1.

The exponential reliability function $e^{-\lambda t}$ in scalar form can be considered a matrix-form reliability function $\boldsymbol{a}e^{\boldsymbol{A}t}\boldsymbol{u}$ with

$$\boldsymbol{a} = [1], \quad \boldsymbol{A} = [-\lambda], \quad \boldsymbol{u} = [1],$$

where [.] represents a vector or matrix depending on its dimensions. ■

This example means that the matrix-form reliability function is a generalization of exponential reliability function by generalizing the scalar parameter λ into a matrix parameter $\boldsymbol{A}$, which is a square matrix in any dimension. If the dimension of $\boldsymbol{A}$ is one, then it becomes the popularly known exponential reliability function.

Example 1.5 The 2-Erlang reliability function in scalar form $R(t) = e^{-\lambda t} + \lambda t e^{-\lambda t}$ can be expressed as a matrix-form reliability function $\boldsymbol{a}e^{\boldsymbol{A}t}\boldsymbol{u}$ with

$$\boldsymbol{a} = [1 \;\; 0], \quad \boldsymbol{A} = \begin{bmatrix} -\lambda & \lambda \\ 0 & -\lambda \end{bmatrix}, \boldsymbol{u} = \begin{bmatrix} 1 \\ 1 \end{bmatrix}.$$

The correspondence between the matrix-form and the scalar-form is from the exponential result of matrix exponential in the Jordan canonical form given by

$$\exp\left(\begin{bmatrix} -\lambda & \lambda \\ 0 & -\lambda \end{bmatrix} t\right) = \begin{bmatrix} e^{-\lambda t} & \lambda t e^{-\lambda t} \\ 0 & e^{-\lambda t} \end{bmatrix}.$$

where $\exp(x) = e^x$.

Hence,

$$\boldsymbol{a}e^{\boldsymbol{A}t}\boldsymbol{u} = \begin{bmatrix} 1 & 0 \end{bmatrix} \begin{bmatrix} e^{-\lambda t} & \lambda t e^{-\lambda t} \\ 0 & e^{-\lambda t} \end{bmatrix} \begin{bmatrix} 1 \\ 1 \end{bmatrix} = \begin{bmatrix} e^{-\lambda t} & \lambda t e^{-\lambda t} \end{bmatrix} \begin{bmatrix} 1 \\ 1 \end{bmatrix} = e^{-\lambda t} + \lambda t e^{-\lambda t}.$$

This result shows that the matrix-form reliability function is identical to the scalar-form reliability function since both express the same 2-Erlang distribution. ■

Expressing the component reliability function in matrix form provides many conveniences in the dynamic analysis of system reliability.

Example 1.6 The reliability of a 3-series system with nonidentical components using a matrix-form reliability function can be found as

$$R_S(t) = R_1(t)R_2(t)R_3(t) = \left(\boldsymbol{a}_1 e^{\boldsymbol{A}_1 t}\boldsymbol{u}\right)\left(\boldsymbol{a}_2 e^{\boldsymbol{A}_2 t}\boldsymbol{u}\right)\left(\boldsymbol{a}_3 e^{\boldsymbol{A}_3 t}\boldsymbol{u}\right),$$

which by using a result in the matrix theory (will be studied in Chapter 2) yields

$$R_S(t) = \boldsymbol{a}_S e^{\boldsymbol{A}_S t}\boldsymbol{u},$$

where $\boldsymbol{a}_S$ is the result of matrix algebra operations on $\boldsymbol{a}_1$, $\boldsymbol{a}_2$, and $\boldsymbol{a}_3$ and $\boldsymbol{A}_S$ is the result of matrix algebra operations on $\boldsymbol{A}_1$, $\boldsymbol{A}_2$, and $\boldsymbol{A}_3$, whereas $\boldsymbol{u}$ remains a unitary column vector whose dimensions are adjusted to the dimensions of $\boldsymbol{A}_S$. ■

Therefore, the operations are to the parameters of the function, instead of to the function itself. Operations to the parameters are much easier to program on a computer. Interestingly, the result is also a matrix-form reliability function whose parameters can be calculated from the parameters of the component matrix-form reliability functions using matrix algebra operators; hence, it eliminates the need for symbolic processing. The same process applies to all system structures with any number of components. This is one good feature of a matrix-form reliability function; that is, it is closed in system reliability analysis, meaning if the component reliability functions are in matrix form, then the resulting system reliability function is also in matrix form.

Example 1.7 The reliability functions of components in Example 1.1 can be expressed in matrix form as:

- Component 1:

$$R_1(t) = 2e^{-t} - e^{-2t} = \begin{bmatrix} 1 & 0 \end{bmatrix} e^{\begin{bmatrix} -1 & 1 \\ 0 & -2 \end{bmatrix} t} \begin{bmatrix} 1 \\ 1 \end{bmatrix} = \boldsymbol{a}_1 e^{\boldsymbol{A}_1 t}\boldsymbol{u}.$$

where $\boldsymbol{a}_1 = \begin{bmatrix} 1 & 0 \end{bmatrix}$ and $\boldsymbol{A}_1 = \begin{bmatrix} -1 & 1 \\ 0 & -2 \end{bmatrix}$.

- Component 2:

$$R_2(t) = 0.8e^{-t} + 0.2e^{-4t} = \begin{bmatrix}0.8 & 0.2\end{bmatrix} \boldsymbol{e}^{\begin{bmatrix}-1 & 0\\ 0 & -4\end{bmatrix}t} \begin{bmatrix}1\\1\end{bmatrix} = \boldsymbol{a}_2 e^{\boldsymbol{A}_2 t}\boldsymbol{u}.$$

where $\boldsymbol{a}_2 = \begin{bmatrix}0.8 & 0.2\end{bmatrix}$ and $\boldsymbol{A}_2 = \begin{bmatrix}-1 & 0\\ 0 & -4\end{bmatrix}$.

The system reliability function can be derived using a property of matrix-form distribution as (will be studied in Chapter 2)

$$R_S(t) = \boldsymbol{a}_S e^{\boldsymbol{A}_S t}\boldsymbol{u} \quad t \geq 0,$$

where

$$\boldsymbol{a}_S = \begin{bmatrix}1 & 0 & 0 & 0\end{bmatrix},$$

$$\boldsymbol{A}_S = \begin{bmatrix}\boldsymbol{A}_1 & -\boldsymbol{A}_1\boldsymbol{u}\boldsymbol{a}_2\\ \boldsymbol{0} & \boldsymbol{A}_2\end{bmatrix} = \begin{bmatrix}\begin{bmatrix}-1 & 1\\ 0 & -2\end{bmatrix} & -\begin{bmatrix}-1 & 1\\ 0 & -2\end{bmatrix}\begin{bmatrix}1\\1\end{bmatrix}\begin{bmatrix}0.8 & 0.2\end{bmatrix}\\ \boldsymbol{0} & \begin{bmatrix}-1 & 0\\ 0 & -4\end{bmatrix}\end{bmatrix} = \begin{bmatrix}-1 & 1 & 0 & 0\\ 0 & -2 & 1.6 & 0.4\\ 0 & 0 & -1 & 0\\ 0 & 0 & 0 & -4\end{bmatrix}, \quad \boldsymbol{u} = \begin{bmatrix}1\\1\\1\\1\end{bmatrix}.$$

As seen, only matrix algebra operation is required to produce the system reliability function; the convolution integral is no longer needed when using matrix-form distribution functions.

To show that the system reliability functions obtained from Examples 1.1 and 1.7 are identical, their graphs are presented in Figure 1.4. ■

Comparison of Examples 1.1 and 1.7 concludes that using a matrix-form reliability function makes the finding of system reliability more systematic and easily programmed.

The hazard function in matrix-form can be derived by substituting Equation (1.4) into Equation (1.2) to yield

$$h(t) = \frac{(\boldsymbol{a}e^{\boldsymbol{A}t}\boldsymbol{u})'}{\boldsymbol{a}e^{\boldsymbol{A}t}\boldsymbol{u}} = (\boldsymbol{a}e^{\boldsymbol{A}t}\boldsymbol{A}\boldsymbol{u})(\boldsymbol{a}e^{\boldsymbol{A}t}\boldsymbol{u})^{-1} \quad \text{or} \quad (\boldsymbol{a}\boldsymbol{A}e^{\boldsymbol{A}t}\boldsymbol{u})(\boldsymbol{a}e^{\boldsymbol{A}t}\boldsymbol{u})^{-1}. \tag{1.5}$$

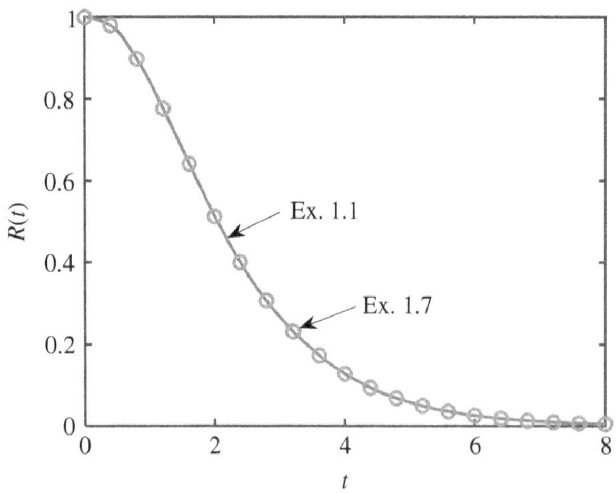

Figure 1.4 System reliability functions from Examples 1.1 and 1.7.

Both the aforementioned forms of system hazard function are identical, so they can be selected according to the needs. Note that the inverted term is scalar, so it does not need matrix inversion to calculate.

Equation (1.5) shows that the system hazard function, although defined using differentials, can be found using only matrix algebra based on the parameters of the matrix-form reliability function. Once the system reliability function is known, determining the system hazard function is straightforward.

In addition to the ease of deriving the system reliability function, finding the system hazard function can also be made programmable.

Likewise, the expected lifetime in matrix-form can be derived by substituting Equation (1.4) into Equation (1.3) and acknowledging the nature of reliability function that approaches a value of 0 as $t \to \infty$ to yield

$$MTTF = \int_0^\infty (\boldsymbol{a}e^{\boldsymbol{A}t}\boldsymbol{u})dt = \boldsymbol{a}\left(\int_0^\infty e^{\boldsymbol{A}t}dt\right)\boldsymbol{u} = \boldsymbol{a}\boldsymbol{A}^{-1}(e^{(\boldsymbol{A})\infty} - e^{(\boldsymbol{A})0})\boldsymbol{u} = -\boldsymbol{a}\boldsymbol{A}^{-1}\boldsymbol{u}. \tag{1.6}$$

Example 1.8 The expected lifetime of the system in Example 1.1 can be found using its matrix-form reliability function derived in Example 1.7 as

$$MTTF_S = -\boldsymbol{a}_S\boldsymbol{A}_S^{-1}\boldsymbol{u} = -\begin{bmatrix}1 & 0 & 0 & 0\end{bmatrix}\begin{bmatrix}-1 & 1 & 0 & 0\\ 0 & -2 & 1.6 & 0.4\\ 0 & 0 & -1 & 0\\ 0 & 0 & 0 & -4\end{bmatrix}^{-1}\begin{bmatrix}1\\1\\1\\1\end{bmatrix} = 2.35.$$

As with the system hazard function, the system expected lifetime defined using integrals can be calculated using matrix algebra, avoiding the needs of using integration. Both scalar and matrix approaches require computer assistance; however, the numerical integration in scalar approach produces an estimate of the real value with its accuracy depending on the step size used in calculating the integral, while matrix inverse in matrix approach produces the exact value. A method to avoid matrix inversion in calculating statistical reliability measures will be discussed in Chapter 3. Furthermore, calculating *MTTF* using matrix approach is easily programmable.

Comparison of calculating *MTTF* using scalar approach in several step sizes and using matrix approach is given in Table 1.3. ■

Table 1.3 Comparison of the system *MTTF* from Examples 1.3 and 1.8.

Step size	1	0.5	0.1	0.01	0.001	Exact value
MTTF	2.8473	2.5998	2.4	2.355	2.3505	2.35

These examples demonstrate that deriving reliability and hazard functions and calculating expected lifetimes of a system can all be performed using matrix algebra if the component reliability functions are expressed in matrix form. The use of matrix algebra is easier to program in computer, so the process of calculating system reliability measures can be carried out with the help of a computer. Equations derived from the definitions of system reliability measures can be utilized to design the algorithms needed for preparing the computer programs. Thus, the task of the system reliability analyst is greatly assisted, not burdened with calculations, so that they can focus on system modeling.

Example 1.6 used a closure property of independent reliability functions in matrix form that are closed under the minimum operation. Likewise, Example 1.7 used a closure property that independent reliability functions in matrix form are closed under the addition operation. Independent reliability functions are also closed under the maximum operation. Minimum, maximum, and addition are basic operations in system reliability modeling and analysis. The closure of independent reliability functions under those operations is convenient in developing algorithms for generating the reliability function of a system from the reliability functions of its components under the assumption that the randomness of component lifetimes follows independent distributions.

Another important feature of a matrix-form reliability function is that such a function can be used for multistate systems. In classical reliability analysis, systems are assumed to have only two states representing two conditions of the systems that either are good for the system to function properly or failed for the system that does not function as expected. The reliability function for such systems can be represented in scalar form. Real-world systems may have more than two conditions. The system may experience slow deterioration or gradual wear-out between being in excellent condition to fail. Systems may be exposed to shocks that do not immediately make the system fail, but each shock will downgrade the system condition. Representing the reliability of such system in scalar reliability function is complicated. It needs as many scalar reliability functions as the number of states the system has. The matrix-form reliability function, on the contrary, can represent the system reliability in a single function.

Example 1.9 The conditions of a radar sensor can be classified based on its capability in detecting a target within a given distance. The deterioration process of radar sensitivity is given in Table 1.4. The sensitivity of radar sensor at the beginning of its operation is more than 400 km, or it is in state 1.

Table 1.4 State and transition for radar sensor example.

Sensitivity	State	1	2	3	4
More than 400 km	1	−0.01	0.008	0.002	0
300–400	2	0	−0.02	0.02	0
200–300	3	0	0	−0.04	0.04
Less than 200 km	4	0	0	0	−0.05

The reliability function of the radar sensor in matrix form is $R(t) = \boldsymbol{a}e^{\boldsymbol{A}t}\boldsymbol{u}$, where $\boldsymbol{a}$ is determined from the probability being in each state in the beginning of its operation and $\boldsymbol{A}$ is determined from the transition among states. Hence,

$$\boldsymbol{a} = \begin{bmatrix} 1 & 0 & 0 & 0 \end{bmatrix}, \quad \boldsymbol{A} = \begin{bmatrix} -0.01 & 0.008 & 0.002 & 0 \\ 0 & -0.02 & 0.02 & 0 \\ 0 & 0 & -0.04 & 0.04 \\ 0 & 0 & 0 & -0.05 \end{bmatrix}.$$

Other reliability measures can be derived from the reliability function. ■

Illustrations given in Examples 1.4–1.9 can be used to conclude that using matrix-form reliability has advantages, including:

- Produce system reliability function, not system reliability values.
- Easily programmable.
- Closed in many system structures, including networks.
- Require only matrix algebra, no calculus.
- Cover many well-known distributions, especially those frequently used in reliability analysis.
- Applicable to multistate systems.

The method can be developed further to enhance currently available simulation models so that they can be used to produce system reliability parameters instead of system reliability values. The parameters can be directly used to construct functional reliability measures and calculate statistical reliability measures. However, it is beyond the scope of this book. Hopefully, this book is able to convince some readers to develop such simulation models.

1.4 Approach and Organization of the Book

This book describes the use of matrix algebra in dynamic analysis of system reliability. The approach is to express component reliability functions in terms of matrix-form reliability functions. This approach is generally applicable because matrix-form reliability functions can be used to express various scalar reliability functions used in system reliability modeling, such as exponential, Erlang, hyperexponential, hypoexponential, Coxian, and any functions derived as the results of mixing, maximum, minimum, or addition operations on those functions. Even other scalar reliability functions, such as Weibull and lognormal, can also be approximated with matrix-form reliability functions with acceptable accuracy.

This approach offers many conveniences in finding various reliability measures, and all can be found simply using matrix algebra, without the need for calculus (differential and integral). The use of matrix algebra allows the equations for reliability measures to be expressed in simple algorithms. The algorithms are then programmed as MATLAB functions as a convenience for the readers. The functions can be used to model the system for its reliability analysis and to produce the outputs of the analysis. As a companion to this book, a MATLAB software package **dysrel-am** is also prepared in executable files under MATLAB license. Translating MATLAB code into other languages, such as Python, is only a small step because everything is already expressed following structured programming. However, at the moment, the best computer language for matrix operations is MATLAB.

The use of matrix-form reliability functions also makes it easier to broaden the definition of component conditions. In scalar reliability functions, component conditions are expressed in two states: good and fail. Reliability functions in matrix form allow modeling the components in multistates, such as new, good, fair, and failed, or more complex states, depending on the needs of real-world applications. This capability makes matrix-form reliability functions more useful in dynamic system reliability analysis.

This book focuses on the process of converting dynamic system reliability analysis using scalar reliability functions into dynamic system reliability analysis using matrix-form reliability functions. Therefore, users of this book are expected to understand the basics of system reliability analysis using the scalar reliability function. Many textbooks already cover these basics, so they need not be discussed again in detail unless they are necessary as the basis for deriving the results in matrix form. Readers should also have a sufficient understanding of the basics of matrix algebra. Advanced matrix algebra that is regularly used in system reliability analysis will still be explained as necessary. Therefore, this book requires conventional reliability analysis (static and scalar) and the basics of matrix algebra as prerequisites. However, they are lightly inserted while explaining the methods in matrix-form reliability functions. From the authors experience, undergraduate students have no difficulties in understanding the book, especially since the book is supported with a software package.

The book is organized as follows. Chapter 2 explains distribution functions in matrix form used to express matrix-form reliability functions, along with their physical interpretation, formulation, and properties. The discussion begins with an explanation of continuous-time Markov chains (CTMC) and their relationship to PH distribution, one of the matrix-form distributions. This chapter also discusses state transition diagrams as a model for representing CTMC. Another matrix-form distribution, the matrix-exponential (ME) distribution, as a generalization of the PH distribution, is also explained, including matrix-similarity transformations to transform PH distributions into ME distributions. Chapter 3 develops reliability modeling and analysis of one-component systems in matrix-form distributions, the derivation of equations for functional and statistical reliability measures, namely reliability function, hazard function, residual reliability function, moments of lifetime, expected residual lifetime, and algorithms for their calculation. Development continues with the use of reliability measures for system replacement, such as calculating the number of replacements and the risk of spare-part stockout, including optimum replacement that minimizes the total of replacement and failure costs. Reliability-based maintenance programs, perfect and imperfect, are also developed. All methods are presented in MATLAB functions that can be utilized as models for various purposes of system analyses.

Chapter 4 discusses the dynamic reliability modeling and analysis of systems with many components, arranged in basic structures, namely series, parallel, and standby with perfect and imperfect switches structures. The algorithms for generating the system reliability function from the component reliability functions for all basic structures are developed. The algorithms are programmed as MATLAB functions that can be used as system models for their reliability analyses. The resulting functions are used as models for complex systems with combinations of basic structures in a hierarchical manner, from the top at the system level, to the subsystem level, all the way down to the bottom at the component level. Chapter 5 starts with developing algorithms for reliability analysis of systems in K-out-of-N structure with any number of primary and backup components. The model for K-out-of-N structure is generalized to general standby structures to include various forms of backups and a method to express them all in a single general standby model. The backup components may have any level of warmness so that they could be in cold standby, warm standby, or hot standby by introducing deceleration matrix that adjusts the transition rates of component states. The model is applied to find the optimum ordering rules of backup components in various settings. The discussion concludes by showing that the general standby model can serve as a unifying model because it can be used to represent systems with any basic structure. An optimization model for finding the optimum warmness level with respect to operation and opportunity costs is added as an application for the general standby structure.

Chapter 6 discusses reliability of networks; that is, systems with bridge components that make the basic structure models discussed in Chapters 4 and 5 no longer applicable. This chapter introduces a combinatorial model to reliability analysis, using minimal paths (or minimum cuts) as a way to derive system structure functions. The process of converting Boolean system structure functions to algebraic system structure functions in the form of sums of disjoint products is also explained. The equations necessary to design algorithms for generating matrix-form network reliability functions from the matrix-form reliability functions of its components are derived. The development continues with deriving algorithms for general networks and capacitated networks. The state transition models defined in Chapters 4 and 5 are applied to network reliability analysis and compared with combinatorial model. Finally, a method to combine the methods of Chapters 4–6 for the complex system with network subsystems is explained. In Chapter 7, all methods and algorithms that have been developed for continuous-time systems are translated into discrete-time systems. In the last chapter, Chapter 8, experiments in using artificial intelligence (AI) for system reliability analysis are reported. Several problems used in Chapters 2–7 are submitted to AI, the AI response to each question is analyzed, and if, in some cases, a clarification is submitted to AI in cases where AI responses are not in line with the book.

For completeness as course material, each chapter provides exercises, applications, problems, and references. Advanced material is marked with stars that can be omitted without disturbing the continuity of the development.

The learning content of this book may take several possible routes depending on the needs and time allocation. Several possible routes are depicted in Figure 1.5.

Following the learning content organization, the book can be used to support the following learning programs:

1) Second course on Reliability: all chapters.
2) First course on Reliability: Chapters 1, 2, 3.1–3.6, 4, 5.1, and 6.1–6.5.
3) Network Reliability: Chapters 1, 2, 3.1–3.6, 4, 5.1, 6, and 7.9.
4) Reliability in Discrete-Time Systems: Chapters 1 and 7.
5) Applied System Reliability: Chapters 1–5.

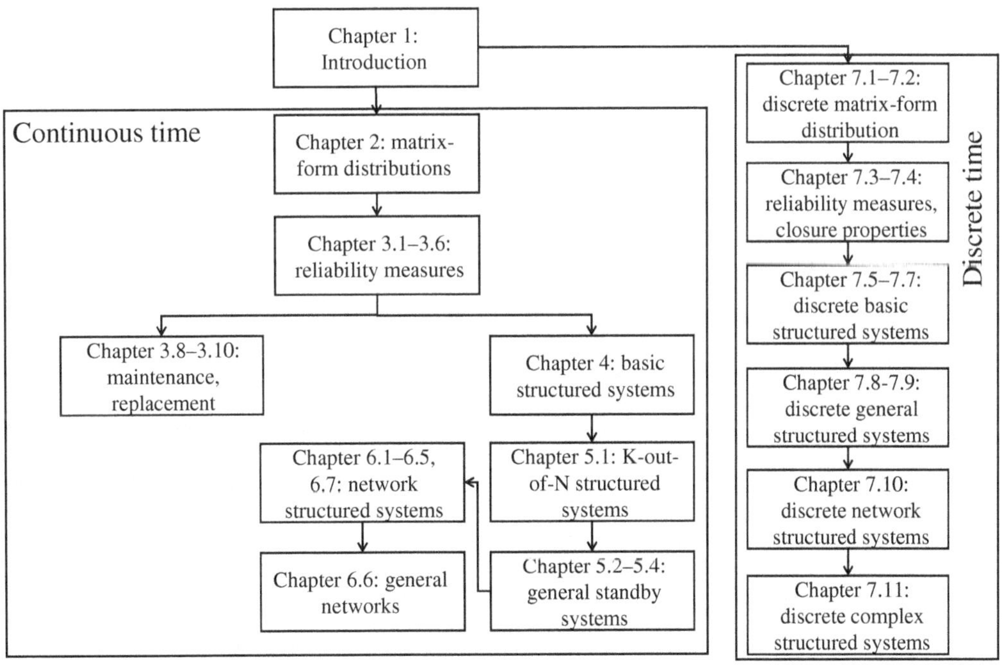

Figure 1.5 Learning content organization of the book.

1.5 Supporting Software

All materials in the book are supported by computer programs available in the software package **dysrel-am (Dynamic System Reliability-Analytical Model)**. The software package contains all algorithms derived in the book prepared as MATLAB functions. The systems being studied can be represented in the available functions as the system models. For example, **phseriessys** function is for modeling continuous-time series systems. Functional and statistical reliability measures are also supported by functions that are called to produce the system reliability measures of the system being studied. For example, **phrelplot** is for plotting system reliability function in continuous time. All functions are given in MATLAB codes so that readers can use them as a basis for their own programs in case they want to translate it to other computer languages or modify them in accordance with their specific needs.

To use the package for analyzing a system, the reader must write a main program to call the appropriate functions for modeling the system and for producing system reliability measures, as well as the input given in the system description. The common inputs are the parameters of reliability function of each component of the system and the number of components in the system. Reliability functions for known distributions are input as the name of the distributions and their parameters. The known distributions are Erlang, hypoexponential, hyperexponential, and Coxian. Unknown distributions must be input as their PH or ME representations. Known distributions that are approximated with PH distribution are input following their PH representation approximations. Other inputs depend on the purposes of the system analysis, such as replacement interval, maintenance interval, cost parameters for optimization, and deceleration factor for warm standby. They can be input through provided dialog boxes in the user interface of the software package.

To help the reader in creating the main program, most examples in the book are supported with main programs available as example files in the software package. The example files are named after the example number in the book, with **examplex_y.m** file is for Example x.y, which is example number y in Chapter x. Likewise, application files are named after the application number in the book following the same rule as for example files. All functions and examples are prepared using MATLAB version 2021a and executed on a PC with Intel i7 processor version 8565U and 8 GB RAM. These specifications are the nominal requirements for executing the programs. Most examples (70%) can still be executed in a PC with Intel i3 processor and 4 GB RAM, but for a few applications, even 8 GB RAM requires memory swapping with SSD. The book provides a link to access the software package repository.

1.6 Problems

1.1 Create a comparative table of the capabilities of three system reliability models in producing system reliability measures, including whether it must be done manually, can use a computer for symbolic processing or number processing, using algebra or calculus.

1.2 Prove that the convolution of exponential reliability function $e^{-\lambda t}$ to itself will yield a 2-Erlang reliability function as given in Equation (1.1).

1.3 Find the reliability function resulting from the convolution of the exponential reliability function $e^{-\lambda t}$ with the 2-Erlang reliability function in Equation (1.1). Note that the result is a 3-Erlang reliability function.

1.4 Determine the expected lifetime of a component with the 2-Erlang reliability function in Equation (1.1) using Equation (1.3). Then express the 2-Erlang in a matrix-form reliability function and use Equation (1.3) to find its expected lifetime. Show that both methods yield the same value.

1.5 Calculate the expected lifetime of a component that has a reliability function in the form of a matrix with parameters

$$\boldsymbol{a} = \begin{bmatrix} 1 & 0 & 0 \end{bmatrix}, \quad \boldsymbol{A} = \begin{bmatrix} -2 & 1 & 1 \\ 0 & -2 & 1 \\ 0 & 0 & -1 \end{bmatrix}.$$

References

1 Ardakan, M.A., Mirzaei, Z., Hamadani, A.Z., Elsayed, E.A., 2017. Reliability optimization by considering time-dependent reliability for components. Qual Reliab Eng Int 33, 1641–1654. https://doi.org/10.1002/qre.2132.

2 Rebello, S., Yu, H., Ma, L., 2018. An integrated approach for system functional reliability assessment using Dynamic Bayesian Network and Hidden Markov Model. Reliab Eng Syst Saf 180, 124–135. https://doi.org/10.1016/j.ress.2018.07.002.

3 Elsayed, E.A., 2021. Reliability Engineering, 3rd ed. Wiley.

4 Navarro, J., 2021. Introduction to System Reliability Theory. Springer International Publishing.

5 Rausand, M., Barros, A., Hoyland, A., 2020. System Reliability Theory: Models, Statistical Methods, and Applications, 3rd ed. Wiley.

6 Bernstein, J.B., Bensoussan, A., Bender, E., 2024. Reliability Prediction for Microelectronics. John Wiley & Sons.

7 Saket, R.K., Sanjeevikumar, P. (Eds.), 2024. Reliability Analysis of Modern Power Systems. Wiley-IEEE Press.

8 Li, Y.-F., Zio, E., Kleyner, A.V., 2022. System Reliability Assessment and Optimization: Methods and Applications. Wiley.

9 Trivedi, K.S., Bobbio, A., 2017. Reliability and Availability Engineering: Modeling, Analysis, and Applications. Cambridge University Press.

10 Eryilmaz, S., 2022. Discrete Stochastic Models and Applications for Reliability Engineering and Statistical Quality Control. CRC Press, Boca Raton. https://doi.org/10.1201/9781003160632.

11 Alkaff, A., Qomarudin, M.N., 2020. Modeling and analysis of system reliability using phase-type distribution closure properties. Appl Stoch Model Bus Ind 36, 548–569. https://doi.org/10.1002/asmb.2509.

12 Horváth, A., Telek, M., 2024. Phase Type Distributions: Theory and Application. Wiley-ISTE.

13 Ncuts, M.F., 1994. Matrix-Geometric Solutions in Stochastic Models: An Algorithmic Approach. Dover Publication, Inc.

2

Matrix-Form Distributions

Learning objectives:

1) Understand the derivations of all equations for functions and moments of PH and ME distributions.
2) Understand the properties of PH and ME distributions and their applications.
3) Able to construct PH representations from random process descriptions and to transform PH distributions into ME distributions in Jordan canonical forms.
4) Able to utilize the software to calculate distribution functions and moments of PH and ME distributions.

Abbreviations

APH	acyclic phase-type
CDF	cumulative distribution function
CPH	cyclic phase-type
CTMC	continuous-time Markov chain
ME	matrix-exponential
PDF	probability density function
PH	phase-type

Notations

t	time, mission duration
n	order, dimension
$\boldsymbol{a}$	initial probability vector, starting vector
$\boldsymbol{A}$	transient matrix, transition matrix
$\boldsymbol{I}$	identity matrix
$\boldsymbol{u}$	unitary vector
$\boldsymbol{v}$	closing vector
$\boldsymbol{b}$	absorbing vector
$(n, \boldsymbol{a}, \boldsymbol{A})$	PH representation
$(n, \boldsymbol{a}, \boldsymbol{A}, \boldsymbol{v})$	ME representation

$\boldsymbol{Q}$	infinitesimal generator
$E(.)$	expected value
$F(t)$	CDF
$f(t)$	PDF
$\otimes$	Kronecker product
$\oplus$	Kronecker sum

Matrix-based distributions generalize exponential distribution by replacing the scalar parameter in the exponential distribution with a matrix. Since the distribution expresses scalar-valued probabilities, the resulting matrix exponential function must be premultiplied by a row vector and postmultiplied by a column vector to produce a scalar. This generalization is related to the assumption that exponential distribution is a special case when the process underlying the randomness of a variable has only one initial state, and with one transition has reached the final state and stopped. The matrix-based distribution assumes that the process underlying the randomness of a variable has many states, requiring many transitions to reach the final state and stop.

The generalization of the process underlying the randomness of a variable from one stage in the exponential distribution to several stages has been widely used. For example, Erlang distribution is a generalization of exponential distribution by allowing the process to have many stages with one transition from one stage to the next. However, the resulting distribution is still expressed in a scalar form to ease its manual calculation. With the development and accessibility of computers, calculations with matrices are no longer an obstacle, which makes representing many stages in a matrix form no longer a problem; thus, matrix-based distributions are becoming increasingly popular. Expressing transitions in a matrix form provides more flexibility to handle not only addition operations such as in Erlang distribution but also merging, minimum, maximum, and all their combinations. All these operations can be expressed in matrix operations that are easy to process with the help of computers. Moreover, the need for many stages in modeling randomness comes naturally from real problems. The system condition cannot be said to be only good or fail. There are many intermediate states from good to fail with many transitions among them, so the randomness needs to be modeled in a matrix.

In matrix-based distributions, all transitions from one state to another are expressed in a transition matrix, instead of a scalar transition as in the exponential distribution. The duration of stay in each state is still assumed to be the same as in the exponential distribution, namely the scalar exponential distribution. This description means that the process underlying the randomness of a variable, which the matrix-based distribution will model, is a continuous-time Markov process. This is because the transition from one state to another can occur at any time, and the duration of stay in each state is exponentially distributed. Given that the states are discrete and limited to only certain values, such a Markov process is known as a continuous-time discrete-state Markov process, also commonly called a continuous-time Markov chain (CTMC). Therefore, understanding the CTMC as the first step to understanding the matrix-based distribution is a necessity.

This chapter discusses CTMC and its properties. A particular interest is a special form of CTMC with one absorbing state and the distribution of absorption time in such CTMC. It is the basis of the development of phase-type (PH) distribution, which is one of the matrix-based distributions. The mathematical expressions used to model PH distribution, along with diagrams to represent its states and transitions, are explained. The PH distribution is shown to be a general class of distributions with known distributions as its members. Several popular distributions of PH distribution class members are exhibited completely with their diagram, scalar, and matrix representations. For other distributions not members of the PH distribution class, methods to approximate them in PH distributions are explained.

The chapter continues by deriving equations for the moments of the PH distribution. The closure properties of the PH distribution under several operations and the derivation of the representation of the resulting PH distribution from these operations are discussed. Next, the ME distribution is proposed as a generalization of the PH distribution, along with transforming the PH distribution into the ME distribution using a matrix similarity transformation. Various transformation results are also given, including a method to reduce the order of the distribution based on the transformation results. The closure properties of the ME distribution are also given as an analogy to the closure properties of the PH distribution. Comparisons of several results of operations using the PH and ME distributions are given.

The development of matrix-form distribution starts with a description of CTMC.

2.1 Continuous-Time Markov Chain

CTMC is a discrete-state random process in continuous time that follows the Markov property. The states can be numbered as state 1, state 2, ... The process stays in a state for some time. Depending on the duration of stay, states in CTMC are classified into transient and absorbing states. A transient state is a state that is occupied for a certain random duration. An absorbing state is a state that is occupied forever; that is, once the process enters an absorbing state, it will never leave that state. CTMC has at least two states, with at least one transient state and does not necessarily have an absorbing state.

As a consequence of the Markov property, the duration of stay of CTMC in each transient state has an exponential distribution with a certain transition rate. CTMC will exit a transient state with a transition rate equal to the rate of the exponential distribution of the duration of stay in that state. The process is assumed to be stationary, meaning the transition rate is constant and does not change over time. If T_i denotes the duration of stay in transient state i, then

$$T_i \sim \exp(\lambda_i), \tag{2.1}$$

where $T_i \sim \exp(\lambda_i)$ denotes T_i has exponential distribution with rate λ_i; thus, its cumulative distribution function (CDF) is given by $F_T(t) = 1 - e^{-\lambda_i t}$ for $t \geq 0$.

The parameter λ_i is the rate of stay in state i, which is also the transition rate out of state i. The CDF of T_i can be obtained from Equation (2.1) as

$$P[T_i \leq t] = F_{T_i}(t) = 1 - e^{-\lambda_i t}. \tag{2.2}$$

An important property of the exponential distribution, which makes it appropriate as the distribution of duration of stay in each state in a CTMC, is the memoryless property, that is,

$$P[T \leq t+x | T > x] = \frac{P[x < T \leq t+x]}{P[T > x]} = \frac{P[T \leq t+x] - P[T \leq x]}{P[T > x]}.$$

Using Equation (2.2) obtains

$$P[T \leq t+x | T > x] = \frac{1 - e^{-\lambda(t+x)} - (1 - e^{-\lambda x})}{e^{-\lambda x}} = 1 - e^{-\lambda t} = P[T \leq t].$$

This property means the duration of time the process has been in a state does not affect the probability the process will exit the state; it is as if the process has just entered that state. Thus, time

$t = 0$ can be taken at any time according to the needs of the analysis; for example, the time the analysis begins.

More comprehensive yet simple explanations on CTMC can be found in stochastic modeling books for beginners, such as Ross [1] for using scalar and Kleinrock [2] for using matrix. A detailed explanation of the transition from the exponential distribution to the PH distribution can be found in He [3].

The next sections describe the methods and models of CTMC.

2.1.1 Modeling CTMC

CTMC is observed when the process leaves a transient state. When leaving a state, the process has a certain probability of going to another state. The sum of all transition probabilities out of a transient state must equal 1 because it represents all possibilities, and there is no probability of staying in the transient state that is being left. If the transition rate out of a state is zero, then the process never leaves that state, and the state is referred to as an absorbing state, not a transient state. Transient states have positive transition rates out of them to indicate that the process will eventually leave transient states.

The CTMC mechanism is expressed in a matrix whose elements are the transition rates from each state to all other states. The transition rate from state i to state j is the product of the transition rate out of state i and the probability of transitioning from that state i to state j. This relationship can be written as

$$\lambda_{ij} = p_{ij}\lambda_i, \tag{2.3}$$

where

- λ_i is the rate of transition out of state i, which is the rate of the exponential distribution of the duration of stay in state i.
- p_{ij} is the probability of making a transition to state $j \neq i$ when the process leaves state i, referred to as the transition probability from state i to state j.
- λ_{ij} is the transition rate from state i to state j.

Because the process is leaving state i, obviously $p_{ii} = 0$. Therefore,

$$\sum_j p_{ij} = \sum_{j\neq i} p_{ij} = 1. \tag{2.4}$$

The matrix representing the CTMC mechanism is constructed with λ_{ij} as its elements. This matrix is called the rate transition matrix or the infinitesimal generator matrix and is denoted as $\boldsymbol{Q}$, where $\boldsymbol{Q} = [q_{ij}] = [\lambda_{ij}, i \neq j]$. Diagonal element of the matrix $\boldsymbol{Q}$ is the transition into state i, which is the negative of the transition out of state i, and the transition out of state i is the sum of all transitions from state i to all other states $j \neq i$. Hence,

$$q_{ii} = -\sum_{j\neq i} \lambda_{ij}. \tag{2.5}$$

Substituting Equation (2.3) into Equation (2.5) yields

$$q_{ii} = -\sum_{j\neq i} p_{ij}\lambda_i = -\left(\sum_{j\neq i} p_{ij}\right)\lambda_i,$$

and using Equation (2.4) yields

$$q_{ii} = -\lambda_i. \tag{2.6}$$

Using Equations (2.3)–(2.6), matrix $\boldsymbol{Q}$ is given as

$$\boldsymbol{Q} = \begin{bmatrix} -\lambda_1 & p_{12}\lambda_1 & \cdots & p_{1n}\lambda_1 \\ p_{21}\lambda_2 & -\lambda_2 & \cdots & p_{2n}\lambda_2 \\ \cdots & \cdots & \cdots & \cdots \\ p_{n1}\lambda_n & p_{n2}\lambda_n & \cdots & -\lambda_n \end{bmatrix} = \begin{bmatrix} -\lambda_1 & \lambda_{12} & \cdots & \lambda_{1n} \\ \lambda_{21} & -\lambda_2 & \cdots & \lambda_{2n} \\ \cdots & \cdots & \cdots & \cdots \\ \lambda_{n1} & \lambda_{n2} & \cdots & -\lambda_n \end{bmatrix}, \tag{2.7}$$

where n is the number of states in CTMC.

The negative value on the diagonal element (i, i) of the matrix $\boldsymbol{Q}$ indicates that the transition is into state i, while a positive value of the off-diagonal element (i, j), $j \neq i$, of the matrix $\boldsymbol{Q}$ indicates the transition is out from state i into state j.

Based on Equation (2.6), the sum of all elements in the row i is equal to $-\lambda_i + \sum_{j \neq i} p_{ij}\lambda_i$. According to Equation (2.4), $\sum_{j \neq i} p_{ij}\lambda_i = \left(\sum_{j \neq i} p_{ij}\right)\lambda_i = \lambda_i$, which means the sum of all elements in each row, or the row sum of the matrix $\boldsymbol{Q}$, is equal to zero. This result is in accordance with the universal law of flow conservation, which states that the amount of inflow is equal to the amount of outflow, so the sum of inflows and outflows is zero. The rate of leaving an absorbing state equals zero because the process never leaves absorbing states. According to Equation (2.3), the transition rate into other states from the absorbing states must also be equal to zero. Thus, all elements in the row corresponding to the absorbing states are 0's.

The characteristics of the infinitesimal generator matrix $\boldsymbol{Q}$ can be summarized as:

- Diagonal elements are nonpositive.
- Off-diagonal elements are nonnegative.
- The row sum is equal to zero.
- All elements in the rows representing absorbing states are zero.

The requirement that the row sum of $\boldsymbol{Q}$ is equal to 0 can be expressed as

$$\boldsymbol{Qu} = \boldsymbol{0}, \tag{2.8}$$

where $\boldsymbol{u}$ is a unitary column vector, a vector whose elements are all 1.

If $\boldsymbol{a}$ is an initial probability vector representing the probabilities that the process starts from all states, with the elements a_i representing the probability that the process is initially in state i, $i = 1, 2, \ldots, n$, then

$$a_i \geq 0, \sum_{i=1}^{n} a_i = 1. \tag{2.9}$$

The matrix $\boldsymbol{Q}$ is called an infinitesimal generator matrix because the probability of being in each state at any future time can be entirely determined from the values $\boldsymbol{Q}$ and probabilities of being in each state at the beginning of the process (or at present). Thus, the probability of being in each state at any time t can be determined from $\boldsymbol{a}$ and $\boldsymbol{Q}$ as $\boldsymbol{a}e^{\boldsymbol{Q}t}$. A more detailed but simple explanation of the CTMC dynamic equations in matrix form can be found in Kleinrock [2].

Example 2.1 A four-state CTMC has the following description. States 1, 2, and 3 are transient states, and state 4 is an absorbing state. The duration of stay in states 1, 2, and 3 follows exponential distribution with rates 1, 2, and 3, respectively. The transition probabilities from state 1 to states 2 and 3, respectively, are 0.6 and 0.4. The transition probabilities from state 2 to states 3 and 4, respectively, are 0.8 and 0.2. The transition probabilities from state 3 to states 2 and 4, respectively, are 0.5 and 0.5. Note that the sum of the transition probabilities from each state equals one, and no information is required about the duration of stay in the absorbing state or the transition probabilities out of the absorbing state, which in this case is state 4.

The CTMC description provides the following information:

- The transition rate from state 1 to state 2 can be calculated using Equation (2.3) as

$$\lambda_{12} = p_{12}\lambda_1 = 0.6 \times 1 = 0.6.$$

Other transition rates can be calculated accordingly.

The infinitesimal generator matrix of this CTMC using Equation (2.7) and the results of calculating λ_{ij}, $i, j = 1, 2, 3, 4$, is

$$\boldsymbol{Q} = [q_{ij}] = \begin{bmatrix} -1 & 0.6 & 0.4 & 0 \\ 0 & -2 & 1.6 & 0.4 \\ 0 & 1.5 & -3 & 1.5 \\ 0 & 0 & 0 & 0 \end{bmatrix},$$

where $q_{ij} = \lambda_{ij}$, $i, j = 1, 2, 3, 4$, $i \neq j$, are obtained using Equation (2.3), and $q_{ii} = -\lambda_i$, $i = 1, 2, 3, 4$, are obtained using Equation (2.6).

Note that the sum of the elements in each row (the row sum) is equal to zero, and all elements in the row corresponding to the absorbing state (state 4) are zero. The requirement that the sum of all elements of each row is equal to zero can be verified using Equation (2.8) as

$$\boldsymbol{Qu} = \begin{bmatrix} -1 & 0.6 & 0.4 & 0 \\ 0 & -2 & 1.6 & 0.4 \\ 0 & 1.5 & -3 & 1.5 \\ 0 & 0 & 0 & 0 \end{bmatrix} \begin{bmatrix} 1 \\ 1 \\ 1 \\ 1 \end{bmatrix} = \begin{bmatrix} 0 \\ 0 \\ 0 \\ 0 \end{bmatrix}.$$

Other information that can be obtained from the process description is:

- The duration of stay in state 1 can be determined using Equation (2.1) as

$$T_1 \sim \exp(1).$$

- When in state 1, the probability that the process is there for more than two time-units can be calculated using Equation (2.2) as

$$P[T_1 > 2] = 1 - P[T_1 \leq 2] = 1 - \left(1 - e^{-\lambda_1 \times 2}\right) = e^{-2}.$$ ■

Exercise 2.1 Calculate other λ_{ij} and show that the results match the elements of $\boldsymbol{Q}$.

Exercise 2.2 Calculate the probability that the process stays in each state for more than 2 time-units. (Hint: The duration of stay in each state is independent of the duration of stays in other states.)

Another way to represent CTMC is using state transition diagram presented next.

2.1.2 State Transition Diagram

To better understand its transition behavior, CTMC is often represented in a state transition diagram, a graph with nodes representing states and arcs representing transition probabilities or transition rates, whichever is considered easier to describe the system. If the arcs represent probabilities, then the nodes must be given information about the transition rate out of the state represented by that node (except for nodes representing absorbing states). If the arcs represent transition rates, then the nodes do not need to be given any information because it is known that the transition rate out of a state is the sum of all transition rates out of that state. The absorbing states are represented using nodes with two concentric circles to distinguish them from transient states.

Example 2.2 The CTMC in Example 2.1 can be exhibited in two different state transition diagrams, which are based on transition probability and transition rate, as shown in Figure 2.1.

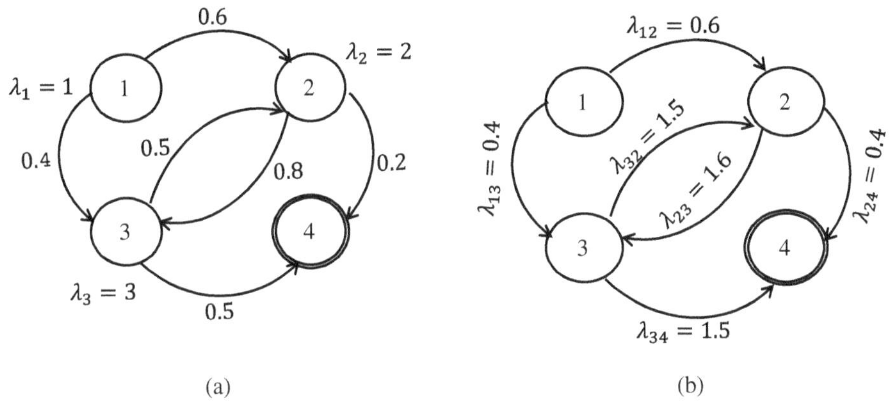

Figure 2.1 State transition diagrams of CTMC in Example 2.1. (a) Using transition probabilities and (b) using transition rates.

These two diagrams are equivalent; one diagram can produce the other. ■

Example 2.3 The transition rate from state 2 to state 3 can be calculated from Figure 2.1a to be $0.8 \times 2 = 1.6$. The transition probability from state 2 to state 3 can be calculated from Figure 2.1b to be $1.6 \div (1.6 + 0.4) = 0.8$. ■

Figure 2.1a follows the results of field observations of the system to be analyzed, while Figure 2.1b is in accordance with the infinitesimal generator matrix that will be used to analyze the system. Thus, both are needed, with Figure 2.1a obtained from field observations must be converted into Figure 2.1b to produce the infinitesimal generator matrix for the analysis.

The state transition diagram does not include the transition rate from one state to itself; therefore, the diagonal elements of the matrix $\boldsymbol{Q}$ are not readily available from the state transition diagram. If the CTMC is represented using Figure 2.1a, then each diagonal element of matrix $\boldsymbol{Q}$ is the negative of the transition rate given in each node. If the CTMC is represented using Figure 2.1b, then each diagonal element of matrix $\boldsymbol{Q}$ is the negative of the sum of all the transition rates out of the state that represents each diagonal element.

Example 2.4 Based on Figure 2.1a, the value of $q_{11} = -1$, $q_{22} = -2$, $q_{33} = -3$, $q_{44} = 0$, which is equal to the transition rate out of states 1, 2, 3, and 4, given as information in nodes 1, 2, 3, and 4, respectively. Whereas if referring to Figure 2.1b, then the value of $q_{11} = -(0.6 + 0.4) = -1$, $q_{22} = -(0.4 + 1.6) = -2$, $q_{33} = -(1.5 + 1.5) = -3$, $q_{44} = 0$, which is the negative of the sum of all transitions out of states 1, 2, 3, and 4, respectively. Since q_{ii} is the negative of the rate of the exponential distribution of the duration of stay in the transient state i, the duration of stay in states 1, 2, and 3 is each exponentially distributed with rates = 1, 2, and 3, which are the values of q_{11}, q_{22}, and q_{33}, respectively. ■

For convenience, even though it is not common in scalar analysis, the negative transition within each state in the infinitesimal generator can be added to the state transition diagram. It will be clear later that such a state transition diagram will be necessary for the state transition diagram using matrix transitions. With this addition, the state transition diagram for CTMC for Example 2.1 becomes as shown in Figure 2.2.

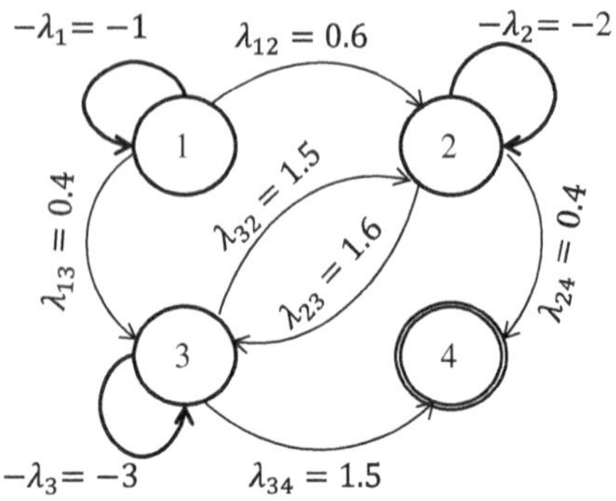

Figure 2.2 State transition diagram of CTMC in Example 2.1 using transition rate.

To be consistent with the state transition diagram using matrix transition discussed later, this book uses the state transition diagram as in Figure 2.2 for the state transition diagram using transition rate.

The next section describes a CTMC with one absorbing state. Such CTMC is important since its distribution of absorption time is directly related to PH distribution, a matrix-form distribution used extensively throughout this book.

2.2 Phase-Type Distribution

The CTMC in Example 2.1 is a CTMC with one absorbing state. CTMC with one absorbing state is the basis of the PH distribution. This does not mean PH distribution can only be used for systems that have CTMC with one absorbing state. The PH distribution can still be used for systems with more than one absorbing state by lumping all absorbing states into one superstate.

2.2.1 CTMC with One Absorbing State

By arranging the state such that the absorbing state is the last and referring to it as the $(n+1)^{th}$ state, CTMC with one absorbing state has an infinitesimal generator matrix with the following structure:

$$Q = \left[\begin{array}{cccc|c} -\lambda_1 & p_{12}\lambda_1 & \cdots & p_{1n}\lambda_1 & p_{1,n+1}\lambda_1 \\ p_{21}\lambda_2 & -\lambda_2 & \cdots & \lambda_{2n} & p_{2,n+1}\lambda_2 \\ \cdots & \cdots & \cdots & \cdots & - \\ p_{n1}\lambda_n & p_{n2}\lambda_n & \cdots & -\lambda_n & p_{n,n+1}\lambda_n \\ \hline 0 & 0 & 0 & 0 & 0 \end{array}\right],$$

if using transition probabilities, or

$$Q = \left[\begin{array}{cccc|c} -\lambda_1 & \lambda_{12} & \cdots & \lambda_{1n} & \lambda_{1,n+1} \\ \lambda_{21} & -\lambda_2 & \cdots & \lambda_{2n} & \lambda_{2,n+1} \\ \cdots & \cdots & \cdots & \cdots & - \\ \lambda_{n1} & \lambda_{n2} & \cdots & -\lambda_n & \lambda_{n,n+1} \\ \hline 0 & 0 & 0 & 0 & 0 \end{array}\right],$$

if using transition rate.

In general, the infinitesimal generator matrix of CTMC with one absorbing state can be written as

$$Q = \begin{bmatrix} \boldsymbol{A} & \boldsymbol{b} \\ \mathbf{0} & 0 \end{bmatrix}. \tag{2.10}$$

The transition matrix $\boldsymbol{A}$ is called the transient matrix, which expresses the transition rates among transient states, and the transition vector $\boldsymbol{b}$ is called the absorbing vector representing the transition rates from all transient states to the absorbing state.

Using Equation (2.8) yields

$$\boldsymbol{Qu} = \begin{bmatrix} \boldsymbol{A} & \boldsymbol{b} \\ \mathbf{0} & 0 \end{bmatrix} \boldsymbol{u} = \begin{bmatrix} \boldsymbol{A} & \boldsymbol{b} \\ \mathbf{0} & 0 \end{bmatrix} \begin{bmatrix} \boldsymbol{u} \\ 1 \end{bmatrix} = \mathbf{0}.$$

Hence,

$$\boldsymbol{Au} + \boldsymbol{b} = \mathbf{0}, \tag{2.11}$$

or

$$\boldsymbol{b} = -\boldsymbol{Au}. \tag{2.12}$$

Example 2.5 The transient matrix and absorbing vector of CTMC in Example 2.1 are

$$\boldsymbol{A} = \begin{bmatrix} -1 & 0.6 & 0.4 \\ 0 & -2 & 1.6 \\ 0 & 1.5 & -3 \end{bmatrix}, \quad \boldsymbol{b} = \begin{bmatrix} 0 \\ 0.4 \\ 1.5 \end{bmatrix}.$$

It can be verified that Equation (2.12) is satisfied,

$$\boldsymbol{Au} = \begin{bmatrix} -1 & 0.6 & 0.4 \\ 0 & -2 & 1.6 \\ 0 & 1.5 & -3 \end{bmatrix} \begin{bmatrix} 1 \\ 1 \\ 1 \end{bmatrix} = \begin{bmatrix} 0 \\ -0.4 \\ -1.5 \end{bmatrix} = -\boldsymbol{b}.$$

■

Based on the properties of $\boldsymbol{Q}$, matrix $\boldsymbol{A}$ is a matrix with nonpositive diagonal elements and nonnegative off-diagonal elements. In addition, the sum of elements in each row (each row sum) is nonpositive because elements of vector $\boldsymbol{b}$ that are always nonnegative are not included as elements of $\boldsymbol{A}$. Transient matrices with these characteristics are always nonsingular (have an inverse) [4].

2.2.2 Distribution of Absorption Time

CTMC with absorbing states will eventually enter the absorbing states. An important variable in a CTMC with one absorbing state is the waiting time for the process starting in any state until it enters the absorbing state. This waiting time is a random variable referred to as the absorption time. The PH distribution is defined as the distribution of absorption times in a CTMC with one absorbing state. If T denotes the absorption time for a CTMC with $\boldsymbol{Q}$ as given by Equation (2.10), then the CDF of T is [4]

$$F_T(t) = P[T \leq t] = 1 - \boldsymbol{a}e^{\boldsymbol{A}t}\boldsymbol{u} \quad t \geq 0, \tag{2.13}$$

where $\boldsymbol{a}$ is the initial probability vector, which represents the probability that the process starts from each transient state. If the CTMC has one state, then $\boldsymbol{a} = \boldsymbol{u} = 1$ and Equation (2.13) reduces to Equation (2.2), the CDF of exponential distribution. This observation is in accordance with the definition of PH distribution, which is a generalization of the exponential distribution for the process underlying the randomness of a variable that requires more than one state to describe.

Since the initial probability starting from the absorbing state is not included as an element of $\boldsymbol{a}$, the sum of elements of $\boldsymbol{a}$ is

$$\boldsymbol{au} = 1 - a_{n+1} \leq 1, \tag{2.14}$$

where a_{n+1} is the probability that the process starts from the absorbing state. The overall initial probability vector of the process is given by the vector $[\boldsymbol{a} \;\; a_{n+1}]$. Starting a process from the absorbing state is indeed rarely encountered in practice; however, for completeness, the PH distribution includes the possibility that the process starts from the absorbing state. Starting from the absorbing state means that the absorption time is zero with a positive probability, equal to the probability that the process starts from the absorbing state. Thus, the absorption time has a discrete component at $t = 0$ as

$$P[T = 0] = a_{n+1}. \tag{2.15}$$

Because of this discrete component, the CDF is not always continuous in the domain of definition $t \geq 0$; it has a jump of size a_{n+1} at $t = 0$. In other words, T is a mixture of discrete and continuous random variables.

PH distribution can be expressed in a state transition diagram with two states, namely, the transient state and the absorbing state. The transient state is a superstate representing many transient states combined. There is a transition from the transient superstate to itself, which is a consolidation of several transitions between pairs of transient states. This transition is expressed as a transition matrix among transient states, which, according to Equation (2.10), is the transient matrix $\boldsymbol{A}$. This is the reason that state transition diagrams with scalar transitions are complemented with a transition rate from each state to itself to make them equivalent to the state transition diagram with matrix transitions. The transient matrix $\boldsymbol{A}$ has a dimension $n \times n$ that represents the transition from n transient states to n transient states. There are transitions from all states in the transient superstate to one absorbing state. These transitions are consolidated in a transition vector, which, according to Equation (2.10), is the absorbing vector $\boldsymbol{b}$. The absorbing vector $\boldsymbol{b}$ has a dimension $n \times 1$ that represents the transition from n transient states to 1 absorbing state.

The PH distribution with transient matrix $\boldsymbol{A}$ and absorbing vector $\boldsymbol{b}$ has a state transition diagram exhibited in Figure 2.3.

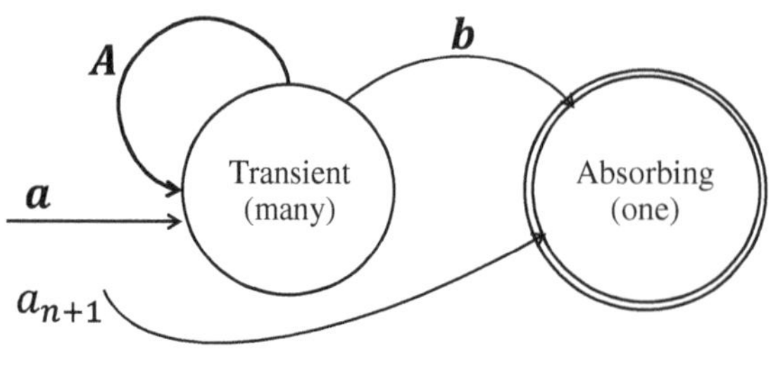

Figure 2.3 State transition diagram of PH distribution.

Example 2.6 The PH distribution for the absorption time of CTMC in Example 2.1 can be depicted in the state transition diagram using matrix transition shown in Figure 2.4 (initial probability vector $\boldsymbol{a}$ is not yet known).

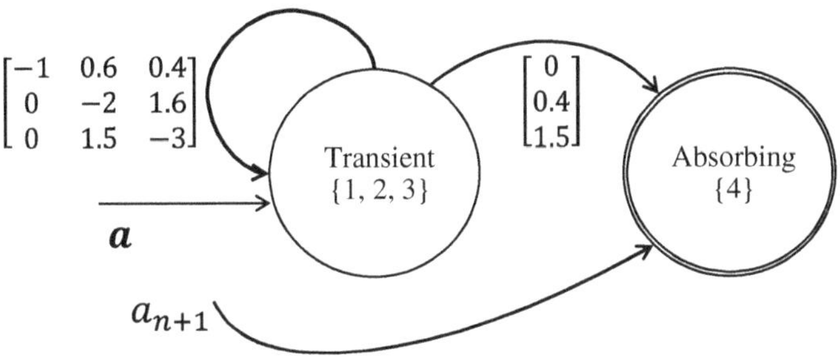

Figure 2.4 PH distribution for the absorption time of CTMC in Example 2.1. ■

Variable T with a distribution given by Equation (2.13) is said to have a PH distribution with representation $(n, \boldsymbol{a}, \boldsymbol{A})$; it is sometimes called a "PH distribution of order n with representation $(\boldsymbol{a}, \boldsymbol{A})$." The number of transient states n, which are the dimensions of the matrix $\boldsymbol{A}$ and vector $\boldsymbol{a}$, is referred to as the order of the PH distribution. Its mathematical notation is $T \sim PH_n(\boldsymbol{a}, \boldsymbol{A})$ or $T \sim PH(n, \boldsymbol{a}, \boldsymbol{A})$. Sometimes, the order is not mentioned, so it is simply written as $T \sim PH(\boldsymbol{a}, \boldsymbol{A})$. Without mentioning the order does not create confusion since such information is implicitly available from the dimensions of $\boldsymbol{A}$ and $\boldsymbol{b}$. This book uses $PH(n, \boldsymbol{a}, \boldsymbol{A})$ as the representation of the PH distribution to inform its order without finding it from the dimension of $\boldsymbol{A}$.

PH distribution has a discontinuity at the origin ($t = 0$) because using Equations (2.13)–(2.14) yields

$$F_T(0) = 1 - \boldsymbol{a}e^{\boldsymbol{0}}\boldsymbol{u} = 1 - \boldsymbol{a}\boldsymbol{u} = a_{n+1}.$$

The magnitude of the discontinuity jump at the origin is equal to the probability of starting from the absorbing state. This value is the probability of the absorption time being zero because if the process starts from the absorbing state, the time to enter the absorbing state is naturally zero. Following Equation (2.15),

$$F_T(0) = P[T \leq 0] = P[T = 0] = a_{n+1}.$$

The probability density function (PDF) is the derivative of the CDF given by Equation (2.13). Hence,

$$f_T(t) = \frac{dF_T(t)}{dt} = -\boldsymbol{a}\boldsymbol{A}e^{\boldsymbol{A}t}\boldsymbol{u} + a_{n+1}\delta(t) \quad t \geq 0, \tag{2.16}$$

where $\delta(t)$ is the impulse function at the point of origin $t = 0$ due to the discontinuity at that point. The magnitude of this impulse is equal to the probability that the absorption time is equal to zero; that is, the probability that the process starts from the absorbing state a_{n+1}. The matrix $\boldsymbol{A}$ in Equation (2.16) can be a post- or premultiplier of $e^{\boldsymbol{A}t}$, whichever is more convenient for the analysis.

Calculating CDF and PDF of PH distributions and plotting their graphs are not possible manually since they involve matrix exponential function. They must be calculated with the help of a computer. The MATLAB functions **phcdfplot** and **phpdfplot** in the software package **dysrel-am** can be used to calculate and plot the graphs of CDF and PDF of any PH distribution.

Depending on the structure of transient matrix $\boldsymbol{A}$, PH distribution is classified into two types as described next.

2.2.3 Types of PH Distribution

There are two types of PH distributions. One is acyclic PH distribution, abbreviated as APH distribution, and the other is cyclic PH distribution, abbreviated as CPH distribution. Given that the absorbing state is the last in the sequence of states, the APH distribution is a PH distribution with CTMC having transitions only from states earlier in sequence to states later in sequence. Its state transition diagram has transitions in the same forward direction, with no transitions in the opposite or backward directions. Therefore, the transient matrix of APH distribution is an upper diagonal matrix, with all elements below the diagonal being zero. This structure can speed up the calculation of its CDF. The APH distribution is found in engineering systems without maintenance intervention, so its condition can only worsen over time, before it finally fails. The transitions in CPH distribution can be to states earlier or later in the sequence. Its state transition diagram has one or more transitions in the opposite direction of the other transitions; hence, its transient matrix is a full matrix. CPH distribution can be found in service systems that require process repetition or systems with direct maintenance while in operation. Because its transition matrix is a full matrix, the computation time for calculating the CDF of CPH distribution is longer than APH distribution.

Example 2.7 The graphs of CDF and PDF for the absorption time of the CTMC in Example 2.1 with two different initial probability vectors are presented in Figure 2.5.

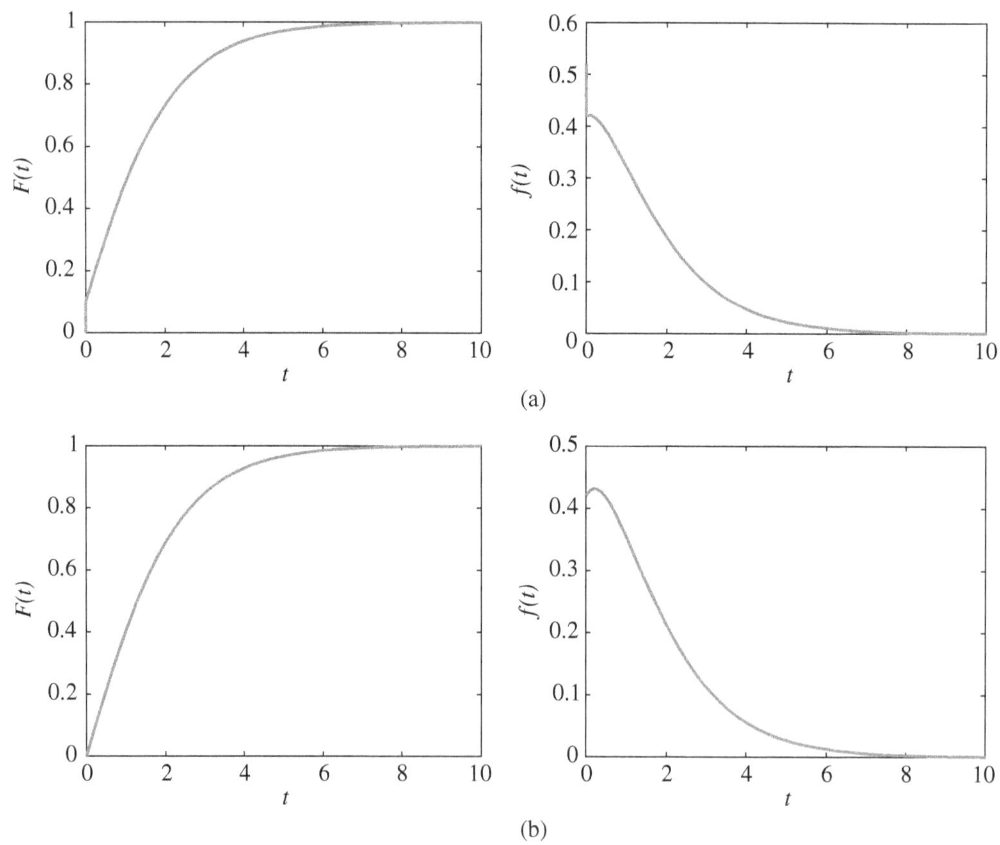

Figure 2.5 CDF and PDF of absorption time of CTMC in Example 2.1. (a) $\boldsymbol{a} = [0.4\ \ 0.3\ \ 0.2]$ and (b) $\boldsymbol{a} = [0.5\ \ 0.3\ \ 0.2]$.

Note that the CDF and PDF in Figure 2.5a have a jump of 0.1 at point $t = 0$, which is the probability that the process starts from the absorbing state. The value of this jump can be calculated from the sum of the probabilities of the process starting from the transient states, which is $\boldsymbol{au} = 0.4+0.3+0.2 = 0.9 = 1-a_4$. The deficiency $1-0.9 = 0.1$ is the value of a_4, which is the probability that the process starts from the absorbing state. Thus, the CDF has a jump of size 0.1 at the origin. ■

The main program to call **phcdfplot** and **phpdfplot** for producing Figure 2.5 is provided in **example2_7.m** file in the software package **dysrel-am**. Its source code is as follows:

MATLAB code for example2_7.m

```
close all;clear
addpath('..\dysrel-am','..\dysrel-am\expokit')
% PH Component
T.a = [0.4 0.3 0.2]; % With a jump
% T.a = [0.5 0.3 0.2]; % Without jump
T.A = [-1 0.6 0.4;0 -2 1.6;0 1.5 -3];
T.n = 3;
% Plot
tmax = 10;
t = linspace(0,tmax,41);
[tt,fs] = phpdfplot(t,T);
[tt,Fs] = phcdfplot(t,T);
```

The main programs for all examples and applications in Chapter 2 are provided in **example2_x.m** or **application2_x** files, where x is the example or application number.

Example 2.8 State transition diagrams using transition probabilities and transition rates in Figure 2.6 represent the same second-order hypoexponential distribution (2-hypoexponential). Since all transitions are in one direction, the hypoexponential distribution is an APH distribution.

The following quantities can be derived from Figure 2.6 and Equation (2.10) as

$$\boldsymbol{Q} = \left[\begin{array}{cc|c} -0.1 & 0.1 & 0 \\ 0 & -0.2 & 0.2 \\ \hline 0 & 0 & 0 \end{array}\right], \quad \boldsymbol{A} = \begin{bmatrix} -0.1 & 0.1 \\ 0 & -0.2 \end{bmatrix}, \quad \boldsymbol{b} = \begin{bmatrix} 0 \\ 0.2 \end{bmatrix}, \quad \boldsymbol{a} = [1 \;\; 0].$$

The CDF and PDF graphs of the absorption time of the process with the state transition diagram in Figure 2.6 are presented in Figure 2.7. Both graphs do not have jumps at $t = 0$ because the probability of starting from the absorbing state is $1 - \boldsymbol{au} = 0$.

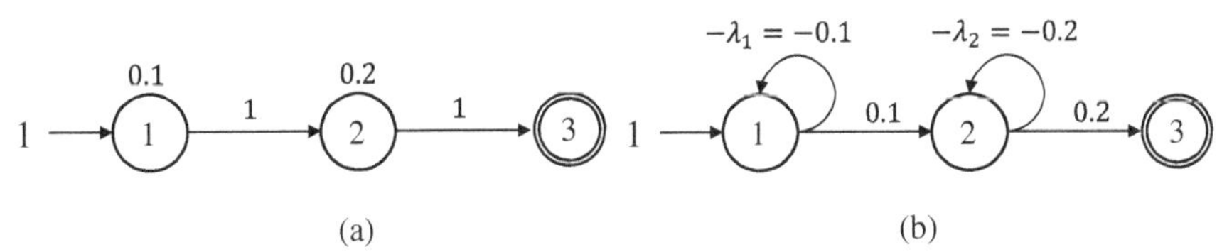

Figure 2.6 State transition diagram of a 2-hypoexponential distribution. (a) Using transition probabilities and (b) using transition rates.

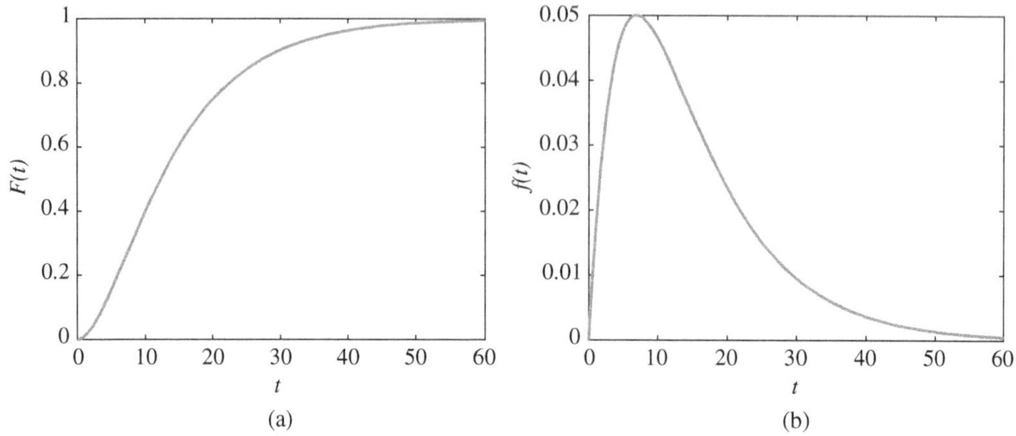

Figure 2.7 CDF and PDF of the APH distribution of Example 2.8. (a) CDF and (b) PDF. ∎

Example 2.9 State transition diagrams using transition probabilities and transition rates in Figure 2.8 contain transitions in the opposite direction. Thus, the absorption time in the underlying CTMC has CPH distribution.

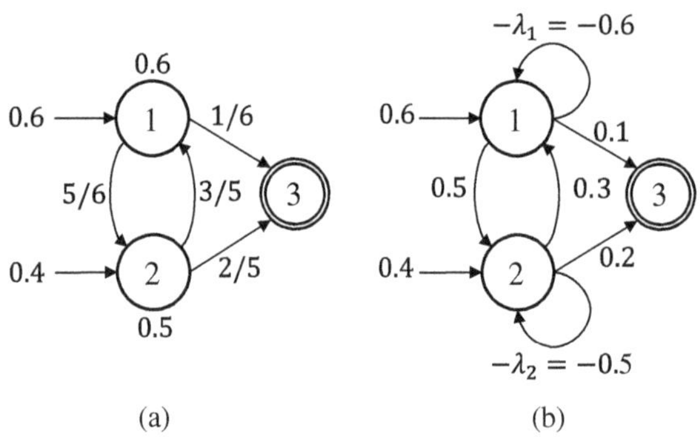

Figure 2.8 State transition diagram for the CPH distribution of Example 2.9. (a) Using transition probabilities and (b) using transition rates.

The following quantities can be derived from Figure 2.8 and Equation (2.10) as

$$\boldsymbol{Q} = \left[\begin{array}{cc|c} -0.6 & 0.5 & 0.1 \\ 0.3 & -0.5 & 0.2 \\ \hline 0 & 0 & 0 \end{array}\right], \quad \boldsymbol{A} = \begin{bmatrix} -0.6 & 0.5 \\ 0.3 & -0.5 \end{bmatrix}, \quad \boldsymbol{b} = \begin{bmatrix} 0.1 \\ 0.2 \end{bmatrix}, \quad \boldsymbol{a} = \begin{bmatrix} 0.6 & 0.4 \end{bmatrix}.$$

The graphs of CDF and PDF of the absorption time of CTMC in Figure 2.8 are presented in Figure 2.9. Both graphs also do not have a jump at the starting point because $1 - \boldsymbol{au} = 0$. ∎

Exercise 2.3 Compare the computing time to produce 100 points of Figures 2.7 and 2.9.

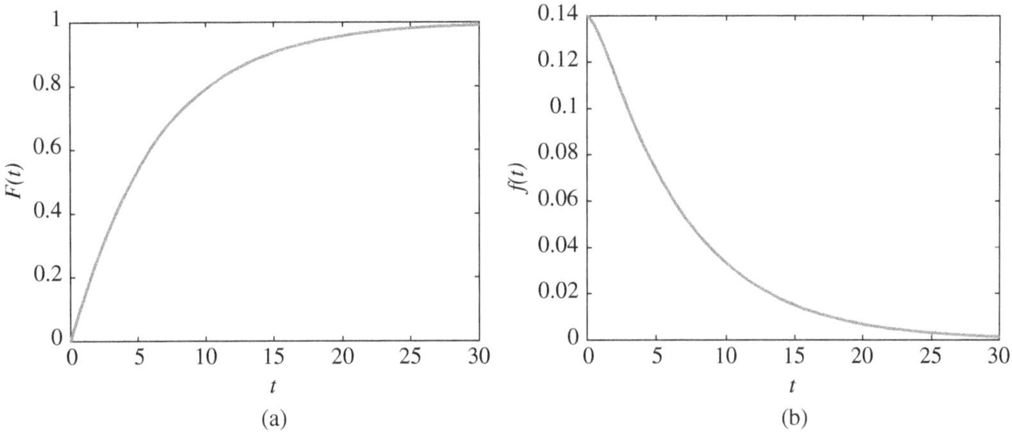

Figure 2.9 CDF and PDF of absorption time on CTMC in Example 2.9. (a) CDF and (b) PDF.

Application 2.1 The condition of the radar sensor is continuously monitored. The monitoring results show that its condition can be classified into four states: Excellent, Decent, Fair, and Broken. The equipment is initially in the Excellent state. The duration of being in the Excellent state has an exponential distribution with a rate of 0.5/year before finally changing to the Decent state. The duration of being in the Decent state has an exponential distribution with a rate of 1/year. Out from the Decent state, the equipment can be in the Broken state with a probability of 0.2; otherwise, its state changes to the Fair state. The duration of being in the Fair state also has an exponential distribution with a rate of 2/year before it finally enters Broken state. Broken equipment will remain broken forever (until it is repaired).

The process underlying the changes in the equipment conditions, as described, is a CTMC with three transient states (Excellent, Decent, and Fair) and one absorbing state (Broken). The CTMC can be represented in a state transition diagram using transition probabilities as given in Figure 2.10a or in a state transition diagram using transition rates as given in Figure 2.10b. It can be seen that the transition in the CTMC is unidirectional, from better states to worse states. Thus, the duration of use of the equipment from the initial state until failure has an APH distribution. Since the equipment is initially known to be in Excellent state, the initial probability of starting from the Excellent state is 1; thus, the initial probability of starting from other states must be 0.

The infinitesimal generator matrix and initial probability vector of the CTMC underlying this equipment states can be obtained using Figure 2.10a or b as

$$\boldsymbol{Q} = \left[\begin{array}{ccc|c} -0.5 & 0.5 & 0 & 0 \\ 0 & -1 & 0.8 & 0.2 \\ 0 & 0 & -2 & 2 \\ \hline 0 & 0 & 0 & 0 \end{array}\right], \quad \boldsymbol{a} = [\boldsymbol{a}\ a_{n+1}] = [[1\ 0\ 0]\ 0].$$

The duration of use until the equipment fails is the same as the absorption time in the CTMC underlying the change in the equipment states. Equation (2.10) tells the duration of use until failure

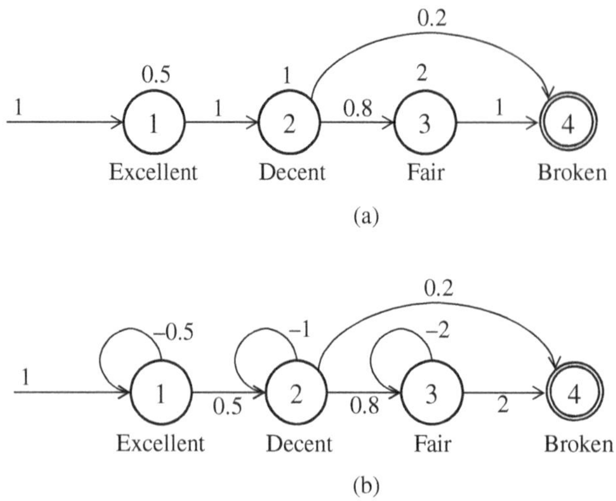

Figure 2.10 State transition diagram of the radar in Application 2.1. (a) Using transition probabilities and (b) using transition rate.

for this equipment has a PH distribution with a representation

$$n = 3, \quad \boldsymbol{a} = \begin{bmatrix} 1 & 0 & 0 \end{bmatrix}, \quad \boldsymbol{A} = \begin{bmatrix} -0.5 & 0.5 & 0 \\ 0 & -1 & 0.8 \\ 0 & 0 & -2 \end{bmatrix}.$$

Suppose that for safety reasons management considers the Fair state as bad as broken and must be repaired. This provision causes the CTMC underlying the radar states to have two absorbing states: the Fair state, which is considered fail, and the Broken state. The state transition diagram for this equipment becomes as shown in Figure 2.11a. To find the PH distribution for the duration of use of the radar from its initial state until it fails, whatever the definition of failure, the two absorbing states representing Fair state and Broken state must be lumped into one absorbing state. Combining

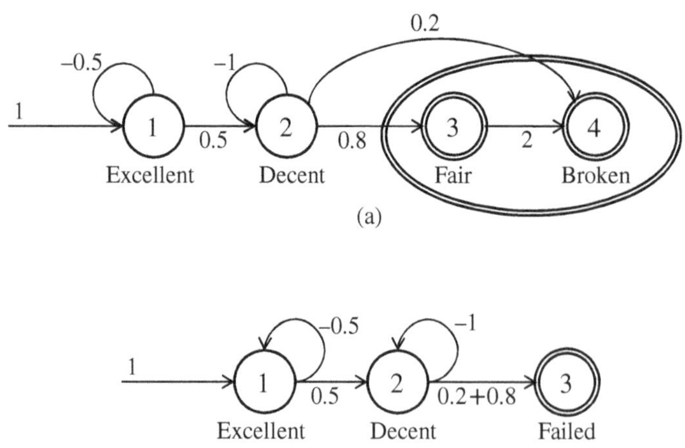

Figure 2.11 State transition diagram of the radar in Application 2.1 with one and two absorbing states. (a) State transition diagram of the radar with two absorbing states and (b) state transition diagram of the radar with one absorbing state.

the two absorbing states into one produces a new state transition diagram given in Figure 2.11b. The transition rate from state 2 to state 3 in Figure 2.11b is the sum of the transition rates from state 2 to states 3 and 4 in Figure 2.11a.

The infinitesimal generator matrix of the CTMC underlying the radar states becomes

$$\boldsymbol{Q} = \left[\begin{array}{cc|c} -0.5 & 0.5 & 0 \\ 0 & -1 & 1 \\ \hline 0 & 0 & 0 \end{array}\right].$$

The PH distribution for the duration of use until the radar fails has a representation

$$n = 2, \quad \boldsymbol{a} = \begin{bmatrix} 1 & 0 \end{bmatrix}, \quad \boldsymbol{A} = \begin{bmatrix} -0.5 & 0.5 \\ 0 & -1 \end{bmatrix}.$$

■

Exercise 2.4 Compare the graphs of the CDFs of absorption times for the cases of one and two absorbing states. Conclude and explain the result.

Application 2.2 The repair process in a workshop is as follows: The service begins with an initial inspection, the duration of which is exponentially distributed at a rate of 10/day. The initial inspection classifies the repair into major and minor repairs with a ratio of 1:4. Major repair requires two stages, with the first stage using a machine that requires exponentially distributed service times at a rate of 2/day. The second stage is sufficiently handled by mechanics who require an exponentially distributed service time at a rate of 5/day. Minor repair only requires the second stage, which has the same distribution of repair time as the second stage of major repair. All items that have completed all stages of repair must go through a final inspection. The duration of the final inspection is exponentially distributed at a rate of 10/day. The result of the final inspection is that 10% of them repeat stage 2 repairs, regardless of its previous inspection result, and the rest are declared finished. The repair work is carried out sequentially without interruption until it is finished; the next item starts after the current one is completed.

This repair process can be expressed in CTMC with four transient states: initial inspection, first stage repair, second stage repair, and final inspection. The finished state is the absorbing state. The state transition diagrams using transition probability and transition rate are given in Figure 2.12, which shows that the service completion time has a CPH distribution. Since all repairs must start from the initial inspection, the initial probability of starting from the initial inspection state is 1 and from other states is 0.

Based on Figure 2.12, the repair time in this workshop follows a CTMC with an infinitesimal generator matrix

$$\boldsymbol{Q} = \left[\begin{array}{cccc|c} -10 & 2 & 8 & 0 & 0 \\ 0 & -2 & 2 & 0 & 0 \\ 0 & 0 & -5 & 5 & 0 \\ 0 & 0 & 1 & -10 & 9 \\ \hline 0 & 0 & 0 & 0 & 0 \end{array}\right].$$

Hence, the PH distribution of the repair time in this workshop has a representation

$$n = 4, \quad \boldsymbol{a} = \begin{bmatrix} 1 & 0 & 0 & 0 \end{bmatrix}, \quad \boldsymbol{A} = \begin{bmatrix} -10 & 2 & 8 & 0 \\ 0 & -2 & 2 & 0 \\ 0 & 0 & -5 & 5 \\ 0 & 0 & 1 & -10 \end{bmatrix}.$$

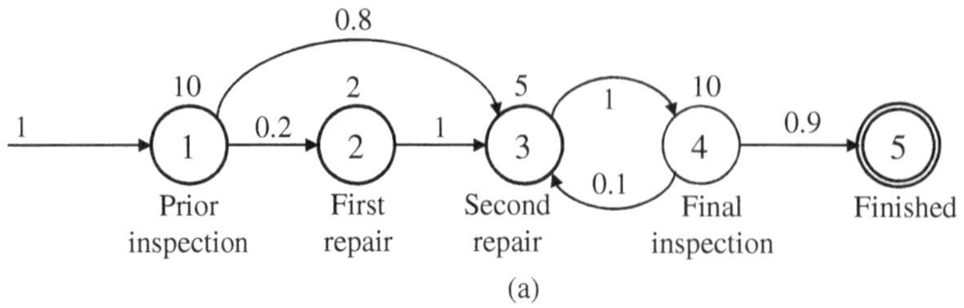

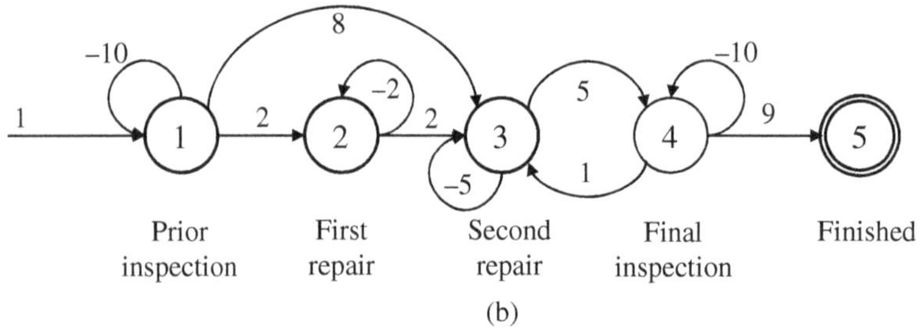

Figure 2.12 State transition diagram of Application 2.2. (a) Using transition probability and (b) using transition rate. ■

Exercise 2.5 The inspection results are 10% must repeat from the first repair, 20% must repeat from the second repair, and the rest do not need to repeat the repair. Calculate the CDF and compare the result with the CDF of Application 2.2. Make a conclusion from the result.

The randomness of absorption time in CTMC can be represented in numbers representing its statistics. These statistics can be calculated from the moments of PH distribution as described in the next section.

2.3 Moments of Random Variables in PH Distribution

Moments are numbers used to measure the randomness of a variable. It is convenient for people with difficulty interpreting CDF. However, the information contained in moments is not as rich as the information contained in the CDF.

Random variables have many moments. The kth-moment of a random variable, $k = 1, 2, \ldots$, is defined as the expectation of the random variable raised to the power k. The expectation of a function of continuous random variable T using PDF or CDF, as PDF is the differential of CDF, is given by

$$E[g(T)] = \int_{-\infty}^{\infty} g(t)f_T(t)dt = \int_{-\infty}^{\infty} g(t)dF_T(t). \tag{2.17}$$

The kth moment, by its definition, can be found using Equation (2.17) as

$$E[T^k] = \int_{-\infty}^{\infty} t^k f_T(t)dt = \int_{-\infty}^{\infty} t^k dF_T(t), \quad k = 1, 2, \ldots. \tag{2.18}$$

The first moment is called the expected value or mean of the random variable, which can be calculated using Equation (2.18) for $k = 1$,

$$E[T] = \int_{-\infty}^{\infty} tf_T(t)dt = \int_{-\infty}^{\infty} tdF_T(t).$$

For $T \sim PH(n, \boldsymbol{a}, \boldsymbol{A})$, using Equation (2.16) yields

$$E[T] = \int_0^{\infty} t(-\boldsymbol{a}\boldsymbol{A}e^{\boldsymbol{A}t}\boldsymbol{u} + a_{n+1}\delta(t))dt. \tag{2.19}$$

The lower bound of the integral becomes 0 because PH distribution is for absorbtion times which is nonnegative random variables. Given that t is a scalar and $\boldsymbol{a}$, $\boldsymbol{A}$, and $\boldsymbol{u}$ are constant, Equation (2.19) can be rewritten as

$$E[T] = -\boldsymbol{a}\left(\int_0^{\infty} t\boldsymbol{A}e^{\boldsymbol{A}t}dt\right)\boldsymbol{u} + a_{n+1}\int_0^{\infty} t\delta(t)dt.$$

The impulse integral in the second term on the right-hand side is equal to the value of the function to be integrated (which is t), at the impulse point $t = 0$; therefore, the result for the second term is equal to 0. Substituting $\boldsymbol{A}e^{\boldsymbol{A}t}dt = de^{\boldsymbol{A}t}$ and using partial integration obtain

$$E[T] = -\boldsymbol{a}\left(\int_0^{\infty} tde^{\boldsymbol{A}t}\right)\boldsymbol{u} = -\boldsymbol{a}\left(te^{\boldsymbol{A}t} - \int_0^{\infty} e^{\boldsymbol{A}t}dt\right)\boldsymbol{u} = -\boldsymbol{a}(te^{\boldsymbol{A}t} - \boldsymbol{A}^{-1}e^{\boldsymbol{A}t})\Bigg|_{t=0}^{t=\infty}\boldsymbol{u}.$$

Since the limit $te^{\boldsymbol{A}t} \to 0$ for $t \to \infty$, the first moment is obtained as

$$E[T] = -\boldsymbol{a}\boldsymbol{A}^{-1}\boldsymbol{u}. \tag{2.20}$$

Finding higher moments requires more effort when using integration. A more efficient way is to use the Laplace–Stieltjes transform $F^*(s)$ of the CDF defined as

$$F_T^*(s) = \int_0^{\infty} e^{-st}dF_T(t). \tag{2.21}$$

Substituting Equation (2.16) into Equation (2.21) yields

$$\begin{aligned}
F_T^*(s) &= \int_0^{\infty} e^{-st}(-\boldsymbol{a}e^{\boldsymbol{A}t}\boldsymbol{A}\boldsymbol{u} + a_{n+1}\delta(t))dt \\
&= -\boldsymbol{a}\left(\int_0^{\infty} e^{-(s\boldsymbol{I}-\boldsymbol{A})t}dt\right)\boldsymbol{A}\boldsymbol{u} + a_{n+1}\int_0^{\infty} e^{-st}\delta(t)dt \\
&= \boldsymbol{a}(s\boldsymbol{I} - \boldsymbol{A})^{-1}\left(e^{-(s\boldsymbol{I}-\boldsymbol{A})t}\Bigg|_{t=0}^{t=\infty}\right)\boldsymbol{A}\boldsymbol{u} + a_{n+1}e^{-st}\big|_{t=0} \\
&= \boldsymbol{u}(s\boldsymbol{I} - \boldsymbol{A})^{-1}(\boldsymbol{0} - \boldsymbol{I})\boldsymbol{A}\boldsymbol{u} \mid a_{n+1}.
\end{aligned}$$

Thus,

$$F_T^*(s) = -\boldsymbol{a}(s\boldsymbol{I} - \boldsymbol{A})^{-1}\boldsymbol{A}\boldsymbol{u} + a_{n+1}. \tag{2.22}$$

Equation (2.21) can be interpreted as the expectations of $g(T) = e^{-sT}$ in Equation (2.17). Thus,

$$E[e^{-sT}] = \int_0^{\infty} e^{-st}dF_T(t) = F_T^*(s). \tag{2.23}$$

Substituting Equation (2.22) into Equation (2.23) yields

$$E[e^{-sT}] = F_T^*(s) = -\boldsymbol{a}(s\boldsymbol{I} - \boldsymbol{A})^{-1}\boldsymbol{A}\boldsymbol{u} + a_{n+1}.$$

Its first derivative with respect to s is

$$\frac{dE[e^{-sT}]}{ds} = E[-Te^{-sT}] = F_T^*(s)' = \boldsymbol{a}(s\boldsymbol{I} - \boldsymbol{A})^{-2}\boldsymbol{A}\boldsymbol{u}.$$

For $s = 0$, $e^{-sT} = 1$. Therefore,

$$E[T] = E[Te^{-sT}]\,|_{s=0} = -F_T^*(0)' = -\boldsymbol{a}(s\boldsymbol{I} - \boldsymbol{A})^{-2}\boldsymbol{A}\boldsymbol{u}\,|_{s=0} = -\boldsymbol{a}\boldsymbol{A}^{-1}\boldsymbol{u},$$

the same result as using integration given by Equation (2.20).

Equations for higher moments can be obtained by differentiating $E[e^{-sT}]$ several times and substituting $s = 0$.

The second moment is

$$E[T^2] = E[(-T)^2 e^{-sT}]\,|_{s=0} = F_T^*(0)'' = -2\boldsymbol{a}(s\boldsymbol{I} - \boldsymbol{A})^{-3}\boldsymbol{A}\boldsymbol{u}\,|_{s=0} = 2\boldsymbol{a}\boldsymbol{A}^{-2}\boldsymbol{u}.$$

Continuing in this manner produces

$$E[T^k] = (-1)^k k!(\boldsymbol{a}\boldsymbol{A}^{-k}\boldsymbol{u}), \quad k = 1, 2, \ldots. \tag{2.24}$$

The variance of T can be found from the first and second moments as

$$var(T) = E[T^2] - E[T]^2 = 2\boldsymbol{a}\boldsymbol{A}^{-2}\boldsymbol{u} - (\boldsymbol{a}\boldsymbol{A}^{-1}\boldsymbol{u})^2. \tag{2.25}$$

Standard deviation $sd(T)$ is the square root of the variance, and the coefficient of variation $cv(T)$ is the standard deviation divided by the mean,

$$sd(T) = \sqrt{var(T)}; \tag{2.26}$$

$$cv(T) = \frac{sd(T)}{E[T]}. \tag{2.27}$$

Manual calculations of statistics of random variables with PH distributions are difficult. For this purpose, a MATLAB function **phstat** is provided in the software package.

Example 2.10 The mean, variance, standard deviation, and coefficient of variation of random variables in Example 2.7b can be calculated using its PH representation as follows:

$$\boldsymbol{a} = \begin{bmatrix} 0.5 & 0.3 & 0.2 \end{bmatrix}, \quad \boldsymbol{A} = \begin{bmatrix} -1 & 0.6 & 0.4 \\ 0 & -2 & 1.6 \\ 0 & 1.5 & -3 \end{bmatrix}.$$

$$E[T] = -\boldsymbol{a}\boldsymbol{A}^{-1}\boldsymbol{u} = -\begin{bmatrix} 0.5 & 0.3 & 0.2 \end{bmatrix} \begin{bmatrix} -1 & 0.6 & 0.4 \\ 0 & -2 & 1.6 \\ 0 & 1.5 & -3 \end{bmatrix}^{-1} \begin{bmatrix} 1 \\ 1 \\ 1 \end{bmatrix} = 1.66;$$

$$E[T^2] = 2\boldsymbol{a}\boldsymbol{A}^{-2}\boldsymbol{u} = 2\begin{bmatrix} 0.5 & 0.3 & 0.2 \end{bmatrix} \begin{bmatrix} -1 & 0.6 & 0.4 \\ 0 & -2 & 1.6 \\ 0 & 1.5 & -3 \end{bmatrix}^{-2} \begin{bmatrix} 1 \\ 1 \\ 1 \end{bmatrix} = 4.81;$$

$$var(T) = E[T^2] - E[T]^2 = 2.07;$$

$$sd(T) = \sqrt{var(T)} = 1.44;$$

$$cv(T) = \frac{sd(T)}{E[T]} = 0.87.$$

■

The main program to call **phstat** for calculating these statistics is given in **example2_10.m** file in the software package **dysrel-am**. Its code is as follows:

MATLAB code for example2_10.m

```
close all;clear
addpath('..\dysrel-am','..\dysrel-am\expokit')
% PH
T.a = [0.5 0.3 0.2];
T.A = [-1 0.6 0.4; 0 -2 1.6; 0 1.5 -3];
T.n = 3;
[mttf,vttf,sdttf,cvttf] = phstat(T)
```

Exercise 2.6 Calculate the mean and variance of the time to failure of the equipment in Application 2.1 using **phstat**. Compare the result with manual calculation using the sum of the mean duration of stay in each state before failing.

The next section shows that PH distribution is a general distribution for which some popular distributions are special cases of PH distribution.

2.4 Popular Members of the PH Distribution Class

The PH distribution is a class of distributions. Several popular distributions are special forms of the PH distribution; therefore, they are members of the PH distribution class. Basically, all distributions formed from the exponential distribution as the results of mixture, addition, minimum, and maximum operations are in the PH distribution class. All distributions in the PH distribution class can be expressed in various representations, such as diagrams, scalar functions, and matrix functions. Table 2.1 provides widely used probability models as members of the PH distribution class in various representations. In this table, the scalar representations are CDFs in scalar functions. The scalar representation for the Coxian distribution is given only for order 3 because its general form is complicated. For practical purposes, the table also provides equations for the mean and variance of each random variable.

It is apparent from Table 2.1 that most scalar representations are more complicated than their matrix representations, especially for distributions that have many transitions, such as Coxian. There are still other distributions in the PH distribution class that are not mentioned in Table 2.1. Basically, distributions derived from the addition, minimum, maximum, mixing, and sorting of several PH distributions will produce PH distributions. Table 2.1 shows that hyperexponential results from mixing exponential distributions, Erlang and hypoexponential are the results of adding exponential distributions, and Coxian is the result of sorting and adding exponential distributions. The processes of sorting, adding, and mixing can be applied to any PH distribution, and the results are still PH distributions. For example, hyper-Erlang and hypo-Erlang, which, respectively, are the results of mixing and adding several Erlang-distributed random variables, are also members of the PH distribution class. The properties of the PH distribution with respect to several operations are explained mathematically in Section 2.7.

Table 2.1 Popular probability models members of the PH distribution class.

No.	Distribution		Representation
1.	Exponential (λ)	Diagram	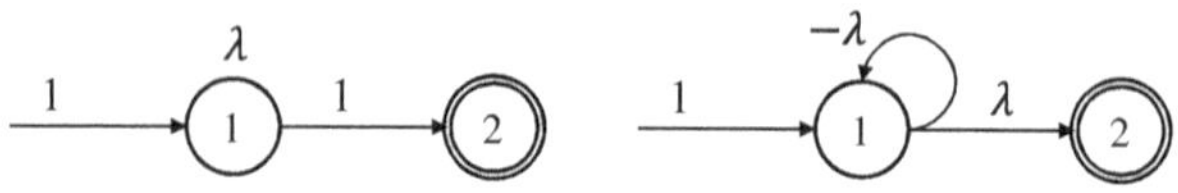
		Scalar	$1 - e^{-\lambda t}$
		Matrix	$n = 1 \quad \boldsymbol{a} = [1] \quad \boldsymbol{A} = [-\lambda]$
		Mean	$\frac{1}{\lambda}$
		Variance	$\frac{1}{\lambda^2}$
2.	n-Hyperexponential (p_i, λ_i) $0 < p_i < 1$ $\sum_{i=1}^{n} p_i = 1$ (mixture exponential)	Diagram	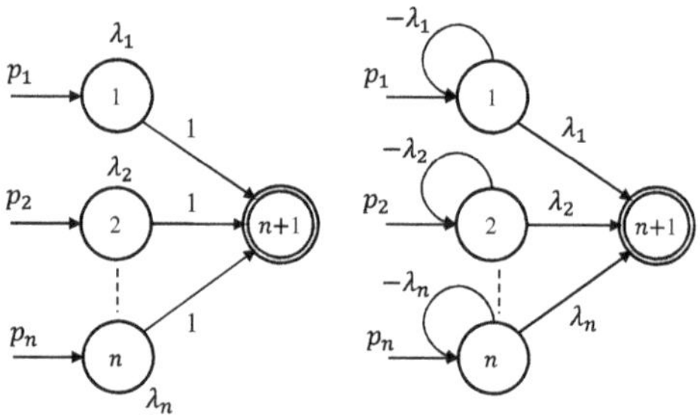
		Scalar	$1 - \sum_{i=1}^{n} p_i e^{-\lambda_i t}$
		Matrix	$n = n \quad \boldsymbol{a} = \begin{bmatrix} p_1 & p_2 \cdots p_n \end{bmatrix} \quad \boldsymbol{A} = \begin{bmatrix} -\lambda_1 & 0 & \dots & 0 \\ 0 & -\lambda_2 & \dots & 0 \\ \vdots & \vdots & \ddots & \vdots \\ 0 & 0 & \dots & -\lambda_n \end{bmatrix}$
		Mean	$\frac{p_1}{\lambda_1} + \frac{p_2}{\lambda_2} + \cdots + \frac{p_n}{\lambda_n}$
3.	n-Erlang (λ) (sum)	Diagram	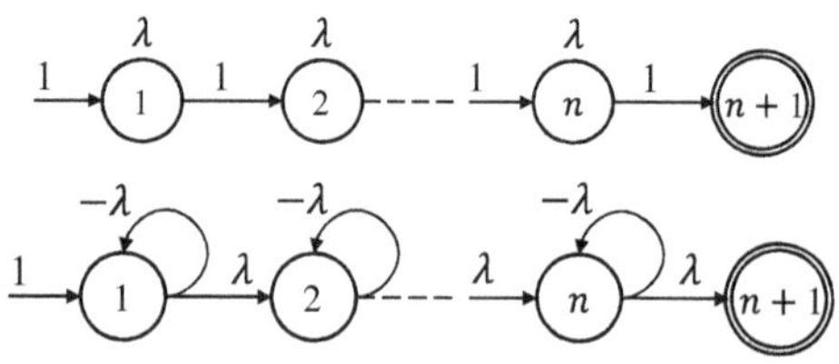
		Scalar	$1 - \sum_{i=0}^{n-1} \frac{1}{i!} (\lambda t)^i e^{-\lambda t}$
		Matrix	$n = n \quad \boldsymbol{a} = \begin{bmatrix} 1 & 0 \dots 0 \end{bmatrix} \quad \boldsymbol{A} = \begin{bmatrix} -\lambda & \lambda & \dots & 0 \\ 0 & -\lambda & \dots & 0 \\ \vdots & \vdots & \ddots & \vdots \\ 0 & 0 & \dots & -\lambda \end{bmatrix}$
		Mean	$\frac{n}{\lambda}$
		Variance	$\frac{n}{\lambda^2}$

Table 2.1 (Continued)

No.	Distribution		Representation
4.	n-Hypo-exponential (λ_i) (sum)	Diagram	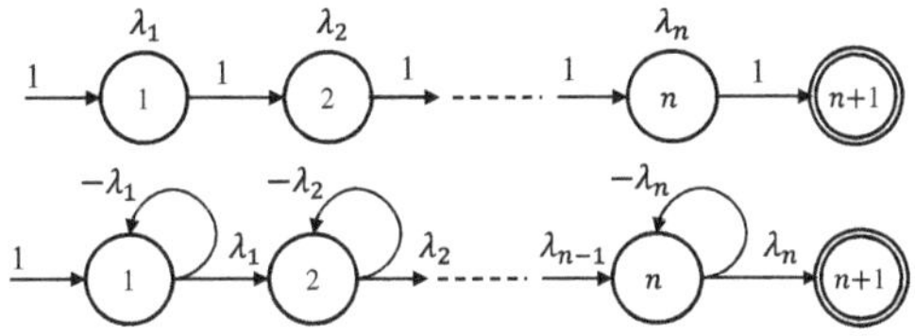
		Scalar	$1-\sum_{i=1}^{n}\frac{e^{-\lambda_i t}}{P_i}$ with $P_i=\prod_{j=1,j\neq i}^{n}\left(1-\frac{\lambda_i}{\lambda_j}\right)$
		Matrix	$n=n$ $\boldsymbol{a}=\begin{bmatrix}1 & 0\ldots 0\end{bmatrix}$ $\boldsymbol{A}=\begin{bmatrix}-\lambda_1 & \lambda_1 & \ldots & 0 & 0\\ 0 & -\lambda_2 & \ldots & 0 & 0\\ \vdots & \vdots & \ddots & \vdots & \vdots\\ 0 & 0 & \ldots & -\lambda_{n-1} & \lambda_{n-1}\\ 0 & 0 & \ldots & 0 & -\lambda_n\end{bmatrix}$
		Mean	$\frac{1}{\lambda_1}+\frac{1}{\lambda_2}+\cdots+\frac{1}{\lambda_n}$
		Variance	$\frac{1}{\lambda_1^2}+\frac{1}{\lambda_2^2}+\cdots+\frac{1}{\lambda_n^2}$
5.	n-Coxian (selection and sum)	Diagram	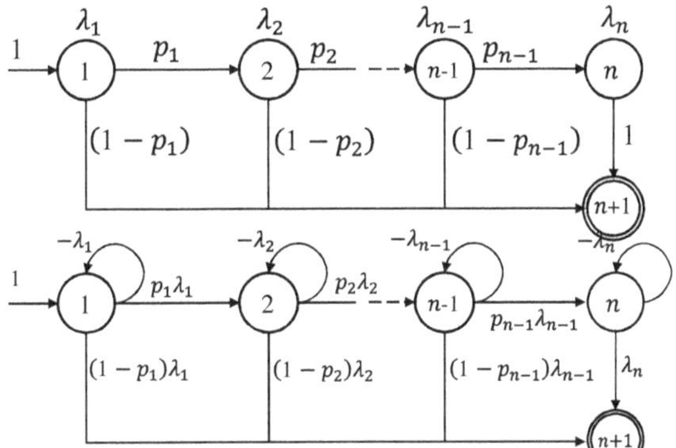
		Scalar (order 3)	$1-p_1p_2\left\{\frac{\left(\lambda_2\lambda_3(\lambda_2-\lambda_3)e^{-\lambda_1 t}+\lambda_1\lambda_3(\lambda_3-\lambda_1)e^{-\lambda_2 t}+\lambda_1\lambda_2(\lambda_1-\lambda_2)e^{-\lambda_3 t}\right)}{(\lambda_1-\lambda_2)(\lambda_1-\lambda_3)(\lambda_2-\lambda_3)}\right\}-$ $p_1(1-p_2)\left(\frac{\lambda_2e^{-\lambda_1 t}-\lambda_1e^{-\lambda_2 t}}{\lambda_1-\lambda_2}\right)-(1-p_1)e^{-\lambda_1 t}$
		Matrix	$n=n$ $\boldsymbol{a}=\begin{bmatrix}1 & 0\ldots 0\end{bmatrix}$ $\boldsymbol{A}=\begin{bmatrix}-\lambda_1 & p_1\lambda_1 & \ldots & 0 & 0\\ 0 & -\lambda_2 & \ldots & 0 & 0\\ \vdots & \vdots & \ddots & \vdots & \vdots\\ 0 & 0 & \ldots & -\lambda_{n-1} & p_{n-1}\lambda_{n-1}\\ 0 & 0 & \ldots & 0 & -\lambda_n\end{bmatrix}$
		Mean	$\frac{1}{\lambda_1}+\frac{p_1}{\lambda_2}+\frac{p_1p_2}{\lambda_3}+\cdots+\frac{p_1p_2\cdots p_n}{\lambda_n}$

Example 2.11 A repair process requires several stages. The process begins with inspecting the incoming item for a duration that follows an exponential distribution with a rate of 5/hour. The inspection results in 20% of items that do not need repair, so they are immediately considered finished, and the rest need minor or major repairs with equal proportion. All items that require repair must undergo an initial repair for a duration that follows an exponential distribution with a rate of 2/hour. Major repair requires further repairs for a duration that follows an exponential distribution with a rate of 1/hour. The repair process for each item is carried out continuously until it is finished without interruption.

The described repair process can be expressed in a state transition diagram in Figure 2.13.

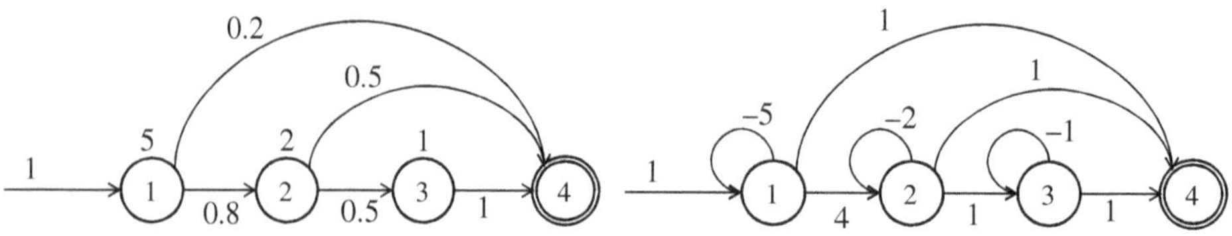

Figure 2.13 State transition diagram of the repair process in Example 2.11. (a) Using transition probabilities and (b) using transition rate.

Since all repairs must first go through inspection in state 1, the initial probability vector is

$$\boldsymbol{a} = \begin{bmatrix} 1 & 0 & 0 \end{bmatrix}, \quad a_4 = 0,$$

which means the PH distribution for overall repair time is without discontinuity at the starting point.

The infinitesimal generator matrix can be derived from Figure 2.13b as

$$\boldsymbol{Q} = \left[\begin{array}{ccc|c} -5 & 4 & 0 & 1 \\ 0 & -2 & 1 & 1 \\ 0 & 0 & -1 & 1 \\ \hline 0 & 0 & 0 & 0 \end{array}\right].$$

The PH distribution of the overall repair time until finished has a matrix representation

$$n = 3, \quad \boldsymbol{a} = \begin{bmatrix} 1 & 0 & 0 \end{bmatrix}, \quad \boldsymbol{A} = \begin{bmatrix} -5 & 4 & 0 \\ 0 & -2 & 1 \\ 0 & 0 & -1 \end{bmatrix},$$

with CDF

$$F(t) = 1 - \boldsymbol{a}e^{\boldsymbol{A}t}\boldsymbol{u}.$$

The graph of $F(t)$ can be generated using the provided **phcdfplot** function with the known values of $\boldsymbol{a}$ and $\boldsymbol{A}$ as inputs to the function.

Scalar representation can also be derived from Figure 2.13. Comparing the diagram with Table 2.1 shows that the repair time for minor repair is two stages, which produces 2-hypoexponential distribution with rates of 5/hour and 2/hour (path 1–2–4), while the repair time for major repair is three stages, which produces 3-hypoexponential with rates of 5/hour, 2/hour, and 1/hour (path 1–2–3–4).

Using the scalar function of n-hypoexponential in Table 2.1, path 1–2–3 for minor repair has a CDF ($\lambda_1 = 5$, $\lambda_2 = 2$)

$$F_1(t) = 1 - \frac{1}{\left(1 - \frac{\lambda_1}{\lambda_2}\right)} e^{-\lambda_1 t} - \frac{1}{\left(1 - \frac{\lambda_2}{\lambda_1}\right)} e^{-\lambda_2 t} = 1 - \frac{5e^{-2t} - 2e^{-5t}}{3},$$

and path 1–2–3–4 for major repair has CDF ($\lambda_1 = 5$, $\lambda_2 = 2$, $\lambda_3 = 1$)

$$F_2(t) = 1 - \frac{1}{\left(1 - \frac{\lambda_1}{\lambda_2}\right)\left(1 - \frac{\lambda_1}{\lambda_3}\right)} e^{-\lambda_1 t} - \frac{1}{\left(1 - \frac{\lambda_2}{\lambda_1}\right)\left(1 - \frac{\lambda_2}{\lambda_3}\right)} e^{-\lambda_2 t} - \frac{1}{\left(1 - \frac{\lambda_3}{\lambda_1}\right)\left(1 - \frac{\lambda_3}{\lambda_2}\right)} e^{-\lambda_3 t}$$
$$= 1 - \frac{2e^{-5t} - 20e^{-2t} + 30e^{-t}}{12}.$$

The CDF of repair time for items that do not require repair follows path 1–4, which has an exponential distribution with rate 5; hence,

$$F_3(t) = 1 - e^{-5t}.$$

The CDF of the overall repair time for all items is the result of mixing the three PH distributions: $F_1(t)$ with a probability of $0.8 \times 0.5 = 0.4$, $F_2(t)$ with a probability of $0.8 \times 0.5 = 0.4$, and $F_3(t)$ with a probability of 0.2. Thus,

$$F(t) = 0.4F_1(t) + 0.4F_2(t) + 0.2F_3(t).$$

The resulting CDF of the overall repair time is a mixture of hypoexponential or hyper-hypoexponential distribution. It involves adding and mixing PH distributions, which, in this case, is the addition and mixing of exponential distributions.

The CDF and PDF graphs of the PH distribution are generated using the obtained scalar and matrix representations. The results are exhibited in Figure 2.14, which shows that they are indeed identical.

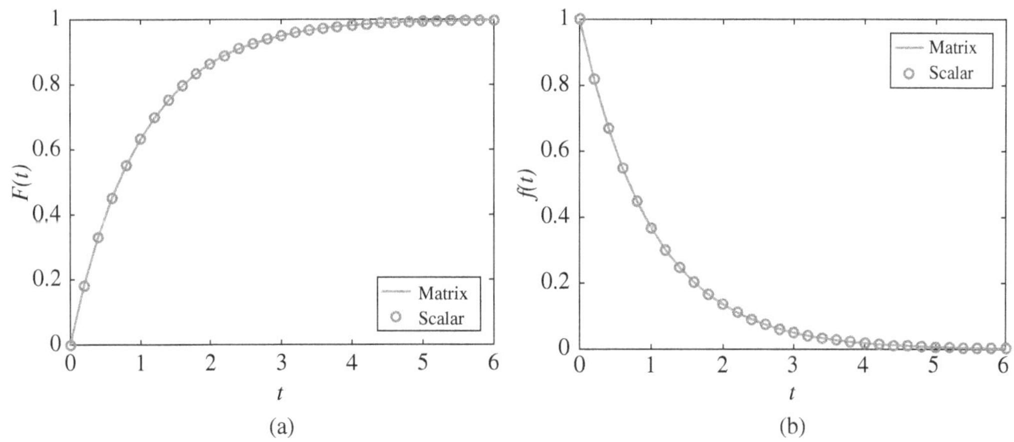

Figure 2.14 CDF and PDF of PH distributions of the repair time in Example 2.11. (a) CDF and (b) PDF. ■

Exercise 2.7 Show that the mean service completion time calculated using the matrix and scalar representations is equal.

In addition to PH distribution accommodating some popular distributions, it can also be used to approximate other popular distributions that are not its members. The next section discusses methods to approximate other distributions with PH distribution.

2.5 PH Distribution as an Approximation of Other Distributions

PH distribution can be used to approximate distributions not naturally included in the PH distribution class. This approximation is possible because for all nonnegative number distributions, there is always a series of PH distributions that converge to the distribution being approximated [5]. In fact, the PH distribution can express all distributions with any desired closeness by making its order sufficiently large [6].

The models in Table 2.1 can be used as an approximation to represent the data collected from the field. The easiest way is to use the two-moment approach, which generally are the mean and the coefficient of variation. The coefficient of variation value is used to select the appropriate model and its order. If the data has a coefficient of variation greater than one, then the simplest model that fits is hyperexponential; a more complex model is mixture-Erlang, a linear combination of several Erlang distributions with different parameters. If the coefficient of variation approaches one, then the most suitable model is exponential.

The Erlang distribution has the smallest coefficient of variation compared to all members of the PH distribution class in the same order [7]. Based on Table 2.1, n-Erlang distribution has a coefficient of variation = $\sqrt{n/\lambda^2}$: $n/\lambda = 1/\sqrt{n}$. Therefore, the smallest PH distribution order that must be used to express data with a coefficient of variation equal to cv is $n_{min} = (1/cv)^2$, rounded up. The larger the value of n, the better the model fits the data because a higher n means more parameters to choose to fit the data. Table 2.2 provides the coefficient of variation values for the PH distributions in Table 2.1.

Table 2.2 Coefficient of variations of some PH distributions.

No.	Distribution	Coefficient of Variation
1	Exponential	$cv = 1$
2	n-Erlang	$cv = \frac{1}{\sqrt{n}}$
3	n-Hyperexponential	$cv > 1$
4	n-Hypoexponential	$\frac{1}{\sqrt{n}} < cv < 1$
5	n-Coxian	$cv > \frac{1}{\sqrt{n}}$

Example 2.12 Data with $cv = 0.65$ must be expressed in a PH distribution with order greater than or equal to $(1/0.65)^2 = 2.367$, rounded up to 3. The distributions that can be used to model the data are hypoexponential, mixture-Erlang, or Coxian distributions. Using 3-Erlang will produce a distribution with $cv = 1/\sqrt{3} = 0.577$, smaller than 0.65. Hence, the 3-Erlang distribution is unsuitable, but the 3-hypoexponential distribution is. If using the PH distribution of order 2, then the value of the coefficient of variation that can be represented is greater than $1/\sqrt{2} = 0.707$, where 0.65 is not included in the range. ■

Example 2.13 Random data with a mean of 11.24 and a standard deviation of 7.56 can be expressed in a PH distribution with order rounded up to

$$n = \left(\frac{1}{cv}\right)^2 = \left(\frac{11.24}{7.56}\right)^2 \approx 3.$$

If a 3-hypoexponential distribution is used, then there are three parameters, λ_1, λ_2, and λ_3, that must be searched to fit the two information given. Hence, one of the parameters can be determined arbitrarily. As given in Table 2.1, the mean and variance of random variables having a hypoexponential distribution of order 3 with parameters λ_1, λ_2, and λ_3 are

$$E[T] = \frac{1}{\lambda_1} + \frac{1}{\lambda_2} + \frac{1}{\lambda_3} = 11.24;$$

$$Var[T] = \frac{1}{\lambda_1^2} + \frac{1}{\lambda_2^2} + \frac{1}{\lambda_3^2} = 7.56^2.$$

Selecting $\lambda_1 = \lambda_2$ results in $\lambda_1 = \lambda_2 = 0.462$ and $\lambda_3 = 0.145$, which will yield a PH representation

$$n = 3, \quad \boldsymbol{a} = \begin{bmatrix} 1 & 0 & 0 \end{bmatrix}, \quad \boldsymbol{A} = \begin{bmatrix} -0.462 & 0.462 & 0 \\ 0 & -0.462 & 0.462 \\ 0 & 0 & -0.145 \end{bmatrix}.$$

This example shows that many 3-hypoexponential distributions fit the two-given information. There would be even more suitable PH distributions if other probability models such as Coxian and mixture-Erlang are considered as candidates. ■

The two-moment approach method is used when field data is insufficient. Horváth and Telek [8] offer a more accurate method using three moments. For sufficient data, a more accurate PH distribution search method (higher agreement with data) can be done with a more sophisticated method as discussed in Buchholz et al. [9] and Horváth and Telek [8]. There are many software for this purpose, such as BuTools (MATLAB), PhFit (Java), Empht (C), and Mapfit (R).

Example 2.14 Weibull distribution with CDF $F_T(t) = 1 - e^{-(0.1t)^{1.5}}$ is approximated with APH distributions using the PhFit software [10]. Since this distribution has $cv^2 = 0.46$, it should be approximated with PH distribution having order not less than $1/0.46 = 2.17$, rounded up to at least 3. Their resulting parameter values of PH distributions of orders 3 and 4 are

$$\text{Order 3: } \boldsymbol{a} = \begin{bmatrix} 1 & 0 & 0 \end{bmatrix}, \quad \boldsymbol{A} = \begin{bmatrix} -0.270 & 0.147 & 0.099 \\ 0 & -0.270 & 0.270 \\ 0 & 0 & -0.270 \end{bmatrix}.$$

$$\text{Order 4: } \boldsymbol{a} = \begin{bmatrix} 1 & 0 & 0 & 0 \end{bmatrix}, \quad \boldsymbol{A} = \begin{bmatrix} -0.339 & 0.169 & 0.009 & 0.154 \\ 0 & -0.323 & 0.323 & 0 \\ 0 & 0 & -0.323 & 0.323 \\ 0 & 0 & 0 & -0.326 \end{bmatrix}.$$

A comparison of the graphs of the CDF of PH distribution approximations with the original Weibull distribution is shown in Figure 2.15, which shows that their CDF graphs are indistinguishable. The largest difference between the Weibull distribution and each PH distribution approximation is given in Table 2.3, including the difference in their means and coefficients of variation. Given those differences, it is not possible to distinguish between the CDF of the original Weibull distribution and its PH distribution approximations. It is apparent that the deviation from the original Weibull distribution is smaller for larger PH orders.

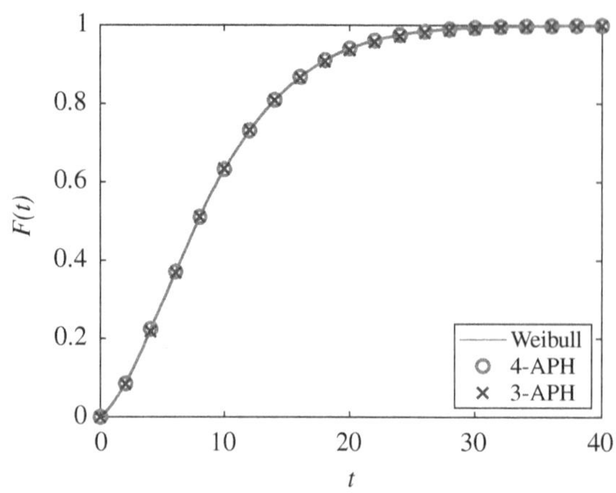

Figure 2.15 CDFs of the Weibull distribution and its PH distribution approximations.

Table 2.3 Weibull distribution statistics and its PH distribution approximations.

Distribution	Mean	cv^2	Maximum difference
Original Weibull	9.03	0.46	–
3-APH approximation	9.07	0.48	0.0051
4-APH approximation	9.03	0.46	0.0008

■

Exercise 2.8 Use two-moment approximation for the Weibull distribution in Example 2.14 and find a PH distribution order 2 that matches the two moments. Compare the CDF of the resulting PH distribution with the original Weibull distribution and conclude the result.

Using cv^2 as a guiding number to determine the order of the PH distribution is not always as clear as the result in Example 2.14.

Example 2.15 Okamura and Dohi [11] approximated a lognormal distribution with PDF $f(t) = \frac{1}{2} - \frac{1}{2}\mathrm{erf}\left(\frac{\ln t-\mu}{\sqrt{2}\sigma}\right)$ with PH distributions using the BuTools software. BuTools uses mixture-Erlang distribution that is not as flexible as Coxian distribution utilized in PhFit regarding the number of parameters that can be selected. The lognormal to be approximated has $cv^2 = 0.63$, which means it can be approximated using PH distribution order 2. However, the PH distribution approximation of order 2 has a significant difference from the original lognormal distribution. Comparisons with several orders of PH distribution are shown in Table 2.4 and Figure 2.16.

- Order 2: $\boldsymbol{a} = [0.491\ 0.509]$, $\boldsymbol{A} = \begin{bmatrix} -0.138 & 0 \\ 0 & -0.164 \end{bmatrix}$.
- Order 4: $\boldsymbol{a} = \begin{bmatrix} 0.119 & 0.881 & 0 & 0 \end{bmatrix}$,

$$\boldsymbol{A} = \begin{bmatrix} -0.103 & 0 & 0 & 0 \\ 0 & -0.607 & 0.607 & 0 \\ 0 & 0 & -0.607 & 0.607 \\ 0 & 0 & 0 & -0.607 \end{bmatrix}.$$

Table 2.4 Statistics of the lognormal distribution and its PH distribution approximations.

Distribution	Mean	cv^2	Maximum difference
Lognormal	5.74	0.63	–
2-APH	6.65	1.01	0.1444
4-APH	5.51	0.68	0.0324
5-APH	5.51	0.59	0.0185

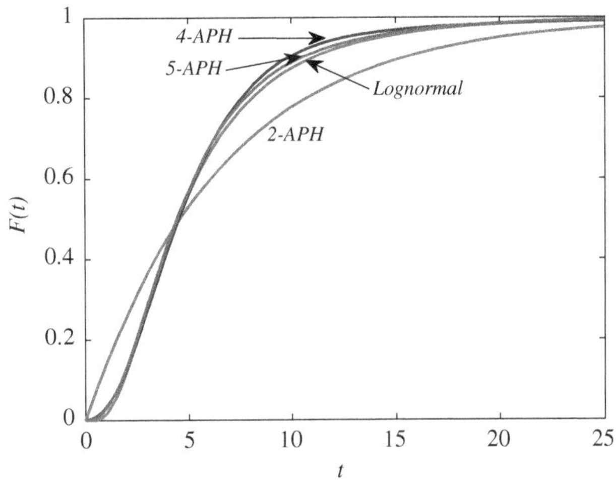

Figure 2.16 The CDFs of the lognormal distribution and its PH distribution approximations.

- Order 5: $\boldsymbol{a} = \begin{bmatrix} 0.269 & 0 & 0.731 & 0 & 0 \end{bmatrix}$,

$$\boldsymbol{A} = \begin{bmatrix} -0.235 & 0.235 & 0 & 0 & 0 \\ 0 & -0.235 & 0 & 0 & 0 \\ 0 & 0 & -0.681 & 0.681 & 0 \\ 0 & 0 & 0 & -0.681 & 0.681 \\ 0 & 0 & 0 & 0 & -0.681 \end{bmatrix}.$$

As apparent from the resulting transient matrices, the PH distribution approximations in this example are mixture-Erlang distributions. For example, the PH distribution approximation of order 4 is a mixture of an exponential distribution (1-Erlang) with $\lambda_1 = 0.103$ and a 3-Erlang distribution with $\lambda_2 = 0.607$ with the mixture probabilities 0.119 and 0.881 as given in its initial probability vector $\boldsymbol{a}$. Therefore, the scalar representation of the PH distribution approximation of order 4 can be derived using Table 2.1 as

$$F(t) = 0.119F_1(t) + 0.881F_2(t),$$

where

$$F_1(t) = 1 - e^{-0.103t},$$

$$F_2(t) = 1 - \left(\sum_{i=0}^{2} \frac{1}{i!}(0.607t)^i e^{-0.607t} \right).$$

Scalar representations for PH distribution approximations of order 5 can be derived accordingly. An advantage of using mixture-Erlang is that, if needed, the scalar representation can be derived easily. ■

Example 2.16 The lognormal distribution in Example 2.15 is attempted to be approximated by a PH distribution of order 4. A moment-by-moment comparison through trial-and-error carries out the search for parameter values. The result is a PH distribution of order 4 with the representation

$$n = 4, \quad \boldsymbol{\alpha} = \begin{bmatrix} 0.3 & 0.7 & 0 & 0 \end{bmatrix}, \quad \boldsymbol{A} = \begin{bmatrix} -0.17 & 0.05 & 0.05 & 0.07 \\ 0 & -0.6 & 0.6 & 0 \\ 0 & 0 & -0.7 & 0.7 \\ 0 & 0 & 0 & -0.75 \end{bmatrix}.$$

A comparison of their CDF graphs is presented in Figure 2.17. This PH distribution approximation has mean = 5.6639, cv^2= 0.6294, and max dev = 0.0147, slightly better than that obtained using BuTools. This example shows that finding a PH distribution approximation can be done by carefully trying out the parameter values until a satisfactory result is achieved. Using the two-moment approach method as an initial guess is certainly helpful. It is also possible by tuning the parameters obtained from using application software.

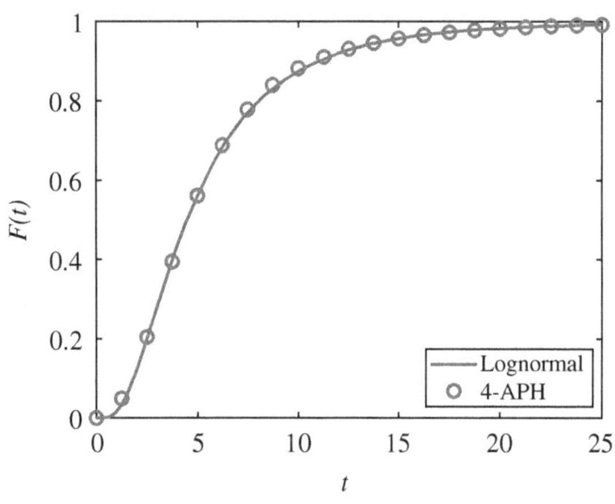

Figure 2.17 CDF of the lognormal and its PH distribution approximation of order 4.

■

Exercise 2.9 Find a better PH approximation order 4 than the results of Examples 2.15 and 2.16. Give a justification that the result is better.

The most important aspect of PH distribution that makes it appealing to be used for system reliability modeling is its closure properties described in the next section.

2.6 Closure Properties of PH Distribution

PH distribution is closed under some operations intensively used in reliability analyses. Closed under an operation means that the operation on the PH distributions will still produce PH distribution, although with a different representation. This definition is a special form of the definition of a closed set. A set A is said to be closed under an operation $\odot$ if and only if $(A, \odot) = A$.

Example 2.17 For the set of exponential functions $E = \{ae^{bt}, a, b \in R\}$,

$$x(t) = ae^{bt} \in E \rightarrow \dot{x}(t) = abe^{bt} \in E;$$

$$x(t) = ae^{bt} \in E \rightarrow \int x(t)dt = \frac{a}{b}e^{bt} \in E;$$

$$(x(t) = ae^{bt} \in E \wedge y(t) = ce^{dt} \in E) \rightarrow x(t)y(t) = ace^{(b+d)t} \in E;$$

$$(x(t) = ae^{bt} \in E \wedge y(t) = ce^{dt} \in E) \rightarrow x(t)/y(t) = a/b\, e^{(b-d)t} \in E.$$

Thus, the set $E = \{ae^{bt}, a, b \in R\}$ is closed under differentiation, integration, multiplication, and division. ■

2.6.1 Kronecker Operation

PH distribution is closed under certain tensor matrix operations. Therefore, it is necessary to understand the tensor algebra operations for matrices, which are better known in probability studies as Kronecker operations. There are two types of Kronecker operations, namely, Kronecker product and Kronecker sum.

The Kronecker product, denoted by $\otimes$, of two matrices $\boldsymbol{A} = [A_{ij}]$ and $\boldsymbol{B} = [B_{ij}]$ with dimensions $m_1 \times m_2$ and $n_1 \times n_2$, respectively, is defined as

$$\boldsymbol{A} \otimes \boldsymbol{B} = \begin{bmatrix} A_{11}\boldsymbol{B} & A_{12}\boldsymbol{B} & \cdots & A_{1m_2}\boldsymbol{B} \\ \vdots & \vdots & \ddots & \vdots \\ A_{m_11}\boldsymbol{B} & A_{m_12}\boldsymbol{B} & \cdots & A_{m_1m_2}\boldsymbol{B} \end{bmatrix}, \tag{2.28}$$

which is a matrix with dimension $m_1n_1 \times m_2n_2$.

The Kronecker sum, denoted by $\oplus$, of two square matrices $\boldsymbol{A} = [A_{ij}]$ and $\boldsymbol{B} = [B_{ij}]$ with dimensions $m \times m$ and $n \times n$ (must be square matrices) is defined as

$$\boldsymbol{A} \oplus \boldsymbol{B} = \boldsymbol{A} \otimes \boldsymbol{I}_n + \boldsymbol{I}_m \otimes \boldsymbol{B}, \tag{2.29}$$

which is a matrix with dimension $mn \times mn$.

Example 2.18 For two matrices $\boldsymbol{A}_{2x2} = \begin{bmatrix} 0 & 3 \\ 2 & 1 \end{bmatrix}$ and $\boldsymbol{B}_{2x1} = \begin{bmatrix} -2 & 1 \end{bmatrix}$,

$$\boldsymbol{A} \otimes \boldsymbol{B} = \begin{bmatrix} 0\begin{bmatrix} -2 & 1 \end{bmatrix} & 3\begin{bmatrix} -2 & 1 \end{bmatrix} \\ 2\begin{bmatrix} -2 & 1 \end{bmatrix} & 1\begin{bmatrix} -2 & 1 \end{bmatrix} \end{bmatrix} = \begin{bmatrix} 0 & 0 & -6 & 3 \\ -4 & 2 & -2 & 1 \end{bmatrix},$$

is a matrix with dimension 2×4 as the result of multiplying their dimensions 2×2 and 1×2, and

$$\boldsymbol{B} \otimes \boldsymbol{A} = \begin{bmatrix} -2\begin{bmatrix} 0 & 3 \\ 2 & 1 \end{bmatrix} & 1\begin{bmatrix} 0 & 3 \\ 2 & 1 \end{bmatrix} \end{bmatrix} = \begin{bmatrix} 0 & -6 & 0 & 3 \\ -4 & -2 & 2 & 1 \end{bmatrix},$$

is a matrix with dimension 2×4 resulting from multiplying their dimensions 1×2 and 2×2.

Note that $\boldsymbol{A} \otimes \boldsymbol{B} \neq \boldsymbol{A} \otimes \boldsymbol{B}$, which means the Kronecker product, is not commutative. ■

Example 2.19 For square matrices $\boldsymbol{A}_{2\times 2} = \begin{bmatrix} 0 & 3 \\ 2 & 1 \end{bmatrix}$ and $\boldsymbol{B}_{1\times 1} = [2]$,

$$\boldsymbol{A} \oplus \boldsymbol{B} = \begin{bmatrix} 0 & 3 \\ 2 & 1 \end{bmatrix} \otimes \boldsymbol{I}_1 + \boldsymbol{I}_2 \otimes [2] = \begin{bmatrix} 0 & 3 \\ 2 & 1 \end{bmatrix} \otimes [1] + \begin{bmatrix} 1 & 0 \\ 0 & 1 \end{bmatrix} \otimes [2]$$

$$= \begin{bmatrix} 0\times 1 & 3\times 1 \\ 2\times 1 & 1\times 1 \end{bmatrix} + \begin{bmatrix} 1\times 2 & 0\times 2 \\ 0\times 2 & 1\times 2 \end{bmatrix} = \begin{bmatrix} 2 & 3 \\ 2 & 3 \end{bmatrix},$$

where $\boldsymbol{I}_1$ and $\boldsymbol{I}_2$ are identity matrices with dimensions 1×1 and 2×2, following the dimensions of $\boldsymbol{B}$ and $\boldsymbol{A}$, respectively.

The resulting matrix has dimension 2×2 as the result of multiplying the dimensions of $\boldsymbol{A}$ and $\boldsymbol{B}$ which are 2×2 and 1×1, respectively.

$$\boldsymbol{B} \oplus \boldsymbol{A} = [2] \otimes \boldsymbol{I}_2 + \boldsymbol{I}_1 \otimes \begin{bmatrix} 0 & 3 \\ 2 & 1 \end{bmatrix} = \begin{bmatrix} 2 & 0 \\ 0 & 2 \end{bmatrix} + \begin{bmatrix} 0 & 3 \\ 2 & 1 \end{bmatrix} = \begin{bmatrix} 2 & 3 \\ 2 & 3 \end{bmatrix},$$

is a matrix with dimension 2×2 resulting from multiplying dimensions 1×1 and 2×2.

The Kronecker sum is also not commutative. This example happened to produce $\boldsymbol{A} \oplus \boldsymbol{B} = \boldsymbol{B} \oplus \boldsymbol{A}$ because $\boldsymbol{B}$ is a scalar, and multiplication with a scalar is commutative. ■

The Kronecker product of a matrix with a scalar or a scalar with a matrix is the same as the multiplication of a matrix with a scalar. It can be seen that the Kronecker operation requires more computation than the matrix multiplication operation. Using Kronecker operations will not be troublesome since they are available as ready-to-use functions in many computer languages.

Some properties of Kronecker operations that are needed in later development are (g and h are scalars, and $\boldsymbol{A}$, $\boldsymbol{B}$, $\boldsymbol{C}$, $\boldsymbol{D}$ are matrices):

Property 1: $g \otimes h = gh$.
Property 2: $g \otimes \boldsymbol{A} = g\boldsymbol{A}$.
Property 3: $\boldsymbol{AB} \otimes \boldsymbol{CD} = (\boldsymbol{A} \otimes \boldsymbol{C})(\boldsymbol{B} \otimes \boldsymbol{D})$.
Property 4: $\boldsymbol{A}^k \otimes \boldsymbol{B}^k = (\boldsymbol{A} \otimes \boldsymbol{B})^k$.
Property 5: $\boldsymbol{A}^{-1} \otimes \boldsymbol{B}^{-1} = (\boldsymbol{A} \otimes \boldsymbol{B})^{-1}$.
Property 6: $e^{\boldsymbol{A}t} \otimes e^{\boldsymbol{B}t} = e^{(\boldsymbol{A} \oplus \boldsymbol{B})t}$.
Property 7: $\boldsymbol{u}_m \otimes \boldsymbol{u}_n = \boldsymbol{u}_{mn}$.

Properties 2–4 apply to matrices of arbitrary dimension, while Properties 5 and 6 apply only to square matrices. All properties can be applied to several operations and matrices by processing them recursively.

Closure properties of PH distribution under some operations commonly encountered in system reliability analyses are developed using Kronecker operations. The developments here are shown using two random variables T_1 and T_2 following independent and not necessarily identical PH distributions with representation $(n_1, \boldsymbol{a}_1, \boldsymbol{A}_1)$ and $(n_2, \boldsymbol{a}_2, \boldsymbol{A}_2)$.

2.6.2 Minimum Operation

The CDF of $T_S = \min(T_1, T_2)$ can be derived from its definition as

$$F_{T_S}(t) = P[T_S \le t] = P[\min(T_1, T_2) \le t] = 1 - P[\min(T_1, T_2) > t].$$

The minimum of two variables being greater than a certain value is satisfied by both variables being greater than that value,

$$P[\min(T_1, T_2) > t] = P[(T_1 > t) \wedge (T_2 > t)],$$

where $\wedge$ is the conjunction, or "and" logical operator.

Since T_1 and T_2 are independent, it follows that

$$F_{T_S}(t) = 1 - P[(T_1 > t) \wedge (T_2 > t)] = 1 - P[T_1 > t]P[T_2 > t]. \tag{2.30}$$

For special case $T_1 \sim \exp(\lambda_1)$, $T_2 \sim \exp(\lambda_2)$,

$$F_{T_S}(t) = 1 - \left(e^{-\lambda_1 t}\right)\left(e^{-\lambda_2 t}\right) = 1 - e^{-(\lambda_1+\lambda_2)t},$$

which means $T_S \sim \exp(\lambda_1 + \lambda_2)$, or T_S has exponential distribution with rate equal to the sum of the rates of T_1 and T_2. The same result should apply for PH distribution as well, but the sum of scalar rates is replaced with the Kronecker sum of matrix rates, as shown in the following derivation.

Use $T_1 \sim PH(n_1, \boldsymbol{a}_1, \boldsymbol{A}_1)$, $T_2 \sim PH(n_2, \boldsymbol{a}_2, \boldsymbol{A}_2)$, and Equation (2.13) to obtain $P[T_1 > t] = 1 - F_{T_1}(t) = \boldsymbol{a}_1 e^{\boldsymbol{A}_1 t}\boldsymbol{u}_1$ and $P[T_2 > t] = 1 - F_{T_2}(t) = \boldsymbol{a}_2 e^{\boldsymbol{A}_2 t}\boldsymbol{u}_2$. Substituting them into Equation (2.30) yields

$$F_{T_S}(t) = 1 - \left(\boldsymbol{a}_1 e^{\boldsymbol{A}_1 t}\boldsymbol{u}_1\right)\left(\boldsymbol{a}_2 e^{\boldsymbol{A}_2 t}\boldsymbol{u}_2\right),$$

which is a multiplication of two scalars.

Using Property 1 of the Kronecker product to replace scalar multiplication with Kronecker product yields

$$F_{T_S}(t) = 1 - \left(\boldsymbol{a}_1 e^{\boldsymbol{A}_1 t}\boldsymbol{u}_1\right) \otimes \left(\boldsymbol{a}_2 e^{\boldsymbol{A}_2 t}\boldsymbol{u}_2\right).$$

Using Property 3 of the Kronecker operations recursively produces

$$F_{T_S}(t) = 1 - (\boldsymbol{a}_1 \otimes \boldsymbol{a}_2)\left(e^{\boldsymbol{A}_1 t} \otimes e^{\boldsymbol{A}_2 t}\right)(\boldsymbol{u}_1 \otimes \boldsymbol{u}_2).$$

Applying Property 6 to the Kronecker product of two exponential functions yields

$$F_{T_S}(t) = 1 - (\boldsymbol{a}_1 \otimes \boldsymbol{a}_2)\left(e^{(\boldsymbol{A}_1 \oplus \boldsymbol{A}_2)t}\right)(\boldsymbol{u}_1 \otimes \boldsymbol{u}_2) = 1 - \boldsymbol{a}_S e^{\boldsymbol{A}_S t}\boldsymbol{u}_S,$$

where $\boldsymbol{a}_S = \boldsymbol{a}_1 \otimes \boldsymbol{a}_2$, $\boldsymbol{A}_S = \boldsymbol{A}_1 \oplus \boldsymbol{A}_2$, $\boldsymbol{u}_S = \boldsymbol{u}_1 \otimes \boldsymbol{u}_2$.

The last equation shows that T_S has a PH distribution with representation $(n_S, \boldsymbol{a}_S, \boldsymbol{A}_S)$. By following the Kronecker product rule, the resulting PH distribution order is $n_S = n_1 n_2$.

This result shows that the minimum operation of two independent random variables having PH distributions produces a random variable having PH distribution. Thus, the PH distribution is closed under the minimum operation, and the following theorem is proven.

Theorem 2.1 [4] If $T_1 \sim PH(n_1, \boldsymbol{a}_1, \boldsymbol{A}_1)$ and $T_2 \sim PH(n_2, \boldsymbol{a}_2, \boldsymbol{A}_2)$ are independent, then $T_S = \min(T_1, T_2) \sim PH(n_S, \boldsymbol{a}_S, \boldsymbol{A}_S)$ with

$$n_S = n_1 n_2, \quad \boldsymbol{a}_S = \boldsymbol{a}_1 \otimes \boldsymbol{a}_2, \quad \boldsymbol{A}_S = \boldsymbol{A}_1 \oplus \boldsymbol{A}_2. \tag{2.31}$$

Example 2.20 For T_1 and T_2 having PH distributions with representations $n_1 = 2, \boldsymbol{a}_1 = \begin{bmatrix}1 & 0\end{bmatrix}$, $\boldsymbol{A}_1 = \begin{bmatrix}-1 & 1 \\ 0 & -1\end{bmatrix}$ and $n_2 = 2, \boldsymbol{a}_2 = \begin{bmatrix}1 & 0\end{bmatrix}, \boldsymbol{A}_2 = \begin{bmatrix}-2 & 2 \\ 0 & -2\end{bmatrix}$, respectively, the minimum of T_1 and T_2 has PH distribution with a representation that can be calculated using Theorem 2.1 as follows (with $\boldsymbol{I}_1$ and $\boldsymbol{I}_2$ being identity matrices having the same dimensions as $\boldsymbol{A}_1$ and $\boldsymbol{A}_2$, respectively):

$$n_S = n_1 n_2 = 4, \quad \boldsymbol{a}_S = \boldsymbol{a}_1 \otimes \boldsymbol{a}_2 = \begin{bmatrix}1 & 0\end{bmatrix} \otimes \begin{bmatrix}1 & 0\end{bmatrix} = \begin{bmatrix}1 & 0 & 0 & 0\end{bmatrix},$$

$$
\begin{aligned}
\boldsymbol{A}_S &= \boldsymbol{A}_1 \oplus \boldsymbol{A}_2 = \boldsymbol{A}_1 \otimes \boldsymbol{I}_2 + \boldsymbol{I}_1 \otimes \boldsymbol{A}_2 \\
&= \begin{bmatrix} -1 & 1 \\ 0 & -1 \end{bmatrix} \otimes \begin{bmatrix} 1 & 0 \\ 0 & 1 \end{bmatrix} + \begin{bmatrix} 1 & 0 \\ 0 & 1 \end{bmatrix} \otimes \begin{bmatrix} -2 & 2 \\ 0 & -2 \end{bmatrix} \\
&= \begin{bmatrix} -1\begin{bmatrix} 1 & 0 \\ 0 & 1 \end{bmatrix} & 1\begin{bmatrix} 1 & 0 \\ 0 & 1 \end{bmatrix} \\ 0\begin{bmatrix} 1 & 0 \\ 0 & 1 \end{bmatrix} & -1\begin{bmatrix} 1 & 0 \\ 0 & 1 \end{bmatrix} \end{bmatrix} + \begin{bmatrix} 1\begin{bmatrix} -2 & 2 \\ 0 & -2 \end{bmatrix} & 0\begin{bmatrix} -2 & 2 \\ 0 & -2 \end{bmatrix} \\ 0\begin{bmatrix} -2 & 2 \\ 0 & 2 \end{bmatrix} & 1\begin{bmatrix} -2 & 2 \\ 0 & -2 \end{bmatrix} \end{bmatrix} \\
&= \begin{bmatrix} \begin{bmatrix} -3 & 2 \\ 0 & -3 \end{bmatrix} & \begin{bmatrix} 1 & 0 \\ 0 & 1 \end{bmatrix} \\ \begin{bmatrix} 0 & 0 \\ 0 & 0 \end{bmatrix} & \begin{bmatrix} -3 & 2 \\ 0 & -3 \end{bmatrix} \end{bmatrix} = \begin{bmatrix} -3 & 2 & 1 & 0 \\ 0 & -3 & 0 & 1 \\ 0 & 0 & -3 & 2 \\ 0 & 0 & 0 & -3 \end{bmatrix}.
\end{aligned}
$$

Note that $\boldsymbol{\alpha}_S$ and $\boldsymbol{A}_S$, respectively, can be verified to satisfy the requirements of being an initial probability vector and a transient matrix; thus, $(n_S, \boldsymbol{\alpha}_S, \boldsymbol{A}_S)$ is valid as a PH representation. ■

2.6.3 Mixing Operation

The mixing of two random variables T_1 and T_2 to produce $T_S = \text{mix}[T_1, T_2]$ is based on the concept of random selection that

$$T_S = T_1 \text{ with probabilities } p_1, \text{and}$$

$$T_S = T_2 \text{ with probabilities } p_2,$$

where $p_1 + p_2 = 1$.

The CDF of T_S can be derived using the law of total probability. For mutually exclusive and exhaustive events B_i, $i = 1, 2, \ldots,$ holds

$$P[A] = P[A|B_1]P[B_1] + P[A|B_2]P[B_2] + \cdots.$$

Substituting $T_S \le t$ for A and $T_S = T_i$ for B_i, $i = 1, 2$ obtains

$$
\begin{aligned}
F_{T_S}(t) &= P[T_S \le t | T_S = T_1]P[T_S = T_1] + P[T_S \le t | T_S = T_2]P[T_S = T_2] \\
&= P[T_1 \le t]p_1 + P[T_2 \le t]p_2.
\end{aligned}
$$

Thus, $F_{T_S}(t)$ is a linear combination of $F_{T_1}(t)$ and $F_{T_2}(t)$ with combination weights p_1 and p_2,

$$F_{T_S}(t) = p_1 F_{T_1}(t) + p_2 F_{T_2}(t). \tag{2.32}$$

For the case of exponential distribution, Equation (2.32) produces

$$F_{T_S}(t) = p_1 F_{T_1}(t) + p_2 F_{T_2}(t) = p_1\left(1 - e^{\lambda_1 t}\right) + p_2\left(1 - e^{\lambda_2 t}\right) = p_1 + p_2 - p_1 e^{\lambda_1 t} - p_2 e^{\lambda_2 t},$$

which can be written in matrix form as

$$F_{T_S}(t) = p_1 + p_2 - p_1 e^{\lambda_1 t} - p_2 e^{\lambda_2 t} = 1 - \begin{bmatrix} p_1 & p_2 \end{bmatrix} \begin{bmatrix} e^{\lambda_1 t} & 0 \\ 0 & e^{\lambda_2 t} \end{bmatrix} \begin{bmatrix} 1 \\ 1 \end{bmatrix}.$$

This result can be extended to PH distribution. Substituting the CDF of T_1 and T_2 into Equation (2.32) obtains

$$F_{T_S}(t) = p_1\left(1 - \boldsymbol{\alpha}_1 e^{\boldsymbol{A}_1 t}\boldsymbol{u}_1\right) + p_2\left(1 - \boldsymbol{\alpha}_2 e^{\boldsymbol{A}_2 t}\boldsymbol{u}_2\right) = p_1 + p_2 - p_1\boldsymbol{\alpha}_1 e^{\boldsymbol{A}_1 t}\boldsymbol{u}_1 - p_2\boldsymbol{\alpha}_2 e^{\boldsymbol{A}_2 t}\boldsymbol{u}_2.$$

Combining the two exponential terms into one by matrix augmentation and substituting $p_1 + p_2 = 1$ yield

$$F_{T_S}(t) = 1 - \begin{bmatrix} p_1\boldsymbol{a}_1 & p_2\boldsymbol{a}_2 \end{bmatrix} \begin{bmatrix} e^{\boldsymbol{A}_1 t} & \boldsymbol{0} \\ \boldsymbol{0} & e^{\boldsymbol{A}_2 t} \end{bmatrix} \begin{bmatrix} \boldsymbol{u}_1 \\ \boldsymbol{u}_2 \end{bmatrix}$$

$$= 1 - \begin{bmatrix} p_1\boldsymbol{a}_1 & p_2\boldsymbol{a}_2 \end{bmatrix} \exp\left(\begin{bmatrix} \boldsymbol{A}_1 & \boldsymbol{0} \\ \boldsymbol{0} & \boldsymbol{A}_2 \end{bmatrix} t \right) \begin{bmatrix} \boldsymbol{u}_1 \\ \boldsymbol{u}_2 \end{bmatrix} = 1 - \boldsymbol{a}_S e^{\boldsymbol{A}_S t} \boldsymbol{u}_S,$$

where $\boldsymbol{a}_S = \begin{bmatrix} p_1\boldsymbol{a}_1 & p_2\boldsymbol{a}_2 \end{bmatrix}, \boldsymbol{A}_S = \begin{bmatrix} \boldsymbol{A}_1 & \boldsymbol{0} \\ \boldsymbol{0} & \boldsymbol{A}_2 \end{bmatrix}$, and $\boldsymbol{u}_S = \begin{bmatrix} \boldsymbol{u}_1 \\ \boldsymbol{u}_2 \end{bmatrix}$.

The resulting CDF is a PH distribution with representations $\boldsymbol{a}_S = \begin{bmatrix} p_1\boldsymbol{a}_1 & p_2\boldsymbol{a}_2 \end{bmatrix}$, $\boldsymbol{A}_S = \begin{bmatrix} \boldsymbol{A}_1 & \boldsymbol{0} \\ \boldsymbol{0} & \boldsymbol{A}_2 \end{bmatrix}$, $\boldsymbol{u}_S = \begin{bmatrix} \boldsymbol{u}_1 \\ \boldsymbol{u}_2 \end{bmatrix}$. Thus, PH distribution is closed under mixing operations. The following theorem formalizes this result.

Theorem 2.2 [4] If $T_1 \sim PH(n_1, \boldsymbol{a}_1, \boldsymbol{A}_1)$ and $T_2 \sim PH(n_2, \boldsymbol{a}_2, \boldsymbol{A}_2)$ are independent, then $T_S = \text{mix}(T_1, T_2) \sim PH(n_S, \boldsymbol{a}_S, \boldsymbol{A}_S)$ with

$$n_S = n_1 + n_2, \quad \boldsymbol{a}_S = \begin{bmatrix} p_1\boldsymbol{a}_1 & p_2\boldsymbol{a}_2 \end{bmatrix}, \quad \boldsymbol{A}_S = \begin{bmatrix} \boldsymbol{A}_1 & \boldsymbol{0} \\ \boldsymbol{0} & \boldsymbol{A}_2 \end{bmatrix}. \tag{2.33}$$

Example 2.21 For T_1 and T_2 as given in Example 2.20, the mixture of the two random variables has PH distribution with a representation that can be derived using Theorem 2.2 as follows:

$$n_S = n_1 + n_2 = 4, \quad \boldsymbol{a}_S = \begin{bmatrix} p_1\boldsymbol{a}_1 & p_2\boldsymbol{a}_2 \end{bmatrix} = \begin{bmatrix} p_1 & 0 & p_2 & 0 \end{bmatrix},$$

$$\boldsymbol{A}_S = \begin{bmatrix} \boldsymbol{A}_1 & \boldsymbol{0} \\ \boldsymbol{0} & \boldsymbol{A}_2 \end{bmatrix} = \begin{bmatrix} \begin{bmatrix} -1 & 1 \\ 0 & -1 \end{bmatrix} & \boldsymbol{0} \\ \boldsymbol{0} & \begin{bmatrix} -2 & 2 \\ 0 & -2 \end{bmatrix} \end{bmatrix} = \begin{bmatrix} -1 & 1 & 0 & 0 \\ 0 & -1 & 0 & 0 \\ 0 & 0 & -2 & 2 \\ 0 & 0 & 0 & -2 \end{bmatrix}.$$

The requirements of being an initial probability vector and a transient matrix can be verified are satisfied by $\boldsymbol{a}_S$ and $\boldsymbol{A}_S$; hence, $(n_S, \boldsymbol{a}_S, \boldsymbol{A}_S)$ is valid as a PH representation. ■

2.6.4 Addition Operation

The distribution function of the sum of two independent random variables is the convolution of their respective distribution functions. Since the PH distribution is based on state transitions, there is another way to find the distribution of the sum of two variables having PH distributions, namely, by using the state transition diagram. The sum of T_1 and T_2, or $T_S = T_1 + T_2$, means that the end of CTMC for T_1 becomes the beginning of CTMC for T_2. That is, there is a transition from the absorbing state of T_1 to the transient state of T_2 and the absorbing vector of T_1 is connected to the initial probability vector of T_2. This process is shown in Figure 2.18, where T denotes the transient superstate, and A denotes the absorbing state.

It is apparent from Figure 2.18 that the transition $\boldsymbol{b}_1$ in the state transition diagram for T_1 is followed by $\boldsymbol{a}_2$ in the state transition diagram for T_2. In the concept of probability, the two probabilities

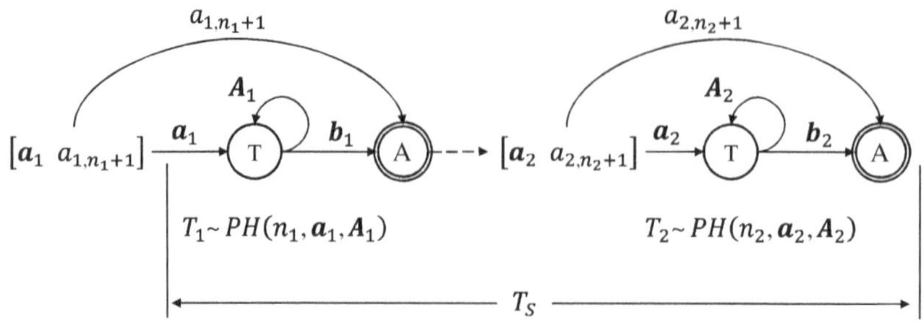

Figure 2.18 State transition diagram of the sum of two PH-distributed variables.

are combined by multiplication because they represent two events that occur simultaneously. Thus, the infinitesimal generator matrix of the sum of two PH-distributed random variables is

$$Q = \left[\begin{array}{cc|c} \boldsymbol{A}_1 & \boldsymbol{b}_1\boldsymbol{a}_2 & \boldsymbol{b}_1 a_{2,n_2+1} \\ \boldsymbol{0} & \boldsymbol{A}_2 & \boldsymbol{b}_2 \\ \hline \boldsymbol{0} & \boldsymbol{0} & 0 \end{array}\right], \quad \boldsymbol{a} = \left[\boldsymbol{a}_1\ a_{1,n_1+1}\boldsymbol{a}_2\ a_{1,n_1+1}a_{2,n_2+1}\right].$$

Exercise 2.10 Show that $\boldsymbol{Qu} = \boldsymbol{0}$ and $\boldsymbol{au} = 1$.

The infinitesimal generator starts from the variable T_1 so that the initial probability vector is $\boldsymbol{a}_1$. However, there is a possibility that it starts from T_2 if T_1 starts from the absorbing state with a probability $a_{1,n+1}$ and then it is continued with the initial probability of T_2, namely, $\boldsymbol{a}_2$. Another possibility is that both T_1 and T_2 start from their respective absorbing states with probability $a_{1,n_1+1}a_{2,n_2+1}$. However, it is not part of the initial probability vector for the PH distribution. Thus, the initial probability vector of the sum of these two variables is $\boldsymbol{a}_S = \left[\boldsymbol{a}_1\ a_{1,n_1+1}\boldsymbol{a}_2\right]$. It can be seen that the order of the PH distribution of $T_S = \text{add}(T_1, T_2)$ is $n_1 + n_2$. This result is summarized in the following theorem.

Theorem 2.3 [4] If $T_1 \sim PH(n_1, \boldsymbol{a}_1, \boldsymbol{A}_1)$ and $T_2 \sim PH(n_2, \boldsymbol{a}_2, \boldsymbol{A}_2)$ are independent, then $T_S = \text{add}(T_1, T_2) = (T_1 + T_2) \sim PH(n_S, \boldsymbol{a}_S, \boldsymbol{A}_S)$ with

$$n_S = n_1 + n_2, \quad \boldsymbol{a}_S = \left[\boldsymbol{a}_1\ a_{1,n_1+1}\boldsymbol{a}_2\right], \quad \boldsymbol{A}_S = \begin{bmatrix} \boldsymbol{A}_1 & \boldsymbol{b}_1\boldsymbol{a}_2 \\ \boldsymbol{0} & \boldsymbol{A}_2 \end{bmatrix}. \tag{2.34}$$

Example 2.22 For T_1 and T_2 as given in Example 2.20, the sum of them has PH distribution with representation that can be found using Theorem 2.3 as follows.

PH distributions for both variables do not have initial probabilities starting from the absorbing states or $a_{i,n_i+1} = 0, i = 1, 2$. Hence,

$$n_S = n_1 + n_2 = 4,$$

$$\boldsymbol{a}_S = \left[\boldsymbol{a}_1\ a_{1,n_1+1}\boldsymbol{a}_2\right] = \left[\boldsymbol{a}_1\ \boldsymbol{0}\right] = \left[[1\ 0]\ [0\ 0]\right] = \left[1\ 0\ 0\ 0\right],$$

$$\boldsymbol{b}_1 = -\boldsymbol{A}_1\boldsymbol{u} = -\begin{bmatrix} -1 & 1 \\ 0 & -1 \end{bmatrix}\begin{bmatrix} 1 \\ 1 \end{bmatrix} = \begin{bmatrix} 0 \\ 1 \end{bmatrix},$$

$$A_S = \begin{bmatrix} A_1 & b_1 a_2 \\ 0 & A_2 \end{bmatrix} = \begin{bmatrix} \begin{bmatrix} -1 & 1 \\ 0 & -1 \end{bmatrix} & \begin{bmatrix} 0 \\ 1 \end{bmatrix} \begin{bmatrix} 1 & 0 \end{bmatrix} \\ 0 & \begin{bmatrix} -2 & 2 \\ 0 & -2 \end{bmatrix} \end{bmatrix} = \begin{bmatrix} -1 & 1 & 0 & 0 \\ 0 & -1 & 1 & 0 \\ 0 & 0 & -2 & 2 \\ 0 & 0 & 0 & -2 \end{bmatrix}.$$

This result shows the advantages of using the matrix-based distribution in solving problems involving convolution operations; that is, there is no need to use convolution integrals, just simply use transitions between states. It can be seen that $\boldsymbol{a}_S$ and $\boldsymbol{A}_S$ satisfy the requirements as an initial probability vector and a transient matrix, respectively; hence, $(n_S, \boldsymbol{a}_S, \boldsymbol{A}_S)$ is valid as a PH representation. ■

A mathematical proof of Theorem 2.3 can be carried out using the Laplace–Stieltjes transform for the CDF given in Equation (2.21),[1]

$$F^*_{T_S}(s) = E\left[e^{-sT_S}\right] = E\left[e^{-s(T_1+T_2)}\right] = E\left[e^{-sT_1}e^{-sT_2}\right].$$

Since T_1 and T_2 are independent, it follows that

$$F^*_{T_S}(s) = E\left[e^{-sT_1}\right] E\left[e^{-sT_2}\right] = F^*_{T_1}(s)F^*_{T_2}(s).$$

Using Equation (2.22) obtains

$$F^*_{T_S}(s) = a_{S,n_S+1} - \boldsymbol{a}_S(s\boldsymbol{I} - \boldsymbol{A}_S)^{-1}\boldsymbol{A}_S\boldsymbol{u}_S. \tag{2.35}$$

Using Equation (2.34), $\boldsymbol{b}_1 = -\boldsymbol{A}_1\boldsymbol{u}_1$, and $1 - \boldsymbol{a}_2\boldsymbol{u}_2 = a_{2,n_2+1}$ obtains

$$\boldsymbol{A}_S\boldsymbol{u}_S = \begin{bmatrix} \boldsymbol{A}_1 & \boldsymbol{b}_1\boldsymbol{a}_2 \\ \boldsymbol{0} & \boldsymbol{A}_2 \end{bmatrix} \begin{bmatrix} \boldsymbol{u}_1 \\ \boldsymbol{u}_2 \end{bmatrix} = \begin{bmatrix} \boldsymbol{A}_1\boldsymbol{u}_1 - \boldsymbol{A}_1\boldsymbol{u}_1\boldsymbol{a}_2\boldsymbol{u}_2 \\ \boldsymbol{A}_2\boldsymbol{u}_2 \end{bmatrix} = \begin{bmatrix} \boldsymbol{A}_1\boldsymbol{u}_1(1 - \boldsymbol{a}_2\boldsymbol{u}_2) \\ \boldsymbol{A}_2\boldsymbol{u}_2 \end{bmatrix} = \begin{bmatrix} a_{2,n_2+1}\boldsymbol{A}_1\boldsymbol{u}_1 \\ \boldsymbol{A}_2\boldsymbol{u}_2 \end{bmatrix}. \tag{2.36}$$

Substituting $\boldsymbol{A}_S$ in Equation (2.34) into Equation (2.35) and using the matrix inverse lemma

$$\begin{bmatrix} \boldsymbol{A} & \boldsymbol{B} \\ \boldsymbol{0} & \boldsymbol{C} \end{bmatrix}^{-1} = \begin{bmatrix} \boldsymbol{A}^{-1} & -\boldsymbol{A}^{-1}\boldsymbol{B}\boldsymbol{C}^{-1} \\ \boldsymbol{0} & \boldsymbol{C}^{-1} \end{bmatrix},$$

produce

$$(s\boldsymbol{I} - \boldsymbol{A}_S)^{-1} = \begin{bmatrix} s\boldsymbol{I} - \boldsymbol{A}_1 & \boldsymbol{A}_1\boldsymbol{u}_1\boldsymbol{a}_2 \\ \boldsymbol{0} & s\boldsymbol{I} - \boldsymbol{A}_2 \end{bmatrix}^{-1} = \begin{bmatrix} (s\boldsymbol{I} - \boldsymbol{A}_1)^{-1} & -(s\boldsymbol{I} - \boldsymbol{A}_1)^{-1}(\boldsymbol{A}_1\boldsymbol{u}_1\boldsymbol{a}_2)(s\boldsymbol{I} - \boldsymbol{A}_2)^{-1} \\ \boldsymbol{0} & (s\boldsymbol{I} - \boldsymbol{A}_2)^{-1} \end{bmatrix}. \tag{2.37}$$

Multiplying Equation (2.37) with Equation (2.36) and $\boldsymbol{a}_S = \begin{bmatrix} \boldsymbol{a}_1 & a_{1,n_1+1}\boldsymbol{a}_2 \end{bmatrix}$ obtains

$$\begin{aligned} \boldsymbol{a}_S(s\boldsymbol{I} - \boldsymbol{A}_S)^{-1}\boldsymbol{A}_S\boldsymbol{u}_S &= \boldsymbol{a}_1(s\boldsymbol{I} - \boldsymbol{A}_1)^{-1}a_{2,n_2+1}\boldsymbol{A}_1\boldsymbol{u}_1 \\ &\quad - \boldsymbol{a}_1(s\boldsymbol{I} - \boldsymbol{A}_1)^{-1}(\boldsymbol{A}_1\boldsymbol{u}_1\boldsymbol{a}_2)(s\boldsymbol{I} - \boldsymbol{A}_2)^{-1}\boldsymbol{A}_2\boldsymbol{u}_2 \\ &\quad + a_{1,n_1+1}\boldsymbol{a}_2(s\boldsymbol{I} - \boldsymbol{A}_2)^{-1}\boldsymbol{A}_2\boldsymbol{u}_2. \end{aligned} \tag{2.38}$$

The term that expresses discontinuity can be broken down into

$$\begin{aligned} a_{S,n_S+1} &= (1 - \boldsymbol{a}_S\boldsymbol{u}_S) = 1 - \begin{bmatrix} \boldsymbol{a}_1 & a_{1,n_1+1}\boldsymbol{a}_2 \end{bmatrix} \begin{bmatrix} \boldsymbol{u}_1 \\ \boldsymbol{u}_2 \end{bmatrix} = 1 - \boldsymbol{a}_1\boldsymbol{u}_1 - (1 - \boldsymbol{a}_1\boldsymbol{u}_1)\boldsymbol{a}_2\boldsymbol{u}_2 \\ &= (1 - \boldsymbol{a}_1\boldsymbol{u}_1)(1 - \boldsymbol{a}_2\boldsymbol{u}_2). \end{aligned}$$

1 Advanced material, can be omitted in lecture.

Hence,

$$a_{S,n_S+1} = a_{1,n_1+1}a_{2,n_2+1}. \tag{2.39}$$

Substituting Equations (2.38) and (2.39) into Equation (2.35) obtains

$$\begin{aligned} F^*_{T_S}(s) &= a_{1,n_1+1}a_{2,n_2+1} - \left(\boldsymbol{a}_1(s\boldsymbol{I} - \boldsymbol{A}_1)^{-1}\boldsymbol{A}_1\boldsymbol{u}_1\right) a_{2,n_2+1} \\ &\quad + \boldsymbol{a}_1(s\boldsymbol{I} - \boldsymbol{A}_1)^{-1}(\boldsymbol{A}_1\boldsymbol{u}_1\boldsymbol{a}_2)(s\boldsymbol{I} - \boldsymbol{A}_2)^{-1}\boldsymbol{A}_2\boldsymbol{u}_2 \\ &\quad - a_{1,n_1+1}\left(\boldsymbol{a}_2(s\boldsymbol{I} - \boldsymbol{A}_2)^{-1}\boldsymbol{A}_2\boldsymbol{u}_2\right). \end{aligned} \tag{2.40}$$

Equation (2.40) can be factorized into

$$F^*_{T_S}(s) = \left(a_{1,n_1+1} - \boldsymbol{a}_1(s\boldsymbol{I} - \boldsymbol{A}_1)^{-1}\boldsymbol{A}_1\boldsymbol{u}_1\right)\left(a_{2,n_2+1} - \boldsymbol{a}_2(s\boldsymbol{I} - \boldsymbol{A}_2)^{-1}\boldsymbol{A}_2\boldsymbol{u}_2\right) = F^*_{T_1}(s)F^*_{T_2}(s),$$

as must be proven.

2.6.5 Maximum Operation

PH distribution of the random variable resulting from the maximum operation of two independent PH-distributed random variables can be derived using the state transition diagram as in the addition operation. The method needs to be explained in detail because it will be used in Chapters 4 and 5.

The maximum of two PH-distributed random variables T_1 and T_2 has four superstates, namely, states where:

- Both T_1 and T_2 are in their transient states. These superstates are denoted as (T,T).
- Both T_1 and T_2 are in their absorbing state. This state is denoted as (A,A).
- T_1 is in its transient states, and T_2 is in its absorbing state. These superstates are denoted as (T,A).
- T_2 is in its transient states, and T_1 is in its absorbing state. These superstates are denoted as (A,T).

The derivation of the distribution of maximum of two independent variables is illustrated with the simplest case where $T_i \sim \exp(\lambda_i)$, $i = 1, 2$. Transition from state (T,T) to state (T,A) is due to T_2 entering its absorbing state, which has transition rate λ_2. Transition from state (T,T) to state (A,T) is due to T_1 entering its absorbing state which has transition rate λ_1. Transition from state (T,A) to state (A,A) is due to T_1 entering its absorbing state, and because of memoryless property of exponential distribution, its transition rate stays the same as λ_1. Similarly, transition rate from state (A,T) to state (A,A) is λ_2. These transitions are presented in a state transition diagram as depicted in Figure 2.19.

These results are extended for PH distribution. From the result of the minimum operation, the superstate (T,T) has a transient matrix $\boldsymbol{A}_1 \oplus \boldsymbol{A}_2$. The duration of stay in superstate (T,A) is until T_1 enters the absorbing state, so the transient matrix for this state is $\boldsymbol{A}_1$. Likewise, the transient matrix in the superstate (A,T) is $\boldsymbol{A}_2$. The transition from superstate (T,T) to superstate (A,T) is due to T_1 entering the absorbing state with a transition rate given by its absorbing vector $\boldsymbol{b}_1$, while T_2 remains in its transient states, which is represented by an identity matrix. Therefore, the transition from superstate (T,T) to superstate (T,A) is the Kronecker product of $\boldsymbol{b}_1$ and $\boldsymbol{I}_2$, where $\boldsymbol{I}_2$ is an identity matrix with the same dimension as $\boldsymbol{A}_2$. The Kronecker product with the identity matrix of the same dimension as $\boldsymbol{A}_2$ when making the transition from superstate (T,T) to superstate (A,T) is to guarantee that the transition can reach all transient states of T_2. Likewise, the transition from superstate (T,T) to superstate (T,A) is the Kronecker product of $\boldsymbol{b}_2$ and $\boldsymbol{I}_1$, where $\boldsymbol{I}_1$ is an identity matrix with the same dimension as $\boldsymbol{A}_1$. The transition rate from superstate (T,A) to superstate (A,A) is the absorbing vector of T_1, which is $\boldsymbol{b}_1$. Likewise, the transition rate from superstate (A,T) to

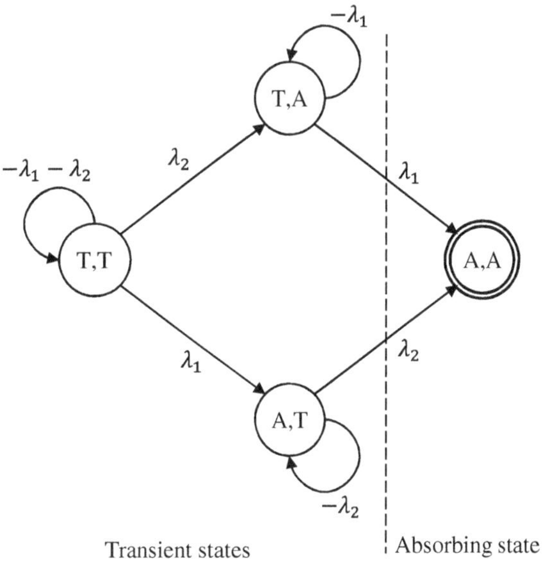

Figure 2.19 State transition diagram of the maximum of two exponentially distributed variables.

superstate (A,A) is the absorbing vector of T_2, which is $\boldsymbol{b}_2$. There is no transition from superstate (T,T) to superstate (A,A) because such a transition can only occur if there are two random events occur simultaneously, something that is impossible in a continuous-time process. However, there is a possibility that the process starts at (A,A) state, namely, if both variables start at their absorbing state. The transition rate directly to (A,A) state is equal to the product of the probabilities each variable starts from its absorbing state, thus equal to $a_{1,n_1+1}a_{2,n_2+1}$. Note that the transient matrix of T_1 in the superstate (T,A) is the same as its transient matrix in the superstate (T,T). This is a nice property of PH distribution that will be derived and explained in Chapter 3.

This explanation is presented as a state transition diagram in Figure 2.20. The diagram includes the possibilities of jumping to the next superstate (T,A) and (A,T) without going through the transient state (T,T) due to the possibility of the process starting from the absorbing state of either one of the variables. There is also a possibility to enter state (A,A) directly, namely, if the CTMCs of the two random variables start from their respective absorbing states.

The infinitesimal generator matrix and initial probability vector for $T_S = \max(T_1, T_2)$ can be derived from Figure 2.20 as

$$
\boldsymbol{Q} = \begin{bmatrix} TT \to TT & TT \to TA & TT \to AT & TT \to AA \\ TA \to TT & TA \to TA & TA \to AT & TA \to AA \\ AT \to TT & AT \to TA & AT \to AT & AT \to AA \\ AA \to TT & AA \to TA & AA \to AT & AA \to AA \end{bmatrix}
= \left[\begin{array}{ccc|c} \boldsymbol{A}_1 \oplus \boldsymbol{A}_2 & \boldsymbol{I}_{n_1} \otimes \boldsymbol{b}_2 & \boldsymbol{b}_1 \otimes \boldsymbol{I}_{n_2} & \boldsymbol{0} \\ \boldsymbol{0} & \boldsymbol{A}_1 & \boldsymbol{0} & \boldsymbol{b}_1 \\ \boldsymbol{0} & \boldsymbol{0} & \boldsymbol{A}_2 & \boldsymbol{b}_2 \\ \hline \boldsymbol{0} & \boldsymbol{0} & \boldsymbol{0} & 0 \end{array}\right],
$$

$$
\boldsymbol{a} = \left[\boldsymbol{a}_1 \otimes \boldsymbol{a}_2 \;\; a_{2,n_2+1}\boldsymbol{a}_1 \;\; a_{1,n_1+1}\boldsymbol{a}_2 \;\; a_{1,n_1+1}a_{2,n_2+1}\right].
$$

Exercise 2.11 Show that $\boldsymbol{Qu} = \boldsymbol{0}$ and $\boldsymbol{au} = 1$.

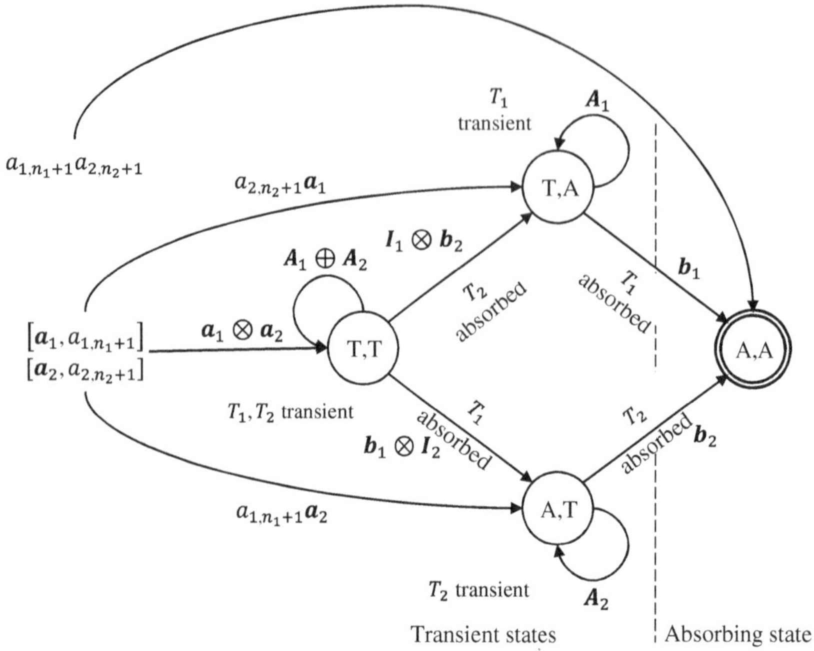

Figure 2.20 State transition diagram of the maximum of two PH-distributed variables.

The initial probability vector, order, and transient matrix for the PH distribution of the maximum of two independent random variables having PH distributions can be derived from $\boldsymbol{Q}$ and $\boldsymbol{a}$. The result is presented as Theorem 2.4.

Theorem 2.4 [4] If $T_1 \sim PH(n_1, \boldsymbol{a}_1, \boldsymbol{A}_1)$ and $T_2 \sim PH(n_2, \boldsymbol{a}_2, \boldsymbol{A}_2)$ are independent, then $T_S = \max(T_1, T_2) \sim PH(n_S, \boldsymbol{a}_S, \boldsymbol{A}_S)$ with

$$n_S = n_1 n_2 + n_1 + n_2, \quad \boldsymbol{a}_S = \begin{bmatrix} \boldsymbol{a}_1 \otimes \boldsymbol{a}_2 & a_{2,n_2+1}\boldsymbol{a}_1 & a_{1,n_1+1}\boldsymbol{a}_2 \end{bmatrix},$$

$$\boldsymbol{A}_S = \begin{bmatrix} \boldsymbol{A}_1 \oplus \boldsymbol{A}_2 & \boldsymbol{I}_1 \otimes \boldsymbol{b}_2 & \boldsymbol{b}_1 \otimes \boldsymbol{I}_2 \\ \boldsymbol{0} & \boldsymbol{A}_1 & \boldsymbol{0} \\ \boldsymbol{0} & \boldsymbol{0} & \boldsymbol{A}_2 \end{bmatrix}. \tag{2.41}$$

Mathematical proof of Theorem 2.4 requires CTMC understanding, which is beyond the scope of this book. A not-so-general proof of this theorem can be read in He [3].

Example 2.23 The maximum of T_1 and T_2 in Example 2.20 has PH distribution with a representation that can be obtained using Theorem 2.4.

Since the PH distributions of T_1 and T_2 have no jumps at the origin, $a_{1,n_1+1} = a_{2,n_1+1} = 0$. Thus,

$$n_S = n_1 n_2 + n_1 + n_2 = 8,$$

$$\boldsymbol{a}_S = \begin{bmatrix} \boldsymbol{a}_1 \otimes \boldsymbol{a}_2 & \boldsymbol{0} & \boldsymbol{0} \end{bmatrix} = \begin{bmatrix} \begin{bmatrix} 1 & 0 & 0 & 0 \end{bmatrix} & \begin{bmatrix} 0 & 0 \end{bmatrix} & \begin{bmatrix} 0 & 0 \end{bmatrix} \end{bmatrix} = \begin{bmatrix} 1 & 0 & 0 & 0 & 0 & 0 & 0 & 0 \end{bmatrix}.$$

From Examples 2.20 and 2.22,

$$\boldsymbol{A}_1 = \begin{bmatrix} -1 & 1 \\ 0 & -1 \end{bmatrix}, \quad \boldsymbol{A}_2 = \begin{bmatrix} -2 & 2 \\ 0 & -2 \end{bmatrix}, \quad \boldsymbol{b}_1 = -\boldsymbol{A}_1\boldsymbol{u} = \begin{bmatrix} 0 \\ 1 \end{bmatrix}, \quad \boldsymbol{b}_2 = -\boldsymbol{A}_2\boldsymbol{u} = \begin{bmatrix} 0 \\ 2 \end{bmatrix},$$

$$\boldsymbol{A}_1 \oplus \boldsymbol{A}_2 = \begin{bmatrix} -3 & 2 & 1 & 0 \\ 0 & -3 & 0 & 1 \\ 0 & 0 & -3 & 2 \\ 0 & 0 & 0 & -3 \end{bmatrix}.$$

Therefore,

$$\boldsymbol{b}_1 \otimes \boldsymbol{I}_2 = \begin{bmatrix} 0 \\ 1 \end{bmatrix} \otimes \begin{bmatrix} 1 & 0 \\ 0 & 1 \end{bmatrix} = \begin{bmatrix} 0 & 0 \\ 0 & 0 \\ 1 & 0 \\ 0 & 1 \end{bmatrix}, \quad \boldsymbol{I}_1 \otimes \boldsymbol{b}_2 = \begin{bmatrix} 1 & 0 \\ 0 & 1 \end{bmatrix} \otimes \begin{bmatrix} 0 \\ 2 \end{bmatrix} = \begin{bmatrix} 0 & 0 \\ 2 & 0 \\ 0 & 0 \\ 0 & 2 \end{bmatrix},$$

$$\boldsymbol{A}_S = \begin{bmatrix} \boldsymbol{A}_1 \oplus \boldsymbol{A}_2 & \boldsymbol{I}_1 \otimes \boldsymbol{b}_2 & \boldsymbol{b}_1 \otimes \boldsymbol{I}_2 \\ \boldsymbol{0} & \boldsymbol{A}_1 & \boldsymbol{0} \\ \boldsymbol{0} & \boldsymbol{0} & \boldsymbol{A}_2 \end{bmatrix} = \left[\begin{array}{cccc|cc|cc} -3 & 2 & 1 & 0 & 0 & 0 & 0 & 0 \\ 0 & -3 & 0 & 1 & 2 & 0 & 0 & 0 \\ 0 & 0 & -3 & 2 & 0 & 0 & 1 & 0 \\ 0 & 0 & 0 & -3 & 0 & 2 & 0 & 1 \\ \hline 0 & 0 & 0 & 0 & -1 & 1 & 0 & 0 \\ 0 & 0 & 0 & 0 & 0 & -1 & 0 & 0 \\ \hline 0 & 0 & 0 & 0 & 0 & 0 & -2 & 2 \\ 0 & 0 & 0 & 0 & 0 & 0 & 0 & -2 \end{array}\right].$$

Note that $\boldsymbol{\alpha}_S$ and $\boldsymbol{A}_S$, respectively, satisfy the requirement of being an initial probability vector and a transient matrix; hence, $(n_S, \boldsymbol{\alpha}_S, \boldsymbol{A}_S)$ is a valid PH representation.

Although the matrix dimension is quite large, many of its elements are zero, and it has an upper block diagonal structure that can be computed quickly. ■

2.6.6 Sorting Operations

Random variables $\{T_i, i = 1, 2, \dots\}$ having identical PH distribution with representation $(n, \boldsymbol{\alpha}, \boldsymbol{A})$ is a sequence of interarrival time of incoming events. The events are selected randomly with a known probability of satisfying a criterion. The interest is to find the distribution of the waiting time T_S until an event that satisfies the criterion occurs. The possible realizations are as follows:

- $T_S = T_1$ if event 1 is selected.
- $T_S = T_1 + T_2$ if event 1 is not selected and event 2 is selected.

…

- $T_S = T_1 + T_2 + \dots T_k$ if events 1, 2, …, $(k-1)$ are not selected and event k is selected.

The random variable T_S can be expressed as

$$T_S = T_0 + T_1 + T_2 + \cdots + T_W,$$

where $T_0 = 0$ and W is a discrete-valued random variable. For sorting operation, W cannot be equal to zero; thus, $P[W = 0] = 0$.

The distribution of T_S is given by the following theorem.

Theorem 2.5 [4] If W has a discrete PH distribution with representation $(m, \boldsymbol{w}, \boldsymbol{W})$, namely, $P[W=k]=\boldsymbol{w}\boldsymbol{W}^{k-1}(\boldsymbol{I}-\boldsymbol{W}\boldsymbol{u})$, $k=1,2,\ldots$, with $P[W=0]=w_{m+1}$, and $T_i \sim PH(n, \boldsymbol{a}, \boldsymbol{A})$, $i=1,2,\ldots$, then $T_S \sim PH(n_S, \boldsymbol{a}_S, \boldsymbol{A}_S)$, where

$$n_S = nm, \quad \boldsymbol{a}_S = \boldsymbol{a} \otimes \boldsymbol{w}(\boldsymbol{I} - a_{n+1}\boldsymbol{W})^{-1},$$

$$\boldsymbol{A}_S = \boldsymbol{A} \otimes \boldsymbol{I} + (1 - a_{n+1})\boldsymbol{b}\boldsymbol{a} \otimes (\boldsymbol{I} - a_{n+1}\boldsymbol{W})^{-1}\boldsymbol{W}. \tag{2.42}$$

The proof of this theorem requires an understanding of CTMC, which is beyond the scope of this book. Those interested can read Neuts [4]. (Note: discrete PH distribution will be discussed in Chapter 7 for discrete-time systems.)

In practice, the selection process assumes that the probability of getting a not suitable choice is a constant scalar q for each sorting. In this case, W is a geometric random variable with $P[W = k] = q^{k-1}(1-q)$, which is a discrete PH distribution with representation $m = 1$, $\boldsymbol{w} = [1]$, $\boldsymbol{W} = q$. Since the interest is in making a suitable selection, define $p = 1-q$ as the probability of making a suitable selection. Based on Theorem 2.5, the following lemma can be derived.

Lemma 2.1 If W is geometrically distributed and $T_i \sim PH(n, \boldsymbol{a}, \boldsymbol{A})$, then $T_S \sim PH(n_S, \boldsymbol{a}_S, \boldsymbol{A}_S)$ with

$$n_S = n, \quad \boldsymbol{a}_S = \boldsymbol{a} \otimes (1 - a_{n+1}(1-p))^{-1} = \frac{\boldsymbol{a}}{(1 - a_{n+1}(1-p))},$$

$$\boldsymbol{A}_S = \boldsymbol{A} \otimes 1 + (1 - a_{n+1})\boldsymbol{b}\boldsymbol{a} \otimes (1 - a_{n+1}(1-p))^{-1}(1-p) = \boldsymbol{A} + \frac{(1 - a_{n+1})(1-p)}{(1 - a_{n+1}(1-p))}\boldsymbol{b}\boldsymbol{a}. \tag{2.43}$$

If T_i has no discontinuity at the origin, namely, $a_{n+1} = 0$, then Equation (2.43) reduces to

$$n_S = n, \quad \boldsymbol{a}_S = \boldsymbol{a}, \quad \boldsymbol{A}_S = \boldsymbol{A} + (1-p)\boldsymbol{b}\boldsymbol{a}. \tag{2.44}$$

Example 2.24 Given an independent sequence of random variables $\{T_i, i = 1, 2, \ldots\}$ having identical PH distribution with a representation

$$n = 2, \quad \boldsymbol{a} = \begin{bmatrix} 1 & 0 \end{bmatrix}, \quad \boldsymbol{A} = \begin{bmatrix} -1 & 1 \\ 0 & -1 \end{bmatrix},$$

selection is carried out with the probability of obtaining the appropriate choice each time a selection is made constantly equal to 0.9.

Since T_S has no discontinuity at the origin, the PH distribution for T_S has a representation that can be obtained from Equation (2.44) as

$$n_S = n = 2, \quad \boldsymbol{a}_S = \boldsymbol{a} = \begin{bmatrix} 1 & 0 \end{bmatrix}, \quad \boldsymbol{b} = -\boldsymbol{A}\boldsymbol{u} = -\begin{bmatrix} -1 & 1 \\ 0 & -1 \end{bmatrix}\begin{bmatrix} 1 \\ 1 \end{bmatrix} = \begin{bmatrix} 0 \\ 1 \end{bmatrix},$$

$$\boldsymbol{A}_S = \boldsymbol{A} + (1-p)\boldsymbol{b}\boldsymbol{a} = \begin{bmatrix} -1 & 1 \\ 0 & -1 \end{bmatrix} + (1-0.9)\begin{bmatrix} 0 \\ 1 \end{bmatrix}\begin{bmatrix} 1 & 0 \end{bmatrix} = \begin{bmatrix} -1 & 1 \\ 0.1 & -1 \end{bmatrix}.$$

■

It is important to note that closure properties of PH distribution always result in nonsingular transient matrices [4, 12]. This nonsingularity property is important in calculating moments, which always involve matrix inversion.

The closure properties of PH distribution can be utilized for system modeling to derive PH distribution of variables related to system operation.

2.6.7 System Modeling Using Closure Properties of PH Distribution

The closure properties of PH distribution can be used to model systems involving PH distribution. The following applications show system modeling using the closure properties of PH distribution.

Application 2.3 A company uses two machines to produce an item. The processing time of each item on machines 1 and 2 has 2-Erlang and 3-Erlang distributions with means of 1 and 3 hours, respectively. The items produced are not all acceptable, with the proportions of acceptable items produced by machines 1 and 2 being 0.5 and 0.7, respectively. An acceptable item from any machine is packed with packaging time that has 4-Erlang distribution with a mean 1 hour. The packed item is shipped directly to the customer. Assume that each product is processed continuously, and the next product starts after the previous product is shipped to the customer. The interest is in finding the distribution of the waiting time until the first item is shipped to the customer.

Let $T_i \sim PH(n_i, \boldsymbol{a}_i, \boldsymbol{A}_i)$, $i = 1, 2$, be the processing time for each item on machines 1 and 2, respectively. The values of $(n_i, \boldsymbol{a}_i, \boldsymbol{A}_i)$, $i = 1, 2$, are available from the distribution of the processing time on each machine as given in Table 2.1:

- Processing time at machine 1 has a 2-Erlang distribution with mean = 1. Use Table 2.1 for 2-Erlang to obtain

$$\text{Mean} = 1 = 2/\lambda \rightarrow \lambda = 2, \quad \boldsymbol{A}_1 = \begin{bmatrix} -\lambda & \lambda \\ 0 & -\lambda \end{bmatrix} = \begin{bmatrix} -2 & 2 \\ 0 & -2 \end{bmatrix}.$$

  The PH representation of processing time at machine 1 is

$$n_1 = 2, \quad \boldsymbol{a}_1 = \begin{bmatrix} 1 & 0 \end{bmatrix}, \quad \boldsymbol{A}_1 = \begin{bmatrix} -2 & 2 \\ 0 & -2 \end{bmatrix}.$$

- Processing time at machine 2 has 3-Erlang distribution with mean = 3. Use Table 2.1 for 3-Erlang to obtain

$$\text{Mean} = 3 = 3/\lambda \rightarrow \lambda = 1, \quad \boldsymbol{A}_2 = \begin{bmatrix} -\lambda & \lambda & 0 \\ 0 & -\lambda & \lambda \\ 0 & 0 & -\lambda \end{bmatrix} = \begin{bmatrix} -1 & 1 & 0 \\ 0 & -1 & 1 \\ 0 & 0 & -1 \end{bmatrix}.$$

  The PH representation of processing time at machine 2 is

$$n_2 = 3, \quad \boldsymbol{a}_2 = \begin{bmatrix} 1 & 0 & 0 \end{bmatrix}, \quad \boldsymbol{A}_2 = \begin{bmatrix} -1 & 1 & 0 \\ 0 & -1 & 1 \\ 0 & 0 & -1 \end{bmatrix}.$$

  The proportion of acceptable items produced by machine 1 is 0.5. According to Lemma 2.1, the PH representation for T_1', the time required by machine 1 to produce an acceptable item, is $(n_1', \boldsymbol{a}_1', \boldsymbol{A}_1')$, where

$$n_1' = n_1 = 2, \quad \boldsymbol{a}_1' = \boldsymbol{a}_1 = \begin{bmatrix} 1 & 0 \end{bmatrix}, \quad \boldsymbol{b}_1 = -\boldsymbol{A}_1 \boldsymbol{u} = \begin{bmatrix} 0 \\ 2 \end{bmatrix},$$

$$\boldsymbol{A}_1' = \boldsymbol{A}_1 + (1 - 0.5)\boldsymbol{b}_1 \boldsymbol{a}_1 = \begin{bmatrix} -2 & 2 \\ 0 & -2 \end{bmatrix} + 0.5 \begin{bmatrix} 0 \\ 2 \end{bmatrix} \begin{bmatrix} 1 & 0 \end{bmatrix} = \begin{bmatrix} -2 & 2 \\ 1 & -2 \end{bmatrix}.$$

Likewise, the PH representation for T_2', the time required by machine 2 to produce an acceptable item, is $(n_2', \boldsymbol{a}_2', \boldsymbol{A}_2')$, where

$$n_2' = n_2 = 3, \quad \boldsymbol{a}_2' = \boldsymbol{a}_2 = \begin{bmatrix} 1 & 0 & 0 \end{bmatrix}, \quad \boldsymbol{b}_2 = -\boldsymbol{A}_2\boldsymbol{u} = \begin{bmatrix} 0 \\ 0 \\ 1 \end{bmatrix},$$

$$\boldsymbol{A}_2' = \boldsymbol{A}_2 + (1-0.7)\boldsymbol{b}_2 a_2 = \begin{bmatrix} -1 & 1 & 0 \\ 0 & -1 & 1 \\ 0 & 0 & -1 \end{bmatrix} + 0.3\begin{bmatrix} 0 \\ 0 \\ 1 \end{bmatrix}\begin{bmatrix} 1 & 0 & 0 \end{bmatrix} = \begin{bmatrix} -1 & 1 & 0 \\ 0 & -1 & 1 \\ 0.3 & 0 & -1 \end{bmatrix}.$$

The processing time to produce one acceptable item from either machine 1 or machine 2 is $T_3 = \min(T_1', T_2')$. The PH representation for T_3 is $(n_3, \boldsymbol{a}_3, \boldsymbol{A}_3)$, which can be found using Theorem 2.1 as

$$n_3 = n_1' n_2' = 2 \times 3 = 6,$$

$$\boldsymbol{a}_3 = \boldsymbol{a}_1' \otimes \boldsymbol{a}_2' = \begin{bmatrix} 1 & 0 \end{bmatrix} \otimes \begin{bmatrix} 1 & 0 & 0 \end{bmatrix} = \begin{bmatrix} 1 & 0 & 0 & 0 & 0 & 0 \end{bmatrix},$$

$$\boldsymbol{A}_3 = \boldsymbol{A}_1' \oplus \boldsymbol{A}_2' = \begin{bmatrix} -2 & 2 \\ 1 & -2 \end{bmatrix} \oplus \begin{bmatrix} -1 & 1 & 0 \\ 0 & -1 & 1 \\ 0.3 & 0 & -1 \end{bmatrix} = \begin{bmatrix} -3 & 1 & 0 & 2 & 0 & 0 \\ 0 & -3 & 1 & 0 & 2 & 0 \\ 0.3 & 0 & -3 & 0 & 0 & 2 \\ 1 & 0 & 0 & -3 & 1 & 0 \\ 0 & 1 & 0 & 0 & -3 & 1 \\ 0 & 0 & 1 & 0.3 & 0 & -3 \end{bmatrix}.$$

Let the packaging time be $T_4 \sim PH(n_4, \boldsymbol{a}_4, \boldsymbol{A}_4)$, which is known to have a 4-Erlang distribution with a mean 1 hour. Its PH representation is

$$n_4 = 4, \quad \boldsymbol{a}_4 = \begin{bmatrix} 1 & 0 & 0 & 0 \end{bmatrix}, \quad \boldsymbol{A}_4 = \begin{bmatrix} -4 & 4 & 0 & 0 \\ 0 & -4 & 4 & 0 \\ 0 & 0 & -4 & 4 \\ 0 & 0 & 0 & -4 \end{bmatrix}.$$

The time to produce one packed item is $T_5 = T_3 + T_4$. The PH representation for T_5 is $(n_4, \boldsymbol{a}_5, \boldsymbol{A}_5)$, which can be found using Theorem 2.3 as

$$n_5 = n_3 + n_4 = 10,$$

$$a_5 = \begin{bmatrix} \boldsymbol{a}_3 & \boldsymbol{0} \end{bmatrix} = \begin{bmatrix} 1 & 0 & 0 & 0 & 0 & 0 & 0 & 0 & 0 & 0 \end{bmatrix},$$

$$\boldsymbol{b}_3 = -\boldsymbol{A}_3\boldsymbol{u} = \begin{bmatrix} 0 & 0 & 0.7 & 1 & 1 & 1.7 \end{bmatrix}^T,$$

$$\boldsymbol{b}_3\boldsymbol{a}_4 = \begin{bmatrix} \boldsymbol{b}_3 & \boldsymbol{0} & \boldsymbol{0} & \boldsymbol{0} \end{bmatrix},$$

$$A_5 = \begin{bmatrix} A_3 & b_3 a_4 \\ 0 & A_4 \end{bmatrix} = \begin{bmatrix} -3 & 1 & 0 & 2 & 0 & 0 & 0 & 0 & 0 & 0 \\ 0 & -3 & 1 & 0 & 2 & 0 & 0 & 0 & 0 & 0 \\ 0.3 & 0 & -3 & 0 & 0 & 2 & 0.7 & 0 & 0 & 0 \\ 1 & 0 & 0 & -3 & 1 & 0 & 1 & 0 & 0 & 0 \\ 0 & 1 & 0 & 0 & -3 & 1 & 1 & 0 & 0 & 0 \\ 0 & 0 & 1 & 0.3 & 0 & -3 & 1.7 & 0 & 0 & 0 \\ 0 & 0 & 0 & 0 & 0 & 0 & -4 & 4 & 0 & 0 \\ 0 & 0 & 0 & 0 & 0 & 0 & 0 & -4 & 4 & 0 \\ 0 & 0 & 0 & 0 & 0 & 0 & 0 & 0 & -4 & 4 \\ 0 & 0 & 0 & 0 & 0 & 0 & 0 & 0 & 0 & -4 \end{bmatrix}.$$

■

Application 2.4 Machines 1 and 2 in Application 2.3 produce components 1 and 2, respectively. The two components are assembled into a product. The time to assemble one product has an exponential distribution with a mean of 1 hour.

The processing time needed to produce two types of components that must be assembled into one product is the maximum of the processing time on each machine to produce one acceptable component. If the time to produce two types of components is T_6, then $T_6 = \max\left(T_1', T_2'\right)$, where T_1' and T_2' are the same as described in Application 2.3. The PH representation for T_6 is $(n_6, \boldsymbol{a}_6, \boldsymbol{A}_6)$, which can be found using Theorem 2.4 as

$$n_6 = n_1' n_2' + n_1' + n_2' = 2 \times 3 + 2 + 3 = 11,$$

$$\boldsymbol{a}_6 = \left[\boldsymbol{a}_1' \otimes \boldsymbol{a}_2' \ \ \boldsymbol{0} \ \ \boldsymbol{0}\right] = \left[\left[1 \ 0\right] \otimes \left[1 \ 0 \ 0\right] \ \left[0 \ 0\right] \ \left[0 \ 0 \ 0\right]\right] = \left[1 \ 0 \ 0 \ 0 \ 0 \ 0 \ 0 \ 0 \ 0 \ 0 \ 0\right],$$

$$\boldsymbol{b}_1' = -\boldsymbol{A}_1'\boldsymbol{u} = \begin{bmatrix} 0 \\ 1 \end{bmatrix}, \quad \boldsymbol{b}_2' = -\boldsymbol{A}_2'\boldsymbol{u} = \begin{bmatrix} 0 \\ 0 \\ 0.7 \end{bmatrix},$$

$$\boldsymbol{A}_6 = \begin{bmatrix} \boldsymbol{A}_1' \oplus \boldsymbol{A}_2' & \boldsymbol{I}_1 \otimes \boldsymbol{b}_2' & \boldsymbol{b}_1' \otimes \boldsymbol{I}_2 \\ \boldsymbol{0} & \boldsymbol{A}_1' & \boldsymbol{0} \\ \boldsymbol{0} & \boldsymbol{0} & \boldsymbol{A}_2' \end{bmatrix} = \begin{bmatrix} -3 & 1 & 0 & 2 & 0 & 0 & 0 & 0 & 0 & 0 & 0 \\ 0 & -3 & 1 & 0 & 2 & 0 & 0 & 0 & 0 & 0 & 0 \\ 0.3 & 0 & -3 & 0 & 0 & 2 & 0.7 & 0 & 0 & 0 & 0 \\ 1 & 0 & 0 & -3 & 1 & 0 & 0 & 0 & 1 & 0 & 0 \\ 0 & 1 & 0 & 0 & -3 & 1 & 0 & 0 & 0 & 1 & 0 \\ 0 & 0 & 1 & 0.3 & 0 & -3 & 0 & 0.7 & 0 & 0 & 1 \\ 0 & 0 & 0 & 0 & 0 & 0 & -2 & 2 & 0 & 0 & 0 \\ 0 & 0 & 0 & 0 & 0 & 0 & 1 & -2 & 0 & 0 & 0 \\ 0 & 0 & 0 & 0 & 0 & 0 & 0 & 0 & -1 & 1 & 0 \\ 0 & 0 & 0 & 0 & 0 & 0 & 0 & 0 & 0 & -1 & 1 \\ 0 & 0 & 0 & 0 & 0 & 0 & 0 & 0 & 0.3 & 0 & -1 \end{bmatrix}.$$

The time to assemble the two acceptable components, T_7, is known to be exponentially distributed with a mean of 1 hour. Thus,

$$n_7 = 1, \quad \boldsymbol{a}_7 = [1], \quad \boldsymbol{A}_7 = [-1].$$

The time to complete the production of one product is $T_8 = T_6 + T_7$. The PH representation of T_8 can be found using Theorem 2.3 as

$$n_8 = n_6 + n_7 = 12,$$

$$\boldsymbol{a}_8 = [\boldsymbol{a}_6 \ 0] = [1\ 0\ 0\ 0\ 0\ 0\ 0\ 0\ 0\ 0\ 0\ 0],$$

$$\boldsymbol{b}_6 = -\boldsymbol{A}_6\boldsymbol{u} = [0\ 0\ 0\ 0\ 0\ 0\ 0\ 1\ 0\ 0\ 0.7]^T,$$

$$\boldsymbol{A}_8 = \begin{bmatrix} \boldsymbol{A}_6 & \boldsymbol{b}_6\boldsymbol{a}_7 \\ \boldsymbol{0} & \boldsymbol{A}_7 \end{bmatrix} = \left[\begin{array}{ccccccccccc|c} -3 & 1 & 0 & 2 & 0 & 0 & 0 & 0 & 0 & 0 & 0 & 0 \\ 0 & -3 & 1 & 0 & 2 & 0 & 0 & 0 & 0 & 0 & 0 & 0 \\ 0.3 & 0 & -3 & 0 & 0 & 2 & 0.7 & 0 & 0 & 0 & 0 & 0 \\ 1 & 0 & 0 & -3 & 1 & 0 & 0 & 0 & 1 & 0 & 0 & 0 \\ 0 & 1 & 0 & 0 & -3 & 1 & 0 & 0 & 0 & 1 & 0 & 0 \\ 0 & 0 & 1 & 0.3 & 0 & -3 & 0 & 0.7 & 0 & 0 & 1 & 0 \\ 0 & 0 & 0 & 0 & 0 & 0 & -2 & 2 & 0 & 0 & 0 & 0 \\ 0 & 0 & 0 & 0 & 0 & 0 & 1 & -2 & 0 & 0 & 0 & 1 \\ 0 & 0 & 0 & 0 & 0 & 0 & 0 & 0 & -1 & 1 & 0 & 0 \\ 0 & 0 & 0 & 0 & 0 & 0 & 0 & 0 & 0 & -1 & 1 & 0 \\ 0 & 0 & 0 & 0 & 0 & 0 & 0 & 0 & 0.3 & 0 & -1 & 0.7 \\ \hline 0 & 0 & 0 & 0 & 0 & 0 & 0 & 0 & 0 & 0 & 0 & -1 \end{array}\right].$$

■

Exercise 2.12 Suppose that the result of assembling is inspected with probability of passing the inspection is 0.8. After passing the inspection, the product must be packed. Inspection and packaging times are exponentially distributed with mean 1 hour and 2 hours, respectively. Determine the PH distribution of the time to produce one packed acceptable product.

Example 2.25 Closure properties of PH distribution are utilized to determine the PH distribution of repair time in Example 2.11.

Let T_i denote the duration of stay at transient state i in CTMC in Figure 2.13. From the transition rate out of state i, it is known that T_i, $i = 1, 2, 3$, has an exponential distribution with rates of 5, 2, and 1, respectively; thus, they have no discontinuity at the origin. The distribution of T_i has PH representation $(n_i, \boldsymbol{a}_i, \boldsymbol{A}_i)$ where

$$n_1 = n_2 = n_3 = 1, \quad \boldsymbol{a}_1 = \boldsymbol{a}_2 = \boldsymbol{a}_3 = [1], \quad \boldsymbol{A}_1 = [-5], \quad \boldsymbol{A}_2 = [-2], \quad \boldsymbol{A}_3 = [-1].$$

Use $\boldsymbol{b}_i = -\boldsymbol{A}_i\boldsymbol{u}$ to obtain

$$\boldsymbol{b}_1 = [5], \quad \boldsymbol{b}_2 = [2], \quad \boldsymbol{b}_3 = [1].$$

The overall repair time, T_S, according to Figure 2.13, has three possibilities:

- $T_{S,1} = T_1$ with probability 0.2.
- $T_{S,2} = T_1 + T_2$ with probability $0.8 \times 0.5 = 0.4$.
- $T_{S,3} = T_1 + T_2 + T_3$ with probability $0.8 \times 0.5 \times 1 = 0.4$,

and T_S is a mixture of $T_{S,1}$, $T_{S,2}$, and $T_{S,3}$ with a mixing probability of 0.2, 0.4, and 0.4, respectively. Since $T_{S,1} = T_1$, the PH representation of the distribution of $T_{S,1}$ is

$$n_{S,1} = 1, \quad \boldsymbol{a}_{S,1} = [1], \quad \boldsymbol{A}_{S,1} = [-5].$$

The PH representation of the distribution of $T_{S,2} = T_1 + T_2$ can be found using Theorem 2.3 as

$$n_{S,2} = n_1 + n_2 = 2, \quad \boldsymbol{a}_{S,2} = \begin{bmatrix} \boldsymbol{a}_1 & 0 \end{bmatrix} = \begin{bmatrix} 1 & 0 \end{bmatrix},$$

$$\boldsymbol{A}_{S,2} = \begin{bmatrix} \boldsymbol{A}_1 & \boldsymbol{b}_1\boldsymbol{a}_2 \\ \boldsymbol{0} & \boldsymbol{A}_2 \end{bmatrix} = \begin{bmatrix} -5 & 5\times 1 \\ 0 & -2 \end{bmatrix} = \begin{bmatrix} -5 & 5 \\ 0 & -2 \end{bmatrix},$$

and

$$\boldsymbol{b}_{S,2} = -\boldsymbol{A}_{S,2}\boldsymbol{u} = \begin{bmatrix} -5 & 5 \\ 0 & -2 \end{bmatrix}\begin{bmatrix} 1 \\ 1 \end{bmatrix} = \begin{bmatrix} 0 \\ 2 \end{bmatrix}.$$

The PH representation of the distribution of $T_{S,3} = T_1 + T_2 + T_3$ can be found using Theorem 2.3 by writing $T_{S,3} = T_{S,2} + T_3$. Thus,

$$n_{S,3} = n_{S,2} + n_3 = 3, \quad \boldsymbol{a}_{S,3} = \begin{bmatrix} \boldsymbol{a}_{S,2} & 0 \end{bmatrix} = \begin{bmatrix} 1 & 0 & 0 \end{bmatrix},$$

$$\boldsymbol{A}_{S,3} = \begin{bmatrix} \boldsymbol{A}_{S,2} & \boldsymbol{b}_{S,2}\boldsymbol{a}_3 \\ \boldsymbol{0} & \boldsymbol{A}_3 \end{bmatrix} = \begin{bmatrix} \begin{bmatrix} -5 & 5 \\ 0 & -2 \end{bmatrix} & \begin{bmatrix} 0 \\ 2 \end{bmatrix}[1] \\ \boldsymbol{0} & -1 \end{bmatrix} = \begin{bmatrix} -5 & 5 & 0 \\ 0 & -2 & 2 \\ 0 & 0 & -1 \end{bmatrix}.$$

Since T_S is a mixture of $T_{S,1}$, $T_{S,2}$, and $T_{S,3}$ with mixing probabilities 0.2, 0.4, and 0.4, respectively, the PH representation $(n_S, \boldsymbol{a}_S, \boldsymbol{A}_S)$ for the distribution of T_S can be found using Theorem 2.2 as

$$n_S = n_{S,1} + n_{S,2} + n_{S,3} = 1 + 2 + 3 = 6,$$

$$\boldsymbol{a}_S = \begin{bmatrix} p_1\boldsymbol{a}_{S,1} & p_2\boldsymbol{a}_{S,2} & p_3\boldsymbol{a}_{S,3} \end{bmatrix} = \begin{bmatrix} p_1[1] & p_2\begin{bmatrix} 1 & 0 \end{bmatrix} & p_3\begin{bmatrix} 1 & 0 & 0 \end{bmatrix} \end{bmatrix} = \begin{bmatrix} 0.2 & 0.4 & 0 & 0.4 & 0 & 0 \end{bmatrix},$$

$$\boldsymbol{A}_S = \begin{bmatrix} \boldsymbol{A}_{S,1} & \boldsymbol{0} & \boldsymbol{0} \\ \boldsymbol{0} & \boldsymbol{A}_{S,2} & \boldsymbol{0} \\ \boldsymbol{0} & \boldsymbol{0} & \boldsymbol{A}_{S,3} \end{bmatrix} = \begin{bmatrix} -5 & \boldsymbol{0} & \boldsymbol{0} \\ \boldsymbol{0} & \begin{bmatrix} -5 & 5 \\ 0 & -2 \end{bmatrix} & \boldsymbol{0} \\ \boldsymbol{0} & \boldsymbol{0} & \begin{bmatrix} -5 & 5 & 0 \\ 0 & -2 & 2 \\ 0 & 0 & -1 \end{bmatrix} \end{bmatrix} = \begin{bmatrix} -5 & 0 & 0 & 0 & 0 & 0 \\ 0 & -5 & 5 & 0 & 0 & 0 \\ 0 & 0 & -2 & 0 & 0 & 0 \\ 0 & 0 & 0 & -5 & 5 & 0 \\ 0 & 0 & 0 & 0 & -2 & 2 \\ 0 & 0 & 0 & 0 & 0 & -1 \end{bmatrix}.$$

The PH distribution of service completion time obtained using the closure properties differs from that obtained using the state transition diagram given in Example 2.11. However, they are identical because they represent the same random variables. To show that they are indeed identical, the CDF of the PH distribution obtained here, denoted by $F_1(t)$, is compared with the CDF of the PH distribution obtained in Example 2.11, denoted by $F_2(t)$. The result is presented in Figure 2.21, which shows that the two CDFs are indeed identical. ■

The use of closure properties to model the repair service provided in Example 2.25 can be generalized to PH distributions of arbitrary order, not just for the PH distribution of order 1.

Example 2.26 Assume T_i, $i = 1, 2, 3$, in Example 2.25 have arbitrary PH distributions with representations $(n_i, \boldsymbol{a}_i, \boldsymbol{A}_i)$, $i = 1, 2, 3$. The PH representations for $T_{S,1}$, $T_{S,2}$, and $T_{S,3}$ can be derived as follows:

Distribution of $T_{S,1}$ is $PH(n_{S,1}, \boldsymbol{a}_{S,1}, \boldsymbol{A}_{S,1})$, where

$$n_{S,1} = n_1, \quad \boldsymbol{a}_{S,1} = \boldsymbol{a}_1, \quad \boldsymbol{A}_{S,1} = \boldsymbol{A}_1.$$

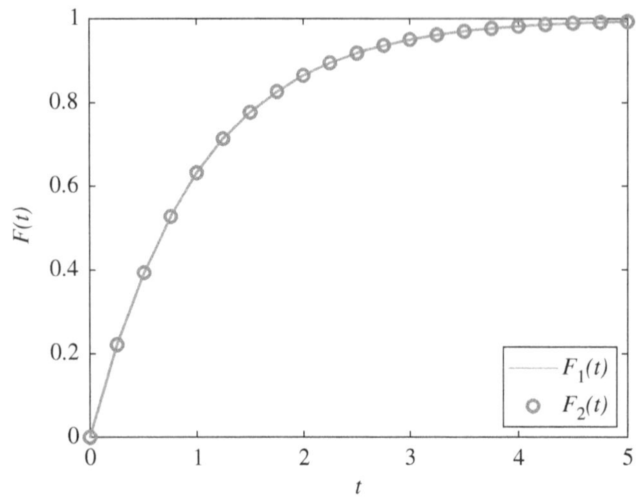

Figure 2.21 Comparison of CDFs generated from Examples 2.11 and 2.25.

Distribution of $T_{S,2}$ is $PH(n_{S,2}, \boldsymbol{a}_{S,2}, \boldsymbol{A}_{S,2})$, where

$$n_{S,2} = n_1 + n_2, \quad \boldsymbol{a}_{S,2} = \begin{bmatrix} \boldsymbol{a}_1 & \boldsymbol{0} \end{bmatrix}, \quad \boldsymbol{A}_{S,2} = \begin{bmatrix} \boldsymbol{A}_1 & \boldsymbol{b}_1\boldsymbol{a}_2 \\ \boldsymbol{0} & \boldsymbol{A}_2 \end{bmatrix},$$

$$\boldsymbol{b}_{S,2} = -\boldsymbol{A}_{S,2}\boldsymbol{u} = -\begin{bmatrix} \boldsymbol{A}_1 & \boldsymbol{b}_1\boldsymbol{a}_2 \\ \boldsymbol{0} & \boldsymbol{A}_2 \end{bmatrix}\boldsymbol{u} = -\begin{bmatrix} \boldsymbol{A}_1\boldsymbol{u} + \boldsymbol{b}_1\boldsymbol{a}_2\boldsymbol{u} \\ \boldsymbol{A}_2\boldsymbol{u} \end{bmatrix} = \begin{bmatrix} \boldsymbol{0} \\ \boldsymbol{b}_2 \end{bmatrix}.$$

Distribution of $T_{S,3}$ is $PH(n_{S,3}, \boldsymbol{a}_{S,3}, \boldsymbol{A}_{S,3})$, where

$$n_{S,3} = n_{S,2} + n_3, \quad \boldsymbol{a}_{S,3} = \begin{bmatrix} \boldsymbol{a}_{S,2} & \boldsymbol{0} \end{bmatrix} = \begin{bmatrix} \boldsymbol{a}_1 & \boldsymbol{0} & \boldsymbol{0} \end{bmatrix},$$

$$\boldsymbol{A}_{S,3} = \begin{bmatrix} \boldsymbol{A}_{S,2} & \boldsymbol{b}_{S,2}\boldsymbol{a}_3 \\ \boldsymbol{0} & \boldsymbol{A}_3 \end{bmatrix} = \begin{bmatrix} \begin{bmatrix} \boldsymbol{A}_1 & \boldsymbol{b}_1\boldsymbol{a}_2 \\ \boldsymbol{0} & \boldsymbol{A}_2 \end{bmatrix} & \boldsymbol{b}_{S,2}\boldsymbol{a}_3 \\ \boldsymbol{0} & \boldsymbol{A}_3 \end{bmatrix} = \begin{bmatrix} \boldsymbol{A}_1 & \boldsymbol{b}_1\boldsymbol{a}_2 & \boldsymbol{0} \\ \boldsymbol{0} & \boldsymbol{A}_2 & \boldsymbol{b}_2\boldsymbol{a}_3 \\ \boldsymbol{0} & \boldsymbol{0} & \boldsymbol{A}_3 \end{bmatrix}.$$

Distribution of T_S is $PH(n_S, \boldsymbol{a}_S, \boldsymbol{A}_S)$, where

$$n_S = n_{S,1} + n_{S,2} + n_{S,3}, \quad \boldsymbol{a}_S = \begin{bmatrix} p_1\boldsymbol{a}_{S,1} & p_2\boldsymbol{a}_{S,2} & p_3\boldsymbol{a}_{S,3} \end{bmatrix},$$

$$\boldsymbol{A}_S = \begin{bmatrix} \boldsymbol{A}_{S,1} & \boldsymbol{0} & \boldsymbol{0} \\ \boldsymbol{0} & \boldsymbol{A}_{S,2} & \boldsymbol{0} \\ \boldsymbol{0} & \boldsymbol{0} & \boldsymbol{A}_{S,3} \end{bmatrix}.$$

The PH representation for T_S can also be found using the state transition diagram for the PH distribution by replacing the scalar transitions in Figure 2.13 with matrix transitions. The result is shown in Figure 2.22.

The PH representation for the distribution of T_S obtained from Figure 2.22 is

$$n_S = n_1 + n_2 + n_3, \quad \boldsymbol{a}_S = \begin{bmatrix} \boldsymbol{a}_1 & \boldsymbol{0} & \boldsymbol{0} \end{bmatrix},$$

$$\boldsymbol{A}_S = \begin{bmatrix} \boldsymbol{A}_1 & (1-p_1)\boldsymbol{b}_1\boldsymbol{a}_2 & \boldsymbol{0} \\ \boldsymbol{0} & \boldsymbol{A}_2 & (1-p_2)\boldsymbol{b}_2\boldsymbol{a}_3 \\ \boldsymbol{0} & \boldsymbol{0} & \boldsymbol{A}_3 \end{bmatrix}.$$

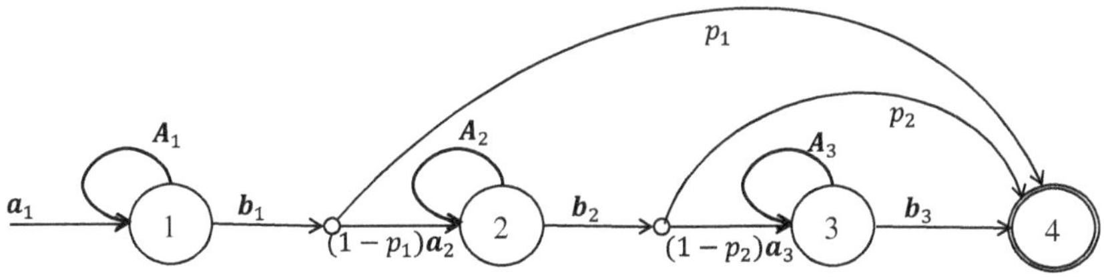

Figure 2.22 State transition diagram of Example 2.11 in PH distribution.

To show that using the properties of the PH distribution and the state transition diagram gives identical results, the CDF of T_S is calculated for $p_1 = 0.2$ and $p_2 = 0.5$ as in Example 2.11; thus, the values of p_1, p_2, and p_3 for the model using the closure properties of the PH distribution are 0.2, $0.5 \times 0.8 = 0.4$, and $(1 - 0.5) \times 0.8 = 0.4$, respectively. The calculations are made for T_i, $i = 1, 2, 3$, has k-Erlang distribution with the same rates as in Example 2.11, namely, 5, 2, and 1. The results for $k = 1, 2, 3$ are presented in Figure 2.23.

Figure 2.23 Comparison of CDF of T_S calculated using the closure properties and the state transition diagram.

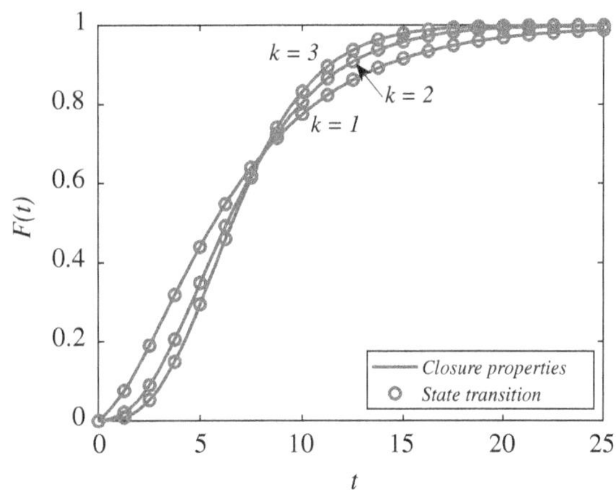

The graphs in Figure 2.23 show that both methods produce identical results; thus, the two resulting PH distributions can be concluded to be identical, even though they have different representations. It is an example that a PH distribution may have several identical representations.

It can be verified manually that for the exponential distributions, namely, by assigning values of $\boldsymbol{A}_1 = [-5]$, $\boldsymbol{A}_2 = [-2]$, $\boldsymbol{A}_3 = [-1]$, $\boldsymbol{b}_1 = 5$, and $\boldsymbol{b}_2 = 2$, the PH distributions generated using the methods in Examples 2.11 and 2.26 are identical. ■

Exercise 2.13 Determine the infinitesimal generator $\boldsymbol{Q}$ of CTMC in Figure 2.22 and show that $\boldsymbol{Qu} = \boldsymbol{0}$.

The next section describes another matrix-form distribution that generalizes PH distribution. This generalization can be utilized to transform PH distribution into ME distribution so that the computation time using the transformed distribution is faster.

2.7 ME Distribution

ME distribution is a generalization of PH distribution. The PDF of ME distribution is given by [7]

$$f_T(t) = \mathbf{s}e^{\mathbf{A}t}\mathbf{c} \quad t \geq 0, \tag{2.45}$$

for any combination of $\mathbf{s}$, $\mathbf{c}$, and $\mathbf{A}$ that makes Equation (2.45) a valid PDF with finite moments. The conditions that must be satisfied by the parameters of $f_T(t)$, in addition to $\mathbf{A}$ must be nonsingular, is $f_T(t)$ must be nonnegative and its integral for all values t is equal to 1. Thus,

$$-\mathbf{s}\mathbf{A}^{-1}\mathbf{c} = 1 \quad \text{dan} \quad \mathbf{s}\mathbf{c} \geq 0. \tag{2.46}$$

Equation (2.45) shows that the ME distribution assumes no discontinuity at the origin. The CDF of ME distribution is obtained by integrating Equation (2.45) to yield

$$F_T(t) = 1 + \mathbf{s}\mathbf{A}^{-1}e^{\mathbf{A}t}\mathbf{c} \quad t \geq 0. \tag{2.47}$$

Equation (2.47) can be changed to [13]

$$F_T(t) = 1 - \mathbf{a}e^{\mathbf{A}t}\mathbf{v} \qquad t \geq 0, \tag{2.48}$$

by defining $-\mathbf{s}\mathbf{A}^{-1} = \mathbf{a}$ and $\mathbf{c} = \mathbf{v}$ to make the CDF of ME distribution have the same form as the CDF of PH distribution.

One advantage of using ME distribution is its flexibility in selecting the values of parameters $\mathbf{a}$, $\mathbf{A}$, and $\mathbf{v}$. For ME distributions:

- $\mathbf{a}$ is not a probability vector, so its elements may be negative or greater than 1, and their sum does not have to be less than or equal to 1.
- $\mathbf{A}$ does not have to have nonpositive diagonal elements, nonnegative off-diagonal elements, and the row sums do not have to be nonpositive.
- $\mathbf{v}$ does not have to be a unitary column vector with all elements equal to 1.

These are required for PH distribution.

A variable T with CDF as given by Equation (2.48) is said to have ME distribution with representation $(n, \mathbf{a}, \mathbf{A}, \mathbf{v})$ and is denoted by $T \sim ME(n, \mathbf{a}, \mathbf{A}, \mathbf{v})$ or $T \sim ME_n(\mathbf{a}, \mathbf{A}, \mathbf{v})$. Different from PH representation, which does not need to include $\mathbf{u}$ in its representation, $\mathbf{v}$ must be included in ME representation because its value is not always the same for all ME distributions. ME distribution is useful for dynamic analysis of system reliability that cannot or is difficult to express in a state transition diagram, such as network structured systems [13].

Example 2.27 A distribution having representation [14]

$$\mathbf{a} = \begin{bmatrix}1 & 0 & 0\end{bmatrix}, \quad \mathbf{A} = \begin{bmatrix}0 & -2 & 2\\ 3 & 2 & -6\\ 2 & 2 & -5\end{bmatrix}, \quad \mathbf{v} = \mathbf{u},$$

is an ME distribution because $\mathbf{A}$ does not qualify as a transient matrix, but $F(t) = 1 - \mathbf{a}e^{\mathbf{A}t}\mathbf{u}$ is a valid CDF. ■

Next, a method to transform PH distribution into an ME distribution is presented.

2.7.1 Similarity Transformation

In addition to the flexibility in its representation, another advantage of using ME distribution is that its representation is not unique; many triplets $\boldsymbol{a}, \boldsymbol{A}$, and $\boldsymbol{v}$ produce equivalent ME distributions. Any matrix similarity transformation to an ME distribution will yield another ME distribution. Applying matrix transformation to $\boldsymbol{A}$ can transform $\boldsymbol{A}$ to the Jordan canonical form, which is generally a diagonal or semi-diagonal matrix so that the computation time becomes much faster.

A matrix $\boldsymbol{P}$ whose columns are generalized eigenvectors of the transient matrix $\boldsymbol{A}$ can transform PH and ME distributions into an ME distribution with a transient matrix in Jordan canonical form. The process is as follows:

$$F_T(t) = 1 - \boldsymbol{a}e^{\boldsymbol{A}t}\boldsymbol{v} = 1 - \boldsymbol{a}\boldsymbol{P}\boldsymbol{P}^{-1}e^{\boldsymbol{A}t}\boldsymbol{P}\boldsymbol{P}^{-1}\boldsymbol{v} = 1 - \boldsymbol{a}\boldsymbol{P}e^{(\boldsymbol{P}^{-1}\boldsymbol{A}\boldsymbol{P})t}\boldsymbol{P}^{-1}\boldsymbol{v}.$$

The resulting CDF can be written as

$$F_T(t) = 1 - \boldsymbol{a}'e^{\boldsymbol{A}'t}\boldsymbol{v}', \tag{2.49}$$

where

$$\boldsymbol{a}' = \boldsymbol{a}\boldsymbol{P}, \quad \boldsymbol{v}' = \boldsymbol{P}^{-1}\boldsymbol{v}, \quad \boldsymbol{A}' = \boldsymbol{P}^{-1}\boldsymbol{A}\boldsymbol{P}. \tag{2.50}$$

Finding $\boldsymbol{P}$ and calculating $\boldsymbol{a}'$, $\boldsymbol{v}'$, and $\boldsymbol{A}'$ from PH representation (n, $\boldsymbol{a}$, $\boldsymbol{A}$) to produce an ME representation (n, $\boldsymbol{a}'$, $\boldsymbol{A}'$, $\boldsymbol{v}'$) can be done using a MATLAB function **phmesimilar** available in the software package **dysrel-am**.

Example 2.28 The ME representation in a Jordan canonical form for the PH distribution used to produce the CDF in Figure 2.5b can be found in the following steps:

- Find the eigenvalues of $\boldsymbol{A}$. The results are $-1, -4.128$, and -0.872.
- Find the eigenvector for each eigenvalue. The results are

$$\boldsymbol{e}_1 = \begin{bmatrix} 1 \\ 0 \\ 0 \end{bmatrix}, \quad \boldsymbol{e}_2 = \begin{bmatrix} 0.0164 \\ -0.752 \\ 1 \end{bmatrix}, \quad \boldsymbol{e}_3 = \begin{bmatrix} 9.784 \\ 1.419 \\ 1 \end{bmatrix}.$$

- Construct the similarity transformation matrix $\boldsymbol{P} = [\boldsymbol{e}_1\ \boldsymbol{e}_2\ \boldsymbol{e}_3]$. The result is

$$\boldsymbol{P} = \begin{bmatrix} 1 & 0.0164 & 9.784 \\ 0 & -0.752 & 1.419 \\ 0 & 1 & 1 \end{bmatrix}.$$

- Use Equation (2.50) to transform the PH representation into an ME representation in a Jordan canonical form. The result is

$$\boldsymbol{a}' = [0.5\ -0.017\ \ 5.517], \quad \boldsymbol{A}' = \begin{bmatrix} -1 & 0 & 0 \\ 0 & -4.128 & 0 \\ 0 & 0 & -0.872 \end{bmatrix}, \quad \boldsymbol{v}' = \begin{bmatrix} -6.9 \\ 0.193 \\ 0.807 \end{bmatrix}.$$

Note that all eigenvalues are negative, indicating $e^{\boldsymbol{A}'t}$ converges to 0 when $t \to \infty$. Thus, $e^{\boldsymbol{A}t}$ is also convergent.

The resulting ME representation is (n, $\boldsymbol{a}'$, $\boldsymbol{A}'$, $\boldsymbol{v}'$), which is not a PH representation because $\boldsymbol{a}'$ contains a negative element, and its sum is greater than 1, which is not an initial probability vector, and $\boldsymbol{v}'$ is not unitary vector. However, the PH and ME representations have identical CDFs, as shown in Figure 2.24.

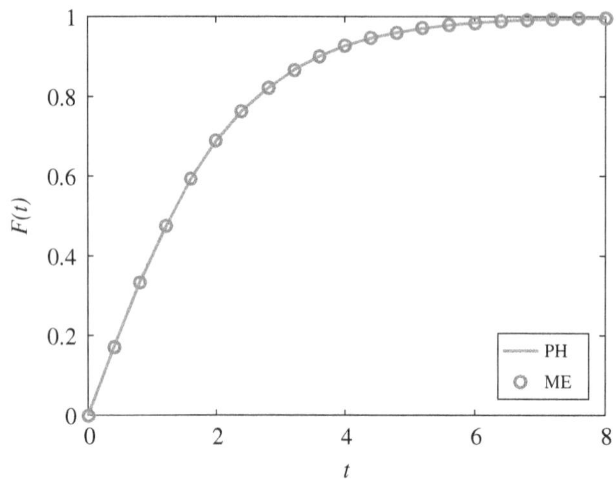

Figure 2.24 CDF with PH and ME representations in Example 2.28. ■

The main program to produce the ME representation and Figure 2.23 is available in **example2_28.m** file in the software package **dysrel-am**. The source code is as follows:

MATLAB code for example2_28.m

```
% CDF of real distinct eigen values
close all;clear
addpath('..\dysrel-am','..\dysrel-am\expokit')
% PH representation
PH.a = [0.5 0.3 0.2];
PH.A = [-1 0.6 0.4;0 -2 1.6;0 1.5 -3];
PH.n = 3;
% CDF from PH
t = linspace(0,8,21);
[tt,Fph] = phcdfplot(t,PH);
% Generate ME representation
ME = phmesimilar(PH);
% CDF from ME
[t,Fme] = mecdfplot(t,ME,0);
close all
figure
plot(tt,Fph,'b',t,Fme,'ro','LineWidth',2,'MarkerSize',6);
set(gca,'FontName','Times New Roman','FontSize',18,'LineWidth',1.5,...
  'FontWeight','bold');
xlabel('\itt');ylabel('\itF(t)');
legend('PH','ME','location','southeast')
```

Example 2.29 The PH distribution in Example 2.28 is a special case with three distinct real eigenvalues. It is possible for matrix $\boldsymbol{A}$ to have complex or identical (repeated) eigenvalues as shown by the results of similarity transformations of the following PH representations of order 3:

- Two complex eigenvalues and one real eigenvalue denoted by $F_1(t)$:

$$\text{PH representation: } \boldsymbol{a}_1 = \begin{bmatrix}0.5 & 0.3 & 0.2\end{bmatrix}, \quad \boldsymbol{A}_1 = \begin{bmatrix}-2 & 1 & 0\\ 0 & -1 & 0.5\\ 0.7 & 0 & -2\end{bmatrix};$$

ME representation: $\boldsymbol{a}_1' = \left[1.729 \;\; 0.007 + 0.320i \;\; 0.007 - 0.320i\right]$,

$$\boldsymbol{A}_1' = \begin{bmatrix} -0.769 & 0 & 0 \\ 0 & -2.116 + 0.521i & 0 \\ 0 & 0 & -2.116 - 0.521i \end{bmatrix}, \quad \boldsymbol{v}_1' = \begin{bmatrix} 0.550 \\ 0.225 - 0.072i \\ 0.225 + 0.072i \end{bmatrix}.$$

- Two repeated eigenvalues with independent eigenvectors denoted by $F_2(t)$:

PH representation: $\boldsymbol{a}_2 = \left[0.5 \;\; 0.3 \;\; 0.2\right]$, $\boldsymbol{A}_2 = \begin{bmatrix} -2 & 1 & 0 \\ 0 & -1 & 0 \\ 0 & 0 & -2 \end{bmatrix}$;

ME representation: $\boldsymbol{a}_2' = \left[0.8 \;\; 0.5 \;\; 0.2\right]$, $\boldsymbol{A}_2' = \begin{bmatrix} -1 & 0 & 0 \\ 0 & -2 & 0 \\ 0 & 0 & -2 \end{bmatrix}$, $\boldsymbol{v}_2' = \begin{bmatrix} 1 \\ 0 \\ 1 \end{bmatrix}$.

- Two repeated eigenvalues with dependent eigenvectors denoted by $F_3(t)$:

PH representation: $\boldsymbol{a}_3 = \left[0.5 \;\; 0.3 \;\; 0.2\right]$, $\boldsymbol{A}_3 = \begin{bmatrix} -1 & 0.6 & 0.4 \\ 0 & -1 & 0.6 \\ 0 & 0 & -3 \end{bmatrix}$;

ME representation: $\boldsymbol{a}_3' = \left[0.055 \;\; -0.11 \;\; -0.055\right]$,

$$\boldsymbol{A}_3' = \begin{bmatrix} -3 & 0 & 0 \\ 0 & -1 & 1 \\ 0 & 0 & -1 \end{bmatrix} \quad \boldsymbol{v}_3' = \begin{bmatrix} 1 \\ -6.818 \\ -3.546 \end{bmatrix}.$$

The similarity between the original PH distributions and their equivalent ME distributions resulting from the similarity transformations is shown through their CDF graphs in Figure 2.25.

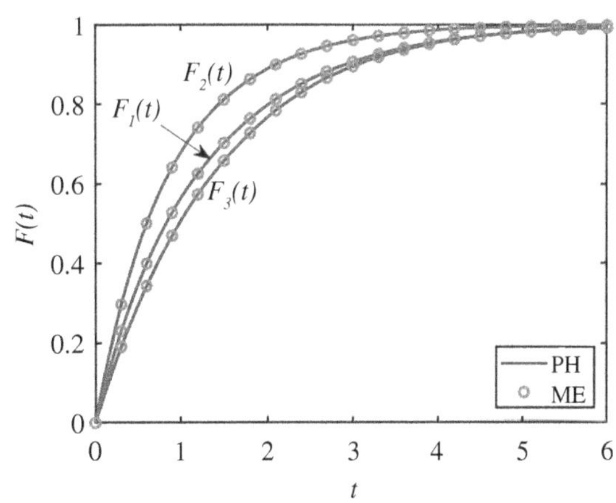

Figure 2.25 CDFs of PH and ME distributions in Example 2.29.

■

The case of repeated eigenvalues with independent eigenvectors can be simplified to order 2 by combining two similar exponential functions into one,

$$\begin{aligned}\boldsymbol{a}'e^{\boldsymbol{A}'t}\boldsymbol{v}' &= \begin{bmatrix}0.8 & 0.5 & 0.2\end{bmatrix} e^{\begin{bmatrix}-1 & 0 & 0\\ 0 & -2 & 0\\ 0 & 0 & -2\end{bmatrix}t} \begin{bmatrix}1\\0\\1\end{bmatrix}\\ &= \begin{bmatrix}0.8 & 0.5 & 0.2\end{bmatrix} \begin{bmatrix}e^{-t} & 0 & 0\\ 0 & e^{-2t} & 0\\ 0 & 0 & e^{-2t}\end{bmatrix} \begin{bmatrix}1\\0\\1\end{bmatrix}\\ &= \begin{bmatrix}0.8e^{-t} & 0.5e^{-2t} & 0.2e^{-2t}\end{bmatrix} \begin{bmatrix}1\\0\\1\end{bmatrix} = 0.8e^{-t} + 0.2e^{-2t}.\end{aligned}$$

The result can be written in the form of $\boldsymbol{a}''e^{\boldsymbol{A}''t}\boldsymbol{v}''$ with

$$\boldsymbol{a}'' = \begin{bmatrix}0.8 & 0.2\end{bmatrix}, \quad \boldsymbol{A}'' = \begin{bmatrix}-1 & 0\\ 0 & -2\end{bmatrix}, \quad \boldsymbol{v}'' = \begin{bmatrix}1\\1\end{bmatrix},$$

which has order 2.

This method of reducing order can be used for all ME distributions that have multiple eigenvalues with independent eigenvectors. This result shows another advantage of ME distribution: it can have a smaller order than its equivalent PH distribution. For this example, the simplification result is still a PH distribution, which is just a coincidence, because the result will generally be an ME distribution.

Matrix $\boldsymbol{A}'$ with diagonal or semi-diagonal structures will speed up the calculation of its exponential value to plot the CDF graph. Likewise, calculating moments becomes faster because calculating the inverse of a diagonal or semi-diagonal matrix is faster than a full matrix. The difference in computation time to generate a CDF graph and calculate moment value using equivalent PH and ME representation is significant for complex systems involving CPH distribution. Experiments on the efficiency of using PH distribution and its equivalent ME distribution will be presented in Chapter 4.

Sometimes, Jordan canonical form obtained from matrix similarity transformation is unstable for complex numerical calculations. However, the numerical calculation and the behavior of CDF are simple, so the possibility of experiencing numerical instability is small. If this happens, a slight perturbation will save the calculation from the instability trap.

2.7.2 Closure Properties of ME Distribution

The closure properties of PH distribution in Theorems 2.1–2.4 also apply to ME distribution [13]. The difference is ME distribution requires an additional equation for $\boldsymbol{v}$, which is not always a unitary vector. The closure properties of ME distribution are given as Theorems 2.6–2.9, while the proofs are given as problems. Since ME distribution assumes no discontinuities at the origin, it follows that $a_{n+1} = 0$. The closure properties of PH distribution when used for ME distribution need to be adjusted by inserting $a_{n+1} = 0$, and $\boldsymbol{b}_i = -\boldsymbol{A}_i\boldsymbol{v}_i$ instead $\boldsymbol{b}_i = -\boldsymbol{A}_i\boldsymbol{u}_i$.

Theorem 2.6 If $T_1 \sim ME(n_1, \boldsymbol{a}_1, \boldsymbol{A}_1, \boldsymbol{v}_1)$ and $T_2 \sim ME(n_2, \boldsymbol{a}_2, \boldsymbol{A}_2, \boldsymbol{v}_2)$ are independent, then $T_S = \min(T_1, T_2) \sim ME(n_S, \boldsymbol{a}_S, \boldsymbol{A}_S, \boldsymbol{v}_S)$ with

$$n_S = n_1 n_2, \quad \boldsymbol{a}_S = \boldsymbol{a}_1 \otimes \boldsymbol{a}_2, \quad \boldsymbol{A}_S = \boldsymbol{A}_1 \oplus \boldsymbol{A}_2, \quad \boldsymbol{v}_S = \boldsymbol{v}_1 \otimes \boldsymbol{v}_2. \tag{2.51}$$

Theorem 2.7 If $T_1 \sim ME(n_1, \boldsymbol{a}_1, \boldsymbol{A}_1, \boldsymbol{v}_1)$ and $T_2 \sim ME(n_2, \boldsymbol{a}_2, \boldsymbol{A}_2, \boldsymbol{v}_2)$ are independent, then $T_S = \text{mix}(T_1, T_2) \sim ME(n_S, \boldsymbol{a}_S, \boldsymbol{A}_S, \boldsymbol{v}_S)$ with

$$n_S = n_1 + n_2, \quad \boldsymbol{a}_S = [p_1\boldsymbol{a}_1 \ \ p_2\boldsymbol{a}_2], \quad \boldsymbol{A}_S = \begin{bmatrix} \boldsymbol{A}_1 & \boldsymbol{0} \\ \boldsymbol{0} & \boldsymbol{A}_2 \end{bmatrix}, \quad \boldsymbol{v}_S = \begin{bmatrix} \boldsymbol{v}_1 \\ \boldsymbol{v}_2 \end{bmatrix}. \tag{2.52}$$

Theorem 2.8 If $T_1 \sim ME(n_1, \boldsymbol{a}_1, \boldsymbol{A}_1, \boldsymbol{v}_1)$ and $T_2 \sim ME(n_2, \boldsymbol{a}_2, \boldsymbol{A}_2, \boldsymbol{v}_2)$ are independent, then $T_S = \text{add}(T_1, T_2) = T_1 + T_2 \sim ME(n_S, \boldsymbol{a}_S, \boldsymbol{A}_S, \boldsymbol{v}_S)$ with

$$n_S = n_1 + n_2, \quad \boldsymbol{a}_S = [\boldsymbol{a}_1 \ \ \boldsymbol{0}], \quad \boldsymbol{A}_S = \begin{bmatrix} \boldsymbol{A}_1 & -\boldsymbol{A}_1\boldsymbol{v}_1\boldsymbol{a}_2 \\ \boldsymbol{0} & \boldsymbol{A}_2 \end{bmatrix}, \quad \boldsymbol{v}_S = \begin{bmatrix} \boldsymbol{v}_1 \\ \boldsymbol{v}_2 \end{bmatrix}. \tag{2.53}$$

Theorem 2.9 If $T_1 \sim ME(n_1, \boldsymbol{a}_1, \boldsymbol{A}_1, \boldsymbol{v}_1)$ and $T_2 \sim ME(n_2, \boldsymbol{a}_2, \boldsymbol{A}_2, \boldsymbol{v}_2)$ are independent, then $T_S = \max(T_1, T_2) \sim ME(n_S, \boldsymbol{a}_S, \boldsymbol{A}_S, \boldsymbol{v}_S)$ with

$$\begin{gathered} n_S = n_1 n_2 + n_1 + n_2, \quad \boldsymbol{a}_S = [\boldsymbol{a}_1 \otimes \boldsymbol{a}_2 \ \ \boldsymbol{0} \ \ \boldsymbol{0}], \\ \boldsymbol{A}_S = \begin{bmatrix} \boldsymbol{A}_1 \oplus \boldsymbol{A}_2 & \boldsymbol{I}_{n_1} \otimes -\boldsymbol{A}_2\boldsymbol{v}_2 & -\boldsymbol{A}_1\boldsymbol{v}_1 \otimes \boldsymbol{I}_{n_2} \\ \boldsymbol{0} & \boldsymbol{A}_1 & \boldsymbol{0} \\ \boldsymbol{0} & \boldsymbol{0} & \boldsymbol{A}_2 \end{bmatrix}, \quad \boldsymbol{v}_S = \begin{bmatrix} \boldsymbol{v}_1 \otimes \boldsymbol{v}_2 \\ \boldsymbol{v}_1 \\ \boldsymbol{v}_2 \end{bmatrix}. \end{gathered} \tag{2.54}$$

The results of the maximum operation of two random variables having ME distributions can be expressed differently from those given by Equation (2.54). The CDF of the maximum operation is given by

$$P[T_S \le t] = P[\max(T_1, T_2) \le t]. \tag{2.55}$$

The maximum of two variables is less than or equal to t is fulfilled by both variables are less than or equal to t. Use the independence of the two variables to obtain

$$P[T_S \le t] = P[T_1 \le t \wedge T_2 \le t] = P[T_1 \le t]P[T_2 \le t]. \tag{2.56}$$

Substituting Equation (2.48) into Equation (2.56) yields

$$P[T_S \le t] = \left(1 - \boldsymbol{a}_1 e^{\boldsymbol{A}_1 t}\boldsymbol{v}_1\right)\left(1 - \boldsymbol{a}_2 e^{\boldsymbol{A}_2 t}\boldsymbol{v}_2\right) = 1 - \boldsymbol{a}_1 e^{\boldsymbol{A}_1 t}\boldsymbol{v}_1 - \boldsymbol{a}_2 e^{\boldsymbol{A}_2 t}\boldsymbol{v}_2 + \left(\boldsymbol{a}_1 e^{\boldsymbol{A}_1 t}\boldsymbol{v}_1\right)\left(\boldsymbol{a}_2 e^{\boldsymbol{A}_2 t}\boldsymbol{v}_2\right).$$

Replacing the multiplication of two scalars in the last term of the right-hand side with a Kronecker product gives

$$P[T_S \le t] = 1 - \boldsymbol{a}_1 e^{\boldsymbol{A}_1 t}\boldsymbol{v}_1 - \boldsymbol{a}_2 e^{\boldsymbol{A}_2 t}\boldsymbol{v}_2 + \left(\boldsymbol{a}_1 e^{\boldsymbol{A}_1 t}\boldsymbol{v}_1\right) \otimes \left(\boldsymbol{a}_2 e^{\boldsymbol{A}_2 t}\boldsymbol{v}_2\right).$$

Using Properties 3 and 6 of Kronecker operations on the last term of the right-hand side yields

$$P[T_S \le t] = 1 - \boldsymbol{a}_1 e^{\boldsymbol{A}_1 t}\boldsymbol{v}_1 - \boldsymbol{a}_2 e^{\boldsymbol{A}_2 t}\boldsymbol{v}_2 + (\boldsymbol{a}_1 \otimes \boldsymbol{a}_2)\left(e^{(\boldsymbol{A}_1 \oplus \boldsymbol{A}_2)t}\right)(\boldsymbol{v}_1 \otimes \boldsymbol{v}_2),$$

which can be expressed using matrix augmentation as

$$P[T_S \le t] = 1 - [-\boldsymbol{a}_1 \otimes \boldsymbol{a}_2 \ \ \boldsymbol{a}_1 \ \ \boldsymbol{a}_2] \begin{bmatrix} \boldsymbol{e}^{(\boldsymbol{A}_1 \oplus \boldsymbol{A}_2)t} & \boldsymbol{0} & \boldsymbol{0} \\ \boldsymbol{0} & e^{\boldsymbol{A}_1 t} & \boldsymbol{0} \\ \boldsymbol{0} & \boldsymbol{0} & e^{\boldsymbol{A}_2 t} \end{bmatrix} \begin{bmatrix} \boldsymbol{v}_1 \otimes \boldsymbol{v}_2 \\ \boldsymbol{v}_1 \\ \boldsymbol{v}_2 \end{bmatrix} = 1 - \boldsymbol{a}_S e^{\boldsymbol{A}_S t}\boldsymbol{v}_S.$$

This result is stated in the following theorem.

Theorem 2.10 If $T_1 \sim ME(n_1, \boldsymbol{a}_1, \boldsymbol{A}_1, \boldsymbol{v}_1)$ and $T_2 \sim ME(n_2, \boldsymbol{a}_2, \boldsymbol{A}_2, \boldsymbol{v}_2)$ are independent, then $T_S = \max(T_1, T_2) \sim ME(n_S, \boldsymbol{a}_S, \boldsymbol{A}_S, \boldsymbol{v}_S)$ with

$$n_S = n_1 n_2 + n_1 + n_2, \boldsymbol{a}_S = [-\boldsymbol{a}_1 \otimes \boldsymbol{a}_2 \;\; \boldsymbol{a}_1 \;\; \boldsymbol{a}_2],$$

$$\boldsymbol{A}_S = \begin{bmatrix} \boldsymbol{A}_1 \oplus \boldsymbol{A}_2 & \boldsymbol{0} & \boldsymbol{0} \\ \boldsymbol{0} & \boldsymbol{A}_1 & \boldsymbol{0} \\ \boldsymbol{0} & \boldsymbol{0} & \boldsymbol{A}_2 \end{bmatrix}, \quad \boldsymbol{v}_S = \begin{bmatrix} \boldsymbol{v}_1 \otimes \boldsymbol{v}_2 \\ \boldsymbol{v}_1 \\ \boldsymbol{v}_2 \end{bmatrix}. \tag{2.57}$$

Theorem 2.10 can also be applied to the PH distribution, but the result is ME distribution, not PH distribution, since $\boldsymbol{a}_S$ is not a probability vector. Consequently, Theorem 2.10 is a closure property of ME distribution, but not of PH distribution.

Example 2.30 T_1 and T_2 are independent and identically distributed ME random variables with representations generated in Example 2.28 as

$$\boldsymbol{a}_1 = \boldsymbol{a}_2 = \left[0.5 \;\; -0.017 \;\; 5.517\right],$$

$$\boldsymbol{A}_1 = \boldsymbol{A}_2 = \begin{bmatrix} -1 & 0 & 0 \\ 0 & -4.128 & 0 \\ 0 & 0 & -0.872 \end{bmatrix}, \quad \boldsymbol{v}_1' = \boldsymbol{v}_2' = \begin{bmatrix} -6.9 \\ 0.193 \\ 0.807 \end{bmatrix}.$$

The ME representation for $\min(T_1, T_2)$ is

$$\begin{aligned} n_S &= n_1 n_2 = 9, \\ \boldsymbol{a}_S &= \boldsymbol{a}_1 \otimes \boldsymbol{a}_2 = \left[0.5 \;\; -0.017 \;\; 5.517\right] \otimes \left[0.5 \;\; -0.017 \;\; 5.517\right] \\ &= \left[0.5\left[0.5 \;\; -0.017 \;\; 5.517\right] - 0.017\left[0.5 \;\; -0.017 \;\; 5.517\right] \;\; 5.517\left[0.5 \;\; -0.017 \;\; 5.517\right]\right], \\ \boldsymbol{v}_S &= \boldsymbol{v}_1 \otimes \boldsymbol{v}_2 = \left[-6.9 \;\; 0.193 \;\; 0.807\right]^T \otimes \left[-6.9 \;\; 0.193 \;\; 0.807\right]^T \\ &= \begin{bmatrix} -6.9\left[-6.9 \;\; 0.193 \;\; 0.807\right]^T \\ 0.193\left[-6.9 \;\; 0.193 \;\; 0.807\right]^T \\ 0.807\left[-6.9 \;\; 0.193 \;\; 0.807\right]^T \end{bmatrix}, \\ \boldsymbol{A}_S &= \boldsymbol{A}_1 \oplus \boldsymbol{A}_2 = \begin{bmatrix} -1 & 0 & 0 \\ 0 & -4.128 & 0 \\ 0 & 0 & -0.872 \end{bmatrix} \oplus \begin{bmatrix} -1 & 0 & 0 \\ 0 & -4.128 & 0 \\ 0 & 0 & -0.872 \end{bmatrix} \\ &= \begin{bmatrix} -1 & 0 & 0 \\ 0 & -4.128 & 0 \\ 0 & 0 & -0.872 \end{bmatrix} \otimes \begin{bmatrix} 1 & 0 & 0 \\ 0 & 1 & 0 \\ 0 & 0 & 1 \end{bmatrix} + \begin{bmatrix} 1 & 0 & 0 \\ 0 & 1 & 0 \\ 0 & 0 & 1 \end{bmatrix} \otimes \begin{bmatrix} -1 & 0 & 0 \\ 0 & -4.128 & 0 \\ 0 & 0 & -0.872 \end{bmatrix} \\ &= diag(-1, -1, -1, -4.128, -4.128, -4.128, -0.872, -0.872, -0.872) \\ &\quad + diag(-1, -4.128, -0.872, -1, -4.128, -0.872, -1, -4.128, -0.872) \\ &= diag(-2, -5.128, -1.872, -5.128, -8.256, -5, -1.872, -5, -1.744). \end{aligned}$$

The ME representation for $\text{add}(T_1, T_2) = T_1 + T_2$ is

$$n_S = n_1 + n_2 = 6,$$

$$\boldsymbol{a}_S = [\boldsymbol{a}_1 \;\; \boldsymbol{0}] = \left[0.5 \;\; -0.017 \;\; 5.517 \;\; 0 \;\; 0 \;\; 0\right],$$

$$-\boldsymbol{A}_1\boldsymbol{v}_1\boldsymbol{a}_2 = -\begin{bmatrix} -1 & 0 & 0 \\ 0 & -4.128 & 0 \\ 0 & 0 & -0.872 \end{bmatrix}\begin{bmatrix} -6.9 \\ 0.193 \\ 0.807 \end{bmatrix}\begin{bmatrix} 0.5 & -0.017 & 5.517 \end{bmatrix} = \begin{bmatrix} -3.45 & 0.12 & -38.07 \\ 0.398 & -0.014 & 4.393 \\ 0.352 & -0.012 & 3.884 \end{bmatrix},$$

$$\boldsymbol{A}_S = \begin{bmatrix} \boldsymbol{A}_1 & -\boldsymbol{A}_1\boldsymbol{v}_1\boldsymbol{a}_2 \\ \boldsymbol{0} & \boldsymbol{A}_2 \end{bmatrix} = \begin{bmatrix} -1 & 0 & 0 & -3.45 & 0.12 & -38.07 \\ 0 & -4.128 & 0 & 0.398 & -0.014 & 4.393 \\ 0 & 0 & -0.872 & 0.352 & -0.012 & 3.884 \\ & & & -1 & 0 & 0 \\ & \boldsymbol{0} & & 0 & -4.128 & 0 \\ & & & 0 & 0 & -0.872 \end{bmatrix},$$

$$\boldsymbol{v}_S = \begin{bmatrix} \boldsymbol{v}_1 \\ \boldsymbol{v}_2 \end{bmatrix} = \begin{bmatrix} -6.9 & 0.193 & 0.807 & -6.9 & 0.193 & 0.807 \end{bmatrix}^T.$$

CDFs of the resulting ME distribution in this example are compared with the result using the original PH distribution. The graphs are presented in Figure 2.26, which shows that the CDFs for the minimum and addition operations using ME distribution obtained from the matrix similarity transformation are identical to those using the original PH distribution. ■

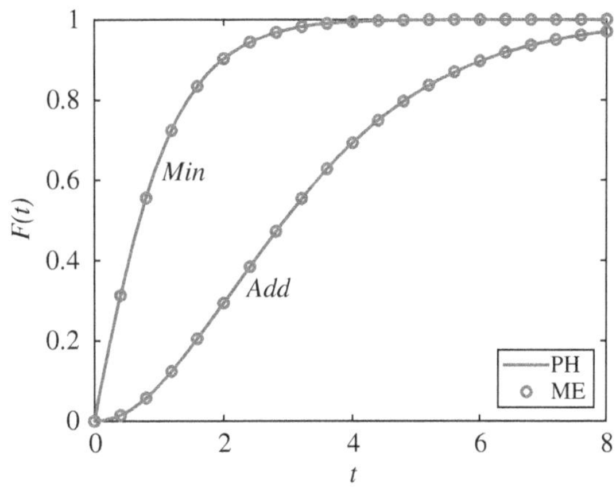

Figure 2.26 Comparison of CDF using PH and ME representations.

Example 2.31 The maximum operation on T_1 and T_2 in Example 2.30 using Theorems 2.9 and 2.10 produce CDFs with graphs as shown in Figure 2.27. It can be seen that both theorems give identical CDFs. The manual calculation is not shown because the result is an ME distribution of order 15. ■

The discussions of PH and ME distributions in this chapter are limited. But they are sufficient to construct the foundation for understanding the materials described in the following chapters. There are other and more advanced aspects of these two distributions that are not discussed in this introductory chapter. Interested readers can find them in the books dedicated to describing the two distributions in great details as mentioned and can be found in the reference list at the end of this chapter.

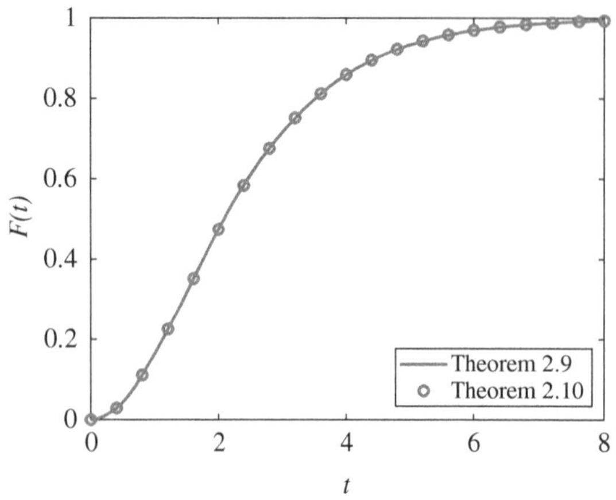

Figure 2.27 Comparison of CDF using Theorems 2.9 and 2.10.

2.8 Problems

2.1 Devise a method to calculate the transition probability of each arc from the transition rate of each arc and vice versa.

2.2 For the system in Application 2.1, compare the cases of one and two absorbing states for
a) The probability will fail in more than 2 months.
b) The means and standard deviations.

2.3 With proper maintenance, the system in Decent state in Application 2.1 will not be broken. Explain the effect of this change using graphs and numbers.

2.4 The system in Application 2.1 is maintained when it is in the Fair state without stopping its operation. The maintenance will bring the system state back from Fair state to Decent state with probability 0.2. Draw the state transition diagram for the system with maintenance and determine its infinitesimal generator, transient matrix, and absorbing vector.

2.5 For the repair process in Application 2.2,
a) Plot the CDF of repair completion time.
b) Determine the probability that the repair completion time will not be more than a half day.
c) Calculate the expected repair time and its standard deviation.

2.6 What is the effect of changing the proportion of repair types in Application 2.2 from 1:4 to 1:3. Explain the effects using graphs and numbers.

2.7 Use Property 3 of the Kronecker operation to prove Property 4.

2.8 Create a comparison table of the results of mixing, addition, minimum, and maximum operations on two independent random variables with PH distributions in terms of order, transient matrix, and initial probability vector.

2.9 Determine the resulting PH distributions from mixing, addition, minimum, and maximum operations on independent random variables with:
a) Two identical exponential distributions.
b) Two identical 2-Erlang distributions.
c) One exponential distribution and one 2-Erlang distribution with the mean of the 2-Erlang distribution is twice the mean of exponential distribution.
Conclude your results.

2.10 Prove that mixing, addition, minimum, and maximum operations are commutative.

2.11 Prove mathematically that the following PH representations are identical:
a) $(\boldsymbol{a}_1 \otimes \boldsymbol{a}_2, \boldsymbol{A}_1 \oplus \boldsymbol{A}_2)$ and $(\boldsymbol{a}_2 \otimes \boldsymbol{a}_1, \boldsymbol{A}_2 \oplus \boldsymbol{A}_1)$
b) $\left(\begin{bmatrix}\boldsymbol{a}_1 & \boldsymbol{0}\end{bmatrix}, \begin{bmatrix}\boldsymbol{A}_1 & \boldsymbol{b}_1\boldsymbol{a}_2 \\ \boldsymbol{0} & \boldsymbol{A}_2\end{bmatrix}\right)$ and $\left(\begin{bmatrix}\boldsymbol{a}_2 & \boldsymbol{0}\end{bmatrix}, \begin{bmatrix}\boldsymbol{A}_2 & \boldsymbol{b}_2\boldsymbol{a}_1 \\ \boldsymbol{0} & \boldsymbol{A}_1\end{bmatrix}\right)$.

2.12 Prove that the following statements are true:
a) Exponential distribution is closed under minimum operation.
b) Hyperexponential distribution is closed under mixing operation.
c) Hypoexponential distribution is closed under addition operation.
d) Erlang distribution is closed under addition operation.
e) Coxian distribution is closed under addition operation.

2.13 Prove that if X and Y are independent PH distributions, then $\max[X - Y, 0]$ is a PH distribution with a discontinuity at point 0. Explain why there is discontinuity at the origin even though X and Y have no discontinuity.

2.14 Find the convolution of the PH distribution in Example 2.8 to itself. Do this using the convolution of two scalar distributions and the closure properties of PH distribution under addition.

2.15 Determine the infinitesimal generator matrices from the results of minimum, mixing, and sorting operations.

2.16 Prove Lemma 2.1 in the case of T without discontinuity.

2.17 Use the analogy with hyperexponentials and hypoexponentials to determine representations of hyper-Erlang and hypo-Erlang distributions.

2.18 The PH distribution with initial probability vectors $\begin{bmatrix}0.5 & 0.5\end{bmatrix}$ and transient matrices $\begin{bmatrix}-2 & 2 \\ 1 & -2\end{bmatrix}$ is to be transformed to the Jordan canonical form.
a) Determine the transformation matrix and the resulting ME representation.
b) Show that the PH and ME representations have the same mean and variance.

2.19 Extend the state transition diagram in Application 2.1 to the case where the duration of stay in each state has an arbitrary PH distribution. Use the diagram to determine the infinitesimal generator and transient matrix.

2.20 Create a PH distribution model using the closure properties of PH distribution for the time required to produce one acceptable product. The production process includes assembling components 1 and 2 into one product, the assembly time has PH distribution, and the assembled product must be tested with a probability of passing the test is p. The manufacturing of component 1 goes through two stages; all stages require identical PH-distributed processing time and, after completion, must be tested with a probability of passing the test is q. The manufacturing of component 2 goes through three stages, all stages require identical PH-distributed processing time and, after completion, must be tested with a probability of passing the test is r. The components that pass the test are sent to the assembly section.

2.21 Example 2.25 involves mixing several dependent random variables. Nevertheless, this method produces the correct distribution. What are the consequences of this result to Theorem 2.2?

2.22 Prove the closure properties of ME distribution under minimum, mixing, and addition operations.

References

1 Ross, S.M., 2023. Introduction to Probability Models. Academic Press.
2 Kleinrock, L., 1975. Queueing Systems. Volume 1: Theory. Wiley-Interscience.
3 He, Q.-M., 2014. Fundamentals of Matrix-Analytic Methods. Vol. 365. Springer New York, New York.
4 Neuts, M.F., 1994. Matrix-Geometric Solutions in Stochastic Models: An Algorithmic Approach. Dover Publication, Inc.
5 Asmussen, S., 2003. Applied Probability and Queues, Stochastic Modelling and Applied Probability. Springer, New York.
6 Cox, D.R., Miller, H.D., 1977. The Theory of Stochastic Processes. Vol. 134. CRC Press.
7 Bladt, M., Nielsen, B.F., 2017. Matrix-Exponential Distributions in Applied Probability. Springer, Boston, MA.
8 Horváth, A., Telek, M., 2024. Phase Type Distributions: Theory and Application. Wiley-ISTE.
9 Buchholz, P., Kriege, J., Felko, I., 2014. Input Modeling with Phase-Type Distributions and Markov Models: Theory and Applications. Springer International Publishing, Cham.
10 Trivedi, K.S., Bobbio, A., 2017. Reliability and Availability Engineering: Modeling, Analysis, and Applications. Cambridge University Press.
11 Okamura, H., Dohi, T., 2016. PH Fitting Algorithm and Its Application to Reliability Engineering. J Oper Res Soc Jpn 59, 72–109. https://doi.org/10.15807/jorsj.59.72.
12 Bellman, R., 1997. Introduction to Matrix Analysis, 2nd ed. Society for Industrial and Applied Mathematics, Philadelphia.
13 Alkaff, A., Qomarudin, M.N., Bilfaqih, Y., 2021. Network reliability analysis: Matrix-exponential approach. Reliab Eng Syst Saf 212, 107591. https://doi.org/10.1016/j.ress.2021.107591.
14 Bladt, M., Neuts, M.F., 2003. Matrix-exponential distributions: Calculus and interpretations via flows. Stoch Model 19, 113–124.

3

One-Component Systems

Learning objectives:

1) Understand the derivations of all equations for system reliability measures in PH distribution.
2) Able to design maintenance and replacement programs.
3) Able to generate PH distribution of system lifetime from shock model descriptions.
4) Able to utilize the software package to calculate system reliability measures, to design perfect and imperfect maintenance programs, and to design replacements with and without possible reuse programs.

Abbreviations

BFR	bathtub failure rate
CDF	cumulative distribution function
CFR	constant failure rate
DFR	decreasing failure rate
IFR	increasing failure rate
ME	matrix-exponential
PDF	probability density function
PH	phase-type
$CVTTF$	coefficient of variations of time to failure
$MTBR$	mean time between replacement
$MTBR_M$	mean time to between replacement of system with maintenance
$MTTF$	mean time to failure
$MTTF_M$	mean time to failure of system with maintenance
$SDTTF$	standard deviation of time to failure
TBR	time between replacement
$VTTF$	variance of time to failure

Notations

T	time to failure
t	time, mission duration
g	age

n	order, dimension
$\boldsymbol{a}$	initial probability vector
$\boldsymbol{A}$	survival matrix
$\boldsymbol{I}$	identity matrix
$\boldsymbol{u}$	unitary vector
s	replacement interval
h	maintenance interval
b	maintenance level
$(n, \boldsymbol{a}, \boldsymbol{A})$	PH representation
$E(\cdot)$	expected value
$R(t)$	reliability function
$RR(t \mid g)$	residual reliability function at age g
$h(t)$	hazard function
$H(t)$	integral hazard function
$MRL(g)$	mean residual life function
$R_G(t, s)$	reliability function with replacement interval s
$R_M(t, h)$	reliability function with maintenance interval h
$N(t)$	number of failures up to time t
$M(t)$	number of replacements up to time t
$P_n(t)$	probability up to time t the system experiences n failures ($P[N(t) = n]$)
$\otimes$	Kronecker product
$\oplus$	Kronecker sum

The PH distributions discussed in Chapter 2 are closely related to reliability. The states in the CTMC can represent the system states in reliability analysis. Transient states are states that represent the many good conditions in which the system is considered still function. An absorbing state is no different from a failed state since a failed system will remain in the failed state forever (unless repair is performed). With these analogies, the reliability of a multistate system can be modeled in one distribution. It is simpler than the conventional approach that uses a separate model for each state, which makes the analysis become complicated. To reduce its complexity, the systems in conventional reliability analysis are usually assumed to have only two states, one good state and one failed state, and as a result, modeling multistate systems requires several reliability models, one model for each state. However, many studies of system reliability modeling required the use of matrix-based distribution due to the failure description is too complex to be represented by scalar distributions that consider the states are either good or failed, one of which is Ranjkesh et al. [1].

Besides its simplicity in dealing with multistate systems, the use of PH distributions as a model for system reliability has other advantages, including [2]:

- PH distributions can be used as models for all types of distributions commonly used in reliability analysis, such as distributions with constant failure rate (CFR), increasing failure rate (IFR), and decreasing failure rate (DFR). Even distributions with a combination of decreasing, constant, and IFR functions in the form of bathtub failure rate (BFR) curves can also be modeled in PH distributions.
- PH distributions can be used to approximate other distributions that are not related to CTMC, as shown in Chapter 2.
- PH distributions have closure properties, as given in Chapter 2, that can be utilized to express the relationship between the system reliability function and the reliability functions of its

components. These properties are useful for dynamic reliability analyses for systems with many components.

- PH distributions can reduce the need to use calculus in calculating reliability measures to only use matrix algebra and even produce closed-form solutions in simple algebraic expressions. Using PH distributions in calculating the moments of lifetimes does not require integrals, and deriving the hazard function no longer requires differentials.

These advantages promise that if the system reliability functions are modeled in PH distributions, then the path to finding a simple method for exact dynamic analyses of system reliability will be smooth.

Systems generally consist of many subsystems, with each subsystem consisting of many components. However, system reliability analysis often views the system as an inseparable entity, so it is unnecessary to recognize its subsystems or components. Such reliability analyses consider the system under study as composed of one large component. System failure is due to the system not functioning, regardless of which subsystem or component causes the system to fail. This approach leads to a simple system reliability analysis. This book develops system reliability analysis starting from a system that is considered to consist of a single component as given in this Chapter 3. Subsequent chapters will discuss the reliability of systems that need to be represented as compositions of many subsystems and components. For one-component systems, system reliability is the same as component reliability; therefore, the terms system and component are used both as needed in this chapter.

This chapter starts with the reliability interpretation of PH distribution by describing the relationship between transient–absorbing states in a CTMC and good–failed states of a system. This way, the similarity between the absorption time in PH distribution and the system lifetime will be clear. Assuming the system lifetime distribution follows a PH distribution, this chapter then develops methods to express functional and statistical system reliability measures in terms of the PH representation of system lifetime distribution. The equations for system reliability measures available in beginner reliability books are developed further to facilitate multistate systems with lifetimes following PH distributions. Functional reliability measures include reliability function, hazard function, residual reliability function, and mean residual life (*MRL*) function, along with the relationship among these functions. Statistical reliability measures include *MTTF*, variance time to failure (*VTTF*), standard deviation of time to failure (*SDTTF*), and coefficient of variations of time to failure (*CVTTF*), along with the relationship between these entities. The reliability measures expressed in PH representation of the system lifetime distribution are shown to be sufficiently calculated using matrix algebra. Numerical methods for efficiently computing reliability measures are also developed.

Development in this chapter continues with the utilization of PH representations of system lifetime distributions to design reliability-based maintenance and replacement programs. Maintenance and replacement problems that are often found in popular reliability engineering books, such as Lewis [3], Trivedi and Bobbio [4], Elsayed [5], are developed further for multistate systems having PH-distributed lifetimes. Periodic maintenance programs are developed to determine the maintenance interval required to maintain the system reliability for the desired mission duration within specified values. The development includes perfect and imperfect maintenance with the level of perfection that can vary between no maintenance and perfect maintenance. The periodic replacement program is studied for its effect on the system reliability and the mean time between replacement (*MTBR*). A method to calculate spare-part requirements to meet the desired level of stockout risk is also given, including determining the optimal replacement interval to minimize the associated replacement costs. Modeling for replacement in the event of failure does not always require replacement is also presented. All resulting equations will be expressed in terms of matrix algebra. MATLAB programs are provided for all algorithms presented in this chapter. At the end of this chapter, various shock and wear models are discussed along with their use to determine PH representations of system lifetime distributions.

The discussion in this and subsequent chapters uses PH distributions with no discontinuities at the origin. This assumption means that no system has already or instantly failed when first used, an assumption that is commonly encountered in practice. Considering the similarity of PH and ME distributions as have been shown in Chapter 2, the results of this chapter are applicable also for ME distribution.

First, the interpretation of PH distribution in reliability context is explained in the next section.

3.1 Reliability Interpretation of PH Distribution

Transient states in CTMC can be used to represent several system functioning levels, while the absorbing state represents the system is failed. Transition from one state to another represents a change in the system conditions. Internal factors, such as aging, fatigue, material wear, and external factors, such as temperature, humidity, dust, shocks, and other operational environmental conditions, can cause the change. In the context of system reliability, Figure 2.2 can be restated by interpreting the transient state as "good" state and the absorbing state as "failed" state. This interpretation results in Figure 3.1, which, unlike Figure 2.2, does not contain the element a_{n+1}, a direct transition to the absorbing state, in accordance with the assumption that the system is not instantly or already failed at the start of operation.

Transitions among transient states indicate that the system is still viable and can operate in acceptable conditions. In this sense, the transient matrix $\boldsymbol{A}$ in CTMC in reliability analysis is referred to as the survival matrix. Transitions from transient states (good states) to the absorbing state (failed states) indicate the system is failing. In reliability analysis, the absorbing vector $\boldsymbol{b}$ is referred to as the failure vector. A failed system will remain in the failed state, so there is no transition out of the failed state. The relationship between $\boldsymbol{A}$ and $\boldsymbol{b}$ is the same as in Chapter 2, namely, $\boldsymbol{b} = -\boldsymbol{Au}$. Like the transient matrix, the system survival matrix must have nonpositive diagonal elements, nonnegative off-diagonal elements, and a nonpositive row sum. Likewise, the failure vector has the same characteristics as the absorbing vector in that all its elements are nonnegative and the sum is less than or equal to 1.

The absorption time in CTMC, which is the time to enter the absorbing state from any initial states, in a reliability context is the time from the system starting to operate until it fails. In reliability theory, this duration is called the lifetime or time to failure. In Chapter 2, it was explained that the absorption time in CTMC with one absorbing state has PH distribution. Thus, the lifetime of a system whose state changes follow a CTMC with one absorbing state has PH distribution, or its distribution can be expressed in PH representation. If T is the system lifetime, then T has a PH

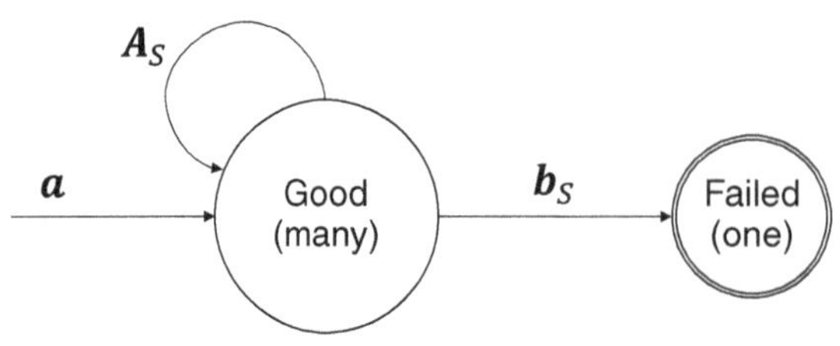

Figure 3.1 State transition diagram for reliability analysis using PH distribution.

distribution, denoted as $T \sim PH(n, \boldsymbol{a}, \boldsymbol{A})$, whose CDF as given in Chapter 2 is

$$F_T(t) = P[T \le t] = 1 - \boldsymbol{a}e^{\boldsymbol{A}t}\boldsymbol{u} \quad t \ge 0. \tag{3.1}$$

The system reliability measures, functional and statistical, will all be expressed in PH representation $(n, \boldsymbol{a}, \boldsymbol{A})$. The following sections develop methods for using Equation (3.1) to express system reliability measures in PH representations. System reliability measures comprise functional reliability measures, which includes reliability function, hazard function, residual reliability function, mean residual reliability function, and statistical reliability measures, which include moments of time to failures. The development starts with the reliability function as the most important reliability measure, which is given in the next section.

3.2 Reliability Function

The basic measure of reliability is of course the reliability function. The reliability function can be said to be the source of all reliability measures. Other reliability measures can be derived from the reliability function. The reliability function, denoted by $R(t)$, is defined as the probability that the system is still functioning, or is still good, at time t. In other words, the reliability function $R(t)$ measures the probability that the system lifetime exceeds the time t. This definition can be expressed mathematically as

$$R(t) = P[T > t] = 1 - P[T \le t] = F_T(t). \tag{3.2}$$

Substituting Equation (3.1) into Equation (3.2) yields

$$R(t) = \boldsymbol{a}e^{\boldsymbol{A}t}\boldsymbol{u} + a_{n+1}\,\delta(t) \quad t \ge 0. \tag{3.3}$$

The additional impulse of a_{n+1} unit is to make the reliability value at $t = 0$ is equal to 1 as it is generally assumed that when the system is new, it must be in good condition, or $R(0) = 1$. If no system has failed from the start, then $a_{n+1} = 0$, and Equation (3.2) can be simplified to

$$R(t) = \boldsymbol{a}e^{\boldsymbol{A}t}\boldsymbol{u} \quad t \ge 0. \tag{3.4}$$

Equation (3.4) guarantees that $R(0) = 1$ since $R(0) = \boldsymbol{a}e^{\boldsymbol{0}}\boldsymbol{u} = \boldsymbol{au}$, and $\boldsymbol{au} = 1 - a_{n+1} = 1 - 0 - 1$. Given that CDF is a monotonic nondecreasing function from 0 at $t = 0$ to 1 at $t \to \infty$, the reliability function is a monotonic nonincreasing function from 1 at $t = 0$ to 0 at $t \to \infty$; thus, $0 \le R(t) \le 1$ and $t_1 > t_2 \to R(t_1) \le R(t_2)$.

The parameter t in the reliability function $R(t)$ states the duration of system usage. The system reliability function shows the change of the system reliability value over time, or dynamic; it is used to show that the system reliability value depends on the duration of its usage. The longer the usage, the lower the system reliability value, because the reliability function is monotonic nonincreasing. As explained in Chapter 1, this book develops dynamic reliability analyses using $R(t)$. Other reliability measures are derived from $R(t)$. When associated with a particular mission whose duration has been determined, for example, during a unit time, the system reliability to carry out the mission is a constant with a value of $R(a)$, often written simply as R. Constant reliability values are used in static reliability analysis. Knowing the system reliability function can determine the system reliability value at any time; but knowing the system reliability value is not sufficient to know the system reliability function. Thus, static reliability analyses can be carried out using the results of the dynamic reliability analyses, but not vice versa.

The following examples illustrate obtaining a PH representation of the system lifetime from observational data on changes in the system state at any time. A method for generating a PH representation of the system lifetime related to the shocks and wear experienced by the system is discussed in Section 3.11.

Example 3.1 A one-component system has several states related to its conditions that can be categorized into several good states and one failed state. The good states are categorized based on the number of defects the system produces in its operation. They are classified into four states: "None," "Few," "Some," and "Many." The observation data show that the duration of the system in a "None" state is exponentially distributed with an average of 10 days, which then enters the "Few" state. The duration of the system in the "Few" state is exponentially distributed with an average of 2.5 days before finally entering the "Some" state, "Many" state, and even failed, each with a probability of 0.5, 0.25, and 0.25, respectively. The duration of the system in the "Some" state is exponentially distributed with an average of 2 days and with a probability of 0.6 entering the "Many" state and the rest becoming failed. The system is in the "Many" state for an exponentially distributed duration with an average of 2.5 days, then fails. The categories and their values can be adjusted to management needs and ease of analysis. The system at the start of operation is in a new condition, with "None" defects produced.

Since the time spent in each state is exponentially distributed, the process that governs the change in system states is CTMC. The information provided can be presented in Table 3.1. The state transition diagram for the CTMC based on the transition probability created using this table is shown in Figure 3.2, with one-circle nodes indicating the good states and two-concentric circle node indicating the failed state. The determination of the good and failed states is in accordance with the management policy, namely, that the management assumes that the system is good if it can still be operated, even though it produces many defects. Information that at the beginning of the operation the system is in a new state is given as an arrow with a value of 1 from the outside

Table 3.1 State, duration of stay, and transition probability of Example 3.1.

State number	State description	Mean duration of stay (day)	Transition probability between states: 1	2	3	4	5
1	None	10	–	1	–	–	–
2	Few	2.5	–	–	0.5	0.25	0.25
3	Some	2	–	–	–	0.6	0.4
4	Many	2.5	–	–	–	–	1
5	Failed	–	–	–	–	–	–

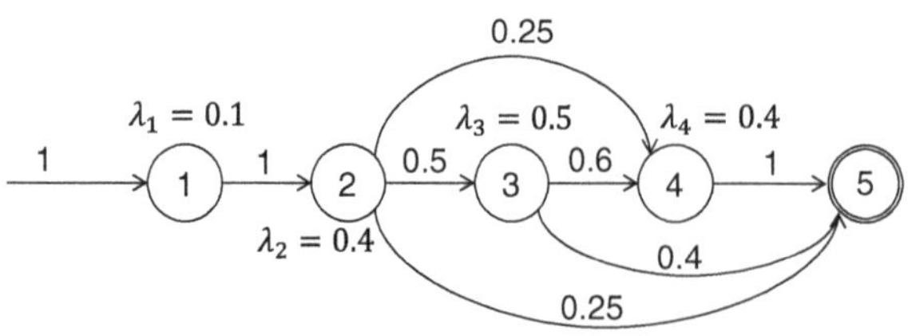

Figure 3.2 State transition diagram in transition probability for Example 3.1.

toward the state 1, which indicates that the system is in the "None" state. Note that the sum of the transition probabilities out of each node is equal to 1.

To determine the PH representation of the system lifetime distribution, the transition probabilities must first be converted into transition rates among states because the survival matrix is determined based on the transition rates among good states. As explained in Chapter 2, since the duration of stay in state 1 has exponential distribution with a mean of 10 days, the exit rate from state 1 is $\lambda_1 = 1/10 = 0.1/\text{day}$, and the transition rate from state 1 to state 2 can be calculated as $\lambda_{12} = p_{12}\lambda_1 = 1 \times 0.1 = 0.1$. For state 2,

$$\lambda_{23} = p_{23}\lambda_2 = 0.5 \times 1/2.5 = 0.2,$$
$$\lambda_{24} = p_{24}\lambda_2 = 0.25 \times 1/2.5 = 0.1,$$
$$\lambda_{25} = p_{25}\lambda_2 = 0.25 \times 1/2.5 = 0.1.$$

The transition rates among other states can be calculated similarly. The results are presented in Table 3.2, and the state transition diagram in transition rate is presented in Figure 3.3.

Table 3.2 State and rate of transition between states of Example 3.1.

State number	State description	Transition rate between states				
		1	2	3	4	5
1	None	–	0.1	–	–	–
2	Few	–	–	0.2	0.1	0.1
3	Some	–	–	–	0.3	0.2
4	Many	–	–	–	–	0.4
5	Failed	–	–	–	–	–

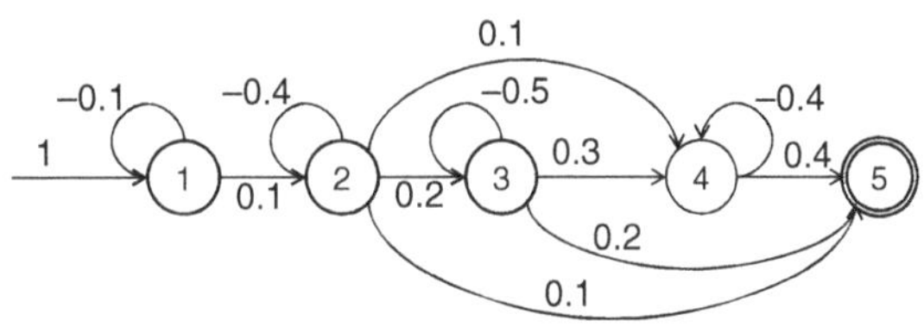

Figure 3.3 State transition diagram in transition rate for Example 3.1.

The infinitesimal generator matrix of this system can be obtained from Table 3.2 by adding diagonal elements, each equal to the negative of the row sum of the off-diagonal elements in that row, to make the row sum equal to 0. Thus,

$$\boldsymbol{Q} = \left[\begin{array}{cccc|c} -0.1 & 0.1 & 0 & 0 & 0 \\ 0 & -0.4 & 0.2 & 0.1 & 0.1 \\ 0 & 0 & -0.5 & 0.3 & 0.2 \\ 0 & 0 & 0 & -0.4 & 0.4 \\ \hline 0 & 0 & 0 & 0 & 0 \end{array}\right].$$

The survival matrix can be obtained from the matrix $\boldsymbol{Q}$ by removing the rows and columns corresponding to the failed state, and the failure vector can be calculated using Equation (2.9) or obtained from the last column of matrix Q without the last row,

$$\boldsymbol{A} = \begin{bmatrix} -0.1 & 0.1 & 0 & 0 \\ 0 & -0.4 & 0.2 & 0.1 \\ 0 & 0 & -0.5 & 0.3 \\ 0 & 0 & 0 & -0.4 \end{bmatrix}, \quad \boldsymbol{b} = -\boldsymbol{A}\boldsymbol{u} = \begin{bmatrix} 0 \\ 0.1 \\ 0.2 \\ 0.4 \end{bmatrix}.$$

The initial probability vector is determined from the information that the system was still new at the start of operation. This information means the probability starting in state 1 is 1, and the probability starting at other states is 0. Hence,

$$\boldsymbol{a} = [1\ \ 0\ \ 0\ \ 0].$$

The order of the PH representation of the system lifetime distribution is $n = 4$, the dimensions of the survival matrix and initial probability vector, or the number of transient states.

Since the duration of stay at each state in CTMC is exponentially distributed, the time the system is in the "None" state, T_1, can be obtained from the value of a_{11}; that is, $T_1 \sim \exp(0.1)$. The probability that the duration of the system in the "None" state is more than 2 days is $e^{-0.1 \times 2} = e^{-0.2}$. ■

Exercise 3.1 Determine the probability that the system stays in the "Few" state more than 1 day.

Example 3.2 Management considers the "Many" state to be failed. In this case, there are two failed states. Since PH distributions only allow one absorbing state, the two failed states must be combined into one failed superstate. The state transition diagram in transition probabilities becomes as shown in Figure 3.4.

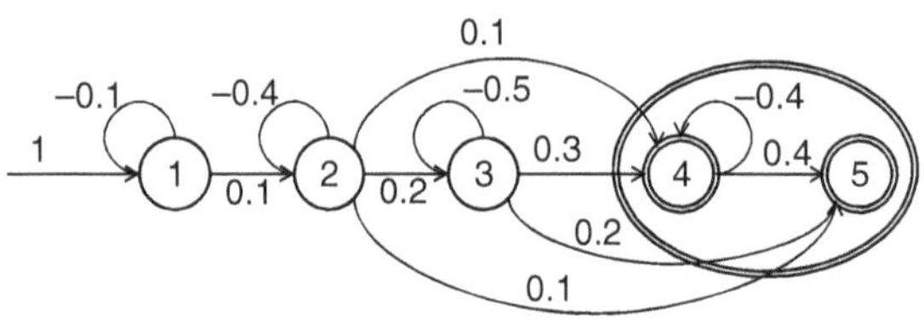

Figure 3.4 State transition diagram of Example 3.2 with two absorbing states.

The transition rate from state 2 to the failed state becomes $q_{24} + q_{25} = 0.1 + 0.1 = 0.2$ and from state 3 to the failed state becomes $q_{34} + q_{35} = 0.3 + 0.2 = 0.5$. The state transition diagram with the two failed states merged into one is presented in Figure 3.5.

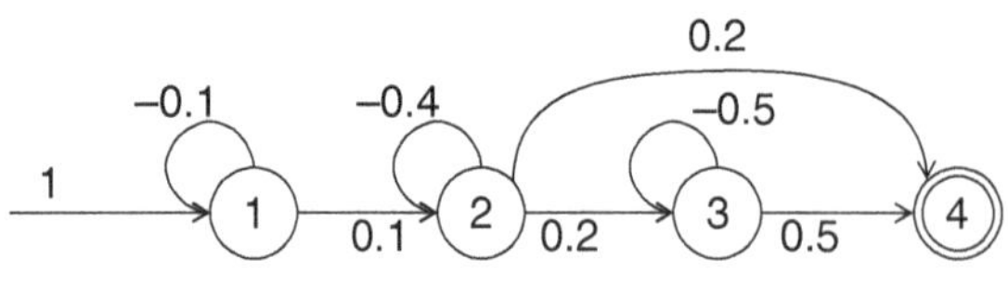

Figure 3.5 State transition diagram of Figure 3.4 after merging the two failed states.

The following quantities can be derived based on Figure 3.5:

$$\boldsymbol{Q} = \begin{bmatrix} -0.1 & 0.1 & 0 & 0 \\ 0 & -0.4 & 0.2 & 0.2 \\ 0 & 0 & -0.5 & 0.5 \\ 0 & 0 & 0 & 0 \end{bmatrix},$$

$$A = \begin{bmatrix} -0.1 & 0.1 & 0 \\ 0 & -0.4 & 0.2 \\ 0 & 0 & -0.5 \end{bmatrix}, \quad b = \begin{bmatrix} 0 \\ 0.2 \\ 0.5 \end{bmatrix}, \quad a = \begin{bmatrix} 1 & 0 & 0 \end{bmatrix}.$$

■

Exercise 3.2 Are the durations of stays in transient states affected by the new definition of failure?

Example 3.3 The system in Example 3.1 is maintained without stopping its operation while in state 4. If this maintenance is successful before the system fails (enters the failed state), then the system condition improves to state 3. The probability that this maintenance will be successful is 0.5. Using transition probabilities, the state transition diagram of this system with maintenance is as presented in Figure 3.6. If maintenance does not affect the duration of stay in state 4, then λ_4 remains equal to 0.4.

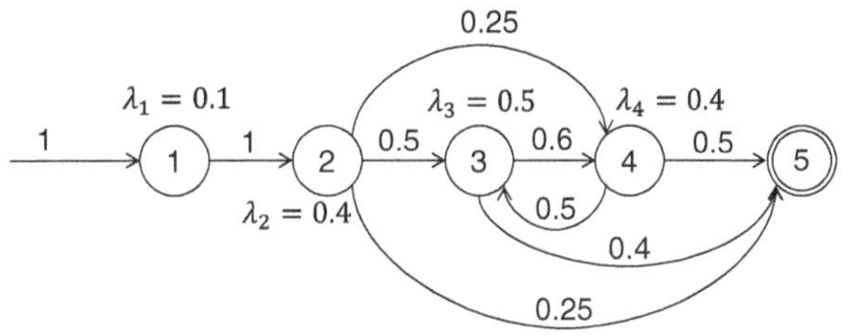

Figure 3.6 State transition diagram in transition probability for Example 3.3.

The transition rate from state 4 to state 3 is $q_{43} = 0.5 \times 0.4 = 0.2$. Since the duration of stay in state 4 remains the same, the addition $q_{43} = 0.2$ from maintenance causes the value of q_{45} to change from 0.4 to $0.5 \times 0.4 = 0.2$ to make the transition rate out of state 4 remain the same equal to 0.4. The state transition diagram in transition rate for Example 3.3 is presented in Figure 3.7. Note that Example 3.3 produces CPH, while Example 3.1 and Example 3.2 produce APH. (The definitions of APH and CPH are given in Chapter 2.)

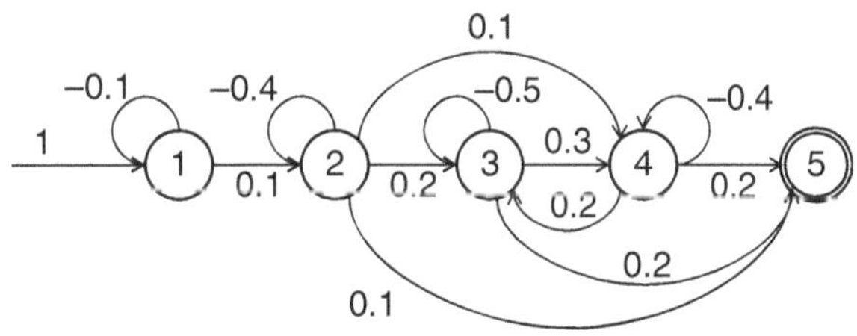

Figure 3.7 State transition diagram in transition rate for Example 3.3.

The following quantities can be derived from Figure 3.7:

$$Q = \begin{bmatrix} -0.1 & 0.1 & 0 & 0 & 0 \\ 0 & -0.4 & 0.2 & 0.1 & 0.1 \\ 0 & 0 & -0.5 & 0.3 & 0.2 \\ 0 & 0 & 0.2 & -0.4 & 0.2 \\ 0 & 0 & 0 & 0 & 0 \end{bmatrix},$$

$$A = \begin{bmatrix} -0.1 & 0.1 & 0 & 0 \\ 0 & -0.4 & 0.2 & 0.1 \\ 0 & 0 & -0.5 & 0.3 \\ 0 & 0 & 0.2 & -0.4 \end{bmatrix}, \quad b = \begin{bmatrix} 0 \\ 0.1 \\ 0.2 \\ 0.2 \end{bmatrix}, \quad a = \begin{bmatrix} 1 & 0 & 0 & 0 \end{bmatrix}.$$

■

By knowing their PH representations, the reliability functions of systems in Examples 3.1–3.3 can be plotted using Equation (3.4). Graphing reliability functions are not easy to do manually because they involve calculations of matrix exponential functions. For this purpose, a MATLAB function **phrelplot** for calculating and plotting reliability function in the software package **dysrel-am** is prepared. The program receives the PH representation of distribution of the system lifetime $(n, \boldsymbol{a}, \boldsymbol{A})$ as inputs.

Example 3.4 A comparison of the reliability functions of systems in Examples 3.1–3.3 is shown in Figure 3.8, with $R_1(t)$, $R_2(t)$, and $R_3(t)$ representing the reliability function of Examples 3.1–3.3, respectively.

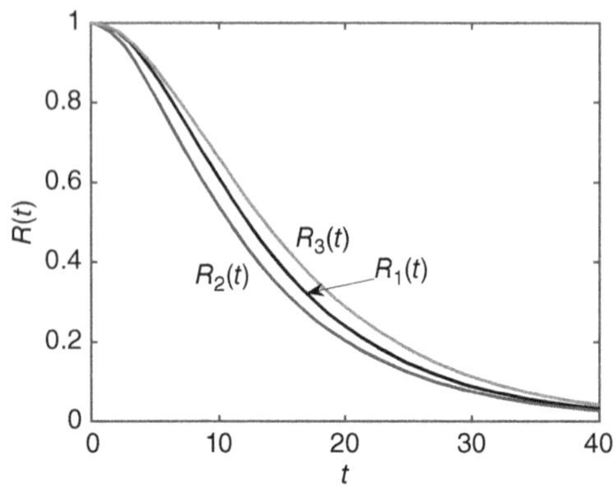

Figure 3.8 Reliability functions of systems in Examples 3.1–3.3.

■

MATLAB code for example3_4.m

```
close all;clear
addpath('..\dysrel-am','..\dysrel-am\expokit');
% Component's lifetime in PH distribution:
% Example 3.1
T(1).a = [1 0 0 0];
T(1).A = 0.1*[-1 1 0 0;0 -4 2 1;0 0 -5 3;0 0 0 -4];
T(1).n = 4;
% Example 3.2
T(2).a = [1 0 0];
T(2).A = 0.1*[-1 1 0;0 -4 2;0 0 -5];
T(2).n = 3;
% Example 3.3
T(3).a = [1 0 0 0];
T(3).A = 0.1*[-1 1 0 0;0 -4 2 1;0 0 -5 3;0 0 2 -4];
T(3).n = 4;
% Time vector
tmax = 40;
t = linspace(0,tmax,21);
for j = 1:3
  [tt,R(:,j)] = phrelplot(t,T(j));
end
close all
% Reliability plot
lgnd = {'R_1(t)','R_2(t)','R_3(t)'};
multiplot(3,tt,R,lgnd,'t','R(t)');
```

The main programs for some examples and applications in Chapter 3 are given in **example3_x.m** files in the software package **dysrel-am**, where x is the example or application number.

Figure 3.8 indicates a behavior that should have been expected from the description of each system; that is,

$$R_2(t) \leq R_1(t) \leq R_3(t) \quad \forall t \geq 0.$$

Based on this reliability function ordering, it can be ascertained that the system in Example 3.2 has a lifetime stochastically shorter than the system in Example 3.1 ([6]). With the notation $\preccurlyeq$ meaning "stochastically shorter or equal," the systems in Examples 3.1–3.3 have lifetime ordering

$$T_2 \preccurlyeq T_1 \preccurlyeq T_3,$$

to state that the lifetime of the system in Example 3.2 is stochastically shorter than or equal to the lifetime of the system in Example 3.1 and that the system in Example 3.1 has a lifetime stochastically shorter than or equal to the lifetime of the system in Example 3.3. This conclusion is realistic and can be obtained intuitively from the operational descriptions of the systems. The system in Example 3.2 has the shortest lifetime because it has more stringent operational requirements; that is, the system in Example 3.2 is considered failed when it is in state 4, while the systems in Examples 3.1 and 3.3 are still considered good and continue to be operated. Likewise, the system in Example 3.3 has the longest lifetime because it is maintained when it is in state 4 and its condition improves back to state 3.

Example 3.5 The reliability values of the systems in Examples 3.1–3.3 for various mission durations are given in Table 3.3. These values are obtained from the reliability function $R(t)$ by substituting the given mission durations as the values of t. Table 3.3 shows that for the same mission duration, the reliability values ordering of these systems is consistent

$$R_2 \leq R_1 \leq R_3,$$

following their reliability functions ordering.

Table 3.3 Comparison of reliability values for several mission durations.

Mission duration	System in example	Reliability: Function	Reliability: Value
5	3.1	$R_1(5)$	0.87
	3.2	$R_2(5)$	0.82
	3.3	$R_3(5)$	0.89
10	3.1	$R_1(10)$	0.61
	3.2	$R_2(10)$	0.54
	3.3	$R_3(10)$	0.66
15	3.1	$R_1(15)$	0.39
	3.2	$R_2(15)$	0.33
	3.3	$R_3(15)$	0.45

■

The reliability function measures system survival. Sometimes the interest is on the system unreliability, or system vulnerability. As the complement of the system reliability, the system vulnerability is given by $1 - R(t) = F_T(t)$. Another important reliability measure is the rate of system failure. The next section discusses this reliability measure, which is popularly known as hazard function.

3.3 Hazard Function

Hazard function (or instantaneous failure function), denoted by $h(t)$, is defined as the instantaneous failure rate for a system that has not failed at time t, or $T > t$, will fail in Δt time later, or $T \leq t + \Delta t$, for $\Delta t \to 0$. The relationship between the reliability and hazard functions can be derived from this definition as follows. ($\wedge$ is the logical conjunction operator.)

$$\begin{aligned} h(t) &= \lim_{\Delta t \to 0} \frac{P[T \leq t + \Delta t | T > t]}{\Delta t} \\ &= \lim_{\Delta t \to 0} \frac{P[T \leq t + \Delta t \wedge T > t]}{\Delta t P[T > t]} \\ &= \lim_{\Delta t \to 0} \frac{P[t < T \leq t + \Delta t]}{\Delta t P[T > t]} \\ &= \lim_{\Delta t \to 0} \frac{P[T > t] - P[T > t + \Delta t]}{\Delta t P[T > t]} \\ &= \frac{-\lim_{\Delta t \to 0} \frac{R(t + \Delta t) - R(t)}{\Delta t}}{R(t)} = -\frac{R'(t)}{R(t)}. \end{aligned}$$

Thus, the relationship between the reliability function and the hazard function is

$$h(t) = -\frac{R'(t)}{R(t)} \quad t \geq 0. \tag{3.5}$$

Substituting Equation (3.4) into Equation (3.5) produces

$$h(t) = -\frac{(\boldsymbol{a}e^{\boldsymbol{A}t}\boldsymbol{u})'}{\boldsymbol{a}e^{\boldsymbol{A}t}\boldsymbol{u}} = -\frac{\boldsymbol{a}e^{\boldsymbol{A}t}\boldsymbol{A}\boldsymbol{u}}{\boldsymbol{a}e^{\boldsymbol{A}t}\boldsymbol{u}} = -\frac{\boldsymbol{a}\boldsymbol{A}e^{\boldsymbol{A}t}\boldsymbol{u}}{\boldsymbol{a}e^{\boldsymbol{A}t}\boldsymbol{u}}.$$

Therefore,

$$h(t) = -(\boldsymbol{a}e^{\boldsymbol{A}t}\boldsymbol{A}\boldsymbol{u})(\boldsymbol{a}e^{\boldsymbol{A}t}\boldsymbol{u})^{-1} = -(\boldsymbol{a}\boldsymbol{A}e^{\boldsymbol{A}t}\boldsymbol{u})(\boldsymbol{a}e^{\boldsymbol{A}t}\boldsymbol{u})^{-1} \quad t \geq 0. \tag{3.6}$$

Note that the inverted term in Equation (3.6) is a scalar, not a matrix, so it does not burden the computation.

Computer assistance is required to calculate and plot the hazard function of a system whose lifetimes follow a high-order PH distribution. A MATLAB function **phhazplot** for calculating and plotting hazard function is provided in the software package **dysrel-am.** The program receives PH representation of system lifetime distribution $(n, \boldsymbol{a}, \boldsymbol{A})$ as input.

Example 3.6 Graphs of the hazard functions of systems in Examples 3.1–3.3 are presented in Figure 3.9. The functions $h_1(t)$, $h_2(t)$, and $h_3(t)$ represent the hazard functions of the systems in Examples 3.1–3.3, respectively.

The graphs show that these hazard functions follow the reverse of their reliability functions ordering, that is,

$$h_3(t) \leq h_1(t) \leq h_2(t) \; \forall t \geq 0.$$

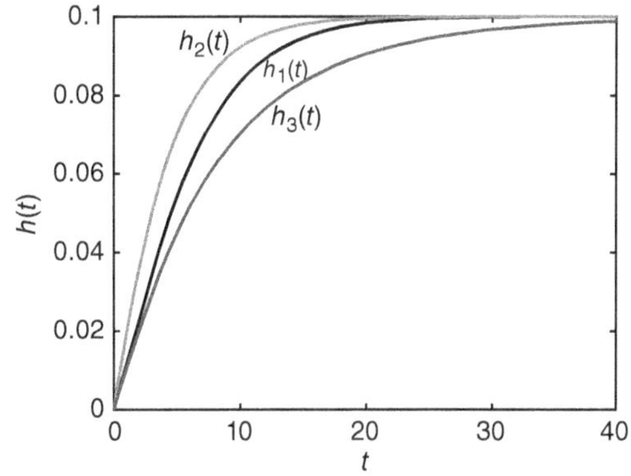

Figure 3.9 Hazard functions of systems in Examples 3.1–3.3.

■

It makes sense that the hazard functions ordering is the reverse of the reliability functions ordering because the higher the reliability, the lower the failure rate. This conclusion is clear from the negative relationship between the reliability and the hazard functions given by Equation (3.5). Consequently, the stochastic order of time to failure follows the reverse of hazard function ordering.

An important use of hazard function is for determining the type of a system based on its failure rate behavior.

3.3.1 Type of System

The hazard function is often used to determine the system type. As mentioned in the introduction in this chapter, there are several types of systems depending on the shapes of their hazard function. DFR, CFR, and IFR systems are systems with monotonically decreasing, constant, and monotonically increasing hazard functions, respectively. Their descriptions are:

- DFR systems are systems whose hazard functions are decreasing over time. The longer they are used, the lower their failure rates.

  $$t_1 > t_2 \rightarrow h(t_1) < h(t_2) \quad \forall t_1,\ t_2 \geq 0.$$

  An example of PH distributions with DFRs is all hyperexponential distributions.

- CFR systems are systems whose hazard functions are constant over time. The failure rates of these systems are not affected by the length of use.

  $$t_1 > t_2 \rightarrow h(t_1) = h(t_2) \quad \forall t_1,\ t_2 \geq 0.$$

  The exponential distribution is the only PH distribution with CFRs.

- IFR systems are systems whose hazard functions are increasing over time. The longer they are used, the higher the failure rates.

  $$t_1 > t_2 \rightarrow h(t_1) > h(t_2) \quad \forall t_1,\ t_2 \geq 0.$$

  An example of PH distributions with IFRs is all Erlang distributions.

The hazard function graphs in Figure 3.9 indicate that the one-component systems in Examples 3.1–3.3 are IFR. The following examples present other types of systems whose lifetimes follow distributions in the class of PH distributions described in Chapter 2.

Example 3.7 The PH distribution order 1, which is exponential distribution, has a representation $(1, [1], \boldsymbol{A})$, where $\boldsymbol{A}$ is a scalar. Therefore, $\boldsymbol{a}e^{\boldsymbol{A}t}\boldsymbol{A}\boldsymbol{u}$ can be rearranged into $\boldsymbol{A}(\boldsymbol{a}e^{\boldsymbol{A}t}\boldsymbol{u})$, and its hazard function is

$$h(t) = -\boldsymbol{A}(\boldsymbol{a}e^{\boldsymbol{A}t}\boldsymbol{u})(\boldsymbol{a}e^{\boldsymbol{A}t}\boldsymbol{u})^{-1} = -\boldsymbol{A},$$

which is a constant. ■

Since the hazard function is constant, systems whose lifetimes follow PH distributions of order 1 are CFR. If $\boldsymbol{A}$ is $[-\lambda]$, then the hazard function is λ, and λ is called the failure rate. The constant hazard function means the failure rate is not affected by the duration of use. A system with a constant hazard rate is a system whose reliability for a certain duration of use is not affected by how long the system has been used previously, often referred to as the system's age. This property has been explained in Section 2.1 as the memoryless property of exponential distribution.

Example 3.8 Graphs of hazard functions of systems having lifetimes that follow n-hyperexponential distributions with PH representations

$$n = n, \quad \boldsymbol{a} = \left[1/n \ 1/n \dots 1/n\right], \quad \boldsymbol{A} = \begin{bmatrix} -1/10 & 0 & \dots & 0 \\ 0 & -2/10 & \dots & 0 \\ \vdots & \vdots & \ddots & \vdots \\ 0 & 0 & \dots & -n/10 \end{bmatrix},$$

for $n = 1, 2, \dots, 5$ is presented in Figure 3.10. These graphs show that systems whose lifetimes follow n-hyperexponential distribution have decreasing hazard functions. Thus, the system type is DFR.

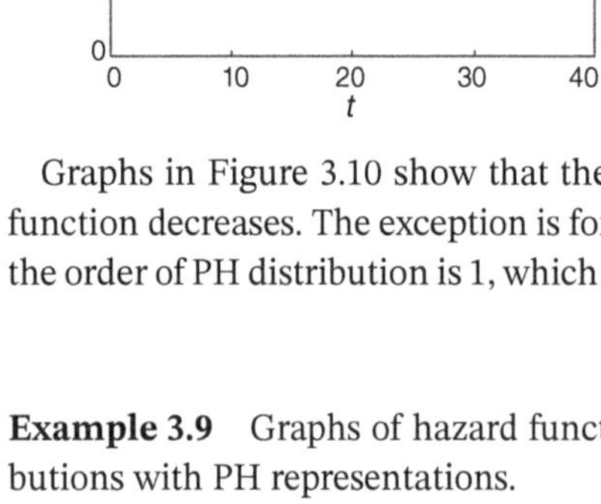

Figure 3.10 Hazard functions of n-hyperexponential distribution, $n = 1, 2, \dots, 5$.

Graphs in Figure 3.10 show that the larger the order of the distribution, the faster the hazard function decreases. The exception is for $n = 1$, for which the hazard function is a constant because the order of PH distribution is 1, which makes the distribution become an exponential distribution. ■

Example 3.9 Graphs of hazard functions of systems having lifetimes following n-Erlang distributions with PH representations.

$$n = n, \quad \boldsymbol{a} = \left[1 \ 0 \dots 0\right], \quad \boldsymbol{A} = \begin{bmatrix} -1/10 & 1/10 & \dots & 0 \\ 0 & -1/10 & \dots & 0 \\ \vdots & \vdots & \ddots & \vdots \\ 0 & 0 & \dots & -1/10 \end{bmatrix},$$

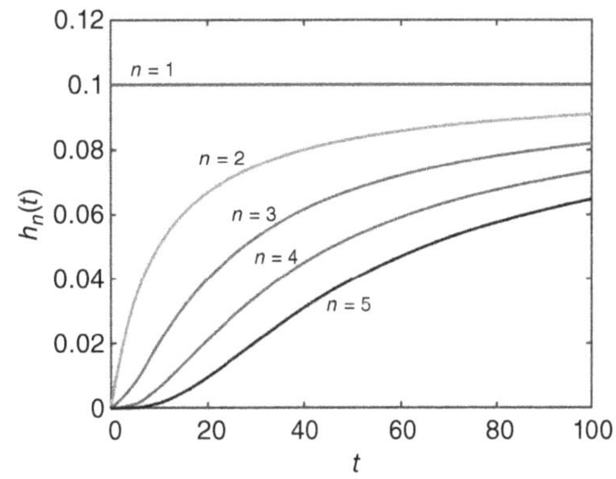

Figure 3.11 Hazard functions of n-Erlang distribution, $n = 1, 2, \ldots, 5$.

for $n = 1, 2, \ldots, 5$ is depicted in Figure 3.11. These graphs show that the type of systems with n-Erlang distributed lifetimes is IFR.

Graphs in Figure 3.11 show that the larger the order of the distribution, the slower the hazard function increases over time, except for $n = 1$, which results in a constant hazard function because the distribution reduces to an exponential distribution. ■

Exercise 3.3 Investigate the type of systems that have lifetimes following n-hypoexponential distribution and the effect of PH order to the hazard function.

Depending on their types, hazard functions have different effects on the reliability function of systems that have been used for some time but have not failed. A detailed explanation of such effects is discussed in Section 3.4.

The system hazard function is not always increasing, constant, or decreasing. Most real-world systems have hazard functions that are decreasing, constant, and increasing over time.

3.3.2 Bathtub Hazard Function

In practice, hazard functions sometimes have curves that are decreasing, constant, and increasing over time. One of them is the bathtub hazard function, which is a hazard function with a curve that initially decreases, then becomes constant, and finally increases. This type of system is the BFR system, which are systems whose hazard functions are in the form of a bathtub as shown in Figure 3.12. An explanation of the reasons for encountering bathtub hazard functions in practice can be found in reliability books such as Trivedi and Bobbio [4] and Lewis [3]. PH distributions with the right choices of representations can produce all types of hazard functions including BFR.

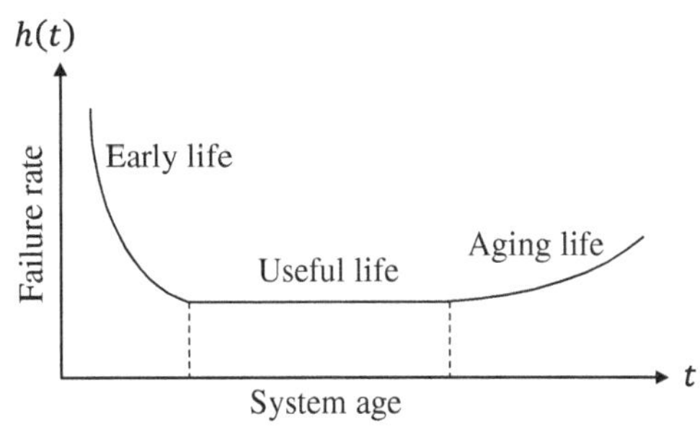

Figure 3.12 Bathtub hazard function.

One member of the PH distribution class that is flexible enough to be used as a model for lifetimes of DFR, IFR, and BFR system is the Coxian distribution as shown in Example 3.10. Although not shown in the graph, a Coxian distribution can also be used as a model for the CFR system because, just like Erlang and hyperexponential, the Coxian distribution of order 1 also reduces to the exponential distribution.

Example 3.10 A PH distribution with proper parameter selection can produce a distribution with complex behaviors such as a bathtub curve as Faddy [7] showed. Following his step, 3-Coxian distributions with appropriate parameters can be used to produce several types of hazard functions. Using the values in Table 3.4 as parameters of 3-Coxian distribution in Table 2.1 will produce Figure 3.13. This figure shows hazard functions of DFR, IFR, and BFR types of systems.

Table 3.4 The 3-Coxian parameters to create graphs on Figure 3.13.

Distribution Type	Parameters				
IFR	$p_1 = 0.96$;	$p_2 = 0.81$;	$\lambda_1 = 0.47$;	$\lambda_2 = 0.31$;	$\lambda_3 = 0.15$
DFR	$p_1 = 0.75$;	$p_2 = 0.81$;	$\lambda_1 = 0.6$;	$\lambda_2 = 0.31$;	$\lambda_3 = 0.02$
BFR	$p_1 = 0.75$;	$p_2 = 0.93$;	$\lambda_1 = 0.6$;	$\lambda_2 = 0.27$;	$\lambda_3 = 0.15$

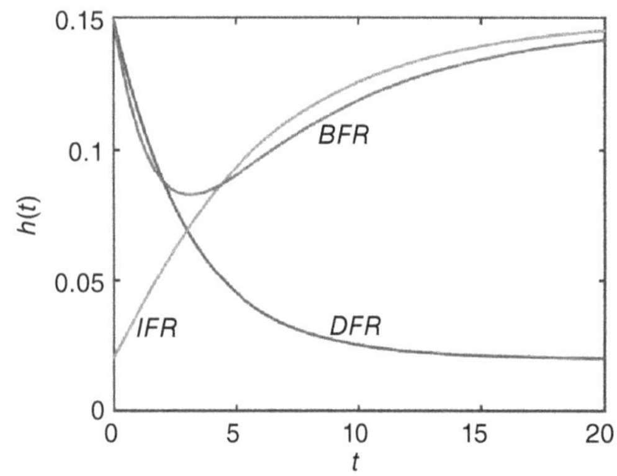

Figure 3.13 Graph of the hazard function using 3-Coxian distribution.

■

Hazard functions with BFR distributions can be generated using several PH distributions by splitting the curve into three segments; each is modeled in an appropriate distribution. For example, the initial segment is modeled using a hyperexponential distribution to accommodate DFR behavior, the middle segment is modeled using an exponential distribution to accommodate CFR behavior, and the final segment is modeled using an Erlang distribution to accommodate IFR behavior.

Example 3.11 The following combination of PH distributions produces a bathtub hazard function curve as shown in Figure 3.14.

- $0 \le t < 10$ using a 3-hyperexponential distribution with representation

$$\boldsymbol{\alpha} = [0.4 \quad 0.3 \quad 0.3], \quad \boldsymbol{A} = \begin{bmatrix} -0.1 & 0 & 0 \\ 0 & -0.2 & 0 \\ 0 & 0 & -0.3 \end{bmatrix}.$$

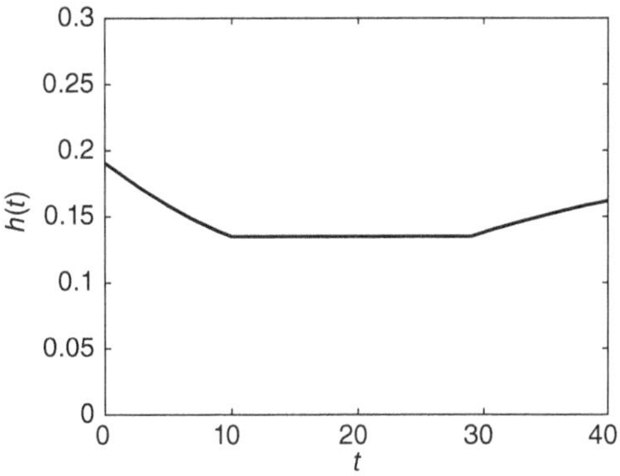

Figure 3.14 Bathtub hazard function of system in Example 3.11.

- $10 \leq t < 30$ using an exponential distribution with representation

$$\boldsymbol{a} = [1], \quad \boldsymbol{A} = \begin{bmatrix} -0.135 \end{bmatrix}.$$

- $t \geq 30$ using a 5-Erlang distribution with representation

$$\boldsymbol{a} = [1 \quad 0 \quad 0 \quad 0 \quad 0],$$

$$\boldsymbol{A} = \begin{bmatrix} -0.2467 & 0.2467 & 0 & 0 & 0 \\ 0 & -0.2467 & 0.2467 & 0 & 0 \\ 0 & 0 & -0.2467 & 0.2467 & 0 \\ 0 & 0 & 0 & -0.2467 & 0.2467 \\ 0 & 0 & 0 & 0 & -0.2467 \end{bmatrix}.$$

■

Given that hazard function can be obtained from reliability function, there must be a way to derive reliability function from hazard function as discussed next.

3.3.3 Finding Reliability Functions from Hazard Functions

There are times when field data produces a hazard function. Equation (3.5) suggests there is a one-to-one relationship between reliability and hazard functions. Therefore, the known hazard function can produce the unknown reliability function.

Integrating both sides of Equation (3.5) yields

$$\int_0^t \frac{R'(x)}{R(x)} dx = -\int_0^t h(x) dx.$$

Defining $H(t) = -\int_0^t h(x)dx$, called the integral hazard function, produces

$$\ln R(t) - \ln R(0) = -H(t).$$

Substituting $\ln R(0) = \ln 1 = 0$ yields

$$\ln R(t) = -H(t).$$

Thus,

$$R(t) = e^{-H(t)} \quad t \geq 0. \tag{3.7}$$

Equation (3.7) is the basis for constructing the standard Weibull distribution, which is very popular in reliability analysis because it can accommodate all types of systems. The standard

Weibull distribution assumes that the hazard function is

$$h(t) = at^b \quad t \geq 0. \tag{3.8}$$

The values of b in Equation (3.8) determine the system types:

- DFR if $b < 0$.
- CFR if $b = 0$.
- IFR if $b > 0$.

The standard Weibull distribution cannot accommodate BFR systems. Producing hazard functions with bathtub curves requires a more complicated function than the simple polynomial function in Equation (3.8). Researchers proposed many mathematical functions to approximate bathtub curves. Using such hazard functions, generalized Weibull distributions, which are an extension of the standard Weibull distribution, can be produced.

The generalized Weibull distribution can also be constructed using piecewise continuous functions, with decreasing values for $0 \leq t < t_1$, constant values for $t_1 \leq t < t_2$, and increasing values for $t \geq t_2$. One way is to use a standard Weibull distribution for each graph segment by selecting values of b that match the characteristics of the curve segment. To use the methods described in this book, each standard Weibull distribution is approximated by a PH distribution as described in Chapter 2.

Example 3.12 Each segment of the hazard function of a BFR system can be modeled using a standard Weibull distribution. One example is using the following hazard functions:

$$h_1(t) = 0.3t^{-0.5} \qquad 0 \leq t < 4$$

$$h_2(t) = 0.15 \qquad 4 \leq t < 25$$

$$h_3(t) = 0.0012t^{1.5} \qquad t \geq 25$$

The system hazard function curve is as shown in Figure 3.15.

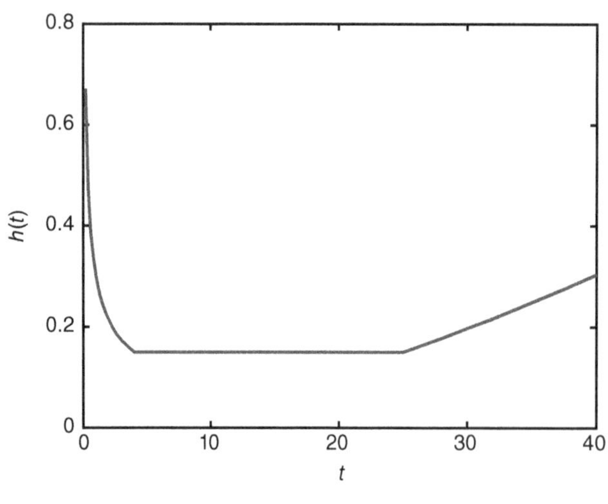

Figure 3.15 Hazard function of the system in Example 3.12. ■

Example 3.13 The reliability function for the BFR system in Example 3.12 can be found using Equation (3.7) by performing piecewise integration to the hazard function pieces.

- $0 \leq t < 4$: $H_1(t) = -\int_0^t h_1(x)dx = 0.6t^{0.5} \rightarrow R_1(t) = e^{-0.6t^{0.5}}$.

- $4 \leq t < 25$: $H_2(t) = -\int_0^4 h_1(x)dx - \int_4^t h_2(x)dx = -1.2 - 0.15(t-4)$
  $\rightarrow R_2(t) = e^{-1.2-0.15(t-4)}$.
- $t \geq 25$: $H_3(t) = -\int_0^4 h_1(x)dx - \int_4^{25} h_2(x)dx - \int_{25}^t h_3(x)dx$
  $= -1.2 - 3.15 - 0.00048(t^{2.5} - 25^{2.5})$
  $\rightarrow R_3(t) = e^{-1.2-3.15-0.00048(t^{2.5}-25^{2.5})}$.

The resulting graph of the system reliability function is shown in Figure 3.16. Even though the system hazard function fluctuates, the system reliability function still meets the monotonically decreasing requirement, starting from a value of 1 at $t = 0$ and reaching a value of 0 at $t \rightarrow \infty$.

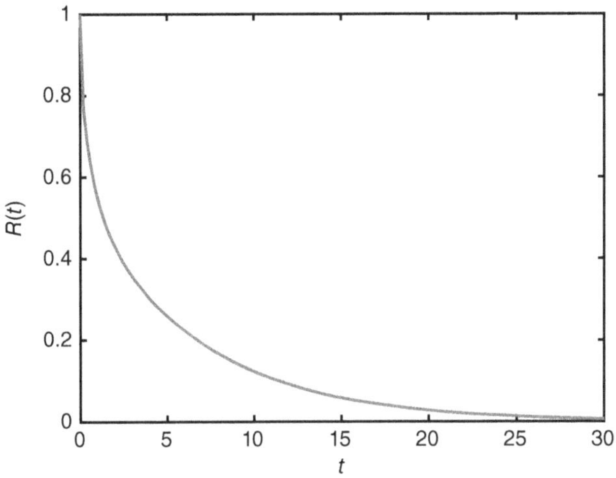

Figure 3.16 The reliability function of BFR system in Example 3.12. ■

Example 3.14 The standard Weibull distribution for each piece of hazard function in Example 3.12 can be approximated by a PH distribution with representation:

- $0 \leq t < 4$: $\boldsymbol{a}_1 = [1 \quad 0]$, $\boldsymbol{A}_1 = \begin{bmatrix} -0.6 & 0.1 \\ 0.47 & -0.1 \end{bmatrix}$.
- $4 \leq t < 25$: $\boldsymbol{a}_3 = [1]$, $\boldsymbol{A}_3 = [-0.18]$
- $t \geq 25$: $\boldsymbol{a}_2 = [1 \quad 0]$, $\boldsymbol{A}_2 = \begin{bmatrix} -0.6 & 0.2 \\ 0.1 & -0.2 \end{bmatrix}$.

The graphs of the system reliability functions generated using the PH distribution approximations and using the original Weibull distributions are presented in Figure 3.17. There is a discrepancy between the two functions. However, this discrepancy can be reduced by using higher orders PH distributions.

The mean value of time to failure can be calculated using the relationship between reliability function and mean value given in Section 1.3 as (will be derived in Section 3.5)

$$E[T] = \int_0^4 R_1(t)dt + \int_4^{25} R_2(t)dt + \int_{25}^{\infty} R_3(t)dt.$$

$E[T]$ obtained using the Weibull distributions and their PH distribution approximations are 386.8363 and 384.2808, respectively, with a difference of 0.66%. This difference can be made smaller by increasing the order of the PH distribution approximations. ■

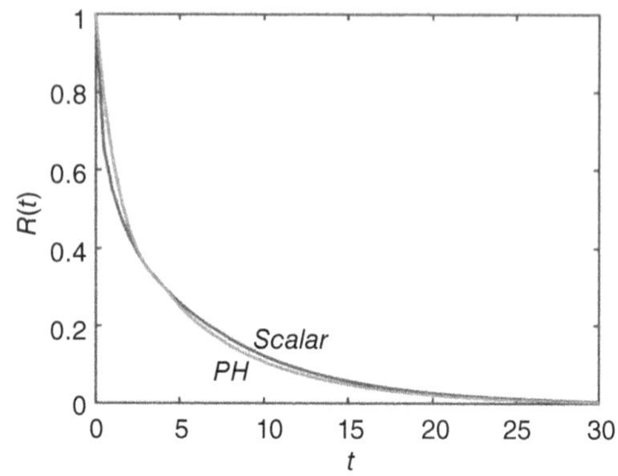

Figure 3.17 Comparison of system reliability functions in Examples 3.12 and 3.14.

Exercise 3.4 Compare the hazard functions using Weibull distribution in Example 3.13 and using PH distribution in Example 3.14. Compare the discrepancies in these hazard functions with the discrepancy in Figure 3.17.

System age, which represents the duration of a system being used and still functioning, greatly influences system reliability. The next section describes the effect of system age to system reliability function.

3.4 Residual Reliability Function

The residual reliability function is the reliability function of a system that has not failed after being used for a certain time to be used for an additional time. The length of time the system has been used until the time of observation, provided that the system has not failed, is called the system age. The residual reliability function for a system with age g, denoted as $RR(t + g \mid g)$, measures the system reliability to be used for t additional time. Residual reliability is a conditional probability because it is for a system known to have a lifetime longer than g, or $T > g$. Following its definition,

$$RR(t+g|g) = P[T > t+g|T > g] \qquad t \geq 0. \tag{3.9}$$

Equation (3.9) can be simplified by using the law of conditional probability,

$$RR(t+g|g) = \frac{P[T > t+g \wedge T > g]}{P[T > g]} = \frac{P[T > t+g]}{P[T > g]} \quad t \geq 0. \tag{3.10}$$

Substituting Equation (3.4) into Equation (3.10) yields

$$RR(t+g|g) = \frac{\boldsymbol{a}e^{\boldsymbol{A}(t+g)}\boldsymbol{u}}{\boldsymbol{a}e^{\boldsymbol{A}g}\boldsymbol{u}} = \frac{\boldsymbol{a}e^{\boldsymbol{A}g}}{\boldsymbol{a}e^{\boldsymbol{A}g}\boldsymbol{u}}e^{\boldsymbol{A}t}\boldsymbol{u} = \boldsymbol{a}\left(\frac{e^{\boldsymbol{A}g}}{\boldsymbol{a}e^{\boldsymbol{A}g}\boldsymbol{u}}\right)e^{\boldsymbol{A}t}\boldsymbol{u}.$$

Thus,

$$RR(t+g|g) = \mathbf{c}(g)e^{\boldsymbol{A}t}\boldsymbol{u} \quad t \geq 0, \tag{3.11}$$

where $\mathbf{c}(g) = \boldsymbol{a}\left(\frac{e^{\boldsymbol{A}g}}{\boldsymbol{a}e^{\boldsymbol{A}g}\boldsymbol{u}}\right)$.

Equation (3.11) shows that the PH distribution of the lifetime of a system at any age has the same survival matrix. The only change is the initial probability vector corrected by $\left(\frac{e^{\boldsymbol{A}g}}{\boldsymbol{a}e^{\boldsymbol{A}g}\boldsymbol{u}}\right)$, from $\boldsymbol{a}$ to become $\boldsymbol{a}\frac{e^{\boldsymbol{A}g}}{\boldsymbol{a}e^{\boldsymbol{A}g}\boldsymbol{u}}$. This correction factor is a function of age. This result has been used and mentioned in Chapter 2.6, and it is essential for dynamic reliability analyses of systems with many components discussed in the following chapters.

Note that $\mathbf{c}(0) = \boldsymbol{a}\left(\frac{e^{A\times 0}}{\boldsymbol{a}e^{A\times 0}\boldsymbol{u}}\right) = \boldsymbol{a}\left(\frac{1}{\boldsymbol{a}\boldsymbol{u}}\right) = \boldsymbol{a}$ (because $\boldsymbol{a}\boldsymbol{u} = 1$). Therefore,

$$RR(t + g|g)\,|_{g=0} = R(t).$$

Example 3.15 The PH distribution of order 1 has $\boldsymbol{a} = \boldsymbol{u} = [1]$ and e^{Ag} is a scalar; hence,

$$\frac{e^{Ag}}{\boldsymbol{a}e^{Ag}\boldsymbol{u}} = \frac{e^{Ag}}{e^{Ag}} = 1.$$

This result means that the initial probability vectors of the reliability functions of CFR systems are not corrected by their age and remain the same $= \boldsymbol{a} = [1]$ as if it were new regardless of its age.

■

Residual reliability function identical to reliability function corresponds to CFR systems having constant hazard functions; no matter how old the system is, the rate of failure is always constant, the same as when it was new, a property popularly referred to as memoryless property. For this reason, the residual system reliability function for CFR system is the same as the system reliability function.

For systems with lifetimes following high-order PH distributions, the residual reliability function must be calculated and plotted with the help of a computer program. For this purpose, a MATLAB function **phresrelplot** for plotting residual reliability function is provided in the software package **dysrel-am.** The function receives the PH representation of the distribution of system lifetime $(n, \boldsymbol{a}, \boldsymbol{A})$ and the system age g.

Example 3.16 Graphs of residual reliability functions of systems in Examples 3.1–3.3 at 10 days old created using **phresrelplot** are shown in Figure 3.18. In this figure, RR_1, RR_2, and RR_3 are the residual reliability for the systems in Examples 3.1, 3.2, and 3.3, respectively. Note that the residual reliability function of these three systems at 10 days old does not differ much from each other. This behavior is different from their reliability function when they were new or their reliability function at $t = 10$. However, the residual reliability functions ordering is the same as their reliability functions ordering, that is,

$$RR_2(t + g|g) \leq RR_1(t + g|g) \leq RR_3(t + g|g)\ \forall t, g.$$

■

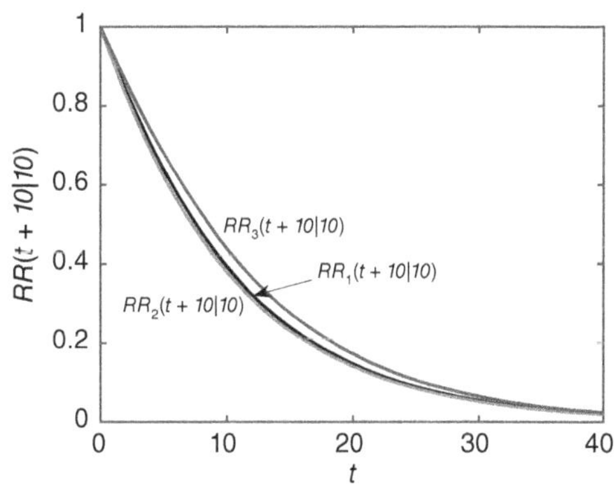

Figure 3.18 Residual reliability functions of systems in Examples 3.1–3.3.

Example 3.17 A comparison of the residual reliability function at 10-day old, $RR(t + 10 \mid 10)$, with the reliability function when it was new, $R(t)$, for the system in Example 3.1 is shown in Figure 3.19.

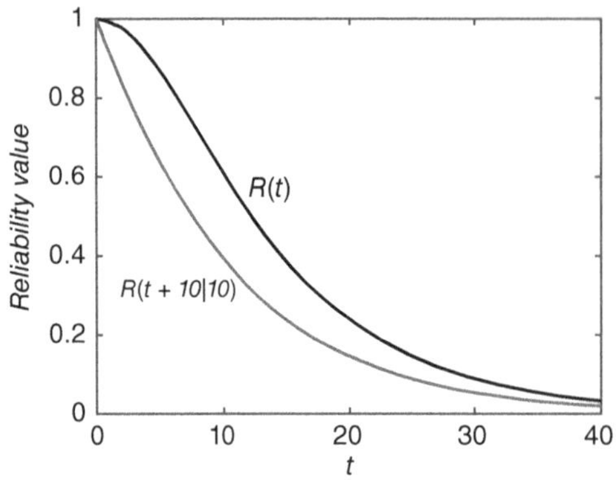

Figure 3.19 Comparison of reliability and residual reliability functions of the system in Example 3.1.

■

The code for the main program of **example3_17.m** is as follows:

MATLAB code for example3_17.m

```
close all;clear
addpath('..\dysrel-am','..\dysrel-am\expokit');
% Component's lifetime in PH distribution:
% Example 3.1
T.a = [1 0 0 0];
T.A = 0.1*[-1 1 0 0;0 -4 2 1;0 0 -5 3;0 0 0 -4];
T.n = 4;
% Time vector
tmax = 40;
g = 10;
t = linspace(0,tmax,21);
% Calculate R(t) and RR(t)
[tt,R] = phrelplot(t,T);
[tt,RR] = phresrelplot(t,T,g);
% Residual Reliability plot
close all
lgnd = {'R(t)',['R(t+',num2str(g),'|',num2str(g),')']};
multiplot(2,tt,[R' RR'],lgnd,'t','Reliability value');
```

The result of Example 3.17 is closely related to the system in Example 3.1 is an IFR system. For this case, the residual reliability function is lower than the reliability function. Considering the hazard function determines whether the system gets better, worse, or just as good as it ages, the hazard function strongly influences the behavior of the residual reliability function relative to the reliability function. Table 3.5 provides the effects of system types on the characteristics of the relationship between the reliability function and its residual reliability function.

Examples 3.18 and 3.19 show graphs of the residual reliability function against the reliability function for DFR and IFR systems. For the CFR systems, the two functions are identical as mathematically shown in Example 3.15.

Example 3.18 A graphical comparison of the reliability and residual reliability functions of a DFR system having a lifetime following a 3-hyperexponential distribution with PH representation

$$\boldsymbol{a} = [0.2 \quad 0.3 \quad 0.5], \quad \boldsymbol{A} = \begin{bmatrix} -0.1 & 0 & 0 \\ 0 & -0.05 & 0 \\ 0 & 0 & -0.02 \end{bmatrix}$$

is presented in Figure 3.20.

Table 3.5 Characteristics of residual reliability functions.

System type	Properties of residual reliability function
DFR	$RR(t+g \mid g) > R(t)$ $g_1 > g_2 \rightarrow RR(t+g_1 \mid g_1) > RR(t+g_2 \mid g_2)$
CFR	$RR(t+g \mid g) = R(t)$ $g_1 > g_2 \rightarrow RR(t+g_1 \mid g_1) = RR(t+g_2 \mid g_2)$
IFR	$RR(t+g \mid g) < R(t)$ $g_1 > g_2 \rightarrow RR(t+g_1 \mid g_1) < RR(t+g_2 \mid g_2)$

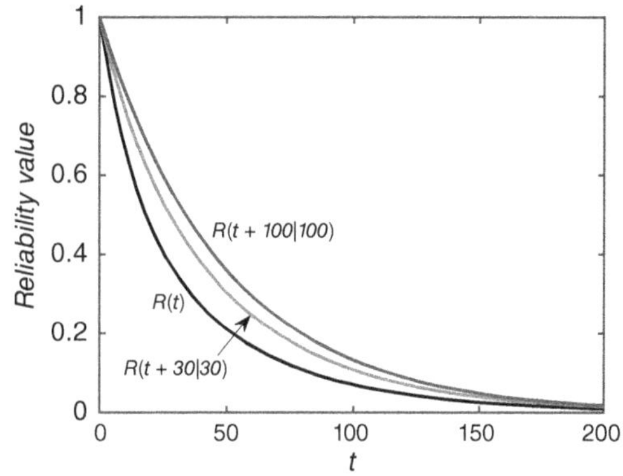

Figure 3.20 A comparison of the reliability and residual reliability functions of the DFR system in Example 3.18.

The graphs show that the system reliability function gets higher, or the system gets better, with age, that is $R(t + 100 \mid 100) > R(t + 30 \mid 30) > R(t)$, in accordance with Table 3.5. ■

Example 3.19 A graphical comparison of reliability and residual reliability functions of an IFR system whose lifetime follows a hypoexponential distribution with PH representation

$$\boldsymbol{a} = [1 \quad 0 \quad 0], \quad \boldsymbol{A} = \begin{bmatrix} -0.1 & 0.1 & 0 \\ 0 & -0.05 & 0.05 \\ 0 & 0 & -0.02 \end{bmatrix}$$

is presented in Figure 3.21.

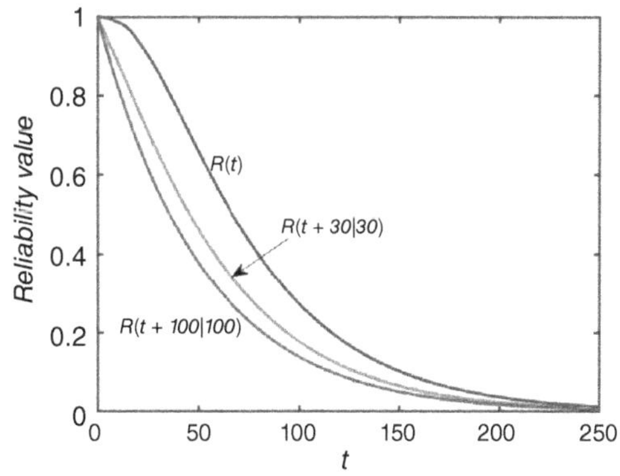

Figure 3.21 A comparison of the reliability and residual reliability functions of the IFR system in Example 3.19.

The graphs show that the system reliability function gets lower, or the system gets worse, with age, that is $R(t+100 \mid 100) < R(t+30 \mid 30) < R(t)$, in accordance with Table 3.5. ■

Exercise 3.5 Compare the residual reliability functions at several ages for a system with lifetime following Erlang distribution and check their consistency with the system type.

Functions are not easily digestible compared to numbers, especially for practical purposes. To provide simple information on system survival in numbers, moments are used as reliability measures of system as described in the next section.

3.5 Moments of Lifetime

The moments of lifetimes are numbers used as statistical measures of reliability. They can be expressed in the PH representation of the system lifetime distribution in closed-form expressions. Such simplicity is one advantage of dynamic system reliability analysis using PH distributions where moments of system lifetimes can be calculated using closed-form expressions involving matrix algebra. In contrast, dynamic system reliability analyses using scalar distribution functions need integral of function to calculate them.

Frequently used moments in reliability analyses are the first and second moments. The first moment of lifetime is the expected value of the random variables of time to failure, popularly known as *MTTF*. The expected lifetime of a system is closely related to the reliability function. The relationship between the two can be derived from the definition of the first moment, which is

$$MTTF = \int_0^\infty t f_T(t)dt = \int_0^\infty t dF_T(t),$$

where $f_T(t)$ and $F_T(t)$ are the PDF and CDF of system lifetime, respectively. Substituting $R(t) = 1 - F_T(t)$ produces

$$MTTF = \int_0^\infty t d(1 - R(t)) = -\int_0^\infty t dR(t) = -0R(0) + \lim_{t\to\infty} tR(t) + \int_0^\infty R(t)dt.$$

The first term of the right-hand side is zero, while the limit of the second term is toward 0 because $R(t)$ decreases exponentially but t increases linearly. Thus,

$$MTTF = \int_0^\infty R(t)dt. \tag{3.12}$$

Moments for PH distributions are given in Equation (2.19) in Chapter 2. Using the equation gives the first and second moments as

$$E[T] = -\boldsymbol{a}\boldsymbol{A}^{-1}\boldsymbol{u}, E[T^2] = 2\boldsymbol{a}\boldsymbol{A}^{-2}\boldsymbol{u}.$$

Since the mean value is the first moment, it follows that

$$MTTF = E[T] = -\boldsymbol{a}\boldsymbol{A}^{-1}\boldsymbol{u}, \tag{3.13}$$

Another important statistical reliability measure is the variance of time to failure (*VTTF*). Following the definition of variance, subtracting the second moment by the square of the first moment yields

$$VTTF = E[T^2] - E[T]^2 = 2\boldsymbol{a}\boldsymbol{A}^{-2}\boldsymbol{u} - (\boldsymbol{a}\boldsymbol{A}^{-1}\boldsymbol{u})^2, \tag{3.14}$$

where *VTTF* is the variance of time to system failure.

$MTTF$ and $VTTF$ calculated using Equations (3.13) and (3.14) are guaranteed will result in finite values because, as explained in Chapter 2, the survival matrix is nonsingular. Other important statistics in reliability analyses are the standard deviation of time to system failure ($SDTTF$) and the coefficient of variation of time to system failure ($CVTTF$), where $SDTTF$ is the square root of $VTTF$ and $CVTTF$ is the ratio of $SDTTF$ to $MTTF$.

Example 3.20 For a system whose lifetime follows a PH distribution of order 1 with representation $\boldsymbol{a} = \boldsymbol{u} = [1]$ and $\boldsymbol{A} = [-\lambda]$, some of its lifetime statistics can be found using Equations (3.13) and (3.14) as:

- $MTTF = -\boldsymbol{a}\boldsymbol{A}^{-1}\boldsymbol{u} = -(-\lambda)^{-1} = 1/\lambda$.
- Second moment of time to failure $E[T^2] = 2\boldsymbol{a}\boldsymbol{A}^{-2}\boldsymbol{u} = 2(-\lambda)^2 = 2/\lambda^2$.
- $VTTF = E[T^2] - E[T]^2 = 1/\lambda^2$.
- $SDTTF = \sqrt{VTTF} = 1/\lambda$.
- $CVTTF = \frac{1/\lambda}{1/\lambda} = 1$.

■

Calculating the first and second moments using Equations (3.13) and (3.14) can be done without matrix inversion. For this purpose, the matrix inversion is converted into a system of linear equations that can be solved using a linear system solution program. Solving a system of linear equations is faster than calculating a matrix inversion. Furthermore, survival matrix that needs to be inverted is a scarce matrix and, in most cases, an upper block diagonal matrix. These special characteristics will disappear when the matrix is inverted. With such characteristics, the complexity of a matrix inverse calculation is proportional to the cube of the matrix dimension, whereas the complexity of solving a system of linear equations is proportional to the square of the matrix dimension. In addition, solving a system of linear equations is more numerically stable than finding matrix inverse.

To convert the matrix inverse term in Equation (3.13) into a system of linear equations, an auxiliary variable $\boldsymbol{c}$ is defined so that

$$\boldsymbol{a}\boldsymbol{A}^{-1} = \boldsymbol{c}. \tag{3.15}$$

Postmultiplying both sides of Equation (3.15) with $\boldsymbol{A}$ yields

$$\boldsymbol{c}\boldsymbol{A} = \boldsymbol{a}, \tag{3.16}$$

which is a system of linear equations with $\boldsymbol{c}$ as the solution. $MTTF$ can be calculated by substituting Equation (3.15) into Equation (3.13) to produce

$$E[T] = -\boldsymbol{c}\boldsymbol{u}. \tag{3.17}$$

To avoid matrix inversion in calculating the second moment, $\boldsymbol{d} = \boldsymbol{A}^{-1}\boldsymbol{u}$ is defined and substituted into Equation (3.14) to produce

$$E[T^2] = 2\boldsymbol{a}\boldsymbol{A}^{-2}\boldsymbol{u} = 2(\boldsymbol{a}\boldsymbol{A}^{-1})(\boldsymbol{A}^{-1}\boldsymbol{u}) = 2\boldsymbol{c}\boldsymbol{d}. \tag{3.18}$$

Convert $\boldsymbol{A}^{-1}\boldsymbol{u} = \boldsymbol{d}$ into a system of linear equations

$$\boldsymbol{A}\boldsymbol{d} = \boldsymbol{u}, \tag{3.19}$$

whose solution is the values of $\boldsymbol{d}$.

Substituting Equations (3.18) and (3.16) into Equation (3.14) yields

$$VTTF = 2\boldsymbol{c}\boldsymbol{d} - (-\boldsymbol{c}\boldsymbol{u})^2, \tag{3.20}$$

where $\boldsymbol{c}$ and $\boldsymbol{d}$ are calculated by solving systems of linear equation in Equations (3.16) and (3.19). The calculations for these statistics are supported by **phstat** function provided in the software package **dysrel-am**. The function receives the PH representation of the distribution of system lifetime $(n, \boldsymbol{a}, \boldsymbol{A})$.

Example 3.21 Statistics for the lifetimes of one-component systems in Examples 3.1–3.3 are calculated using **phstat** function. The results are presented in Table 3.6.

Table 3.6 Some statistics of the lifetimes of systems in Examples 3.1–3.3.

	Example		
Statistics	**3.1**	**3.2**	**3.3**
MTTF	14.87	13.50	16.25
VTTF	114.48	109.25	129.69
SDTTF	10.70	10.45	11.39
CVTTF	0.72	0.77	0.70

Table 3.6 shows that the system in Example 3.3 has the longest life expectancy, followed by the system in Example 3.1, and the shortest is the system in Example 3.2. This conclusion should be predictable from the description of each system. It is also consistent with their reliability and lifetime stochastic ordering.

The code for the main program of **example3_21.m** is as follows:

MATLAB code for example3_21.m

```
close all;clear
addpath('..\dysrel-am','..\dysrel-am\expokit');

% Component's lifetime in PH distribution:
% Example 3.1
T(1).a = [1 0 0 0];
T(1).A = 0.1*[-1 1 0 0;0 -4 2 1;0 0 -5 3;0 0 0 -4];
T(1).n = 4;
% Example 3.2
T(2).a = [1 0 0];
T(2).A = 0.1*[-1 1 0;0 -4 2;0 0 -5];
T(2).n = 3;
% Example 3.3
T(3).a = [1 0 0 0];
T(3).A = 0.1*[-1 1 0 0;0 -4 2 1;0 0 -5 3;0 0 2 -4];
T(3).n = 4;
% Calculate statistics
for j = 1:3
  [mttf(j),vttf(j),sdttf(j),cvttf(j)] = phstat(T(j));
end
% Display results
mttf
vttf
sdttf
cvttf
```

■

Exercise 3.6 Compare the computation times to make Table 3.6 by inverting matrices and solving linear equations.

The value of *CVTTF* is closely related to the system type. The following examples illustrate this relationship.

Example 3.22 The lifetime with a hypoexponential distribution has

$$MTTF = 1/\lambda_1 + 1/\lambda_2 + \cdots + 1/\lambda_n,$$
$$VTTF = 1/\lambda_1^2 + 1/\lambda_2^2 + \cdots + 1/\lambda_n^2.$$

It can be proven that

$$\frac{1}{\lambda_1^2} + \frac{1}{\lambda_2^2} + \cdots + \frac{1}{\lambda_n^2} < \left(\frac{1}{\lambda_1} + \frac{1}{\lambda_2} + \cdots + \frac{1}{\lambda_n}\right)^2.$$

Hence,

$$VTTF < MTTF^2$$

or

$$\frac{VTTF}{MTTF^2} = \left(\frac{SDTTF}{MTTF}\right)^2 < 1.$$

Thus,

$$CVTTF = \frac{SDTTF}{MTTF} < 1.$$

Since systems with hypoexponentially distributed lifetimes are already known from Example 3.19 to belong to IFR types, this result suggests that IFR systems have $CVTTF < 1$. This conclusion also holds for Erlang distributed lifetime for which $\lambda_1 = \lambda_2 = \cdots = \lambda_n = \lambda$; hence, its $CVTTF = 1/\sqrt{n} < 1$. ■

CFR components have exponentially distributed lifetimes. Its *CVTTF* from Example 3.20 is equal to 1. Likewise, components with hyperexponentially distributed lifetimes are already known in Example 3.8 to be DFR components. Hyperexponential can be shown to have $CVTTF > 1$. These results are summarized as follows:

- The system is DFR ⇔ the system lifetime distribution has $CVTTF > 1$.
- The system is CFR ⇔ the system lifetime distribution has $CVTTF = 1$.
- The system is IFR ⇔ the system lifetime distribution has $CVTTF < 1$.

Example 3.23 The n-hyperexponentially distributed lifetimes in Example 3.8 have *CVTTF* as given in Table 3.7. The graphs of their hazard function are shown in Figure 3.22 for several values of n.

Table 3.7 and Figure 3.22 show that for DFR systems, the larger the *CVTTF*, the faster the hazard function decreases.

Table 3.7 *CVTTF* of n-hyperexponentially distributed lifetime.

n	1	2	3	5	10	15
CVTTF	1	1.11	1.20	1.34	1.62	1.82

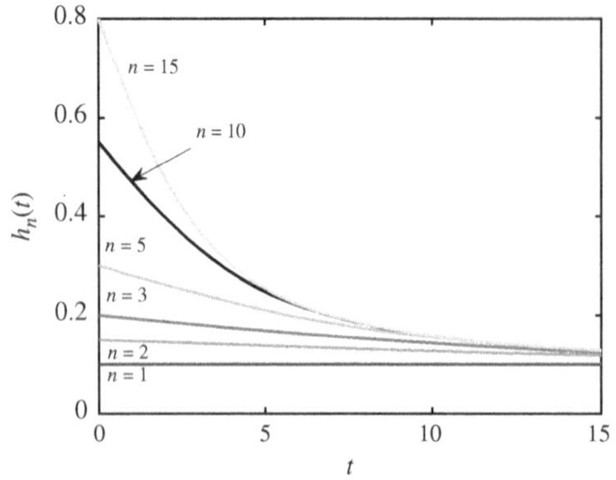

Figure 3.22 Hazard functions of n-hyperexponential distributions.

■

Example 3.24 Random variables with n-Erlang distribution are known to have $CVTTF = 1/\sqrt{n} < 1$. Graphs of the hazard functions of systems with lifetimes following n-Erlang distribution given in Example 3.9 are presented in Figure 3.23 for several n.

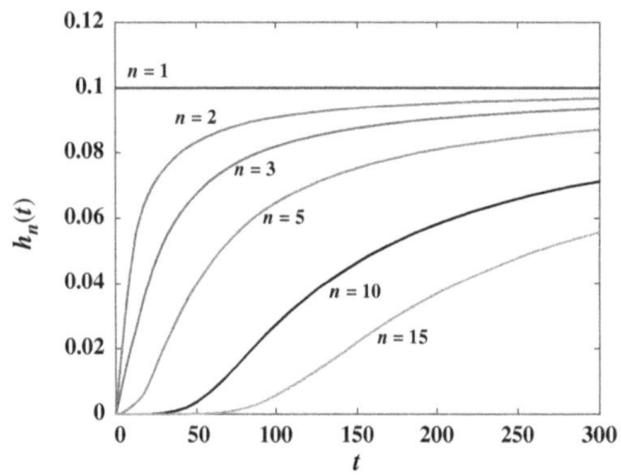

Figure 3.23 Hazard functions of n-Erlang distributed lifetimes.

These graphs show that for IFR systems, the larger the $CVTTF$, the faster the hazard function increases. ■

For a system that has been in operation and still functioning, its residual lifetime is affected by its age. The next section discusses a method to represent MRL as a function of system age.

3.6 Mean Residual Life (MRL) Function

Another important functional reliability measure is the MRL function, which measures the expected residual lifetime of a system as a function of its age. This function, denoted as $MRL(g)$, measures the expected residual lifetime of a system of age g. MRL is a conditional expectation of the remaining lifetime as it is subject to the system not failing after being used for g duration. By its definition,

$$MRL(g) = E[T - g|T > g], \tag{3.21}$$

that can be calculated using Equation (3.12) by replacing the reliability function with the residual reliability function in Equation (3.11) and the integration started at $t = g$,

$$MRL(g) = E[T - g|T > g] = \int_g^{\infty} RR(t + g - g|T > g)dt = \int_g^{\infty} RR(t|T > g)dt. \tag{3.22}$$

Substituting Equation (3.11) into Equation (3.22) produces

$$MRL(g) = \int_g^{\infty} \mathbf{c}(g)e^{\boldsymbol{A}(t-g)}\boldsymbol{u}dt. \tag{3.23}$$

Defining $v = t - g$, then $dv = dt$ and substituting them into Equation (3.23) produce

$$MRL(g) = \int_0^{\infty} \mathbf{c}(g)e^{\boldsymbol{A}v}\boldsymbol{u}dv. \tag{3.24}$$

Analogy to Equations (3.12) and (3.13) will give the same result as $MTTF$ by replacing the initial probability vector $\boldsymbol{a}$ with $\mathbf{c}(g)$. Thus,

$$MRL(g) = -\mathbf{c}(g)\boldsymbol{A}^{-1}\boldsymbol{u}, \tag{3.25}$$

where $\mathbf{c}(g) = \boldsymbol{a}\left(\frac{e^{Ag}}{ae^{Ag}\boldsymbol{u}}\right)$.

For $g = 0$, Equation (3.25) becomes

$$MRL(0) = -\mathbf{c}(0)\boldsymbol{A}^{-1}\boldsymbol{u} = \boldsymbol{a}\left(\frac{e^0}{ae^0\boldsymbol{u}}\right)\boldsymbol{A}^{-1}\boldsymbol{u} = \boldsymbol{a}\boldsymbol{A}^{-1}\boldsymbol{u} = MTTF,$$

in accordance with its definition that the MRL of an unused system is its expected lifetime.

Example 3.25 Based on the result of Example 3.15, the PH distribution of order 1 has $\mathbf{c}(g) = \boldsymbol{a}$, independent of g. The MRL for a system whose lifetime follows a PH distribution of order 1 is obtained by substituting $\mathbf{c}(g) = \boldsymbol{a}$ into Equation (3.25) to produce

$$MRL(g) = -\boldsymbol{a}\boldsymbol{A}^{-1}\boldsymbol{u} = MTTF,$$

a constant regardless of its age. ■

The calculation of MRL as a function of age for systems with high-order PH distributed lifetimes requires computer support. A MATLAB function **phmrlplot** is prepared as a part of the software package **dysrel-am**. The program receives the PH representation (n, $\boldsymbol{a}$, $\boldsymbol{A}$) of the distribution of system lifetime and the system age g.

Example 3.26 Graphs of MRL functions of systems in Examples 3.1–3.3, denoted as $MRL_1(g)$, $MRL_2(g)$, and $MRL_3(g)$, respectively, are shown in Figure 3.24. These graphs were generated using **phmrlplot**.

The graphs in Figure 3.24 show that MRL functions follow the same ordering as their reliability function, that is,

$$MRL_2(g) \leq MRL_1(g) \leq MRL_3(g)\ \forall g.$$

■

Example 3.27 MRL functions of DFR systems whose lifetimes follow n-hyperexponential distributions, $n = 1, 2, \ldots, 5$, given in Example 3.8 are depicted in Figure 3.25.

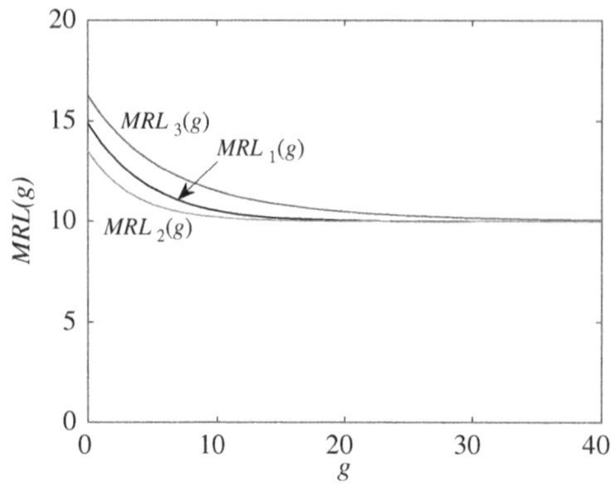

Figure 3.24 *MRL* functions for systems in Examples 3.1–3.3.

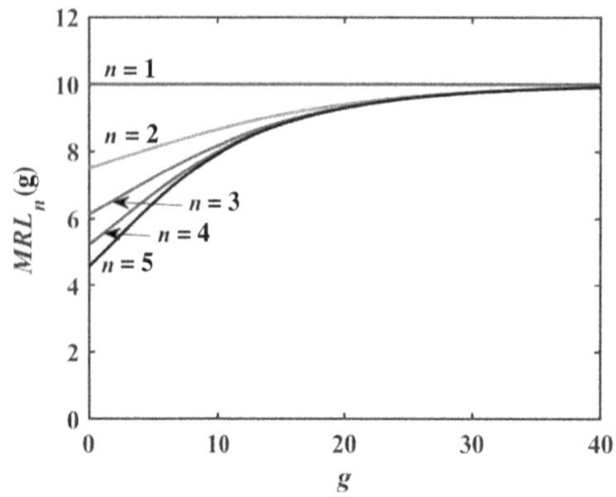

Figure 3.25 *MRL* functions for systems with *n*-hyperexponential lifetime distributions.

Figure 3.25 shows that *MRL* functions for DFR systems are increasing with age, which means that the longer they are used, the longer the expected remaining lifetime will be. This conclusion is in accordance with the nature of the decreasing hazard function, in which the failure rate decreases over time. Although monotonically increasing, these functions do not exceed a specific limit, which is the *MRL* value for $n = 1$, a constant equal to its *MTTF*. Using the result of Example 3.23, *MRL* functions in Figure 3.25 become lower with an increase in *CVTTF*. ■

Example 3.28 *MRL* functions of IFR systems with *n*-Erlang distributed lifetimes given in Example 3.9 for $n = 1, 2, \ldots, 5$ are depicted in Figure 3.26.

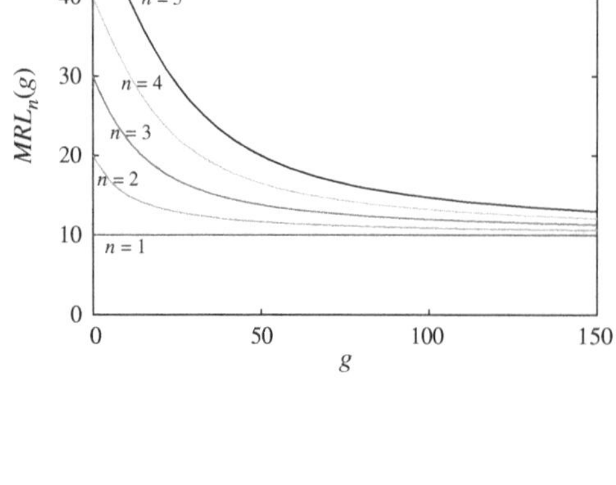

Figure 3.26 *MRL* functions for systems with *n*-Erlang lifetime distributions.

Figure 3.26 shows that *MRL* functions of IFR systems decrease with age, which means that the longer they are used, the shorter their remaining life expectancies will be. This conclusion is in accordance with the nature of the increasing hazard function, in which the failure rate increases over time. Like Example 3.27, although their *MRL* functions are monotonically decreasing, they eventually converge to the *MRL* value for $n = 1$, a constant equal to *MTTF*. Using the result of Example 3.24, the graphs in Figure 3.26 show that the lower the *CVTTF*, the higher the *MRL*. ■

Example 3.29 The remaining life expectancies, or *MRL*, of systems with exponential (CFR), 3-Erlang (IFR), and 3-hyperexponential (DFR) distributed lifetimes, all have *MTTF* = 10 days, are compared when their ages are 5, 10, and 15 days. The results are presented in Table 3.8.

Table 3.8 *MRL*(*g*) for systems of different types and ages.

	***MRL*(*g*) with *MTTF* = 10.00**		
Age (*g*)	**DFR system**	**CFR system**	**IFR system**
0	10.00	10.00	10.00
5	15.58	10.00	6.55
10	17.81	10.00	5.29
15	18.90	10.00	4.72

■

The numbers in Table 3.8 can be used to explain the characteristics of *MRL* functions regarding the system types. The results are summarized in Table 3.9.

Table 3.9 Characteristics of *MRL* functions.

System type	***MRL* function characteristic**
DFR	$MRL(g) \geq MTTF = MRL(0)$ $g_1 < g_2 \rightarrow MRL(g_1) < MRL(g_2)$
CFR	$MRL(g) = MTTF = MRL(0)$ $g_1 < g_2 \rightarrow MRL(g_1) = MRL(g_2)$
IFR	$MRL(g) \leq MTTF = MRL(0)$ $g_1 < g_2 \rightarrow MRL(g_1) > MRL(g_2)$

Exercise 3.7 Compare *MRL* functions of two systems, both have lifetimes following PH distributions with the same orders and mean values, but one follow Erlang distribution and the other follow hypoexponential distribution.

So far, system reliability is measured in a deterministic time, which means its duration of use is certain. However, there are cases where mission duration is uncertain. The next section discusses the effect of uncertainty in mission duration to system reliability.

3.7 Mission Reliability with Uncertain Mission Time[1]

Mission reliability is the reliability of a system to carry out a specific mission. Mission reliability certainly depends on the duration of the mission. If the mission duration is fixed, for example t_1, then the mission reliability can be determined by simply substituting the mission duration carried out by the system into the system reliability function $R(t)$. The problem is that the mission duration is not always certain. Sometimes, tolerance must be given, and the mission reliability value is sought to carry out the mission within the given time span.

Mission reliability when mission duration is a random variable V can be calculated using conditional probabilities

$$R(V) = \int_b^c R(V|V = v) f_V(v) dv, \tag{3.26}$$

where $f_V(v)$ is the PDF of random variable V defined in the range $[b, c]$. Substituting Equation (3.4) into Equation (3.26) and using the fact that $f_V(v)$ is a scalar produce

$$R(V) = \int_b^c (\boldsymbol{a}e^{\boldsymbol{A}v}\boldsymbol{u}) f_V(v) dv = \boldsymbol{a}\left(\int_b^c e^{\boldsymbol{A}v} f_V(v) dv\right)\boldsymbol{u}. \tag{3.27}$$

Example 3.30 A system that has lifetimes following a PH distribution of order 1 with $MTTF = 20$ days is assigned to carry a mission with random duration following a uniform distribution between 10 and 15 days.

PDF of uniform distribution between 10 and 15 days is

$$\begin{aligned} f_V(v) &= \frac{1}{15-10} = \frac{1}{5} \quad 10 \le v \le 15, \\ &= 0 \qquad\qquad\quad \text{other } v. \end{aligned}$$

PH distribution of order 1 with $MTTF = 20$ has a PH representation

$$n = 1, \quad \boldsymbol{a} = [1], \quad \boldsymbol{A} = \left[-\frac{1}{20}\right] = [-0.05].$$

The system reliability function is

$$R(t) = \boldsymbol{a}e^{\boldsymbol{A}t}\boldsymbol{u} = e^{-\frac{1}{20}t}.$$

Substituting $R(t)$ and $f_V(v)$ into Equation (3.27) yields

$$\begin{aligned} R(V) &= \left(\int_{10}^{15} e^{-0.05v}\left(\frac{1}{5}\right) dv\right) = \left(\frac{1}{5}\right)\left(-\frac{1}{0.05}\right)(e^{-(0.05)15} - e^{-(0.05)10}) \\ &= 4(e^{-0.5} - e^{-0.75}) = 0.54. \end{aligned}$$

Mission reliability can also be calculated using the expectation of the mission duration, which is $(10 + 15)/2 = 12.5$. This approach results in mission reliability $R(12.5) = 0.53$. Hence, using the expected value for the random mission duration gives an approximate of the actual value obtained using the distribution of the random mission duration. ■

One particular interest is understanding the effect of random variations in the mission duration on mission reliability. If S is the random variation of the mission duration t with PDF $f_S(s)$ in the range $[b, c]$, then the mission reliability function considering the random variation S is

$$R(t + S) = \int_{s=b}^c (\boldsymbol{a}e^{\boldsymbol{A}(t+s)}\boldsymbol{u}) f_S(s) ds. \tag{3.28}$$

1 Advanced material, can be omitted in lecture.

Often, the variation is expressed as a percentage of the mission duration t. In this case, Equation (3.28) can be made more general by making variation S depending on the mission duration t, that is,

$$R(t+S(t)) = \int_{s=b(t)}^{c(t)} (\boldsymbol{a}e^{\boldsymbol{A}(t+s(t))}\boldsymbol{u}) f_{S(t)}(s)ds. \tag{3.29}$$

If $b(t)$ and $c(t)$ are both nonnegatives, then $S(t)$ expresses the random delay in the mission duration t; if $b(t)$ and $c(t)$ are both nonpositives, then $S(t)$ expresses the uncertain acceleration in mission duration; if $b(t)$ is negative and $c(t)$ is positive, then the mission can be completed sooner or later than its planned mission duration.

Example 3.31 Suppose the mission duration in Example 3.30 experiences a uniformly distributed delay [b%, c%] from the planned mission duration t. Given $S(t)$ is uniformly distributed between b% and c% of t, its PDF is

$$f_{S(t)}(s) = \frac{1}{\frac{ct}{100} - \frac{bt}{100}} = \frac{100}{(c-b)t}, \quad 0 < b \leq s \leq c.$$

Substituting $f_{S(t)}(s)$ into Equation (3.29) obtains

$$R(t+S(t)) = \int_{s=bt/100}^{ct/100} (\boldsymbol{a}e^{\boldsymbol{A}(t+s(t))}\boldsymbol{u}) \frac{100}{(c-b)t} ds.$$

Since the integration variable is s, which means t is a constant, it follows that

$$R(t+S(t)) = \frac{\boldsymbol{a}}{(c-b)t}(e^{\boldsymbol{A}(t+ct/100)} - e^{\boldsymbol{A}(t+bt/100)})\boldsymbol{A}^{-1}\boldsymbol{u}.$$

If $\boldsymbol{a}$ and $\boldsymbol{A}$ are as given in Example 3.30, and $S(t)$ is uniform [5%, 10%], then

$$\begin{aligned} R(t+S(t)) &= \frac{1}{(0.1-(0.05))t}\left(e^{-\frac{1}{20}(t+0.1t)} - e^{-\frac{1}{20}(t+0.05t)}\right)(-20)(1) \\ &= \frac{-133.3333}{t}\left(e^{-\frac{1.1}{20}t} - e^{-\frac{1.05}{20}t}\right) \quad t > 0. \end{aligned}$$

The mission reliability for a mission duration of 10 days is

$$R(t+S(t)) = \frac{-200}{10}\left(e^{-\frac{1.1}{20}10} - e^{-\frac{1.05}{20}10}\right) = 0.58.$$

A graph that shows the difference between the mission reliability function without random delay in mission duration, $R(t)$, and with random delay in mission duration, $R(t+S(t))$, is presented in Figure 3.27. ■

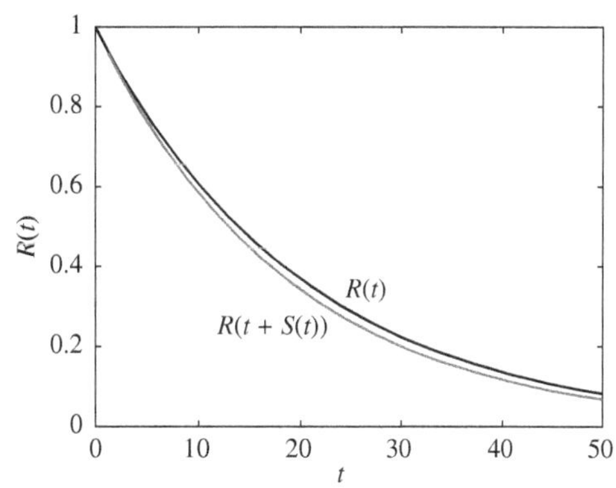

Figure 3.27 Mission reliability function of the system in Example 3.31 with and without random variation in mission duration.

For random mission duration or mission duration variations having a PH distribution, deriving mission reliability equations requires using Kronecker operation properties. Suppose the mission duration is a random variable following PH distribution with representation (n, $\boldsymbol{d}$, $\boldsymbol{D}$). Use Equation (3.26) to obtain

$$R(V) = \int_{v=0}^{\infty} R(V|V=v)f_V(v)dv = \int_{v=0}^{\infty} (\boldsymbol{a}e^{\boldsymbol{A}v}\boldsymbol{u})(-\boldsymbol{d}\boldsymbol{D}e^{\boldsymbol{D}v}\boldsymbol{u})dv. \tag{3.30}$$

Applying the appropriate properties of Kronecker operations obtains

$$R(V) = \int_{v=0}^{\infty} (\boldsymbol{a}e^{\boldsymbol{A}v}\boldsymbol{u}) \otimes (-\boldsymbol{d}\boldsymbol{D}e^{\boldsymbol{D}v}\boldsymbol{u})dv = \boldsymbol{a} \otimes (-\boldsymbol{d}\boldsymbol{D}) \left(\int_{v=0}^{\infty} (e^{\boldsymbol{A}v}) \otimes (e^{\boldsymbol{D}v})dv \right) \boldsymbol{u}$$

$$= \boldsymbol{a} \otimes (-\boldsymbol{d}\boldsymbol{D}) \left(\int_{v=0}^{\infty} e^{(\boldsymbol{A}\oplus\boldsymbol{D})v}dv \right) \boldsymbol{u} = (\boldsymbol{a} \otimes (-\boldsymbol{d}\boldsymbol{D}))(-(\boldsymbol{A} \oplus \boldsymbol{D})^{-1})\boldsymbol{u}.$$

Thus,

$$R(V) = (\boldsymbol{a} \otimes \boldsymbol{d}\boldsymbol{D})(\boldsymbol{A} \oplus \boldsymbol{D})^{-1}\boldsymbol{u}. \tag{3.31}$$

Example 3.32 A system having an exponentially distributed lifetime with $MTTF = 10$ days is assigned for a mission with a random duration following 2-Erlang distribution with a mean of 10 days. The mission reliability can be calculated using Equation (3.31) as follows.

PH representations of system's lifetime (1, $\boldsymbol{a}$, $\boldsymbol{A}$) and mission duration (2, $\boldsymbol{d}$, $\boldsymbol{D}$) are known as

$$\boldsymbol{a} = [1], \quad \boldsymbol{A} = [-0.1], \quad \boldsymbol{d} = [1 \;\; 0], \quad \boldsymbol{D} = \begin{bmatrix} -0.2 & 0.2 \\ 0 & -0.2 \end{bmatrix}.$$

$$\boldsymbol{a} \otimes (\boldsymbol{d}\boldsymbol{D}) = [1] \otimes [-0.2 \;\; 0.2] = [-0.2 \;\; 0.2].$$

$$\boldsymbol{A} \oplus \boldsymbol{D} = [-0.1] \oplus \begin{bmatrix} -0.2 & 0.2 \\ 0 & -0.2 \end{bmatrix} = \begin{bmatrix} -0.1 & 0 \\ 0 & -0.1 \end{bmatrix} + \begin{bmatrix} -0.2 & 0.2 \\ 0 & -0.2 \end{bmatrix} = \begin{bmatrix} -0.3 & 0.2 \\ 0 & -0.3 \end{bmatrix}.$$

Substituting these results into Equation (3.31) obtains

$$R(V) = [-0.2 \;\; 0.2] \begin{bmatrix} -0.3 & 0.2 \\ 0 & -0.3 \end{bmatrix}^{-1} \begin{bmatrix} 1 \\ 1 \end{bmatrix} = 0.44.$$

If the mission reliability is calculated for the average duration of the mission, then the mission reliability is obtained as

$$R(V) = e^{-0.1E[V]} = e^{-1} = 0.37.$$

The difference between the two is significant; hence, using the average value for a random duration produced an inaccurate approximation for mission reliability. ■

The mission reliability function with a PH distributed delay can be found from Equation (3.28) by using a PH representation (n, $\boldsymbol{d}$, $\boldsymbol{D}$) for $f_S(s)$,

$$R(t+S) = \int_{s=0}^{\infty} (\boldsymbol{a}e^{\boldsymbol{A}(t+s)}\boldsymbol{u})(-\boldsymbol{d}\boldsymbol{D}e^{\boldsymbol{D}s}\boldsymbol{u})ds. \tag{3.32}$$

Using the appropriate Kronecker product properties by first adding $\boldsymbol{I}_D$, an identity matrix with dimension equal to the dimensions of the matrix $\boldsymbol{D}$, produces

$$R(t+S) = \int_{s=0}^{\infty} (\boldsymbol{a}e^{\boldsymbol{A}s}e^{\boldsymbol{A}t}\boldsymbol{u}) \otimes (-\boldsymbol{d}\boldsymbol{D}e^{\boldsymbol{D}s}\boldsymbol{I}_D\boldsymbol{u})ds$$

$$= (\boldsymbol{a} \otimes (-\boldsymbol{d}\boldsymbol{D})) \left(\int_{0}^{\infty} (e^{\boldsymbol{A}s} \otimes e^{\boldsymbol{D}s})ds \right) (e^{\boldsymbol{A}t} \otimes \boldsymbol{I}_D)\boldsymbol{u}$$

$$= (\boldsymbol{a} \otimes (-\boldsymbol{d}\boldsymbol{D}))(-(\boldsymbol{A} \oplus \boldsymbol{D})^{-1}) \left(e^{\boldsymbol{A}t} \otimes e^{\boldsymbol{0}_D}\right) \boldsymbol{u} = (\boldsymbol{a} \otimes (\boldsymbol{d}\boldsymbol{D}))(\boldsymbol{A} \oplus \boldsymbol{D})^{-1} \left(e^{(\boldsymbol{A}\oplus\boldsymbol{0}_D)t}\right) \boldsymbol{u}.$$

Hence,

$$R(t+S) = \boldsymbol{a}' e^{\boldsymbol{A}'t} \boldsymbol{u}, \tag{3.33}$$

with $\boldsymbol{a}' = (\boldsymbol{a} \otimes (\boldsymbol{dD}))(\boldsymbol{A} \oplus \boldsymbol{D})^{-1}$ and $\boldsymbol{A}' = \boldsymbol{A} \oplus \boldsymbol{0}_D = \boldsymbol{A} \otimes \boldsymbol{I}_D$.

Example 3.33 The PH representation for the mission reliability function for the system in Example 3.32 can be found using Equation (3.33) as

$$\boldsymbol{a}' = (\boldsymbol{a} \otimes (\boldsymbol{dD}))(\boldsymbol{A} \oplus \boldsymbol{D})^{-1} = [-0.2\ \ 0.2] \begin{bmatrix} -0.3 & 0.2 \\ 0 & -0.3 \end{bmatrix}^{-1} = [0.6667 - 0.2222],$$

$$\boldsymbol{A}' = \boldsymbol{A} \otimes \boldsymbol{I}_D = \begin{bmatrix} -0.1 & 0 \\ 0 & -0.1 \end{bmatrix}.$$

The mission reliability value for a given mission duration can be found using this representation. A comparison between the reliability function of a mission with no delay and with a PH distributed delay is presented in Figure 3.28.

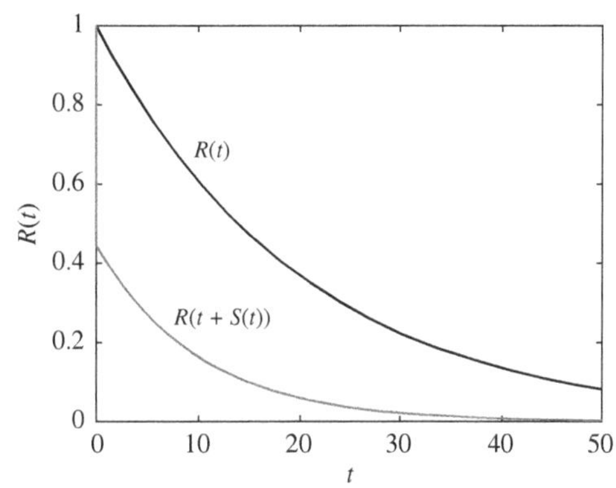

Figure 3.28 A comparison of mission reliability of the system in Example 3.33 without and with PH-distributed mission delay.

The graphs show that the $R(t+S)$ starts from $R(S)$ when $t = 0$, whose value has been calculated in Example 3.32, which is 0.44. This number means there is a jump or discontinuity of 0.56 at $t = 0$ due to the positive probability that the mission is accomplished in zero duration as the mission duration has an Erlang distribution. To make the problem more realistic, S should be a function of t as in Example 3.31, which makes the problem more complicated. This problem is left as a challenge for interested readers. Note that Equation (3.33) may produce reliability function in ME distribution (see Chapter 2) because $\boldsymbol{a}'$ may have negative value as shown in Example 3.33; hence, it is not a valid initial probability vector for PH distribution. ■

Information on system reliability function can be utilized to design programs to effectively improve mission reliability of a system. The next section discusses a way to improve system reliability through maintenance program. Maintenance can be carried out with different maintenance intervals and maintenance levels. It will show the effect of maintenance interval and maintenance level to system reliability measures, specifically to reliability function and *MTTF*.

3.8 Maintenance

Systems that have not failed should be maintained periodically to improve their conditions, thus lengthening their lifetimes stochastically. It is sometimes called preventive maintenance, in contrast to corrective maintenance when the system is repaired, overhauled, or replaced. This book refers to preventive maintenance simply as maintenance, and corrective maintenance as replacement. Maintenance requirements are closely related to reliability. Therefore, maintenance programs are often designed based on the reliability of the system being maintained, and the origin of the reliability-based maintenance term is coined. There are two types of maintenance: perfect and imperfect maintenance. Perfect maintenance is a maintenance program that is able to restore the system condition to be as good as new, while imperfect maintenance is only able to make the system condition better, but not as good as new. As explained in Sections 3.4 and 3.6 on residual reliability and *MRL*, the age of a system only affects the initial probability vector of the PH distribution of the system lifetime, while its survival matrix is not affected by age, remaining the same for all ages. Therefore, the maintenance program for a system with lifetime following PH distribution only affects and is aimed at the initial probability vector.

The discussion on maintenance starts with perfect maintenance that is easier to formulate and then follows with a more complex imperfect maintenance involving an additional variable representing the maintenance level.

3.8.1 Perfect Maintenance

Define h as the maintenance interval. If, in use during a maintenance interval h, the system has not failed, then the system is maintained at time h. This maintenance occurs with a probability equal to the system reliability for use during h, $R(h)$. Perfect maintenance can restore the initial probability vector of the system with age h, $\mathbf{c}(h)$ given by Equation (3.11), back to the initial probability vector $\boldsymbol{a}$ as a new system. In the second usage interval, that is, $h \le t \le 2h$, the system reliability function becomes the same as in the interval $0 \le t \le h$, corrected by $R(h)$, because the second interval is only served by the system that has not failed at time h. Because the survival matrix is $\boldsymbol{A}$ all the time and the initial probability vector return to $\boldsymbol{a}$, the reliability function in the interval $h \le t \le 2h$ is the same as the reliability function in the interval $0 \le t \le h$, which is $R(t)$, shifted h unit time to the right to become $R(t-h)$. Therefore, the system reliability in the interval $h \le t \le 2h$ is

$$R(t) = P[T > t|T > h]P[T > h] = R(t-h)R(h).$$

The system reliability in the interval probability $2h \le t \le 3h$ is

$$R(t) = P[T > t|T > 2h]P[T > 2h] = R(t-2h)[R(t-h)R(h)]\,|_{t=2h} = R(t-2h)R^2(h).$$

Continuing in this manner yields the reliability function of the system with periodic perfect maintenance in general as

$$R_M^{(k)}(t,h) = R^k(h)R(t-kh) \text{ for } kh \le t \le (k+1)h, \quad k = 0, 1, \ldots, \tag{3.34}$$

where $R_M^{(k)}(t,h)$ is the reliability function of a system under perfect maintenance with interval h in the interval $kh \le t \le (k+1)h$.

For a system with lifetime following PH distribution with representation $(n, \boldsymbol{a}, \boldsymbol{A})$, Equation (3.34) becomes

$$R_M^{(k)}(t,h) = (\boldsymbol{a}e^{\boldsymbol{A}h}\boldsymbol{u})^k\boldsymbol{a}e^{\boldsymbol{A}(t-kh)}\boldsymbol{u} \text{ for } kh \le t \le (k+1)h. \tag{3.35}$$

The reliability function of a system with perfect maintenance interval h is denoted as $R_M(t,h)$.

Example 3.34 The reliability without maintenance for a system with a lifetime following a PH distribution of order 1, or exponential, at time t is $R(t) = e^{-\lambda t}$. Its reliability with perfect maintenance every interval h, assuming t lies within the $(k+1)^{th}$ maintenance interval, $kh < t \leq (k+1)h$, can be obtained from Equation (3.35) as

$$R^k(h)R(t-kh) = (e^{-\lambda h})^k e^{-\lambda(t-kh)} = e^{-k\lambda h}e^{-\lambda t + k\lambda h} = e^{-\lambda t},$$

equal to the reliability function without maintenance, regardless of the maintenance interval. This result indicates that the CFR system does not require periodic maintenance because its reliability cannot be improved through maintenance. ■

Calculating reliability values and plotting graphs of reliability functions with perfect maintenance for systems with a general PH-distributed lifetime require computer assistance. For this purpose, a MATLAB function **phmaintrel** is prepared as part of **dysrel-am** software package. This function accepts the PH representation of the system lifetime $(n, \boldsymbol{a}, \boldsymbol{A})$ and the maintenance interval h as inputs. This program can also be used for replacement programs as will be described in Section 3.9.

Example 3.35 The reliability function of the one-component system in Example 3.1 without and with perfect maintenance every 5 days is shown in Figure 3.29.

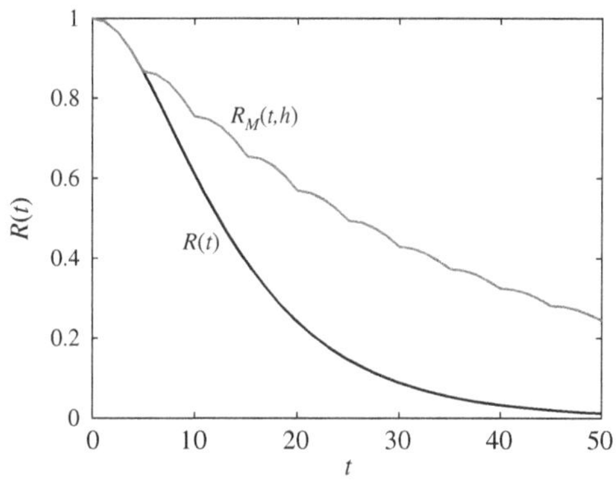

Figure 3.29 Reliability function of the system in Example 3.1 without and with perfect maintenance.

The main program to create Figure 3.29 can be found in **example3_35.m** file. Its MATLAB code is as follows:

MATLAB code of example3_35.m

```
clear;close all
addpath('..\dysrel-am','..\dysrel-am\expokit')
% Ex 3.1
T.a = [1 0 0 0];
T.A = 0.1*[-1 1 0 0;0 -4 2 1;0 0 -5 3;0 0 0 -4];
T.n = 4;
% Maintenance interval
h = 5;
% Time vector
tmax = 50;
t = linspace(0,tmax,tmax+1);
% Reliability without maintenance
Rt   = phrel(t,T);
% Reliability with maintenance
Rtwr = phmaintrel(t,T,h);
```

```
% Plot
lgnd = {'R(t)','R_M(t,h)'};
multiplot(2,t,[Rt' Rtwr'],lgnd,'t','R(t)');
```

The graphs show that reliability with maintenance is higher than without maintenance. The reliability values without maintenance and with maintenance every 5 days for 20 days of use are 0.24 and 0.57, respectively, an increase in the reliability value of more than twice. Note that the reliability function with maintenance is not a continuous function. This is consistent with Equation (3.34), a mathematical expression of a piecewise continuous function. Nevertheless, the requirement for being a monotonic nonincreasing function is satisfied. ■

The system in Example 3.1 is an IFR system. Examples 3.34 and 3.35 show that maintenance can increase the reliability of IFR systems but have no effect on CFR systems. The effect on DFR systems is investigated in the next example.

Example 3.36 The reliability function of the DFR system in Example 3.8 without and with perfect maintenance every 5 days is calculated and plotted using **phmaintplot** function. The results are presented in Figure 3.30, which shows that maintenance lowers reliability functions of DFR systems. This conclusion is in accordance with the nature of DFR systems, which become better the longer they are used.

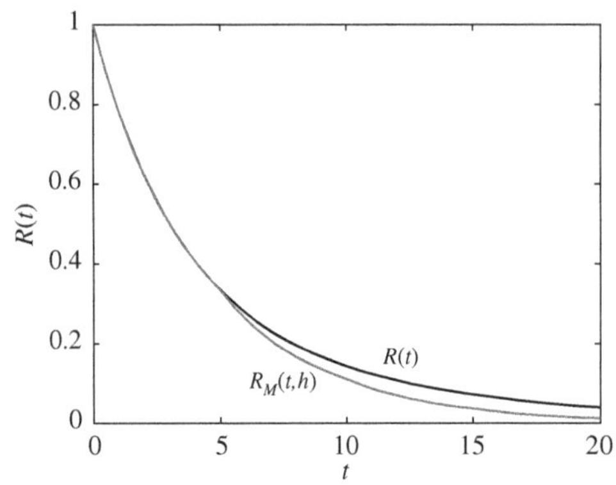

Figure 3.30 Reliability function of the system in Example 3.8 with and without maintenance.

■

The results from Examples 3.34–3.36 can be summarized as:

- DFR systems $\rightarrow R_M(t,h) < R(t)$.
- CFR systems $\rightarrow R_M(t,h) = R(t)$.
- IFR systems $\rightarrow R_M(t,h) > R(t)$.

Since the reliability function with maintenance, as given in Equation (3.34), is discontinuous, the expected lifetime of a system with maintenance can be obtained by piecewise integration of the reliability function with maintenance.

$$
\begin{aligned}
MTTF_M(h) &= \int_0^{\infty} R_M(t,h)dt \\
&= \int_0^{h} R(t)dt + R(h)\int_h^{2h} R(t-h)dt + R^2(h)\int_{2h}^{3h} R(t-2h)dt + \cdots .
\end{aligned}
$$

Substitute $t - h = t$ for the second integral so that the integral limits become 0 and h, substitute $t - 2h = t$ for the third integral, and so on, which ultimately results in

$$\begin{aligned} MTTF_M(h) &= \int_0^h R(t)dt + R(h)\int_0^h R(t)dt + R^2(h)\int_0^h R(t)dt + \cdots \\ &= (1 + R(h) + R^2(h) + \cdots)\int_0^h R(t)dt. \end{aligned}$$

The terms in front of the integral on the right-hand side are an infinite geometric series sum to $1/(1 - R(h))$. Thus,

$$MTTF_M(h) = \frac{1}{1 - R(h)}\int_0^h R(t)dt. \tag{3.36}$$

Example 3.37 For a system with an exponentially distributed lifetime with rate λ, it is known that $R(t) = e^{-\lambda t}$. Substituting this into Equation (3.36) obtains

$$MTTF_M(h) = \frac{1}{1 - e^{-\lambda h}}\int_0^h e^{-\lambda t}dt = \frac{1(-\lambda)^{-1}(e^{-\lambda h} - 1)}{1 - e^{-\lambda h}} = \frac{1}{\lambda},$$

which is the same as $MTTF$ without maintenance, regardless of the maintenance interval used. ■

This result is in accordance with the result of Example 3.34, which states that the reliability functions of CFR systems are not affected by maintenance, so their expected lifetimes are certainly not affected by maintenance. The conclusion of Examples 3.34 and 3.37 cannot be separated from the nature of the exponential distribution that does not recognize age; therefore, no maintenance is needed.

Example 3.38 A system whose lifetime follows a 2-Erlang distribution with $MTTF = 20$ days is perfectly maintained every 3 days. The system reliability without maintenance in a scalar representation is given by

$$R(t) = (1 + 0.1t)e^{-0.1t}t \geq 0.$$

Its reliability for 10 days of use without maintenance is

$$R(10) = (1 + 0.1 \times 10)e^{-0.1\times 10} = 0.736.$$

If the system is maintained every 3 days, the 10th day of use is between 3×3 and 4×3, which is in the range of the third and fourth maintenance, or $k = 3$. Its reliability with perfect maintenance every 3 days can be calculated using Equation (3.34) to yield

$$R(10) = R(3)^3 R(10 - 3 \times 3) = ((1 + 0.1 \times 3)e^{-0.1\times 3})^3(1 + 0.1 \times 1)e^{-0.1\times 1} = 0.889,$$

an increase in 20% from the reliability value without maintenance. The increase will be higher as the maintenance interval becomes smaller, or the maintenance becomes more frequent.

The expected lifetime can be calculated using Equation (3.36) as

$$\begin{aligned} MTTF_M(3) &= \frac{1}{1 - R(3)}\int_0^3 R(t)dt = \frac{1}{1 - (1 + 0.1 \times 3)e^{-0.1\times 3}}\int_0^3 (1 + 0.1t)e^{-0.1t}dt \\ &= \frac{1}{0.0369}\left(-\frac{1}{0.1}e^{-0.1t} - te^{-0.1t} - \frac{1}{0.1}e^{-0.1t}\right)\Bigg|_{t=0}^{t=3} = 80.25 \text{ days}. \end{aligned}$$

Compared with an expected lifetime of 20 days for no maintenance, a perfect maintenance program every 3 days increases the system life expectancy fourfold. ■

For a system whose lifetime follows a PH distribution with representation $(n, \boldsymbol{a}, \boldsymbol{A})$, Equation (3.36) becomes

$$MTTF_M(h) = \frac{1}{1 - \boldsymbol{a}e^{\boldsymbol{A}h}\boldsymbol{u}} \int_0^h \boldsymbol{a}e^{\boldsymbol{A}t}\boldsymbol{u}dt = \frac{\boldsymbol{a}\boldsymbol{A}^{-1}(e^{\boldsymbol{A}h} - \boldsymbol{I})\boldsymbol{u}}{1 - \boldsymbol{a}e^{\boldsymbol{A}h}\boldsymbol{u}}. \tag{3.37}$$

Calculating expected lifetimes with maintenance for systems with PH-distributed lifetimes is supported by a MATLAB function **phmaintmean**. The function receives PH representation of system lifetime distribution $(n, \boldsymbol{a}, \boldsymbol{A})$ and maintenance interval h. The expected lifetime of the one-component system in Example 3.1 with perfect maintenance every 3 days calculated using **phmaintmean** is 58.97 days. Compared to its expected lifetime without maintenance, which is 14.87 days, with perfect maintenance every 3 days increases the expected lifetime 3.5 folds.

Example 3.39 Graphs of the expected lifetime with perfect maintenance, $MTTF_M(h)$, as a function of the maintenance interval h for a system having a k-Erlang distributed lifetime, $k = 1, 2, \dots, 4$, all with $MTTF = 10$, are given in Figure 3.31.

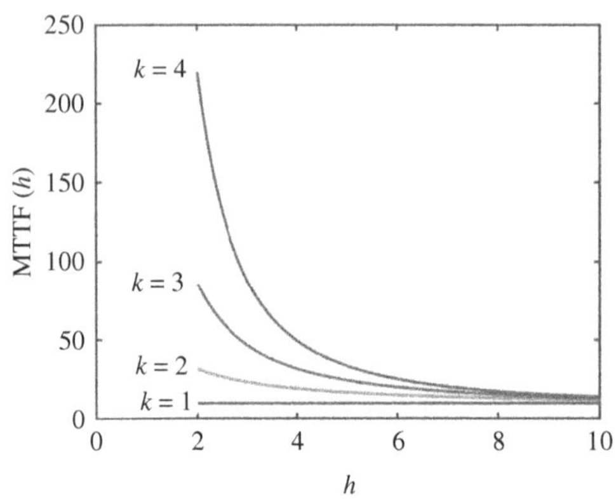

Figure 3.31 Expected lifetime as a function of maintenance interval.

The effect of maintenance in extending the expected lifetime is stronger as k is bigger, or $CVTTF = 1/\sqrt{k}$ is smaller. For sufficiently large maintenance intervals, the value of k, or $CVTTF$, has almost no effect on $MTTF_M(h)$. For maintenance interval h approaching $MTTF = 10$, the effect of maintenance is negligible, or $MTTF_M(h) \to MTTF$. The $MTTF_M(h)$ increases exponentially with decreasing maintenance intervals and becomes infinite for maintenance intervals approaching 0 due to the probability of a system failure is very small. ■

A decision in maintenance programs is to determine a maintenance interval that can make the system under maintenance satisfy predetermined reliability or $MTTF$ values. An add-on program is used to search for such a maintenance interval by trial and error.

Example 3.40 The one-component system in Example 3.1 is perfectly maintained to make its $MTTF$ longer than 30 days, and its reliability for a 10-day mission is higher than 0.7. The search for a minimum maintenance interval that satisfies these two requirements produces Table 3.10.

The table shows that the system must be maintained at least every 6 days to meet the two operational requirements. $R_M(10,15) = R_M(10,10)$ because the system at $t = 10$ has not been maintained yet. ■

The next section extends the discussion of perfect maintenance in previous section into imperfect maintenance by incorporating a variable that represents maintenance level. Maintenance level may vary from 0 for without maintenance to 1 for perfect maintenance.

Table 3.10 Maintenance interval with reliability and *MTTF* requirements.

Maintenance interval h	15	10	8	7	6	5
Reliability value $R_M(10, h)$	0.61	0.61	0.70	0.73	0.75	0.75
$MTTF_M(h)$	17.86	21.65	24.95	27.49	31.04	36.26

3.8.2 Imperfect Maintenance

Maintenance has some limitations, such as time and cost, so it cannot always be expected to be perfect. Imperfect maintenance is unable to restore the system to be as good as new. Pham and Wang [8] described several models for imperfect maintenance, including models that were proposed by Nakagawa [9], Abdel-Hameed [10], and other later models. Two basic models for imperfect maintenance are models that incorporate maintenance success probability and that incorporate age-rejuvenating factors. The following development follows a different model.

As for perfect maintenance, imperfect maintenance affects only the initial probability vector of the PH distribution of the system lifetime, not the survival matrix. An imperfect maintenance can be modeled by assuming that the result of an imperfect maintenance is an initial probability vector whose value is between the initial probability vector at the time of maintenance, $\mathbf{c}(h)$, and the initial probability vector when it was new, $\boldsymbol{a}$. This intermediate value can be determined as a convex combination of $\mathbf{c}(h)$ and $\boldsymbol{a}$ so that the initial probability vector after the first maintenance is given by

$$\mathbf{d}(h) = b\boldsymbol{a} + (1 - b)\mathbf{c}(h), \tag{3.38}$$

where b is a number between 0 and 1 that represents the level of maintenance perfection, with $b = 1$ denotes perfect maintenance and $b = 0$ denotes no maintenance. The derivation of the system reliability function equation on each maintenance interval is as follows.

Define $R_M^k(t, h, b)$ as the reliability function of the system maintained with maintenance intervals h and a maintenance level b in the interval $kh \leq t < (k + 1)h$, $k = 0, 1, 2, \ldots$. For each maintenance interval, it follows that

1) $0 \leq t < h$: There is no maintenance yet, so $R_M^0(t, h, b) = R(t) = \boldsymbol{a}e^{At}\boldsymbol{u}$.
2) $h \leq t < 2h$: Maintenance is carried out at time $t - h$:
   a) If there is no maintenance, then the resulting reliability function is $R(t) = \boldsymbol{a}e^{At}\boldsymbol{u}$, which can be written as

   $$R(h)RR(t - h|h) = (\boldsymbol{a}e^{Ah}\boldsymbol{u})\left(\frac{\boldsymbol{a}e^{Ah}}{\boldsymbol{a}e^{Ah}\boldsymbol{u}}e^{A(t-h)}\boldsymbol{u}\right) = (\boldsymbol{a}e^{Ah}\boldsymbol{u})(\mathbf{c}(\boldsymbol{a}, h)e^{A(t-h)}\boldsymbol{u}).$$

   b) If it is perfectly maintained, the reliability function after maintenance becomes the same as when it was new. Maintenance is carried out if, at time $t = h$, the system has not failed, which happens with probability $R(h)$. The reliability function of the system in this interval is

   $$R(h)R(t - h) = (\boldsymbol{a}e^{Ah}\boldsymbol{u})(\boldsymbol{a}e^{A(t-h)}\boldsymbol{u}).$$

   c) If it is not perfectly maintained, then the reliability function after maintenance becomes between unmaintained and perfectly maintained. The two differ in the initial probability vector for the reliability function in the interval $h \leq t < 2h$, which becomes between $\boldsymbol{a}$ and $\mathbf{c}(\boldsymbol{a}, h)$. Following Equation (3.38) produces

   $$\mathbf{d}(h) = b\boldsymbol{a} + (1 - b)\mathbf{c}(\boldsymbol{a}, h), \tag{3.39}$$

with $0 \leq b \leq 1$. Using $\boldsymbol{d}(h)$ as the initial probability vector, the reliability function with imperfect maintenance in this interval becomes

$$R_M^1(t, h, b) = R(h)R(t-h, b) = (\boldsymbol{a}e^{\boldsymbol{A}h}\boldsymbol{u})(\boldsymbol{d}(h)e^{\boldsymbol{A}(t-h)}\boldsymbol{u}). \tag{3.40}$$

d) Equation (3.40) is the general form of the reliability function of a system with arbitrary maintenance in the interval $h \leq t < 2h$.
  i) If $b = 1$, then $\boldsymbol{d}(h) = \boldsymbol{a}$, and

$$R_M^1(t, h, b) = R(h)R(t-h, 1) = (\boldsymbol{a}e^{\boldsymbol{A}h}\boldsymbol{u})(\boldsymbol{a}e^{\boldsymbol{A}(t-h)}\boldsymbol{u}),$$

  the same as in point b for perfect maintenance.
  ii) If $b = 0$, then $\boldsymbol{d}(h) = \boldsymbol{c}(\boldsymbol{a}, h)$, and

$$R_M(t, h, b) = R(h)R(t-h, 0) = (\boldsymbol{a}e^{\boldsymbol{A}h}\boldsymbol{u})(\boldsymbol{c}(\boldsymbol{a}, h)e^{\boldsymbol{A}(t-h)}\boldsymbol{u}),$$

  the same as in point a for without maintenance.

3) $2h \leq t < 3h$: Maintenance is carried out at time $t = 2h$. Use the result of 2c to produce

$$\begin{aligned} R_M^2(t, h, b) &= R_M^1(2h, h, b)R(t-2h, b) = R(h)R(2h-h, b)R(t-2h, b) \\ &= (\boldsymbol{a}e^{\boldsymbol{A}h}\boldsymbol{u})(\boldsymbol{d}(h)e^{\boldsymbol{A}(2h-h)}\boldsymbol{u})(\boldsymbol{d}(2h)e^{\boldsymbol{A}(t-2h)}\boldsymbol{u}). \end{aligned}$$

As for $\boldsymbol{d}(h)$, $\boldsymbol{d}(2h)$ is also between the initial probability vectors without maintenance and perfect maintenance. Perfect maintenance will return the initial probability vector as at time $t = h$,that is $\boldsymbol{d}(h)$, while without maintenance will produce the initial probability vector of the residual reliability function at time $t = 2h$, that is $\frac{\boldsymbol{d}(h)e^{\boldsymbol{A}h}}{\boldsymbol{d}(h)e^{\boldsymbol{A}h}\boldsymbol{u}} = \boldsymbol{c}(\boldsymbol{d}(h), h)$. Thus, it follows that

$$\boldsymbol{d}(2h) = b\boldsymbol{d}(h) + (1-b)\boldsymbol{c}(\boldsymbol{d}(h), h). \tag{3.41}$$

Following the same pattern produces a recursive equation

$$\boldsymbol{d}(kh) = b\boldsymbol{d}((k-1)h) + (1-b)\boldsymbol{c}(\boldsymbol{d}((k-1)h), h) \quad k = 1, 2, \dots, \tag{3.42}$$

where $\boldsymbol{d}(0) = \boldsymbol{a}$, and the system reliability function in the interval $kh \leq t < (k+1)h$ is given by

$$\begin{aligned} R_M^{(k)}(t, h, b) &= (\boldsymbol{a}e^{\boldsymbol{A}h}\boldsymbol{u})(\boldsymbol{d}(h)e^{\boldsymbol{A}(2h-h)}\boldsymbol{u})(\boldsymbol{d}(2h)e^{\boldsymbol{A}(3h-2h)}\boldsymbol{u}) \dots (\boldsymbol{d}(kh)e^{\boldsymbol{A}(t-kh)}\boldsymbol{u}) \\ &= (\boldsymbol{a}e^{\boldsymbol{A}h}\boldsymbol{u})(\boldsymbol{d}(h)e^{\boldsymbol{A}h}\boldsymbol{u})(\boldsymbol{d}(2h)e^{\boldsymbol{A}h}\boldsymbol{u}) \dots (\boldsymbol{d}(kh)e^{\boldsymbol{A}(t-kh)}\boldsymbol{u}) \quad k = 1, 2, \dots. \end{aligned} \tag{3.43}$$

Equation (3.43) can be written as

$$\begin{aligned} R_M^{(k)}(t, h, b) &= (\boldsymbol{d}(kh)e^{\boldsymbol{A}(t-kh)}\boldsymbol{u}), \quad k = 0 \\ &= (\boldsymbol{d}(kh)e^{\boldsymbol{A}(t-kh)}\boldsymbol{u})\prod_{j=0}^{k-1}(\boldsymbol{d}(jh)e^{\boldsymbol{A}h}\boldsymbol{u}), \quad k = 1, 2, \dots. \end{aligned} \tag{3.44}$$

The system *MTTF* with imperfect maintenance can be calculated using piecewise integration to the reliability function under imperfect maintenance,

$$MTTF_M(h, b) = \int_0^h R_M^{(0)}(t, h, b)dt + \int_h^{2h} R_M^{(1)}(t, h, b)dt + \int_{2h}^{3h} R_M^{(2)}(t, h, b)dt + \cdots.$$

Substituting Equation (3.44) with some manipulation obtains

$$MTTF_M(h, b) = \left(\boldsymbol{a} + R_M^0(h)\boldsymbol{d}(h) + R_M^1(2h)\boldsymbol{d}(2h) + R_M^2(3h)\boldsymbol{d}(3h) + \cdots\right)\boldsymbol{A}^{-1}(e^{\boldsymbol{A}h} - \boldsymbol{I})\boldsymbol{u}. \tag{3.45}$$

For simplicity, the available **phmaintrel** and **phmaintmean** can be employed to calculate reliability functions and expected lifetime of systems with imperfect maintenance having

PH-distributed lifetime. An additional input maintenance level b is required for this purpose, with $b = 1$ is for perfect maintenance.

Example 3.41 The function **phmaintrel** is used to calculate reliability value and plot reliability functions of system in Example 3.1 for maintenance interval 5 days and several maintenance levels. The results are shown in Figure 3.32.

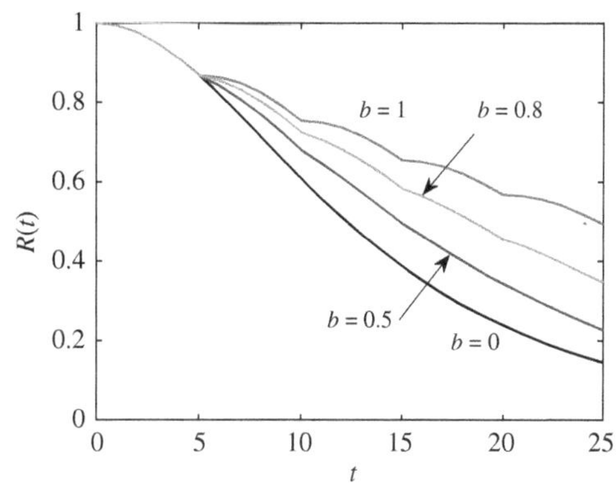

Figure 3.32 Reliability function of Example 3.1 with various levels of maintenance.

Figure 3.32 shows the system reliability is higher for a higher maintenance level as it should be intuitively, and for $b = 0$, the reliability function is identical to the reliability function without maintenance. ■

Example 3.42 The effects of maintenance level and interval on 20-day mission reliability and expected lifetime for the one-component system in Example 3.1 calculated using **phmaintrel** and **phmaintmean** are given in Tables 3.11 and 3.12, respectively.

Table 3.11 Reliability of the system in Example 3.1 for 20-day mission with maintenance.

Reliability value for 20-day mission $R_M(20, h, b)$				
	Maintenance interval h			
Maintenance level b	4	6	8	10
0.0	0.24	0.24	0.24	0.24
0.2	0.27	0.27	0.27	0.27
0.4	0.32	0.31	0.31	0.29
0.6	0.39	0.37	0.35	0.32
0.8	0.49	0.44	0.40	0.34
1.0	0.63	0.54	0.46	0.37

The results show the expected behavior that the system reliability and the system expected lifetime increase with a lower maintenance interval or higher maintenance level. The maintenance interval for maintenance level = 0 does not affect the expected lifetime and reliability value since there is no maintenance.

Table 3.12 *MTTF* of the system in Example 3.1 with maintenance.

	$MTTF_M(h, b)$				
	Maintenance Interval h				
Maintenance level b	5	10	15	20	25
0.0	14.88	14.88	14.88	14.88	14.88
0.2	15.72	15.48	15.26	15.11	15.02
0.4	16.87	16.22	15.71	15.37	15.17
0.6	18.64	17.21	16.23	15.66	15.33
0.8	21.93	18.68	16.91	15.99	15.51
1.0	36.26	21.64	17.86	16.39	15.70

■

Using a convex combination of the initial probability vectors of PH distribution of lifetime at the time of maintenance and when the system was new as a model for imperfect maintenance is equivalent to the imperfect maintenance model with probability of success. If b is the probability that the maintenance succeeds in bringing the system back to be as good as new, then $(1 - b)$ is the probability that the maintenance failed, which causes the system to stay at the same age at the time of maintenance. Hence,

$$R_M(t, h, b) = b(\boldsymbol{a}e^{\boldsymbol{A}t}\boldsymbol{u}) + (1 - b)(\boldsymbol{c}(h)e^{\boldsymbol{A}t}\boldsymbol{u}) = (b\boldsymbol{a} + (1 - b)\boldsymbol{c}(h))e^{\boldsymbol{A}t}\boldsymbol{u} = \boldsymbol{d}(h)e^{\boldsymbol{A}t}\boldsymbol{u},$$

where $\boldsymbol{d}(h) = b\boldsymbol{a} + (1 - b)\boldsymbol{c}(h)$, the same as given by Equation (3.38).

Thus, b in the imperfect maintenance model based on a convex combination of initial probability vectors can be interpreted as the probability of success in the imperfect maintenance model that incorporates maintenance success probability. The difference between the two is in the model based on convex combination the value of b is chosen intentionally (controllable), while in the model based on probability of success, the value of b follows the rule of chance (uncontrollable).

Another imperfect maintenance model is maintenance that rejuvenates the system to age between new, age = 0, and the time of maintenance, age = h. Let the system age after maintenance is $(1 - b)h$, where b is the rejuvenating factor. If the maintenance is perfect, then $b = 1$ and the system age after maintenance becomes 0; if the maintenance failed, then $b = 0$ and the system age after maintenance stays equal to h, which is the system age at the time of maintenance. For a system with a PH-distributed lifetime, this means the initial probability vector after the first maintenance is $\boldsymbol{c}((1 - b)h)$. It follows that the system reliability after the first maintenance is

$$R_M^{(1)}(t, h, b) = R(h)RR(t - h, b) = (\boldsymbol{a}e^{\boldsymbol{A}h}\boldsymbol{u})(\boldsymbol{c}((1 - b)h)e^{\boldsymbol{A}(t-h)}\boldsymbol{u}),$$

and after the second maintenance is

$$\begin{aligned} R_M^{(2)}(t, h, b) &= R_M^{(1)}(2h, h, b)RR(t - 2h, b) \\ &= (\boldsymbol{a}e^{\boldsymbol{A}h}\boldsymbol{u})(\boldsymbol{c}((1 - b)h)e^{\boldsymbol{A}(2h-h)}\boldsymbol{u})(\boldsymbol{c}((1 - b)2h)e^{\boldsymbol{A}(t-2h)}\boldsymbol{u}) \\ &= (\boldsymbol{a}e^{\boldsymbol{A}h}\boldsymbol{u})(\boldsymbol{c}((1 - b)h)e^{\boldsymbol{A}h}\boldsymbol{u})(\boldsymbol{c}((1 - b)2h)e^{\boldsymbol{A}(t-2h)}\boldsymbol{u}). \end{aligned}$$

In general, the system reliability function at any interval can be calculated recursively as

$$R_M^{(k)}(t, h, b) = R_M^{(k-1)}(kh, h, b)RR(t - kh, b), \quad k = 1, 2, \dots, \tag{3.46}$$

with initial condition $R_M^{(0)}(0, h, b) = R(t)$.

Note that b as the weight of convex combination and as a rejuvenating factor is equivalent, but not identical. The same value of b will produce a different reliability function when it is used as a weight of convex combination and as a rejuvenating factor. The **phmaintrel** and **phmaintmean** can be modified slightly to change from a probability of success to a rejuvenating factor.

Example 3.43 The system in Example 3.41 is imperfectly maintained to rejuvenate its age. The graphs of system reliability functions for several maintenance levels that represent rejuvenating factors are shown in Figure 3.33.

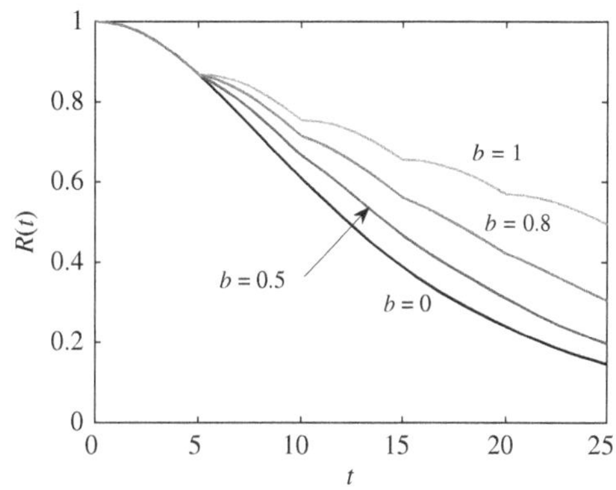

Figure 3.33 Reliability functions with imperfect maintenance to rejuvenate the system in Example 3.41.

The main program to create Figure 3.33 is available in **example3_43.m** file. Its MATLAB code is given below.

MATLAB code of example3_43.m

```
close all;clear
addpath('..\dysrel-am','..\dysrel-am\expokit');
% PH Ex 3.1
T.a = [1 0 0 0];
T.A = 0.1*[-1 1 0 0;0 -4 2 1;0 0 -5 3;0 0 0 -4];
T.n = 4;
% Time vector
tmax = 25;
t = linspace(0,tmax,41);
% Repair interval
h = 5;
% Reliability without maintenance
Rt    = phrel(t,T);
% Reliability with perfect maintenance
[R,Rb,Rr] = phmaintrel(t,T,h,0,1);
Rtwr = Rr;
% Reliability with rejuvenate maintenance
r = 0.8;
[R,Rb,Rr] = phmaintrel(t,T,h,0,r);
Rtwr2 = Rr;
% Reliability with rejuvenate maintenance
r = 0.5;
```

```
[R,Rb,Rr] = phmaintrel(t,T,h,0,r);
Rtwr3 = Rr;
% Plots
lgnd = {'b = 0','b = 1','b = 0.8','b = 0.5'};
multiplot(4,t,[Rt' Rtwr' Rtwr2' Rtwr3'],lgnd,'t','R(t)');
```

■

Combining perfect–imperfect maintenance or using different maintenance levels for different maintenance intervals is possible. In this case, b depends on k to become b_k. For example, an alternating imperfect–perfect maintenance is accommodated using $b_2 = b_4 = \cdots = 1$. Using maintenance level b_k can also be used to make maintenance level higher for larger k, such as $b_k = (k - 1)/k$. Models for imperfect maintenance can be made more complex by making b as a function of maintenance interval to become $b(h)$. For example, the longer the interval requires, the higher value of b can be achieved by using $b(h) = (h - 1)/h$. Experiments using these approaches are given in Problems.

Maintenance cannot be carried out indefinitely to prolong the system lifetime. At one time, the system must be replaced, either because it failed (unplanned replacement) or because it is better to be replaced (planned replacement). The next section discusses the replacement programs and a combination of maintenance-replacement program.

3.9 Replacement Program

A system eventually fails after being used for some time. The duration of use until it fails is random; as a result, the time to system failure cannot be known in advance. A system that fails while in operation will cause losses because it can decrease the quality of services and finished products that utilize the system and even interfere with the operation of other systems. To avoid failure in the middle of operation, it is sometimes better to replace the system periodically, even though it has not failed yet. Replacement can be done when the system is at rest, outside of system operating time, so it does not incur the cost of missing the production opportunity. Periodic replacement is referred to as planned replacement, while replacement when the system fails is referred to as unplanned replacement. Unplanned replacements are executed at random due to the randomness of the system lifetime. Planned replacements are carried out periodically in a deterministic time interval.

The replacement interval should be related to the system reliability function. High reliability systems certainly do not need to be replaced frequently, or their planned replacement intervals are long. In contrast, low reliability systems must be replaced frequently, or the replacement intervals are short. In addition, the cost factor plays an important role in determining the replacement interval. If the losses due to a system failure in the middle of operation are large, the replacement interval must be shorter. Likewise, if the cost of a planned replacement is high, then the replacement interval must be longer. Thus, a replacement program needs to balance the system reliability and the costs associated with the replacement. A replacement plan based on system reliability is called a reliability-based replacement plan.

Keep in mind that even if the replacement is carried out in a planned manner, the system may fail within the planned replacement interval. However, a good replacement program reduces the possibility of unplanned replacement so that losses due to unexpected failure are kept sufficiently small.

3.9.1 Mean Time Between Replacement

Suppose the planned replacement interval is s. One possible realization of replacements is as given in Figure 3.34. Define the ith system is the system used before the ith replacement. Figure 3.34 shows that the first system did not fail until the time of the planned replacement on $t = s$; likewise, the second system did not fail until the time of the planned replacement on $t = 2s$. However, the third system failed before the planned replacement was supposed to be on $t = 3s$. When failed, it was replaced at the time of failure on $t = 2s + T_3$. The fourth system did not fail until the time of the planned replacement on $t = 3s + T_3$, and so on.

The relationship between time between replacement (TBR) and time to failure can be determined from this observation. If the system does not fail within a replacement interval, then TBR equals the replacement interval. But, if the system fails within a replacement interval, then TBR equals time to system failure. A system does not fail in a replacement interval when its lifetime is longer than the replacement interval. On the contrary, a system fails within a replacement interval when its lifetime is less than or equal to the replacement interval. The mathematical expression for this description is

$$T > s \rightarrow TBR = s,$$
$$T \leq s \rightarrow TBR = T,$$

which can be summarized to

$$TBR = \min(T, s). \tag{3.47}$$

Equation (3.47) states that TBR cannot be greater than s and $TBR = T$ for $T < s$. Hence,

$$\begin{aligned} P[TBR > t] &= 0 \text{ for } t > s, \\ &= P[T > t] = R(t) \text{ for } t \leq s. \end{aligned} \tag{3.48}$$

The expected interval of replacement, referred to as mean time between replacement ($MTBR$), can be calculated using Equation (3.12) for $MTTF$ by replacing $R(t) = P[T > t]$ with $P[TBR > t]$ to yield

$$MTBR = E[TBR] = \int_0^\infty P[TBR > t]dt. \tag{3.49}$$

Substituting Equation (3.48) into Equation (3.49) produces

$$MTBR(s) = \int_0^s R(t)dt + \int_s^\infty 0dt = \int_0^s R(t)dt, \tag{3.50}$$

where $MTBR(s)$ is the mean time between replacement, or the expected interval of replacement, if the planned replacement interval is s.

Since $0 \leq R(t) \leq 1$, Equation (3.50) implies that

$$MTBR(s) \leq \int_0^s dt = s,$$

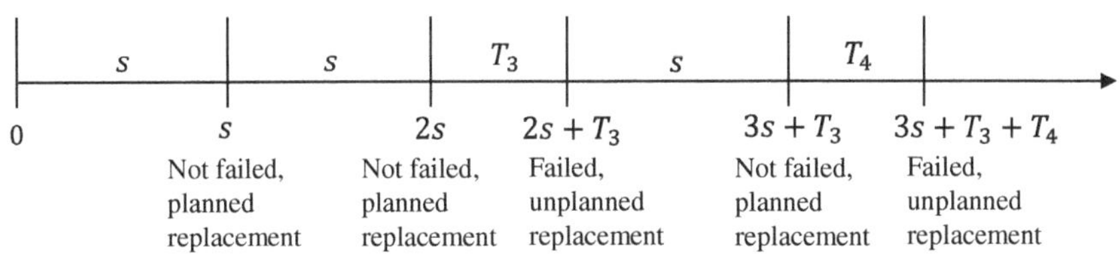

Figure 3.34 One possible realization of replacements.

which means *MTBR* will never be greater than the replacement interval. This result can be explained by the fact that the TBR is never greater than the replacement interval and that there is a possibility that the TBR is less than the replacement interval; this means the system failed before the planned replacement.

If the system is not replaced regularly, then the replacement interval is infinite. Substituting $s \to \infty$ into Equation (3.50) results in Equation (3.12). Hence,

$$MTBR(\infty) = MTTF.$$

This means that without planned replacement, the system is expected to be replaced every its expected lifetime.

For a system with lifetime following PH distribution with representation $(n, \boldsymbol{a}, \boldsymbol{A})$, Equation (3.50) becomes

$$MTBR(s) = \int_0^s \boldsymbol{a}e^{\boldsymbol{A}t}\boldsymbol{u}dt = \boldsymbol{a}\left(\int_0^s e^{\boldsymbol{A}t}dt\right)\boldsymbol{u} = \boldsymbol{a}\boldsymbol{A}^{-1}(e^{\boldsymbol{A}s} - \boldsymbol{I})\boldsymbol{u}. \tag{3.51}$$

Since $\lim_{s\to\infty} e^{\boldsymbol{A}s} \to 0$, Equation (3.51) for $s \to \infty$ produces

$$MTBR(\infty) = \boldsymbol{a}\boldsymbol{A}^{-1}(\boldsymbol{0} - \boldsymbol{I})\boldsymbol{u} = -\boldsymbol{a}\boldsymbol{A}^{-1}\boldsymbol{u} = MTTF,$$

in accordance with its definition.

Example 3.44 A system with a PH-distributed lifetime of order 1 with $MTTF = 10$ days is replaced periodically every its expected lifetime if until that time the system has not failed. The system *MTBR* can be calculated using Equation (3.51) by substituting $\boldsymbol{a} = [1]$, $\boldsymbol{A} = [-0.1]$, and $s = MTTF = 10$ to obtain

$$MTBR(10) = \boldsymbol{a}\boldsymbol{A}^{-1}(e^{\boldsymbol{A}s} - \boldsymbol{I})\boldsymbol{u} = (1)(-0.1)^{-1}(e^{-0.1\times10} - 1) = 6.32,$$

which is less than the replacement interval $s = 10$ as it should be. ■

Calculating and plotting *MTBR(s)* as a function of replacement interval *s* for systems having lifetimes following PH distributions of high order need computer support. For this purpose, **phmtbrplot** is provided as a part of the software package **dysrel-am**. The function accepts the PH representation of the distribution of system lifetime $(n, \boldsymbol{a}, \boldsymbol{A})$ and replacement interval *s* as inputs.

Example 3.45 Graphs of *MTBR(s)* as a function of the replacement interval *s* for the systems in Examples 3.1–3.3 are generated using **phmtbrplot** function. The results are presented in Figure 3.35, with $MTBR_1$, $MTBR_2$, and $MTBR_3$ being *MTBR* of systems in Examples 3.1–3.3, respectively.

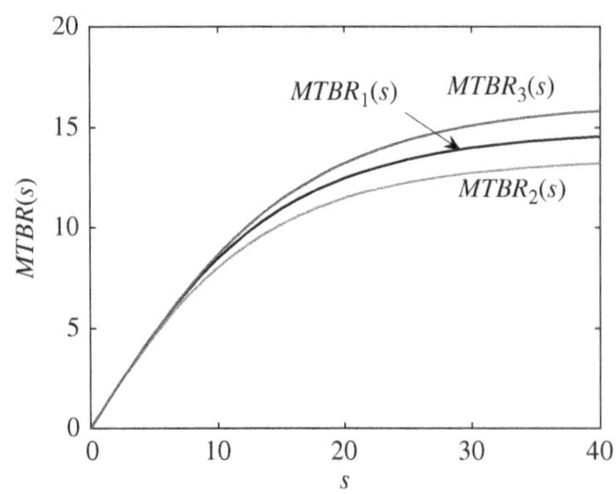

Figure 3.35 *MTBR(s)* for systems in Examples 3.1–3.3.

The main program for producing Figure 3.35 is given in **example3_45.m** file in the software package **dysrel-am**. Its source code is as follows:

MATLAB code for example3_45.m

```
close all;clear
addpath('..\dysrel-am','..\dysrel-am\expokit');

% Component's lifetime in PH distribution:
% Example 3.1
T(1).a = [1 0 0 0];
T(1).A = 0.1*[-1 1 0 0;0 -4 2 1;0 0 -5 3;0 0 0 -4];
T(1).n = 4;
% Example 3.2
T(2).a = [1 0 0];
T(2).A = 0.1*[-1 1 0;0 -4 2;0 0 -5];
T(2).n = 3;
% Example 3.3
T(3).a = [1 0 0 0];
T(3).A = 0.1*[-1 1 0 0;0 -4 2 1;0 0 -5 3;0 0 2 -4];
T(3).n = 4;
% Time vector
smax = 40;
s = linspace(0,smax,21);
for j = 1:3
  [ss,MTBR(:,j)] = phmtbrplot(s,T(j));
end
close all
% MTBR plot
lgnd = {'MTBR_1(s)','MTBR_2(s)','MTBR_3(s)'};
multiplot(3,ss,MTBR,lgnd,'s','MTBR(s)');
```

Graphs in Figure 3.35 show that $MTBR(s)$ for these three systems for the same value of s follow the same ordering as their reliability functions, that is,

$$MTBR_2(s) < MTBR_1(s) < MTBR_3(s).$$

In addition, $MTBR(s)$ of each system converges to a certain number, which is its respective $MTTF$ as the limit of $MTBR(s)$ for $s \to \infty$. The graphs of $MTBR(s)$ for all systems start from zero, since zero replacement interval makes the TBR become zero, or $s \to 0$ implies $MTBR(s) \to 0$. ∎

Example 3.46 An investigation of the effect of $CVTTF$ on $MTBR(s)$ is conducted using n-Erlang-distributed lifetime that all have $MTTF = 10$.

For n-Erlang distributed lifetime, it is known to have $CVTTF = 1/\sqrt{n}$. The graphs of $MTBR(s)$ as a function of s for some n are presented in Figure 3.36. For $CVTTF = 0$, or deterministic lifetime equal to 10, holds

$$\begin{aligned} MTBR(s) &= s \text{ for } s \leq 10, \\ &= 10 \text{ for } s > 10. \end{aligned}$$

Figure 3.36 shows that $MTBR(s)$ is decreasing with $CVTTF$. The reason is thc greater the $CVTTF$, the more likely the system will fail before the planned replacement, which makes the replacement time more likely shorter than the lifetime. ∎

Exercise 3.8 Investigate the effect of $CVTTF > 1$ to $MTBR$. Draw a conclusion from the result.

Incorporating cost functions in replacement program can produce replacement interval that minimizes the total cost, as discussed in the next section.

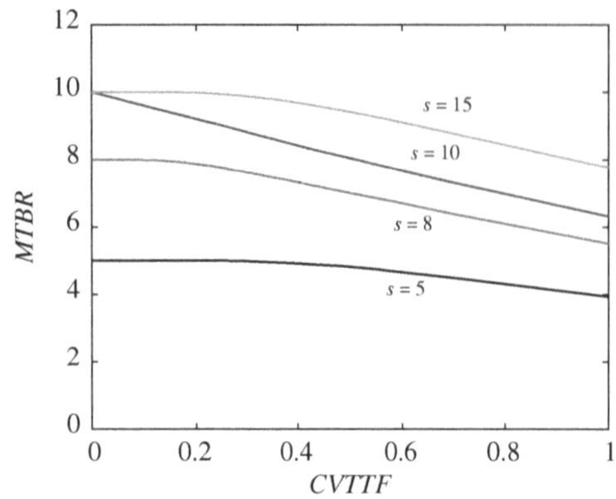

Figure 3.36 The effect of *CVTTF* on *MTBR*(*s*).

3.9.2 Optimum Replacement Interval

As mentioned in the introduction to Section 3.9, there are costs associated with replacement. Planned replacement costs are the costs of replacing the system according to its replacement schedule, which covers the price of the replacement system plus its installation cost. Unplanned replacement costs are planned replacement costs plus losses due to operational disruptions and possibly disruptions to the operations of other systems. The unplanned replacement costs are always greater, and even much greater, than planned replacement. The system may be replaced frequently to reduce unplanned replacement costs. However, systems that are over-replaced (replaced too often) make planned replacement costs expensive. On the other hand, if it is under-replaced (rarely replaced,) it will result in a large amount of additional costs covering operational losses due to system malfunctions. Hence, finding the optimum replacement interval that minimizes the total of planned and unplanned replacement costs is necessary. This optimum value depends on the system reliability. Systems with higher reliability will have longer optimum replacement intervals than systems with lower reliability for the same cost value.

Define

C_U = Replacement costs *due* to unplanned system failure, and

C_P = Replacement cost for replacing the system as planned.

It follows that

$$C_U \gg C_P.$$

The expected total cost of replacement is

$$C_T = C_U \times P_U + C_P \times P_P, \tag{3.52}$$

where P_U and P_P are the probability of unplanned replacement (probability of system failure before replacement schedule) and the probability of planned replacement (probability of system survival until replacement schedule), respectively.

It follows from their definitions that

$$P_P = P(T > s) = R(s), \tag{3.53}$$

and

$$P_U = 1 - R(s). \tag{3.54}$$

Substituting Equations (3.53) and (3.54) into Equation (3.52) yields

$$C_T = C_U(1 - R(s)) + C_P R(s) = C_U - (C_U - C_P)R(s), \tag{3.55}$$

where C_T is the total expected replacement cost. Since this cost is for one replacement interval, *TBR*, the average total replacement cost per unit time, C_A, is

$$C_A = \frac{C_T}{MTBR}. \tag{3.56}$$

Substituting Equations (3.50) and (3.55) into Equation (3.56) yields

$$C_A = \frac{C_U - (C_U - C_P)R(s)}{\int_0^s R(t)dt}. \tag{3.57}$$

Equation (3.57) can be simplified into

$$G(s) = \frac{C_A}{C_U} = \frac{1 - (1 - C)R(s)}{\int_0^s R(t)dt}, \tag{3.58}$$

where $C = \frac{C_P}{C_U}$, which is less than 1.

Knowing the cost coefficient C and the system reliability function $R(t)$, the graph of $G(s)$ as a function of s can be generated. Furthermore, the value of s that minimizes $G(s)$, denoted as s_{opt}, can be searched numerically.

Example 3.47 The lifetime of a system has a PH distribution of order 1 with a rate of failure λ. The system reliability function is

$$R(t) = \boldsymbol{a}e^{\boldsymbol{A}t}\boldsymbol{u} = e^{-\lambda t},$$

$$\int_0^s R(t)dt = -\frac{1}{\lambda}e^{-\lambda s} + \frac{1}{\lambda},$$

$$G(s) = \frac{1 - (1 - C)R(s)}{\int_0^s R(t)dt} = \frac{1 - (1 - C)e^{-\lambda s}}{\frac{1}{\lambda}(1 - e^{-\lambda s})}.$$

The optimum value of replacement interval can be obtained by setting $dG(s)/ds = 0$,

$$\frac{dG(s)}{ds} = \frac{(\lambda(1 - C)e^{-\lambda s})\left(\frac{1}{\lambda}(1 - e^{-\lambda s})\right) - (1 - (1 - C)e^{-\lambda s})e^{-\lambda s}}{\left(\frac{1}{\lambda}(1 - e^{-\lambda s})\right)^2} = 0.$$

An algebraic manipulation yields

$$(1 - C)e^{-\lambda s} - (1 - C)e^{-2\lambda s} - e^{-\lambda s} + (1 - C)e^{-2\lambda s} = 0,$$

or

$$Ce^{-\lambda s} = 0.$$

Since C is not zero, and $e^{-\lambda s} \to 0$ for $s \to \infty$, it follows that $s_{opt} \to \infty$. Thus, systems with exponentially distributed lifetimes, or CFR systems, have no optimum replacement interval. They should be used indefinitely until they fail. ■

Plotting G(s) and finding s_{opt} is tedious for PH distribution of high order. For this purpose, a MATLAB function **phtbroptplot** is prepared as a part of **dysrel-am** to plot $G(s)$ and determine s_{opt}. The function receives the PH representation of the distribution of system lifetime and cost parameter C as inputs.

Example 3.48 The graphs of $G(s)$ for IFR and DFR systems are produced using n-Erlang distributions for IFR systems having $MTTF = 10$ and n-hyperexponential distributions in Example 3.8 normalized to also have $MTTF = 10$ for DFR systems. The resulting graphs obtained using **phtbroptplot** for $C = 0.1$ are shown in Figure 3.37.

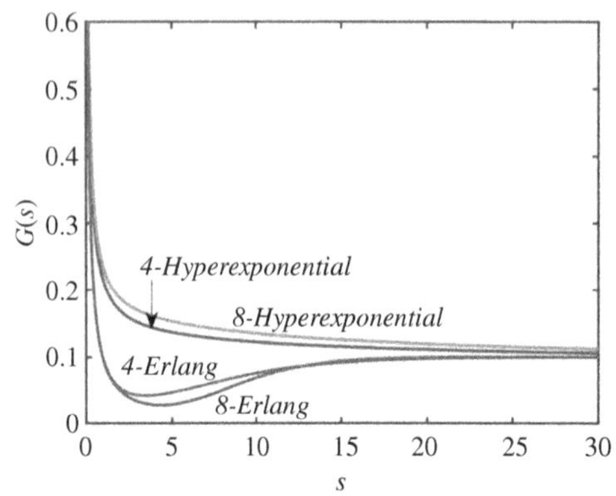

Figure 3.37 $G(s)$ for DFR and IFR systems.

Graphs in Figure 3.37 show that IFR systems have optimum replacement intervals $s_{opt} = 3.5$ for the 4-Erlang distribution and $s_{opt} = 4.4$ for the 8-Erlang distribution. In contrast, DFR systems do not have optimum replacement intervals, which is consistent with the definition of DFR systems that get better with longer use.

The main program to create Figure 3.32 is prepared as **example3_48.m** file in the software package **dysrel-am**. Its source code is as follows:

MATLAB code of example3_48.m

```
clear;close all
addpath('..\dysrel-am','..\dysrel-am\expokit')
% Components PH
PH(1) = phnhyperexp(4);
PH(1).A = (phmttf(PH(1))/10)*PH(1).A;
PH(2) = phnhyperexp(8);
PH(2).A = (phmttf(PH(2))/10)*PH(2).A;
PH(3) = phnErlang (4,10);
PH(4) = phnErlang(8,10);
% Calculate G(s) plot and optimal value
C = 0.1;
smax = 30;
for i = 1:length(PH)
  [s,G(i,:),sopt(i),Gopt(i)] = phtbroptplot(PH(i),C,smax);
end
% Plot
close all
lgnd = {'4-Hyper exponential','8-Hyper exponential','4-Erlang',...
```

```
'8-Erlang'};
multiplot(3,s,G',lgnd,'s','G(s)');
axis([0 30 0 0.6]);
```

■

Example 3.49 The s_{opt} values for one-component systems in Examples 3.1–3.3 as a function of C are given in Table 3.13. This table shows that higher C, which means a weaker effect of failure on overall cost, results in longer s_{opt}.

Table 3.13 Values of s_{opt} as a function of C for systems in Examples 3.1–3.3.

	s_{opt}		
C	3.1	3.2	3.3
0.02	1.84	1.48	2.00
0.05	3.00	2.52	3.32
0.10	4.56	4.10	5.19
0.20	7.99	8.53	9.29

■

Exercise 3.9 Plot s_{opt} as a function of C for systems in Example 3.49.

Application 3.1 A factory uses lathes to produce parts of a product. Lathe chisels wear out easily and eventually break. Worn lathe chisels will produce defective parts, which will cost the factory. The more worn the chisel, the more defective parts are produced and the more losses the factory suffers. So far, the factory has implemented a policy of replacing chisels only when they break, and this policy has resulted in many defective parts being produced, resulting in large losses. Using the result of this section, chisels will be replaced periodically at certain optimum replacement intervals. The optimum replacement interval for each chisel is determined to be that minimize the total cost of replacing the chisel and the loss of producing defective parts due to chisel wear. Data is collected on the lifetime of each chisel used in production as a function of the number of parts that have been produced. In addition, planned replacement costs, such as the price of the chisel and installation costs, and unplanned replacement costs, such as the planned replacement costs plus losses due to producing defective products, were also collected. Data processing was carried out using a widely available scalar statistical program package. The results are presented in Table 3.14.

Table 3.14 Data for Application 3.1.

	Chisel lifetimes (in the number of pieces produced)				
Lathe chisel	Distribution	Shape/scale parameters	cv	C_U	C_P
Part A	Weibull	2.38/8118.72	0.447	835	55
Part B	Weibull	2.38/7818.28	0.448	835	55
Part C	Weibull	2.86/5935.52	0.379	622	35
Part D	Weibull	1.63/5443.28	0.629	1016	52

The resulting Weibull distributions are approximated by PH distributions; in this case are using hypoexponential distributions. First, the order of the PH distribution that meets the requirement is determined from the coefficient of variation value of each chisel lifetime distribution using the relationship $n \geq 1/cv^2$. Because of the use of hypoexponential distributions, all initial probability vectors have elements of one in the first state and zero for all other states. Representations of the resulting PH distribution approximations are presented in Table 3.15. Comparison of the reliability

Table 3.15 PH distribution approximations for Weibull Distributions in Table 3.14.

Chisel	Survival matrix
Part A	$n = 5, \quad \boldsymbol{a} = [1\,0\,0\,0\,0]$, $\boldsymbol{A} = \begin{bmatrix} -0.0006 & 0.0006 & 0 & 0 & 0 \\ 0 & -0.0005 & 0.0005 & 0 & 0 \\ 0 & 0 & -0.0006 & 0.0006 & 0 \\ 0 & 0 & 0 & -0.0026 & 0.0026 \\ 0 & 0 & 0 & 0 & -0.0007 \end{bmatrix}$
Part B	$n = 5, \quad \boldsymbol{a} = [1\,0\,0\,0\,0]$, $\boldsymbol{A} = \begin{bmatrix} -0.0006 & 0.0006 & 0 & 0 & 0 \\ 0 & -0.0006 & 0.0006 & 0 & 0 \\ 0 & 0 & -0.0008 & 0.0008 & 0 \\ 0 & 0 & 0 & -0.0006 & 0.0006 \\ 0 & 0 & 0 & 0 & -0.0014 \end{bmatrix}$
Part C	$n = 7, \quad \boldsymbol{a} = [1\,0\,0\,0\,0\,0\,0]$, $\boldsymbol{A} = \begin{bmatrix} -0.0011 & 0.0011 & 0 & 0 & 0 & 0 & 0 \\ 0 & -0.0013 & 0.0013 & 0 & 0 & 0 & 0 \\ 0 & 0 & -0.0011 & 0.0011 & 0 & 0 & 0 \\ 0 & 0 & 0 & -0.0016 & 0.0016 & 0 & 0 \\ 0 & 0 & 0 & 0 & -0.0018 & 0.0018 & 0 \\ 0 & 0 & 0 & 0 & 0 & -0.0017 & 0.0017 \\ 0 & 0 & 0 & 0 & 0 & 0 & -0.0011 \end{bmatrix}$
Part D	$n = 4, \quad \boldsymbol{a} = [1\,0\,0\,0]$, $\boldsymbol{A} = \begin{bmatrix} -0.0004 & 0.0004 & 0 & 0 \\ 0 & -0.0024 & 0.0024 & 0 \\ 0 & 0 & -0.0009 & 0.0009 \\ 0 & 0 & 0 & -0.0019 \end{bmatrix}$

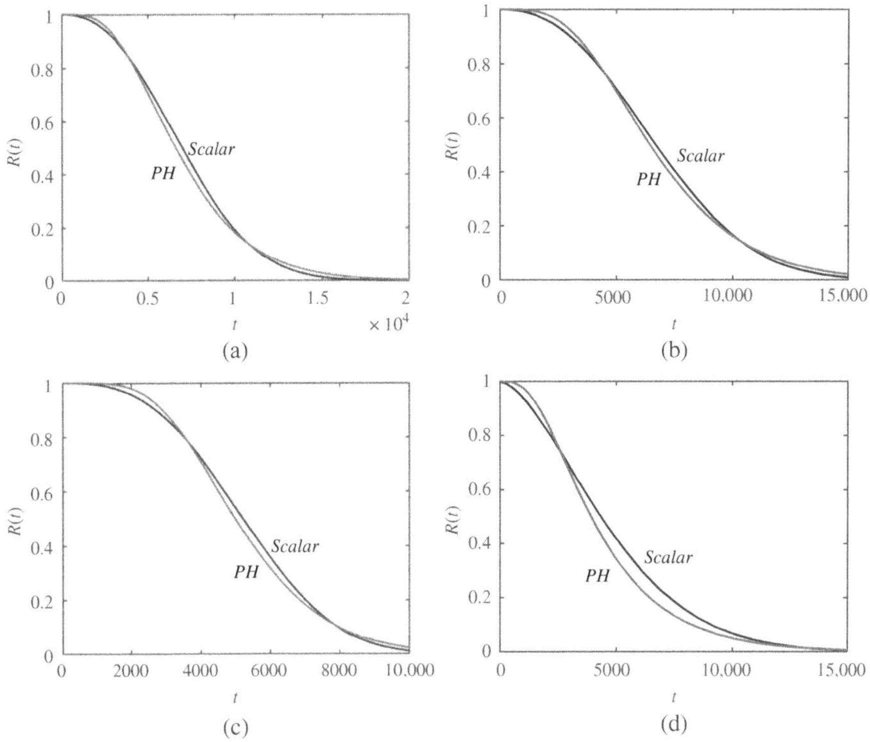

Figure 3.38 Reliability function of each chisel calculated using Weibull and approximate PH distribution.

function calculated using the original Weibull distribution and its PH distribution approximation for each chisel is presented in Figure 3.38.

The cost coefficient is calculated as the ratio of planned replacement cost to unplanned replacement cost. Equation (3.58) is used to produce the cost function for each chisel as given in Figure 3.39. The optimal replacement interval for each chisel can be determined from the graphs. The results are presented in Table 3.16. The replacement interval is measured in terms of the number of parts that have been produced using the chisel. ■

Considering that replacement and perfect maintenance are equivalent, the problems of finding the optimum perfect maintenance interval can be formulated similarly to finding the optimum replacement interval. The more interesting problem would be finding the optimum imperfect

Table 3.16 The optimum replacement interval for each chisel.

Chisel	C	s_{opt} (pieces)
Part A	0.0660	2201
Part B	0.0777	2338
Part C	0.0562	1882
Part D	0.0512	1034

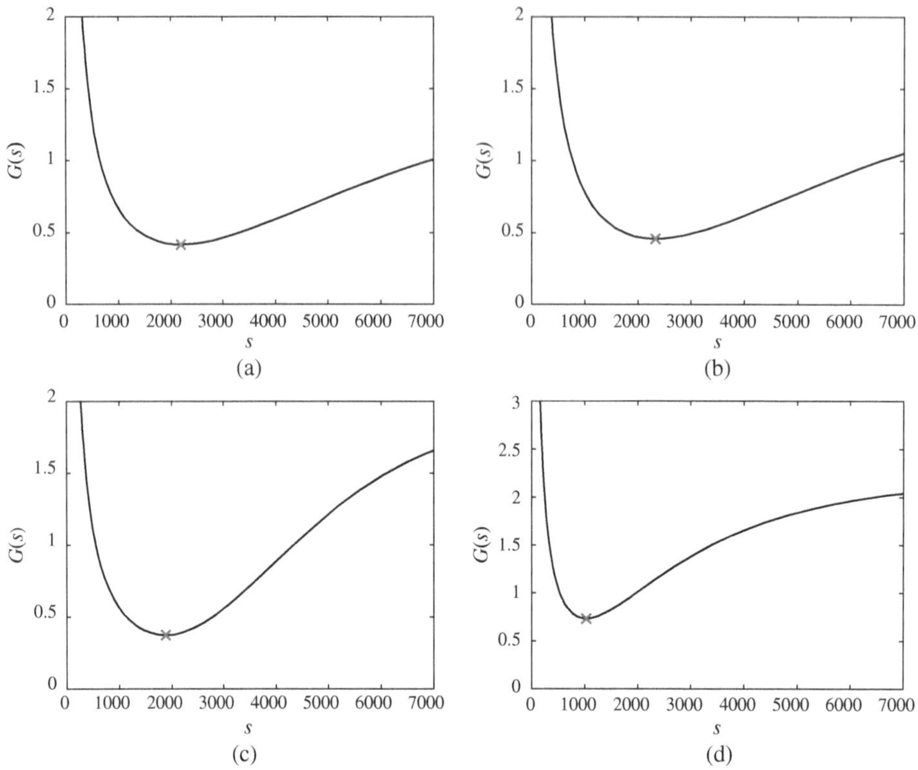

Figure 3.39 Cost function of each chisel.

maintenance as a function of maintenance level. It is left as a research challenge to interested readers.

The next section discusses the effect of replacement to system reliability in carrying a mission. The system is replaced periodically during the mission, even though the system is not failed to reduce the probability that the system is unexpectedly failed during the mission.

3.9.3 Mission Reliability with Planned Replacement

The probability of system failure while in operation can be reduced with planned replacement. In other words, mission reliability can be improved with planned replacement. The mission reliability for duration h is defined as the probability that the system never fails up to time h, including when it is replaced. Therefore, the mission reliability for duration h with a planned replacement is a multiplication of the probability that the mission survived at the end of each replacement interval s in the mission duration $[0, h]$ and the probability that the mission survived at time t in the last replacement interval of its mission duration. Since the system is replaced at the end of each replacement interval s, its reliability when replaced is $R(s)$. This value represents the probability that the system is not fail when replaced. If the system is replaced k times during the mission duration, then the probability that the mission survived until the last replacement is $R^k(s)$. The probability that the system survived in the last interval up to its mission duration h is its reliability to be used for the remaining $(h - ks)$ time units, or $R(h - ks)$. Therefore, the mission reliability for a duration

h with replacement interval s, denoted as $R_G(h, s)$, is

$$R_G(h, s) = R(s)^k R(h - ks). \tag{3.59}$$

For a system with a PH-distributed lifetime with representation $(n, \boldsymbol{a}, \boldsymbol{A})$, Equation (3.59) becomes

$$R_G(h, s) = (\boldsymbol{a}e^{\boldsymbol{A}s}\boldsymbol{u})^k(\boldsymbol{a}e^{\boldsymbol{A}(h-ks)}\boldsymbol{u}). \tag{3.60}$$

Note that Equation (3.60) is identical to Equation (3.35) for system reliability with perfect maintenance. In principle, perfect maintenance is identical to planned replacement since perfect maintenance makes the system as good as new, in that the system is replaced by a new system. However, the same system is used again after maintenance, while in replacement program, the system is replaced by a new one. The two are identical mathematically but different managerially, which causes the focus of maintenance and replacement programs to be different.

Example 3.50 A CFR system has reliability function $R(t) = e^{-\lambda t}$. The reliability of this system for mission duration h without planned replacement is $e^{-\lambda h}$. Its mission reliability with planned replacement interval s calculated using Equation (3.59) is

$$R_G(h, s) = (e^{-\lambda s})^k e^{-\lambda(h-ks)} = e^{-k\lambda s}e^{-\lambda(h-ks)} = e^{-\lambda h},$$

the same as the mission reliability without planned replacement. Thus, CFR systems are not worth replacing periodically as they will not improve their reliability. ■

Example 3.51 Systems with lifetimes following n-Erlang distribution, $n = 1, 2, 5$, all with $MTTF = 20$ days are used to carry out a mission for a duration of 28 days. A comparison of their mission reliability without and with planned replacements is given in Table 3.17. For $s = 28$ days, mission reliability with and without planned replacement is equal because the replacement interval is equal to their mission duration, which means there is no replacement during the mission. The table shows that for smaller n, which means larger $CVTTF(= 1/\sqrt{n})$, the difference between mission reliability with and without planned replacement is smaller, and there is no difference between the two for $n = 1$ which has $CVTTF = 1$ (that is, the system is CFR).

Table 3.17 Comparison of mission reliability without and with planned replacements for Example 3.51.

	$n = 1$		$n = 2$		$n = 5$	
s	$R(28)$	$R_G(28, s)$	$R(28)$	$R_G(28, s)$	$R(28)$	$R_G(28, s)$
5	0.2466	0.2466	0.2311	0.6003	0.1730	0.9542
10		0.2466		0.4378		0.7524
20		0.2466		0.3284		0.4173
28		0.2466		0.2311		0.1730

■

Exercise 3.10 Prove that $R_G(h, s) = R(h)$ for $s > h$.

Maintenance and replacement programs can be combined. The next section discusses a method to combine maintenance and replacement programs.

3.9.4 Replacement with Maintenance

It is possible to design a program that combines maintenance and replacement. In this case, the system is maintained every h units of time and replaced every s units of time with $h < s$. The development of a combined maintenance-replacement program basically combines the results of the replacement program in this section with the results of the maintenance program in Section 3.8.

Let $MTBR_M(h, s)$ be the mean time between replacement with maintenance interval h and replacement intervals s. Substituting the reliability function with maintenance into Equation (3.50) yields

$$MTBR_M(h, s) = \int_0^s R_M(t, h)dt. \tag{3.61}$$

For practical purpose and efficiency, without any loss of generality, the replacement interval is chosen to be a multiple of the maintenance interval, say $s = Kh$, so that the Kth maintenance coincides with replacement. Since $R_M(t, h)$ is piecewise continuous function with discontinuity at $t = h, 2h, \ldots$, Equation (3.61) can be written as

$$MTBR_M(h, K) = \int_0^h R_M^{(0)}(t, h)dt + \int_h^{2h} R_M^{(1)}(t, h)dt + \cdots + \int_{(K-1)h}^{Kh} R_M^{(K-1)}(t, h)dt. \tag{3.62}$$

Substitute Equation (3.35) for perfect maintenance into Equation (3.62) and use suitable changes in integration variables to obtain

$$MTBR_M(h, K) = \int_0^h (\boldsymbol{a}e^{\boldsymbol{A}t}\boldsymbol{u})dt + (\boldsymbol{a}e^{\boldsymbol{A}h}\boldsymbol{u})^1 \int_0^h (\boldsymbol{a}e^{\boldsymbol{A}t}\boldsymbol{u})dt + \cdots + (\boldsymbol{a}e^{\boldsymbol{A}h}\boldsymbol{u})^{K-1} \int_0^h (\boldsymbol{a}e^{\boldsymbol{A}t}\boldsymbol{u})dt.$$

Hence,

$$MTBR_M(h, K) = (1 + (\boldsymbol{a}e^{\boldsymbol{A}h}\boldsymbol{u})^1 + \cdots + (\boldsymbol{a}e^{\boldsymbol{A}h}\boldsymbol{u})^{K-1}) \int_0^h (\boldsymbol{a}e^{\boldsymbol{A}t}\boldsymbol{u})dt. \tag{3.63}$$

Performing the integration leads to

$$MTBR_M(h, K) = (1 + (\boldsymbol{a}e^{\boldsymbol{A}h}\boldsymbol{u})^1 + \cdots + (\boldsymbol{a}e^{\boldsymbol{A}h}\boldsymbol{u})^{K-1})(\boldsymbol{a}\boldsymbol{A}^{-1}(e^{\boldsymbol{A}h} - \boldsymbol{I})\boldsymbol{u}). \tag{3.64}$$

If the system is not replaced periodically, then $K \to \infty$ and Equation (3.64) reduces into Equation (3.37) as an expression for expected lifetime with perfect maintenance. Likewise, if the system is not maintained periodically, then $h = s$ and $K = 1$, which makes Equation (3.64) reduces to Equation (3.51), an expression for the $MTBR$ without maintenance. It can be verified that Equation (3.64) also reduces to Equation (3.51) for $s \leq h$.

Substitute Equation (3.44) or (3.45) for imperfect maintenance into Equation (3.62) and use Equation (3.64), which results in

$$\begin{aligned} &MTBR(b, h, s) \\ &= \left(\boldsymbol{a} + R_M^0(h)\boldsymbol{d}(h) + R_M^1(2h)\boldsymbol{d}(2h) + \cdots + R_M^{K-2}((K-1)h)\boldsymbol{d}((K-1)h)\right) \boldsymbol{A}^{-1}(e^{\boldsymbol{A}h} - \boldsymbol{I})\boldsymbol{u}. \end{aligned} \tag{3.65}$$

If $b = 1$, then $\boldsymbol{d}(kh) = \boldsymbol{a}$ and Equation (3.65) reduces Equation (3.64). Likewise, if $b = 0$, then the system is not maintained, and Equation (3.65) reduces to Equation (3.51). It is left as an exercise to prove these propositions.

A computer program for calculating system $MTBR$ with integrated maintenance-replacement program is given as **phmaintreplmean.**

Example 3.52 The system in Example 3.1 is maintained and replaced periodically. Its *MTBR* as a function of maintenance and replacement intervals is given in Table 3.18.

Table 3.18 *MTBR* of the system in Example 3.1 with maintenance.

Maintenance		Replacement interval *s*			
Interval *h*	Level *b*	4	8	12	None
2	0.0	3.88	7.15	9.59	14.88
	0.5	3.90	7.40	10.30	18.27
	1.0	3.93	7.68	11.27	89.46
4	0.0	3.88	7.15	9.59	14.88
	0.5	3.88	7.29	10.09	17.86
	1.0	3.88	7.46	10.66	44.51

Table 3.18 shows that when $b = 0$, which means the system is not maintained, maintenance intervals do not affect *MTBR*, that is maintenance intervals 2 and 4 yield the same *MTBR*. Likewise, when maintenance interval is equal to replacement interval, the system *MTBR* is unaffected by maintenance level since the system has not been maintained when replaced.

A combination of maintenance and replacement can be considered a combination of imperfect and perfect maintenance since replacement is the same as perfect maintenance. For example, maintenance level = 0.5 with interval = 2 days and replacement interval = 4 days can be reinterpreted as the system is imperfectly maintained with level = 0.5 every 2 days and perfectly maintained every 4 days. Thus, the maintenance program uses alternating imperfect–perfect maintenance. However, the interest will not be in the values of *MTBR*; instead, it is about calculating reliability for a certain mission duration. It is left as a challenge for the interested readers to derive the equation for the reliability function under the maintenance-replacement program. ■

Exercise 3.11 Repeat Example 3.52 using n-Erlang distribution and investigate the effect of n to *MTBR*.

An important random variable in replacement program is the number of replacements. It is important because it is related to the number of stocks that must be provided for a certain level of stockout risk that management is willing to take. The next section describes methods for calculating measures of the number of replacements.

3.10 Number of Replacements

Failed systems must be replaced. If the system lifetime is random, then the number of replacements in a time interval is also random. Define $N(t)$ as the number of replacements in the interval $[0, t]$ and

$$P_n(t) = P[N(t) = n], \quad n = 0, 1, 2, \dots, \tag{3.66}$$

as the probability that the number of replacements in the interval $[0, t]$ is n.

The number of replacements in a time interval depends on the system reliability. For example, from the definition of $P_n(t)$, $P_0(t)$ is the probability that there is no replacement up to time t, which is equal to the probability of the system lifetime that is greater than t. Hence,

$$P_0(t) = P[T > t] = R(t). \tag{3.67}$$

If the system lifetime is PH-distributed with representation $(n, \boldsymbol{a}, \boldsymbol{A})$, then Equation (3.67) becomes

$$P_0(t) = \boldsymbol{a}e^{\boldsymbol{A}t}\boldsymbol{u}. \tag{3.68}$$

The interest is to develop a method to find $P_n(t)$, $n = 1, 2, \dots$, when the system lifetime has PH distribution. Given the number of replacements closely related to the system lifetime with a known distribution, a method for calculating $P_n(t)$, $n = 1, 2, \dots$, can be developed using its relation to the waiting time until the nth replacement. If T_i is the system lifetime between the $(i-1)^{th}$ and i^{th} replacements, $i = 1, 2, \dots, n$, and the 0th replacement is at $t = 0$, then the following relationship applies

$$S_n = T_1 + T_2 + \cdots + T_n, \quad n = 1, 2, \dots, \tag{3.69}$$

where S_n is the waiting time until the nth replacement. By definition, T_1 is the lifetime of the first system before being replaced by the second system, T_2 is the lifetime of the second system before being replaced by the third system, and so on. In general, failed systems are replaced with similar systems; thus, T_i, $i = 1, 2, \dots, n$, have independent and identical distributions.

The development of the methods for calculating measures for the number of replacements starts with the simplest case, namely, exponential distribution.

3.10.1 Exponentially Distributed Lifetimes

If T_i, $i = 1, 2, \dots, n$, follow an independent exponential distribution with identical rate λ, then the replacement process is a CTMC with a state transition diagram given in Figure 3.40, where state i represents the ith replacement.

Referring to Table 2.1, the random variable S_n having state transition diagram in Figure 3.40 follows n-Erlang distribution with CDF in a scalar representation given by

$$F_{S_n}(t) = P[S_n \le t] = 1 - \sum_{i=0}^{n-1} \frac{(\lambda t)^i}{i!} e^{-\lambda t}, \quad n = 1, 2, \dots. \tag{3.70}$$

The relationship between S_n and $N(t)$ can be derived from the fact that the event $S_n \le t$ is equivalent to the number of replacements up to time t is equal to or more than n. Hence,

$$P[S_n \le t] = P[N(t) \ge n], \quad n = 1, 2, \dots. \tag{3.71}$$

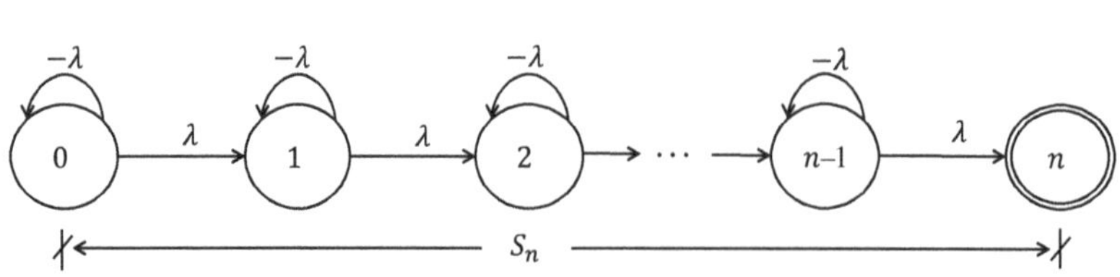

Figure 3.40 State transition diagram for the waiting time until the nth replacement for a system with exponentially distributed lifetime.

The probability that there are n replacements up to time t can be calculated as

$$P_n(t) = P[N(t) = n] = P[N(t) \geq n] - P[N(t) \geq n+1], \quad n = 0, 1, \ldots. \tag{3.72}$$

Substituting Equation (3.71) into Equation (3.72) yields

$$P_n(t) = P[S_n \leq t] - P[S_{n+1} \leq t], \quad n = 0, 1, 2, \ldots. \tag{3.73}$$

For $n = 0$, Equation (3.73) becomes $P_0(t) = P[S_0 \leq t] - P[S_1 \leq t]$. Because $S_0 = 0$, which makes $P[S_0 \leq t] = 1$, $P_0(t) = 1 - P[T_1 \leq t] = P[T_1 > t] = R(t)$, the same as given by Equation (3.67).

Substituting Equation (3.70) into Equation (3.73) yields

$$P_n(t) = 1 - \sum_{i=0}^{n-1} \frac{(\lambda t)^i}{i!} e^{-\lambda t} - \left(1 - \sum_{i=0}^{n} \frac{(\lambda t)^i}{i!} e^{-\lambda t}\right).$$

The first sum negates each other with the second sum, leaving only one term for $i = n$. Therefore,

$$P_n(t) = \frac{(\lambda t)^n}{n!} e^{-\lambda t}, \quad n = 0, 1, \ldots, \tag{3.74}$$

which is a probability mass function of Poisson distribution. This result is widely known that if the time between events follows exponential distribution, then the number of events follows Poisson distribution, and vice versa.

The expected number of replacements up to time t is

$$E[N(t)] = \sum_{n=0}^{\infty} n P_n(t). \tag{3.75}$$

Substituting Equation (3.74) into Equation (3.75) and ignoring the term $n = 0$ (because it is equal to 0) yield

$$E[N(t)] = \sum_{n=1}^{\infty} n \frac{(\lambda t)^n}{n!} e^{-\lambda t} = \lambda t e^{-\lambda t} \sum_{n=1}^{\infty} \frac{(\lambda t)^{n-1}}{(n-1)!}.$$

The sum on the last term in the right-hand side is the Taylor series for $e^{\lambda t}$ that cancelled out with $e^{-\lambda t}$ in front of the summation. Thus,

$$E[N(t)] = \lambda t. \tag{3.76}$$

For CFR systems, it is known that $MTTF = 1/\lambda$. Thus, Equation (3.76) can be written as

$$E[N(t)] = \frac{t}{MTTF}. \tag{3.77}$$

Example 3.53 For CFR systems that have $MTTF$ = 10 days, or λ = 0.1/day, the probability that the third replacement occurs in no more than 1 month (30 days) can be calculated using Equation (3.70) as

$$P[S_3 \leq 30] = F_{S_3}(30) = 1 - \sum_{i=0}^{2} \frac{(0.1 \times 30)^i}{i!} e^{-0.1\times 30} = 1 - (1 + +3 + 4.5)e^{-3} = 0.58.$$

The probability that the system will be replaced twice in 1 month can be calculated using Equation (3.74) as

$$P[N(30) = 2] = P_2(30) = \frac{(0.1 \times 30)^2}{2!} e^{-0.1\times 30} = 4.5e^{-3} = 0.22.$$

The expected number of replacements in 1 month for this system can be found using Equation (3.76) as

$$E[N(30)] = 0.1 \times 30 = 3.$$

This number can also be calculated using Equation (3.77) with the same result. ■

A replacement program often wants to ensure the risk of running out of replacement stocks during a procurement cycle is within a predetermined value. Risk is generally measured in terms of probability.

Example 3.54 If there are currently two replacement stocks, and those being ordered will arrive in 1 month, then the risk of stockout can be calculated as the probability that the system will fail more than three times in 1 month. The reason is that the system that is currently being used, coupled with the two available stocks, can still overcome three system failures; the fourth system failure onwards does not have a replacement. Therefore, stockout occurs when the system requires more than three replacements.

$$\text{Risk of stockout} = P[N(30) > 3] = 1 - P[N(30) \leq 3] = 1 - \sum_{i=0}^{3} \frac{(0.1 \times 30)^i}{i!} e^{-0.1 \times 30}$$

$$= 1 - (1 + 3 + 4.5 + 4.5)e^{-3} = 0.403.$$

■

Exercise 3.12 Show that the risk of out-of-stock for the same planning period is higher for the lower *MTTF*, and for the same *MTTF* is higher for the longer planning period.

Next, the result for exponential distribution is extended to general PH distribution.

3.10.2 PH-Distributed Lifetimes

The state transition diagram for the waiting time until the nth replacement, S_n, for the case of PH distribution is similar to Figure 3.40 by replacing the scalar transition λ with matrix transitions $\boldsymbol{a}, \boldsymbol{A}$, and $\boldsymbol{b}$, which represent initial probability vector, survival matrix, and failure vector of the PH distribution of system lifetime, respectively. The result is Figure 3.41.

There is no closed formula available to calculate $P_n(t)$ for T_i, $i = 1, 2, \ldots, n$, following PH distributions of order greater than 1. However, the algorithm for calculating $P_n(t)$ can be designed in the same manner as deriving Equation (3.73). First, PH representation for S_n must be sought. Theorem 2.3 can be used recursively to produce Lemma 3.1.

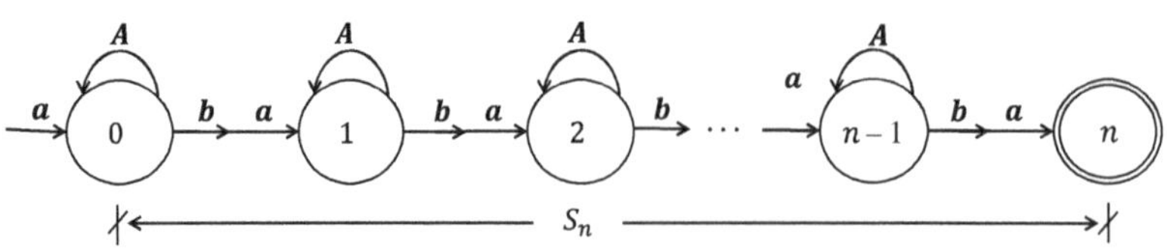

Figure 3.41 State transition diagram for the waiting time until the nth replacement for system with PH-distributed lifetime.

Lemma 3.1 If $T_i \sim PH(k, \boldsymbol{a}, \boldsymbol{A})$, $i = 1, 2, \dots, n$, then $S_n \sim PH(k_{S_n}, \boldsymbol{a}_{S_n}, \boldsymbol{A}_{S_n})$ with $(k_{S_n}, \boldsymbol{a}_{S_n}, \boldsymbol{A}_{S_n})$ can be calculated using the following recursion equations:

$$k_{S_i} = k_{S_{i-1}} + k, \tag{3.78}$$

$$\boldsymbol{a}_{S_i} = [\boldsymbol{a}_{S_{i-1}} \quad \boldsymbol{0}], \tag{3.79}$$

$$\boldsymbol{A}_{S_i} = \begin{bmatrix} \boldsymbol{A}_{S_{i-1}} & \boldsymbol{b}_{S_{i-1}}\boldsymbol{a} \\ \boldsymbol{0} & \boldsymbol{A} \end{bmatrix}, \tag{3.80}$$

for $i = 2, 3, \dots, n$ and $\boldsymbol{b}_{S_{i-1}} = -\boldsymbol{A}_{S_{i-1}}\boldsymbol{u}$, with initial conditions are given by

$$k_{S_1} = k, \quad \boldsymbol{a}_{S_1} = \boldsymbol{a}, \quad \boldsymbol{A}_{S_1} = \boldsymbol{A}. \tag{3.81}$$

Lemma 3.1 can be proved by writing Equation (3.69) recursively as

$$S_i = S_{i-1} + T_i \qquad i = 2, 3, \dots, n, \tag{3.82}$$

starting from the initial condition $S_1 = T_1$ that is known to have a PH distribution. Theorem 2.3 guarantees that $S_2 = S_1 + T_2 = T_1 + T_2$ has a PH distribution, which, in turn, $S_3 = S_2 + T_3$ also has a PH distribution. Continuing in this manner proves $S_n = T_1 + T_2 + \cdots + T_n$ has a PH distribution.

A function **phstandsys** is provided to generate the PH representation of S_n, $(k_{S_n}, \boldsymbol{a}_{S_n}, \boldsymbol{A}_{S_n})$, from the PH representation of T_i, $(k, \boldsymbol{a}, \boldsymbol{A})$, $i = 1, 2, \dots, n$.

Example 3.55 The use of Lemma 3.1 for a CFR system is as follows:

The PH representation of the system lifetime is

$$k = 1, \quad \boldsymbol{a} = [1], \quad \boldsymbol{A} = [-\lambda].$$

- The initial value is obtained using Equation (3.81) as

$$k_{S_1} = 1, \quad \boldsymbol{a}_{S_1} = [1], \quad \boldsymbol{A}_{S_1} = [-\lambda], \quad \boldsymbol{b}_{S_1} = -\boldsymbol{A}_{S_1}\boldsymbol{u} = [\lambda].$$

- Iteration 1 using Equations (3.78–3.80) results in

$$k_{S_2} = k_{S_1} + 1 = 2, \quad \boldsymbol{a}_{S_2} = [\boldsymbol{a}_{S_1} \quad 0] = [1 \quad 0],$$

$$\boldsymbol{A}_{S_2} = \begin{bmatrix} \boldsymbol{A}_{S_1} & \boldsymbol{b}_{S_1}\boldsymbol{a} \\ \boldsymbol{0} & \boldsymbol{A} \end{bmatrix} = \begin{bmatrix} -\lambda & [\lambda][1] \\ 0 & [-\lambda] \end{bmatrix} = \begin{bmatrix} -\lambda & \lambda \\ 0 & -\lambda \end{bmatrix}, \quad \boldsymbol{b}_{S_2} = -\boldsymbol{A}_{S_2}\boldsymbol{u} = \begin{bmatrix} 0 \\ \lambda \end{bmatrix}.$$

- Iteration 2 using Equations (3.78–3.80) results in

$$k_{S_3} = k_{S_2} + 1 = 3, \quad \boldsymbol{a}_{S_3} = [\boldsymbol{a}_{S_2} \quad 0] = [[1 \quad 0] \quad 0] = [1 \quad 0 \quad 0],$$

$$\boldsymbol{A}_{S_3} = \begin{bmatrix} \boldsymbol{A}_{S_2} & \boldsymbol{b}_{S_2}\boldsymbol{a} \\ \boldsymbol{0} & \boldsymbol{A} \end{bmatrix} = \begin{bmatrix} \begin{bmatrix} -\lambda & \lambda \\ 0 & -\lambda \end{bmatrix} & \begin{bmatrix} 0 \\ \lambda \end{bmatrix}[1] \\ \boldsymbol{0} & [-\lambda] \end{bmatrix} = \begin{bmatrix} -\lambda & \lambda & 0 \\ 0 & -\lambda & \lambda \\ 0 & 0 & -\lambda \end{bmatrix},$$

$$\boldsymbol{b}_{S_3} = -\boldsymbol{A}_{S_3}\boldsymbol{u} = \begin{bmatrix} 0 \\ 0 \\ \lambda \end{bmatrix}.$$

Continuing in this manner for iterations 3, 4, ... produces the following PH representation for S_n

$$k_{S_n} = n, \quad \boldsymbol{a}_{S_n} = [1 \quad 0 \quad 0 \dots \quad 0], \quad \boldsymbol{A}_{S_n} = \begin{bmatrix} -\lambda & \lambda & \dots & 0 \\ 0 & -\lambda & \dots & 0 \\ \vdots & \vdots & \ddots & \vdots \\ 0 & 0 & \dots & -\lambda \end{bmatrix},$$

which is the matrix representation of the n-Erlang distribution as given in Table 2.1. This result is in accordance with the result of Subsection 3.10.1 but in a matrix representation, namely, n-Erlang distribution. ■

Use Lemma 3.1 to generate

$$P[S_n \leq t] = F_{S_n}(t) = 1 - \boldsymbol{a}_{S_n} e^{\boldsymbol{A}_{S_n} t} \boldsymbol{u}. \tag{3.83}$$

Equations (3.73) and (3.83) can be used to calculate $P_n(t)$ as

$$P_n(t) = F_{S_n}(t) - F_{S_{n+1}}(t) = 1 - \boldsymbol{a}_{S_n} e^{\boldsymbol{A}_{S_n} t} \boldsymbol{u} - \left(1 - \boldsymbol{a}_{S_{n+1}} e^{\boldsymbol{A}_{S_{n+1}} t} \boldsymbol{u}\right).$$

Thus,

$$P_n(t) = \boldsymbol{a}_{S_{n+1}} e^{\boldsymbol{A}_{S_{n+1}} t} \boldsymbol{u} - \boldsymbol{a}_{S_n} e^{\boldsymbol{A}_{S_n} t} \boldsymbol{u}. \tag{3.84}$$

The expected number of replacements up to time t can be found numerically using Equations (3.75) and (3.84). Unlike exponential distribution, currently, there is no simple closed-form expression available for calculating $E[N(t)]$ for a general class of PH distributions.

Calculation of $F_{S_n}(t), P_n(t)$, and $E[N(t)]$ for PH-distributed lifetime with order greater than 1 is difficult to do manually. For this purpose, MATLAB functions **phFSn, phPn,** and **phEN** are prepared. The program accepts PH representation of system lifetime (n, $\boldsymbol{a}$, $\boldsymbol{A}$) as input.

Example 3.56 The use of Equation (3.83) in the form of **phFSn** function to produce graphs of CDFs of the time until the third replacement, S_3, for one-component systems in Examples 3.1–3.3 results in Figure 3.42, with $F_1(t)$, $F_2(t)$, and $F_3(t)$ being CDFs of S_3 for Examples 3.1–3.3, respectively.

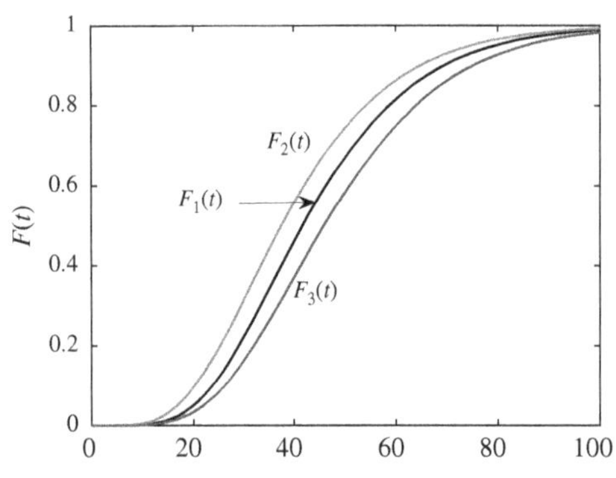

Figure 3.42 Graphs of $F_{S_3}(t)$ for systems in Examples 3.1–3.3.

As expected, the CDFs follow a stochastic ordering

$$F_2(t) > F_1(t) > F_3(t),$$

the opposite of their reliability function ordering.

The probability that no more than two replacements in 1 month (30 days) for systems in Examples 3.1–3.3 can be calculated as

$$P[N(30) \leq 2] = P[S_3 > 30] = 1 - F_{S_3}(30).$$

This equation expresses the events $[N(30) \leq 2]$ and $[S_3 > 30]$ are equivalent. Because the third failure occurs after 30 days, or $S_3 > 30$, it means that until the 30th day, there are at most two failures, or $N(30) \leq 2$.

$P[N(30) \le 2]$ for Examples 3.1–3.3 are obtained using Equation (3.84) as $P[N(30) = 0] + P[N(30) = 1] + P[N(30) = 2]$, which is equal to 0.778, 0.685, and 0.837, respectively.

The graphs of $P_n(30)$ as a function of n for systems in Examples 3.1–3.3 generated using Equation (3.84) in the form of **phPn** are shown in Figure 3.43.

Figure 3.43 $P_n(30)$ for Examples 3.1–3.3.

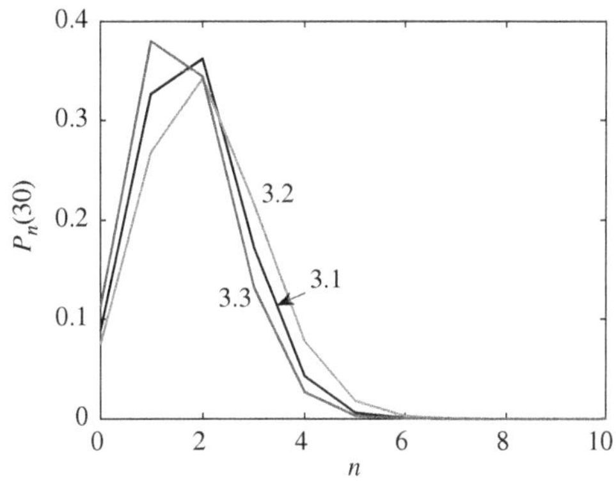

The probability that exactly three replacements are required in 1 month for systems in Examples 3.1–3.3 can be obtained from the graphs in Figure 3.43 for $n = 3$, or from the results of the calculations, which are 0.172, 0.216, and 0.132, respectively.

The expected number of replacements in 1 month for systems in Examples 3.1–3.3 can also be calculated using Equations (3.75) and (3.84) provided as **phEN** function. The results are 1.776, 2.022, and 1.592, respectively. Hence,

$$E[N_2(t)] > E[N_1(t)] > E[N_3(t)],$$

which is the opposite of their reliability function ordering, the higher the reliability, the lower the expected number of replacements.

The main program for this example calls **phFSn, phPn,** and **phEN** functions for calculating $F_{S_n}(t)$, $P_n(t)$, and $E[N(t)]$, respectively. It is available in **example3_56.m** file in the software package **dysrel am**. Its source code is as follows:

MATLAB code for example3_56.m

```
clear;close all
addpath('..\dysrel-am','..\dysrel-am\expokit')
% PH representation of Components lifetime
% Ex 3.1
T(1).a = [1 0 0 0];
T(1).A = 0.1*[-1 1 0 0;0 -4 2 1;0 0 -5 3;0 0 0 -4];
T(1).n = 4;
% Ex 3.2
T(2).a = [1 0 0];
T(2).A = 0.1*[-1 1 0;0 -4 2;0 0 -5];
T(2).n = 3;
% Ex 3.3
T(3).a = [1 0 0 0];
T(3).A = 0.1*[-1 1 0 0;0 -4 2 1;0 0 -5 3;0 0 2 -4];
T(3).n = 4;
% Plot FS3
```

```
tmax = 100;
t = linspace(0,tmax,tmax+1);
N = 3;
U = [];
for i = 1:3
  FS3 = phFSn(t,T(i),N,1);
  U = [U FS3'];
end
lgnd = {'F_1(t)','F_2(t)','F_3(t)'};
multiplot(3,t,U,lgnd,'t','F(t)');
% Plot Pn(30)
t = 30;
n = 0:10;
for i = 1:3
  u = ones(T(i).n,1);
  eAtu = expv(t,T(i).A,u);
  for j = 1:length(n)
    if j == 1
      Pn(i,j) = T(i).a*eAtu;
    else
      Pn(i,j) = phPn(T(i),n(j),t,1);
    end
  end
end
lgnd = {'3.1','3.2','3.3'};
multiplot(3,n,Pn',lgnd,'n','P_n(30)');
% Calculate P3(30)
t = 30;
for i = 1:3
  FS3(i) = phFSn(30,T(i),3,1);
  P3(i)  = 1-FS3(i);
end
disp(['P3(30) = ',num2str(P3)])
% Calculate EN(30)
for i = 1:3
  EN(i) = phEN(T(i),30,1);
end
disp(['EN(30) = ',num2str(EN)])
```

Table 3.6 provides the *MTTF* values of one-component system in Examples 3.1–3.3 as 14.87, 13.50, and 16.25, respectively. Hence, $t/MTTF$ values with $t = 30$ for them are 2.017, 2.222, and 1.846, respectively, while the exact values from Example 3.56 are 1.776, 2.022, and 1.592, respectively. This result suggests that the expected number of replacements for non-CFR systems cannot be calculated as such for CFR systems; Equation (3.77) only produces approximate values for non-CFR systems. The expected number of replacements for one-component systems in Examples 3.1–3.3 that have *CVTTF* less than 1 is less than that calculated using Equation (3.77). These results show that replacement performances, such as $E[N(t)]$, $F_{S_n}(t)$, and $P_n(t)$, in addition to being affected by *MTTF*, are also affected by higher lifetime moments, such as *CVTTF*. To understand the effect of *CVTTF* on replacement performance, an experiment is carried out using distributions with different *CVTTF* but the same *MTTF*. ■

Example 3.57 The effect of *CVTTF* on $E[N(30)]$ is investigated using systems that have lifetimes following n-Erlang distributions for several n that all have $MTTF = 10$. The result is presented in Figure 3.44.

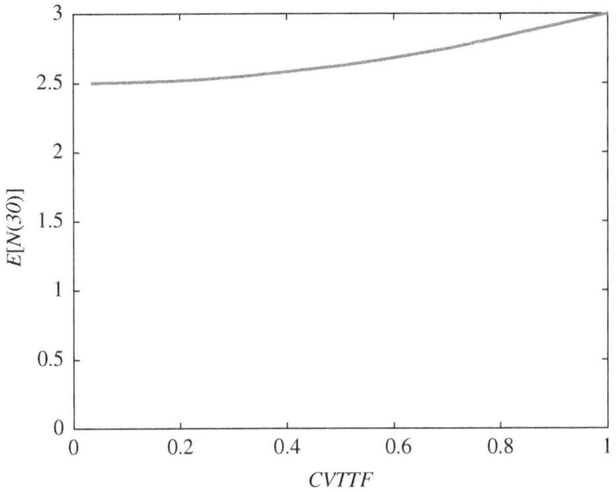

Figure 3.44 Effect of *CVTTF* on $E[N(30)]$.

This graph shows that $E[N(30)] = 3$ for $CVTTF = 1$, the same as calculated using Equation (3.76). The value of $E[N(30)]$ decreases with the decrease in *CVTTF*. For a deterministic system lifetime = 10, which must have $CVTTF = 0$, the number of replacements in 30 days should be exactly equal to 3. This point is unattainable since it requires an Erlang distribution with near-infinite orders that are difficult to achieve with a computer. However, the graph shows that for a very small *CVTTF*, $E[N(30)]$ is still around 2.5. This indicates the influence of higher moments of lifetime on the expected number of replacements. ■

Replacement is closely related to stock inventory and the risk of stockout, as given in Example 3.58.

Example 3.58 The risk of running out of stocks for a one-component system in Example 3.1 in 1 month as a function of the number of stocks currently available can be calculated as

$$P[N(30) > w + 1] = 1 - P[N(30) \leq w + 1],$$

where w is the number of available stocks in inventory. The results are given in Table 3.19. The table indicates the risk of running out of stock in 1 month when currently no spare available is 0.91, and the risk is lower for higher available stocks.

Table 3.19 Risk of running out of stock for the system in Example 3.1.

w	0	1	2	3
Risk of stockout	0.91	0.58	0.22	0.05

■

The problem in a stock procurement program is the opposite of the description in Example 3.58. In general, management wants to determine the minimum number of stocks, popularly refers to as safety stocks, that must be provided to meet the needs over a period of procurement program t so that the risk of running out of stocks does not exceed a certain tolerated value, for example, $\alpha\%$. If $n_{min}(\alpha)$ is the minimum number of stocks that must be reserved to maintain a maximum risk of running out of α, then $n_{min}(\alpha)$ can be searched recursively so that

$$(P[N(t) > n_{min}(\alpha) - 1] < (1 - \alpha)) \wedge (P[N(t) > n_{min}(\alpha)] \geq (1 - \alpha)) \tag{3.85}$$

is satisfied. Equation (3.85) can be expressed in CDF as

$$\left(F_{S_{n_{min}(\alpha)-1}}(t) < (1 - \alpha)\right) \wedge \left(F_{S_{n_{min}(\alpha)}}(t) \geq (1 - \alpha)\right). \tag{3.86}$$

Example 3.59 Management wants to determine the safety stocks for the one-component system in Example 3.1 for 1 month of operation so that the risk of running out of stocks is not more than 5%. The problem can be stated as finding n that satisfy $F_{n-1}(30) < 0.95$ and $F_n(30) \geq 0.95$.

By referring to Table 3.19, the safety stocks to satisfy the requirement of no more than 5% risk of running out of stocks in a month is

$$n_{min}(0.05) = 3,$$

because the risks of stockout for $w = 3$ is 0.0497 (less than 5%), while for $w = 2$, it is 0.2220 (greater than 5%). ■

Exercise 3.13 Determine the safety stocks for 1 month planning as a function of n if the system lifetime follows n-Erlang distribution with MTTF = 10 days.

The next section discusses a method to incorporate probability of reuse of failed system in calculating the statistics of the number of replacements. Because a failed system can be repaired and reused, the number of replacements is different than the number of failures; thus, it is necessary to derive equations for the number of replacements.

3.10.3 Number of Replacements with Possible Reuse

A failed system is possible to be repaired and reused. If it is repaired and reused, then no replacement is required, and the number of replacements is no longer equal to the number of failures. Define $M(t)$ and $N(t)$ as the number of replacements and failures in an interval $(0, t)$, respectively. Section 3.10.2 considers the case where $M(t) = N(t)$; that is, the system must be replaced every time it fails. Possible reuses make $M(t) \leq N(t)$ since some failed systems can still be reused. Let p be the proportion of, or the probability that, failed systems that cannot be reused (or must be replaced). Assumed that p is a constant for all failures, independent of the system lifetime and history (whether it has been reused). It is also assumed that the repair time is negligible and that repairs restore the system to its original state; that is, after being repaired, it is as good as new.

The solution to this problem requires a distribution of waiting times until replacement. This distribution can be found from the distribution of the system lifetime and the probability that the failed system can be reused. Define W_1 as the waiting time until the first replacement. Possible realizations are as follows:

- $W_1 = T_1$ if the first failure causes replacement. This realization occurs with probability p.
- $W_1 = T_1 + T_2$ if the first failure can be repaired, and the second failure causes replacement. This realization occurs with probability $(1 - p)p$.
- $W_1 = T_1 + T_2 + T_3$ if the first and second failures can be repaired and the third failure causes replacement. This realization occurs with probability $(1 - p)^2 p$.

Continuing in this manner arrives at

$$W_1 = T_1 + T_2 + \cdots + T_K, \tag{3.87}$$

where K is a discrete random variable with $P[K = k] = (1 - p)^{k-1}p$. This probability value shows that K is a geometric random variable with parameter p. After the first replacement, the process repeats from the beginning; thus, $W_2, W_3, \ldots$ follow the same pattern as W_1 and Equation (3.87). Thus, Equation (3.87) expresses the replacement interval W in general, not just for the first replacement.

For a CFR system having an exponentially distributed lifetime with rate λ, the PDF of W can be found using the law of total probability,

$$f_W(t) = \sum_{k=1}^{\infty} f_W(t|K = k)P[K = k], \tag{3.88}$$

where $f_W(t)$ is the PDF of W.

If $K = k$, then W in Equation (3.87) is the sum of k identical exponential random variables. Hence, W has a k-Erlang distribution with PDF

$$f_W(t|K = k) = \frac{\lambda(\lambda t)^{k-1}}{(k-1)!}e^{-\lambda t}, \quad k = 1, 2, \ldots. \tag{3.89}$$

Substituting Equation (3.89) and $P[K = k] = p(1 - p)^{k-1}$ into Equation (3.88) yields

$$\begin{aligned} f_W(t) &= \sum_{k=1}^{\infty} \frac{\lambda(\lambda t)^{k-1}}{(k-1)!}e^{-\lambda t}p(1-p)^{k-1} \\ &= p\lambda e^{-\lambda t}\sum_{k=1}^{\infty} \frac{(\lambda t)^{k-1}}{(k-1)!}(1-p)^{k-1} \\ &= p\lambda e^{-\lambda t}\sum_{k=1}^{\infty} \frac{((1-p)\lambda t)^{k-1}}{(k-1)!}. \end{aligned}$$

The summation on the right hand is the Taylor series of $e^{(1-p)\lambda t}$; hence,

$$f_W(t) = p\lambda e^{-\lambda t}e^{(1-p)\lambda t} = p\lambda e^{-p\lambda t}. \tag{3.90}$$

Thus, the replacement interval W has an exponential distribution with rate $p\lambda$. The replacement rate is certainly lower than the failure rate due to a correction factor $0 \leq p \leq 1$ as the probability that a failure requires replacement. The expected replacement interval can be obtained from Equation (3.90) as

$$MTBR = \frac{1}{p\lambda}, \tag{3.91}$$

or

$$MTBR = \frac{MTTF}{p}. \tag{3.92}$$

Since p is a number between 0 and 1, Equation (3.92) implies

$$MTBR \geq MTTF.$$

By knowing the distribution of the replacement interval, other quantities related to the replacement of CFR systems can be obtained by replacing λ with $p\lambda$:

- The probability that the number of replacements up to time t equal to m is

$$P[M(t) = m] = \frac{(p\lambda t)^m}{m!}e^{-p\lambda t}. \tag{3.93}$$

- The expected number of replacements up to time t is

$$E[M(t)] = p\lambda t = pE[N(t)]. \tag{3.94}$$

- The waiting time until the mth replacement is

$$F_{S_m}(t) = P[S_m \le t] = 1 - \sum_{i=0}^{m-1} \frac{(p\lambda t)^i}{i!} e^{-p\lambda t}. \tag{3.95}$$

Example 3.60 If failed systems in Example 3.53 can be reused with a probability of 0.2, then the probability that there will be three replacements in 1 month can be calculated using Equation (3.93) with $p = 1 - 0.2 = 0.8$ as

$$P[M(t) = 3] = \frac{(0.1 \times 0.8 \times 30)^3}{3!} e^{-0.1 \times 0.8 \times 30} = 0.21.$$

The expected number of replacements in 1 month can be calculated using Equation (3.94) as

$$E[M(30)] = 0.1 \times 0.8 \times 30 = 2.4 = 0.8E[N(t)].$$

The probability that the third replacement will occur in no more than 1 month (30 days) can be calculated using Equation (3.95) as

$$P[S_3 \le 30] = F_{S_3}(30) = 1 - \sum_{i=0}^{2} \frac{(0.1 \times 0.8 \times 30)^i}{i!} e^{-0.1 \times 0.8 \times 30} = 0.43.$$

■

The distribution of replacement interval for a system with a PH-distributed lifetime can be found using Theorem 2.5 and Lemma 2.1. If the PH representation of the system lifetime distribution is $(n, \boldsymbol{a}, \boldsymbol{A})$, and the probability that a failed system must be replaced is p, then the replacement interval of the system has a PH distribution with representation $(n, \boldsymbol{a}, \boldsymbol{A}')$ where

$$\boldsymbol{A}' = \boldsymbol{A} + (1 - p)\boldsymbol{ba} = \boldsymbol{A} - (1 - p)\boldsymbol{Aua}. \tag{3.96}$$

Example 3.61 For a system that has a lifetime following PH distribution of order 1 with representation $(1, [1], [-\lambda])$ and failures require replacement with probability p, the replacement interval has a PH distribution with representation $(1, [1], \boldsymbol{A}')$, where $\boldsymbol{A}'$ can be calculated using Equation (3.96) as

$$\boldsymbol{A}' = [-\lambda] - (1 - p)[-\lambda][1][1] = [-p\lambda],$$

the same as obtained using Equation (3.90), which is an exponential distribution with rate $p\lambda$. ■

Example 3.62 The proportion of the one-component system in Example 3.1 that can be reused when failed is 40%. The PH distribution of replacement intervals has a survival matrix calculated using Equation (3.96) with $p = 1 - 0.4 = 0.6$ as

$$\begin{aligned}\boldsymbol{A}' &= \boldsymbol{A} + (1 - p)\boldsymbol{ba} \\ &= \begin{bmatrix} -0.1 & 0.1 & 0 & 0 \\ 0 & -0.4 & 0.2 & 0.1 \\ 0 & 0 & -0.5 & 0.3 \\ 0 & 0 & 0 & -0.4 \end{bmatrix} + (1 - 0.6)\begin{bmatrix} 0 \\ 0.1 \\ 0.2 \\ 0.4 \end{bmatrix} \begin{bmatrix} 1 & 0 & 0 & 0 \end{bmatrix}\end{aligned}$$

$$= \begin{bmatrix} -0.1 & 0.1 & 0 & 0 \\ 0 & -0.4 & 0.2 & 0.1 \\ 0 & 0 & -0.5 & 0.3 \\ 0 & 0 & 0 & -0.4 \end{bmatrix} + \begin{bmatrix} 0 & 0 & 0 & 0 \\ 0.04 & 0 & 0 & 0 \\ 0.08 & 0 & 0 & 0 \\ 0.16 & 0 & 0 & 0 \end{bmatrix}$$

$$= \begin{bmatrix} -0.1 & 0.1 & 0 & 0 \\ 0.04 & -0.4 & 0.2 & 0.1 \\ 0.08 & 0 & -0.5 & 0.3 \\ 0.16 & 0 & 0 & -0.4 \end{bmatrix}.$$

■

With the PH representation of the replacement interval being known, quantities related to replacements with possible reuse can be calculated using the formulas and methods discussed for replacement without possible repair given in Section 3.10.2. The functions **phFSn, phPn,** and **phEN** for replacement without possible reuse can be modified slightly to accommodate replacements with possible reuse. An additional input of the probability of being reused p is required. For simplicity, **phFSn, phPn,** and **phEN** are made for replacement with possible reuse. Replacement without reuse is a special case for $p = 1$. The following examples were calculated using **phFSn, phPn,** and **phEN.**

Example 3.63 Comparison of $F_{S_3}(t)$ for the one-component system in Example 3.1 for $p = 0, 0.2, 0.5, 0.8, 1.0$ is given in Figure 3.45.

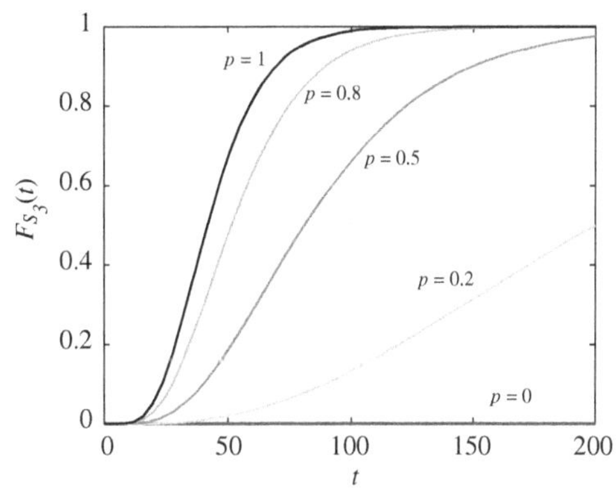

Figure 3.45 $F_{S_3}(t)$ of Example 3.1 for several p.

Graphs in Figure 3.45 show that the greater the p, the higher the $F_{S_3}(t)$. For $p = 0$, all failed systems can always be reused; thus, no system replacement is required, which makes $F_{S_3}(t) = 0$. ■

Example 3.64 The probability that the system in Example 3.1 requires n replacement in one month, $P_m(30)$, is given in Table 3.20 for several p. ■

Example 3.65 The expected number of replacements in 30 days for the one-component system in Example 3.1, $E[M(30)]$, as a function of the probability of failure requiring replacement, p, is given in Figure 3.46.

Table 3.20 Probability of the number of replacements in 1 month for the system in Example 3.1 for several reuse probabilities.

Probability must be replaced (p)	Probability of the number of replacements in 1 month, $P_m(30)$				
	0	1	2	3	4
0.0	1.00	0.00	0.00	0.00	0.00
0.2	0.69	0.27	0.04	0.00	0.00
0.4	0.46	0.40	0.12	0.02	0.00
1.0	0.09	0.33	0.36	0.17	0.04

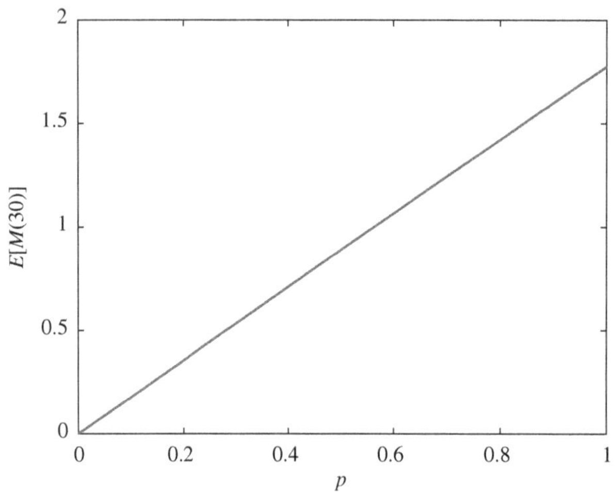

Figure 3.46 $E[M(30)]$ as a function p. ■

Figure 3.46 shows that the expected number of replacements increases linearly with the probability of replacement. This result is clear for the PH distribution of order 1 as shown by Equation (3.94). This linear relationship also applies to high-order PH distributions; this is interesting to investigate further, considering that Equation (3.96) does not clearly show that the proportion of reuse linearly affects the survival matrix of the replacement interval.

The gradient of straight line in Figure 3.46 is certainly the same as $E[N(30)]$ because $M(t) = N(t)$ for $p = 1$. Thus, it can be concluded that systems with a general PH distribution lifetime apply

$$E[M(t)] = pE[N(t)]. \tag{3.97}$$

Equation (3.97) can be proved by defining a Bernoulli random variable,

$$X_i = 1 \text{ if } i^{th} \text{ failure requires replacement}$$
$$= 0 \text{ if } i^{th} \text{ failure requires no replacement}$$

Define

$$Y_{N(t)} = X_1 + X_2 + \cdots + X_{N(t)}, \tag{3.98}$$

as the number of replacements in interval $[0, t]$, which is $M(t)$.

The probability mass function of $M(t)$ can be derived by using the law of total probability with respect to the conditional probability of $N(t)$,

$$\begin{aligned} P[M(t)=m] &= \sum_{n=1}^{\infty} P[Y_{N(t)}=m|N(t)=n]P[N(t)=n] \\ &= \sum_{n=1}^{\infty} P[Y_n=m]P[N(t)=n]. \end{aligned} \tag{3.99}$$

Since X_i is a Bernoulli random variable, Y_n denotes the number of successes of n Bernoulli trials with the probability of success per trial being p. It is well known that Y_n has a binomial distribution with the probability mass function given by

$$P[Y_n=m] = \frac{n!}{m!(n-m)!}p^m(1-p)^{n-m}, \quad m=0,1,\ldots,n, \tag{3.100}$$

with expected value

$$E[Y_n] = \sum_{m=1}^{n} mP[Y_n=m] = np. \tag{3.101}$$

Use Equations (3.98–3.101) to produce

$$\begin{aligned} E[M(t)] &= \sum_{m=1}^{\infty} mP[M(t)=m] = \sum_{m=1}^{\infty} m \sum_{n=1}^{\infty} P[Y_{N(t)}=m|N(t)=n]P[N(t)=n] \\ &= \sum_{n=1}^{\infty}\left(\sum_{m=1}^{n} mP[Y_n=m]\right)P[N(t)=n] = \sum_{n=1}^{\infty} npP[N(t)=n] \\ &= p\sum_{n=1}^{\infty} nP[N(t)=n] = pE[N(t)], \end{aligned}$$

as given by Equation (3.97) to be proven.

Results of Section 3.10 obviously can be employed to enhance the results of Section 3.9. For example, finding the optimum replacement interval for systems with possible reuse can use the same optimization model by replacing the survival matrix without possible reuse in the formulation with the survival matrix with possible reuse. The problem will be simple if there is no cost associated with reuse. The more interesting problem would be adding a cost for reuse with a lower value than the replacement cost. Likewise, calculating mission reliability with replacement can be enhanced with a combination of replacement and reuse. These interesting problems are left as challenges for interested readers.

Exercise 3.14 Determine the effect of probability of reuse to the number of safety stocks required for a given risk of stockout. Use any IFR distributions as the system lifetime distribution.

The next section discusses a method to derive the PH distribution of system lifetime by assuming that the system failed due to the random shocks experience by the system.

3.11 Using Shock Models to Determine PH Distribution[2]

Shock models are widely used in reliability analyses to model the system lifetime distribution. Shock models are collections of models to describe the randomness of shocks experienced

2 Optional material, can be omitted in lecture.

by a system. These models include shock interarrival models, shock strength models, and shock consequence models [11]. Shocks experienced by the system are described according to the shock arrival interval distribution, each shock has a random strength following a shock strength distribution, and each shock has a different effect on the system states. With certain criteria, shocks experienced by a system can cause the system to fail. These criteria can be related to the arrival interval, strength, number of shocks, and shock consequences. The time until the system fails, or the system lifetime, is a random variable whose randomness depends on the randomness of shocks experienced by the system.

Many variations of shock models can be used to describe the randomness of the system lifetime. Generally, the random quantities associated with shocks are assumed to be independent. For more complex models, these quantities depend on each other, such as the dependence of interarrival and strength. Simple shock models assume an exponential or Erlang distribution of interarrival time that allows them to be analyzed as Markov processes [12], and they can be extended to handle cases with PH distributions [13]. Such shock models can be expressed in the state transition diagrams, and the system lifetime distribution can be derived from them.

In addition to shocks, failure can also occur due to wear. The difference between the two is that the shock model describes the influence of external factors, such as pressure, voltage, temperature, and humidity, while the wear model describes the influence of internal factors on system conditions, such as aging and shrinkage [14]. The two are often undistinguishable in systems that are not too sensitive and complex.

The PH distribution of system lifetime can be derived from the random shocks experienced by the system that may cause to degraded and failed. The shock models can be expressed in the state transition diagrams. The discussion in this section starts with a scalar transition and then followed by the matrix transition.

3.11.1 Scalar Transition

State transition diagrams with scalar transitions can represent shock models with interarrival times following distributions that can be represented in scalar transition, such as exponential, Erlang, and hypoexponential.

Example 3.66 A system experiences two types of shocks that arrive randomly and independently. The interarrival time of shock type 1 is exponentially distributed with rate λ_1 and with probability p_1 causing the system state to degrade by one level. The interarrival time of shock type 2 is exponentially distributed with rate λ_2 and with probability p_2 causing the system state to degrade by two levels. It can be said that the magnitude of shock type 1 is one unit because it causes the system state to degrade by one level, and the magnitude of shock type 2 is two units because it causes the system state to degrade by two levels. The system fails if the system state degrades by at least three levels.

Since the interarrival times of shocks that cause the system to undergo state changes have exponential distribution, the process underlying the changes in the system states is CTMC. The state transition diagram using the transition rates for this CTMC is given in Figure 3.47.

The system has four states comprising of three transient states and one absorbing state because it requires at least three-level deterioration to cause it to fail. The states can be represented by $k = 0$, 1, 2, 3+. State 3+ is the failed state to represent the system experiencing three or more levels of deterioration. The deterioration due to shock type 1 is one level, while shock type 2 causes an instant deterioration of two levels. Using Equation (3.90), the rate of deterioration due to shock

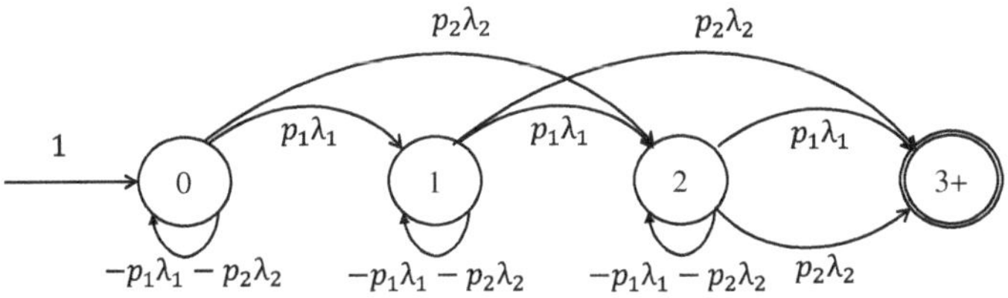

Figure 3.47 State transition diagram of the system in Example 3.66.

type 1 is $p_1\lambda_1$, which is the product of the rate of shock type 1 and the probability that this shock causes a deterioration. Likewise, the rate of deterioration due to shock type 2 is $p_2\lambda_2$. Shock type 2 causes the system to fail when the system is in state 2, because it makes the system state drop two levels, more than necessary to make it fail. The system lifetime can be determined from this state transition diagram to have a PH distribution with representation

$$n = 3, \quad \boldsymbol{a} = [1\ \ 0\ \ 0], \quad \boldsymbol{A} = \begin{bmatrix} -p_1\lambda_1 - p_2\lambda_2 & p_1\lambda_1 & p_2\lambda_2 \\ 0 & -p_1\lambda_1 - p_2\lambda_2 & p_1\lambda_1 \\ 0 & 0 & -p_1\lambda_1 - p_2\lambda_2 \end{bmatrix}.$$

Knowing the system lifetime distribution, all system reliability measures including replacement and maintenance can be determined. ∎

Example 3.67 Data on shock type 1 and shock type 2 in Example 3.66 are $\lambda_1 = 0.2$/day, $\lambda_2 = 0.1$/day, $p_1 = 0.4$, and $p_2 = 0.2$.

Substituting these parameter values into the survival matrix resulting from Example 3.66 obtains

$$\boldsymbol{A} = \begin{bmatrix} -0.1 & 0.08 & 0.02 \\ 0 & -0.1 & 0.08 \\ 0 & 0 & 0.1 \end{bmatrix}.$$

If the system must be replaced with a probability of 0.5 when it fails, then the survival matrix of the PH distribution of the replacement interval can be obtained from Equation (3.96) as

$$\begin{aligned} \boldsymbol{A} - (1 - 0.5)\boldsymbol{Aua} &= \begin{bmatrix} -0.1 & 0.08 & 0.02 \\ 0 & -0.1 & 0.08 \\ 0 & 0 & 0.1 \end{bmatrix} - 0.5\begin{bmatrix} 0 \\ -0.02 \\ -0.1 \end{bmatrix}[1\ \ 0\ \ 0] \\ &= \begin{bmatrix} -0.1 & 0.08 & 0.02 \\ 0.01 & -0.1 & 0.08 \\ 0.05 & 0 & -0.1 \end{bmatrix}. \end{aligned}$$

∎

It is possible to incorporate random shock magnitude into the model as Example 3.68 shows.

Example 3.68 A system receives a shock that has random interarrival time and magnitude. The shock interarrival time is exponentially distributed with rate λ. The shock magnitude has random discrete values equal to 1 with probability p_1 and equal to 2 with probability p_2. There is a possibility that the shock does not cause any change to the system state. The system fails when the cumulative shock magnitudes received are three or more.

This shock can be modeled using a state transition diagram in transition probability. The system state k is for the cumulative shock magnitude received by the system. Hence, $k = 0, 1, 2, 3+$, with $k = 0$ for the new system and $k = 3+$ for a failed system since the system fails when the cumulative shocks received are three or more. The probability that the shock does not cause any deterioration to the system state is $1 - p_1 - p_2$ so that the total probability of leaving any state is equal to 1. The resulting state transition diagram in transition probability based on the shock characteristics is shown in Figure 3.48.

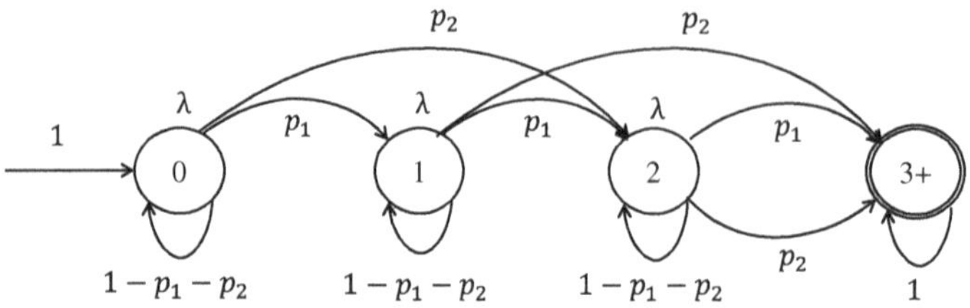

Figure 3.48 State transition diagram in transition probability for Example 3.68.

The state transition diagram using transition probability in Figure 3.48 is different than the common state transition diagram for CTMC, which has no transition from one node to itself. It is because CTMC is observed when the process is making a transition out from each state, while the state transition diagram given in Figure 3.48 is based on the transitions due to the arrival of shocks. However, the state transition diagram in transition rate is the same as for CTMC.

The PH representation of the system lifetime distribution can be derived from its state transition diagram but in transition rate. Convert Figure 3.48 to the state transition diagram in transition rate to yield Figure 3.49, which is equivalent to the state transition diagram in transition rate for CTMC.

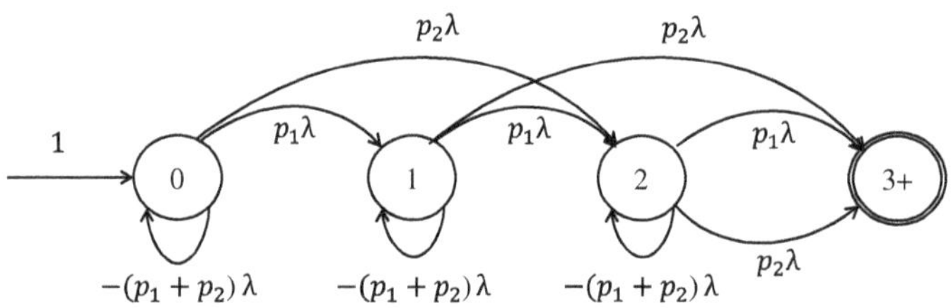

Figure 3.49 State transition diagram in transition rate for Example 3.68.

The similarity between Figures 3.47 and 3.49 shows that shocks with independent random discrete magnitude can be treated as several independent shocks, one for each magnitude value.

■

It is possible to extend the scalar model for interarrival time following nonexponential distribution as long as they have a scalar CTMC representation.

Example 3.69 A system receives two types of shocks with the same magnitude but different interarrival time distributions. The interarrival times of shocks type 1 follow the exponential distribution with mean interarrival time of 25 days, while shocks type 2 follows the 2-Erlang distribution with mean interarrival time 20 days. The system fails when exposed to two shocks, irrespective of their types.

The system states can be described as k, $k = 0, 1, 2$. The transition rate for shock type 1 is $1/25 = 0.04$/day. To model the transitions due to shock type 2, introduce dummy states to accommodate the two transitions inside the 2-Erlang distribution. Each transition rate inside the 2-Erlang

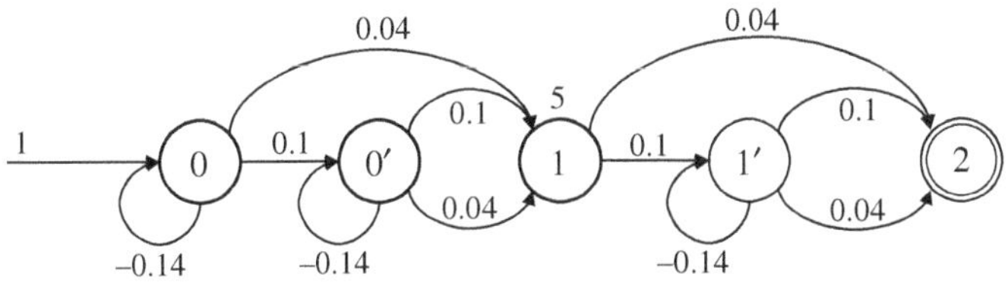

Figure 3.50 State transition diagram for the system in Example 3.69.

distribution is 2/20 days = 0.1/day. Shock type 1 may arrive while shock type 2 is in the dummy state. The resulting state transition diagram in transition rate is presented as Figure 3.50, where states 0′ and 1′ are dummy states.

The system has five states, four transient states, and one absorbing state, including two dummy states 0′ and 1′ for representing 2-Erlang distribution. The PH representation for the system lifetime distribution can be generated from Figure 3.50 as

$$n = 4, \quad \boldsymbol{a} = [1 \;\; 0 \;\; 0 \;\; 0], \quad \boldsymbol{A} = \begin{bmatrix} -0.14 & 0.1 & 0.04 & 0 \\ 0 & -0.14 & 0.14 & 0 \\ 0 & 0 & -0.14 & 0.1 \\ 0 & 0 & 0 & -0.14 \end{bmatrix}.$$

The second and fourth rows in the survival matrix $\boldsymbol{A}$ are for the dummy states. ∎

Although possible, this example shows the difficulty in using scalar transitions to describe the state transition diagram of a system experiencing shocks with nonexponential interarrival distribution. If the interarrival time of shock type 2 has n-Erlang distribution, the number of system states becomes $2n + 1$, and if the system fails at the kth shocks, the number of system states becomes $kn + 1$. Using a matrix transition will eliminate this difficulty, as shown in the next section.

3.11.2 Matrix Transition

If shock interarrival time follows PH distribution, then the shock model can be described using matrix transitions in a state transition diagram. For a simple shock model that can be described using two system states, good and failed, the state transition diagram with matrix transitions as given in Figure 3.1 is sufficient. Systems with simple shock models in PH distributions are versatile. Below are several shock models with independent shock interarrival time following PH distributions $V \sim PH(n,\, \boldsymbol{a},\, \boldsymbol{A})$. Closure properties of PH distributions are utilized to determine the distribution of the system lifetime from the shock models.

1) The system immediately fails whenever it receives a shock.
   In this case, the system lifetime is identical to the shock interarrival time,

$$T \sim PH(n, \boldsymbol{a}, \boldsymbol{A}). \tag{3.102}$$

2) The system fails with probability p every time it receives a shock.
   From Lemma 2.1, the system lifetime distribution is

$$T \sim PH(n, \boldsymbol{a}, \boldsymbol{A} - (1 - p)\boldsymbol{Aua}). \tag{3.103}$$

3) The system experiences N various types of shocks, the interarrival time for shock type i is $V_i \sim PH(n_i, \boldsymbol{a}_i, \boldsymbol{A}_i)$, $i = 1, 2, \ldots, N$. The system fails whenever it receives any type of shock.

The system lifetime is equal to the time of the first arrival of the shocks, which is the minimum of the interarrival times of all the shocks. Thus,

$$T = \min(V_1, V_2, \dots, V_N).$$

The system lifetime distribution can be determined using Theorem 2.1 as $T \sim PH(n_T, \boldsymbol{a}_T, \boldsymbol{A}_T)$, where

$$n_T = n_1 n_2 \dots n_N, \quad \boldsymbol{a}_T = \boldsymbol{a}_1 \otimes \boldsymbol{a}_2 \otimes \dots \otimes \boldsymbol{a}_N,$$
$$\boldsymbol{A}_T = \boldsymbol{A}_1 \oplus \boldsymbol{A}_2 \oplus \dots \oplus \boldsymbol{A}_N. \tag{3.104}$$

4) The system experiences N types of shocks with shock type i has an interarrival time $V_i \sim PH(n_i, \boldsymbol{a}_i, \boldsymbol{A}_i)$, $i = 1, 2, \dots, N$. The shock type i causes the system to fail with probability p_i.
   Using a combination of models in points 2 and 3 yields $T \sim PH(n_T, \boldsymbol{a}_T, \boldsymbol{A}_T)$, where

$$n_T = n_1 n_2 \dots n_N, \quad \boldsymbol{a}_T = \boldsymbol{a}_1 \otimes \boldsymbol{a}_2 \otimes \dots \otimes \boldsymbol{a}_N, \quad q_i = 1 - p_i,$$

$$\boldsymbol{A}_T = (\boldsymbol{A}_1 - q_1 \boldsymbol{A}_1 \boldsymbol{u}\boldsymbol{a}_1) \oplus (\boldsymbol{A}_2 - q_2 \boldsymbol{A}_2 \boldsymbol{u}\boldsymbol{a}_2) \oplus \dots \oplus (\boldsymbol{A}_N - q_N \boldsymbol{A}_N \boldsymbol{u}\boldsymbol{a}_N). \tag{3.105}$$

5) The system receives shocks, and it fails at the time it receives the Nth shocks.
   The system lifetime is the sum of the interarrival times of the first N shocks,

$$T = V_1 + V_2 + \dots + V_N.$$

The system lifetime distribution $T \sim PH(n_T, \boldsymbol{a}_T, \boldsymbol{A}_T)$ can be determined using Lemma 3.1 recursively.

Example 3.70 A system receives two types of shocks. The interarrival times of shocks type 1 and type 2 have 2-Erlang distributions with mean interarrival time of 20 days and 40 days, respectively. Shocks type 1 and shock type 2 cause the system to fail with probabilities 0.2 and 0.6, respectively.

Let $V_1 \sim PH(n_1, \boldsymbol{a}_1, \boldsymbol{A}_1)$ and $V_2 \sim PH(n_2, \boldsymbol{a}_2, \boldsymbol{A}_2)$ be the interarrival times of shock type 1 and 2, respectively. Their PH representations are

$$n_1 = n_2 = 2, \quad \boldsymbol{a}_1 = \boldsymbol{a}_2 = [1 \ \ 0], \quad \boldsymbol{A}_1 = \begin{bmatrix} -0.1 & 0.1 \\ 0 & -0.1 \end{bmatrix}, \quad \boldsymbol{A}_2 = \begin{bmatrix} -0.05 & 0.05 \\ 0 & -0.05 \end{bmatrix}.$$

Use Equation (3.105) from the shock model in point 4) with $q_1 = 0.8$ and $q_2 = 0.4$ to generate a PH representation of the system lifetime distribution.

$$n_T = 2 + 2 = 4, \quad \boldsymbol{a}_T = \boldsymbol{a}_1 \otimes \boldsymbol{a}_2 = [1 \ \ 0] \otimes [1 \ \ 0] = [1 \ \ 0 \ \ 0 \ \ 0].$$

$$\begin{aligned}
\boldsymbol{A}_1 - q_1 \boldsymbol{A}_1 \boldsymbol{u}_1 \boldsymbol{a}_1 &= \begin{bmatrix} -0.1 & 0.1 \\ 0 & -0.1 \end{bmatrix} - 0.8 \begin{bmatrix} -0.1 & 0.1 \\ 0 & -0.1 \end{bmatrix} \begin{bmatrix} 1 \\ 1 \end{bmatrix} [1 \ \ 0] \\
&= \begin{bmatrix} -0.1 & 0.1 \\ 0 & -0.1 \end{bmatrix} - 0.8 \begin{bmatrix} 0 & 0 \\ -0.1 & 0 \end{bmatrix} = \begin{bmatrix} -0.1 & 0.1 \\ 0.08 & -0.1 \end{bmatrix}.
\end{aligned}$$

$$\begin{aligned}
\boldsymbol{A}_2 - q_2 \boldsymbol{A}_2 \boldsymbol{u}_2 \boldsymbol{a}_2 &= \begin{bmatrix} -0.05 & 0.05 \\ 0 & -0.05 \end{bmatrix} - 0.4 \begin{bmatrix} -0.05 & 0.05 \\ 0 & -0.05 \end{bmatrix} \begin{bmatrix} 1 \\ 1 \end{bmatrix} [1 \ \ 0] \\
&= \begin{bmatrix} -0.05 & 0.05 \\ 0 & -0.05 \end{bmatrix} - 0.4 \begin{bmatrix} 0 & 0 \\ -0.05 & 0 \end{bmatrix} = \begin{bmatrix} -0.05 & 0.05 \\ 0.02 & -0.05 \end{bmatrix}.
\end{aligned}$$

$$\begin{aligned}
\boldsymbol{A}_T &= (\boldsymbol{A}_1 - q_1 \boldsymbol{A}_1 \boldsymbol{u}_1 \boldsymbol{a}_1) \oplus (\boldsymbol{A}_2 - q_2 \boldsymbol{A}_2 \boldsymbol{u}_2 \boldsymbol{a}_2) \\
&= \begin{bmatrix} -0.1 & 0.1 \\ 0.08 & -0.1 \end{bmatrix} \oplus \begin{bmatrix} -0.05 & 0.05 \\ 0.02 & -0.05 \end{bmatrix}
\end{aligned}$$

$$= \begin{bmatrix} -0.1 & 0.1 \\ 0.08 & -0.1 \end{bmatrix} \otimes \begin{bmatrix} 1 & 0 \\ 0 & 1 \end{bmatrix} + \begin{bmatrix} 1 & 0 \\ 0 & 1 \end{bmatrix} \otimes \begin{bmatrix} -0.05 & 0.05 \\ 0.02 & -0.05 \end{bmatrix}$$

$$= \begin{bmatrix} -0.1\begin{bmatrix} 1 & 0 \\ 0 & 1 \end{bmatrix} & 0.1\begin{bmatrix} 1 & 0 \\ 0 & 1 \end{bmatrix} \\ 0.08\begin{bmatrix} 1 & 0 \\ 0 & 1 \end{bmatrix} & -0.1\begin{bmatrix} 1 & 0 \\ 0 & 1 \end{bmatrix} \end{bmatrix} + \begin{bmatrix} 1\begin{bmatrix} -0.05 & 0.05 \\ 0.02 & -0.05 \end{bmatrix} & 0\begin{bmatrix} -0.05 & 0.05 \\ 0.02 & -0.05 \end{bmatrix} \\ 0\begin{bmatrix} -0.05 & 0.05 \\ 0.02 & -0.05 \end{bmatrix} & 1\begin{bmatrix} -0.05 & 0.05 \\ 0.02 & -0.05 \end{bmatrix} \end{bmatrix}$$

$$= \begin{bmatrix} -0.15 & 0.05 & 0.1 & 0 \\ 0.02 & -0.15 & 0 & 0.1 \\ 0.08 & 0 & -0.15 & 0.05 \\ 0 & 0.08 & 0.02 & -0.15 \end{bmatrix}.$$

The system lifetime has a PH distribution with representation $(n_T, \boldsymbol{a}_T, \boldsymbol{A}_T)$.

If the system can be repaired and reused with a probability of 0.8, then the PH representation of the replacement interval for this system is

$$n_S = 4, \quad \boldsymbol{a}_S = [1 \ \ 0 \ \ 0 \ \ 0],$$

$$\boldsymbol{A}_S = \boldsymbol{A}_T - q\boldsymbol{A}_T\boldsymbol{ua} = \boldsymbol{A}_T - 0.8\boldsymbol{A}_T\boldsymbol{ua}$$

$$= \begin{bmatrix} -0.15 & 0.05 & 0.1 & 0 \\ 0.02 & -0.15 & 0 & 0.1 \\ 0.08 & 0 & -0.15 & 0.05 \\ 0 & 0.08 & 0.02 & -0.15 \end{bmatrix}$$

$$- 0.8\begin{bmatrix} -0.15 & 0.05 & 0.1 & 0 \\ 0.02 & -0.15 & 0 & 0.1 \\ 0.08 & 0 & -0.15 & 0.05 \\ 0 & 0.08 & 0.02 & -0.15 \end{bmatrix}\begin{bmatrix} 1 \\ 1 \\ 1 \\ 1 \end{bmatrix}[1 \ \ 0 \ \ 0 \ \ 0]$$

$$= \begin{bmatrix} -0.15 & 0.05 & 0.1 & 0 \\ 0.044 & -0.15 & 0 & 0.1 \\ 0.096 & 0 & -0.15 & 0.05 \\ 0.04 & 0.08 & 0.02 & -0.15 \end{bmatrix}.$$

■

The PH distribution of the system lifetime in Example 3.69 can be derived directly from the interarrival time distribution of each shock type and the probability that each shock type may cause system failure. The state transition diagram is unnecessary to derive the PH representation. More complicated shock models need to be represented in a state transition diagram to find the PH distribution of the system lifetime. One of them is shocks that cause the system to undergo several states before it fails. However, the models mentioned earlier are still applicable to simplify the state transition diagram.

Example 3.71 Matrix transition can be applied to Example 3.69. The distributions of interarrival times for shock types 1 and 2 have PH representations $(n_1, \boldsymbol{a}_1, \boldsymbol{A}_1)$ and $(n_2, \boldsymbol{a}_2, \boldsymbol{A}_2)$, respectively, with

$$n_1 = 1, \quad \boldsymbol{a}_1 = [1], \quad \boldsymbol{A}_1 = [-0.04], \quad \boldsymbol{b}_1 = -\boldsymbol{A}_1\boldsymbol{u} = [0.04].$$

$$n_2 = 2, \quad \boldsymbol{a}_2 = [1 \ \ 0], \quad \boldsymbol{A}_2 = \begin{bmatrix} -0.1 & 0.1 \\ 0 & -0.1 \end{bmatrix}, \quad \boldsymbol{b}_2 = -\boldsymbol{A}_2\boldsymbol{u} = \begin{bmatrix} 0 \\ 0.1 \end{bmatrix}.$$

Using the model in point 3, the time S until the system state degraded by one level is the minimum of the interarrival times of the two shocks. Hence, S has PH distribution with representation

$$n_S = n_1 n_2 = 2, \quad \boldsymbol{a}_S = \boldsymbol{a}_1 \otimes \boldsymbol{a}_2 = [1] \otimes \begin{bmatrix} 1 & 0 \end{bmatrix} = \begin{bmatrix} 1 & 0 \end{bmatrix},$$

$$\boldsymbol{A}_S = \boldsymbol{A}_1 \oplus \boldsymbol{A}_2 = [-0.04] \oplus \begin{bmatrix} -0.1 & 0.1 \\ 0 & -0.1 \end{bmatrix} = \begin{bmatrix} -0.14 & 0.1 \\ 0 & -0.14 \end{bmatrix}.$$

Since the system fails when it is degraded two levels, the system lifetime T is the sum of S_1 and S_2, two identical and independent PH-distributed variables with PH representation $(n_S, \boldsymbol{a}_S, \boldsymbol{A}_S)$. PH representation of T can be calculated using Lemma 3.1 as

$$n_T = n_S + n_S = 4, \quad \boldsymbol{a}_T = [\boldsymbol{a}_S\, \boldsymbol{0}] = \begin{bmatrix} 1 & 0 & 0 & 0 \end{bmatrix},$$

$$\boldsymbol{b}_S \boldsymbol{a}_S = -\boldsymbol{A}_S \boldsymbol{u} \boldsymbol{a}_S = -\begin{bmatrix} -0.14 & 0.1 \\ 0 & -0.14 \end{bmatrix} \begin{bmatrix} 1 \\ 1 \end{bmatrix} \begin{bmatrix} 1 & 0 \end{bmatrix} = \begin{bmatrix} 0.04 & 0 \\ 0.14 & 0 \end{bmatrix},$$

$$\boldsymbol{A}_T = \begin{bmatrix} \boldsymbol{A}_S & \boldsymbol{b}_S \boldsymbol{a}_S \\ 0 & \boldsymbol{A}_S \end{bmatrix} = \begin{bmatrix} -0.14 & 0.1 & 0.04 & 0 \\ 0 & -0.14 & 0.14 & 0 \\ 0 & 0 & -0.14 & -0.1 \\ 0 & 0 & 0 & -0.14 \end{bmatrix}.$$

The resulting PH representation is the same as obtained using the scalar transition given in Example 3.69. However, using matrix transition offers simplicity when shocks involved have a higher-order PH representation or more complicated distribution than Erlang, such as Coxian. ■

Research related to shock and wear models is still being developed to produce more complex models and include dependencies between variables in the model. Some recent studies related to this problem are Goyal and Xie [15], Geng et al. [16], Shamstabar et al. [17], and Ranjkesh et al. [1], among others.

Reliability measures that have been developed for one-component systems in this chapter will be used extensively for systems having several components arranged in several structures in the following chapters.

3.12 Problems

3.1 Determine the PH representation of Example 3.1 if the duration of stay in state 1 (“None” state) has 2-Erlang distribution with the same mean.

3.2 Prove numerically that if the maintenance rate in Example 3.3 can restore the “Many” defects state to the “Few” defect state, then the reliability and hazard functions are stochastically higher and lower, respectively, than in Example 3.3.

3.3 Prove that the system *MTTF* in Problem 3.2 is longer than in Table 3.6.

3.4 Derive an equation for the residual hazard function. Use the resulting equation to make a comparison between the residual hazard function and the hazard function of the system in Example 3.1. What conclusions can be drawn from this? Try several ages.

3.5 Calculate the average duration of stay in each state for the one-component system in Example 3.1. Compare the total value of the expected duration of stay in all states with the expected lifetime. Explain your result. Devise a method to use the information on the expected duration of stay in each state to find the expected lifetime.

3.6 A system has lifetime following PH distribution as a mixture of an exponential distribution with a mean value of 1 with a probability of 0.8 and a 2-Erlang distribution with a mean value of 1. Determine the PH distribution of system lifetime and use its representation to plot the hazard and reliability functions and their residuals. What conclusions (many) can be drawn from these graphs?

3.7 Prove that the 2-hyperexponential distribution has $CVTTF > 1$. Extend your proof to n-hyperexponentials.

3.8 Calculate the age of a system with a 2-Erlang distributed lifetime when its residual life expectancy is 3/4 of its expected lifetime. Check your result by using *MRL* graph obtained using the provided *MRL* function.

3.9 Prove mathematically that the residual life expectancy of a system with a 2-Erlang distributed lifetime is never less than a half of its expected lifetime.

3.10 The graphs of the *MRL* functions of the systems in Examples 3.1–3.3 converge to a certain value. Compare these values with the convergent values of their hazard functions.

3.11 Find the relationship between the limits of the hazard and *MRL* functions as $t \to \infty$. Are they related to the *MTTF*?

3.12 PH distributions are used as models for lifetime data with mean =10 and coefficient of variation = 0.8. Determine the minimum value of n (PH order) and the appropriate parameter values if the model is

a) n-Hypoexponential
b) n-Coxian
c) n-Erlang

(The values are not unique; many satisfy the two requirements.)

3.13 Determine the mission reliability of a system having 2-Erlang distributed lifetime with $MTTF = 10$ days to execute a mission with random duration having the following distributions:

a) Uniformly distributed between 0 and 2 days.
b) Exponentially distributed with mean value 1 day.
c) 2-Erlang distributed with mean value 1 day

Which one of the three has the highest mission reliability? Give a plausible justification for the result.

3.14 Graph the relationship between the reliability of the system in Example 3.1 for a 20-day mission and several maintenance intervals if the maintenance is perfect. Use the maintenance intervals as legends for the graphs.

3.15 Create a graph of *MTTF* of system in Example 3.1 as a function of maintenance interval with maintenance levels as the legends.

3.16 Create a graph of $MTTF$ of the system in Example 3.1 as a function of maintenance level with maintenance intervals as the legends.

3.17 Modify the available program to calculate the system reliability function if the maintenance levels are not constant, instead the levels are
a) Alternating with perfect after every two imperfect maintenances.
b) Higher for later maintenance (define your function).
c) Higher for longer maintenance interval (define your function).
Use your programs to calculate and compare the system $MTTF$ under several maintenance level policies given above.

3.18 Compare mission reliability over 25 days using the system in Example 3.3 without and with planned replacements every 10 and 15 days.

3.19 Create a main program to call the program for $MTBR$ so that it can be used to create a matrix $MTBR(s, n)$, where s is the replacement interval and n is the order of the Erlang distribution of system lifetime with $MTTF = 100$.

3.20 Determine the optimal replacement interval for a system with a reliability function $R(t) = e^{-0.0001t^2}$ and a ratio of replacement cost to failure cost of 0.1. Compare the results of calculations using the Weibull distribution and the PH distribution approximation for several orders.

3.21 Brake pad in a car must be replaced regularly to protect the brake clutch. The price of a brake clutch is 20 times more expensive than a brake pad. The lifetime of brake pad has 6-Erlang distribution with mean 10 000 km. Determine the optimum replacement interval for bake pad. Plot the values of optimum replacement interval with respect to the ratio of the price of brake clutch to brake pad.

3.22 Prove that if the replacement interval is shorter than the maintenance interval then the maintenance program does not affect system performance.

3.23 Derive the expression for the reliability function of the system under an integrated maintenance-replacement program.

3.24 Prove that if the maintenance level is zero then the maintenance interval does not affect system reliability.

3.25 Determine the PH representation for the shock model of Example 3.69 if the interarrival times of shock type 1 have a 2-Erlang distribution with the same mean value as Example 3.69.

3.26 Determine the PH representation for the system lifetime distribution in Example 3.69 if shock type 1 has random magnitude with probability 0.6 is equal to 1, and the rest is equal to 2.

References

1 Ranjkesh, S.H., Hamadani, A.Z., Mahmoodi, S., 2019. A new cumulative shock model with damage and inter-arrival time dependency. Reliab Eng Syst Saf 192, 106 047. https://doi.org/10.1016/j.ress.2018.01.006.

2 Alkaff, A., Qomarudin, M.N., 2020. Modeling and analysis of system reliability using phase-type distribution closure properties. Appl Stoch Models Bus Ind 36, 548–569. https://doi.org/10.1002/asmb.2509.

3 Lewis, E.E., 1987. Introduction to Reliability Engineering. Wiley, New York.

4 Trivedi, K.S., Bobbio, A., 2017. Reliability and Availability Engineering: Modeling, Analysis, and Applications. Cambridge University Press, Cambridge.

5 Elsayed, E.A., 2021. Reliability Engineering, 3rd ed. Wiley, Hoboken, NJ.

6 Shaked, M., Shanthikumar, J.G., 2007. Stochastic Orders. Springer, New York.

7 Faddy, M.J., 1995. Phase-type distributions for failure times. Math Comput Model 22, 63–70. https://doi.org/10.1016/0895-7177(95)00181-Z.

8 Pham, H., Wang, H., 1996. Imperfect maintenance. Eur J Oper Res 94, 425–438. https://doi.org/10.1016/S0377-2217(96)00099-9.

9 Nakagawa, T., 1988. Sequential imperfect preventive maintenance policies. IEEE Trans Reliab 37, 295–298. https://doi.org/10.1109/24.3758.

10 Abdel-Hameed, M., 1987. Inspection and maintenance policies of devices subject to deterioration. Adv Appl Probab 19, 917–931. https://doi.org/10.2307/1427108.

11 Eryilmaz, S., Tekin, M., 2019. Reliability evaluation of a system under a mixed shock model. J Comput Appl Math 352, 255–261. https://doi.org/10.1016/j.cam.2018.12.011.

12 Pérez-Ocón, R., Segovia, M., 2009. Shock models under a Markovian arrival process. Math Comput Model 50, 879–884. https://doi.org/10.1016/j.mcm.2008.12.020.

13 Riascos-Ochoa, J., Sanchez-Silva, M., Akhavan-Tabatabaei, R., 2014. Reliability analysis of shock-based deterioration using phase-type distributions. Probabilistic Engineering Mechanics 38, 88–101. https://doi.org/10.1016/j.probengmech.2014.09.004.

14 Montoro-Cazorla, D., Pérez-Ocón, R., 2015. A reliability system under cumulative shocks governed by a BMAP. Appl Math Model 39, 7620–7629. https://doi.org/10.1016/j.apm.2015.03.066.

15 Goyal, D., Xie, M., 2024. On the compound Poisson phase-type process and its application in shock models. J Comput Appl Math 446, 115 852. https://doi.org/10.1016/j.cam.2024.115852.

16 Geng, Y., Wang, S., Shi, J., Zhang, Y., Wang, W., 2023. Reliability modeling of phased degradation under external shocks. Reliab Eng Syst Saf 239, 109 524. https://doi.org/10.1016/j.ress.2023.109524.

17 Shamstabar, Y., Shahriari, H., Samimi, Y., 2021. Reliability monitoring of systems with cumulative shock-based deterioration process. Reliab Eng Syst Saf 216, 107 937. https://doi.org/10.1016/j.ress.2021.107937.

4

Basic Structured Systems

Learning objectives:

1) Understand the derivations and proofs of all equations used for modeling series, parallel, and perfect and imperfect switch standby systems.
2) Understand the characteristics of each basic structure and combinations of them.
3) Able to construct reliability models of complex systems composed of many basic structures.
4) Able to utilize the software for reliability analyses of complex systems by generating and evaluating functional and statistical system reliability measures.

Abbreviations

BFR	bathtub failure rate
CFR	constant failure rate
CVTTF	coefficient of variations of time to failure
DFR	decreasing failure rate
IFR	increasing failure rate
ME	matrix-exponential
MRL	mean residual life
MTTF	mean time to failure
PH	phase-type
SDTTF	standard deviation of time to failure
VTTF	variance of time to failure

Notations

T	time to failure
t	time, mission duration
g	age
n	order, dimension
$\boldsymbol{a}$	initial probability vector
$\boldsymbol{A}$	survival matrix

$\boldsymbol{I}$	identity matrix
$\boldsymbol{u}$	unitary vector
$\boldsymbol{b}$	failure probability vector
$\boldsymbol{Q}$	infinitesimal generator
N	number of components in a system
$(n_i, \boldsymbol{a}_i, \boldsymbol{A}_i)$	PH representation of lifetime distribution of component i
$(n_i, \boldsymbol{a}_i, \boldsymbol{A}_i, \boldsymbol{v}_i)$	ME representation of lifetime distribution of component i
$(n_S, \boldsymbol{a}_S, \boldsymbol{A}_S)$	PH representation of system lifetime distribution
$(n_S, \boldsymbol{a}_S, \boldsymbol{A}_S, \boldsymbol{v}_S)$	ME representation of system lifetime distribution
$E(.)$	expected value
$R(t)$	reliability function
$h(t)$	hazard function
$RR(t \mid g)$	residual reliability function at age g
$MRL(g)$	mean residual life function
$\otimes$	Kronecker product
$\oplus$	Kronecker sum

The ideal way to analyze the reliability of a system is to use the system reliability function. As shown in Chapter 3, all other reliability measures, such as hazard function, residual reliability function, lifetime moments, and *MRL* function, can be derived or calculated easily once the system reliability function is obtained. The use of reliability functions also allows dynamic analyses of system reliability to show the system reliability value changes over time. There are system reliability measures that can be derived only from reliability function and cannot be obtained from static reliability analysis. If desired, static analysis can be performed simply by inserting certain time values into the system reliability function obtained. The urgency of utilizing system reliability functions in dynamic analyses of system reliability in a real application, which cannot be done using reliability values in static analyses of system reliability, is shown in the study of Ardakan et al. [1].

Deriving the system reliability function from its component reliability functions is a classic problem of reliability theory and engineering. Equations expressing the relationship between the system reliability function and the component reliability functions are found in many reliability textbooks, such as Elsayed [2] and Trivedi and Bobbio [3]. However, these equations require manual derivation to obtain the system reliability function from the reliability functions of its components. With the advancement of computer technology and its software, it is now possible to produce the system reliability function as an output from component reliability functions as inputs by utilizing symbolic processing utilities available in many software packages such as MATLAB. However, the result is a long, complicated, and unstructured mathematical expression that is difficult to digest. To overcome this difficulty, the system reliability function is often derived using approximations.

This chapter explains the use of matrix-form distributions in Chapter 2 for exact dynamic analyses of system reliability. The purpose of using matrix-form distributions is to facilitate the design of algorithms that can be easily implemented as a computer program to generate the mathematical expression of the system reliability function directly from the mathematical expression of the component reliability functions. The algorithms generate a PH, or ME, representation of the system reliability function from the PH, or ME, representations of the components reliability functions using matrix algebra. Once the PH representation of the system reliability function is obtained, all system reliability measures can also be found using only matrix algebra, without the need for calculus; thus, it is easier to implement in a simple computer program. The use of PH distribution for some system reliability analyses has been widely found in the literature, such as Zhao et al. [4], Lyu et al. [5], Gong et al. [6, 7], Kim and Kim [8], Liu et al. [9], and Riascos-Ochoa et al. [10]. Alkaff and Qomarudin [11] used PH distribution closure properties to devise algorithms for the reliability analysis of systems in several structures. On the contrary, ME distributions are rarely applied for system reliability analysis; one is Alkaff et al. [12] for network reliability analysis. But, no book, let alone textbook, is currently available that summarizes the methods of system reliability analyses using PH or ME distribution in one source that is easy to understand and covers various system structures. Trivedi and Bobbio [3] briefly discuss PH distribution and its application for reliability analysis of systems with simple structures in one of its chapters, but not until providing algorithms to generate PH representations of reliability functions of complex systems. Eryilmaz [13] focuses on discrete time and not intended for devising simple models and algorithms for complex system reliability analyses. This chapter and the remaining of this book are attempts to fill this gap.

This chapter employs the closure properties of PH distribution given in Chapter 2 to design algorithms for generating PH representations of the reliability functions of systems with basic structures, which cover series, parallel, and standby structures with perfect and imperfect switches. Other structures will be developed in later chapters. The development assumes that the system components are independent, meaning that the failure of one component does not affect the reliability of other components that have not failed, an assumption commonly found and acceptable in practice. With this assumption, the closure properties of PH distribution ensure that if all components have lifetimes following not necessarily identical PH distributions, then the system will have a lifetime following a PH distribution. Equations that express the relationship between the PH representation of the system reliability function and the PH representation of the component reliability functions are derived for each structure using the appropriate closure property. Algorithms are designed based on these properties to produce PH representations of the system reliability functions as output. The inputs are the number of components and the PH representations of the component reliability functions. These algorithms are programmed as ready-to-use MATLAB functions. The properties of the system reliability measures for each structure, both functional and statistical, are described using the relationship between the system lifetime and the component lifetimes. Comparisons of system reliability measures in various structures are also discussed. These discussions are supported by the results of manual calculations and computer calculations using MATLAB functions that have been prepared. The algorithms and the programs are then developed further to make them applicable to ME distributions through their analogy to PH distribution. The consistency between PH distributions and ME distributions resulting from similarity transformations is also demonstrated for all basic structures, including comparing their computational times.

This chapter continues with developing models for systems that have a combination of several basic structures. The technique of modeling such systems through hierarchies is explained through examples, including manual calculations for simple cases. MATLAB functions created for basic structures are used as system models. Their uses are exemplified for systems with simple and complex compound structures, starting from creating a hierarchical system reliability model, preparing a main program to call the appropriate MATLAB functions, and running the program to produce statistical and functional reliability measures of compound structured systems. The properties of system reliability measures with simple compound structures are also given. In cases where the complexity of the resulting PH representation exceeds the computer capability, a combined scalar-matrix model through decomposing the system into several subsystems is also explained and exemplified.

In a system that has many components, the component may be said to be identical or nonidentical. Identicality of components could be based on their physical characteristics, or their statistical characteristics that, in reliability analysis, are related to their lifetime distributions. A system with identical components could mean the system uses several of the same components or different components but has identical lifetime distributions. This book uses the same term for both, which should be clear from its context. Likewise, nonidentical components mean different components or components with different lifetime distributions.

This chapter starts with the descriptions of several types of basic structures in system reliability as the basis for constructing the mathematical model of each structure in later sections.

4.1 Basic Structures

System structure is a logical relationship between the system and its components that relates a good or failed system to good or failed components. The system and the components are said to be good if they function properly to satisfy the operational requirements; otherwise, they are considered failed. A logical relationship can differ from its physical connection, which generally describes the physical arrangement of components to construct the system. Although not the same, a logical relationship must refer to its physical connection. However, the logical relationship is also influenced by the system operational requirements, while the system operational requirements highly depend on immediate needs. Thus, a system with a certain physical connection may have several logical relationships that can change over time to meet immediate needs. A logical relationship is manifested in a reliability block diagram for ease of understanding. The relationship between the functioning or failure of the components and the functioning or failure of the system can be clearly observed through this diagram.

The basic structures in system reliability analyses are series, parallel, and standby structures. The reliability block diagram of a series structured system is presented in Figure 4.1. This diagram shows that if one of the components fails, the input will be disconnected from the output. As a result, the system fails to produce output, and the system is said to have failed. This diagram also shows that for the system to be good, all components must be good. Thus, a series structured system is good when all its components are good; conversely, the system fails when one of its components fails.

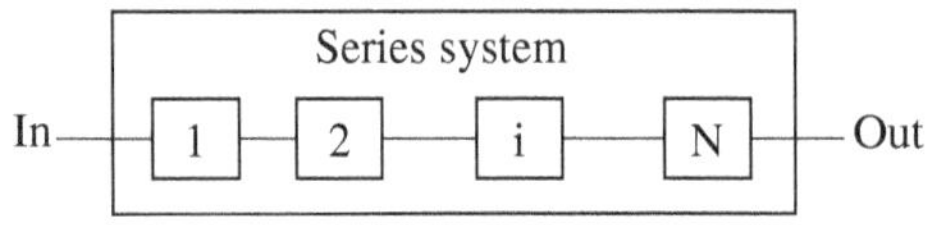

Figure 4.1 Reliability block diagram of a series structured system.

Figure 4.2 shows a reliability block diagram of a parallel structured system. Input to a parallel structured system can produce output as long as one of its components is functioning. Thus, a parallel structured system is good when at least one of its components is good. Conversely, the system fails when all components fail.

A standby structured system is a system consisting of several components, but only one of them is actively operated. The other components are not operated; they are passive as backups to replace the one active component when it fails. A switch is needed to activate one of the passive backup components when the current active component fails. A standby structured system fails if there is no

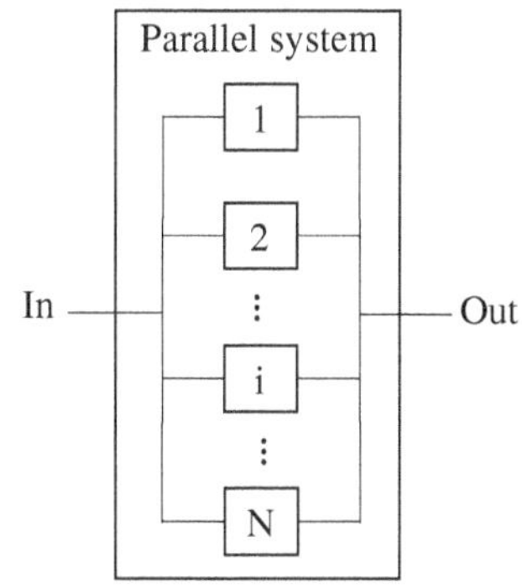

Figure 4.2 Reliability block diagram of a parallel structured system.

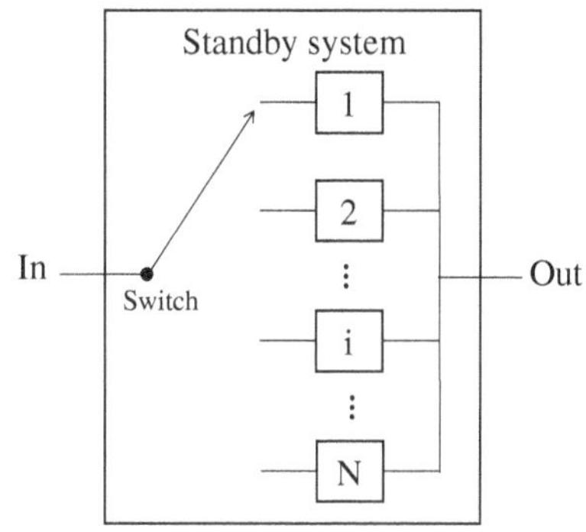

Figure 4.3 Reliability block diagram of a standby structured system.

active component. The possibilities are that the active component fails and no backup component is available to replace it, or the switch fails to put a backup component into an active component to replace the failing active component. Parallel and standby structured systems both require only one good active component. The difference between the two is that all components are actively operated in a parallel structured system; in contrast, in a standby structured system, only one component is active, and the rest are passive. The reliability block diagram of a system with a standby structure is similar to a parallel structure with the addition of a switch as presented in Figure 4.3.

Standby structured systems are the most complicated structure compared to series and parallel structured systems. It is not possible to analyze standby systems using static system reliability method. For this reason, this structure is not discussed in static analysis, even though standby systems are quite common in practice. Dynamic system reliability analysis is also not easy for standby structured systems if the reliability functions are expressed in scalar form as shown in Chapter 1. However, as will be clear later, using matrix-form reliability functions will simplify the analysis of standby systems.

In practice, there are systems with complex structures that, if observed carefully, actually are structures composed of several basic structures. Such systems can be analyzed per subsystem using the appropriate basic structure for each subsystem. The reliability of a system with a complex structure can be calculated hierarchically from the component level to the subsystem level up to the system level. The more complex the system, the more levels must be passed to connect the system to its components.

The following examples illustrate that a single physical connection of several components can have several logical relationships depending on their purpose and needs. Consequently, the reliability block diagram for the system can also change according to its purpose and needs.

Example 4.1 A factory uses a power supply system with three generators (3-generator system) operated simultaneously to meet its power requirement. Generators G_1, G_2, and G_3 have capacities of 100, 90, and 80 kVA, respectively. The factory requires 240 kVA for its maximum operation.

To meet the maximum power requirement of 240 kVA, all generators must function properly. If one of the generators fails, the 3-generator system fails to meet the power requirements. Thus, the logical relationship of the 3-generator system is a three-component in series or a 3-series structure, although the physical relationship is parallel. (In electrical engineering, three generators operated simultaneously are always in a parallel physically.) The reliability block diagram of the 3-generator system to meet a power requirement of 240 kVA is shown in Figure 4.4. ∎

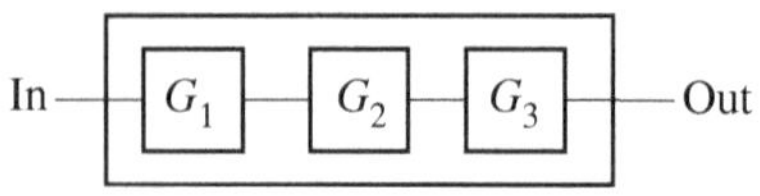

Figure 4.4 Reliability block diagram of Example 4.1.

Example 4.2 The factory in Example 4.1 requires a power supply of 80 kVA for minimum operation. This requirement is met by one generator out of the three available.

This 3-generator system can be designed with three active generators, or one generator is active and the other two are passive. The first design has a 3-parallel structure, and the second has a 3-standby structure. The selection of one of these two structures requires technical and cost considerations [14]. The reliability block diagrams of these two designs are presented in Figure 4.5. ∎

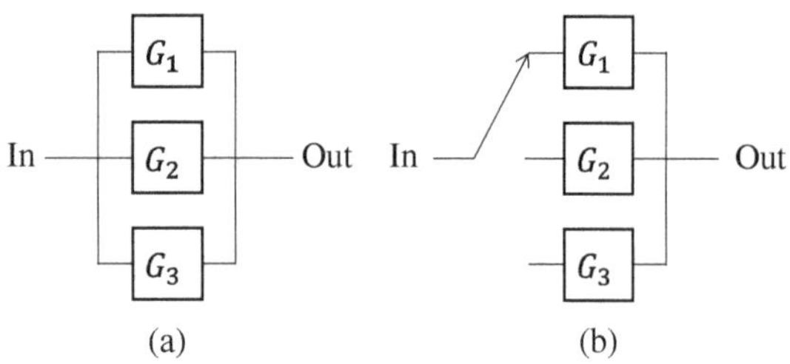

Figure 4.5 Two alternative reliability block diagrams of Example 4.2. (a) Three actives (3-parallel) and (b) one active, two passive (3-standby).

Example 4.3 The factory in Example 4.1 requires nominal power of 175 kVA. This requirement can be met by G_1 and one of G_2 or G_3 must function properly. The logical relationship of this 3-generator system is that G_1 is in a series structure with G_2 and G_3 that are in a parallel structure. The alternative is G_2 and G_3 are operated in a standby structure; for example, in the beginning, G_2 is active and G_3 is passive. Selecting one of these two possibilities also requires technical and cost considerations. The reliability block diagram of the 3-generator system to meet the power requirement of 175 kVA is one of those given in Figure 4.6. ∎

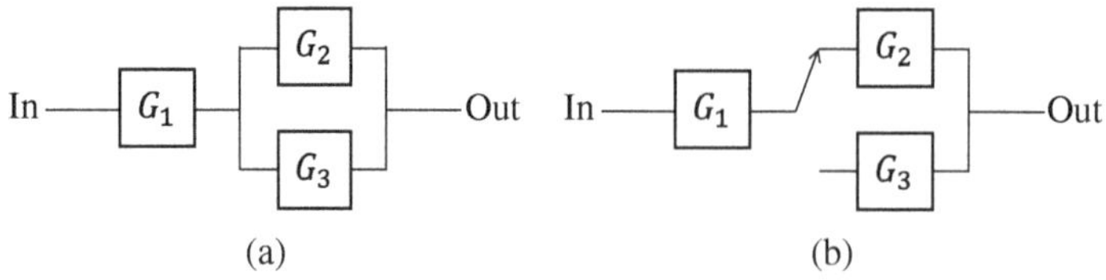

Figure 4.6 Two alternative reliability block diagrams of Example 4.3.

Exercise 4.1 Determine the minimum and maximum power requirements that the structure in Example 4.3 is still able to satisfy.

Exercise 4.2 Determine the system structure to meet a demand of 185 kVA. Explain the result.

Example 4.4 Systems with the same physical components and circuits can have different logical relationships if the components are operated differently or have different purposes, as shown in Figure 4.7.

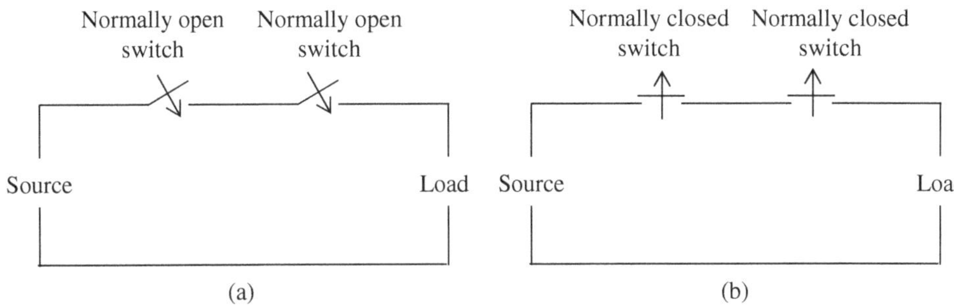

Figure 4.7 Two power supply systems with two types of switches. (a) using a normally open switch and (b) using a normally closed switch.

For the system in Figure 4.7a, the two normally open switches must be closed when there is a need to supply power from the source to the load. If one of the switches fails to close when needed, then this system fails to supply power to the load. Therefore, these two switches have a series structure. For the system in Figure 4.7b with normally closed switches, only one needs to be opened to cut the power supply to the load. If one of the switches is good, the system successfully cuts the supply to the load. Therefore, these two switches have a parallel structure. Figure 4.8 shows the reliability block diagrams of the systems in Figure 4.7. ■

Figure 4.8 Reliability block diagram of systems in Example 4.4. (a) Two normally open switches to connect the source to the load and (b) two normally closed switches to cut the source from the load.

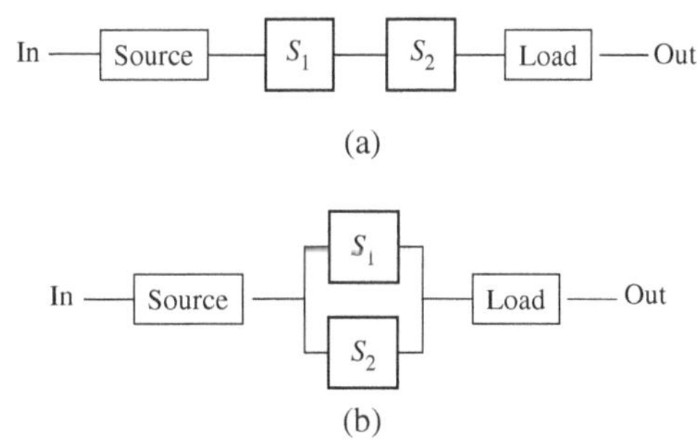

The next section describes ways to express system structures in the mathematical function known as a system structure function, which is simply referred to as structure function.

4.2 Structure Function

A structure function is a mathematical function to express the structure of a system. This function relates the reliability variables of the components to the reliability variable of the system. There are two popular reliability variables: the state variable and the lifetime variable. The system reliability model depends on the reliability variable chosen. Different models may result in different structure functions.

The first model is to express the relationship between the states of the components and the system state. Let $X_i(t), i = 1, 2, \dots, N$, be the state of component i at time t, N be the number of components, and $X_S(t)$ be the system state at time t. The states of each component and system are either good or failed, and the values of $X_i(t)$ and $X_S(t)$ are defined to be either 0 for failed or 1 for good. The relationship between $X_i(t)$ and $X_S(t)$ is expressed as a structure function φ, which is given by

$$X_S(t) = \varphi(X_1(t), X_2(t), \dots, X_N(t)). \tag{4.1}$$

The structure function φ is obtained by finding all combinations of good components that assure the system to be good (or all combinations of failed components that assure the system to be failed). This model is called the combinatorial model. Static and dynamic reliability analyses can be performed using the combinatorial model. Static reliability analysis is performed by eliminating t from Equation (4.1). Since the states of the components and the system are Boolean variables due to their values are 0 or 1, the structure function obtained by the combinatorial model is a Boolean function, a function whose operations and variables are all Boolean.

The second model is based on the relationship between the lifetimes of the components and the system lifetime. Let $T_i, i = 1, 2, \dots, N$, be the lifetime of component i, N be the number of components, and T_S be the system lifetime. The relationship between T_i and T_S can be expressed in a structure function, but in a different form than the structure function of the first model. If ψ is the structure function of the second model, then

$$T_S = \psi(T_1, T_2, \dots, T_N). \tag{4.2}$$

The second model is called the state transition model. This model uses the state transition diagram discussed in Chapter 2. Since the state transition diagram observes the state changes over time, this model can only be used for dynamic reliability analyses. Static reliability analysis using the state transition model starts with a dynamic reliability analysis to produce the system reliability function, then substitute mission duration as the value of t in the resulting system reliability function to yield the system reliability value.

These two models are related to each other because the state $X(t)$ is related to the lifetime T. Their relationship is based on the fact that the event $X_i(t) = 1$ is equivalent to the event that the component i is still good at time t, or the lifetime of component i is longer than t. Therefore,

$$\{X_i(t) = 1\} \Leftrightarrow \{T_i > t\},$$
$$\{X_i(t) = 0\} \Leftrightarrow \{T_i \le t\}.$$

Since the probability for equivalent events are equal, it holds

$$P[X_i(t) = 1] = P[T_i > t] = R_i(t),$$
$$P[X_i(t) = 0] = P[T_i \le t] = 1 - R_i(t).$$

Example 4.5 The structure function in combinatorial model for the series structured system can be derived from the definition that a system with series structure is good if all its components are good. Hence, $X_S(t) = 1$ is satisfied by $X_i(t) = 1, \forall i = 1, 2, \dots, N$, and Equation (4.1) can be expressed in a Boolean function as

$$X_S(t) = \varphi(X_1(t), X_2(t), \dots, X_N(t)) = X_1(t) \wedge X_2(t) \wedge \dots \wedge X_N(t),$$

where $\wedge$ is the Boolean operation "conjunction" or "and." Since the Boolean operation "and" is equivalent to the arithmetic operation "multiplication," a series structure has a structure function

$$X_S(t) = \prod_{i=1}^{N} X_i(t).$$

The structure function of a series structured system in state transition model can be derived from the definition that a series structured system fails when one of its components fails. Hence, the time of system failure, T_S, is the same as the shortest lifetimes of all its components, and Equation (4.2) becomes

$$T_S = \psi(T_1, T_2, \dots, T_N) = \min(T_1, T_2, \dots, T_N) = \min_i(T_i).$$

■

Chapters 4 and 5 employ the state transition model, while Chapters 6 and 7 employ both the combinatorial and state transition models. Both models have advantages and disadvantages. Some structures are easier to analyze using the state transition model, and others are easier to analyze using the combinatorial model. The more appropriate model for each structure will become clear by understanding both models.

Exercise 4.3 Determine a mathematical expression for structure function of parallel structure from its definition that the system fails when all components fail.

The following development assumes all components have lifetimes following independent and not necessarily identical PH distributions. The PH distribution closure properties in Chapter 2 will be employed to guarantee that the lifetimes of basic structured systems also have PH distributions. The PH representation of the distribution of the system lifetime, T_S, is denoted by $(n_S, \boldsymbol{a}_S, \boldsymbol{A}_S)$, while $(n_i, \boldsymbol{a}_i, \boldsymbol{A}_i)$ denotes the PH representation of the distribution of the lifetime of component i, T_i. As shown in Chapter 3, the PH representation of the system lifetime distribution is the same as that of the system reliability function.

The development of reliability models of basic structured systems starts with series structured systems, the simplest and most widely used structure in system design.

4.3 Series Structured System

A series structured system is a system whose components are arranged in a series structure. For simplicity, it will be referred to as a series system, and a series system with N components is referred to as an N-series system. Series systems are the most important systems in system reliability analysis because they are always found in all real systems. Systems generally have many components, and systems most likely fail whenever one of their components fails, which is the definition of a series structured system.

The relationship between the lifetime of an N-series system and the lifetime of its components has been modeled in Example 4.5 with a structure function

$$T_S = \min(T_1, T_2, \dots, T_N). \tag{4.3}$$

Using Equation (4.3) yields

$$R_S(t) = P[T_S > t] = P[\min(T_1, T_2, \dots, T_N) > t].$$

The minimum of several variables is greater than a certain value if all the variables are greater than that certain value. Thus,

$$P[\min(T_1, T_2, \dots, T_N) > t] = P[T_1 > t \wedge T_2 > t \wedge \dots \wedge T_N > t].$$

Using independence of $T_1, T_2, \dots, T_N$ and $R_i(t) = P[T_i > t]$ obtains

$$R_S(t) = R_1(t)R_2(t)\dots R_N(t) = \prod_{i=1}^{N} R_i(t). \tag{4.4}$$

Equation (4.4) is available in classical system reliability textbooks, and it is also applicable for static reliability analysis. For example, a series system constructed using three components with reliability values $R_1 = 0.9$, $R_2 = 0.8$, and $R_3 = 0.5$ has reliability

$$R_S = R_1R_2R_3 = 0.9 \times 0.8 \times 0.5 = 0.36.$$

However, finding the system reliability function from the component reliability function in scalar form, as shown in Chapter 1, is not simple and is difficult to program. An easy-to-program method to generate the reliability of series systems using matrix-form distribution is needed as described next.

4.3.1 Algorithm for Calculating PH Representation in Series Systems

A PH representation for the distribution of T_S can be obtained from Equation (4.3) by using the closure property of the PH distribution under the minimum operation given by Theorem 2.1. Since this theorem is for two variables, its use for many variables must be done recursively. By defining

$$T_S^{(i)} = \min(T_1, T_2, \dots, T_i), i = 2, 3, \dots N.$$

Equation (4.3) can be expressed recursively as [11]

$$T_S^{(i)} = \min\left(T_S^{(i-1)}, T_i\right) \quad i = 2, 3, \dots N, \tag{4.5}$$

with initial condition $T_S^{(1)} = T_1$, which has a PH distribution.

Use Equation (4.5) and Theorem 2.1 recursively as follows:

- For $i = 2$: $T_S^{(2)} = \min(T_S^{(1)}, T_2) = \min(T_1, T_2)$, which according to Theorem 2.1 has a PH distribution with representation $n_S^{(2)} = n_1n_2, \boldsymbol{a}_S^{(2)} = \boldsymbol{a}_1 \otimes \boldsymbol{a}_2$ and $\boldsymbol{A}_S^{(2)} = \boldsymbol{A}_1 \oplus \boldsymbol{A}_2$.
- For $i = 3$: $T_S^{(3)} = \min(T_S^{(2)}, T_3)$. Since $T_S^{(2)}$ has been proven to have a PH distribution, according to Theorem 2.1 $T_S^{(3)}$ has a PH distribution with representation $n_S^{(3)} = n_S^{(2)}n_3, \boldsymbol{a}_S^{(3)} = \boldsymbol{a}_S^{(2)} \otimes \boldsymbol{a}_3$ and $\boldsymbol{A}_S^{(3)} = \boldsymbol{A}_S^{(2)} \oplus \boldsymbol{A}_3$.

Continuing in this manner for $i = 4, 5, \dots, N$ produces the following lemma.

Lemma 4.1 [11] If $T_i \backsim PH(n_i, \boldsymbol{a}_i, \boldsymbol{A}_i)$, $i = 1, 2, \dots, N$, then T_S of a series system follows a PH distribution with representation $(n_S, \boldsymbol{a}_S, \boldsymbol{A}_S)$ that can be generated using recursion equations

$$n_S^{(i)} = n_S^{(i-1)}n_i, \tag{4.6}$$

$$\boldsymbol{a}_S^{(i)} = \boldsymbol{a}_S^{(i-1)} \otimes \boldsymbol{a}_i, \tag{4.7}$$

$$\boldsymbol{A}_S^{(i)} = \boldsymbol{A}_S^{(i-1)} \oplus \boldsymbol{A}_i, \tag{4.8}$$

for $i = 2, 3, \dots, N$, with initial condition

$$n_S^{(1)} = n_1, \quad \boldsymbol{a}_S^{(1)} = \boldsymbol{a}_1, \quad \boldsymbol{A}_S^{(1)} = \boldsymbol{A}_1, \tag{4.9}$$

where the superscript (i) indicates the result of the recursion for the first i components $(1, 2, \ldots, i)$ and the subscript i indicates component i (this definition will be used hereafter).

Recursion Equations (4.6)–(4.8) with the initial conditions given by Equation (4.9) form an algorithm to generate a PH representation of the distribution of lifetime of a series system. This algorithm is called Algorithm 4.1 and is programmed as MATLAB function **phseriessys** available in the software package **dysrel-am**. This function is called by a main program to return $(n_S, \boldsymbol{a}_S, \boldsymbol{A}_S)$, the PH representation of the distribution of the system lifetime, as its output. The inputs are the number of components, N, and the PH representations of the distributions of the lifetime of components, $(n_i, \boldsymbol{a}_i, \boldsymbol{A}_i)$, $i = 1, 2, \ldots, N$. Following Equations (4.6)–(4.9), the pseudocode of **phseriessys** is as follows.

Algorithm 4.1

---

input $N, (n_i, \boldsymbol{a}_i, \boldsymbol{A}_i), i = 1, 2, \ldots, N$
$n_S = n_1$
$\boldsymbol{a}_S = \boldsymbol{a}_1$
$\boldsymbol{A}_S = \boldsymbol{A}_1$
for $i = 2$ to N
  $n_S = n_S n_i$
  $\boldsymbol{a}_S = \boldsymbol{a}_S \otimes \boldsymbol{a}_i$
  $\boldsymbol{A}_S = \boldsymbol{A}_S \oplus \boldsymbol{A}_i$
  $i = i + 1$
end
end

The PH distribution of the lifetime of a series system has an order that can be deduced from Equation (4.6) to be equal to $n_1 n_2 \ldots n_N$. Thus, the complexity of an N-series system is n^N, which is the order of the PH distribution of the lifetimes of the components raised to the power of the number of components. The resulting system survival matrix is nonsingular because the Kronecker sum of nonsingular matrices is nonsingular [15]. Thus, the moments of the system lifetime are guaranteed to be finite.

Algorithm 4.1 will produce a PH representation $(n_S, \boldsymbol{a}_S, \boldsymbol{A}_S)$ of the distribution of the lifetime of an N-series system as

$$n_S = n_1 n_2 \ldots n_N, \tag{4.10}$$

$$\boldsymbol{a}_S = \boldsymbol{a}_1 \otimes \boldsymbol{a}_2 \otimes \ldots \otimes \boldsymbol{a}_N, \tag{4.11}$$

$$\boldsymbol{A}_S = \boldsymbol{A}_1 \oplus \boldsymbol{A}_2 \oplus \ldots \oplus \boldsymbol{A}_N. \tag{4.12}$$

The state transition diagram of an N-series system can be drawn using Equations (4.11) and (4.12) as in Figure 4.9, where, as given in Equation (2.12), $\boldsymbol{b}_S = -\boldsymbol{A}_S \boldsymbol{u}$.

A good component is represented by state g, and a failed component is represented by state f. The system states are expressed in N-tuples of the states of its components. A series system is in a good state if all components are good and in failed states if any of its components fail. Therefore, in Figure 4.9, the good system state is $(g, g, \ldots, g)$, and the failed system states are all states that contain at least one f element in the N-tuples. However, since PH distribution requires the

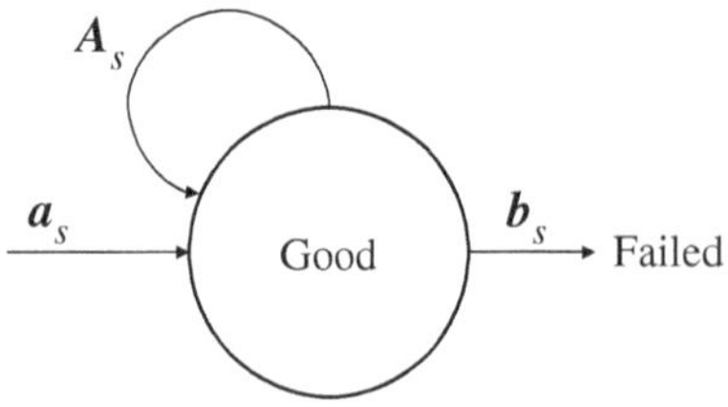

Figure 4.9 State transition diagram of an N-series system.

process underlying the system failure to have only one absorbing state, all those failed states must be lumped into one failed (or absorbing) state to represent the system lifetime in a PH distribution.

Equations (4.11) and (4.12) can also be derived without recursion. For $T_i \backsim PH(n_i, \boldsymbol{a}_i, \boldsymbol{A}_i)$, $i = 1, 2, \ldots, N$, the system reliability function can be obtained using Equation (4.4) and properties of Kronecker product as:

$$R_S(t) = R_1(t)R_2(t)\ldots R_N(t) = \left(\boldsymbol{a}_1 e^{\boldsymbol{A}_1 t}\boldsymbol{u}_1\right)\left(\boldsymbol{a}_2 e^{\boldsymbol{A}_2 t}\boldsymbol{u}_2\right)\ldots\left(\boldsymbol{a}_N e^{\boldsymbol{A}_N t}\boldsymbol{u}_N\right).$$

Multiplication of scalars is equivalent to Kronecker product of scalars,

$$R_S(t) = \left(\boldsymbol{a}_1 e^{\boldsymbol{A}_1 t}\boldsymbol{u}_1\right)\otimes\left(\boldsymbol{a}_2 e^{\boldsymbol{A}_2 t}\boldsymbol{u}_2\right)\ldots\otimes\left(\boldsymbol{a}_N e^{\boldsymbol{A}_N t}\boldsymbol{u}_N\right).$$

Using $\boldsymbol{AC}\otimes\boldsymbol{BD} = (\boldsymbol{A}\otimes\boldsymbol{B})(\boldsymbol{C}\otimes\boldsymbol{D})$ and $e^{\boldsymbol{A}_1 t}\otimes e^{\boldsymbol{A}_2 t} = e^{(\boldsymbol{A}_1 t)\oplus(\boldsymbol{A}_2 t)}$ recursively gives

$$\begin{aligned} R_S(t) &= (\boldsymbol{a}_1\otimes\boldsymbol{a}_2\ldots\otimes\boldsymbol{a}_N)\left(e^{\boldsymbol{A}_1 t}\otimes e^{\boldsymbol{A}_2 t}\ldots e^{\boldsymbol{A}_N t}\right)\ldots(\boldsymbol{u}_1\otimes\boldsymbol{u}_2\ldots\otimes\boldsymbol{u}_N)\\ &= (\boldsymbol{a}_1\otimes\boldsymbol{a}_2\ldots\otimes\boldsymbol{a}_N)\left(e^{(\boldsymbol{A}_1 t)\oplus(\boldsymbol{A}_2 t)\ldots(\boldsymbol{A}_N t)}\right)\ldots(\boldsymbol{u}_1\otimes\boldsymbol{u}_2\ldots\boldsymbol{u}_N) = \boldsymbol{a}_S e^{\boldsymbol{A}_S t}\boldsymbol{u}_S, \end{aligned}$$

where $\boldsymbol{a}_S = \boldsymbol{a}_1\otimes\boldsymbol{a}_2\ldots\otimes\boldsymbol{a}_N, \boldsymbol{A}_S = \boldsymbol{A}_1\oplus\boldsymbol{A}_2\ldots\oplus\boldsymbol{A}_N$, and $\boldsymbol{u}_S = \boldsymbol{u}_1\otimes\boldsymbol{u}_2\ldots\otimes\boldsymbol{u}_N$, the same as will be obtained if using Equations (4.11) and (4.12).

Next, the algorithm is illustrated with the simplest case where all components have lifetimes following nonidentical exponential distributions.

4.3.2 Series Systems with CFR Components

A CFR component has a lifetime following the exponential distribution, which is the simplest PH distribution because its order is equal to one; hence, its representation is a scalar. For this simple case, the calculation of the PH representation of the distribution of the lifetime of a series system can be performed manually. The manual calculation using Algorithm 4.1 is illustrated with a 3-series system as given by the following example.

Example 4.6 A 3-series system uses three CFR components with lifetimes following exponential distributions with $MTTF = 1$, 2, and 5 months.

Given that all component lifetimes have exponential distribution, which is a PH distribution of order 1, their PH representations, as given in Table 2.1, are

$$\begin{aligned} &n_1 = n_2 = n_3 = 1, \quad \boldsymbol{a}_1 = \boldsymbol{a}_2 = \boldsymbol{a}_3 = [1],\\ &\boldsymbol{A}_1 = [-1], \quad \boldsymbol{A}_2 = [-0.5], \quad \boldsymbol{A}_3 = [-0.2]. \end{aligned}$$

The calculation of the PH representation for the lifetime distribution of this series system is carried out recursively using Equations (4.6)–(4.9) as follows:

- The algorithm starts with the initial conditions for $(i) = 1$:

$$\boldsymbol{a}_S^{(1)} = \boldsymbol{a}_1 = [1], \quad \boldsymbol{A}_S^{(1)} = \boldsymbol{A}_1 = [-1], \quad n_S^{(1)} = n_1 = 1.$$

- For $(i) = 2$:

$$n_S^{(2)} = n_S^{(1)} n_2 = 1, \quad \boldsymbol{a}_S^{(2)} = \boldsymbol{a}_S^{(1)} \otimes \boldsymbol{a}_2 = [1] \otimes [1] = [1],$$
$$\boldsymbol{A}_S^{(2)} = \boldsymbol{A}_S^{(1)} \oplus \boldsymbol{A}_2 = [-1] \oplus [-0.5] = [-1.5].$$

- For $(i) = 3$:

$$n_S^{(3)} = n_S^{(2)} n_3 = 1, \quad \boldsymbol{a}_S^{(3)} = \boldsymbol{a}_S^{(2)} \otimes \boldsymbol{a}_3 = [1] \otimes [1] = [1].$$
$$\boldsymbol{A}_S^{(3)} = \boldsymbol{A}_S^{(2)} \oplus \boldsymbol{A}_3 = [-1.5] \oplus [-0.2] = [-1.7].$$

- Because $i = 3 = N$, the algorithm stops.

Thus, the PH representation for the lifetime distribution of this 3-series system is $(n_S, \boldsymbol{a}_S, \boldsymbol{A}_S) = (1, [1], [-1.7])$. Since its order is 1, it is an exponential distribution. The system reliability function in a scalar form as given in Table 2.1 is

$$R_S(t) = e^{-\lambda_S t}\ t \geq 0.$$

The survival matrix is obtained as $\boldsymbol{A}_S = [-\lambda_S]$, or $\lambda_S = 1.7$, which is the system failure rate. The expected lifetime of the system is $MTTF_S = 1/\lambda_S = 1/1.7$. Note that $\boldsymbol{A}_S = [-1.7] = [-1] + [-0.5] + [-0.2]$, which means the system survival matrix is the arithmetic sum of the survival matrices of its components. It is due to its order being 1, which makes the Kronecker sum $\boldsymbol{A}_1 \oplus \boldsymbol{A}_2 \oplus \boldsymbol{A}_3$ becomes an arithmetic sum $\boldsymbol{A}_1 + \boldsymbol{A}_2 + \boldsymbol{A}_3$.

The same result will be obtained using Equations (4.10)–(4.12),

$$n_S = 1 \times 1 \times 1 = 1, \boldsymbol{a}_S = [1] \otimes [1] \otimes [1] = [1],$$
$$\boldsymbol{A}_S = [-1] \oplus [-0.5] \oplus [-0.2] = [-1.7].$$

■

Exercise 4.4 Determine the PH distribution of the lifetime of an N-series system with identical CFR components having $MTTF = b$ months.

Example 4.6 provides some properties of a series system with CFR components:

- The system is CFR because its lifetime has PH distribution of order $n_S = 1$.
- The failure rate of the system is the sum of failure rates of all its components,

$$\boldsymbol{A}_S = \boldsymbol{A}_1 + \boldsymbol{A}_2 + \cdots + \boldsymbol{A}_N \to \lambda_S = \lambda_1 + \lambda_2 + \cdots + \lambda_N = \sum_{i=1}^{N} \lambda_i. \tag{4.13}$$

Equation (4.13) can also be written in terms of hazard functions as

$$h_S(t) = h_1(t) + h_2(t) + \ldots h_N(t) = \sum_{i=1}^{N} h_i(t).$$

- The relationship between $MTTF_S$ and $MTTF_i$, $i = 1, 2, \ldots, N$, can be obtained using Equation (4.13) and $MTTF = 1/\lambda$ or $\lambda = 1/MTTF$,

$$\frac{1}{MTTF_S} = \frac{1}{MTTF_1} + \frac{1}{MTTF_2} + \cdots + \frac{1}{MTTF_N} = \sum_{i=1}^{N} \frac{1}{MTTF_i}. \tag{4.14}$$

Knowing that the system lifetime is exponentially distributed with rate λ_S, the variance of the system lifetime is

$$VTTF_S = \frac{1}{\lambda_S^2} = MTTF_S^2. \tag{4.15}$$

Exercise 4.5 Calculate *MTTF*, *VTTF*, *CVTTF* of the system in Exercise 4.4.

The development is continued in the next section for systems with components having lifetimes following PH distributions of high order.

4.3.3 Series Systems with Non-CFR Components

Non-CFR components have lifetimes following PH distributions of order greater than 1. A manual calculation to produce a PH representation of the lifetime distribution of such a system is troublesome as shown in the following example.

Example 4.7 A 2-series system is constructed using two components having 2-Erlang distributed lifetimes with $MTTF = 2$ and 1, respectively.

PH representations of components 1 and 2 are

$$n_1 = n_2 = 2, \quad \boldsymbol{a}_1 = [1\ \ 0], \quad \boldsymbol{A}_1 = \begin{bmatrix} -1 & 1 \\ 0 & -1 \end{bmatrix}, \quad \boldsymbol{a}_2 = \begin{bmatrix} 1 & 0 \end{bmatrix}, \quad \boldsymbol{A}_2 = \begin{bmatrix} -2 & 2 \\ 0 & -2 \end{bmatrix}.$$

Using Algorithm 4.1 yields

$$n_S = n_1 n_2 = 2 \times 2 = 4.$$

$$\boldsymbol{a}_S = \boldsymbol{a}_1 \otimes \boldsymbol{a}_2 = \begin{bmatrix} 1 & 0 \end{bmatrix} \otimes \begin{bmatrix} 1 & 0 \end{bmatrix} = \begin{bmatrix} 1 & 0 & 0 & 0 \end{bmatrix}.$$

$$\boldsymbol{A}_S = \boldsymbol{A}_1 \oplus \boldsymbol{A}_2 = \begin{bmatrix} -1 & 1 \\ 0 & -1 \end{bmatrix} \oplus \begin{bmatrix} -2 & 2 \\ 0 & -2 \end{bmatrix}$$

$$= \begin{bmatrix} -1\begin{bmatrix} 1 & 0 \\ 0 & 1 \end{bmatrix} & 1\begin{bmatrix} 1 & 0 \\ 0 & 1 \end{bmatrix} \\ 0\begin{bmatrix} 1 & 0 \\ 0 & 1 \end{bmatrix} & -1\begin{bmatrix} 1 & 0 \\ 0 & 1 \end{bmatrix} \end{bmatrix} + \begin{bmatrix} 1\begin{bmatrix} -2 & 2 \\ 0 & -2 \end{bmatrix} & 0\begin{bmatrix} -2 & 2 \\ 0 & -2 \end{bmatrix} \\ 0\begin{bmatrix} -2 & 2 \\ 0 & -2 \end{bmatrix} & 1\begin{bmatrix} -2 & 2 \\ 0 & -2 \end{bmatrix} \end{bmatrix}$$

$$= \begin{bmatrix} -3 & 2 & 1 & 0 \\ 0 & -3 & 0 & 1 \\ 0 & 0 & -3 & 2 \\ 0 & 0 & 0 & -3 \end{bmatrix}.$$

Thus, the lifetime of this series system has PH distribution with representation $(n_S, \boldsymbol{a}_S, \boldsymbol{A}_S)$ as given in the above calculations.

For comparison, the PH representation obtained using the **phseriessys** function is

```
≫ example4_7
ns =
     4
as =
     1     0     0     0
As =
    -3     2     1     0
     0    -3     0     1
     0     0    -3     2
     0     0     0    -3
```

the same result as manual calculations. ■

The difficulty in calculating PH representation of the lifetime distribution of series systems can be avoided by using **phseriessys**. Once the PH representation is obtained, system reliability measures can be generated as needed using the results of Chapter 3.

Example 4.8 The **phseriessys** function is used to calculate the PH representation of the lifetime distribution of a 5-series system. The kth component of this system has lifetime following k-Erlang distributions with $MTTF = 100$, $k = 1, 2, \ldots, 5$. The PH representation obtained from **phseriessys** is then used to calculate and plot the system reliability and *MRL* functions and to calculate the system *MTTF* and *SDTTF*.

The main program for this purpose is given in **example4_8.m** file in the software package **dysrel-am**. Its MATLAB code is as follows:

MATLAB code for example4_8.m

```
clear;close all
addpath('..\dysrel-am','..\dysrel-am\expokit')
N = 5;                                                  (inputs)
% PH Components
for k = 1:N
  T(k) = phnErlang(k,100);
end
% PH Series system                                      (system model)
Ts = phseriessys(T);
% Result in command window                              (statistical outputs)
mttf = phmttf(Ts)
sdttf = phsdttf(Ts)
% Time vector
tmax = 100;
t = linspace(0,tmax,21);
% Reliability plot                                      (graphical outputs)
phrelplot(t,Ts);
% MRL plot
phmrlplot(t,Ts);
```

Executing this file produces Figure 4.10.

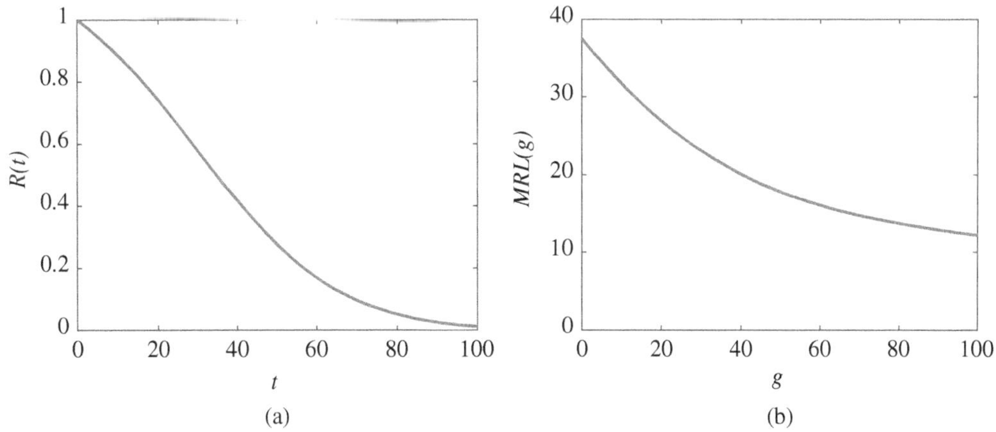

Figure 4.10 Outputs of example4_8.m.

```
≫ example4_8
mttf =
   37.4996
sdttf =
   23.4401
```

■

The main program of each example and application in Chapter 4 is available in **example4_x.m** and **application 4_x.m** files in the software package **dysrel-am**, where x is the example or application number.

Example 4.9 A series system is constructed using N identical IFR components in Example 3.2. The PH representation of the system lifetime distribution is calculated using **phseriessys**. The results are used to plot graphs of the system reliability measures for several values of N, as shown in Figure 4.11. The values of some reliability measures are given in Table 4.1.

The increasing system hazard functions in Figure 4.11b show that the system is IFR.

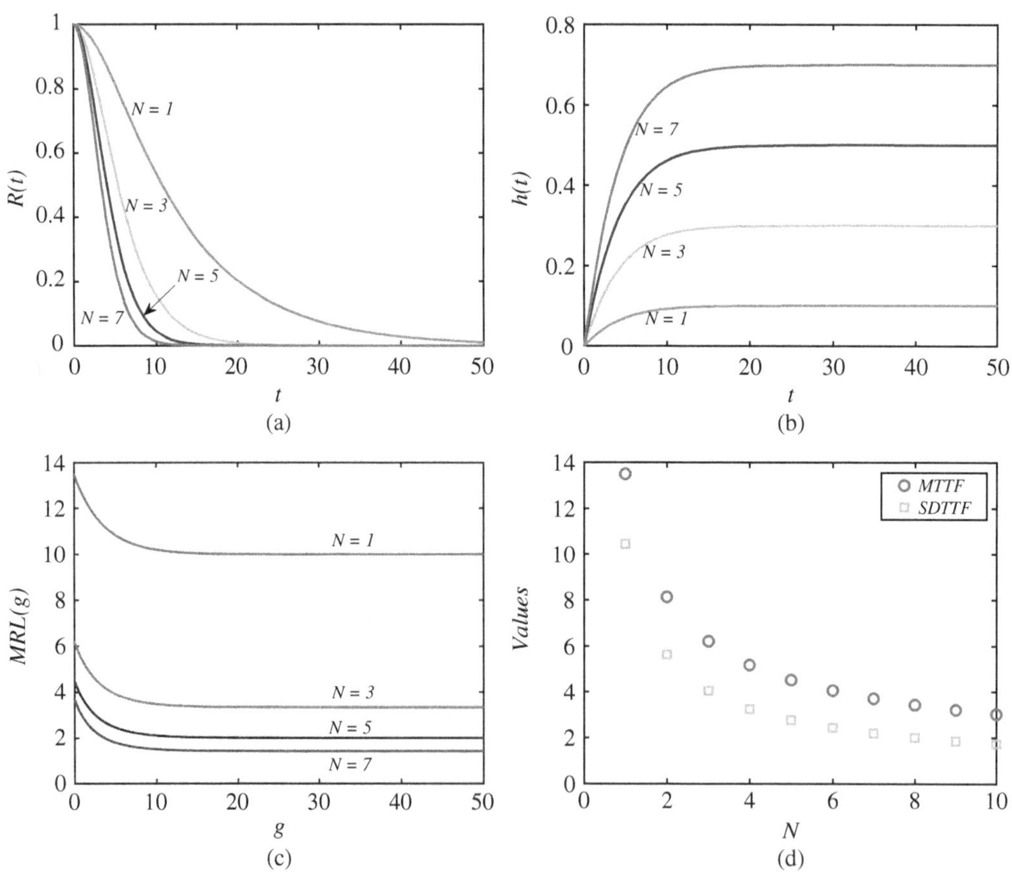

Figure 4.11 Reliability measures of series systems in Example 4.9. (a) System reliability function, (b) system hazard function, (c) system *MRL* function, and (d) system *MTTF* and *SDTTF*.

Table 4.1 The reliability and *MRL* values at $t = 5$ days as functions of *N*.

	N						
Measures	**1**	**2**	**3**	**4**	**5**	**6**	**7**
Reliability	0.82	0.67	0.54	0.44	0.36	0.29	0.24
MRL	10.84	5.69	3.92	3.01	2.45	2.07	1.80

These graphs and table can assist in assessing the system reliability performances. For example, if the system has three components, its reliability for a mission of 5 days is less than 0.6. With that many components, its residual life expectancy after being used for a mission of 5 days is less than 4 days. ■

Exercise 4.6 Determine the maximum number of components in the system in Example 4.9 for a given mission time if the mission reliability must be higher than 0.5 and the expected residual life after the mission is not less than 3 days.

Example 4.10 The **phseriessys** function is also utilized to generate graphs of system reliability measures of N-series systems constructed using identical DFR components whose lifetimes have a hyperexponential distribution with a PH representation

$$\boldsymbol{a} = [0.2 \quad 0.3 \quad 0.5], \quad \boldsymbol{A} = \begin{bmatrix} -0.1 & 0 & 0 \\ 0 & -0.05 & 0 \\ 0 & 0 & -0.02 \end{bmatrix}.$$

The results are presented in Figure 4.12.

The graphs of system hazard and *MRL* functions show that the system is DFR as evidence by these functions is increasing over time. ■

Exercise 4.7 Describe the relationship between the graphs of *MTTF* and *SDTTF* of series systems with CFR components.

Exercise 4.8 Describe the behavior of the *CVTTF* of the systems in Examples 4.9 and 4.10 based on the graphs of their *MTTF* and *SDTTF*.

Next, the properties of series systems are examined.

4.3.4 Properties of Series Structured Systems

Examples 4.7–4.10 show that the type of a series system with the same type of components is the same as the type of its components, as shown in Example 4.6, Figure 4.11b, and Figure 4.12b. This property can be stated as follows:

- All components are DFR $\Rightarrow$ the series system is DFR.
- All components are CFR $\Rightarrow$ the series system is CFR.
- All components are IFR $\Rightarrow$ the series system is IFR.

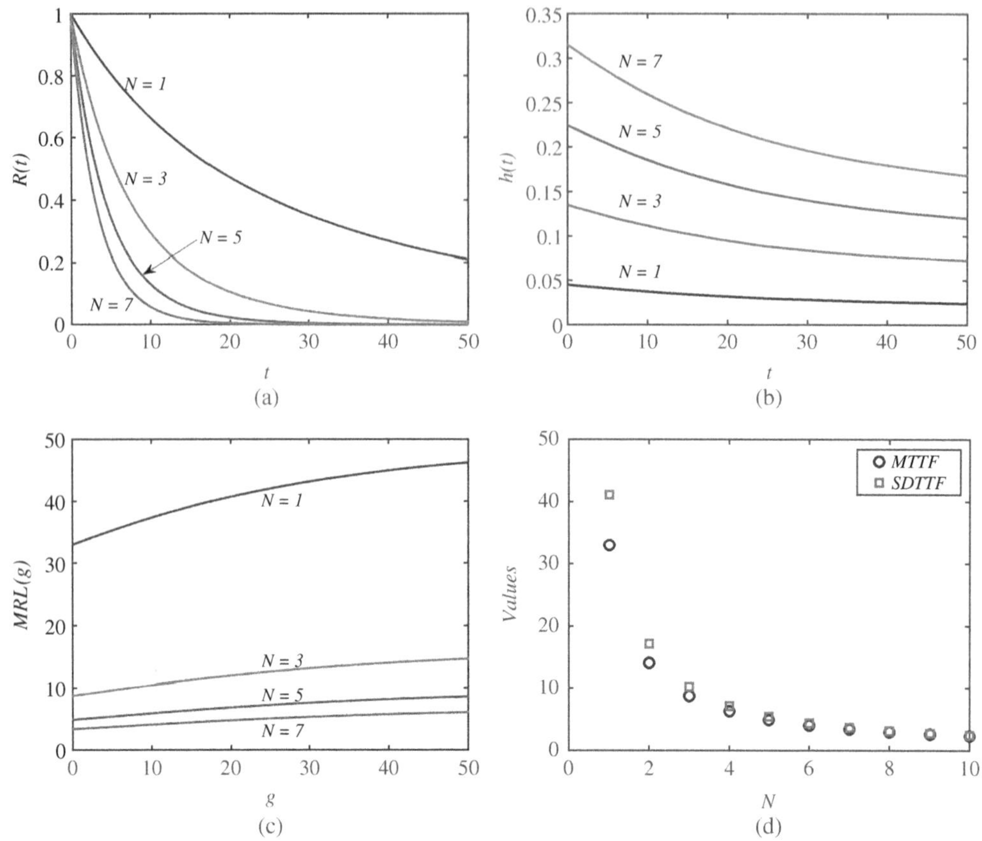

Figure 4.12 Reliability measures of series systems in Example 4.10. (a) System reliability function, (b) system hazard function, (c) system *MRL* function, and (d) system *MTTF* and *SDTTF*.

The abovementioned property is referred to as Property 1 of the series system. Other properties of the series system can be examined from its definition.

Given $0 \leq R_i(t) \leq 1$, the value of the right-hand side of Equation (4.4) will be smaller with an increase in the value of N. It means that the reliability function of a series system will never be higher with the addition of components in the system, even though the added component has a higher reliability than all existing components. Define $R_S(t, N)$ as the reliability function of an N-series system and $R_S(t, N + 1)$ as the reliability function of the N-series system by adding one component. It follows from Equation (4.4) that

$$R_S(t, N+1) = \prod_{i=1}^{N+1} R_i(t) = \left(\prod_{i=1}^{N} R_i(t)\right) R_{N+1}(t) = R_S(t, N) R_{N+1}(t) \leq R_S(t, N).$$

Thus, Property 2 holds:

$$R_S(t, N+1) \leq R_S(t, N). \tag{4.16}$$

Figures 4.11a and 4.12a clearly show that Property 2 is satisfied. Equation (4.16) shows that the addition of components makes the lifetime of a series system stochastically smaller based on its reliability function. Therefore,

$$T_S(N+1) \preccurlyeq T_S(N).$$

In addition, since $0 \le R_i(t) \le 1$, the right-hand side of Equation (4.4) is smaller than the smallest term in the multiplication in the left-hand side,

$$R_S(t) \le \min_i [R_i(t)]. \tag{4.17}$$

Equation (4.17) is Property 3 of series systems, which states that the reliability function of a series system at any time is not higher than the lowest reliability function of its components at that time. In a reliability sense, the system will never be better than its worst component. Consequently, the lifetime of a series system is stochastically smaller than the lifetime of its worst component,

$$T_S \preccurlyeq \min_i (T_i).$$

Hence,

$$MTTF_S \le \min_i [MTTF_i]. \tag{4.18}$$

Equation (4.18) is Property 4 of a series system, which means that the expected lifetime of a series system is shorter than the expected lifetime of its worst component.

Properties 3 and 4 provide upper limits on the reliability and expected lifetime of a series system. These limits are important in series system design for its reliability: avoid components with too low reliability or expected lifetimes. For example, it is better to use two components with reliability of 0.8 and 0.7 than two components with reliability of 0.9 and 0.6 or with reliability 1.0 and 0.5.

The hazard function of a series system can be investigated by substituting Equation (3.7) into Equation (4.4), yielding

$$R_S(t) = e^{-H_S(t)} = \prod_{i=1}^{N} R_i(t) = \prod_{i=1}^{N} e^{-H_i(t)} = e^{-\sum_{i=1}^{N} H_i(t)},$$

where $H_S(t)$ and $H_i(t)$, $i = 1, 2, \ldots, N$, are the integral hazard functions of the system and its components, respectively.

Thus,

$$H_S(t) = \sum_{i=1}^{N} H_i(t). \tag{4.19}$$

Since $h(t)$ is the differential of $H(t)$, it follows that

$$h_S(t) = \sum_{i=1}^{N} h_i(t), \tag{4.20}$$

where $h_S(t)$ and $h_i(t)$, $i = 1, 2, \ldots, N$, are the hazard functions of the system and its components, respectively.

Since the hazard function is nonnegative, Equation (4.20) yields Property 5 of a series system, which states that the addition of components to a series system will increase the system hazard function,

$$h_S(t, N+1) \ge h_S(t, N). \tag{4.21}$$

Equation (4.20) can also be used to produce Property 6 of a series system,

$$h_S(t) \ge \max_i [h_i(t)], \tag{4.22}$$

which can be proven following the proof of Equation (4.17). Equation (4.22) means that the system hazard function is higher than the highest hazard function of its component.

The properties of the series system are shown by the following examples:

Example 4.11 Three components with PH-distributed lifetimes are arranged in a series structure. The component types and their PH representations are

- Component 1 (DFR): $\boldsymbol{a} = [0.5\ \ 0.3\ \ 0.2]$, $\boldsymbol{A} = \begin{bmatrix} -0.02 & 0 & 0 \\ 0 & -0.2 & 0 \\ 0 & 0 & -0.4 \end{bmatrix}$.
- Component 2 (CFR): $\boldsymbol{a} = [1]$, $\boldsymbol{A} = [-0.07]$.
- Component 3 (IFR): $\boldsymbol{a} = [1\ \ 0\ \ 0]$, $\boldsymbol{A} = \begin{bmatrix} -0.18 & 0.18 & 0 \\ 0 & -0.18 & 0.18 \\ 0 & 0 & -0.18 \end{bmatrix}$.

The reliability and hazard functions of the system and all its components are shown in Figure 4.13.

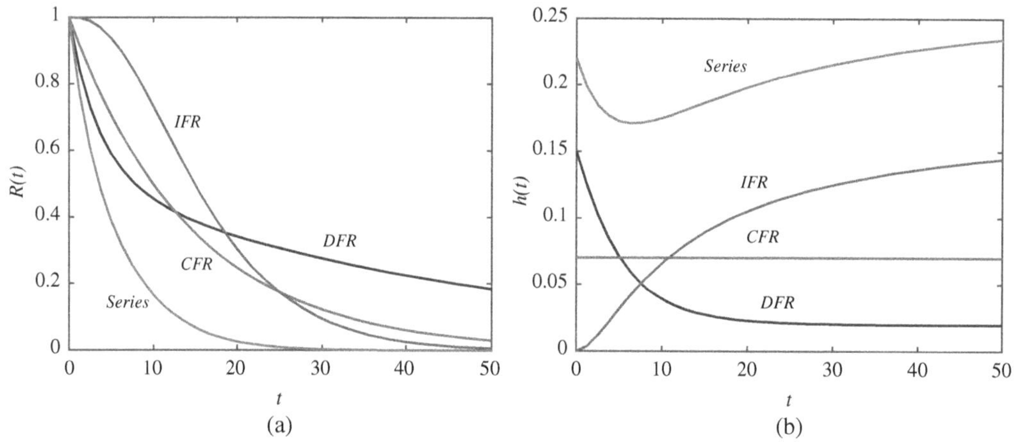

Figure 4.13 Reliability and hazard functions of the system and components in Example 4.11. (a) Reliability functions and (b) hazard functions.

Figure 4.13a shows that the system reliability function at any time is always lower than the lowest reliability function of its components at that time in accordance with Property 3. On the contrary, the system hazard function at any time is higher than the highest hazard function of its components at that time in accordance with Property 6 as shown in Figure 4.13b. The expected lifetimes of the system and all components calculated using the method in Chapter 3 are $MTTF_1 = 27.00$, $MTTF_2 = 14.29$, $MTTF_3 = 16.67$, and $MTTF_S = 5.37$. Hence, the system $MTTF$ is smaller than the smallest component $MTTF$ in accordance with Property 4. ■

Example 4.12 DFR, IFR, and BFR components in Example 3.10 are arranged in a series structure. The reliability and hazard functions of the system and its components are presented in Figure 4.14.

It is apparent from Figure 4.14 that the system reliability function is lower than the reliability functions of its components, while the system hazard function is higher than the hazard functions of its components, consistent with the results of using only DFR or IFR components. ■

Exercise 4.9 Calculate the *MTTFs* of the system and its components in Example 4.12 and show that Property 4 is satisfied.

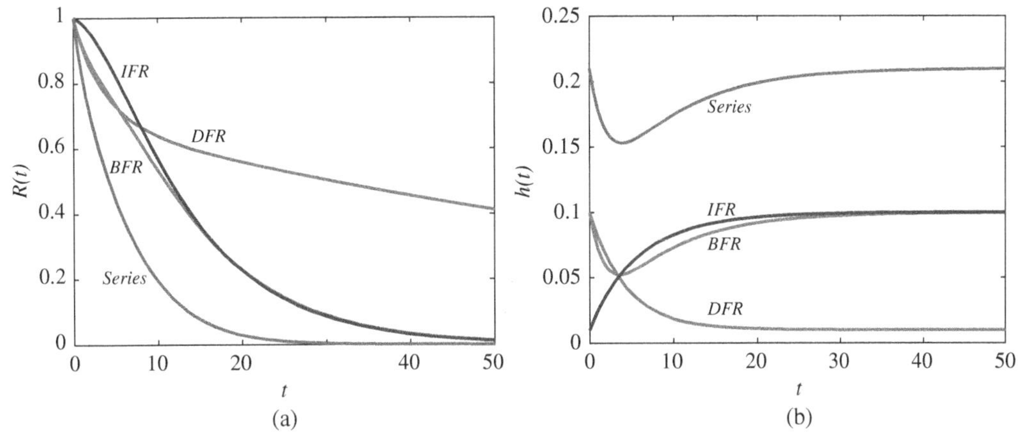

Figure 4.14 Reliability and hazard functions of the system and components in Example 4.12. (a) Reliability functions and (b) hazard functions.

The next section develops reliability models of parallel structured systems. While the reliability of series structured systems is lower than the reliability of their components, the reliability of parallel structured systems is higher than the reliability of their components. Parallel structure is a way to increase system reliability.

4.4 Parallel Structured Systems

A parallel structured system, simply referred to as a parallel system, is a system that fails when all its components have failed. The system continues to function as long as there is still one good component. By this definition, the lifetime of a parallel system is equal to the lifetime of the component that fails last, which is the component with the longest lifetime. Therefore, the lifetime of a parallel system is the maximum of the lifetimes of its components. Parallel systems are often found in real systems that require high reliability. To improve the system reliability and the speed of replacing a failed component with a good one, the system is equipped with a number of backup components that play an active role as the main components; thus, the main and backup components are indistinguishable. A parallel system with N components is referred to as an N-parallel system.

The relationship between the lifetime of an N-parallel system and the lifetime of its components can be modeled in a structure function based on its definition that the system fails when all of its components have failed,

$$T_S = \max(T_1, T_2, \dots, T_N). \tag{4.23}$$

Using Equation (4.23) yields

$$R_S(t) = P[T_S > t] = P[\max(T_1, T_2, \dots, T_N) > t] = 1 - P[\max(T_1, T_2, \dots, T_N) \leq t].$$

The maximum of several variables, which is less than or equal to a certain value, is satisfied if all variables are less than or equal to that certain value. Thus,

$$R_S(t) = 1 - P[T_1 \leq t \wedge T_2 \leq t \wedge \dots \wedge T_N \leq t].$$

Since $T_1, T_2, \ldots, T_N$ are independent and $P[T_i \leq t] = 1 - R_i(t)$, it follows that

$$R_S(t) = 1 - (1 - R_1(t))(1 - R_2(t)) \ldots (1 - R_N(t)) = 1 - \prod_{i=1}^{N}(1 - R_i(t)), \tag{4.24}$$

where $R_i(t)$ is the reliability of component $i, i = 1, 2, \ldots, N$.

Equation (4.24) is available in classical system reliability textbooks, and it is also applicable for static reliability. For example, a 3-parallel system constructed using three components with reliability $R_1 = 0.9$, $R_2 = 0.8$, and $R_3 = 0.5$ has reliability

$$R_S = 1 - (1 - R_1)(1 - R_2)(1 - R_3) = 1 - 0.1 \times 0.2 \times 0.5 = 0.99.$$

However, deriving the parameters of system reliability function from the parameters of reliability functions using Equation (4.24) is not simple, even with the help of computers, if the reliability functions are expressed in scalar form as shown in Chapter 1. Therefore, it is necessary to devise an easy-to-programmed method, which can be accomplished by using reliability functions in matrix form, as presented next.

4.4.1 Algorithm for Calculating PH Representations of Parallel Systems

Equation (4.23) suggests that the distribution of T_S can be found using Theorem 2.4 for the maximum operation on two independent and not necessarily identical random variables having PH distributions. By defining $T_S^{(i)} = \max(T_1, T_2, \ldots, T_i)$, Equation (4.23) can be expressed recursively as [11]

$$T_S^{(i)} = \max\left(T_S^{(i-1)}, T_i\right) \quad i = 2, 3, \ldots, N, \tag{4.25}$$

with initial condition $T_S^{(1)} = T_1$, which has a PH distribution.

Applying Theorem 2.4 recursively to Equation (4.25), similar to the minimum operation in Section 4.3, and acknowledging that there is no discontinuity at the starting point yields the following lemma.

Lemma 4.2 [11] If $T_i \backsim PH(n_i, \boldsymbol{a}_i, \boldsymbol{A}_i)$, $i = 1, 2, \ldots, N$, then T_S of a parallel system has a PH distribution with representation $(n_S, \boldsymbol{a}_S, \boldsymbol{A}_S)$ that can be generated recursively using recursion equations

$$n_S^{(i)} = n_S^{(i-1)} n_i + n_S^{(i-1)} + n_i, \tag{4.26}$$

$$\boldsymbol{a}_S^{(i)} = \left[\boldsymbol{a}_S^{(i-1)} \otimes \boldsymbol{a}_i \;\; \boldsymbol{0} \;\; \boldsymbol{0}\right], \tag{4.27}$$

$$\boldsymbol{A}_S^{(i)} = \begin{bmatrix} \boldsymbol{A}_S^{(i-1)} \oplus \boldsymbol{A}_i & \boldsymbol{I}_S^{(i-1)} \otimes \boldsymbol{b}_i & \boldsymbol{b}_S^{(i-1)} \otimes \boldsymbol{I}_i \\ \boldsymbol{0} & \boldsymbol{A}_S^{(i-1)} & \boldsymbol{0} \\ \boldsymbol{0} & \boldsymbol{0} & \boldsymbol{A}_i \end{bmatrix}, \tag{4.28}$$

for $i = 2, 3, \ldots, N$, with the initial conditions given by Equations (4.9), $\boldsymbol{b}_i = -\boldsymbol{A}_i\boldsymbol{u}$, $\boldsymbol{b}_S^{(i-1)} = -\boldsymbol{A}_S^{(i-1)}\boldsymbol{u}$, and $\boldsymbol{I}_i$ and $\boldsymbol{I}_S^{(i-1)}$ are identity matrices with the same dimensions as $\boldsymbol{A}_i$ and $\boldsymbol{A}_S^{(i-1)}$, respectively.

Equation (4.28) indicates that the system survival matrix is an upper block diagonal matrix. This structure greatly speeds up numerical calculations.

Recursion equations (4.26)–(4.28) with the initial conditions of Equation (4.9) is an algorithm to generate PH representations of the distributions of the lifetimes of parallel systems. This algorithm is called Algorithm 4.2, and it is programmed in MATLAB as **phparallelsys** function using the Kronecker function provided by MATLAB. The MATLAB code is available in the software package **dysrel-am**. The function generates $(n_S, \boldsymbol{a}_S, \boldsymbol{A}_S)$, the PH representation of the distribution of the system lifetime, as its output. The inputs are the number of components, N, and the PH representations of the distributions of the component lifetimes, $(n_i, \boldsymbol{a}_i, \boldsymbol{A}_i)$, $i = 1, 2, \ldots, N$. The pseudo-code for **phparallelsys** is as follows:

Algorithm 4.2

---

input $N, (n_i, \boldsymbol{a}_i, \boldsymbol{A}_i), i = 1, 2, \ldots, N$
$n_S = n_1$
$\boldsymbol{a}_S = \boldsymbol{a}_1$
$\boldsymbol{A}_S = \boldsymbol{A}_1$
for $i = 2$ to N
  $n_S = n_S n_i + n_S + n_i$
  $\boldsymbol{a}_S = [\boldsymbol{a}_S \otimes \boldsymbol{a}_i \;\; \boldsymbol{0} \;\; \boldsymbol{0}]$
  $\boldsymbol{b}_S = -\boldsymbol{A}_S \boldsymbol{u}$
  $\boldsymbol{b}_i = -\boldsymbol{A}_i \boldsymbol{u}$

$$\boldsymbol{A}_S = \begin{bmatrix} \boldsymbol{A}_S \oplus \boldsymbol{A}_i & \boldsymbol{I}_S \otimes \boldsymbol{b}_i & \boldsymbol{b}_S \otimes \boldsymbol{I}_i \\ \boldsymbol{0} & \boldsymbol{A}_S & \boldsymbol{0} \\ \boldsymbol{0} & \boldsymbol{0} & \boldsymbol{A}_i \end{bmatrix}$$

  $i = i + 1$
end
end

It is too complicated to write down the order of the resulting PH representation for parallel systems with nonidentical components. For identical components, the order of the resulting PH representation can be deduced from Equation (4.28) equal to $\sum_{j=1}^{N} C_N^j n^j = (n+1)^N - 1$, where $C_N^j = \frac{N!}{j!(N-j)!}$. Thus, its complexity, or the order of the resulting PH distribution, is $(n + 1)^N$. This complexity is higher than the complexity of a series system whose magnitude is n^N. Since the Kronecker sum operation on nonsingular matrices is nonsingular [15], the system survival matrix is an upper block diagonal matrix whose block diagonals are nonsingular [16]. Thus, the system survival matrix is nonsingular, and the moments of the system lifetime are guaranteed to be finite.

Unlike series systems, the general expression for the PH distribution of the lifetime of an N-parallel system is difficult to summarize in a simple equation. The difficulty is that the number of good system states grows exponentially with the number of components. It is still possible to derive an expression of the PH representation for a small N by recursively using Equations (4.27) and (4.28).

For $N = 3$, Equation (4.28) gives

$$\boldsymbol{A}_S^{(3)} = \begin{bmatrix} \boldsymbol{A}_S^{(2)} \oplus \boldsymbol{A}_3 & \boldsymbol{I}_S^{(2)} \otimes \boldsymbol{b}_3 & \boldsymbol{b}_S^{(2)} \otimes \boldsymbol{I}_3 \\ \boldsymbol{0} & \boldsymbol{A}_S^{(2)} & \boldsymbol{0} \\ \boldsymbol{0} & \boldsymbol{0} & \boldsymbol{A}_3 \end{bmatrix}, \tag{4.29}$$

where

$$\boldsymbol{A}_S^{(2)} = \begin{bmatrix} \boldsymbol{A}_S^{(1)} \oplus \boldsymbol{A}_2 & \boldsymbol{I}_1 \otimes \boldsymbol{b}_2 & \boldsymbol{b}_S^{(1)} \otimes \boldsymbol{I}_2 \\ \boldsymbol{0} & \boldsymbol{A}_S^{(1)} & \boldsymbol{0} \\ \boldsymbol{0} & \boldsymbol{0} & \boldsymbol{A}_2 \end{bmatrix} = \begin{bmatrix} \boldsymbol{A}_1 \oplus \boldsymbol{A}_2 & \boldsymbol{I}_1 \otimes \boldsymbol{b}_2 & \boldsymbol{b}_1 \otimes \boldsymbol{I}_2 \\ \boldsymbol{0} & \boldsymbol{A}_1 & \boldsymbol{0} \\ \boldsymbol{0} & \boldsymbol{0} & \boldsymbol{A}_2 \end{bmatrix}, \tag{4.30}$$

and

$$\boldsymbol{b}_S^{(2)} = -\boldsymbol{A}_S^{(2)}\boldsymbol{u} = -\begin{bmatrix} \boldsymbol{A}_1 \oplus \boldsymbol{A}_2 & \boldsymbol{I}_1 \otimes \boldsymbol{b}_2 & \boldsymbol{b}_1 \otimes \boldsymbol{I}_2 \\ \boldsymbol{0} & \boldsymbol{A}_1 & \boldsymbol{0} \\ \boldsymbol{0} & \boldsymbol{0} & \boldsymbol{A}_2 \end{bmatrix} \begin{bmatrix} \boldsymbol{u}_1 \otimes \boldsymbol{u}_2 \\ \boldsymbol{u}_1 \\ \boldsymbol{u}_2 \end{bmatrix}$$
$$= -\begin{bmatrix} (\boldsymbol{A}_1 \oplus \boldsymbol{A}_2)(\boldsymbol{u}_1 \otimes \boldsymbol{u}_2) + (\boldsymbol{I}_1 \otimes \boldsymbol{b}_2)\boldsymbol{u}_1 + (\boldsymbol{b}_1 \otimes \boldsymbol{I}_2)\boldsymbol{u}_2 \\ \boldsymbol{A}_1\boldsymbol{u}_1 \\ \boldsymbol{A}_2\boldsymbol{u}_2 \end{bmatrix}. \tag{4.31}$$

The first row of the unitary vector $\boldsymbol{u}$ in Equation (4.31) is $\boldsymbol{u}_1 \otimes \boldsymbol{u}_2$ because the first block row consists of $n_1 n_2$ rows, the number of rows in that block row.

The first term in the first row of $\boldsymbol{b}_S^{(2)}$ can be expanded into

$$(\boldsymbol{A}_1 \oplus \boldsymbol{A}_2)(\boldsymbol{u}_1 \otimes \boldsymbol{u}_2) = (\boldsymbol{A}_1 \otimes \boldsymbol{I}_2 + \boldsymbol{I}_1 \otimes \boldsymbol{A}_2)(\boldsymbol{u}_1 \otimes \boldsymbol{u}_2) = (\boldsymbol{A}_1 \otimes \boldsymbol{I}_2)(\boldsymbol{u}_1 \otimes \boldsymbol{u}_2) + (\boldsymbol{I}_1 \otimes \boldsymbol{A}_2)(\boldsymbol{u}_1 \otimes \boldsymbol{u}_2).$$

Using the Kronecker product property $(\boldsymbol{A} \otimes \boldsymbol{B})(\boldsymbol{C} \otimes \boldsymbol{D}) = (\boldsymbol{AC}) \otimes (\boldsymbol{BD})$ for the two terms on the right-hand side produces

$$(\boldsymbol{A}_1 \oplus \boldsymbol{A}_2)(\boldsymbol{u}_1 \otimes \boldsymbol{u}_2) = (\boldsymbol{A}_1\boldsymbol{u}_1 \otimes \boldsymbol{I}_2\boldsymbol{u}_2) + (\boldsymbol{I}_1\boldsymbol{u}_1 \otimes \boldsymbol{A}_2\boldsymbol{u}_2).$$

Using $\boldsymbol{A}_i\boldsymbol{u}_i = -\boldsymbol{b}_i$ produces

$$(\boldsymbol{A}_1 \oplus \boldsymbol{A}_2)(\boldsymbol{u}_1 \otimes \boldsymbol{u}_2) = (-\boldsymbol{b}_1 \otimes \boldsymbol{I}_2\boldsymbol{u}_2) + (\boldsymbol{I}_1\boldsymbol{u}_1 \otimes -\boldsymbol{b}_2). \tag{4.32}$$

Substituting Equation (4.32) into Equation (4.31) makes the first row of $\boldsymbol{b}_S^{(2)}$ in Equation (4.31) become zero. Hence,

$$\boldsymbol{b}_S^{(2)} = \begin{bmatrix} \boldsymbol{0} \\ \boldsymbol{b}_1 \\ \boldsymbol{b}_2 \end{bmatrix}. \tag{4.33}$$

The first block row of the failure vector $\boldsymbol{b}_S^{(2)}$ in Equation (4.33) represents components 1 and 2 failing simultaneously. Its zero value is due to the impossibility of two components failing simultaneously in a continuous-time system.

Following the dimension of $\boldsymbol{A}_S^{(2)}$ results in

$$\boldsymbol{I}_S^{(2)} = \begin{bmatrix} \boldsymbol{I}_1 \otimes \boldsymbol{I}_2 & \boldsymbol{0} & \boldsymbol{0} \\ \boldsymbol{0} & \boldsymbol{I}_1 & \boldsymbol{0} \\ \boldsymbol{0} & \boldsymbol{0} & \boldsymbol{I}_2 \end{bmatrix}. \tag{4.34}$$

Substituting $\boldsymbol{A}_S^{(2)}$, $\boldsymbol{b}_S^{(2)}$, and $\boldsymbol{I}_S^{(2)}$ in Equations (4.30), (4.33), and (4.34) into Equation (4.29) yields

$$\boldsymbol{A}_S^{(3)} = \begin{bmatrix} \begin{bmatrix} \boldsymbol{A}_1 \oplus \boldsymbol{A}_2 & \boldsymbol{I}_1 \otimes \boldsymbol{b}_2 & \boldsymbol{b}_1 \otimes \boldsymbol{I}_2 \\ \boldsymbol{0} & \boldsymbol{A}_1 & \boldsymbol{0} \\ \boldsymbol{0} & \boldsymbol{0} & \boldsymbol{A}_2 \end{bmatrix} \oplus \boldsymbol{A}_3 & \begin{bmatrix} \boldsymbol{I}_1 \otimes \boldsymbol{I}_2 & \boldsymbol{0} & \boldsymbol{0} \\ \boldsymbol{0} & \boldsymbol{I}_1 & \boldsymbol{0} \\ \boldsymbol{0} & \boldsymbol{0} & \boldsymbol{I}_2 \end{bmatrix} \otimes \boldsymbol{b}_3 & \begin{bmatrix} \boldsymbol{0} \\ \boldsymbol{b}_1 \\ \boldsymbol{b}_2 \end{bmatrix} \otimes \boldsymbol{I}_3 \\ \boldsymbol{0} & \begin{bmatrix} \boldsymbol{A}_1 \oplus \boldsymbol{A}_2 & \boldsymbol{I}_1 \otimes \boldsymbol{b}_2 & \boldsymbol{b}_1 \otimes \boldsymbol{I}_2 \\ \boldsymbol{0} & \boldsymbol{A}_1 & \boldsymbol{0} \\ \boldsymbol{0} & \boldsymbol{0} & \boldsymbol{A}_2 \end{bmatrix} & \boldsymbol{0} \\ \boldsymbol{0} & \boldsymbol{0} & \boldsymbol{A}_3 \end{bmatrix}. \tag{4.35}$$

Simplifying the block matrix in the first block row and column of $\boldsymbol{A}_S^{(3)}$ yields

$$
\begin{aligned}
&\begin{bmatrix} \boldsymbol{A}_1 \oplus \boldsymbol{A}_2 & \boldsymbol{I}_1 \otimes \boldsymbol{b}_2 & \boldsymbol{b}_1 \otimes \boldsymbol{I}_2 \\ \boldsymbol{0} & \boldsymbol{A}_1 & \boldsymbol{0} \\ \boldsymbol{0} & \boldsymbol{0} & \boldsymbol{A}_2 \end{bmatrix} \oplus \boldsymbol{A}_3 \\
&= \begin{bmatrix} \boldsymbol{A}_1 \oplus \boldsymbol{A}_2 & \boldsymbol{I}_1 \otimes \boldsymbol{b}_2 & \boldsymbol{b}_1 \otimes \boldsymbol{I}_2 \\ \boldsymbol{0} & \boldsymbol{A}_1 & \boldsymbol{0} \\ \boldsymbol{0} & \boldsymbol{0} & \boldsymbol{A}_2 \end{bmatrix} \otimes \boldsymbol{I}_3 + \begin{bmatrix} \boldsymbol{I}_1 \oplus \boldsymbol{I}_2 & \boldsymbol{0} & \boldsymbol{0} \\ \boldsymbol{0} & \boldsymbol{I}_1 & \boldsymbol{0} \\ \boldsymbol{0} & \boldsymbol{0} & \boldsymbol{I}_2 \end{bmatrix} \otimes \boldsymbol{A}_3 \\
&= \begin{bmatrix} (\boldsymbol{A}_1 \oplus \boldsymbol{A}_2) \otimes \boldsymbol{I}_3 + (\boldsymbol{I}_1 \oplus \boldsymbol{I}_2) \otimes \boldsymbol{A}_3 & \boldsymbol{I}_1 \otimes \boldsymbol{b}_2 \otimes \boldsymbol{I}_3 & \boldsymbol{b}_1 \otimes \boldsymbol{I}_2 \otimes \boldsymbol{I}_3 \\ \boldsymbol{0} & \boldsymbol{A}_1 \otimes \boldsymbol{I}_3 + \boldsymbol{I}_1 \otimes \boldsymbol{A}_3 & \boldsymbol{0} \\ \boldsymbol{0} & \boldsymbol{0} & \boldsymbol{A}_2 \otimes \boldsymbol{I}_3 + \boldsymbol{I}_2 \otimes \boldsymbol{A}_3 \end{bmatrix} \\
&= \begin{bmatrix} \boldsymbol{A}_1 \oplus \boldsymbol{A}_2 \oplus \boldsymbol{A}_3 & \boldsymbol{I}_1 \otimes \boldsymbol{b}_2 \otimes \boldsymbol{I}_3 & \boldsymbol{b}_1 \otimes \boldsymbol{I}_2 \otimes \boldsymbol{I}_3 \\ \boldsymbol{0} & \boldsymbol{A}_1 \oplus \boldsymbol{A}_3 & \boldsymbol{0} \\ \boldsymbol{0} & \boldsymbol{0} & \boldsymbol{A}_2 \oplus \boldsymbol{A}_3 \end{bmatrix}.
\end{aligned}
$$

Substituting this result back to $\boldsymbol{A}_S^{(3)}$ in Equation (4.35) produces

$$
\boldsymbol{A}_S^{(3)} = \begin{bmatrix}
\boldsymbol{A}_1 \oplus \boldsymbol{A}_2 \oplus \boldsymbol{A}_3 & \boldsymbol{I}_1 \otimes \boldsymbol{b}_2 \otimes \boldsymbol{I}_3 & \boldsymbol{b}_1 \otimes \boldsymbol{I}_2 \otimes \boldsymbol{I}_3 & \boldsymbol{I}_1 \otimes \boldsymbol{I}_2 \otimes \boldsymbol{b}_3 & \boldsymbol{0} & \boldsymbol{0} & \boldsymbol{0} \\
\boldsymbol{0} & \boldsymbol{A}_1 \oplus \boldsymbol{A}_3 & \boldsymbol{0} & \boldsymbol{0} & \boldsymbol{I}_1 \otimes \boldsymbol{b}_3 & \boldsymbol{0} & \boldsymbol{b}_1 \ 0 \ \boldsymbol{I}_3 \\
\boldsymbol{0} & \boldsymbol{0} & \boldsymbol{A}_2 \oplus \boldsymbol{A}_3 & \boldsymbol{0} & \boldsymbol{0} & \boldsymbol{I}_2 \otimes \boldsymbol{b}_3 & \boldsymbol{b}_2 \otimes \boldsymbol{I}_3 \\
\boldsymbol{0} & \boldsymbol{0} & \boldsymbol{0} & \boldsymbol{A}_1 \oplus \boldsymbol{A}_2 & \boldsymbol{I}_1 \otimes \boldsymbol{b}_2 & \boldsymbol{b}1 \otimes \boldsymbol{I}_2 & \boldsymbol{0} \\
\boldsymbol{0} & \boldsymbol{0} & \boldsymbol{0} & \boldsymbol{0} & \boldsymbol{A}_1 & \boldsymbol{0} & \boldsymbol{0} \\
\boldsymbol{0} & \boldsymbol{0} & \boldsymbol{0} & \boldsymbol{0} & \boldsymbol{0} & \boldsymbol{A}_2 & \boldsymbol{0} \\
\boldsymbol{0} & \boldsymbol{0} & \boldsymbol{0} & \boldsymbol{0} & \boldsymbol{0} & \boldsymbol{0} & \boldsymbol{A}_3
\end{bmatrix}. \tag{4.36}
$$

Failure vector $\boldsymbol{b}_S^{(3)} = -\boldsymbol{A}_S^{(3)}\boldsymbol{u}$ can be derived the same way as $\boldsymbol{b}_S^{(2)}$, but it requires longer steps. The result intuitively should be

$$
\boldsymbol{b}_S^{(3)} = \begin{bmatrix} \boldsymbol{0} & \boldsymbol{0} & \boldsymbol{0} & \boldsymbol{0} & \boldsymbol{b}_1 & \boldsymbol{b}_2 & \boldsymbol{b}_3 \end{bmatrix}^T, \tag{4.37}
$$

because there is only one failure that may happen at a time. The proof of Equation (4.37) is given in Problems.

The resulting infinitesimal generator matrix of 3-parallel system is

$$
\begin{aligned}
\boldsymbol{Q}_S^{(3)} &= \begin{bmatrix} \boldsymbol{A}_S^{(3)} & \boldsymbol{b}_S^{(3)} \\ \boldsymbol{0} & 0 \end{bmatrix} \\
&= \begin{bmatrix}
\boldsymbol{A}_1 \oplus \boldsymbol{A}_2 \oplus \boldsymbol{A}_3 & \boldsymbol{I}_1 \otimes \boldsymbol{b}_2 \otimes \boldsymbol{I}_3 & \boldsymbol{b}_1 \otimes \boldsymbol{I}_2 \otimes \boldsymbol{I}_3 & \boldsymbol{I}_1 \otimes \boldsymbol{I}_2 \otimes \boldsymbol{b}_3 & \boldsymbol{0} & \boldsymbol{0} & \boldsymbol{0} & \boldsymbol{0} \\
\boldsymbol{0} & \boldsymbol{A}_1 \oplus \boldsymbol{A}_3 & \boldsymbol{0} & \boldsymbol{0} & \boldsymbol{I}_1 \otimes \boldsymbol{b}_3 & \boldsymbol{0} & \boldsymbol{b}_1 \otimes \boldsymbol{I}_3 & \boldsymbol{0} \\
\boldsymbol{0} & \boldsymbol{0} & \boldsymbol{A}_2 \oplus \boldsymbol{A}_3 & \boldsymbol{0} & \boldsymbol{0} & \boldsymbol{I}_2 \otimes \boldsymbol{b}_3 & \boldsymbol{b}_2 \otimes \boldsymbol{I}_3 & \boldsymbol{0} \\
\boldsymbol{0} & \boldsymbol{0} & \boldsymbol{0} & \boldsymbol{A}_1 \oplus \boldsymbol{A}_2 & \boldsymbol{I}_1 \otimes \boldsymbol{b}_2 & \boldsymbol{b}_1 \otimes \boldsymbol{I}_2 & \boldsymbol{0} & \boldsymbol{0} \\
\boldsymbol{0} & \boldsymbol{0} & \boldsymbol{0} & \boldsymbol{0} & \boldsymbol{A}_1 & \boldsymbol{0} & \boldsymbol{0} & \boldsymbol{b}_1 \\
\boldsymbol{0} & \boldsymbol{0} & \boldsymbol{0} & \boldsymbol{0} & \boldsymbol{0} & \boldsymbol{A}_2 & \boldsymbol{0} & \boldsymbol{b}_2 \\
\boldsymbol{0} & \boldsymbol{0} & \boldsymbol{0} & \boldsymbol{0} & \boldsymbol{0} & \boldsymbol{0} & \boldsymbol{A}_3 & \boldsymbol{b}_3 \\
\boldsymbol{0} & \boldsymbol{0} & \boldsymbol{0} & \boldsymbol{0} & \boldsymbol{0} & \boldsymbol{0} & \boldsymbol{0} & 0
\end{bmatrix}.
\end{aligned} \tag{4.38}
$$

The state transition diagram of a 3-parallel system constructed using $\boldsymbol{Q}_S^{(3)}$ in Equation (4.38) is shown in Figure 4.15. In this diagram, the system states are expressed as combinations of

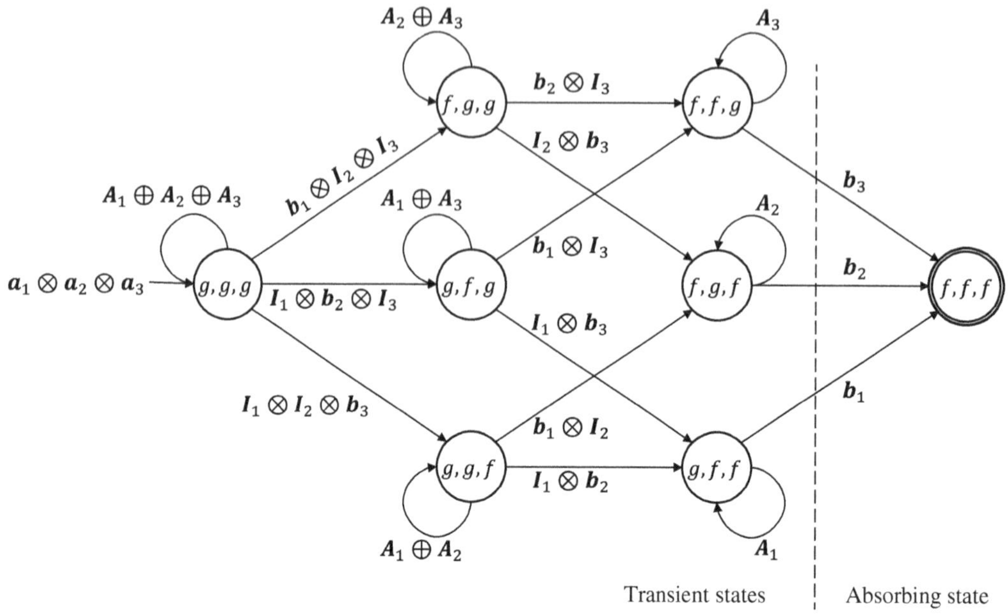

Figure 4.15 State transition diagram of a 3-parallel system.

component states, with g for good and f for failed. For example, the state (f, g, f) indicates that component 1 failed, component 2 is good, and component 3 failed. The diagram starts with zero failed components, followed by a group of states with one failed component, a group of states with two failed components, until a state with all failed components. The sequence of states within a group is arbitrary, but the weight of each arc must be determined or adjusted accordingly. By the definition of parallel structure, all states with at least one good component are transient states, and the state with no good (or all failed) component is the absorbing state. Drawing the state transition diagram of N-parallel systems is too complicated for an arbitrary N.

The algorithm is first illustrated with the simplest case where all components have lifetimes following nonidentical exponential distributions.

4.4.2 Parallel Systems with CFR Components

A manual calculation of the PH representation of the distribution of lifetime of a parallel system using Algorithm 4.2 is still possible for parallel systems with CFR components.

Example 4.13 The PH representation of the lifetime distribution of a 3-parallel system with the same components as Example 4.6 is derived.

Using n_i, $\boldsymbol{a}_i$, and $\boldsymbol{A}_i$ in Example 4.6 determines

$$\boldsymbol{b}_1 = -\boldsymbol{A}_1\boldsymbol{u} = -[-1][1] = [1], \boldsymbol{b}_2 = -\boldsymbol{A}_2\boldsymbol{u} = -[-0.5][1] = [0.5],$$
$$\boldsymbol{b}_3 = -\boldsymbol{A}_3\boldsymbol{u} = -[-0.2][1] = [0.2].$$

The iteration using Algorithm 4.2 is as follows:

- Initial conditions: $n_S^{(1)} = n_1 = 1, \boldsymbol{a}_S^{(1)} = \boldsymbol{a}_1 = [1], \boldsymbol{A}_S^{(1)} = \boldsymbol{A}_1 = [-1]$.

- Second and third iterations to find n_S:

$$n_S^{(2)} = n_S^{(1)} n_2 + n_S^{(1)} + n_2 = 1 \times 1 + 1 + 1 = 3,$$
$$n_S^{(3)} = n_S^{(2)} n_3 + n_S^{(2)} + n_3 = 3 \times 1 + 3 + 1 = 7.$$

- Second and third iterations to find $\boldsymbol{a}_S$:

$$\boldsymbol{a}_S^{(2)} = \begin{bmatrix} \boldsymbol{a}_S^{(1)} \otimes \boldsymbol{a}_2 & \boldsymbol{0} & \boldsymbol{0} \end{bmatrix} = \begin{bmatrix} 1 \otimes 1 & 0 & 0 \end{bmatrix} = \begin{bmatrix} 1 & 0 & 0 \end{bmatrix}.$$
$$\boldsymbol{a}_S^{(3)} = \begin{bmatrix} \boldsymbol{a}_S^{(2)} \otimes \boldsymbol{a}_3 & \boldsymbol{0} & \boldsymbol{0} \end{bmatrix} = \begin{bmatrix} \begin{bmatrix} 1 & 0 & 0 \end{bmatrix} \otimes [1] & \begin{bmatrix} 0 & 0 & 0 \end{bmatrix} & 0 \end{bmatrix} = \begin{bmatrix} 1 & 0 & 0 & 0 & 0 & 0 & 0 \end{bmatrix}.$$

- Second and third iterations to find $\boldsymbol{A}_S$:

$$\boldsymbol{A}_S^{(2)} = \begin{bmatrix} \boldsymbol{A}_S^{(1)} \oplus \boldsymbol{A}_2 & \boldsymbol{I}_S^{(1)} \otimes \boldsymbol{b}_2 & \boldsymbol{b}_S^{(1)} \otimes \boldsymbol{I}_2 \\ \boldsymbol{0} & \boldsymbol{A}_S^{(1)} & \boldsymbol{0} \\ \boldsymbol{0} & \boldsymbol{0} & \boldsymbol{A}_2 \end{bmatrix} = \begin{bmatrix} [-1] \oplus [-0.5] & 1 \otimes [0.5] & [1] \otimes [1] \\ 0 & [-1] & 0 \\ 0 & 0 & [-0.5] \end{bmatrix}$$
$$= \begin{bmatrix} -1.5 & 0.5 & 1 \\ 0 & -1 & 0 \\ 0 & 0 & -0.5 \end{bmatrix},$$

$$\boldsymbol{b}_S^{(2)} = -\begin{bmatrix} -1.5 & 0.5 & 1 \\ 0 & -1 & 0 \\ 0 & 0 & -0.5 \end{bmatrix} \begin{bmatrix} 1 \\ 1 \\ 1 \end{bmatrix} = \begin{bmatrix} 0 \\ 1 \\ 0.5 \end{bmatrix}.$$

$$\boldsymbol{A}_S^{(3)} = \begin{bmatrix} \boldsymbol{A}_S^{(2)} \oplus \boldsymbol{A}_3 & \boldsymbol{I}_S^{(2)} \otimes \boldsymbol{b}_3 & \boldsymbol{b}_S^{(2)} \otimes \boldsymbol{I}_3 \\ \boldsymbol{0} & \boldsymbol{A}_S^{(2)} & \boldsymbol{0} \\ \boldsymbol{0} & \boldsymbol{0} & \boldsymbol{A}_3 \end{bmatrix}$$
$$= \begin{bmatrix} \begin{bmatrix} -1.5 & 0.5 & 1 \\ 0 & -1 & 0 \\ 0 & 0 & -0.5 \end{bmatrix} \oplus [-0.2] & \begin{bmatrix} 1 & 0 & 0 \\ 0 & 1 & 0 \\ 0 & 0 & 1 \end{bmatrix} \otimes [0.2] & \begin{bmatrix} 0 \\ 1 \\ 0.5 \end{bmatrix} \otimes [1] \\ \boldsymbol{0} & \begin{bmatrix} -1.5 & 0.5 & 1 \\ 0 & -1 & 0 \\ 0 & 0 & -0.5 \end{bmatrix} & \boldsymbol{0} \\ \boldsymbol{0} & \boldsymbol{0} & [-0.2] \end{bmatrix}$$
$$= \begin{bmatrix} \begin{bmatrix} -1.7 & 0.5 & 1 \\ 0 & -1.2 & 0 \\ 0 & 0 & -0.7 \end{bmatrix} & \begin{bmatrix} 0.2 & 0 & 0 \\ 0 & 0.2 & 0 \\ 0 & 0 & 0.2 \end{bmatrix} & \begin{bmatrix} 0 \\ 1 \\ 0.5 \end{bmatrix} \\ \begin{bmatrix} 0 & 0 & 0 \\ 0 & 0 & 0 \\ 0 & 0 & 0 \end{bmatrix} & \begin{bmatrix} -1.5 & 0.5 & 1 \\ 0 & -1 & 0 \\ 0 & 0 & -0.5 \end{bmatrix} & \begin{bmatrix} 0 \\ 0 \\ 0 \end{bmatrix} \\ \begin{bmatrix} 0 & 0 & 0 \end{bmatrix} & \begin{bmatrix} 0 & 0 & 0 \end{bmatrix} & [-0.2] \end{bmatrix}$$

$$
= \begin{bmatrix} -1.7 & 0.5 & 1 & 0.2 & 0 & 0 & 0 \\ 0 & -1.2 & 0 & 0 & 0.2 & 0 & 1 \\ 0 & 0 & -0.7 & 0 & 0 & 0.2 & 0.5 \\ 0 & 0 & 0 & -1.5 & 0.5 & 1 & 0 \\ 0 & 0 & 0 & 0 & -1 & 0 & 0 \\ 0 & 0 & 0 & 0 & 0 & -0.5 & 0 \\ 0 & 0 & 0 & 0 & 0 & 0 & -0.2 \end{bmatrix}.
$$

$$
\boldsymbol{b}_S^{(3)} = -\begin{bmatrix} -1.7 & 0.5 & 1 & 0.2 & 0 & 0 & 0 \\ 0 & -1.2 & 0 & 0 & 0.2 & 0 & 1 \\ 0 & 0 & -0.7 & 0 & 0 & 0.2 & 0.5 \\ 0 & 0 & 0 & -1.5 & 0.5 & 1 & 0 \\ 0 & 0 & 0 & 0 & -1 & 0 & 0 \\ 0 & 0 & 0 & 0 & 0 & -0.5 & 0 \\ 0 & 0 & 0 & 0 & 0 & 0 & -0.2 \end{bmatrix} \begin{bmatrix} 1 \\ 1 \\ 1 \\ 1 \\ 1 \\ 1 \\ 1 \end{bmatrix} = \begin{bmatrix} 0 \\ 0 \\ 0 \\ 0 \\ 1 \\ 0.5 \\ 0.2 \end{bmatrix}.
$$

Thus, the lifetime of this 3-parallel system has a PH distribution of order 7 with the survival matrix and initial probability vector as given above. Although still possible, this example illustrates the difficulties faced by manual calculation.

Using Equation (4.36) yields

$$
\boldsymbol{A}_S^{(3)} = \begin{bmatrix} -1-0.5-0.2 & (1)(0.5)(1) & (1)(1)(1) & (1)(1)(0.2) & 0 & 0 & 0 \\ 0 & -1-0.2 & 0 & 0 & (1)(0.2) & 0 & (1)(1) \\ 0 & 0 & -0.5-0.2 & 0 & 0 & (1)(0.2) & (0.5)(1) \\ 0 & 0 & 0 & -1-0.5 & (1)(0.5) & (1)(1) & 0 \\ 0 & 0 & 0 & 0 & -1 & 0 & 0 \\ 0 & 0 & 0 & 0 & 0 & -0.5 & 0 \\ 0 & 0 & 0 & 0 & 0 & 0 & -0.2 \end{bmatrix},
$$

the same as produced by recursive calculation using Algorithm 4.2.

For comparison, the survival matrix of the PH distribution of the system lifetime produced using the **phparallelsys** function is

```
As =
   -1.7000    0.5000    1.0000    0.2000         0         0         0
         0   -1.2000         0         0    0.2000         0    1.0000
         0         0   -0.7000         0         0    0.2000    0.5000
         0         0         0   -1.5000    0.5000    1.0000         0
         0         0         0         0   -1.0000         0         0
         0         0         0         0         0   -0.5000         0
         0         0         0         0         0         0   -0.2000
```

the same result as the manual calculations. ∎

Unlike series systems, parallel systems with CFR components are not CFR systems because the order of PH distribution of the system lifetime distribution is 7.

State transition diagram of an N-parallel system is still possible to be presented for arbitrary N if the components are identical. For a parallel system with all CFR components having identical failure rates λ, the system states can be defined as the numbers of good components with values N, $N-1, \ldots, 1, 0$. The transition from state N to state $N-1$ is due to one of the N good components failing, which makes the duration of stay in state N is the minimum of N component lifetimes. Thus, the transition rate from state N to state $N-1$ is the sum of the transition rates of N components, which is $N\lambda$. Because of the memoryless property of exponential distribution, other states follow the same argument. The resulting state transition diagram is Figure 4.16. Table 2.1 shows that this diagram is for N-hypoexponential distribution $(N\lambda, (N-1)\lambda, \ldots, \lambda)$. This result means the lifetime of an N-parallel system with identical components has N-hypoexponential distribution. Its representations can be deduced from Table 2.1 as follows:

- Diagram representation:

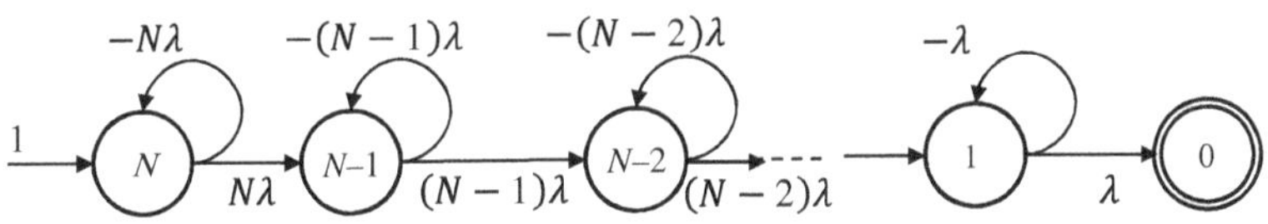

Figure 4.16 State transition diagram of a parallel system with identical CFR components.

- Matrix representation:

$$n_S = N, \quad \boldsymbol{a}_S = \begin{bmatrix} 1 & 0 \ldots 0 \end{bmatrix}, \quad \boldsymbol{A}_S = \begin{bmatrix} -N\lambda & N\lambda & \ldots & 0 & 0 \\ 0 & -(N-1)\lambda & \ldots & 0 & 0 \\ \vdots & \vdots & \ddots & \vdots & \vdots \\ 0 & 0 & \ldots & -2\lambda & 2\lambda \\ 0 & 0 & \ldots & 0 & -\lambda \end{bmatrix}. \tag{4.39}$$

- Scalar representation:

$$R_S(t) = \sum_{i=1}^{N} \frac{e^{-\lambda_i t}}{P_i} \text{ with } P_i = \prod_{j=1, j\neq i}^{N} \left(1 - \frac{\lambda_i}{\lambda_j}\right), \tag{4.40}$$

where $\lambda_i = (N+1-i)\lambda$.

The system lifetime T_S according to Figure 4.16 is the sum of the duration of stay in each state,

$$T_S = V_N + V_{N-1} + \cdots + V_1, \tag{4.41}$$

where V_i is the duration of staying in state $i, i = N, N-1, \ldots, 1$. From the memoryless property of exponential distributions, V_i, $i = N, N-1, \ldots, 1$, are independent random variables having exponential distribution with rate $i\lambda$, or $E[V_i] = 1/i\lambda$, $i = 1, 2, \ldots, N$.

Independence of V_i, $i = N, N-1, \ldots, 1$, in Equation (4.41) leads to

$$E[T_S] = E[V_N] + E[V_{N-1}] + \cdots + E[V_1].$$

$$Var[T_S] = Var[V_N] + Var[V_{N-1}] + \cdots + Var[V_1].$$

Thus, the system $MTTF$ and $VTTF$ can be calculated as

$$MTTF_S = \frac{1}{N\lambda} + \frac{1}{(N-1)\lambda} + \cdots + \frac{1}{\lambda} = \frac{1}{\lambda}\sum_{i=1}^{N}\frac{1}{i}. \tag{4.42}$$

$$VTTF_S = \frac{1}{(N\lambda)^2} + \frac{1}{((N-1)\lambda)^2} + \cdots + \frac{1}{\lambda^2} = \frac{1}{\lambda^2}\sum_{i=1}^{N}\frac{1}{i^2}. \tag{4.43}$$

Since $MTTF_{comp} = 1/\lambda$ and $VTTF_{comp} = 1/\lambda^2 = (MTTF_{comp})^2$, Equations (4.42) and (4.43) can be rewritten as

$$MTTF_S = MTTF_{comp}\sum_{i=1}^{N}\frac{1}{i}. \tag{4.44}$$

$$VTTF_S = (MTTF_{comp})^2\sum_{i=1}^{N}\frac{1}{i^2}. \tag{4.45}$$

Example 4.14 Three identical CFR components with $MTTF = 1$ are arranged in parallel. The PH representation of the distribution of the system lifetime calculated using Equation (4.39) for $N = 3$ and $\lambda = 1$ is

$$n_S = 3, \boldsymbol{a}_S = \begin{bmatrix}1 & 0 & 0\end{bmatrix}, \boldsymbol{A}_S = \begin{bmatrix}-3 & 3 & 0\\ 0 & -2 & 2\\ 0 & 0 & -1\end{bmatrix}.$$

Calculated using Algorithm 4.2 or Equation (4.36) results in

$$n_S = 7, \quad \boldsymbol{a}_S = [1 \; 0 \; 0 \; 0 \; 0 \; 0 \; 0],$$

$$\boldsymbol{A}_S = \begin{bmatrix}-1-1-1 & 1 & 1 & 1 & 0 & 0 & 0\\ 0 & -1-1 & 0 & 0 & 1 & 0 & 1\\ 0 & 0 & -1-1 & 0 & 0 & 1 & 1\\ 0 & 0 & 0 & -1-1 & 1 & 1 & 0\\ 0 & 0 & 0 & 0 & -1 & 0 & 0\\ 0 & 0 & 0 & 0 & 0 & -1 & 0\\ 0 & 0 & 0 & 0 & 0 & 0 & -1\end{bmatrix} = \begin{bmatrix}-3 & 1 & 1 & 1 & 0 & 0 & 0\\ 0 & -2 & 0 & 0 & 1 & 0 & 1\\ 0 & 0 & -2 & 0 & 0 & 1 & 1\\ 0 & 0 & 0 & -2 & 1 & 1 & 0\\ 0 & 0 & 0 & 0 & -1 & 0 & 0\\ 0 & 0 & 0 & 0 & 0 & -1 & 0\\ 0 & 0 & 0 & 0 & 0 & 0 & -1\end{bmatrix}.$$

These two seemingly different results are actually identical if the three identical states (g, g, f), (g, f, g), and (f, g, g) are lumped into one state 2 and identical states (g, f, f), (f, g, f), and (f, f, g) are lumped into one state 1. The elements of the survival matrix associated with those states are then combined, as shown with rectangle blocks in $\boldsymbol{A}_S$. Lumping several states into one super-state will reduce the order of the PH representation; this example shows that the order is reduced from 7 to 3.

The system $MTTF$ and $VTTF$ calculated using Equations (4.44) and (4.45) are

$$MTTF_S = 1\sum_{i=1}^{3}\frac{1}{i} = 1 + \frac{1}{2} + \frac{1}{3} = \frac{11}{6}.$$

$$VTTF_S = (1)^2\sum_{i=1}^{N}\frac{1}{i^2} = 1 + \frac{1}{4} + \frac{1}{9} = \frac{49}{9}.$$

■

Exercise 4.10 Use **phmttf** and **phvttf** functions to calculate the system $MTTF$ and $VTTF$ from the two resulting PH representations in Example 4.14. Show that they are equal, and also equal to the results obtained using Equations (4.44) and (4.45).

Next, the algorithm is applied to parallel systems with components having lifetimes following higher order PH distributions.

4.4.3 Parallel Systems with Non-CFR Components

It is difficult to obtain a simple expression for the PH representation of the lifetime distribution of a parallel system in general. However, as for CFR components, it is still possible to derive a simple expression for parallel systems with identical non-CFR components. For systems with identical components, the system states are sufficiently represented by the number of good components. The new defined system states makes Figure 4.15 can be simplified by lumping the states (g, g, f), (g, f, g), and (f, g, g) into state 2 and the states (g, f, f), (f, g, f), and (f, f, g) into state 1. For identical components, $\boldsymbol{A}_1 = \boldsymbol{A}_2 = \boldsymbol{A}_3 = \boldsymbol{A}$, and $\boldsymbol{b}_1 = \boldsymbol{b}_2 = \boldsymbol{b}_3 = \boldsymbol{b}$. Thus, Figure 4.15 reduces to Figure 4.17.

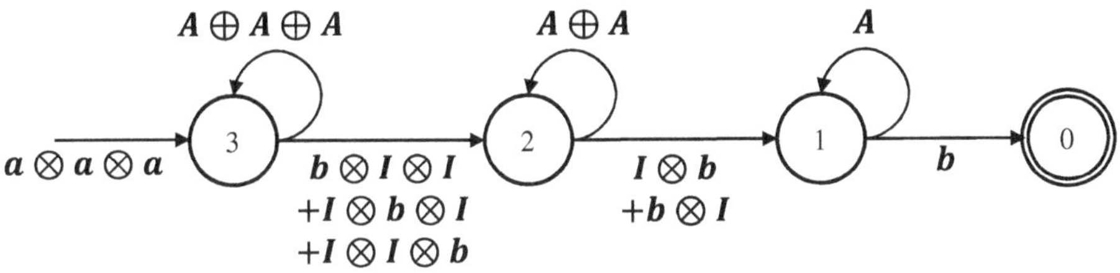

Figure 4.17 State transition diagram of a 3-parallel system with identical components.

The transition from state 3 to state 2 is due to a failure of one of the three components, either component 1, component 2, or component 3. The transition due to the failure of component 1, which means components 2 and 3 are good, has a failure vector $\boldsymbol{b} \otimes \boldsymbol{I} \otimes \boldsymbol{I}$. Likewise, the failure probability vector due to the failure of component 2 is $\boldsymbol{I} \otimes \boldsymbol{b} \otimes \boldsymbol{I}$ and that of due to the failure of component 3 is $\boldsymbol{I} \otimes \boldsymbol{I} \otimes \boldsymbol{b}$. Therefore, the transition vector from state 3 to state 2 is the sum of these three probability vectors, which is $\boldsymbol{b} \otimes \boldsymbol{I} \otimes \boldsymbol{I} + \boldsymbol{I} \otimes \boldsymbol{b} \otimes \boldsymbol{I} + \boldsymbol{I} \otimes \boldsymbol{I} \otimes \boldsymbol{b}$. To simplify the expressions, write $\boldsymbol{A} \oplus \boldsymbol{A} \oplus \boldsymbol{A}$ as $(\oplus \boldsymbol{A})^3$, $\boldsymbol{a} \otimes \boldsymbol{a} \otimes \boldsymbol{a}$ as $(\otimes \boldsymbol{a})^3$, and $\boldsymbol{b} \otimes \boldsymbol{I} \otimes \boldsymbol{I} + \boldsymbol{I} \otimes \boldsymbol{b} \otimes \boldsymbol{I} + \boldsymbol{I} \otimes \boldsymbol{I} \otimes \boldsymbol{b}$ as $\sum \otimes (\boldsymbol{I}^2, \boldsymbol{b})$. The state transition diagram of an N-parallel system with identical components can be constructed using the simplified notations as presented in Figure 4.18.

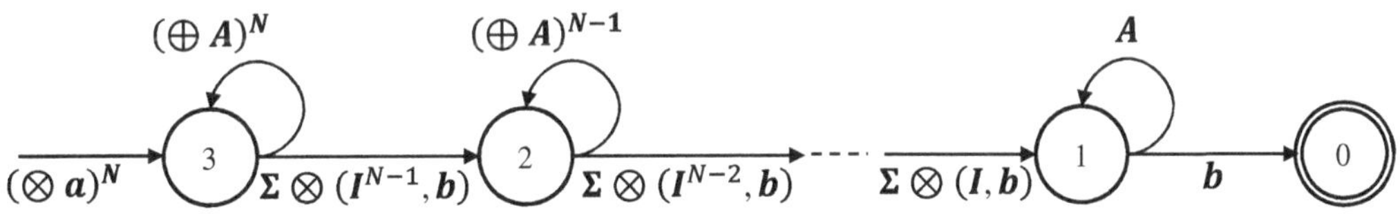

Figure 4.18 State transition diagram of N-parallel system with identical components.

The PH distribution of the system lifetime has a representation that can be derived from Figure 4.18 as

$$n_S = n^N + n^{N-1} + \cdots + n, \boldsymbol{a}_S = \left[(\otimes \boldsymbol{a})^N \ \boldsymbol{0} \ \ldots \ \boldsymbol{0} \right],$$

$$\boldsymbol{A}_S = \begin{bmatrix} (\oplus \boldsymbol{A})^N & \sum \otimes (\boldsymbol{I}^{N-1}, \boldsymbol{b}) & \ldots & \boldsymbol{0} & \boldsymbol{0} \\ \boldsymbol{0} & (\oplus \boldsymbol{A})^{N-1} & \ldots & \boldsymbol{0} & \boldsymbol{0} \\ \vdots & \vdots & \ddots & \vdots & \vdots \\ \boldsymbol{0} & \boldsymbol{0} & \ldots & (\oplus \boldsymbol{A})^2 & \sum \otimes (\boldsymbol{I}, \boldsymbol{b}) \\ \boldsymbol{0} & \boldsymbol{0} & \ldots & \boldsymbol{0} & \boldsymbol{A} \end{bmatrix}. \tag{4.46}$$

Example 4.15 The PH representation of a 4-parallel system with identical components can be derived using Equation (4.46) for $N = 4$ to yield

$$n_S = n^4 + n^3 + n^2 + n, \quad \boldsymbol{a}_S = \left[(\otimes\boldsymbol{a})^4 \ \boldsymbol{0} \ \boldsymbol{0} \ \boldsymbol{0}\right],$$

$$\boldsymbol{A}_S = \begin{bmatrix} (\oplus\boldsymbol{A})^4 & \sum\otimes(\boldsymbol{I}^3, \boldsymbol{b}) & \boldsymbol{0} & \boldsymbol{0} \\ \boldsymbol{0} & (\oplus\boldsymbol{A})^3 & \sum\otimes(\boldsymbol{I}^2, \boldsymbol{b}) & \boldsymbol{0} \\ \boldsymbol{0} & \boldsymbol{0} & (\oplus\boldsymbol{A})^2 & \sum\otimes(\boldsymbol{I}, \boldsymbol{b}) \\ \boldsymbol{0} & \boldsymbol{0} & \boldsymbol{0} & \boldsymbol{A} \end{bmatrix},$$

where

$$(\oplus\boldsymbol{A})^4 = \boldsymbol{A} \oplus \boldsymbol{A} \oplus \boldsymbol{A} \oplus \boldsymbol{A}, (\oplus\boldsymbol{A})^3 = \boldsymbol{A} \oplus \boldsymbol{A} \oplus \boldsymbol{A}, (\oplus\boldsymbol{A})^2 = \boldsymbol{A} \oplus \boldsymbol{A},$$

$$\sum\otimes(\boldsymbol{I}^3, \boldsymbol{b}) = \boldsymbol{b} \otimes \boldsymbol{I} \otimes \boldsymbol{I} \otimes \boldsymbol{I} + \boldsymbol{I} \otimes \boldsymbol{b} \otimes \boldsymbol{I} \otimes \boldsymbol{I} + \boldsymbol{I} \otimes \boldsymbol{I} \otimes \boldsymbol{b} \otimes \boldsymbol{I} + \boldsymbol{I} \otimes \boldsymbol{I} \otimes \boldsymbol{I} \otimes \boldsymbol{b},$$

$$\sum\otimes(\boldsymbol{I}^2, \boldsymbol{b}) = \boldsymbol{b} \otimes \boldsymbol{I} \otimes \boldsymbol{I} + \boldsymbol{I} \otimes \boldsymbol{b} \otimes \boldsymbol{I} + \boldsymbol{I} \otimes \boldsymbol{I} \otimes \boldsymbol{b}, \sum\otimes(\boldsymbol{I}, \boldsymbol{b}) = \boldsymbol{b} \otimes \boldsymbol{I} + \boldsymbol{I} \otimes \boldsymbol{b}.$$

■

The survival matrix in Equation (4.46) can be generated recursively starting from the last column until the first column using backward recursion equations

$$n_S = n_S n + n, \tag{4.47}$$

$$\boldsymbol{a}_S = \left[\boldsymbol{a}_S \otimes \boldsymbol{a} \ \boldsymbol{0}\right], \tag{4.48}$$

$$\boldsymbol{I}_S = \boldsymbol{I}_S \otimes \boldsymbol{I}, \tag{4.49}$$

$$\boldsymbol{A}_S(N-k, N-k) = \boldsymbol{A}_S(N-k+1, N-k+1) \oplus \boldsymbol{A}, \tag{4.50}$$

$$\boldsymbol{A}_S(N-k-1, N-k) = \boldsymbol{A}_S(N-k, N-k+1) \otimes \boldsymbol{I} + \boldsymbol{I}_S \otimes \boldsymbol{b}, \tag{4.51}$$

for $k = 1, 2, \dots, N-1$, except Equation (4.51), is until $k = N - 2$, with initial conditions $n_S = n$, $\boldsymbol{a}_S = \boldsymbol{a}$, $\boldsymbol{A}_S(N, N) = \boldsymbol{A}$, $\boldsymbol{A}_S(N-1, N) = \boldsymbol{b} \otimes \boldsymbol{I} + \boldsymbol{I} \otimes \boldsymbol{b}$, $\boldsymbol{I}_S = \boldsymbol{I}$.

The algorithm is referred to as Algorithm 4.3 and is presented as MATLAB function **phparallelidnsys** available in the software package **dysrel-am**. Its pseudo-code is as follows:

Algorithm 4.3

input $N, (n, \boldsymbol{a}, \boldsymbol{A})$
$n_S = n$
$\boldsymbol{a}_S = \boldsymbol{a}$
$\boldsymbol{I}_S = \boldsymbol{I}$
$\boldsymbol{b} = -\boldsymbol{A}\boldsymbol{u}$
$\boldsymbol{A}_S(N, N) = \boldsymbol{A}$
$\boldsymbol{A}_S(N-1, N) = \boldsymbol{b} \otimes \boldsymbol{I} + \boldsymbol{I} \otimes \boldsymbol{b}$
for $k = 1$ to $N - 1$
  $n_S = n_S n + n$
  $\boldsymbol{a}_S = \left[\boldsymbol{a}_S \otimes \boldsymbol{a} \ \boldsymbol{0}\right]$
  $\boldsymbol{I}_S = \boldsymbol{I}_S \otimes \boldsymbol{I}$
    $\boldsymbol{A}_S(N-k, N-k) = \boldsymbol{A}_S(N-k+1, N-k+1) \oplus \boldsymbol{A}$
  if $k < N - 1$ then
    $\boldsymbol{A}_S(N-k-1, N-k) = \boldsymbol{A}_S(N-k, N-k+1) \otimes \boldsymbol{I} + \boldsymbol{I}_S \otimes \boldsymbol{b}$
    $k = k + 1$
end
end

It is clear from Example 4.15 that a parallel system with identical components treated using Algorithm 4.3 will have a smaller order PH distribution than using Algorithm 4.2 for nonidentical components. The order of the PH distribution of the lifetime of an N-parallel system with identical components having a lifetime following PH distribution of order n is $n^N + n^{N-1} + \cdots + n$, while the order for the same system with nonidentical components is $C_N^N n^N + C_N^{N-1} n^{N-1} + C_N^{N-2} n^{N-2} \ldots + C_N^1 n$, a significant reduction. Therefore, it is necessary to use **phparallelidnsys** instead of **phparallelsys** whenever possible, especially considering that, in most cases, parallel systems are made of identical components.

Example 4.16 A 2-parallel system has identical components with lifetimes following 2-Erlang distributions with $MTTF = 2$.

PH representations of components 1 and 2 are

$$n_1 = n_2 = 1, \quad \boldsymbol{a}_1 = \boldsymbol{a}_2 = \begin{bmatrix} 1 & 0 \end{bmatrix}, \quad \boldsymbol{A}_1 = \boldsymbol{A}_2 = \begin{bmatrix} -1 & 1 \\ 0 & -1 \end{bmatrix}, \quad \boldsymbol{b}_1 = \boldsymbol{b}_2 = \begin{bmatrix} 0 \\ 1 \end{bmatrix}.$$

Using Algorithm 4.2 yields

$$n_S = n \times n + n + n = 2 \times 2 + 2 + 2 = 8.$$

$$\boldsymbol{a}_S = \begin{bmatrix} \boldsymbol{a}_1 \otimes \boldsymbol{a}_2 & \boldsymbol{0} & \boldsymbol{0} \end{bmatrix} = \begin{bmatrix} [1\ 0] \otimes [1\ 0] & [0\ 0] & [0\ 0] \end{bmatrix} = \begin{bmatrix} 1 & 0 & 0 & 0 & 0 & 0 & 0 & 0 \end{bmatrix}.$$

$$\boldsymbol{A}_S = \begin{bmatrix} \boldsymbol{A}_1 \oplus \boldsymbol{A}_2 & \boldsymbol{I} \otimes \boldsymbol{b}_2 & \boldsymbol{b}_1 \otimes \boldsymbol{I} \\ \boldsymbol{0} & \boldsymbol{A}_1 & \boldsymbol{0} \\ \boldsymbol{0} & \boldsymbol{0} & \boldsymbol{A}_2 \end{bmatrix} = \begin{bmatrix} \begin{bmatrix} -1 & 1 \\ 0 & -1 \end{bmatrix} \oplus \begin{bmatrix} -1 & 1 \\ 0 & -1 \end{bmatrix} & \begin{bmatrix} 1 & 0 \\ 0 & 1 \end{bmatrix} \otimes \begin{bmatrix} 0 \\ 1 \end{bmatrix} & \begin{bmatrix} 0 \\ 1 \end{bmatrix} \otimes \begin{bmatrix} 1 & 0 \\ 0 & 1 \end{bmatrix} \\ \boldsymbol{0} & \begin{bmatrix} -1 & 1 \\ 0 & -1 \end{bmatrix} & \boldsymbol{0} \\ \boldsymbol{0} & \boldsymbol{0} & \begin{bmatrix} -1 & 1 \\ 0 & -1 \end{bmatrix} \end{bmatrix}$$

$$= \begin{bmatrix} \begin{bmatrix} -2 & 1 & 1 & 0 \\ 0 & -2 & 0 & 1 \\ 0 & 0 & -2 & 1 \\ 0 & 0 & 0 & -2 \end{bmatrix} & \begin{bmatrix} \begin{bmatrix} 0 \\ 1 \end{bmatrix} & \begin{bmatrix} 0 \\ 0 \end{bmatrix} \\ \begin{bmatrix} 0 \\ 0 \end{bmatrix} & \begin{bmatrix} 0 \\ 1 \end{bmatrix} \end{bmatrix} & \begin{bmatrix} \begin{bmatrix} 0 & 0 \\ 0 & 0 \end{bmatrix} \\ \begin{bmatrix} 1 & 0 \\ 0 & 1 \end{bmatrix} \end{bmatrix} \\ \boldsymbol{0} & \begin{bmatrix} -1 & 1 \\ 0 & -1 \end{bmatrix} & \boldsymbol{0} \\ \boldsymbol{0} & \boldsymbol{0} & \begin{bmatrix} -2 & 2 \\ 0 & -2 \end{bmatrix} \end{bmatrix} = \begin{bmatrix} -2 & 1 & 1 & 0 & 0 & 0 & 0 & 0 \\ 0 & -2 & 0 & 1 & 1 & 0 & 0 & 0 \\ 0 & 0 & -2 & 1 & 0 & 0 & 1 & 0 \\ 0 & 0 & 0 & -2 & 0 & 1 & 0 & 1 \\ 0 & 0 & 0 & 0 & -1 & 1 & 0 & 0 \\ 0 & 0 & 0 & 0 & 0 & -1 & 0 & 0 \\ 0 & 0 & 0 & 0 & 0 & 0 & -1 & 1 \\ 0 & 0 & 0 & 0 & 0 & 0 & 0 & -1 \end{bmatrix}.$$

Using Algorithm 4.3 yields

$$n_S = n \times n + n = 4 + 2 = 6.$$

$$\boldsymbol{a}_S = \begin{bmatrix} \boldsymbol{a}_1 \otimes \boldsymbol{a}_2 & \boldsymbol{0} \end{bmatrix} = \begin{bmatrix} [1\ 0] \otimes [1\ 0] & [0\ 0] \end{bmatrix} = \begin{bmatrix} 1 & 0 & 0 & 0 & 0 & 0 \end{bmatrix}.$$

$$\boldsymbol{A}_S(2,2) = \boldsymbol{A} = \begin{bmatrix} -1 & 1 \\ 0 & -1 \end{bmatrix},$$

$$\boldsymbol{A}_S(1,2) = \boldsymbol{b} \otimes \boldsymbol{I} + \boldsymbol{I} \otimes \boldsymbol{b} = \begin{bmatrix} 0 \\ 1 \end{bmatrix} \otimes \begin{bmatrix} 1 & 0 \\ 0 & 1 \end{bmatrix} + \begin{bmatrix} 1 & 0 \\ 0 & 1 \end{bmatrix} \otimes \begin{bmatrix} 0 \\ 1 \end{bmatrix} = \begin{bmatrix} 0 & 0 \\ 0 & 0 \\ 1 & 0 \\ 0 & 1 \end{bmatrix} + \begin{bmatrix} 0 & 0 \\ 1 & 0 \\ 0 & 0 \\ 0 & 1 \end{bmatrix} = \begin{bmatrix} 0 & 0 \\ 1 & 0 \\ 1 & 0 \\ 0 & 2 \end{bmatrix},$$

$$\boldsymbol{A}_S(1,1) = \boldsymbol{A}_S(2,2) \oplus \boldsymbol{A} = \begin{bmatrix} -1 & 1 \\ 0 & -1 \end{bmatrix} \oplus \begin{bmatrix} -1 & 1 \\ 0 & -1 \end{bmatrix} = \begin{bmatrix} -2 & 1 & 1 & 0 \\ 0 & -2 & 0 & 1 \\ 0 & 0 & -2 & 1 \\ 0 & 0 & 0 & -2 \end{bmatrix},$$

$$\boldsymbol{A}_S = \begin{bmatrix} \boldsymbol{A}_S(1,1) & \boldsymbol{A}_S(1,2) \\ 0 & \boldsymbol{A}_S(2,2) \end{bmatrix} = \begin{bmatrix} \begin{bmatrix} -2 & 1 & 1 & 0 \\ 0 & -2 & 0 & 1 \\ 0 & 0 & -2 & 1 \\ 0 & 0 & 0 & -2 \end{bmatrix} & \begin{bmatrix} 0 & 0 \\ 1 & 0 \\ 1 & 0 \\ 0 & 2 \end{bmatrix} \\ \mathbf{0} & \begin{bmatrix} -1 & 1 \\ 0 & -1 \end{bmatrix} \end{bmatrix} = \begin{bmatrix} -2 & 1 & 1 & 0 & 0 & 0 \\ 0 & -2 & 0 & 1 & 1 & 0 \\ 0 & 0 & -2 & 1 & 1 & 0 \\ 0 & 0 & 0 & -2 & 0 & 2 \\ 0 & 0 & 0 & 0 & -1 & 1 \\ 0 & 0 & 0 & 0 & 0 & -1 \end{bmatrix}.$$

For comparison, the PH representations obtained using **phparallelsys** is

```
ns =
     8
as =
     1     0     0     0     0     0     0     0
As =
    -2     1     1     0     0     0     0     0
     0    -2     0     1     1     0     0     0
     0     0    -2     1     0     0     1     0
     0     0     0    -2     0     1     0     1
     0     0     0     0    -1     1     0     0
     0     0     0     0     0    -1     0     0
     0     0     0     0     0     0    -1     1
     0     0     0     0     0     0     0    -1
```

and using **phparallelidnsys** is

```
ns =
     6
as =
     1     0     0     0     0     0
As =
    -2     1     1     0     0     0
     0    -2     0     1     1     0
     0     0    -2     1     1     0
     0     0     0    -2     0     2
     0     0     0     0    -1     1
     0     0     0     0     0    -1
```

which are equal to the results of using manual calculations. ■

It is apparent from Example 4.16 that using **phparallelidnsys** leads to a simpler PH representation than using **phparallelsys**; hence, **phparallelidnsys** should be employed whenever possible. This example also shows that manual calculations are difficult for N-parallel systems with components having a higher order PH-distributed lifetime, even for systems with identical components. Using **phparallelsys** or **phparallelidnsys** is the only viable way to analyze N-parallel systems whose components have a high-order PH-distributed lifetime.

Example 4.17 The **phparallelsys** function is used to calculate the PH representation of the lifetime distribution of a 5-parallel system. The kth component of this system has lifetime following k-Erlang distributions with $MTTF$ = 100, $k = 1, 2, \ldots, 5$. The PH representation obtained from **phparallelsys** is used to calculate and plot the system reliability and MRL functions and also the system $MTTF$ and $SDTTF$.

The main program for this purpose is given in **example4_17.m** file in the software package **dysrel-am**. Its MATLAB code is as follows:

MATLAB code for example4_17.m

```
clear;close all;
addpath('..\dysrel-am','..\dysrel-am\expokit')
N = 5;                                          (inputs)
% PH Components
for k = 1:N
  T(k) = phnErlang(k,100);
end
% PH Parallel system                            (system model)
Ts = phparallelsys(T);
% Result in command window                      (statistical outputs)
mttf = phmttf(Ts)
sdttf = phsdttf(Ts)
% Time vector
tmax = 500;
t = linspace(0,tmax,21);
% Reliability plot                              (graphical outputs)
phrelplot(t,Ts);
% MRL plot
phmrlplot(t,Ts);
```

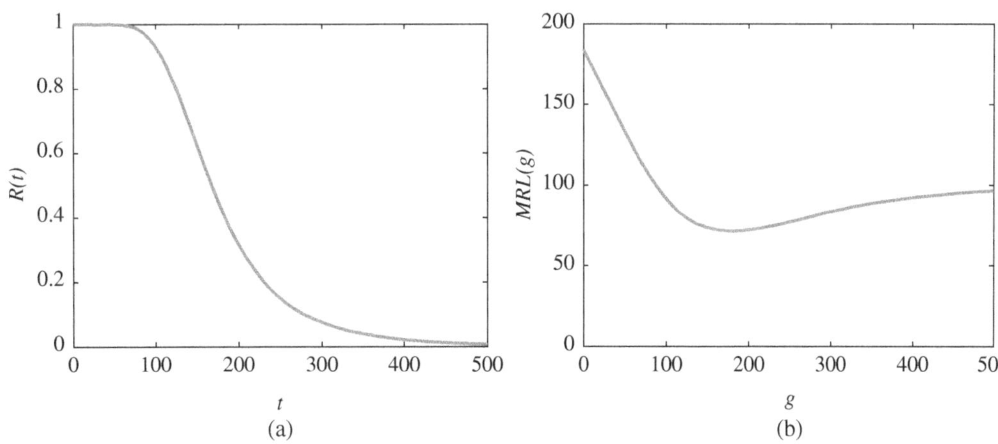

Figure 4.19 Outputs of example4_17.m.

Executing this program produces Figure 4.19.

```
≫ example4_17
mttf =
   184.2439
sdttf =
    79.2095
```

■

Seemingly different results obtained using Algorithms 4.2 and 4.3 in Example 4.16 actually are identical, as shown in Example 4.18.

Example 4.18 Functions **phparallelsys** and **phparallelidnsys** are compared for generating the PH representations of N-parallel systems whose component lifetimes follow an identical 4-Erlang distribution with $MTTF = 10$. The resulting PH distributions are used to generate the system reliability and hazard functions as shown in Figure 4.20. Table 4.2 presents the PH distribution order and computation times for producing Figure 4.20.

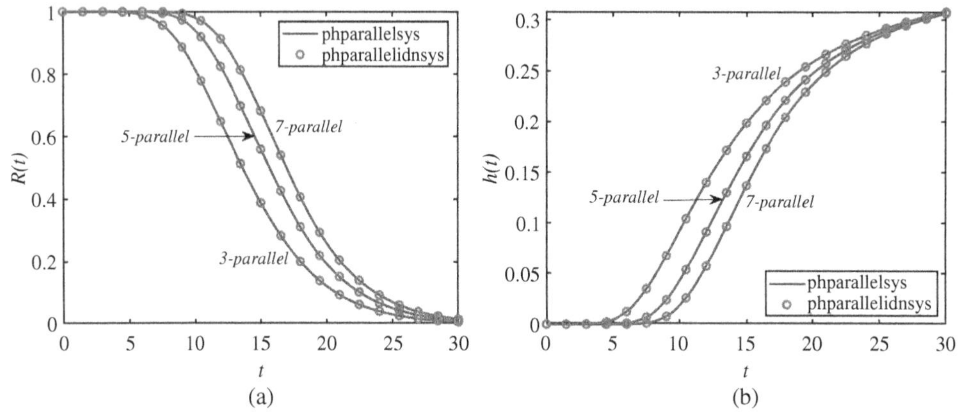

Figure 4.20 System reliability and hazard functions of N-parallel systems generated using phparallelsys and phparallelidnsys. (a) System reliability function and (b) system hazard function.

Table 4.2 Comparisons of PH distribution orders and computation times using phaparallelsys and phparallelidnsys.

	PH order		Computation time (s)	
N	**phparallelsys**	**phparallelidnsys**	**phparallelsys**	**phparallelidnsys**
2	24	20	0.0071	0.0100
3	124	84	0.0376	0.0332
4	624	340	0.0892	0.0563
5	3,124	1,364	0.4800	0.1469
6	15,624	5,460	1.9762	0.9040
7	78,124	21,844	9.4018	2.9006
8	390,624	87,380	106.9242	11.5435

Figure 4.20 shows that both functions produce the same result. However, according to Table 4.2, using **phparallelidnsys** for $N > 2$ saves computer memory and time, and the saving is higher for larger N. ■

Example 4.19 N identical IFR components from Example 3.2 are arranged in a parallel structure. The PH representations obtained using **phparallelsys** or **phparallelidnsys** are used to plot the system reliability measures for several values of N. The results are shown in Figure 4.21. The values of some reliability measures are given in Table 4.3.

These graphs and table can be used to determine the number of components in a parallel system to achieve certain design requirements. If the system must have reliability greater than 0.5 for a

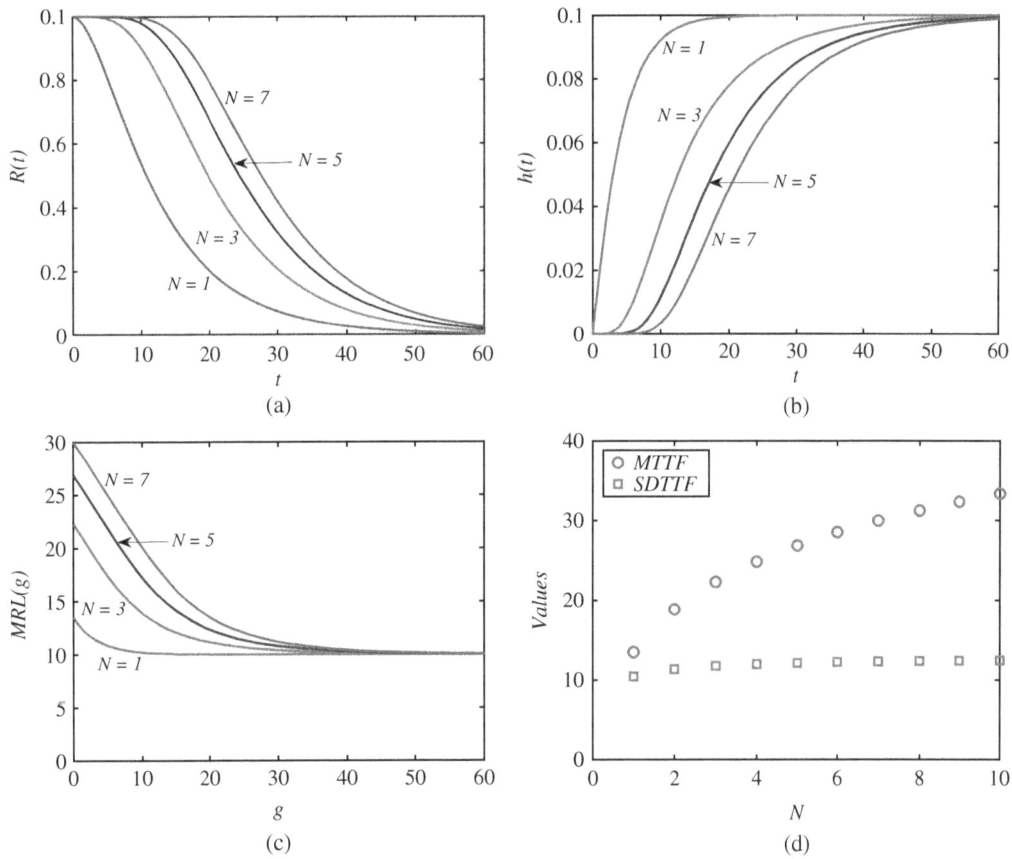

Figure 4.21 Reliability measures of the system in Example 4.19. (a) System reliability function, (b) system hazard function, (c) system *MRL* function, and (d) system *MTTF* and *SDTTF*.

Table 4.3 The system reliability and *MRL* values at $t = 20$ as functions of N.

	N						
Measures	**1**	**2**	**3**	**4**	**5**	**6**	**7**
Reliability	0.20	0.36	0.49	0.60	0.68	0.74	0.80
MRL	10.01	10.58	11.15	11.73	12.31	12.90	13.48

mission of 20 days, then N must be at least 4, or the system must be equipped with at least three active backup components. If the requirement is the system after being used in a mission for 20 days must have residual life expectancy higher than 11 days, then N must be at least 3, or the number of backup components must be at least 2. To satisfy the two requirements, the system must have a 4-parallel structure or have three active backups. ∎

Exercise 4.11 Determine the number of components, N, to make the system have reliability at least 0.8 and *MRL* at least 15 days for a 10-day mission.

Example 4.20 Example 4.19 is repeated for an N-parallel system having N identical DFR components in Example 4.10. The resulting reliability measures are presented in Figure 4.22 for several values of N. ∎

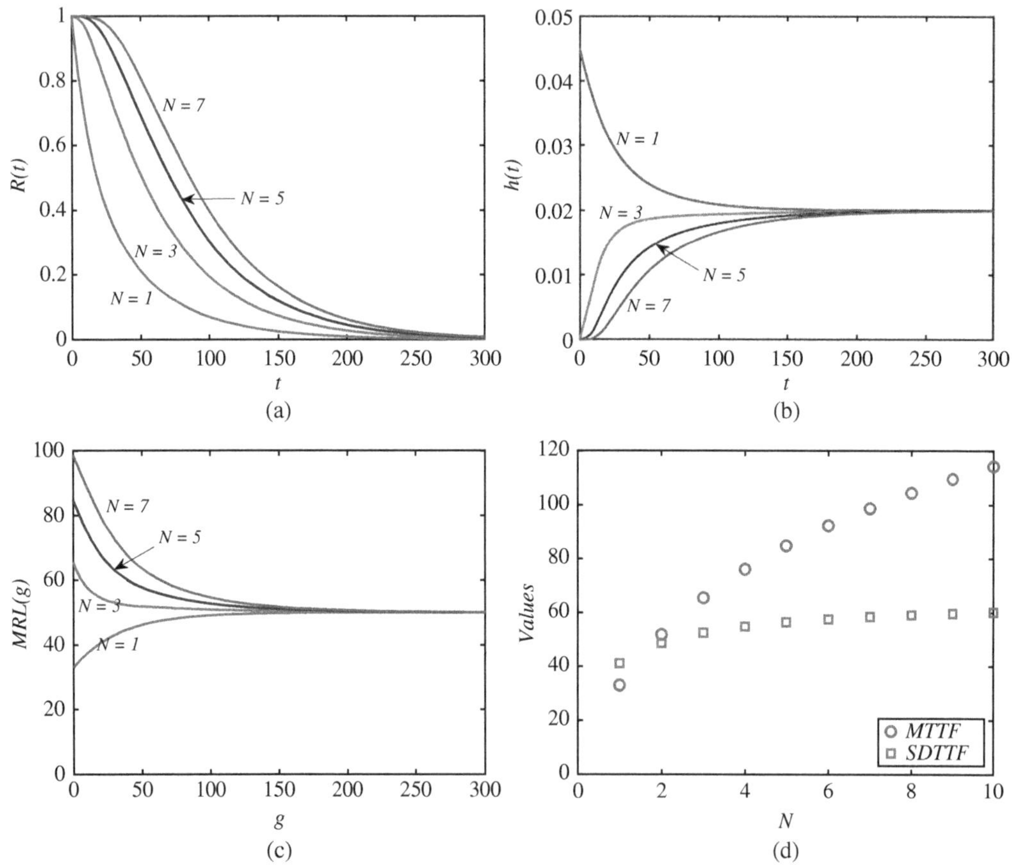

Figure 4.22 System reliability measures of Example 4.20. (a) System reliability function, (b) system hazard function, (c) system *MRL* function, and (d) system *MTTF* and *SDTTF*.

Examples 4.18–4.20 show that the system type is not always the same as the component type. When all components are IFR, the system is also IFR. However, when all components are DFR, the system can be IFR as shown in Figure 4.22b. When all components are identical CFR, the system is also IFR because the system lifetime for $N > 1$ has a hypoexponential distribution. This behavior is different from a series system where the system is always the same type as the components. Interestingly, *SDTTF* is constant after a certain value of N while *MTTF* keeps increasing, and thus *CVTTF* is decreasing, with N. Consequently, depending on the value of *SDTTF* of its components, all parallel systems eventually will have *CVTTF* less than 1, or the system becomes IFR. The value of *CVTTF* as a function of N for the system in Example 4.20 is given in Table 4.4, which shows that the system becomes IFR for $N > 1$.

Table 4.4 *CVTTF* of systems in Example 4.20 as a function of *N*.

N	1	2	3	4	5	6	7
CVTTF	1.2461	0.9387	0.8030	0.7218	0.6661	0.6248	0.5925

Exercise 4.12 Determine the largest value of component *CVTTF* so that N-parallel system constructed using this component is IFR for $N = 2$.

The next section discusses the properties of parallel systems.

4.4.4 Properties of Parallel Systems

The abovementioned result is referred to as Property 1 of parallel systems:

- All components are CFR ⇒ the parallel system is IFR.
- All components are IFR ⇒ the parallel system is IFR.
- All components are DFR ⇒ the parallel system is IFR for a sufficiently large N.

Parallel systems also have several other properties useful in designing parallel systems for reliability.

Using Equation (4.24) for component i, $i = 1, 2, \dots, N$, having lifetimes following PH distribution with representation $(n_i, \boldsymbol{a}_i, \boldsymbol{A}_i)$ results in

$$R_S(t) = 1 - \left(1 - \boldsymbol{a}_1 e^{\boldsymbol{A}_1 t}\boldsymbol{u}\right)\left(1 - \boldsymbol{a}_2 e^{\boldsymbol{A}_2 t}\boldsymbol{u}\right) \dots \left(1 - \boldsymbol{a}_N e^{\boldsymbol{A}_N t}\boldsymbol{u}\right),$$

which is not easy to simplify. Therefore, Equation (4.24) is not a practical way to find the reliability function of a parallel system even with components having PH-distributed lifetimes order 1.

Given $0 \le 1 - R_i(t) \le 1$, the multiplication value on the right side of Equation (4.24) is smaller for larger N. Since the subtracting term gets smaller, the value of the right-hand side becomes larger. Thus, the reliability function of the parallel system becomes higher as the number of components is increased. This statement is in accordance with Figures 4.21a and 4.22a. Define $R_S(t, N)$ as the reliability function of the N-parallel system, and $R_S(t, N+1)$ as the reliability function of the N-parallel system with the addition of one component. Using Equation (4.24) yields

$$\begin{aligned} 1 - R_S(t, N+1) &= \prod_{i=1}^{N+1}(1 - R_i(t)) = \left(\prod_{i=1}^{N}(1 - R_i(t))\right)(1 - R_{N+1}(t)) \\ &= (1 - R_S(t, N))(1 - R_{N+1}(t)) \le 1 - R_S(t, N). \end{aligned}$$

Hence,

$$R_S(t, N+1) \ge R_S(t, N). \tag{4.52}$$

Equation (4.52) is referred to as Property 2 of parallel systems, which states that the reliability of a parallel system increases with the number of its components.

Following the logic for the series system, Equation (4.52) implies

$$R_S(t) \geq \max_i [R_i(t)]. \tag{4.53}$$

Equation (4.53) is referred to as Property 3 of parallel systems, which states that the reliability function of a parallel system at any time is higher than the reliability function of its best component at that time. In reliability sense, a parallel system at any time is always better than its best components. Equation (4.52) also suggests that having a less good parallel backup is still better than having no backup.

Since *MTTF* is the integral of the reliability function, Equation (4.53) implies that *MTTF* of a parallel system is larger than the largest *MTTF* of its components, that is

$$MTTF_S \geq \max_i [MTTF_i]. \tag{4.54}$$

Equation (4.54) is referred to as Property 4 of parallel systems, which states that the expected lifetime of a parallel system is longer than the expected lifetime of its best component, which is the component with the longest expected lifetime.

Finding a simple mathematical function that expresses the relationship between the system hazard function and the hazard functions of its components for a parallel system is not easy. Therefore, it is also not easy to mathematically ensure the existence of similar to Properties 5 and 6 in series systems for a parallel system. Figures 4.21b and 4.22b show that the more components there are, the smaller the hazard function of a parallel system will be. However, these two examples are for parallel systems with identical components.

Example 4.21 Three components in Example 4.11 are arranged in a parallel structure. The graphs of the reliability and hazard functions of the system and all its components are presented in Figure 4.23.

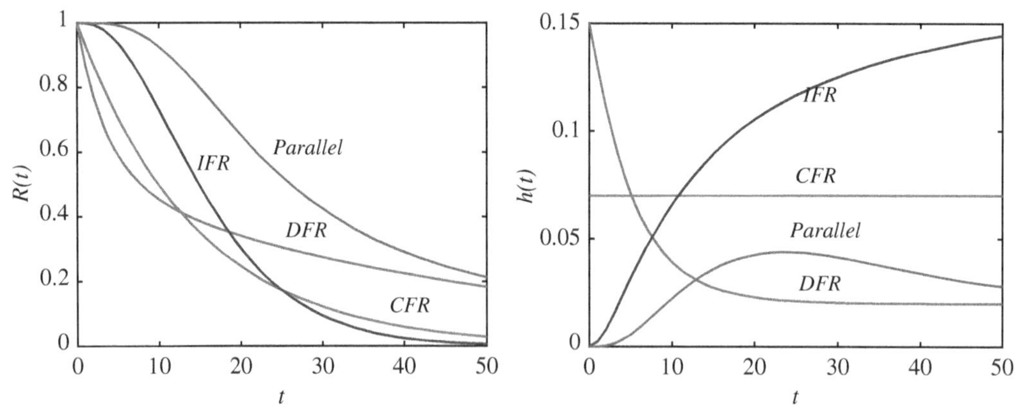

Figure 4.23 Reliability and hazard functions of Example 4.21. (a) Reliability functions and (b) hazard functions.

Figure 4.23a shows that the system reliability function at any time is always greater than the highest component reliability function at that time, in accordance with Property 2. The system *MTTF* is 38.6719, which is greater than the longest *MTTF* of its components, in accordance with

Property 4. However, Figure 4.23b shows that the system hazard function at any time is not always smaller than the smallest hazard function of its component at that time. ∎

Example 4.22 Example 4.12 is repeated for the components that are arranged in a parallel structure. The reliability and hazard functions of the parallel system and its components are presented in Figure 4.24, which shows that the system reliability function is greater than the reliability functions of its components. Still, the hazard function of the system at any time is not always smaller than the hazard function of its components.

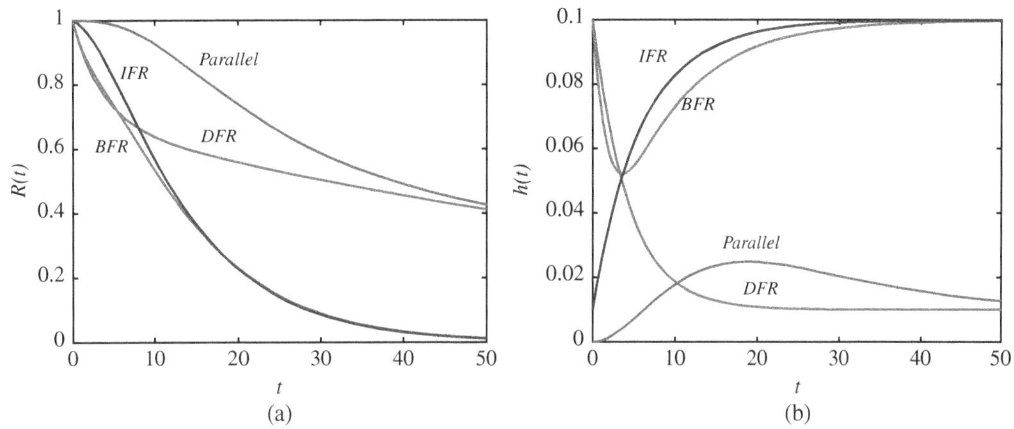

Figure 4.24 Reliability and hazard functions of Example 4.22. (a) Reliability function and (b) hazard function.

∎

The last two examples show that parallel systems with nonidentical components do not follow Properties 5 and 6 for the hazard function of series systems. Thus, Properties 5 and 6 for series systems are satisfied by parallel systems with identical components:

If a parallel system has identical components, then

$$h_S(t, N+1) \leq h_S(t, N). \tag{4.55}$$

$$h_S(t) \leq \min_i [h_i(t)]. \tag{4.56}$$

Determining the number of components that should be used to form a parallel system is a classic reliability optimization problem [17]. One simple problem involving reliability function is to find the value of N that minimizes the total cost of a mission. The costs involved are the system procurement that grows with N, and the cost due to mission failure that decreases with N. These costs depend on the mission time and so does the optimum value of N.

Application 4.1 A system with a lifetime following 4-Erlang distribution with $MTTF = 20$ days is employed to carry out a mission for s days. The mission may use several identical systems arranged in a parallel structure to increase the mission reliability. The interest is to find the number of such systems in parallel to minimize the expected mission cost. The costs associated with the mission are composed of the procurement cost $P = 10$ per system and the penalty cost due to mission failure $F = 200$ per mission. The cost of operating the systems is negligible compared to procurement and penalty costs.

The penalty cost is a random variable with a probability equal to the probability of system failure during the mission of s days, which is $1 - R(s)$. The expected mission total cost TC can be calculated as the sum of procurement cost and expected failure cost, which can be formulated as

$$TC = NP + F\,(1 - R(s)) = 10N + 200(1 - R(s)),$$

where N is the number of systems in a parallel structure for the mission.

The optimum values of N for several mission times are summarized in Table 4.5.

Table 4.5 The expected mission cost as a function of mission time and N.

	Mission time			
N	10	15	20	25
1	38.58	80.55	123.31	156.99
2	**24.08**	44.89	84.19	128.04
3	30.58	**38.78**	66.37	109.4
4	40.08	43.10	**60.60**	98.36
5	50.01	51.09	61.67	92.89
6	60	60.39	66.61	**91.53**
7	70	70.14	73.75	93.17

The optimal values are given in bold.

Table 4.5 shows that the optimum number of systems in parallel structure for the mission is higher for longer mission times. ■

The optimization problem in Application 4.1 becomes complicated if the operating cost is not negligible. This more general problem is a challenge for the interested reader.

Exercise 4.13 Investigate the effect of $MTTF$ of each system to the optimum number of systems in parallel structure.

The next section discusses systems with standby structure. This type of basic structure is seldom discussed in classic reliability books since it cannot be analyzed using static models based on reliability values and requires integral convolutions to derive the system reliability function from its component reliability functions. However, this structure is simpler than parallel structure if matrix-form distribution is used. Standby structure is also intended to make system reliability higher than the reliability of its components.

4.5 Standby Structured Systems

A standby structured system is a system whose components are arranged in a structure with one component actively operating (connected/online) and supported by several passive components that are not operating (disconnected/offline). The active component is called the primary component, and the passive components are called the backup components. When the primary component fails, the backup component at the top of the list is activated to take over the failed component and become the primary component. This process is repeated; the next failing primary component is replaced by the next backup component until there is no backup component that can replace the failing primary component. At that point, the system fails. Only active components can fail, while passive components cannot because they are not operated. The system needs a switch to activate an offline passive component. This switch is often assumed to be perfect and always functions properly whenever required to connect an offline backup component to become active as the primary component. A standby structured system is referred to simply as a standby system. A standby system with a perfect switch is simply called a standby system; in other words, the switch is generally assumed to be perfect.

Although both require only one good component to ensure the system is good, the standby structure differs from the parallel structure. The difference is that in the parallel structure, all components are operated simultaneously, while in the standby structure, only one is actively operated while the rest are not. The parallel structure is sometimes called a hot standby structure because the backup components are operated the same as the primary component; in contrast, the standby structure is sometimes called a cold standby structure because the backup components are not operated. This distinction is necessary when defining the general standby structure discussed in Chapter 5. A system with a standby structure has been represented as a reliability block diagram in Figure 4.3. A standby system with N components, which means that initially there is one primary component and N-1 backup components, is denoted as an N-standby system. Standby systems are found in practice as systems that require high reliability and the process of replacing a failing primary component with a backup does not take a long time to wait for the readiness of the replacement component. Cost considerations are important in determining whether to arrange the components in a parallel structure or a standby structure [14].

Since the components are operated one by one, backup components cannot fail, and the system fails when all components fail, the lifetime of an N-standby system is the sum of the lifetimes of its components,

$$T_S = T_1 + T_2 + \cdots + T_N = \text{add}(T_1, T_2, \dots, T_N). \tag{4.57}$$

Using Equation (4.57) yields

$$R_S(t) = P[T_S > t] = P[T_1 + T_2 + \cdots + T_N > t] = R_1(t) * R_2(t) * \dots * R_N(t),$$

where * is the convolution integral operator.

Unlike series and parallel systems, there is no algebraic expression that relates the system reliability function to component reliability functions. Therefore, it is not possible to determine the reliability value of a standby system directly from the reliability values of its components. It must be calculated through the system reliability function derived using N-fold convolution integrals of the reliability functions of its components, which is complicated conducting manually even for the simplest reliability function and even with the help of a computer. It is necessary to devise a simple programmable method to derive the parameters of reliability function of a standby system from the parameters of its components. Using matrix-form reliability function will make this task achievable.

A standby system with imperfect switches means the switch can fail when needed, resulting in system failure. Equation (4.57) must be modified by including a variable representing the switch status. For a 2-standby system with an imperfect switch, the system lifetime is

$$\begin{aligned} T_S &= T_1 && \text{if the switch fails} \\ &= T_1 + T_2 && \text{if the switch does not fail} \end{aligned}$$

Let X_1 be a Bernoulli random variable representing the switch status when component 1 fails, with values

$$X_1 = 1 \qquad \text{if the switch works}$$
$$= 0 \qquad \text{if the switch does not work}$$

Using X_1 yields

$$T_S = T_1 + X_1 T_2. \tag{4.58}$$

Assume that the switch is good with probability $p_1 = P[X_1 = 1]$. The value p_1 represents the reliability of the switch when component 1 fails, and the switch must place component 2 to become active. The state transition diagram for this system is given in Figure 4.25. The transition rate from component 1 to component 2 is corrected by p_1, and there is a direct transition to the absorbing state with a probability $1 - p_1$ when component 1 fails. This direct transition to the absorbing state is due to the fact the switch fails to activate component 2, which makes the system fail. The transition rate from component 1 to component 2 in Figure 4.25 is the product of $\boldsymbol{b}_1$, p_1, and $\boldsymbol{a}_2$. Thus, the survival matrix of a 2-standby system can be obtained from Theorem 2.3 with the correction p_1 to yield

$$\boldsymbol{a}_S = \begin{bmatrix} \boldsymbol{a}_1 & \boldsymbol{0} \end{bmatrix}, \quad \boldsymbol{A}_S = \begin{bmatrix} \boldsymbol{A}_1 & p_1 \boldsymbol{b}_1 \boldsymbol{a}_2 \\ \boldsymbol{0} & \boldsymbol{A}_2 \end{bmatrix}. \tag{4.59}$$

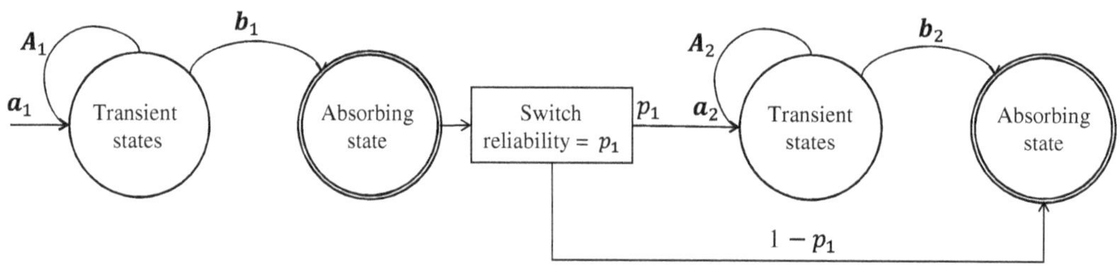

Figure 4.25 State transition diagram of a 2-standby system with imperfect switch.

If the switch is perfect, the state transition diagram for a 2-standby system becomes the same as in Figure 2.18 given to explain Theorem 2.3 without discontinuities.

For an N-standby system with an imperfect switch, Equations (4.57) and (4.58) can be developed recursively to yield [11]

$$T_S = T_1 + X_1(T_2 + X_2(T_3 + X_3(T_4 + \cdots + X_{N-1}T_N))). \tag{4.60}$$

If the switch is perfect, then $X_i = 1$, and Equation (4.60) becomes Equation (4.57).

An N-standby system is modeled as

$$T_S = \text{add}(T_1, T_2, \dots, T_N; p_1, p_2, \dots p_{N-1}), \tag{4.61}$$

where p_i is the switch reliability when component i is failing, and the switch must activate component $(i+1)$. For an N-standby system with the perfect switch, $p_i = 1$, $i = 1, 2, \dots, N$.

Exercise 4.14 Determine the infinitesimal generator of a 2-standby system.

Next, Equations (4.57) and (4.60) are translated into a systematic way to calculate the PH representation of standby system.

4.5.1 Algorithm for Calculating PH Representation of Standby Systems

The design of the algorithm for calculating the PH representation of the lifetime distribution of standby systems starts from a standby system with a perfect switch. The results are then developed further to imperfect switches. From Equation (4.57), it is apparent that the closure property of the PH distribution under the addition operation given by Theorem 2.3 can be used to design the algorithm. Since this theorem is for two variables, extending it to many variables recursively is necessary. By defining

$$T_S^{(i)} = T_1 + T_2 + \cdots + T_i,$$

Equation (4.57) can be expressed recursively as [11]

$$T_S^{(i)} = T_S^{(i-1)} + T_i \quad i = 2, 3, \ldots, N, \tag{4.62}$$

starting from the initial conditions $T_S^{(1)} = T_1$ that have a PH distribution. By repeatedly using Theorem 2.3 and assuming no discontinuity at the origin, or $a_i(n_i + 1) = 0$, $i = 1, 2, \ldots, N$, yield the following lemma.

Lemma 4.3 [11] If $T_i \backsim PH(n_i, \boldsymbol{a}_i, \boldsymbol{A}_i)$, $i = 1, 2, \ldots, N$, then T_S of an N-standby system with perfect switch has a PH distribution with representation $(n_S, \boldsymbol{a}_S, \boldsymbol{A}_S)$ given by the recursion equations

$$n_S^{(i)} = n_S^{(i-1)} + n_i, \tag{4.63}$$

$$\boldsymbol{a}_S^{(i)} = \begin{bmatrix}\boldsymbol{a}_S^{(i-1)} & \boldsymbol{0}\end{bmatrix}, \tag{4.64}$$

$$\boldsymbol{A}_S^{(i)} = \begin{bmatrix}\boldsymbol{A}_S^{(i-1)} & \boldsymbol{b}_S^{(i-1)}\boldsymbol{a}_i \\ \boldsymbol{0} & \boldsymbol{A}_i\end{bmatrix}, \tag{4.65}$$

for $i = 1, 2, 3 \ldots, N$, with the initial conditions given by Equation (4.9), and $\boldsymbol{b}_S^{(i-1)} = -\boldsymbol{A}_s^{(i-1)}\boldsymbol{u}$.

Lemma 4.3 is a generalization of Lemma 3.1 for identical T_i, $i = 1, 2, \ldots, N$.
Multiplying Equation (4.65) for $i - 1$ with $\boldsymbol{u}$ yields

$$\boldsymbol{b}_S^{(i-1)} = -\boldsymbol{A}_S^{(i-1)}\boldsymbol{u} = \begin{bmatrix} -\left(\boldsymbol{A}_S^{(i-2)}\boldsymbol{u} + \boldsymbol{b}_S^{(i-2)}\boldsymbol{a}_{i-1}\boldsymbol{u}\right) \\ -\boldsymbol{A}_{i-1}\boldsymbol{u}\end{bmatrix}. \tag{4.66}$$

Without discontinuity at the origin makes $\boldsymbol{a}_{i-1}\boldsymbol{u} = 1$ and

$$\boldsymbol{b}_S^{(i-2)}\boldsymbol{a}_{i-1}\boldsymbol{u} = \boldsymbol{b}_S^{(i-2)}, \quad \boldsymbol{A}_S^{(i-2)}\boldsymbol{u} = -\boldsymbol{b}_S^{(i-2)}. \tag{4.67}$$

Substituting Equation (4.67) into the first row of Equation (4.66) yields

$$\boldsymbol{A}_S^{(i-2)}\boldsymbol{u} + \boldsymbol{b}_S^{(i-2)}\boldsymbol{a}_{i-1}\boldsymbol{u} = -\boldsymbol{b}_S^{(i-2)} + \boldsymbol{b}_S^{(i-2)} = \boldsymbol{0}.$$

Multiply Equation (4.66) by $\boldsymbol{a}_i$ and using $-\boldsymbol{A}_{i-1}\boldsymbol{u} = \boldsymbol{b}_{i-1}$ yield

$$\boldsymbol{b}_S^{(i-1)}\boldsymbol{a}_i = \begin{bmatrix}\boldsymbol{0} \\ \boldsymbol{b}_{i-1}\boldsymbol{a}_i\end{bmatrix}. \tag{4.68}$$

Substituting Equation (4.68) into Equation (4.65) yields

$$\boldsymbol{A}_S^{(i)} = \left[\begin{array}{c|c}\boldsymbol{A}_S^{(i-1)} & \begin{array}{c}\boldsymbol{0} \\ \boldsymbol{b}_{i-1}\boldsymbol{a}_i\end{array} \\ \hline \boldsymbol{0} & \boldsymbol{A}_i\end{array}\right]. \tag{4.69}$$

Using Equation (4.69) recursively for $i = 1, 2, 3, 4, \ldots, N$, and Equation (4.59) with $p_i = 1$ for $i = 2$ produce the survival matrix of an N-standby system,

$$A_S = \begin{bmatrix} A_1 & b_1a_2 & 0 & 0 & 0 & 0 \\ 0 & A_2 & b_2a_3 & 0 & 0 & 0 \\ \vdots & \vdots & \ddots & \ddots & \vdots & \vdots \\ 0 & 0 & 0 & A_i & b_ia_{i+1} & 0 \\ \vdots & \vdots & \vdots & 0 & \ddots & \ddots \\ 0 & 0 & 0 & 0 & 0 & A_N \end{bmatrix}. \tag{4.70}$$

The order and initial probability vector can be calculated recursively using Equations (4.63) and (4.64) to yield

$$n_S = n_1 + n_2 + \cdots + n_N, \tag{4.71}$$

$$a_S = \left[a\ 0\ \ldots\ 0\right]. \tag{4.72}$$

The system survival matrix in Equation (4.70) is nonsingular because it is an upper block diagonal matrix with nonsingular block diagonal matrices.

Using Equations (4.70) and (4.72) produces the state transition diagram of an N-standby system as given in Figure 4.26.

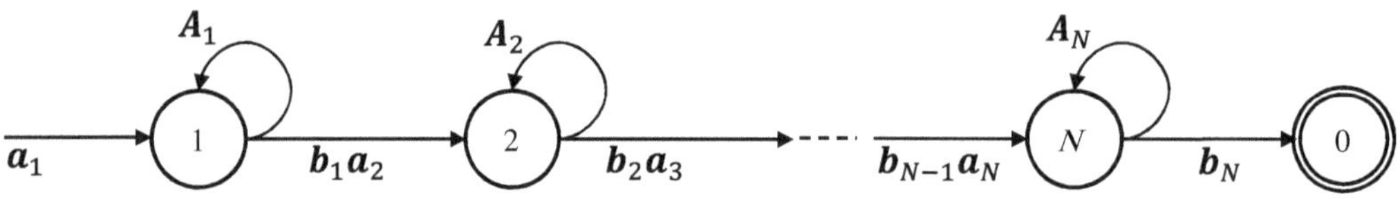

Figure 4.26 State transition diagram of an N-standby system with perfect switch.

The PH representation of a standby system with imperfect switches can be found using Equation (4.60), which can be expressed in a recursion equation

$$T_S^{(i)} = T_S^{(i-1)} + X_{i-1}T_i, \tag{4.73}$$

for $i = 2, 3, \ldots, N$, with the initial condition $T_S^{(1)} = T_1$ that has a PH distribution. Thus, the survival matrix of an N-standby system with imperfect switches can be obtained recursively using Equation (4.69) corrected by the switch reliability. The result is stated in Lemma 4.4.

Lemma 4.4 [11] If $T_i \backsim PH(n_i, a_i, A_i)$, $i = 1, 2, \ldots, N$, then the distribution of the lifetime T_S of an N-standby system with an imperfect switch has a PH representation (n_S, a_S, A_S) given by the recursion equations

$$n_S^{(i)} = n_S^{(i-1)} + n_i, \tag{4.74}$$

$$a_S^{(i)} = \left[a_S^{(i-1)}\ 0\right], \tag{4.75}$$

$$A_S^{(i)} = \left[\begin{array}{c|c} A_S^{(i-1)} & \begin{array}{c} 0 \\ p_{i-1}b_{i-1}a_i \end{array} \\ \hline 0 & A_i \end{array}\right], \tag{4.76}$$

for $i = 3, 4, \ldots, N$, with initial conditions for $i = 1$ and $i = 2$ are given by Equations (4.9) and (4.59), respectively.

If $p_i = 1$, then Equation (4.76) becomes Equation (4.69). Therefore, Equation (4.76) is applicable for any switch, perfect and imperfect. The algorithm for calculating the PH representation for the lifetime distribution of standby systems is made only for imperfect switches because the perfect switch is a special case of imperfect switches. The algorithm for calculating the PH representation of the distribution of the lifetime of N-standby system with an arbitrary switch is designed using Equations (4.74)–(4.76) and is called Algorithm 4.4. This algorithm is programmed as **phstandsys** function in MATLAB, available in the software package **dysrel-am**. This function accepts N, $(n_i, \boldsymbol{a}_i, \boldsymbol{A}_i)$, $i = 1, 2, \ldots, N$, and p_i, $i = 1, 2, \ldots, N-1$, as inputs. The pseudo-code for **phstandsys** is as follows:

Algorithm 4.4

---

input $N, p_i, i = 1, 2, \ldots, N-1, (n_i, \boldsymbol{a}_i, \boldsymbol{A}_i), i = 1, 2, \ldots, N$

$n_S = n_1$

$\boldsymbol{a}_S = \boldsymbol{a}_1$

$\boldsymbol{A}_S = \boldsymbol{A}_1$

for $i = 2$ to N

  $n_S = n_S + n_i$

  $\boldsymbol{a}_S = \begin{bmatrix}\boldsymbol{a}_S & \boldsymbol{0}\end{bmatrix}$

  $\boldsymbol{b}_i = -\boldsymbol{A}_i \boldsymbol{u}$

  if $i = 2$ then $\boldsymbol{A}_S = \begin{bmatrix}\boldsymbol{A}_S & p_{i-1}\boldsymbol{b}_{i-1}\boldsymbol{a}_i \\ \boldsymbol{0} & \boldsymbol{A}_i\end{bmatrix}$

  if $i > 2$ then $\boldsymbol{A}_S = \begin{bmatrix}\boldsymbol{A}_S & \begin{matrix}\boldsymbol{0} \\ p_{i-1}\boldsymbol{b}_{i-1}\boldsymbol{a}_i\end{matrix} \\ \boldsymbol{0} & \boldsymbol{A}_i\end{bmatrix}$

  $i = i + 1$

end

end

The recursion Equation (4.74) implies that the PH distribution for the lifetime of an N-standby system with an arbitrary switch has order equal to $n_1 + n_2 + \cdots + n_N$; hence, its complexity is nN, which is the order of the PH distribution of the lifetimes of the components multiply by the number of components. As for series and parallel systems, the survival matrix of standby systems with components having lifetimes following PH distributions is always nonsingular.

Next, the algorithm is first illustrated using standby systems with components having lifetimes following nonidentical exponential distributions.

4.5.2 Standby Systems with CFR Components

Algorithm 4.4 is illustrated by manually calculating PH representation of the distribution of lifetime of a standby system having nonidentical CFR components.

Example 4.23 A 3-standby system with a perfect switch is constructed using the same components as in Example 4.6. The PH representation of the distribution of its lifetime is derived using Equations (4.63)–(4.65).

The lifetime distributions of the components have PH representations:

$$n_1 = n_2 = n_3 = 1, \quad \boldsymbol{a}_1 = \boldsymbol{a}_2 = \boldsymbol{a}_3 = [1],$$

$$\boldsymbol{A}_1 = [-1], \quad \boldsymbol{A}_2 = [-0.5], \quad \boldsymbol{A}_3 = [-0.2], \quad \boldsymbol{b}_1 = [1], \quad \boldsymbol{b}_2 = [0.5], \quad \boldsymbol{b}_3 = [0.2].$$

Iterations using Equation (4.63)–(4.65) are as follows:

- Initial conditions: $n_S^{(1)} = n_1 = 1, \boldsymbol{a}_S^{(1)} = \boldsymbol{a}_1 = [1], \boldsymbol{A}_S^{(1)} = \boldsymbol{A}_1 = [-1]$.
- Second and third iterations for n_S using Equation (4.63):

$$n_S^{(2)} = n_S^{(1)} + n_2 = 1 + 1 = 2, n_S^{(3)} = n_S^{(2)} + n_3 = 2 + 1 = 3.$$

- Second and third iterations for $\boldsymbol{a}_S$ using Equation (4.64):

$$\boldsymbol{a}_S^{(2)} = \begin{bmatrix}\boldsymbol{a}_S^{(1)} & \boldsymbol{0}\end{bmatrix} = \begin{bmatrix}[1] & 0\end{bmatrix} = \begin{bmatrix}1 & 0\end{bmatrix},$$
$$\boldsymbol{a}_S^{(3)} = \begin{bmatrix}\boldsymbol{a}_S^{(2)} & \boldsymbol{0}\end{bmatrix} = \begin{bmatrix}[1\ 0] & 0\end{bmatrix} = \begin{bmatrix}1 & 0 & 0\end{bmatrix}.$$

- Second and third iterations for $\boldsymbol{A}_S$ using Equation (4.65):

$$\boldsymbol{A}_S^{(2)} = \begin{bmatrix}\boldsymbol{A}_S^{(1)} & \boldsymbol{b}_S^{(1)}\boldsymbol{a}_2 \\ \boldsymbol{0} & \boldsymbol{A}_2\end{bmatrix} = \begin{bmatrix}[-1] & [1][1] \\ 0 & [-0.5]\end{bmatrix} = \begin{bmatrix}-1 & 1 \\ 0 & -0.5\end{bmatrix},$$

$$\boldsymbol{b}_S^{(2)} = -\boldsymbol{A}_S^2\boldsymbol{u} = \begin{bmatrix}-1 & 1 \\ 0 & -0.5\end{bmatrix}\begin{bmatrix}1 \\ 1\end{bmatrix} = \begin{bmatrix}0 \\ 0.5\end{bmatrix},$$

$$\boldsymbol{A}_S^{(3)} = \begin{bmatrix}\boldsymbol{A}_S^{(2)} & \boldsymbol{b}_S^{(2)}\boldsymbol{a}_3 \\ \boldsymbol{0} & \boldsymbol{A}_3\end{bmatrix} = \begin{bmatrix}\begin{bmatrix}-1 & 1 \\ 0 & -0.5\end{bmatrix} & \begin{bmatrix}0 \\ 0.5\end{bmatrix}[1] \\ \boldsymbol{0} & [-0.2]\end{bmatrix} = \begin{bmatrix}-1 & 1 & 0 \\ 0 & -0.5 & 0.5 \\ 0 & 0 & -0.2\end{bmatrix}.$$

This result shows that a standby system with CFR components is not a CFR system, as seen from the order of its PH representation is 3. ■

The resulting values of $\boldsymbol{a}_S$ and $\boldsymbol{A}_S$ match Table 2.1 for a 3-hypoexponential distribution with $\lambda_1 = 1$, $\lambda_2 = 0.5$, $\lambda_3 = 0.2$. This result can be generalized to N-standby systems. If component i of an N-standby system has a lifetime following an exponential distribution with a rate λ_i, $i = 1, 2, \dots, N$, then the system lifetime has an N-hypo-exponential distribution with parameter λ_i, $i = 1, 2, \dots, N$.

Various PH representations of N-hypo-exponential distribution can be obtained from Table 2.1. Figure 4.27 is a state transition diagram of an N-standby system with the state indicating the component number that is currently active, and 0 represents no component active. This diagram is equivalent to Figure 4.26 with $\boldsymbol{A}_i = -\lambda_i$, $\boldsymbol{b}_i = \lambda_i$, and $\boldsymbol{a}_i = 1$.

- Diagram representation:

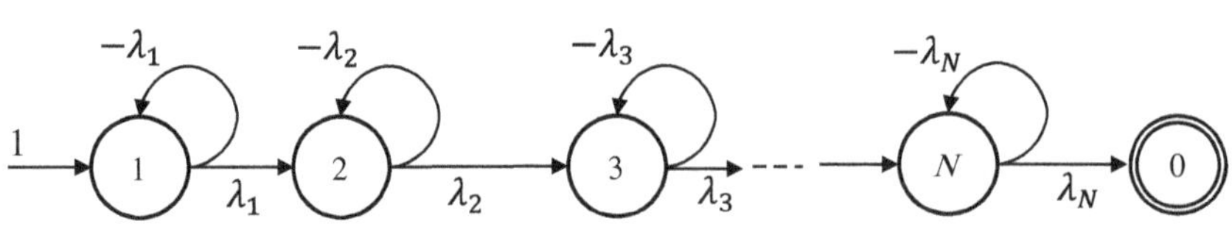

Figure 4.27 State transition diagram of an N-standby system with CFR components.

- Matrix representation:

$$n = N, \quad \boldsymbol{a} = \begin{bmatrix}1 & 0 \dots 0\end{bmatrix}, \quad \boldsymbol{A} = \begin{bmatrix}-\lambda_1 & \lambda_1 & \dots & 0 & 0 \\ 0 & -\lambda_2 & \dots & 0 & 0 \\ \vdots & \vdots & \ddots & \vdots & \vdots \\ 0 & 0 & \dots & -\lambda_{N-1} & \lambda_{N-1} \\ 0 & 0 & \dots & 0 & -\lambda_N\end{bmatrix}. \tag{4.77}$$

- Scalar representation for $\lambda_i \neq \lambda_j$ for $i \neq j$:

$$R_S(t) = \sum_{i=1}^{N} \frac{e^{-\lambda_i t}}{P_i} \text{ with } P_i = \prod_{j=1, j \neq i}^{N} \left(1 - \frac{\lambda_i}{\lambda_j}\right). \tag{4.78}$$

If all components are identical, $\lambda_1 = \lambda_2 = \ldots = \lambda_N = \lambda$, then the lifetime distribution becomes N-Erlang, with a scalar representation of its reliability function as given in Table 2.1 is

$$R_S(t) = \sum_{i=0}^{N-1} \frac{1}{i!} (\lambda t)^i e^{-\lambda t}. \tag{4.79}$$

Using Equation (4.57) yields $E[T_S] = E[T_1] + E[T_2] + \cdots + E[T_N]$ and $Var[T_S] = Var[T_1] + Var[T_2] + \cdots + Var[T_N]$. Hence,

$$MTTF_S = MTTF_1 + MTTF_2 + \cdots + MTTF_N = \sum_{i=1}^{N} MTTF_i, \tag{4.80}$$

$$VTTF_S = VTTF_1 + VTTF_2 + \cdots + VTTF_N = \sum_{i=1}^{N} VTTF_i. \tag{4.81}$$

Example 4.24 The mean and standard deviation of the system in Example 4.23 calculated using Equations (4.80) and (4.81) are

$$MTTF_S = 1 + 2 + 5 = 8.$$
$$SDTTF_S = (VTTF_S)^{0.5} = (1^2 + 2^2 + 5^2)^{0.5} = 5.48.$$
$$CVTTF_S = \frac{SDTTF}{MTTF} = \frac{5.48}{8} = 0.68 < 1.$$

■

From these results, a standby system with CFR components can be concluded to have the following properties:

- The system lifetime has a hypoexponential distribution.
- The system is IFR.
- The system *MTTF* and *VTTF* are the sums of *MTTF* and *VTTF* of its components, respectively.

Algorithm 4.4 can also be applied to imperfect switch standby systems with CFR components as given by Example 4.25.

Example 4.25 The system in Example 4.23 uses a switch with reliability = 0.8. The PH representation of the system lifetime distribution can be derived using Equations (4.74)–(4.76).

The PH representations of the component lifetime distributions are the same as given in Example 4.23.

Iterations using Algorithm 4.4 following Equations (4.74)–(4.76) are:

- Initial conditions: $n_S^{(1)} = n_1 = 1, \boldsymbol{a}_S^{(1)} = \boldsymbol{a}_1 = [1], \boldsymbol{A}_S^{(1)} = \boldsymbol{A}_1 = [-1]$.
- Second and third iterations for n_S using Equation (4.74) yields

$$n_S^{(2)} = n_S^{(1)} + n_2 = 2, \quad n_S^{(3)} = n_S^{(2)} + n_3 = 3.$$

- Second and third iterations to generate $\boldsymbol{a}_S$ using Equation (4.75) yields

$$\boldsymbol{a}_S^{(2)} = \left[\boldsymbol{a}_S^{(1)} \ \mathbf{0}\right] = \left[[1] \ 0\right] = \left[1 \ 0\right],$$
$$\boldsymbol{a}_S^{(3)} = \left[\boldsymbol{a}_S^{(2)} \ \mathbf{0}\right] = \left[[1 \ 0] \ 0\right] = \left[1 \ 0 \ 0\right].$$

- Second and third iterations to generate $\boldsymbol{A}_S$ using Equation (4.76) yields

$$\boldsymbol{A}_S^{(1)} = [-1]; \quad \boldsymbol{b}_S^{(1)} = -\boldsymbol{A}_S^{(1)}\boldsymbol{u} = -[-1][1] = [1],$$

$$\boldsymbol{A}_S^{(2)} = \begin{bmatrix} \boldsymbol{A}_S^{(1)} & p_1\boldsymbol{b}_1\boldsymbol{a}_2 \\ \boldsymbol{0} & \boldsymbol{A}_2 \end{bmatrix} = \begin{bmatrix} [-1] & 0.8[1][1] \\ 0 & [-0.5] \end{bmatrix} = \begin{bmatrix} -1 & 0.8 \\ 0 & -0.5 \end{bmatrix},$$

$$\boldsymbol{b}_2 = -\boldsymbol{A}_2\boldsymbol{u} = [0.5], \quad \boldsymbol{a}_3 = [1],$$

$$\boldsymbol{A}_S^{(3)} = \begin{bmatrix} \boldsymbol{A}_S^2 & \boldsymbol{0} \\ & p_2\boldsymbol{b}_2\boldsymbol{a}_3 \\ \boldsymbol{0} & \boldsymbol{A}_3 \end{bmatrix} = \begin{bmatrix} \begin{bmatrix} -1 & 0.8 \\ 0 & -0.5 \end{bmatrix} & \begin{bmatrix} 0 \\ 0.8[0.5][1] \end{bmatrix} \\ 0 & [-0.2] \end{bmatrix} = \begin{bmatrix} -1 & 0.8 & 0 \\ 0 & -0.5 & 0.4 \\ 0 & 0 & -0.2 \end{bmatrix}.$$

Thus, the PH representation of a 3-standby system with imperfect switches has order equal to 3, which means it is also not a CFR. ■

Referring to Figure 4.25, the state transition diagram of an N-standby system with imperfect switch having CFR components is as shown in Figure 4.28.

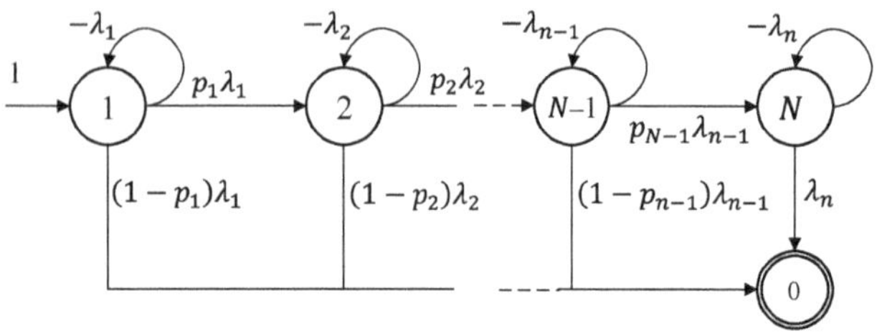

Figure 4.28 State transition diagram of an N-standby system with CFR components and imperfect switch.

The distribution of the lifetime of an N-standby system with CFR components using an imperfect switch can be deduced from Figure 4.28 as an N-Coxian $(\lambda_1, \lambda_2, \dots, \lambda_N; p_1, p_2, \dots, p_{N-1})$. If $p_i = 1$, $i = 1, 2, \dots, N$, Figure 4.28 reduces to Figure 4.27.

The expected lifetime of an imperfect switch standby system can be calculated using Figure 4.28 as the sum of the expected duration of stay at each state. Hence,

$$MTTF_S = MTTF_1 + p_1(MTTF_2 + p_2(MTTF_3 + \cdots + p_{N-1}MTTF_N)). \tag{4.82}$$

$$VTTF_S = VTTF_1 + p_1(VTTF_2 + p_2(VTTF_3 + \cdots + p_{N-1}VTTF_N)). \tag{4.83}$$

Equations (4.82) and (4.83) can also be derived from Equation (4.60).

Example 4.26 The system in Example 4.25 has

$$MTTF_S = 1 + 0.8(2 + 0.8(5)) = 5.8.$$

$$SDTTF_S = (VTTF_S)^{0.5} = (1 + 0.8(2^2 + 0.8(5^2)))^{0.5} = 4.49.$$

$$CVTTF_S = \frac{SDTTF}{MTTF} = \frac{4.49}{5.8} = 0.77 < 1.$$

Obviously, the *MTTF* and *SDTTF* of the system lifetime are smaller than the same system with perfect switch. The lower the switch reliability, the lower the systems *MTTF* and *VTTF* as apparent from Equations (4.82) and (4.83). ■

Exercise 4.15 Use **phmttf** and **phsdttf** functions in **dysrel-am** to calculate the system *MTTF* and *SDTTF* from the PH representation obtained in Example 4.25 and show that the results are the same as obtained in Example 4.26.

Next, the algorithm is illustrated with standby systems with components having lifetimes following not necessarily identical higher order PH distributions.

4.5.3 Standby Systems with Non-CFR Components

The state transition diagram of N-standby systems with non-CFR components is an extension of Figure 4.25 to N components as given in Figure 4.29. This diagram is applicable for any switch by acknowledging that the perfect switch has $1 - p_i = 0$, and deleting all transitions with zero value. If the switch is perfect, then Figure 4.29 reduces to Figure 4.26.

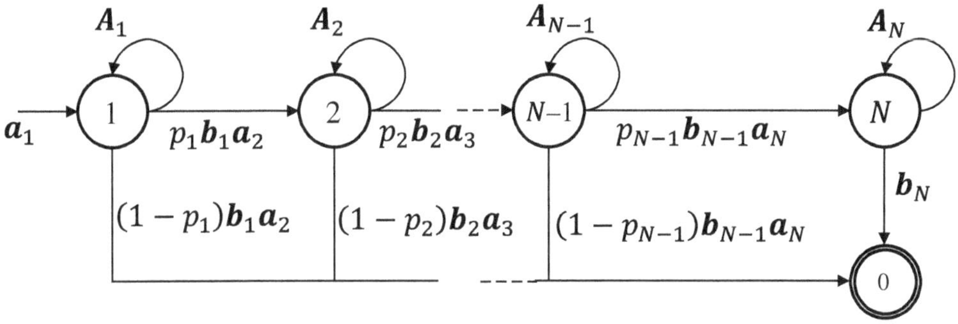

Figure 4.29 The state transition diagram of N-standby system with an imperfect switch.

Based on Figure 4.29, Equations (4.82) and (4.83) are applicable for any component, not necessarily CFR components.

The survival matrix of N-standby systems can be obtained from Figure 4.29 as

$$A_S = \begin{bmatrix} A_1 & p_1 b_1 a_2 & 0 & 0 & 0 & 0 \\ 0 & A_2 & p_2 b_2 a_3 & 0 & 0 & 0 \\ \vdots & \vdots & \ddots & \ddots & \vdots & \vdots \\ 0 & 0 & 0 & A_i & p_i b_i a_{i+1} & 0 \\ \vdots & \vdots & \vdots & 0 & \ddots & \ddots \\ 0 & 0 & 0 & 0 & 0 & p_{N-1} A_N \end{bmatrix}. \tag{4.84}$$

The order and initial probability vector are the same as for the perfect switch given by Equations (4.71) and (4.72).

Example 4.27 An imperfect switch standby system uses two components with lifetimes following 2-Erlang distribution with $MTTF = 2$ and 1. The switch reliability is 0.8.

PH representations of components 1 and 2 are

$$a_1 = \begin{bmatrix} 1 & 0 \end{bmatrix}, \quad A_1 = \begin{bmatrix} -1 & 1 \\ 0 & -1 \end{bmatrix}, \quad b_1 = \begin{bmatrix} 0 \\ 1 \end{bmatrix} \quad a_2 = \begin{bmatrix} 1 & 0 \end{bmatrix}, \quad A_2 = \begin{bmatrix} -2 & 2 \\ 0 & -2 \end{bmatrix}.$$

Using Algorithm 4.4 yields

$$n_S = n_1 + n_2 = 2 + 2 = 4.$$

$$a_S = a_1 = \begin{bmatrix} 1 & 0 \end{bmatrix}.$$

$$A_S = \begin{bmatrix} A_1 & p b_1 a_2 \\ 0 & A_2 \end{bmatrix} = \begin{bmatrix} \begin{bmatrix} -1 & 1 \\ 0 & -1 \end{bmatrix} & 0.8 \begin{bmatrix} 0 \\ 1 \end{bmatrix} \begin{bmatrix} 1 & 0 \end{bmatrix} \\ \mathbf{0} & \begin{bmatrix} -2 & 2 \\ 0 & -2 \end{bmatrix} \end{bmatrix} = + \begin{bmatrix} \begin{bmatrix} -1 & 1 \\ 0 & -1 \end{bmatrix} & \begin{bmatrix} 0 & 0 \\ 0.8 & 0 \end{bmatrix} \\ \mathbf{0} & \begin{bmatrix} -2 & 2 \\ 0 & -2 \end{bmatrix} \end{bmatrix} = \begin{bmatrix} -1 & 1 & 0 & 0 \\ 0 & -1 & 0.8 & 0 \\ 0 & 0 & -2 & 2 \\ 0 & 0 & 0 & -3 \end{bmatrix}.$$

Thus, the lifetime of this standby system has a PH distribution with representation (n_S, a_S, A_S) as given in the above calculations. ■

As for series and parallel systems, manual calculations seem complicated and inefficient for standby systems whose components have lifetimes following PH distributions of high orders.

Instead, it must use the **phstandsys** function. The following examples use the **phstandsys** function.

Example 4.28 The **phstandsys** function is used to calculate the PH representation of the distribution of the lifetime of a 5-standby system with a perfect switch. The kth component of this system has lifetime following k-Erlang distributions with $MTTF = 100$, $k = 1, 2, \ldots, 5$. The PH representation obtained from **phstandsys** is used to calculate and plot the system reliability and *MRL* functions and also the system *MTTF* and *SDTTF*.

The main program for this purpose is given in **example4_28.m** file in the software package **dysrel-am**. Its MATLAB code is as follows:

MATLAB code for example4_28.m

```
clear;close all
addpath('..\dysrel-am','..\dysrel-am\expokit')
N = 5;                                    (inputs)
% PH Components
for k = 1:N
  T(k) = phnErlang(k,100);
end
% PH Cold Standby system                  (system model)
Ts = phstandsys(T,ones(N-1,1));
% Result in command window                (statistical outputs)
mttf = phmttf(Ts)
sdttf = phsdttf(Ts)
% Time vector
tmax = 1000;
t = linspace(0,tmax,21);
% Reliability plot                        (graphical outputs)
phrelplot(t,Ts);
% MRL plot
phmrlplot(t,Ts);
```

Executing this program produces Figure 4.30.

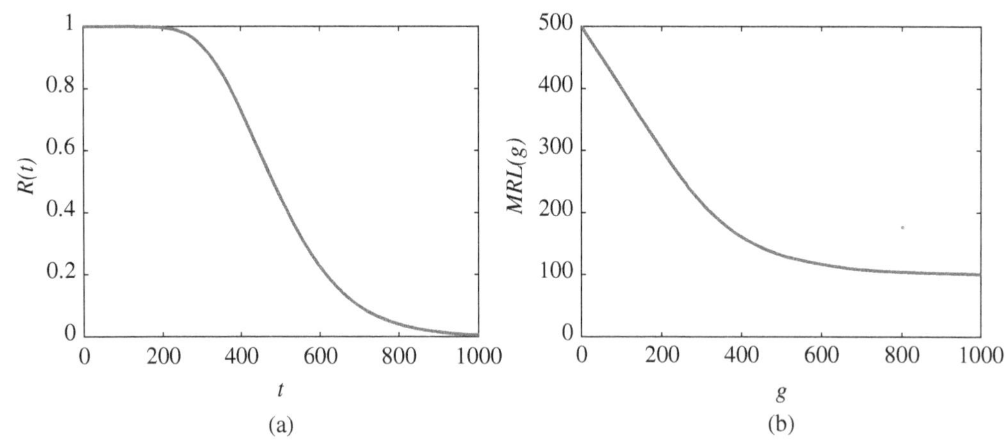

Figure 4.30 Outputs of example4_28.m.

```
>> example4_28
mttf =
   500
sdttf =
  151.1070
```

■

Example 4.29 An N-standby system with a perfect switch ($p = 1$) is constructed using N identical IFR components in Example 3.2. The **phstandsys** function generates the PH representation of the system lifetime distribution. The result is used to calculate the system reliability measures for several values of N shown in Figure 4.31 and Table 4.6.

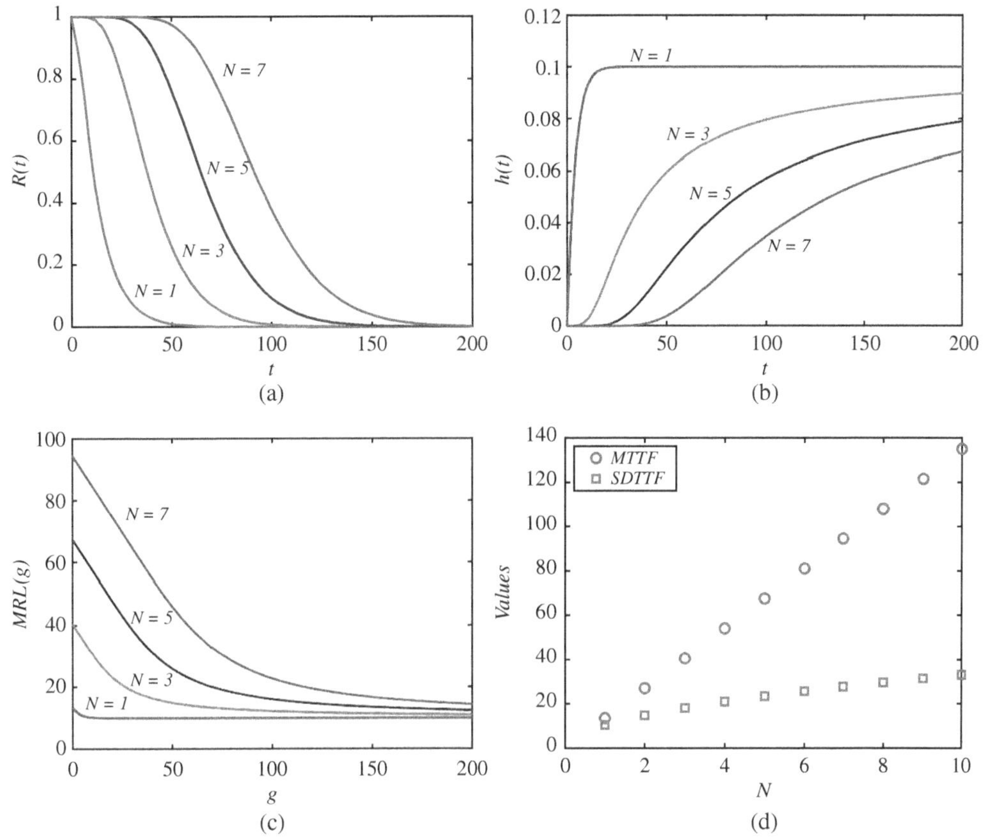

Figure 4.31 Reliability measures of systems in Example 4.29. (a) System reliability function, (b) system hazard function, (c) system *MRL* function, and (d) system *MTTF* and *SDTTF*.

Table 4.6 Reliability and *MRL* values at t = 20 days as functions of N.

	N						
Measures	**1**	**2**	**3**	**4**	**5**	**6**	**7**
Reliability	0.20	0.63	0.90	0.99	1.00	1.00	1.00
MRL	10.01	14.83	23.10	34.55	47.57	61.01	74.50

■

The graphs and table can be used to determine the number of backups in a standby system to meet a certain design requirement. For a mission of 20 days, the number of backups that must be provided to make the system reliability higher than 0.7 is $N - 1 > 3 - 1 = 2$. If the remaining life expectancy after being used for 20 days must be longer than 20 days, then N must be at least equal to 3, or the number of backup components must be at least 2.

Exercise 4.16 Determine the value of N to make the system have reliability not less than 0.9 and MRL not less than 20 days for a 25-day mission.

Application 4.2 Application 4.1 is repeated for the systems to carry out the mission arranged in a standby structure with a perfect switch. The resulting total cost as a function of N and mission duration is presented in Table 4.7.

Table 4.7 Expected mission cost as a function of mission duration and N.

	Mission duration			
N	**20**	**30**	**40**	**50**
1	123.31	179.76	201.52	207.93
2	30.23	71.20	129.41	175.96
3	**30.18**	**34.02**	52.38	90.64
4	40	40.10	**41.65**	**49.75**
5	50	50	50.05	50.69

The optimum values are in bold.

As for the parallel structure in Application 4.1, Table 4.7 shows that the optimum number of systems in perfect switch standby structure to carry out the mission is higher for longer mission duration. However, the optimum number of systems in standby structure is less than that in parallel structure. It does make sense, since only one system is active in a standby structure while all systems are active in a parallel structure. ■

Exercise 4.17 Investigate the effect of cost parameters to the optimum number of systems in standby structure to carry out the mission.

Example 4.30 The experiment in Example 4.29 is repeated using DFR components in Example 4.10. The results are presented in Figure 4.32. ■

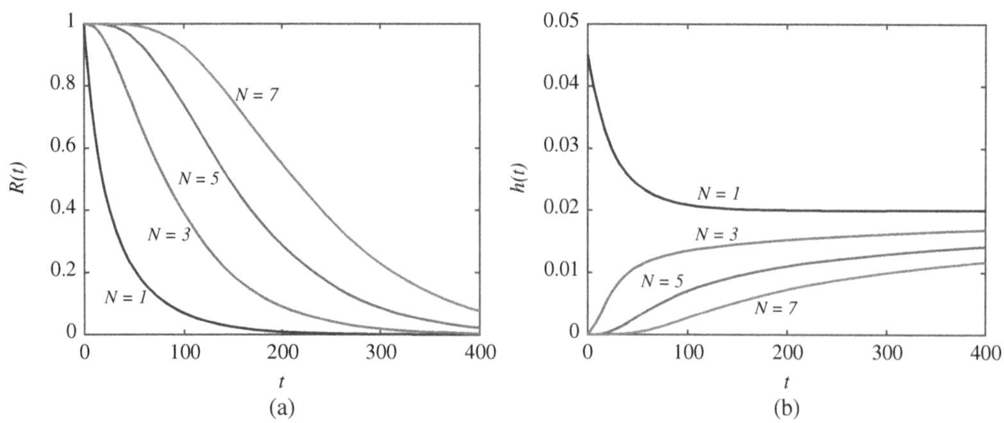

Figure 4.32 Reliability measures of systems in Example 4.30. (a) System reliability function, (b) system hazard function, (c) system *MRL* function, and (d) system *MTTF* and *SDTTF*.

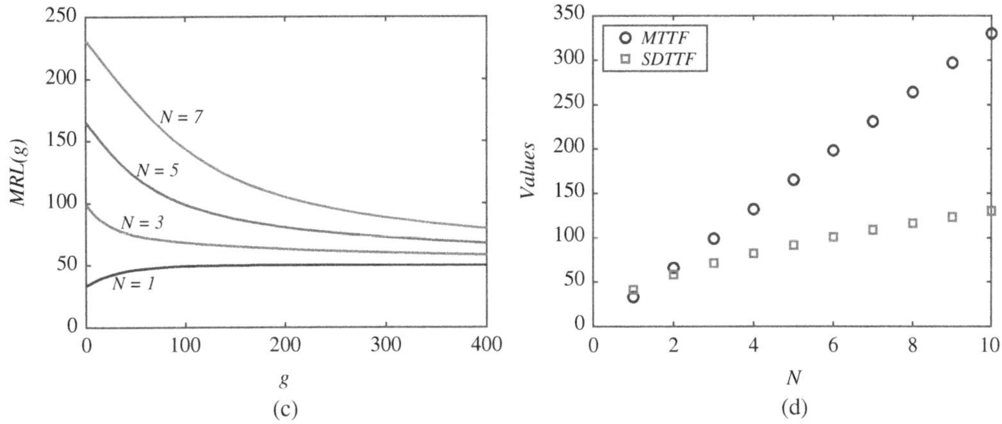

Figure 4.32 *(Continued)*

Example 4.31 The experiment in Example 4.29 is repeated for a standby system with an imperfect switch having reliability = 0.8. The results are presented in Figure 4.33. ■

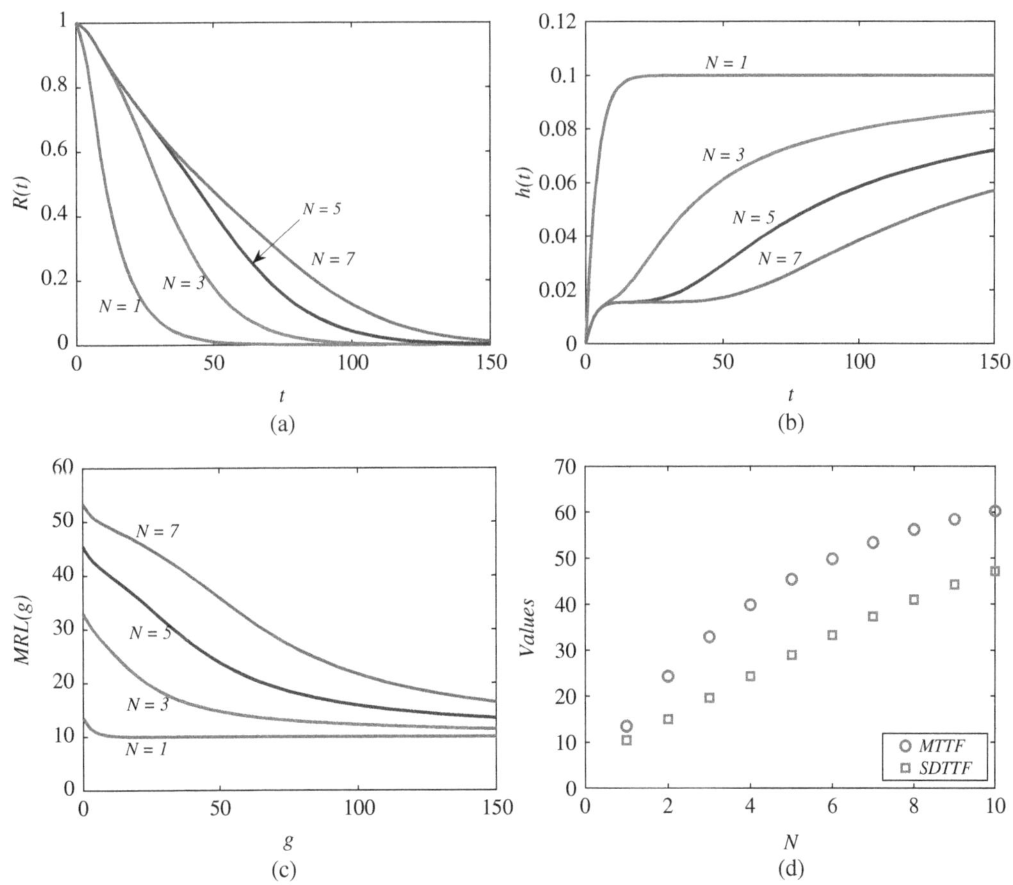

Figure 4.33 Reliability measures of systems in Example 4.31. (a) System reliability function, (b) system hazard function, (c) system *MRL* function, and (d) system *MTTF* and *SDTTF*.

Example 4.32 A 3-standby system is constructed using an IFR component in Example 3.2, a DFR component in Example 4.10, and a CFR component with $MTTF = 15$. The effects of switch reliability on system reliability at $t = 20$, $MTTF$, and $SDTTF$ are presented as functions of switch reliability in Figure 4.34.

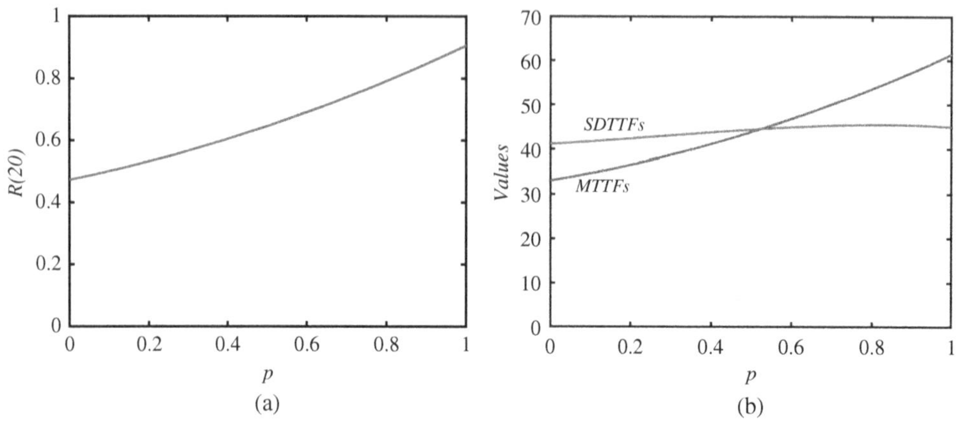

Figure 4.34 System reliability, *MTTF*, and *SDTTF* as functions of switch reliability. (a) System reliability at $t = 20$ and (b) system *MTTF* and *SDTTF*. ■

Examples 4.30–4.32 show that standby systems have the same behavior as parallel systems and differ from series systems in that the type of standby system is not always the same as the type of its components. Although the components are DFR, the system could be an IFR. Standby systems have the same type as their components only if the components are IFR. Standby systems with CFR components are IFR systems because their lifetimes have a hypoexponential distribution, which is IFR.

A standby system with a perfect switch constructed using DFR components can be an IFR system if the number of backups is sufficient. Suppose the system is using identical components. Then, using Equations (4.57) and (4.81) yields

$$MTTF_S = N(MTTF_{comp}) \text{ and } VTTF_S = N(VTTF_{comp}).$$

Thus,

$$CVTTF_S = SDTTF_S/MTTF_S = CVTTF_{comp}/\sqrt{N}.$$

Since DFR components have $CVTTF_{comp} > 1$, N must be greater than $CVTTF^2_{comp}$ to produce $CVTTF_S < 1$ as a condition for an IFR system.

Example 4.33 For a standby system with identical DFR components having $CVTTF_{comp} = 1.5$, the number of components $N > 1.5^2$, rounded to 3, will make the system to be IFR. ■

Determining the number of components to make an imperfect switch standby system become IFR is more complicated. For example, 3-standby systems with nonidentical CFR components and imperfect switches may have system lifetime distributions following Coxian distributions that, as shown in Chapter 3.5, can be IFR or DFR and may even be BFR. By comparing the value of *SDTTF* to $MTTF$ in Figure 4.34, the system in Example 4.32 is DFR for lower switch reliability and IFR for higher switch reliability, and at certain value of switch reliability, the system is CFR when $SDTTF = MTTF$.

Next, the definition of standby systems is used to derive properties of such systems.

4.5.4 Properties of Standby Systems

As have been shown, standby systems have Property 1:

- All components are IFR ⇒ the standby system is IFR.
- All components are CFR ⇒ the standby system is IFR.
- All components are DFR ⇒ the standby system is IFR for a sufficiently large N.

Other properties of standby systems can be derived using its system definition. The following derivation of the properties of the standby system is for the perfect switch. However, the results apply to any switch.

Using Equation (4.57) yields

$$T_S(N+1) = T_1 + T_2 + \cdots + T_{N+1} = T_S(N) + T_{N+1}.$$

Since T_i is nonnegative, it holds that

$$T_S(N+1) \succcurlyeq T_S(N),$$

and

$$T_S \succcurlyeq \max_i T_i.$$

Therefore, standby systems have Property 2 that adding a standby component will increase the system reliability function,

$$R_S(t, N+1) \geq R_S(t, N), \tag{4.85}$$

and Property 3 that the reliability function of a standby system is greater than the reliability function of its best component,

$$R_S(t) \geq \max_i R_i(t). \tag{4.86}$$

Equation (4.80) is referred to as Property 4 of standby systems regarding their $MTTF$. It is stricter than Property 4 of parallel systems such that $MTTF_S \geq \max_i [MTTF_i]$.

As with parallel systems, the relationship between the system hazard function and the component hazard functions cannot be formulated in a simple term. For a standby system with identical components, Figures 4.31b and 4.32b show that the system hazard function decreases with the increasing number of components, and the system hazard function is smaller than the component hazard function. However, this behavior does not apply to a standby system with nonidentical components as shown by the following examples.

Example 4.34 The three components in Example 4.11 are connected in a standby structure. The graphs of the reliability and hazard functions of the system and all its components are presented in Figure 4.35. The graphs in Figure 4.35a show that the system reliability function at any time is always greater than the highest component reliability function at that time in accordance with

Property 3. The expected system lifetime $MTTF_S = 57.953 = 27.000 + 14.286 + 16.667$, the sum of *MTTFs* of all its components in accordance with Property 4. However, the system hazard function is not always lower than the component hazard function as shown in Figure 4.35b. ■

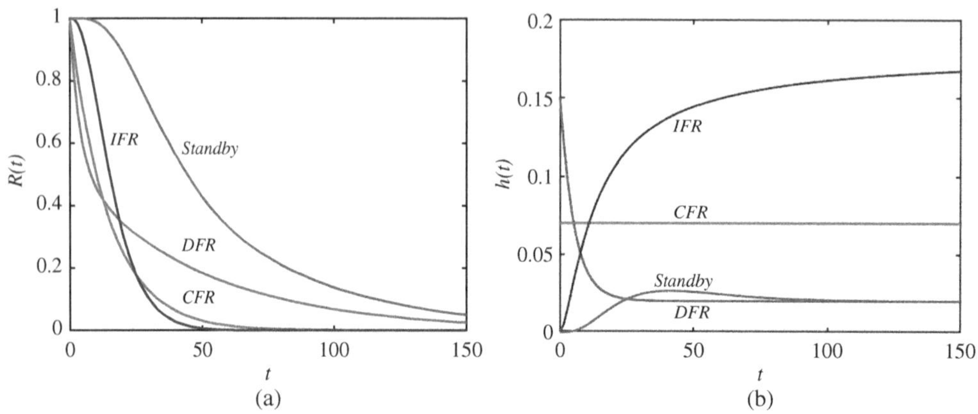

Figure 4.35 Reliability and hazard functions of Example 4.34. (a) System reliability function and (b) system hazard function.

Example 4.35 Example 4.12 is repeated as a standby system. The reliability and hazard functions of the components and the system are shown in Figure 4.36. It can be seen that the reliability function of the system at any time is greater than the reliability functions of each component, while the hazard function of the system at any time is not always smaller than the hazard function of each component. ■

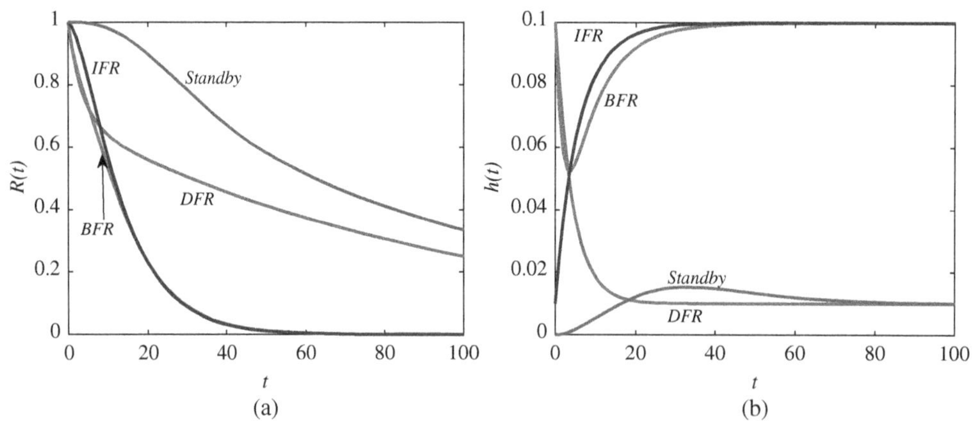

Figure 4.36 Reliability and hazard functions of Example 4.35. (a) Reliability function and (b) hazard function.

Examples 4.34 and 4.35 show that the system hazard function can be higher than the component hazard function. As with parallel systems, Properties 5 and 6 given by Equation (4.56) that the system hazard function is lower than the hazard functions of its components only apply to standby systems with identical components.

The absence of an algebraic expression relating the reliability function of a standby system to the reliability functions of its components causes the calculation of the reliability value of a standby system from the reliability values of its components impossible. The only possible way is finding the system reliability functions from reliability functions of its components, then substituting a value of time to find the system reliability value. The standby system can be analyzed using scalar reliability functions; however, it requires repeated convolution integrals as many as the number of components, which is certainly not easy to do manually or even with the help of a computer. On the other hand, matrix distribution allows dynamic analyses of the standby system reliability using matrix algebra, which is even simpler than parallel systems. It is the most significant advantage of using the matrix-form distribution model in system reliability analysis.

The next section uses Algorithm 4.4 to show that the ordering of nonidentical components in imperfect switch standby systems affects system reliability measures.

4.5.5 Component Ordering in Imperfect Switch Standby Systems

A simple logic will verify that the minimum, maximum, and addition operations give results independent of the sequence of their operands. Equation (4.3) will provide the same value for T_S regardless of the sequence of T_i, $i = 1, 2, \dots, N$. In reliability context, the lifetime of a series system is independent of the sequence of its components. Likewise, Equations (4.23) and (4.57) assert that the sequence of the components does not affect the lifetime of the parallel and standby systems. The sequence of the components does not affect the lifetime of the series, parallel, and standby systems, implying that the reliability measures of these systems are independent of the sequence in which their components are arranged. However, it is different with the imperfect switch standby system. For such a system, component 1 is used for certain, component 2 is used if the switch is working when component 1 fails, component 3 is used if the switch is working when component 1 and component 2 fail, and so on. A component in the later sequence has a smaller probability of being used.

Figure 4.29 and Equation (4.60) can be used to deduce that the lifetime of an imperfect switch standby system is the sum of the lifetimes of its components corrected by the probability of the components being used, that is

$$T_S = T_1 + p_1 T_2 + p_1 p_2 T_3 + \cdots + p_1 p_2 \dots p_{N-1} T_N. \tag{4.87}$$

Equation (4.87) provides guidance that the system lifetime can be made longer by placing better components earlier in the sequence. Components with longer lifetimes have higher reliability functions or lower hazard functions. Therefore, component-ordering strategies for standby systems with imperfect switches can be formulated as:

- If $R_i(t) < R_j(t)\ \forall\ t$, then component i is placed after component j.
- If $h_i(t) < h_j(t)\ \forall\ t$, then component i is placed earlier than component j.

Component ordering to improve system reliability can only be achieved if the reliability and hazard functions of the components can be ordered. Often, the reliability or hazard functions of the components do not follow the same order at any time, such that one is larger at one time and smaller at another time. In this case, the abovementioned rules are not valid. Ordering is still possible only for a certain period when their hazard functions can be ordered, not at all times.

The problem of finding the optimum component ordering in standby systems has been widely studied in the literature. However, it mainly dealt with component ordering in general standby systems, such as Zhai et al. [18], Roy and Gupta [19], and Alkaff et al. [20], that are discussed in the next chapter. In those studies, component ordering is based on the component hazard functions.

Example 4.36 A 3-standby system with switch reliability = 0.8 is constructed using three IFR components whose lifetimes follow 2-Erlang distribution with components 1, 2, and 3 having $MTTF = 5$, 10, and 15, respectively. The component hazard functions are exhibited in Figure 4.37a, which shows that

$$h_1(t) > h_2(t) > h_3(t),$$

where $h_1(t)$, $h_2(t)$, and $h_3(t)$ denote the hazard functions of the components with $MTTF = 5$, 10, and 15, respectively.

Various component orderings were examined, and the resulting system reliability function graphs are shown in Figure 4.37b.

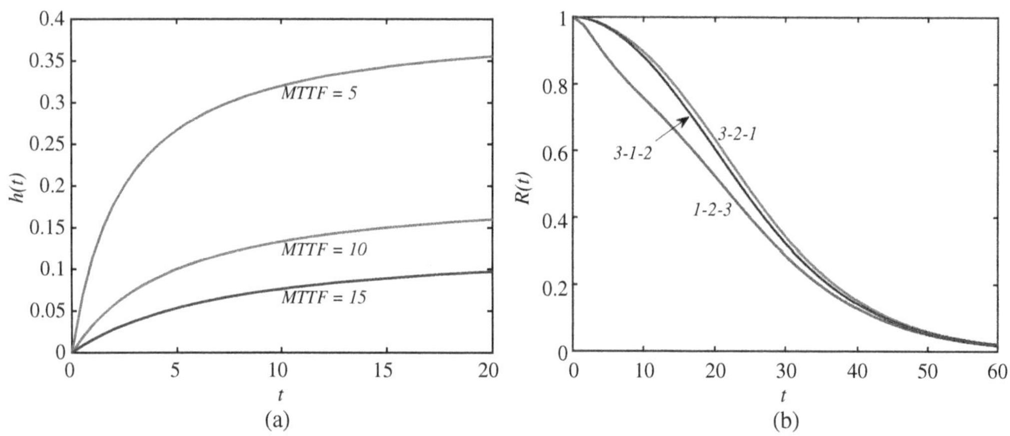

Figure 4.37 Effect of component ordering in imperfect switch standby systems with IFR components. (a) Component hazard function and (b) system reliability function.

This result confirms that components with lower hazard functions should be placed earlier in the backup sequence to increase the system reliability function. The component sequences 3–2-1, 1–2-3, and 3–1-2 produced systems with $MTTF = 26.2$, 22.6, and 23.4, respectively, indicating that the increasing hazard function component ordering, which is 3–2-1 sequence, produced the highest system $MTTF$ as well as the highest system reliability function. On the contrary, the decreasing hazard function component ordering, which is 1–2-3 sequence, produced the lowest $MTTF$ and reliability function. ∎

Example 4.37 Example 4.36 is repeated using three DFR components having 2-Coxian lifetime distributions with the same $CVTTF = 1.2$ and $MTTF = 5$, 10, and 15. The component hazard functions are given in Figure 4.38a, which shows that $h_1(t) < h_2(t) < h_3(t)$. Experiments with several component ordering show the same results as in Example 4.36 namely, increasing hazard function ordering produced the highest reliability function as shown in Figure 4.38b.

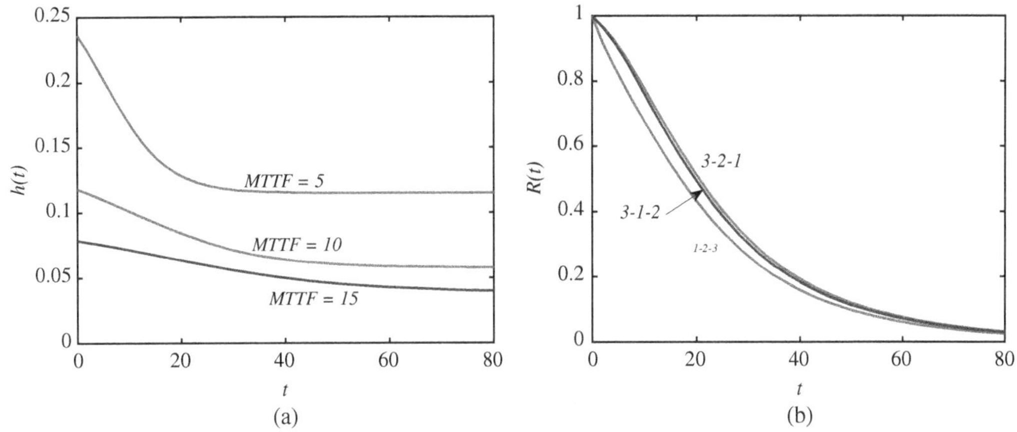

Figure 4.38 The effect of component ordering in an imperfect switch standby system with DFR components. (a) Component hazard function and (b) system reliability function. ■

Examples 4.36 and 4.37 show that the optimum ordering following increasing component hazard functions is applicable for all component types.

Exercise 4.18 Show numerically that the effect of ordering is diminishing with an increase in switch reliability.

Exercise 4.19 Investigate the effect of ordering for a mixture of DFR, CFR, and IFR components having strict hazard function ordering.

Basic structures have different characteristics. Comparison of their reliability measures is discussed in the next section.

4.6 Comparisons of Reliability Measures of Several Basic Structures

The following stochastic ordering is logical for nonnegative random variables,

$$\min[T_1, T_2, \dots, T_N] \preccurlyeq \max[T_1, T_2, \dots, T_N] \preccurlyeq \text{add}[T_1, T_2, \dots, T_N]. \tag{4.88}$$

Operations min, max, and add in Equation (4.88) are mathematical expressions for the structure functions of series, parallel, and standby systems, respectively. For the same distribution and number of components, Equation (4.88) implies

$$T_{series} \preccurlyeq T_{parallel} \preccurlyeq T_{standby}, \tag{4.89}$$

where T_{series}, $T_{parallel}$, and $T_{standby}$ are the lifetimes of the series, parallel, and standby systems, respectively.

Using Equation (4.89) to determine the ordering of reliability and hazard functions of series, parallel, and standby systems yields

$$R_{series}(t) \le R_{parallel}(t) \le R_{standby}(t), \tag{4.90}$$

$$h_{series}(t) \ge h_{parallel}(t) \ge h_{standby}(t), \tag{4.91}$$

where $R_{series}(t)$, $R_{parallel}(t)$, $R_{standby}(t)$, $h_{series}(t)$, $h_{parallel}(t)$, and $h_{standby}(t)$ are the reliability and hazard functions of series, parallel, and standby systems, respectively.

Exercise 4.20 Determine the ordering of the *MRL* functions.

The ordering of the expected system lifetimes can also be inferred from the stochastic ordering of their lifetimes given in Equation (4.89) to yield

$$MTTF_{series} \leq MTTF_{parallel} \leq MTTF_{standby}, \tag{4.92}$$

where $MTTF_{series}$, $MTTF_{parallel}$, and $MTTF_{standby}$ denote the expected lifetime of series, parallel, and standby systems, respectively.

The position of the imperfect switch standby system in the ordering can only be partially ascertained. For T_{sis} is the lifetime of the imperfect switch standby system, the following stochastic ordering can be ascertained from the definition of each structure,

$$T_{series} \preccurlyeq T_{sis} \preccurlyeq T_{standby}. \tag{4.93}$$

It is known that the order of the reliability function follows the stochastic order of the lifetime, while the order of the hazard function is the opposite. Equation (4.93) implies

$$R_{series}(t) \leq R_{sis}(t) \leq R_{standby}(t), \tag{4.94}$$

$$h_{series}(t) \geq h_{sis}(t) \geq h_{standby}(t), \tag{4.95}$$

where $R_{sis}(t)$ and $h_{sis}(t)$, respectively, are the reliability and hazard functions of the imperfect switch standby system. However, which one of the reliability functions of parallel or imperfect switched standby systems is higher cannot be ascertained, as shown in Examples 4.38 and 4.39.

Example 4.38 The IFR components in Example 3.2 are used to construct systems with various structures with the number of components equal to 4. Their reliability and failure functions are presented in Figure 4.39, which strengthens the concluded reliability measures ordering, and that the ordering of parallel and imperfect switch standby structures cannot be ascertained. ∎

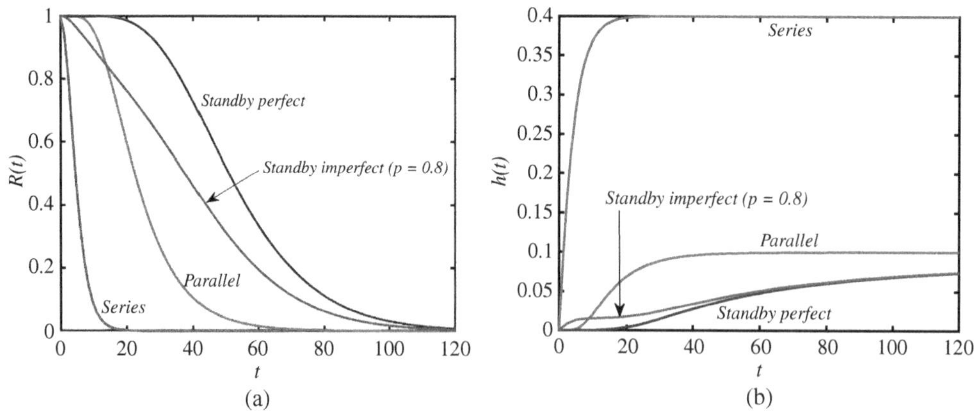

Figure 4.39 Comparisons of system reliability measures with IFR components. (a) IFR components with switch reliability = 0.8 and (b) IFR components with switch reliability = 0.6.

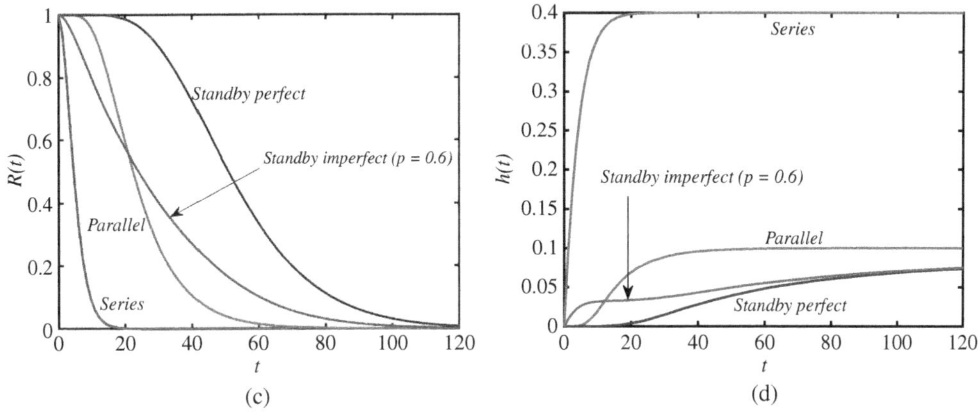

Figure 4.39 *(Continued)*

Example 4.39 Example 4.38 is repeated using the DFR components in Example 4.10. The results are presented in Figure 4.40, all in accordance with the same conclusions as using IFR components in Example 4.38. ■

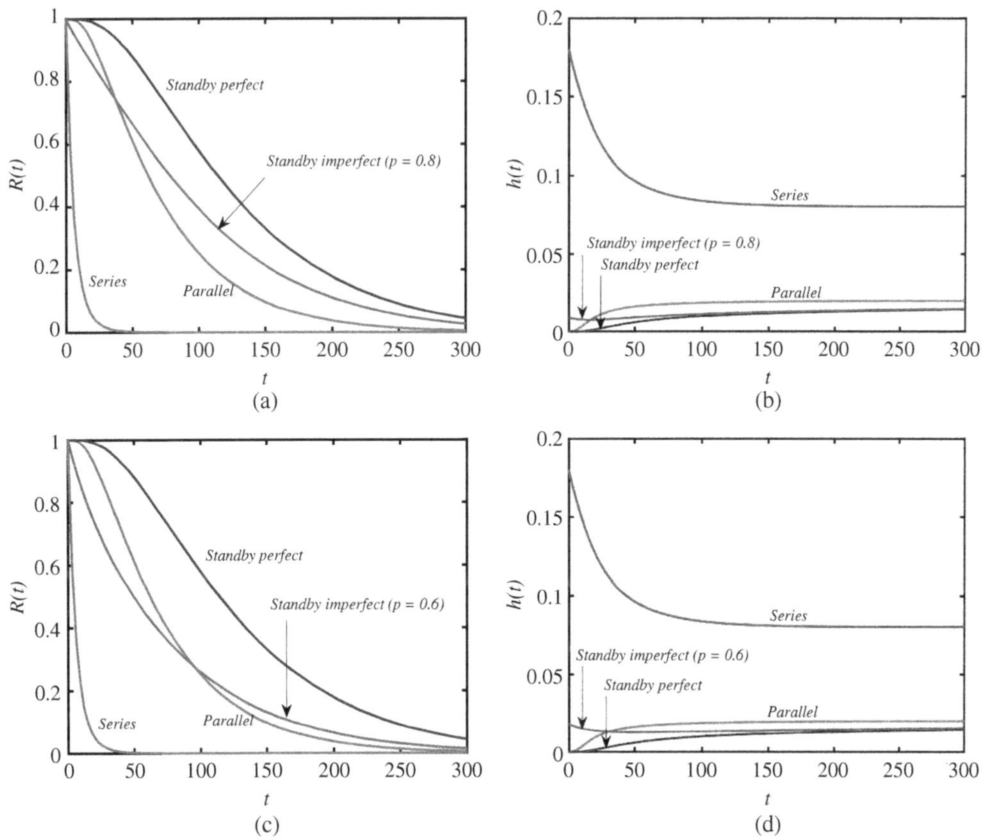

Figure 4.40 Comparison of system reliability measures with DFR components. (a) DFR components and switch reliability = 0.8 and (b) DFR components and switch reliability = 0.6.

Examples 4.38 and 4.39 show that the ordering of system reliability measures in several structures applies to systems with IFR and DFR components. Although not shown, the same conclusion certainly applies to CFR components.

Next, the discussion of basic structures is extended for ME distribution. ME distribution shares the same form and properties as PH distribution. Therefore, algorithms for calculating PH representation of system reliability function from PH representations of its component reliability functions are applicable for directly calculating ME representation of system reliability function from ME representations of its component reliability functions with only a slight modification. Using ME distribution is attractive because, in most cases, it can reduce the computation time.

4.7 Algorithms for ME Distribution

The algorithms for systems with components having lifetimes following PH distributions can be extended to ME distributions using the closure properties of ME distribution presented as Theorems 2.6–2.9. Algorithms for the ME distribution are the same as for the PH distribution with the addition of calculating the closing vector $\boldsymbol{v}_S$ [12]. The following lemmas are derived from Theorems 2.6–2.9; they can be used to design algorithms for ME distribution. The proof follows the same manner as for the PH distribution in Sections 4.3–4.5. As before, let T_i be the lifetime of component i, $i = 1, 2, \ldots, N$, and T_S be the system lifetime.

Lemma 4.5 If $T_i \backsim ME(n_i, \boldsymbol{a}_i, \boldsymbol{A}_i, \boldsymbol{v}_i)$, $i = 1, 2, \ldots, N$, then T_S of an N-series system has an ME distribution with a representation $(n_S, \boldsymbol{a}_S, \boldsymbol{A}_S, \boldsymbol{v}_S)$ given by recursion equations

$$n_S^{(i)} = n_S^{(i-1)} n_i, \tag{4.96}$$

$$\boldsymbol{a}_S^{(i)} = \boldsymbol{a}_S^{(i-1)} \otimes \boldsymbol{a}_i, \tag{4.97}$$

$$\boldsymbol{A}_S^{(i)} = \boldsymbol{A}_S^{(i-1)} \oplus \boldsymbol{A}_i, \tag{4.98}$$

$$\boldsymbol{v}_S^{(i)} = \boldsymbol{v}_S^{(i-1)} \otimes \boldsymbol{v}_i, \tag{4.99}$$

for $i = 2, 3, \ldots, N$, with initial conditions

$$n_S^{(1)} = n_1, \quad \boldsymbol{a}_S^{(1)} = \boldsymbol{a}_1, \quad \boldsymbol{A}_S^{(1)} = \boldsymbol{A}_1, \quad \boldsymbol{v}_S^{(1)} = \boldsymbol{v}_1. \tag{4.100}$$

Lemma 4.5 is utilized to construct Algorithm 4.5, which is then programmed in MATLAB as a **meseriessys** function, available in the software package **dysrel-am**. Its pseudo-code, which is derived from the pseudo-code of **phseriessys** by adding lines to calculate $\boldsymbol{v}_S$ recursively, is as follows:

Algorithm 4.5

---

input N, $(n_i, \boldsymbol{a}_i, \boldsymbol{A}_i, \boldsymbol{v}_i)$, $i = 1, 2, \ldots, N$
$n_S = n_1$
$\boldsymbol{a}_S = \boldsymbol{a}_1$
$\boldsymbol{A}_S = \boldsymbol{A}_1$
$\boldsymbol{v}_S = \boldsymbol{v}_1$
for $i = 2$ to N

$n_S = n_S n_i$
$\boldsymbol{a}_S = \boldsymbol{a}_S \otimes \boldsymbol{a}_i$
$\boldsymbol{A}_S = \boldsymbol{A}_S \oplus \boldsymbol{A}_i$
$\boldsymbol{v}_S = \boldsymbol{v}_S \otimes \boldsymbol{v}_i$
$i = i + 1$
end
end

Lemma 4.6 If $T_i \backsim ME(n_i, \boldsymbol{a}_i, \boldsymbol{A}_i, \boldsymbol{v}_i)$, $i = 1, 2, \ldots, N$, then T_S of an N-parallel system has ME distributions with representation $(n_S, \boldsymbol{a}_S, \boldsymbol{A}_S, \boldsymbol{v}_S)$ given by recursion equations

$$n_S^{(i)} = n_S^{(i-1)} n_i + n_S^{(i-1)} + n_i, \tag{4.101}$$

$$\boldsymbol{a}_S^{(i)} = \left[\boldsymbol{a}_S^{(i-1)} \otimes \boldsymbol{a}_i \;\; \boldsymbol{0} \;\; \boldsymbol{0}\right], \tag{4.102}$$

$$\boldsymbol{A}_S^{(i)} = \begin{bmatrix} \boldsymbol{A}_S^{(i-1)} \oplus \boldsymbol{A}_i & \boldsymbol{I}_S^{(i-1)} \otimes \boldsymbol{b}_i & \boldsymbol{b}_S^{(i-1)} \otimes \boldsymbol{I}_i \\ \boldsymbol{0} & \boldsymbol{A}_S^{(i-1)} & \boldsymbol{0} \\ \boldsymbol{0} & \boldsymbol{0} & \boldsymbol{A}_i \end{bmatrix}, \tag{4.103}$$

$$\boldsymbol{v}_S^{(i)} = \left[\boldsymbol{v}_S^{(i-1)} \otimes \boldsymbol{v}_i \;\; \boldsymbol{v}_S^{(i-1)} \;\; \boldsymbol{v}_i\right]^T, \tag{4.104}$$

for $i = 2, 3, \ldots, N$, with the initial conditions given by Equation (4.100), $\boldsymbol{b}_i = -\boldsymbol{A}_i \boldsymbol{v}_i$, and $\boldsymbol{b}_S^{(i-1)} = -\boldsymbol{A}_S^{(i-1)} \boldsymbol{v}_S$.

As for Lemma 4.5, Lemma 4.6 is utilized to construct Algorithm 4.6 and programmed in MATLAB as **meparallelsys** function, available in the software package **dysrel-am**. Its pseudo-code is as follows:

Algorithm 4.6

---

input N, $(n_i, \boldsymbol{a}_i, \boldsymbol{A}_i, \boldsymbol{v}_i)$, $i = 1, 2, \ldots, N$
$n_S = n_1$
$\boldsymbol{a}_S = \boldsymbol{a}_1$
$\boldsymbol{A}_S = \boldsymbol{A}_1$
$\boldsymbol{v}_S = \boldsymbol{v}_1$
for $i = 2$ to N
$n_S = n_S n_i + n_S + n_i$
$\boldsymbol{a}_S = \left[\boldsymbol{a}_S \otimes \boldsymbol{a}_i \;\; \boldsymbol{0} \;\; \boldsymbol{0}\right]$
$\boldsymbol{b}_S = -\boldsymbol{A}_S \boldsymbol{v}_i$
$\boldsymbol{b}_i = -\boldsymbol{A}_i \boldsymbol{v}_i$
$$\boldsymbol{A}_S = \begin{bmatrix} \boldsymbol{A}_S \oplus \boldsymbol{A}_i & \boldsymbol{I}_S \otimes \boldsymbol{b}_i & \boldsymbol{b}_S \otimes \boldsymbol{I}_i \\ \boldsymbol{0} & \boldsymbol{A}_S & \boldsymbol{0} \\ \boldsymbol{0} & \boldsymbol{0} & \boldsymbol{A}_i \end{bmatrix}$$
$\boldsymbol{v}_S = \left[\boldsymbol{v}_S \otimes \boldsymbol{v}_i \;\; \boldsymbol{v}_S \;\; \boldsymbol{v}_i\right]^T$
$i = i + 1$
end
end

Lemma 4.7 If $T_i \backsim ME(n_i,\, \boldsymbol{a}_i,\, \boldsymbol{A}_i,\, \boldsymbol{v}_i)$, $i = 1, 2, \ldots, N$, then T_S of an N-standby system with an arbitrary switch has an ME distribution with representation $(n_S,\, \boldsymbol{a}_S,\, \boldsymbol{A}_S,\, \boldsymbol{v}_S)$ given by recursion equations

$$n_S^{(i)} = n_S^{(i-1)} + n_i, \tag{4.105}$$

$$\boldsymbol{a}_S^{(i)} = \left[\boldsymbol{a}_S^{(i-1)}\ \boldsymbol{0}\right], \tag{4.106}$$

$$\boldsymbol{A}_S^{(i)} = \left[\begin{array}{c|c} \boldsymbol{A}_S^{(i-1)} & \begin{array}{c}\boldsymbol{0}\\ p_{i-1}\boldsymbol{b}_{i-1}\boldsymbol{a}_i\end{array} \\ \hline \boldsymbol{0} & \boldsymbol{A}_i \end{array}\right], \tag{4.107}$$

$$\boldsymbol{v}_S^{(i)} = \left[\boldsymbol{v}_S^{(i-1)}\ \boldsymbol{v}_i\right]^T, \tag{4.108}$$

for $i = 3, 4, \ldots, N$, with the initial conditions for $i = 1$ is given by Equation (4.100), and for $i = 2$ is given by

$$\boldsymbol{a}_S^{(2)} = \left[\boldsymbol{a}_1\ \boldsymbol{0}\right], \quad \boldsymbol{A}_S^{(2)} = \begin{bmatrix} \boldsymbol{A}_1 & p_1\boldsymbol{b}_1\boldsymbol{a}_2 \\ \boldsymbol{0} & \boldsymbol{A}_2 \end{bmatrix}, \quad \boldsymbol{v}_S^{(2)} = \begin{bmatrix} \boldsymbol{v}_1 \\ \boldsymbol{v}_2 \end{bmatrix}. \tag{4.109}$$

Lemma 4.7 is utilized to construct Algorithm 4.7 and programmed in MATLAB as **mestandsys** function, available in the software package **dysrel-am**. Its pseudo-code is as follows.

Algorithm 4.7

---

input $N, p_i, i = 1, 2, \ldots, N-1, (n_i, \boldsymbol{a}_i, \boldsymbol{A}_i, \boldsymbol{v}_i), i = 1, 2, \ldots, N$
$n_S = n_1$
$\boldsymbol{a}_S = \boldsymbol{a}_1$
$\boldsymbol{A}_S = \boldsymbol{A}_1$
$\boldsymbol{v}_S = \boldsymbol{v}_1$
for $i = 2$ to N
  $n_S = n_S + n_i$
  $\boldsymbol{a}_S = \left[\boldsymbol{a}_S\ \boldsymbol{0}\right]$
  $\boldsymbol{b}_i = -\boldsymbol{A}_i\boldsymbol{v}_i$
  $\boldsymbol{v}_S = \left[\boldsymbol{v}_S\ \boldsymbol{v}_i\right]^T$
  if $i = 2$ then $\boldsymbol{A}_S = \begin{bmatrix} \boldsymbol{A}_S & p_{i-1}\boldsymbol{b}_{i-1}\boldsymbol{a}_i \\ \boldsymbol{0} & \boldsymbol{A}_i \end{bmatrix}$
  if $i > 2$ then $\boldsymbol{A}_S = \begin{bmatrix} \boldsymbol{A}_S & \begin{array}{c}\boldsymbol{0}\\ p_{i-1}\boldsymbol{b}_{i-1}\boldsymbol{a}_i\end{array} \\ \boldsymbol{0} & \boldsymbol{A}_i \end{bmatrix}$
  $i = i + 1$
end
end

Example 4.40 Consider two components, with each having a lifetime following 2-hypoexponential distribution with PH representations

$$n_1 = n_2 = 2, \quad \boldsymbol{a}_1 = \boldsymbol{a}_2 = \left[1\ 0\right], \quad \boldsymbol{A}_1 = \begin{bmatrix} -1 & 1 \\ 0 & -2 \end{bmatrix}, \quad \boldsymbol{A}_2 = \begin{bmatrix} -2 & 2 \\ 0 & -3 \end{bmatrix}.$$

Their ME representations in Jordan canonical forms obtained using matrix similarity transformations are

$$n_1 = n_2 = 2,\quad \boldsymbol{a}_1 = \begin{bmatrix}-1 & 1\end{bmatrix},\quad \boldsymbol{a}_2 = \begin{bmatrix}-2 & 1\end{bmatrix},$$

$$\boldsymbol{A}_1 = \begin{bmatrix}-2 & 0\\ 0 & -1\end{bmatrix},\quad \boldsymbol{A}_2 = \begin{bmatrix}-3 & 0\\ 0 & -2\end{bmatrix},\quad \boldsymbol{v}_1 = \begin{bmatrix}1\\ 2\end{bmatrix},\quad \boldsymbol{v}_2 = \begin{bmatrix}1\\ 3\end{bmatrix}.$$

Note that the ME representation obtained using matrix similarity transformation is not unique; any arrangement of the diagonal elements of matrix $\boldsymbol{A}$ is allowed if the initial and closing vectors are adjusted accordingly.

The two components are used to construct systems in several structures. The PH and ME representations of each system are derived using their respective equations. The results are as follows.

1) 2-series system:
   PH representation:

$$n_S = 4,\quad \boldsymbol{a}_S = \begin{bmatrix}1 & 0 & 0 & 0\end{bmatrix},\quad \boldsymbol{A}_S = \begin{bmatrix}-3 & 2 & 1 & 0\\ 0 & -4 & 0 & 1\\ 0 & 0 & -4 & 2\\ 0 & 0 & 0 & -5\end{bmatrix}.$$

   ME representation:

$$n_S = 4,\quad \boldsymbol{a}_S = \begin{bmatrix}2 & -1 & -2 & 1\end{bmatrix},\quad \boldsymbol{A}_S = \begin{bmatrix}-5 & 0 & 0 & 0\\ 0 & -4 & 0 & 0\\ 0 & 0 & -4 & 0\\ 0 & 0 & 0 & -3\end{bmatrix},\quad \boldsymbol{v}_S = \begin{bmatrix}1\\ 3\\ 2\\ 6\end{bmatrix}.$$

2) 2-parallel system:
   PH representation:

$$n_S = 8,\quad \boldsymbol{a}_S = \begin{bmatrix}1 & 0 & 0 & 0 & 0 & 0 & 0 & 0\end{bmatrix},$$

$$\boldsymbol{A}_S = \begin{bmatrix}-3 & 2 & 1 & 0 & 0 & 0 & 0 & 0\\ 0 & -4 & 0 & 1 & 3 & 0 & 0 & 0\\ 0 & 0 & -4 & 2 & 0 & 0 & 2 & 0\\ 0 & 0 & 0 & -5 & 0 & 3 & 0 & 2\\ 0 & 0 & 0 & 0 & -1 & 1 & 0 & 0\\ 0 & 0 & 0 & 0 & 0 & -2 & 0 & 0\\ 0 & 0 & 0 & 0 & 0 & 0 & -2 & 2\\ 0 & 0 & 0 & 0 & 0 & 0 & 0 & -3\end{bmatrix}.$$

   ME representation:

$$n_S = 8,\quad \boldsymbol{a}_S = \begin{bmatrix}-2 & 1 & 2 & -1 & -1 & 1 & -2 & 1\end{bmatrix},$$

$$\boldsymbol{A}_S = \begin{bmatrix}-5 & 0 & 0 & 0 & 0 & 0 & 0 & 0\\ 0 & -4 & 0 & 0 & 0 & 0 & 0 & 0\\ 0 & 0 & -4 & 0 & 0 & 0 & 0 & 0\\ 0 & 0 & 0 & -3 & 0 & 0 & 0 & 0\\ 0 & 0 & 0 & 0 & -2 & 0 & 0 & 0\\ 0 & 0 & 0 & 0 & 0 & -1 & 0 & 0\\ 0 & 0 & 0 & 0 & 0 & 0 & -3 & 0\\ 0 & 0 & 0 & 0 & 0 & 0 & 0 & -2\end{bmatrix},\quad \boldsymbol{v}_S = \begin{bmatrix}1\\ 3\\ 2\\ 6\\ 1\\ 2\\ 1\\ 3\end{bmatrix}.$$

3) 2-standby system:
PH representation:

$$n_S = 4, \quad \boldsymbol{a}_S = \begin{bmatrix} 1 & 0 & 0 & 0 \end{bmatrix}, \quad \boldsymbol{A}_S = \begin{bmatrix} -1 & 1 & 0 & 0 \\ 0 & -2 & 2 & 0 \\ 0 & 0 & -2 & 2 \\ 0 & 0 & 0 & -3 \end{bmatrix}.$$

ME representation:

$$n_S = 4, \quad \boldsymbol{a}_S = \begin{bmatrix} -1 & 1 & 0 & 0 \end{bmatrix}, \quad \boldsymbol{A}_S = \begin{bmatrix} -2 & 0 & -4 & 2 \\ 0 & -1 & -4 & 2 \\ 0 & 0 & -3 & 0 \\ 0 & 0 & 0 & -2 \end{bmatrix}, \quad \boldsymbol{v}_S = \begin{bmatrix} 1 \\ 2 \\ 1 \\ 3 \end{bmatrix}.$$

Even though their representations are different, their reliability and hazard functions obtained using the two representations are identical for each system as shown in Figure 4.41. ■

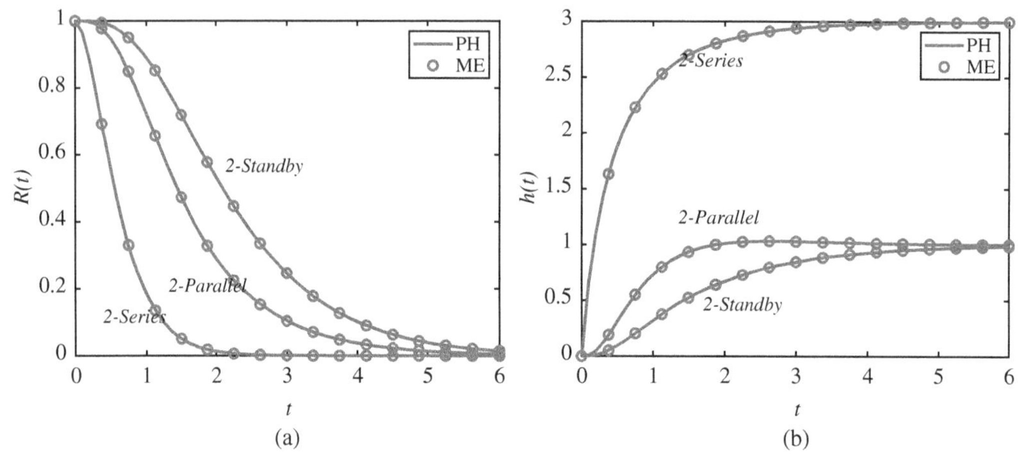

Figure 4.41 Reliability and hazard functions of systems in Example 4.40 calculated using PH and ME distributions. (a) System reliability functions and (b) system hazard functions.

Considering that ME distribution obtained from the PH distribution using a matrix similarity transformation is in Jordan canonical form, the computation time using ME distribution may be expected to be faster than using PH distribution. To compare the computation time using PH and ME distributions, it is important to understand the numerical methods for calculating matrix exponential function in the form of $\boldsymbol{a}e^{\boldsymbol{A}t}\boldsymbol{b}$.

Numerical technique for calculating $e^{\boldsymbol{A}}$ is by expressing the exponential function as Taylor series

$$e^{\boldsymbol{A}} = \boldsymbol{I} + \boldsymbol{A} + \frac{\boldsymbol{A}^2}{2!} + \frac{\boldsymbol{A}^3}{3!} + \cdots \tag{4.110}$$

Method 1 for calculating $\boldsymbol{a}e^{\boldsymbol{A}t}\boldsymbol{b}$ is by calculating $e^{\boldsymbol{A}}$ using Equation (4.110), and the result is premultiplied by $\boldsymbol{a}$ and postmultiplied by $\boldsymbol{b}$. Method 2 is by premultiplied and postmultiplied each term in Equation (4.110) by $\boldsymbol{a}$ and $\boldsymbol{b}$, respectively, and sum the result,

$$\boldsymbol{a}e^{\boldsymbol{A}}\boldsymbol{b} = \boldsymbol{ab} + \boldsymbol{aAb} + \frac{\boldsymbol{aA}^2\boldsymbol{b}}{2!} + \frac{\boldsymbol{aA}^3\boldsymbol{b}}{3!} + \cdots \tag{4.111}$$

The terms added in Equation (4.110) are matrices, while in Equation (4.111) are scalars. Method 2 is supported by an add-on in MATLAB provided at https://www.maths.uq.edu.au/expokit/ [21].

Example 4.41 Systems in several structures are constructed using N identical to either component in Example 3.1 or Example 3.3. The component lifetime in Example 3.1 has an APH distribution whose survival matrix is an upper diagonal matrix. The component lifetime in Example 3.3 has a CPH distribution, whose survival matrix is a full matrix. These PH distributions are transformed into ME distributions in Jordan canonical form using matrix similarity transformations. Reliability measures for these systems are calculated using component lifetime distributions expressed in PH and ME distributions. System reliability measures cover reliability function, hazard function, MRL function, and residual reliability function, with 21 points each. The calculations are using method 1 and method 2. Comparisons of the resulting reliability functions are presented in Figure 4.42, and comparisons of the computation times are presented in Table 4.8. ■

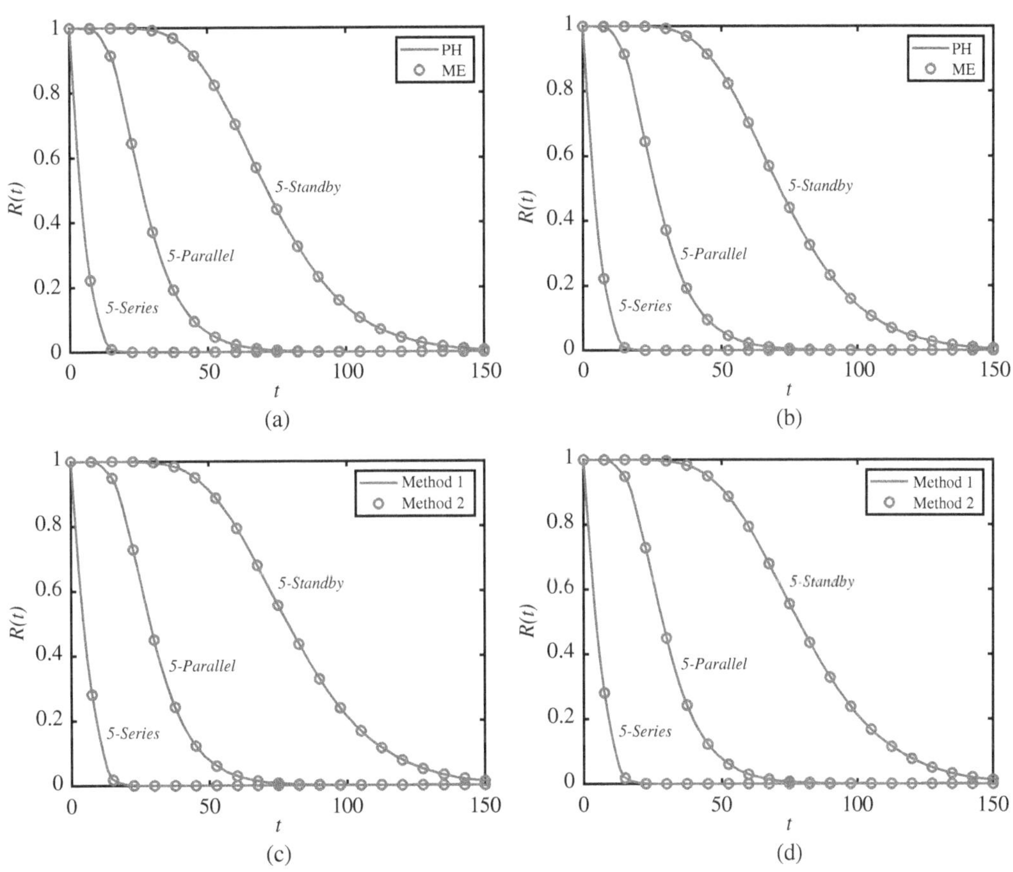

Figure 4.42 Reliability functions of systems in Example 4.36 calculated using PH and ME distributions and methods 1 and 2. (a) APH component, method 1, (b) APH component, method 2, (c) CPH component, PH distribution, and (d) CPH component, ME distribution.

Figure 4.42 indicates that PH distribution and its similar ME distribution calculated using method 1 or method 2 produces the same results for all cases. The entries in Table 4.8 show that

Table 4.8 Computation time for plotting $R(t)$ and $h(t)$ using PH and ME representations.

System			Computation time (s)			
			Method 1		Method 2	
Component	Structure	N	PH	ME	PH	ME
Example 3.1 (APH with ME having off-diagonal elements)	Series	3	0.02	0.12	0.08	0.14
		4	0.09	0.15	0.13	0.21
		5	0.73	0.20	0.39	0.33
		6	11.84	0.30	1.08	1.22
		7	290.94	0.69	3.17	3.48
	Parallel	3	0.05	0.10	0.10	0.20
		4	0.38	0.20	0.19	0.36
		5	10.13	0.78	1.29	1.17
		6	473.59	8.70	3.71	4.16
		7	Hours	148.28	23.67	29.22
	Standby	3	0.01	0.13	0.01	0.15
		4	0.01	0.14	0.02	0.16
		5	0.01	0.17	0.02	0.18
		6	0.01	0.19	0.03	0.21
		7	0.01	0.22	0.03	0.23
Example 3.3 (CPH with ME in diagonal form)	Series	3	0.04	0.06	0.01	0.07
		4	0.36	0.08	0.12	0.09
		5	8.38	0.10	0.25	0.17
		6	456.90	0.15	0.80	0.86
		7	Hours	0.23	3.66	1.05
	Parallel	3	0.11	0.07	0.02	0.10
		4	2.45	0.11	0.20	0.13
		5	200.42	0.29	1.03	0.31
		6	Hours	2.09	6.29	1.90
		7	Hours	18.34	3,444.50	13.21
	Standby	3	0.01	0.09	0.02	0.11
		4	0.01	0.10	0.02	0.12
		5	0.01	0.12	0.02	0.14
		6	0.01	0.13	0.03	0.16
		7	0.01	0.16	0.05	0.18

PH distribution calculated using method 1 is much slower than that calculated using method 2 and that the computation difference between the two is more pronounced for larger PH distribution orders. ME distribution calculated using methods 1 and 2 is not significantly different in their computation times. Therefore, PH distribution should not be calculated using method 1, while ME distribution can be calculated using method 1 or 2. Assuming method 2 is utilized, the only significant difference in computing time between using PH distribution and its similar ME distribution is for the 7-parallel system with CPH distribution that has order around 390,000; for other cases, the computation time difference between the two is not significant. In the case of APH distribution, transforming PH distribution into ME distribution yielded an upper diagonal survival matrix like the original survival matrix of the APH distribution. Therefore, using ME distribution in this case made the computation time slower. This experiment leads to conclusions that method 1 provided by MATLAB should not be used, and transforming PH distribution into ME distribution is necessary when the resulting PH order is very high and the distribution is CPH.

Computation time comparisons between using PH distributions in matrix form and scalar forms using symbolic processing were given by Alkaff and Qomarudin [11]. Their results showed the superiority of using PH distribution and matrix algebra compared to scalar distribution and calculus.

The development of methods for basic structured systems is extended to systems composed of subsystems in different basic structures. Such systems are frequently encountered in real-world applications. The next section develops a method to model and analyze compound basic structured systems.

4.8 Compound Basic Structured Systems

Systems may need to be assembled in combinations of several basic structures to meet their design objectives. Some simple combinations of basic structures frequently encountered in practice are series–parallel, parallel–series, series–standby, and standby–series. Some systems involve more combinations to form complex compound basic structures. As long as the components can still be grouped into several subsystems, each subsystem has a certain basic structure, and the overall system consists of several subsystems that also follow a certain basic structure, the methods explained in Sections 4.3–4.5 can be used in stages. Each subsystem reliability function is generated from its component reliability functions, and the system reliability function is generated from its subsystems reliability functions. Several stages may be needed from the system to the components, not just the system–subsystem–component, depending on the size and complexity of the system being analyzed is. Fortunately, as long as the system can still be staged in a hierarchy and each stage can be represented in a basic structure, the methods explained in Sections 4.3–4.5 are still applicable.

4.8.1 Hierarchical Basic Structured Systems

Real-world large-scale systems consist of many subsystems with various structures. Such systems can be decomposed into several subsystems, each of which can be decomposed into several sub-subsystems. This decomposition process continues to the component level. For example, a system with a lifetime T_S can be decomposed into M subsystems with subsystem m having a lifetime S_m, $m = 1, 2, \ldots, M$. In turn, subsystems m can be decomposed into K_m sub-subsystems with sub-subsystem mk having a lifetime V_{mk}, $k = 1, 2, \ldots, K_m$. Assume that at the last decomposition, sub-subsystem mk contains L_{mk} components with components mkl having a lifetime T_{mkl}, $l = 1, 2, \dots, L_{mk}$. The system lifetime can be represented as a function of the lifetimes of its subsystems using an appropriate structure function. Likewise, subsystems to sub-subsystems and sub-subsystems to components. Therefore, the system structure function can be expressed hierarchically as

$$T_S = \varphi(S_1, S_2, \dots, S_M), \tag{4.112}$$

$$S_m = \varphi_m(V_{m,1}, V_{m,2}, \dots, V_{m,K_m}), \tag{4.113}$$

$$V_{m,k} = \varphi_{m,k}(T_{m,k,1}, T_{m,k,2}, \dots, T_{m,k,L_{mk}}), \tag{4.114}$$

where φ is the system structure function, φ_m is the structure function of subsystem m, and $\varphi_{m,k}$ is the structure function of sub-subsystems mk. The hierarchical model for this system is shown in Figure 4.43, alongside the structure function that relates one level to the next upper level.

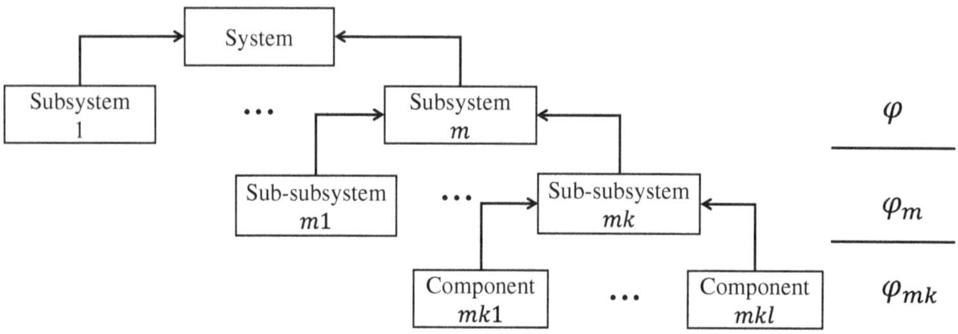

Figure 4.43 A hierarchical structured system model.

By using the closure properties of the PH distribution, if all components have PH-distributed lifetimes, then all sub-subsystems have PH-distributed lifetimes. In the same manner, recursively all the way to the system also has PH-distributed lifetime. The PH representation for the distribution of the system lifetime can be calculated recursively from the lowest level, the component, to the highest level, the system, by using a combination of algorithms that have been derived in Sections 4.3–4.5 in accordance with the structure of each level.

The development of methods for modeling and analyzing compound basic structured systems starts with simple compound basic structures that can be described in two levels of hierarchy.

4.8.2 Simple Compound Basic Structures

Simple compound basic structures are combinations of two basic structures. Such structures require two levels of hierarchy to express system reliability as a function of the reliability of its components. Some popular simple compound basic structures include series–parallel, parallel–series, series–standby, and standby–series structures. It is possible in theory to have parallel–standby and standby–parallel structures; however, they are not popular and can be represented as general standby systems that will be discussed in Chapter 5.

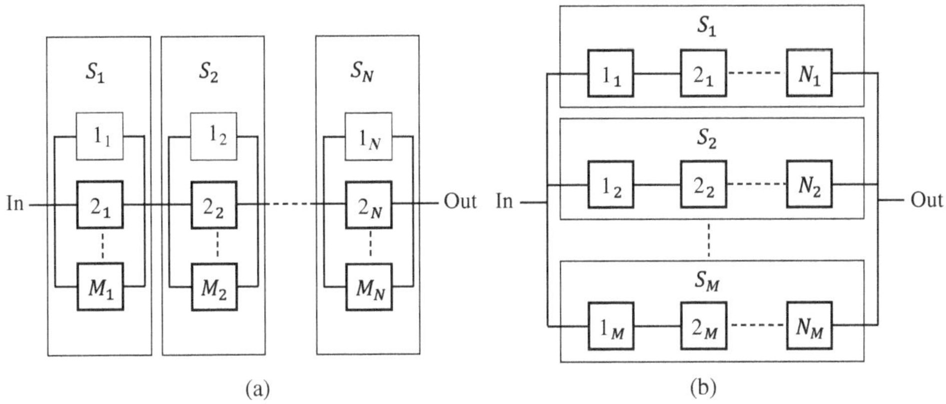

Figure 4.44 Series–parallel and parallel–series systems. (a) Series–parallel system and (b) parallel–series system.

The series–parallel structured system, simply referred to as the series–parallel system, is composed of several subsystems with parallel structures assembled into one system in a series structure. The parallel–series system is several subsystems with series structures assembled into one system in a parallel structure. The reliability block diagrams of a series–parallel system and a parallel–series system are shown in Figure 4.44a,b, respectively. The series–standby and standby–series structured systems follow the same pattern by replacing the parallel structure with the standby structure.

The series–parallel and parallel–series systems can be modeled using combinations of the **phseriessys** and **phparallelsys** functions. The parallel structure portion is modeled using **phparallelsys**, and the series structure portion is modeled using **phseriessys**. To model a series–parallel system in Figure 4.44a with N subsystem in series and the ith subsystem has M_i components in parallel, define:

- T_S is the system lifetime.
- T_{S_i} is the lifetime of subsystem $i, i = 1, 2, \dots, N$.
- T_{j_i} is the lifetime of component j in subsystem $i, j = 1, 2, \dots M_i$.

The system reliability model in these variables is

$$T_{S_i} = \text{phparallelsys}\,(T_{1_i}, T_{2_i}, \dots, T_{M_i}) \quad i = 1, 2, \dots, N,$$
$$T_S = \text{phseriessys}\,(T_{S_1}, T_{S_2}, \dots, T_{S_N}).$$

To model the parallel–series system in Figure 4.44b with M subsystem in parallel and the ith subsystem has N_i components in series, define:

- T_{S_i} is the lifetime of subsystem $i, i = 1, 2, \dots, M$.
- T_{j_i} is the lifetime of component j in the subsystem $i, j = 1, 2, \dots, N_i$.

The system reliability model in these variables is

$$T_{S_i} = \text{phseriessys}\,(T_{1_i}, T_{2_i}, \dots, T_{N_i}) \quad i = 1, 2, \dots, M,$$
$$T_S = \text{phparallelsys}\,(T_{S_1}, T_{S_2}, \dots, T_{S_M}).$$

A main program that calls functions that represent these models must be prepared to generate the PH distribution of the system lifetime.

Series–parallel systems are found in systems with several subsystems, each of which has several components, with components belonging to one subsystem can be utilized by other subsystems. In other words, each subsystem is modular, and each module in one subsystem is compatible with similar modules in other subsystems (modular compatible). In parallel–series systems, the subsystems are not modular, or their modules cannot be removed and used in other subsystems (compact or incompatible).

Example 4.42 A system comprises three devices, each of which has four modules. Each device requires all four modules to perform its function, and the failure of one module causes the device to fail. Other devices can use modules in one device. The three devices are operated simultaneously, even though only one is needed.

This description means this system has four modular subsystems connected in series and each subsystem has three components (modules in each equipment) connected in parallel. The overall system is connected in a series–parallel structure. The reliability block diagram of this system is presented in Figure 4.45.

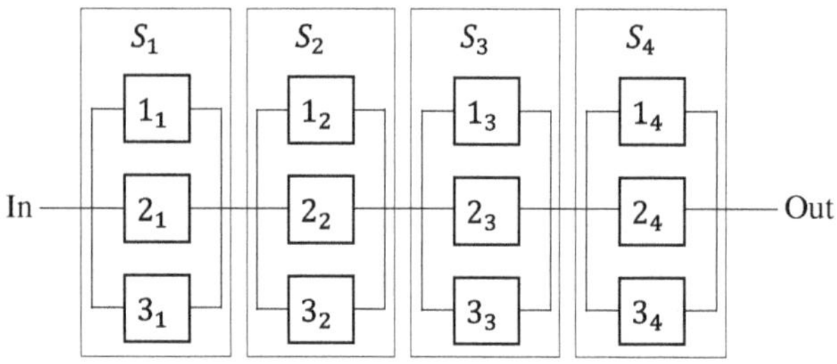

Figure 4.45 Reliability block diagram of the system in Example 4.42.

The system can be modeled using MATLAB functions in two levels: the system level uses the **phseriessys** function because it has a series structure, and the subsystem level uses **phparallelsys** because it has a parallel structure. The system reliability model in the prepared MATLAB functions is

$$T_{S_1} = \text{phparallelsys}\,(T_{1_1}, T_{2_1}, T_{3_1}),$$

$$T_{S_2} = \text{phparallelsys}\,(T_{1_2}, T_{2_2}, T_{3_2}),$$

$$T_{S_3} = \text{phparallelsys}\,(T_{1_3}, T_{2_3}, T_{3_3}),$$

$$T_{S_4} = \text{phparallelsys}\,(T_{1_4}, T_{2_4}, T_{3_4}),$$

$$T_S = \text{phseriessys}\,(T_{S_1}, T_{S_2}, T_{S_3}, T_{S_4}),$$

where T_{S_i} is the lifetime of all modules i, $i = 1, 2, 3, 4$, in the three devices, and T_{j_i} is the lifetime of module i in device j, $j = 1, 2, 3$. If the same modules in different devices have identical lifetime distributions, then **phparallelsys** function to model the modules can be replaced by **phparallelidnsys**. ∎

Example 4.43 Each device in Example 4.42 is compact; its modules cannot be removed to be installed on other devices. If one device fails because one of its modules failed, then all modules in that device cannot be utilized, including those are still good.

This system with three devices has a parallel–series structure in such operational condition. The subsystems are the devices connected in parallel, while the components are the modules of each device connected in series. The reliability block diagram of such a system is given in Figure 4.46.

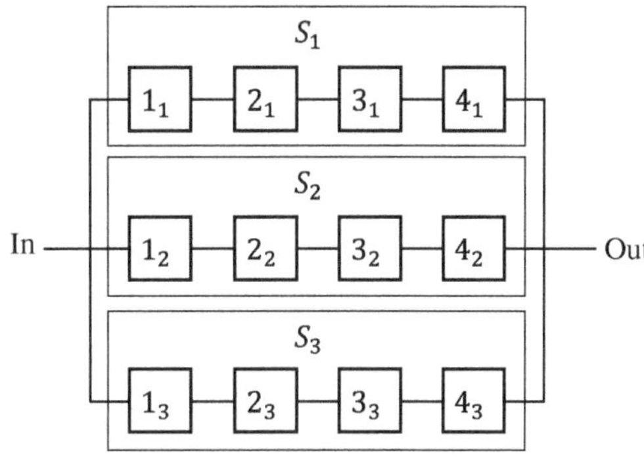

Figure 4.46 Reliability block diagram of the system in Example 4.43.

The system reliability model in the prepared MATLAB functions is

$$T_{S_1} = \text{phseriessys}\,(T_{1_1}, T_{2_1}, T_{3_1}, T_{41}),$$

$$T_{S_2} = \text{phseriessys}\,(T_{1_2}, T_{2_2}, T_{3_2}, T_{4_2}),$$

$$T_{S_3} = \text{phseriessys}\,(T_{1_3}, T_{2_3}, T_{3_3}, T_{4_3}),$$

$$T_S = \text{phparallelsys}\,(T_{S_1}, T_{S_2}, T_{S_3}),$$

where T_{S_i} is the lifetime of device i, $i = 1, 2, 3$, and T_{j_i} is the lifetime of modules j, $j = 1, 2, 3, 4$, in device i.

If the same modules in different devices are identical, then the devices are identical and **phparallelsys** function for the system model can be replaced by **phparallelidnsys**. ■

Example 4.44 If only one required device is operated, while others are backups that are not operated, then the system in Example 4.43 has a standby–series structure. Its reliability block diagram is shown in Figure 4.47.

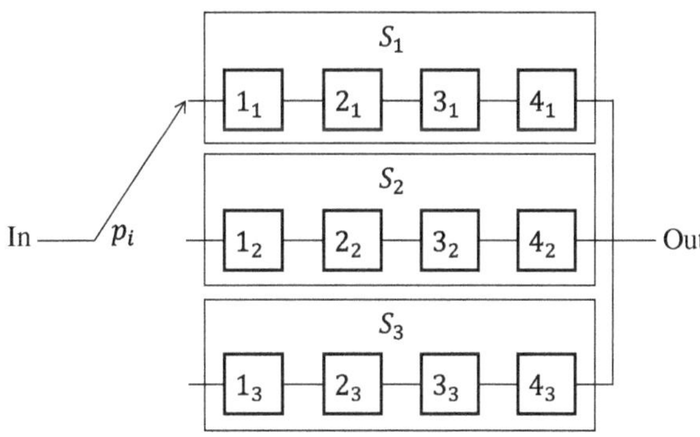

Figure 4.47 Reliability block diagram of the system in Example 4.44.

The reliability model of this system can be obtained from the system reliability model of Example 4.43 by replacing **phparallelsys** with **phstandsys**. The overall system model in the prepared MATLAB functions is

$$T_{S_1} = \text{phseriessys}\,(T_{1_1}, T_{2_1}, T_{3_1}, T_{4_1}),$$
$$T_{S_2} = \text{phseriessys}\,(T_{1_2}, T_{2_2}, T_{3_2}, T_{4_2}),$$
$$T_{S_3} = \text{phseriessys}\,(T_{1_3}, T_{2_3}, T_{3_3}, T_{4_3}),$$
$$T_S = \text{phstandsys}\,(S_1, S_2, S_3, p_1, p_2).$$

where p_1 and p_2, respectively, are the reliability of the switch to replace the active subsystem with the first backup subsystems in the event of first failure and with the second backup subsystems in the event of second failure. ■

Series–standby structures are often found in manufacturing systems that utilize production lines. A production line comprises several workstations arranged in series. To increase its reliability, each workstation is equipped with more than one machine, even though only one is actively operated. The unoperated machines are backup machines to replace the current primary machine when it fails.

Example 4.45 A manufacturing system requires four machines that are used sequentially to produce a product. Each machine is equipped with several passive backups as replacements if the primary fails. Based on technical and cost considerations, machines 1, 2, 3, and 4 have 1, 2, 2, and 3 backups, respectively. The time to activate the backups is assumed negligible, and the switch to activate a passive backup machine is assumed to be perfect.

This manufacturing system has a series–standby structure, which comprises several primary machines operate in series, and each primary machine has backup machines. The reliability block diagram of this system is presented in Figure 4.48.

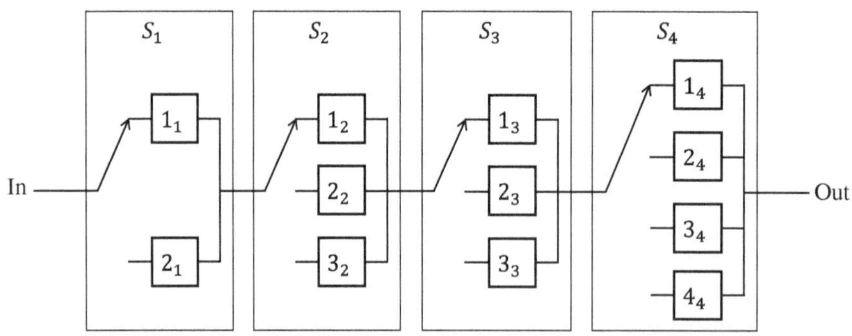

Figure 4.48 Reliability block diagram of a system in Example 4.45.

The system reliability model in the prepared MATLAB functions is

$$T_{S_1} = \text{phstandsys}\,(T_{1_1}, T_{2_1}),$$
$$T_{S_2} = \text{phstandsys}\,(T_{1_2}, T_{2_2}, T_{3_2}),$$
$$T_{S_3} = \text{phstandsys}\,(T_{1_3}, T_{2_3}, T_{3_3}),$$

$$T_{S_4} = \text{phstandsys}\,(T_{1_4}, T_{2_4}, T_{3_4}, T_{4_4}),$$

$$T_S = \text{phseriessys}\,(T_{S_1}, T_{S_2}, T_{S_3}, T_{S_4}).$$

■

The method to derive PH distributions of compound basic structured systems manually is illustrated using systems with components having exponentially distributed lifetimes.

4.8.3 Simple Compound Basic Structured Systems with CFR Components

A manual calculation to find the PH distribution of the lifetime of a compound basic structured system is shown for a system with CFR components.

Example 4.46 A system uses two devices, each of which has two modules. Modules i in device j have an exponentially distributed lifetime with a rate λ_{ij}, where $\lambda_{11} = 1$, $\lambda_{12} = 2$, $\lambda_{21} = 3$, $\lambda_{22} = 4$.

Define M_i, $i = 1, 2$ as modules i in both devices, M_{ij}, $i = 1, 2; j = 1, 2$, as module i in device j, and $D_j, j = 1, 2$ as device j.

The lifetimes of modules M_{ij}, $i = 1, 2; j = 1, 2$, have exponential distributions with PH representation $(n_{M_{ij}}, \boldsymbol{a}_{M_{ij}}, \boldsymbol{A}_{M_{ij}})$, where

$$n_{M_{11}} = n_{M_{12}} = n_{M_{21}} = n_{M_{22}} = 1, \quad \boldsymbol{a}_{M_{11}} = \boldsymbol{a}_{M_{12}} = \boldsymbol{a}_{M_{21}} = \boldsymbol{a}_{M_{22}} = [1],$$

$$\boldsymbol{A}_{M_{11}} = [-1], \quad \boldsymbol{A}_{M_{12}} = [-2], \quad \boldsymbol{A}_{M_{21}} = [-3], \quad \boldsymbol{A}_{M_{22}} = [-4],$$

$$\boldsymbol{b}_{M_{11}} = [1], \quad \boldsymbol{b}_{M_{12}} = [2], \quad \boldsymbol{b}_{M_{21}} = [3], \quad \boldsymbol{b}_{M_{22}} = [4].$$

1) Series–parallel structure:
 - Modules in parallel structure:
   - Module 1 in devices 1 and 2 are connected in parallel:

$$\boldsymbol{a}_{M_1} = \left[\boldsymbol{a}_{M_{11}} \otimes \boldsymbol{a}_{M_{12}}\ 0\ 0\right] = \left[[1] \otimes [1]\ 0\ 0\right] = \left[1\ 0\ 0\right].$$

$$\boldsymbol{A}_{M_1} = \begin{bmatrix} \boldsymbol{A}_{M_{11}} \oplus \boldsymbol{A}_{M_{12}} & \boldsymbol{I}_{M_{11}} \otimes \boldsymbol{b}_{M_{12}} & \boldsymbol{b}_{M_{11}} \otimes \boldsymbol{I}_{M_{12}} \\ \boldsymbol{0} & \boldsymbol{A}_{M_{11}} & \boldsymbol{0} \\ \boldsymbol{0} & \boldsymbol{0} & \boldsymbol{A}_{M_{12}} \end{bmatrix} = \begin{bmatrix} [-1] \oplus [-2] & [1] \otimes [2] & [1] \otimes [1] \\ 0 & [-1] & 0 \\ 0 & 0 & [-2] \end{bmatrix}$$

$$= \begin{bmatrix} -3 & 2 & 1 \\ 0 & -1 & 0 \\ 0 & 0 & -2 \end{bmatrix}.$$

   - Module 2 in devices 1 and 2 are connected in parallel:

$$\boldsymbol{a}_{M_2} = \left[\boldsymbol{a}_{M_{21}} \otimes \boldsymbol{a}_{M_{22}}\ 0\ 0\right] = \left[[1] \otimes [1]\ 0\ 0\right] = \left[1\ 0\ 0\right].$$

$$\boldsymbol{A}_{M_2} = \begin{bmatrix} \boldsymbol{A}_{M_{21}} \oplus \boldsymbol{A}_{M_{22}} & \boldsymbol{I}_{M_{21}} \otimes \boldsymbol{b}_{M_{22}} & \boldsymbol{b}_{M_{21}} \otimes \boldsymbol{I}_{M_{22}} \\ \boldsymbol{0} & \boldsymbol{A}_{M_{21}} & \boldsymbol{0} \\ \boldsymbol{0} & \boldsymbol{0} & \boldsymbol{A}_{M_{22}} \end{bmatrix} = \begin{bmatrix} [-3] \oplus [-4] & [1] \otimes [4] & [3] \otimes [1] \\ 0 & [-3] & 0 \\ 0 & 0 & [-4] \end{bmatrix}$$

$$= \begin{bmatrix} -7 & 4 & 3 \\ 0 & -3 & 0 \\ 0 & 0 & -4 \end{bmatrix}.$$

 - Modules 1 and 2 in a series structure:

$$\boldsymbol{a}_S = \left[1\ 0\ 0\right] \otimes \left[1\ 0\ 0\right] = \left[1\ 0\ 0\ 0\ 0\ 0\ 0\ 0\ 0\right].$$

$$\boldsymbol{A}_S = \boldsymbol{A}_{M_1} \oplus \boldsymbol{A}_{M_2} = \boldsymbol{A}_{M_1} \otimes \boldsymbol{I}_{M_2} + \boldsymbol{I}_{M_1} \otimes \boldsymbol{A}_{M_2}$$

$$= \begin{bmatrix} -3\begin{bmatrix}1&0&0\\0&1&0\\0&0&1\end{bmatrix} & 2\begin{bmatrix}1&0&0\\0&1&0\\0&0&1\end{bmatrix} & 1\begin{bmatrix}1&0&0\\0&1&0\\0&0&1\end{bmatrix} \\ \mathbf{0} & -1\begin{bmatrix}1&0&0\\0&1&0\\0&0&1\end{bmatrix} & \mathbf{0} \\ \mathbf{0} & \mathbf{0} & -2\begin{bmatrix}1&0&0\\0&1&0\\0&0&1\end{bmatrix} \end{bmatrix} + \begin{bmatrix} \begin{bmatrix}-7&4&3\\0&-3&0\\0&0&-4\end{bmatrix} & \mathbf{0} & \mathbf{0} \\ \mathbf{0} & \begin{bmatrix}-7&4&3\\0&-3&0\\0&0&-4\end{bmatrix} & \mathbf{0} \\ \mathbf{0} & \mathbf{0} & \begin{bmatrix}-7&4&3\\0&-3&0\\0&0&-4\end{bmatrix} \end{bmatrix}$$

$$= \begin{bmatrix} \begin{bmatrix}-10&4&3\\0&-4&0\\0&0&-7\end{bmatrix} & \begin{bmatrix}2&0&0\\0&2&0\\0&0&2\end{bmatrix} & \begin{bmatrix}1&0&0\\0&1&0\\0&0&1\end{bmatrix} \\ \mathbf{0} & \begin{bmatrix}-8&4&3\\0&-4&0\\0&0&-5\end{bmatrix} & \mathbf{0} \\ \mathbf{0} & \mathbf{0} & \begin{bmatrix}-9&4&3\\0&-5&0\\0&0&-6\end{bmatrix} \end{bmatrix}$$

$$= \begin{bmatrix} -10&4&3&2&0&0&1&0&0\\ 0&-4&0&0&2&0&0&1&0\\ 0&0&-7&0&0&2&0&0&1\\ 0&0&0&-8&4&3&0&0&0\\ 0&0&0&0&-4&0&0&0&0\\ 0&0&0&0&0&-5&0&0&0\\ 0&0&0&0&0&0&-9&4&3\\ 0&0&0&0&0&0&0&-5&0\\ 0&0&0&0&0&0&0&0&-6 \end{bmatrix}.$$

The order of the resulting PH distribution is $(n_{M_{11}} n_{M_{12}} + n_{M_{11}} + n_{M_{12}})(n_{M_{21}} n_{M_{22}} + n_{M_{21}} + n_{M_{22}}) = 9$.

2) Parallel–series structure:
 - Modules in series structures:
   - Modules 1 and 2 in device 1 in a series structure:

$$\boldsymbol{a}_{D_1} = \boldsymbol{a}_{M_{11}} \otimes \boldsymbol{a}_{M_{21}} = [1] \otimes [1] = [1],$$
$$\boldsymbol{A}_{D_1} = \boldsymbol{A}_{M_{11}} \oplus \boldsymbol{A}_{M_{21}} = [-1] \oplus [-3] = [-4].$$

   - Modules 1 and 2 in device 2 in a series structure:

$$\boldsymbol{a}_{D_2} = \boldsymbol{a}_{M_{12}} \otimes \boldsymbol{a}_{M_{22}} = [1] \otimes [1] = [1],$$
$$\boldsymbol{A}_{D_2} = \boldsymbol{A}_{M_{12}} \oplus \boldsymbol{A}_{M_{22}} = [-2] \oplus [-4] = [-6].$$

- Devices 1 and 2 in a parallel structure:

$$\boldsymbol{a}_S = \left[\boldsymbol{a}_{D_1} \otimes \boldsymbol{a}_{D_2}\ 0\ 0\right] = \left[[1] \otimes [1]\ 0\ 0\right] = \left[1\ 0\ 0\right],$$

$$\boldsymbol{A}_S = \begin{bmatrix} \boldsymbol{A}_{D_1} \oplus \boldsymbol{A}_{D_2} & \boldsymbol{I}_{D_1} \otimes \boldsymbol{b}_{D_2} & \boldsymbol{b}_{D_1} \otimes \boldsymbol{I}_{D_2} \\ \boldsymbol{0} & \boldsymbol{A}_{D_1} & \boldsymbol{0} \\ \boldsymbol{0} & \boldsymbol{0} & \boldsymbol{A}_{D_2} \end{bmatrix} = \begin{bmatrix} [-4] \oplus [-6] & [1] \otimes [6] & [4] \otimes [1] \\ 0 & [-4] & 0 \\ 0 & 0 & [-6] \end{bmatrix}$$

$$= \begin{bmatrix} -10 & 6 & 4 \\ 0 & -4 & 0 \\ 0 & 0 & -6 \end{bmatrix}.$$

The order of the resulting PH distribution is $(n_{M_{11}} n_{M_{21}})(n_{M_{12}} n_{M_{22}}) + (n_{M_{11}} n_{M_{21}}) + (n_{M_{12}} n_{M_{22}}) = 3$.

3) Series–standby structure:
   - Modules in a standby structure:
     - Modules 1 in devices 1 and 2 in a standby structure:

$$\boldsymbol{a}_{M_1} = \left[\boldsymbol{a}_{M_{11}}\ 0\right] = \left[[1]\ 0\right] = \left[1\ 0\right].$$

$$\boldsymbol{A}_{M_1} = \begin{bmatrix} \boldsymbol{A}_{M_{11}} & \boldsymbol{b}_{M_{11}} \boldsymbol{a}_{M_{12}} \\ \boldsymbol{0} & \boldsymbol{A}_{M_{12}} \end{bmatrix} = \begin{bmatrix} -1 & [1][1] \\ 0 & -2 \end{bmatrix} = \begin{bmatrix} -1 & 1 \\ 0 & -2 \end{bmatrix}.$$

     - Modules 2 in devices 1 and 2 in a standby structure:

$$\boldsymbol{a}_{M_2} = \left[\boldsymbol{a}_{M_{21}}\ 0\right] = \left[[1]\ 0\right] = \left[1\ 0\right].$$

$$\boldsymbol{A}_{M_2} = \begin{bmatrix} \boldsymbol{A}_{M_{21}} & \boldsymbol{b}_{M_{21}} \boldsymbol{a}_{M_{22}} \\ \boldsymbol{0} & \boldsymbol{A}_{M_{22}} \end{bmatrix} = \begin{bmatrix} -3 & [3][1] \\ 0 & -4 \end{bmatrix} = \begin{bmatrix} -3 & 3 \\ 0 & -4 \end{bmatrix}.$$

   - Modules 1 and 2 in a series structure:

$$\boldsymbol{a}_S = \boldsymbol{a}_{M_1} \otimes \boldsymbol{a}_{M_2} = \left[1\ 0\right] \otimes \left[1\ 0\right] = \left[1\ 0\ 0\ 0\right].$$

$$\boldsymbol{A}_S = \boldsymbol{A}_{M_1} \oplus \boldsymbol{A}_{M_2} = \begin{bmatrix} -1 & 1 \\ 0 & -2 \end{bmatrix} \oplus \begin{bmatrix} -3 & 3 \\ 0 & -4 \end{bmatrix} = \begin{bmatrix} -1\begin{bmatrix} 1 & 0 \\ 0 & 1 \end{bmatrix} & 1\begin{bmatrix} 1 & 0 \\ 0 & 1 \end{bmatrix} \\ 0\begin{bmatrix} 1 & 0 \\ 0 & 1 \end{bmatrix} & -2\begin{bmatrix} 1 & 0 \\ 0 & 1 \end{bmatrix} \end{bmatrix}$$

$$+ \begin{bmatrix} 1\begin{bmatrix} -3 & 3 \\ 0 & -4 \end{bmatrix} & 0\begin{bmatrix} -3 & 3 \\ 0 & -4 \end{bmatrix} \\ 0\begin{bmatrix} -3 & 3 \\ 0 & -4 \end{bmatrix} & 1\begin{bmatrix} -3 & 3 \\ 0 & -4 \end{bmatrix} \end{bmatrix} = \begin{bmatrix} -4 & 3 & 1 & 0 \\ 0 & -5 & 0 & 1 \\ 0 & 0 & -5 & 3 \\ 0 & 0 & 0 & -6 \end{bmatrix}.$$

The order of the resulting PH distribution is $(n_{M_{11}} + n_{M_{12}})(n_{M_{21}} + n_{M_{22}}) = 4$.

4) Standby–series structure:
   - Modules in series structures:
     - Modules 1 and 2 in device 1 in a series structure:

$$\boldsymbol{a}_{D_1} = \boldsymbol{a}_{M_{11}} \otimes \boldsymbol{a}_{M_{21}} = [1] \otimes [1] = [1],$$

$$\boldsymbol{A}_{D_1} = \boldsymbol{A}_{M_{11}} \oplus \boldsymbol{A}_{M_{21}} = [-1] \oplus [-3] = [-4].$$

– Modules 1 and 2 in device 2 in a series structure:

$$\boldsymbol{a}_{D_2} = \boldsymbol{a}_{M_{12}} \otimes \boldsymbol{a}_{M_{22}} = [1] \otimes [1] = [1],$$

$$\boldsymbol{A}_{D_2} = \boldsymbol{A}_{M_{12}} \oplus \boldsymbol{A}_{M_{22}} = [-2] \oplus [-4] = [-6].$$

- Devices 1 and 2 in a standby structure:

$$\boldsymbol{a}_S = \left[\boldsymbol{a}_{D_1}\ \boldsymbol{0}\right] = \left[[1]\ 0\right] = \left[1\ 0\right],$$

$$\boldsymbol{A}_S = \begin{bmatrix} \boldsymbol{A}_{D_1} & \boldsymbol{b}_{D_1}\boldsymbol{a}_{D_2} \\ \boldsymbol{0} & \boldsymbol{A}_{D_2} \end{bmatrix} = \begin{bmatrix} [-4] & [4][1] \\ 0 & [-6] \end{bmatrix} = \begin{bmatrix} -4 & 4 \\ 0 & -6 \end{bmatrix}.$$

The order of the resulting PH distribution is $n_{M_{11}} n_{M_{21}} + n_{M_{21}} n_{M_{22}} = 2$.
Using the provided MATLAB functions yields the following PH representations for the distributions of the system lifetimes:

- System in a series–parallel structure:

```
As =
   -10     4     3     2     0     0     1     0     0
     0    -6     0     0     2     0     0     1     0
     0     0    -7     0     0     2     0     0     1
     0     0     0    -8     4     3     0     0     0
     0     0     0     0    -4     0     0     0     0
     0     0     0     0     0    -5     0     0     0
     0     0     0     0     0     0    -9     4     3
     0     0     0     0     0     0     0    -5     0
     0     0     0     0     0     0     0     0    -6
```

- System in a parallel–series structure:

```
As =
   -10     6     4
     0    -4     0
     0     0    -6
```

- System in a series–standby structure:

```
As =
    -4     3     1     0
     0    -5     0     1
     0     0    -5     3
     0     0     0    -6
```

- System in a standby–series structure:

```
As =
    -4     4
     0    -6
```

As can be seen, manual calculations and MATLAB functions produce the same results. ∎

Next, properties of simple compound basic structured systems are examined.

4.8.4 Properties of Simple Compound Basic Structured Systems

Consider the parallel–series system in Example 4.43. If module 1 in device 1, module 2 in device 2, and module 3 in device 3 fail, then the system fails. However, such failures do not cause the series–parallel system in Example 4.42 to fail, because device 1 can use module 1 from device 2, allowing device 1 to continue operation and the whole system to function. If the two systems use identical devices and modules, then the lifetime of the series–parallel system in Example 4.42 is stochastically greater than or equal to the lifetime of the parallel–series system in Example 4.43. This reasoning applies to arbitrary series–parallel and parallel–series systems, resulting in

$$T_{s-p} \succcurlyeq T_{p-s},$$

where T_{s-p} and T_{p-s} are the lifetimes of the series–parallel and parallel–series systems, respectively. Thus,

$$R_{s-p}(t) \geq R_{p-s}(t), \tag{4.115}$$

$$h_{s-p}(t) \leq h_{p-s}(t), \tag{4.116}$$

$$MTTF_{s-p} \geq MTTF_{p-s}, \tag{4.117}$$

where $R_{s-p}(t)$, $R_{p-s}(t)$, $h_{s-p}(t)$, and $h_{p-s}(t)$, are the reliability and hazard functions of the series–parallel and parallel–series systems, respectively, while $MTTF_{s-p}$ and $MTTF_{p-s}$ are the expected lifetimes of the series–parallel and parallel–series systems, respectively.

The same reasoning applied to series and standby results in

$$T_{s-st} \succcurlyeq T_{st-s},$$

where T_{s-st} and T_{st-s} are the lifetimes of the series–standby and standby–series systems, respectively. Thus,

$$R_{s-st}(t) \geq R_{st-s}(t), \tag{4.118}$$

$$h_{s-st}(t) \leq h_{st-s}(t), \tag{4.119}$$

$$MTTF_{s-st} \geq MTTF_{st-s}, \tag{4.120}$$

where $R_{s-st}(t)$ and $R_{st-s}(t)$ are the reliability functions of the series–standby system and the standby–series system, respectively, while $MTTF_{s-st}$ and $MTTF_{st-s}$ are the expected lifetimes of the series–standby and standby–series systems, respectively.

As previously shown in Section 4.6 that for parallel and standby systems having the same components, the lifetime of the standby system is stochastically longer than the lifetime of the parallel system or $T_{s-p} \preccurlyeq T_{s-st}$ and $T_{p-s} \preccurlyeq T_{st-s}$. Thus,

$$R_{s-p}(t) \leq R_{s-st}(t), \tag{4.121}$$

$$R_{p-s}(t) \leq R_{st-s}(t), \tag{4.122}$$

$$MTTF_{s-p} \leq MTTF_{s-st}, \tag{4.123}$$

$$MTTF_{p-s} \leq MTTF_{st-s}. \tag{4.124}$$

The ordering of the system hazard functions can be derived as the reverse of the ordering of the system reliability functions, while the ordering of the system's *MRL* functions follows the ordering of the system reliability functions. For standby systems with imperfect switches, as discussed in Section 4.6, Equations (4.118) and (4.120) are still certain to hold, while Equations (4.121)–(4.124) are not.

Example 4.47 Three similar devices, each having four modules, in various combinations of structures are compared for their system reliability and hazard functions. The lifetimes of the modules i, $i = 1, 2, 3, 4$, in each device have independent and identical Erlang distributions with parameters ($n_1 = 1$, $\lambda_1 = 0.1$), ($n_2 = 2$, $\lambda_2 = 0.2$), ($n_3 = 3$, $\lambda_3 = 0.6$), and ($n_4 = 1$, $\lambda_4 = 0.2$). The graphs of the system reliability and hazard functions for various combinations of structures are shown in Figures 4.49–4.51. As seen, the ordering of system reliability and hazard functions are in accordance with the properties already described. A comparison of the *MTTF* and *SDTTF* is presented in Table 4.9, which shows its conformity with the derived properties.

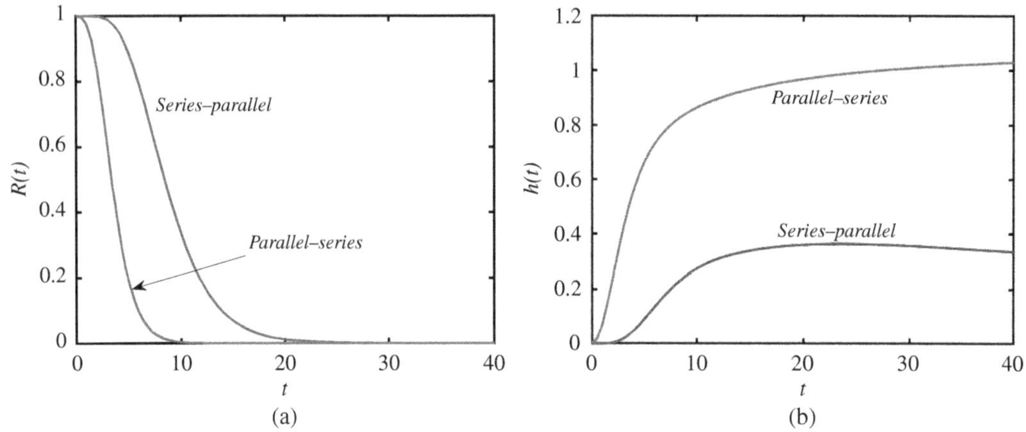

Figure 4.49 Comparisons of reliability and hazard functions of systems in Example 4.47 in series–parallel and parallel–series structures. (a) Reliability functions and (b) hazard functions.

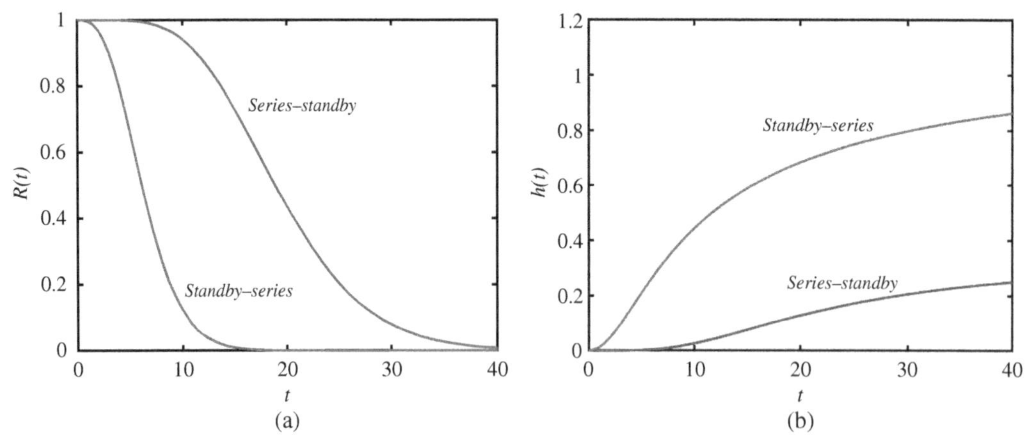

Figure 4.50 Comparison of reliability and hazard functions of systems in Example 4.47 in series–standby and standby–series structures. (a) Reliability functions and (b) hazard functions.

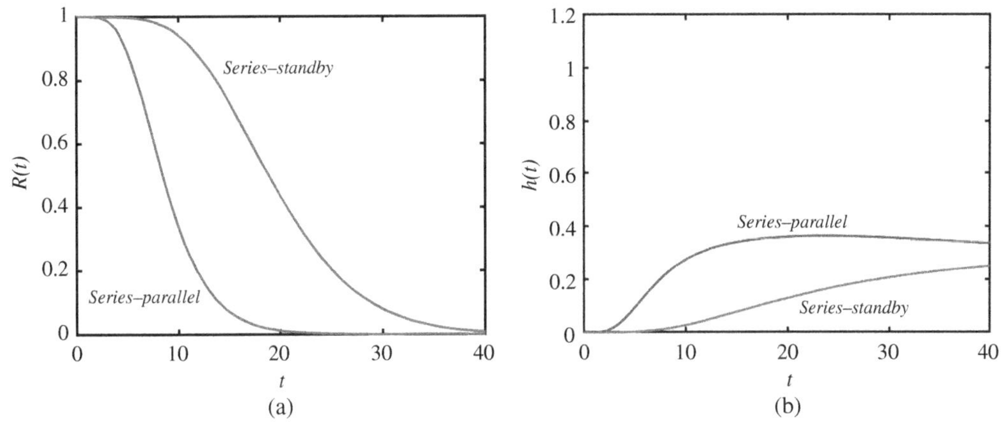

Figure 4.51 Comparison of reliability and hazard functions of systems in Example 4.47 in series–parallel and series–standby structures. (a) Reliability functions and (b) hazard functions.

Table 4.9 Comparison of *MTTF* and *SDTF* of systems in Example 4.47.

Structure	*MTTF*	*SDTTF*
Series–parallel	9.04	3.73
Parallel–series	3.64	1.74
Series–standby	19.63	6.93
Standby–series	6.51	2.94

■

The development of a method for modeling and analyzing compound basic structured systems continues in the next section for complex compound basic structures having more than two levels of hierarchy.

4.8.5 Complex Compound Basic Structures

Systems with complex compound basic structures require longer hierarchies. Such systems can still be modeled in a combination of basic structures, and their PH representations can be calculated recursively using the appropriate models. The modeling of systems with complex compound basic structures is illustrated in Application 4.3.

Application 4.3 The power requirement of a factory is supplied by three identical generators operated simultaneously. The power from one generator is sufficient to meet the needs of the loads. The power generated by the generators is distributed to the load using one of the two available distribution channels, one is active, and the other is passive. The switch to place the passive channel to replace the active one when failed is imperfect with reliability = 0.9. The loads are three identical production lines operated simultaneously, although one production line is sufficient to meet the minimum production target. Each production line uses two machines that carry out sequential processes to produce a certain type of product. Similar machines on both production lines are compatible so that they can be used by any production line that needs them.

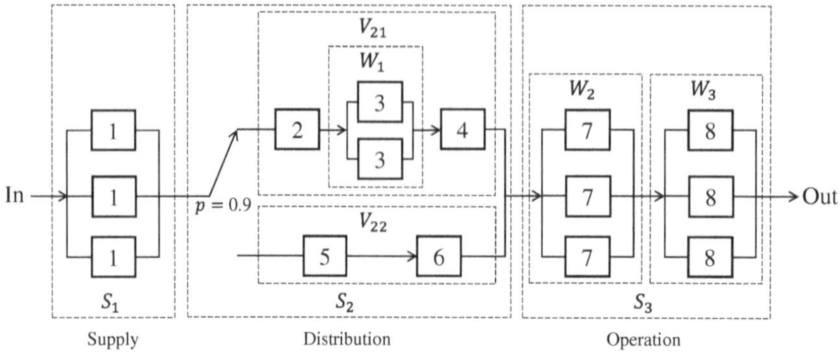

Figure 4.52 Reliability block diagram of the factory system in Application 4.3.

This system can be represented in the reliability block diagram given in Figure 4.52. Components that are indicated by the same number are identical.

Referring to its reliability block diagram, the system can be modeled in the prepared MATLAB functions as

$$T_{S_1} = \text{phparallelsys}(T_1, T_1, T_1),$$

$$W_1 = \text{phparallelsys}(T_3, T_3),$$

$$V_{21} = \text{phseriessys}(T_2, W_1, T_4),$$

$$V_{22} = \text{phseriessys}(T_5, T_6),$$

$$T_{S_2} = \text{phstandsys}(V_{21}, V_{22}; 0.9),$$

$$W_2 = \text{phparallelsys}(T_7, T_7, T_7),$$

$$\mathrm{W}_3 = \text{phparallelsys}(T_8, T_8, T_8),$$

$$T_{S_3} = \text{phseriessys}(W_2, W_3),$$

$$T_S = \text{phseriessys}(T_{S_1}, T_{S_2}, T_{S_3}),$$

where $T_S, T_{S_i}, V_j, W_i, T_i$ are the lifetime of the system, subsystems S_i, subsystems V_j, subsystems W_i, and components i, respectively.

The components of this system have the following PH distributions. All generators have identical hazard function $0.006t^{0.5}$. All components in the main distribution channel have exponentially distributed lifetimes with $MTTF = 300$, and those in the backup distribution channel have exponentially distributed lifetimes with $MTTF = 200$. The first machine in the three production lines has an identical 3-Erlang distributed lifetime with $MTTF = 100$, while the second machine has a lifetime with the same distribution as the first machine but with $MTTF = 150$. The interest is to investigate the effect of the number generator on the system reliability measures.

The reliability function of the generators is approximated with PH distribution using BuTools from MATLAB, as mentioned in Chapter 2. One of the possible PH distribution approximations has a representation

$$\boldsymbol{a} = [0.001 \quad 0.999 \quad 0], \quad \boldsymbol{A} = \begin{bmatrix} -0.0244 & 0 & 0 \\ 0 & -0.0558 & 0.0558 \\ 0 & 0 & -0.0558 \end{bmatrix}.$$

A comparison of graphs of the original generator reliability function and its PH distribution approximation is shown in Figure 4.53a. The graphs of system reliability measures for several numbers of generators are presented in Figure 4.53b–d. The system *MTTF* and *CVTTF* are given in Table 4.10. The system reliability model to produce the graph for $G = 1$, $G = 2$, and $G = 3$ is obtained by changing only the model for T_{S_1}.

Table 4.10 Comparison of *MTTF* and *CVTTF* of the system in Application 4.3.

Number of Generators	*MTTF*	*CVTTF*
1	33.90	0.68
2	45.90	0.52
3	53.01	0.46

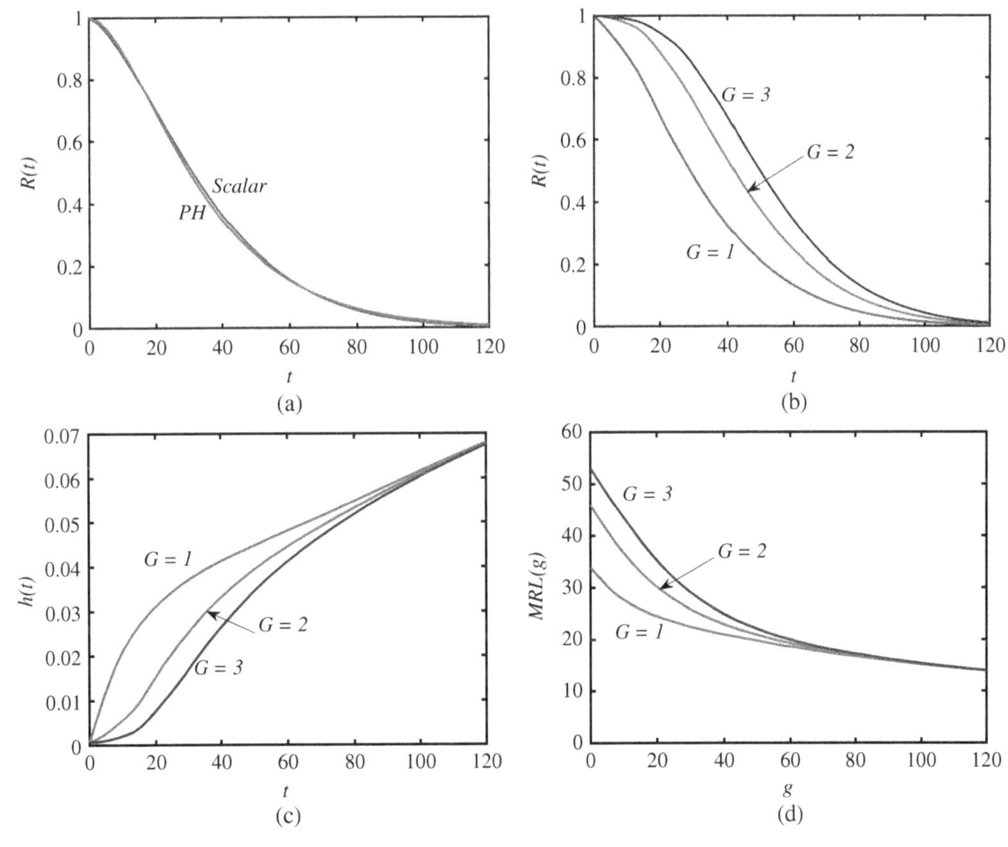

Figure 4.53 The generator reliability function and the effect of the number of generators on reliability measures of the factory system in Application 4.3. (a) Generator reliability functions, (b) system reliability function, (c) system hazard function, and (d) system MRL functions.

The main program to call the required functions for Application 4.3 is available in **application4_3.m** file in the software package **dysrel-am**. Its code is as follows:

MATLAB code for application4_3.m

```
close all;clear
addpath('..\dysrel-am','..\dysrel-am\expokit')
% PH Components                                               (inputs)
% Weibull from phfitting
T(1).a = [0.001 0.999 0];
T(1).A = [-0.0244 0 0;0 -0.0558 0.0558;0 0 -0.0558];
T(1).n = 3;
% n-Erlang
T(2)  = phnErlang(1,300);
T(3)  = phnErlang(1,300);
T(4)  = phnErlang(1,300);
T(5)  = phnErlang(1,200);
T(6)  = phnErlang(1,200);
T(7)  = phnErlang(3,100);
T(8)  = phnErlang(3,150);
tmax = 120;
s     = 21;
t     = linspace(0,tmax,s);
W = []; X = []; Z = [];
% Loop for Number of Generators
for n = 1:3
tic
% Generator
Tg = [];
for i = 1:n
  Tg = [Tg T(1)];
end
% System representation using basic structures                (system model)
Tss(1)  = phparallelsys(Tg);
W1 = phparallelsys([T(3) T(3)]);
V21 = phseriessys([T(2),W1,T(4)]);
V22 = phseriessys([T(5),T(6)]);
Tss(2)  = phstandsys([V21 V22],0.9);
W2 = phparallelsys([T(7) T(7) T(7)]);
W3 = phparallelsys([T(8) T(8) T(8)]);
Tss(3)  = phseriessys([W2 W3]);
Ts = phseriessys(Tss);
tgensys = toc
order = Ts.n
% Statistics Functions                                        (statistical outputs)
tic
mttfs(n)  = phmttf(Ts)
tmttfs = tgensys+toc
sdttfs(n)  = phsdttf(Ts)
cvttfs(n)  = phcvttf(Ts)
% Plot Functions                                              (graphical outputs)
[tt,Rs]  = phrelplot(t,Ts);
[tt,hs]  = phhazplot(t,Ts);
[tt,MRLs]  = phmrlplot(t,Ts);
% Save
W = [W Rs'];
X = [X hs'];
Z = [Z MRLs'];
end
```

```
% PH plot of Generator
[tt,Rg] = phrelplot(t,T(1));
[tt,hg] = phhazplot(t,T(1));
% Scalar plot of Generator
for i = 1:length(tt)
  h(i) = 0.006*tt(i)^0.5;
  R(i) = exp(-0.004*tt(i)^1.5);
end
% Plots                                         (combining graphs)
close all
lgnd = {'\itG = 1';'\itG = 2';'\itG = 3'};
multiplot(2,tt,[R' Rg'],{'Scalar','PH'},'t','R(t)');
multiplot(2,tt,[h' hg'],{'Scalar','PH'},'t','h(t)');
multiplot(3,tt,W,lgnd,'t','R(t)');
multiplot(3,tt,X,lgnd,'t','h(t)');
multiplot(3,tt,Z,lgnd,'g','MRL(g)');
```

Since all parallel subsystems are each using identical components, they can be represented in **phparrallelsys** or **phparallelidnsys** function. The order of the resulting PH representations and computing time using **phparallelsys** are 1,000,188 and 211.16 seconds while using **phaparallelidnsys** are 177,957 and 31.02 seconds, around 80% reduction in the order and computation time. These results show the effectiveness of using **phparallelidnsys** whenever possible. Even though the order of the PH distribution and the number of components are not too large, their multiplier effect can significantly affect the overall system. ■

Application 4.4 The block diagram given in Figure 4.52 is based on assumptions that the power of one generator is sufficient to meet the requirements of the three production lines and the output of one production line is sufficient to meet the demand. However, if the demand for the products increases so that they must be satisfied by all three production lines simultaneously, then subsystem S_3 has a series–series structure and the model changes from

$$W_2 = \text{phparallelidnsys}(T_7, T_7, T_7),$$

$$\text{W}_3 = \text{phparallelidnsys}(T_8, T_8, T_8),$$

$$T_{S_3} = \text{phseriessys}(W_2, W_3),$$

to

$$W_2 = \text{phseriessys}(T_7, T_7, T_7),$$

$$\text{W}_3 = \text{phseriessys}(T_8, T_8, T_8),$$

$$T_{S_3} = \text{phseriessys}(W_2, W_3).$$

If the three production lines are in full capacity to meet the increase in demand, then they consume power that must be supplied simultaneously by the three generators. In this case, the three generators are in a series structure, and the model for S_1 changes from

$$T_{S_1} = \text{phparallelsys}(T_1, T_1, T_1)$$

to

$$T_{S_1} = \text{phseriessys}(T_1, T_1, T_1).$$

Comparison between the system reliability and hazard functions before and after demand increased is shown in Figure 4.54.

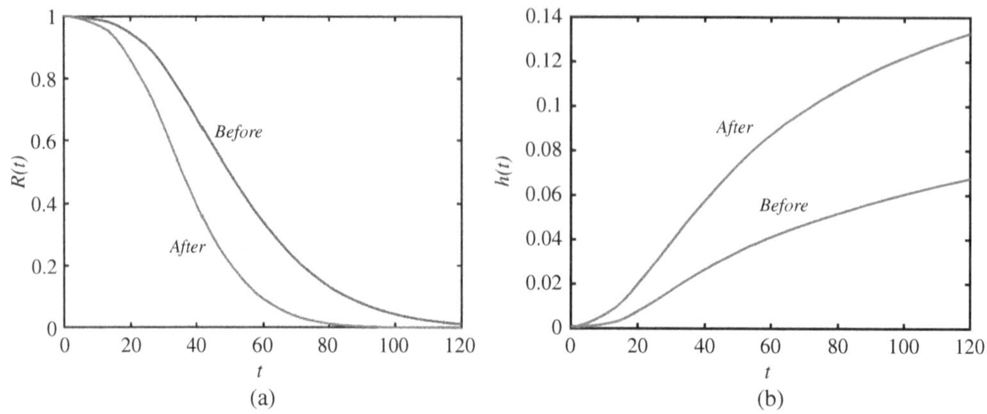

Figure 4.54 Comparisons of system reliability and hazard functions before and after demand increased.

Obviously, the system reliability functions down and the hazard functions up when the demand increases because the system is now more vulnerable to failure, as Figure 4.54 shows. ■

Exercise 4.21 Generate the system reliability and hazard functions if the increase in demand can still be met by one generator.

In case the order of the resulting PH representation for system reliability is too big for a personal computer, it is necessary to decompose the system into a few subsystems. The PH representation of each subsystem is calculated using the algorithms provided in this chapter, and the reliability measures of the overall system are calculated using the conventional scalar method. The next section discusses a method to decompose the system and calculate reliability measures for overall system from the reliability functions of its subsystem.

4.9 Mixed Matrix-Scalar Model

Using matrix-based distributions simplifies the process of finding the system reliability function from the reliability functions of its components. However, using matrix-based distribution has a drawback in that the order of the resulting PH distribution of the system lifetime grows exponentially with the number of components. Even though the survival matrix is sparse, it still needs a vast storage space beyond the capacity of ordinary personal computers. A large-scale system usually has many components in series, while there are few components in parallel or standby structures. For example, a series system with eight subsystems, each subsystem consisting of two components in a parallel structure, and each component having a PH-distributed lifetime of order 4, has a PH-distributed lifetime of order $((4+1)^2-1)^8 = 110{,}075{,}314{,}176$, too large for a personal computer to handle. Such a problem can be solved by decomposing the system into several subsystems. If the system is decomposed into two subsystems in a series structure, each subsystem consisting of four series sub-subsystems, then the order of each subsystem is $((4+1)^2-1)^4 = 331.376$, which a personal computer can still handle. The reliability function of each subsystem is modeled using a matrix distribution, while the system reliability function is the product of reliability function of each subsystem as used in scalar functions. The system reliability measures are sought using the resulted system reliability function.

Suppose that the system is decomposed into two subsystems. Use $R_{S_1}(t)$ and $R_{S_2}(t)$ to express the reliability functions of subsystem 1 and subsystem 2 resulting from the decomposition. Since both are connected in series, the system reliability function is given by

$$R_S(t) = R_{S_1}(t)R_{S_2}(t).$$

By defining $h_{S_1}(t)$ and $h_{S_2}(t)$ as the hazard functions of subsystem 1 and subsystem 2, the system hazard function can be obtained as

$$h_S(t) = -\frac{R'_S(t)}{R_S(t)} = -\frac{R_{S_1}(t)R'_{S_2}(t) + R'_{S_1}(t)R_{S_2}(t)}{R_{S_1}(t)R_{S_2}(t)} = h_{S_1}(t) + h_{S_2}(t),$$

in accordance with **Property 5** of series systems.

For more complex systems, they can be decomposed into K subsystems. The system reliability function expressed as a function of the reliability functions of its subsystems is

$$R_S(t) = \prod_{k=1}^{K} R_{S_k}(t), \tag{4.125}$$

$$h_S(t) = \sum_{k=1}^{K} h_{S_k}(t), \tag{4.126}$$

where $R_{S_k}(t)$ and $h_{S_k}(t)$ are the reliability and hazard functions of subsystem k, respectively, where $k = 1, 2, \ldots, K$.

Application 4.5 The reliability measures of the system in Application 4.3 are calculated by decomposing the system into two large subsystems. The first large subsystem consists of subsystems S_1 and S_2, and the second large subsystem is subsystem S_3. The generators and machines in production lines are assumed to have lifetimes following n-Erlang distribution with the same means as in Application 4.3. All subsystems in parallel structures with identical components are modeled using **phparallelidnsys** to reduce the PH distribution order.

A comparison of the graphs of the system reliability and hazard functions calculated without and with decomposition for 2-Erlang distribution is given in Figure 4.55. It is apparent that the graphs produced with and without decomposition are identical.

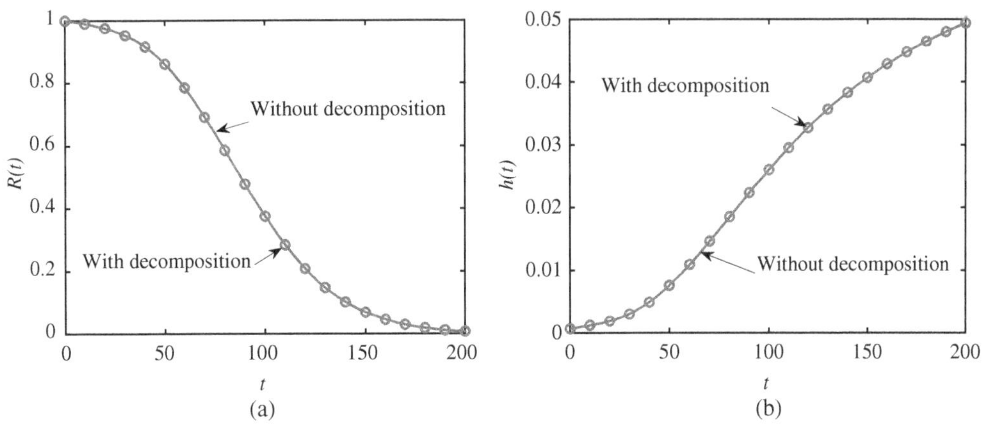

Figure 4.55 Reliability and hazard functions calculated with and without decomposition. (a) System reliability functions and (b) system hazard functions.

Table 4.11 PH order and computation time without and with decomposition.

	PH order		Computation time (s)	
n-Erlang	Without decomposition	With decomposition	Without decomposition	With decomposition
1	81	[9;9]	0.09	0.02
2	8,232	[42;196]	0.54	0.04
3	177,957	[117;1,521]	19.98	0.13
4	1,778,112	[252;7,056]	372.42	0.69

Comparisons of the orders of PH distributions of the system lifetimes and the computation times to produce 21 points of reliability and hazard functions without and with decomposition for several values of n in n-Erlang distribution are presented in Table 4.11. The orders of PH distributions with decomposition are given in two numbers, one for each subsystem. ■

This illustration shows the usefulness of performing decomposition for a system whose lifetime has too large PH representation order. The larger the PH representation order, the greater the efficiency obtained, in terms of computer storage and computation time. The large order is mostly because the system has many subsystems arranged in series structure. Therefore, the process always decomposes systems with long series structures into several subsystems in series structures, and the method in this illustration is always applicable. The mixed matrix-scalar models have a disadvantage: even though they can be used to generate the overall system reliability and hazard functions, they can only be used to calculate the expected lifetime per subsystem resulting from the decomposition and cannot be used to calculate the expected lifetime of the entire system. If necessary, the expected lifetime of the system can be approximated from the expected lifetime per subsystem using Equation (4.14) or calculated using numerical integration.

Exercise 4.22 Should the system in Application 4.5 be decomposed into three subsystems? Explain your reason(s)!

Basic structures discussed in this chapter are special cases of the more general structure that will be discussed in Chapter 5. The general structure covers all basic structures, K-out-of-N structure, and warm standby structure, among others.

4.10 Problems

4.1 Create a comparison table for PH representations of all basic structures with identical and nonidentical components. Use this table to conclude that the parallel structure is the most complicated one.

4.2 For the 3-Generator system in Example 4.1, determine its structure if the power requirement is

a) 185 kVA.
b) 95 kVA.

4.3 Give another example of a system with a physical connection in a series structure, but its logical connection is in a parallel structure or vice versa.

4.4 Determine the basic structure of a system with identical components that have the longest system *MTTF*. Provide mathematical and qualitative reasons for the answer.

4.5 A system uses three identical components. Each component has an exponentially distributed lifetime with rate λ. Derive the system lifetime distribution in matrix and scalar forms if the system has (a) series, (b) parallel, (c) standby, and (d) imperfect switch standby structures.

4.6 Plot and compare the *MRL* functions of all basic structures with N identical components having 2-Erlang distributed lifetimes with $MTTF = 100$ for several values of N. Conclude the effect of N on the limit value of *MRL*.

4.7 Calculate the order of the PH distribution of the lifetime of a system having four components with PH-distributed lifetimes, each of which has order 2, 3, 4, 5, if the system structure is:
a) Series.
b) Parallel.
c) Standby.

4.8 A system has two components, and each has a 2-Erlang distributed lifetime with *MTTF* of 10 and 20, respectively. Manually derive PH distributions for the system lifetime if the system is:
a) 2-Series.
b) 2-Parallel.
c) 2-standby with perfect switch.
d) 2-standby with imperfect switch with reliability = 0.8.
Compare the results with the results using **dysrel-am**.

4.9 Use the state transition diagram of N-standby system to prove that the lifetime of N-standby system with components having identical n-Erlang distributed lifetimes has Nn-Erlang distribution.

4.10 For systems with N identical CFR components, prove that:

$$CVTTF_{series} = CVTTF_{comp}.$$
$$CVTTF_{parallel} < CVTTF_{comp}.$$
$$CVTTF_{standby} < CVTTF_{comp}.$$

Are the same equality–inequality hold for non-CFR components?

4.11 Using the reliability function convolution equation given in Chapter 1, derive the reliability function expression for a 2-standby system whose components have identical 2-Erlang distributed lifetimes if the switch is
a) Perfect.
b) Imperfect.

4.12 Demonstrate by an experiment that a parallel system can have a higher reliability function than an imperfect switch standby system. Use the experiment to determine the largest switch reliability value so that the parallel system's reliability function is still higher than the standby system.

4.13 Prove mathematically that the lower the switch reliability, the lower the $MTTF$ of 2-standby systems with imperfect switch. Generalize the proof to N-standby systems.

4.14 Use components with a 2-Erlang distributed lifetime.
a) Show that $CVTTF$ of a series system gets larger as the number of components increases, but the value never exceeds 1.
b) Show that $CVTTF$ of parallel and standby systems become smaller as the number of components increases.
c) What is the conclusion of this experiment?
d) Repeat questions a–c for components with a 2-hyperexponentially distributed lifetime.

4.15 Prove Equation (4.37).

4.16 Show that it is not necessary to have **phseriesidnsys** or **phstandidnsys**.

4.17 Use the system in Application 4.3 to make a comparison of the computation times for plotting the reliability function and calculating $MTTF$ using the PH distribution and the Jordan ME distribution if all generators and machines have the same PH distribution as the component in:
a) Example 3.1
b) Example 3.2
c) Example 3.3

4.18 Construct a 3-standby system with an imperfect switch using the components in Examples 3.1–3.3 and a switch with reliability = 0.8.
a) Show that there is an optimal ordering of these three components.
b) Find the optimal ordering of the components. Calculate the efficiency gain between using the best and the worst ordering.
c) Plot the percentage of improvement between the best and the worst ordering as a function of switch reliability.

4.19 Prove that the order of PH distribution of the lifetime of a series–parallel system is greater than that of a parallel–series system with the same components.

4.20 Think of a quantitative measure to express the benefit of using the compatible devices in Example 4.42 compared to the incompatible devices in Example 4.43.
a) Calculate the benefit if all modules have lifetimes following identical exponential distribution with $MTTF = 100$.
b) Repeat question a. for 2-Elang distribution.
c) Draw a conclusion from these two results and give an explanation.

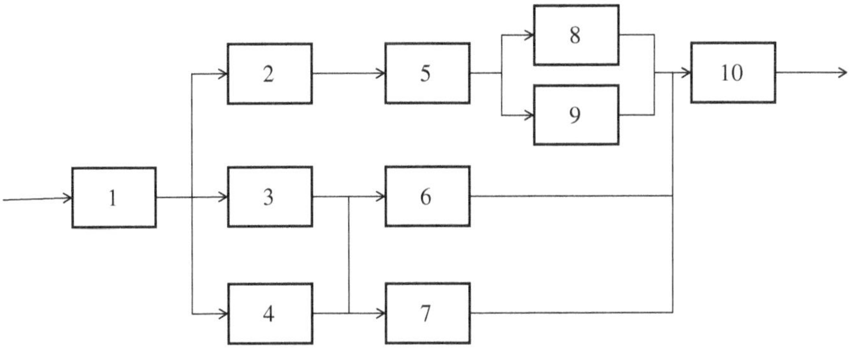

Figure P.4.1 Reliability block diagram for the system in Problems 4.22–4.24.

4.21 Devise an algorithm for parallel systems with identical components having ME-distributed lifetimes. Use the algorithm to create **meparallelidnsys**.

4.22 A system has a reliability block diagram as given in Figure P.4.1.
a) Model the system in the provided MATLAB functions.
b) Compare the system order, reliability, and hazard functions, *MTTF*, and *SDTTF* between all components having exponential distribution and having 2-Erlang distribution.

4.23 Repeat Problem 4.22 for components 3 and 4 as well as components 8 and 9 are arranged in a standby structure, both using imperfect switch with the same reliability.

4.24 Assume that all components in Problems 4.22 and 4.23 have lifetimes following identical 2-Erlang distribution with $MTTF = 100$ days.
a) Compare their system reliability and MRL functions for switch reliability = 0.8.
b) Plot the graph of system *MTTF* as a function of switch reliability.

4.25 A power system connects two power generation sites to two consumption regions through high-voltage DC transmission has a reliability block diagram given in Figure P.4.2. The generators are identical with lifetimes following 4-Erlang distribution with $MTTF = 800$ days. Modules 2 and 3 in the AC/DC and DC/AC converters have lifetimes following exponential distributions with *MTTF* equal to 1000 and 1500 days, respectively. The lifetimes of protection units in the transmission lines each has 2-Erlang distribution with $MTTF = 1000$ days. When the consumptions are low, power generated by one generator can satisfy the demand. However, power generated by both generators is required when the consumptions are high.
a) Compare the reliability, hazard, and MRL functions, and MTTF of this system to meet the low and high demand.
b) Suppose that all standby subsystems (DC/AC and AC/DC converters and protections) use identical unreliable switches. Plot the system MTTF as a function of switch reliability.

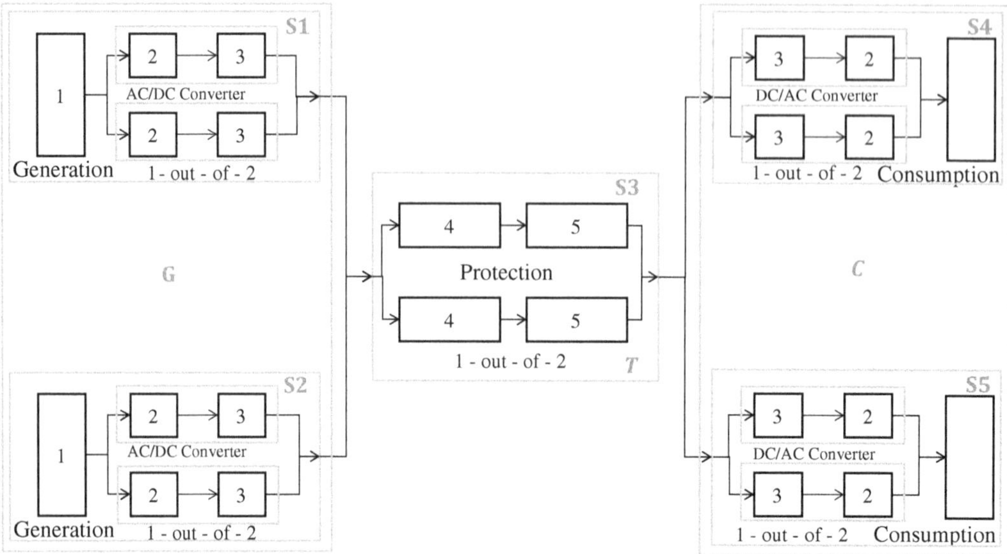

Figure P.4.2 Power system in Problem 4.25.

References

1 Ardakan, M.A., Mirzaei, Z., Hamadani, A.Z., Elsayed, E.A., 2017. Reliability optimization by considering time-dependent reliability for components. Qual Reliab Eng Int 33, 1641–1654. https://doi.org/10.1002/qre.2132.

2 Elsayed, E.A., 2021. Reliability Engineering. Wiley, Hoboken, NJ.

3 Trivedi, K.S., Bobbio, A., 2017. Reliability and Availability Engineering: Modeling, Analysis, and Applications. Cambridge University Press, Cambridge, UK.

4 Zhao, X., Li, R., Fan, Y., Qiu, Q., 2022. Reliability modeling for multi-state systems with a protective device considering multiple triggering mechanism. Proc Inst Mech Eng O J Risk Reliab 236, 173–193. https://doi.org/10.1177/1748006X211013325.

5 Lyu, H., Wang, S., Yang, Z., Qu, H., Ma, L., 2023. Reliability modeling for multi-component system subject to dependent competing failure processes with phase-type distribution considering multiple shock sources. Qual Eng 35, 95–109. https://doi.org/10.1080/08982112.2022.2098043.

6 Gong, M., Liu, H., Peng, R., 2020. Redundancy allocation of mixed warm and cold standby components in repairable K -out-of- N systems. Proc Inst Mech Eng O J Risk Reliab 234, 696–707. https://doi.org/10.1177/1748006X20921306.

7 Gong, M., Xie, M., Yang, Y., 2018. Reliability assessment of system under a generalized run shock model. J Appl Probab 55, 1249–1260. https://doi.org/10.1017/jpr.2018.83.

8 Kim, H., Kim, P., 2017. Reliability models for a nonrepairable system with heterogeneous components having a phase-type time-to-failure distribution. Reliab Eng Syst Saf 159, 37–46. https://doi.org/10.1016/j.ress.2016.10.019.

9 Liu, B., Cui, L., Wen, Y., Guo, F., 2016. A cold standby repairable system with the repairman having multiple vacations and operational, repair, and vacation times following phase-type distributions. Commun Stat Theory Methods 45, 850–858. https://doi.org/10.1080/03610926.2013.851239.

10 Riascos-Ochoa, J., Sanchez-Silva, M., Akhavan-Tabatabaei, R., 2014. Reliability analysis of shock-based deterioration using phase-type distributions. Probab Eng Mech 38, 88–101. https://doi.org/10.1016/j.probengmech.2014.09.004.

11 Alkaff, A., Qomarudin, M.N., 2020. Modeling and analysis of system reliability using phase-type distribution closure properties. Appl Stoch Models Bus Ind 36, 548–569. https://doi.org/10.1002/asmb.2509.

12 Alkaff, A., Qomarudin, M.N., Bilfaqih, Y., 2021a. Network reliability analysis: Matrix-exponential approach. Reliab Eng Syst Saf 212, 107591. https://doi.org/10.1016/j.ress.2021.107591.

13 Eryilmaz, S., 2022. Discrete Stochastic Models and Applications for Reliability Engineering and Statistical Quality Control. CRC Press, Boca Raton. https://doi.org/10.1201/9781003160632.

14 Alkaff, A., 2023. Optimum warmness levels in general standby systems. Reliab Eng Syst Saf 237, 109366. https://doi.org/10.1016/j.ress.2023.109366.

15 Bellman, R., 1997. Introduction to Matrix Analysis, 2nd ed. Society for Industrial and Applied Mathematics, Philadelphia.

16 Neuts, M.F., 1994. Matrix-Geometric Solutions in Stochastic Models: An Algorithmic Approach. Dover Publication, Inc.

17 Coit, D.W., Zio, E., 2019. The evolution of system reliability optimization. Reliab Eng Syst Saf 192, 106259. https://doi.org/10.1016/j.ress.2018.09.008.

18 Zhai, Q., Yang, J., Peng, R., Zhao, Y., 2015. A study of optimal component order in a general 1-out-of-N warm standby system. IEEE Trans Reliab 64, 349–358. https://doi.org/10.1109/TR.2014.2337073.

19 Roy, A., Gupta, N., 2020. Reliability function of k -out-of- n system equipped with two cold standby components. Commun Stat Theory Methods 50, 1–20. https://doi.org/10.1080/03610926.2020.1737122.

20 Alkaff, A., Qomarudin, M.N., Purwantini, E., Wiratno, S.E., 2021b. Dynamic reliability modeling for general standby systems. Comput Ind Eng 161, 107615. https://doi.org/10.1016/j.cie.2021.107615.

21 Sidje, R.B., 1998. Expokit: A software package for computing matrix exponentials. ACM Transactions on Mathematical Software (TOMS) 24(1), 130–156.

5

General Standby Structured Systems

Learning objectives:

1) Understand the derivations of equations for modeling *K*-out-of-*N* general standby systems.
2) Able to construct reliability models of complex systems comprised many types of structures.
3) Able to provide the best system design for standby systems with multistate components having multiple warmness levels.
4) Able to utilize the software for reliability modeling and analyses of complex systems by generating and evaluating their functional and statistical reliability measures.

Abbreviations

BFR	bathtub failure rate
CFR	constant failure rate
DFR	decreasing failure rate
IFR	increasing failure rate
ME	matrix-exponential
PH	phase-type
CVTTF	coefficient of variations of time to failure
MRL	mean residual life
MTTF	mean time to failure
SDTTF	standard deviation of time to failure
VTTF	variance of time to failure

Notations

T	time to failure
t	time, mission duration
g	age
n	order, dimension
$\boldsymbol{a}$	initial probability vector
$\boldsymbol{A}$	survival matrix
$\boldsymbol{I}$	identity matrix

$\boldsymbol{u}$	unitary vector
$\boldsymbol{b}$	failure probability vector
K	number of primary components in a system
N	number of components in a system
$\boldsymbol{D}$	deceleration matrix
d	deceleration factor
OC	operating cost
UOC	unit operating cost
$LOS(S)$	duration of stay in state S
$U(S)$	expected duration of stay in state S
EOC	expected operating cost
TOC	total operation cost
OL	opportunity loss
EOL	expected opportunity loss
TOL	total opportunity loss
TC	total cost
AC	average cost per unit time
$(n_i, \boldsymbol{a}_i, \boldsymbol{A}_i)$	PH representation of lifetime distribution of component i
$(n_i, \boldsymbol{a}_i, \boldsymbol{A}_i, \boldsymbol{v}_i)$	ME representation of lifetime distribution of component i
$(n_S, \boldsymbol{a}_S, \boldsymbol{A}_S)$	PH representation of system lifetime distribution
$(n_S, \boldsymbol{a}_S, \boldsymbol{A}_S, \boldsymbol{v}_S)$	ME representation of system lifetime distribution
$R(t)$	reliability function
$h(t)$	hazard function
$MRL(g)$	mean residual function
$\otimes$	Kronecker product
$\oplus$	Kronecker sum

The basic structures discussed in Chapter 4 have some limitations. The standby and parallel structures require only one component in good condition to ensure the system is good even though the system has many components. The one required component is referred to as the primary component, and all other components are referred to as backup components. The backup components may be actively operated as the primary component; thus, primary and backup components are indistinguishable. Such backup components are said to be in hot standby, and their structure is referred to in Chapter 4 as the parallel structure. The backup components can also be idle or passive. When the primary component fails, the first backup component in the backup list is activated to replace the failed primary component. Activating a backup component to become the primary component requires a switch, usually assumed to be perfect. The backup components in such a structure are said to be in cold standby, and the structure is referred to in Chapter 4 as the standby structure.

In many applications, systems need to be arranged in a more complex structure to meet their design objectives. One increase in structural complexity is the number of primary components, which could be more than one as needed. Another increase in complexity is that the backup components do not have to be all in cold or all in hot standby. They can be operated in warm standby with varying degrees of warmness, from cold to hot, as needed. This flexibility produces a general standby structure that contains parallel and standby structures discussed in Chapter 4 as its special forms. In addition, structural complexity can be increased by adding components that act as bridges to connect one subsystem to another. The presence of these bridge components will produce a network structure. This chapter discusses general standby structured systems, and Chapter 6 will discuss network structured systems.

In a general standby structure, some backup components can be in cold standby because they are not operated, and others can be in hot standby because they are operated like the primary components. Some others can also be operated in warm standby because they are not operated at full power. The number of primary components can be one or more as needed. The number of primary components is denoted as K, with K varying from 1 to N, where N is the number of components in the system. One popular type of general standby structure is the so-called K-out-of-N structure. This structure initially has K primary components and N-K backups in hot standby. The primary components are always maintained, as many as K, while the backup components will decrease from N-K until none are left. The primary and backup components are indistinguishable since both are operated simultaneously. The system fails when the number of good components is less than K.

The development of general standby structures starts with the K-out-of-N structure, which is a generalization of the parallel structure by allowing the number of primary components to be K, any number less than N, while the parallel structure only requires one primary component. This generalization leads to a method for analyzing the reliability of systems in K-out-of-N structures can be constructed by modifying the method for the N-parallel structure. In the K-out-of-N structure, a system in a state with less than K good components has failed, or system states with less than K good components are absorbing states. Therefore, system states with less than K good components in the survival matrix of an N-parallel structure are not included in the survival matrix of a K-out-of-N structure. An effective way to produce the survival matrix of a K-out-of-N structure is to generate the system states in which the system is good, which are the system states that contain at least K good components, and to determine the block transition matrix within and between pairs of these states. The equations to determine the block transition matrices within and between pairs of system states will be described and utilized to generate a PH representation of the system lifetime distribution.

The general standby structure is a generalization of the K-out-of-N structure by allowing the backup components to be not only in hot standby but also to have varying degrees of warmness. This flexibility can be obtained by formulating a mathematical model for warm standby with any degree of warmness from cold to hot. The model for warm standby is incorporated into the model of K-out-of-N structure to create a model for general standby structures. It is necessary to mathematically prove that the model is valid for all compositions of backup components with varying degrees of warmness from cold to hot. This flexibility makes the general standby system become a unifying model for many structures, such as:

- An N-parallel structure is a 1-out-of-N general standby structure with all backup components being in hot standby.
- An N-standby structure is a 1-out-of-N general standby structure with all backup components being in cold standby.
- An N-series structure is an N-out-of-N general standby structure.
- A K-out-of-N structure is a K-out-of-N general standby structure with all backup components being in hot standby.

This unifying model greatly facilitates system reliability analysts in modeling the system to be studied and using the supporting software. The models discussed in previous publications can be said to be special forms of this unifying model. The N-series system with one warm backup component discussed by Wells [1] is an N-out-of-($N + 1$) general standby system. The K-out-of-N system supported by one cold backup component discussed in Eryilmaz [2] and by one warm backup component discussed in Eryilmaz [3] are K-out-of-($N+1$) general standby systems. Likewise, a K-out-of-N system supported by two cold backup components, as discussed in Roy and Gupta [4], is a K-out-of-($N+2$) general standby system. The 1-out-of-N general standby systems with dual-state components investigated in Levitin et al. [5], Ma et al. [6], and Yen and Wang [7] are simple forms of the K-out-of-N general standby system for $K = 1$. The model proposed in this book, in addition to unifying the specific models in those mentioned publications, is also applicable to systems with multistate components by utilizing the PH or ME distributions.

This chapter begins by developing a method to model a system with a K-out-of-N structure. Algorithms for generating block transition matrices within and between pairs of system states in the system survival matrix are discussed and presented. The resulting model and method for the K-out-of-N structure are extended to the general standby structure by adding a factor expressing the warmness levels of the backup components. The standby model with warmness levels will be proven valid for all types of backup components by showing that the PH distribution is closed in a standby system with the backup components having arbitrary warmness levels. The result is used to construct a general standby system model for any combination of standby types. Algorithms for adjusting the survival matrix of a K-out-of-N system by adding warmness levels at appropriate positions are discussed and presented. Complex systems with multiple structures shown can be modeled in a hierarchical manner using the general standby model as a unifying model.

The development continues by determining the sequence of backup components that maximize the reliability of general standby systems. Unlike backup components in an all-hot system, whose sequences do not affect system reliability measures, the sequence of warm standby components with different failure rates or different warmness levels will be shown to have significant effects on system reliability measures. By relating the warmness levels of the backup components to costs, an optimization model formulation for finding the optimum warmness levels to minimize a total cost is derived. Finally, a model with dynamic warmness levels that is more practical in operating general standby systems is developed.

As in Chapter 4, the development in this chapter assumes that component i, $i = 1, 2, \ldots, N$, has PH-distributed lifetimes with representation (n_i, $\boldsymbol{\alpha}_i$, $\boldsymbol{A}_i$), and (n_S, $\boldsymbol{\alpha}_S$, $\boldsymbol{A}_S$) is the PH representation of the distribution of the system lifetime distribution. The results are then extended for ME distributions with representation (n_i, $\boldsymbol{\alpha}_i$, $\boldsymbol{A}_i$, $\boldsymbol{\nu}_i$) for the component lifetime distributions and (n_S, $\boldsymbol{\alpha}_S$, $\boldsymbol{A}_S$, $\boldsymbol{\nu}_S$) for the system lifetime distribution.

First, a method of modeling K-out-of-N structured systems is discussed.

5.1 Modeling *K*-Out-of-*N* Structured System

A K-out-of-N structured system, simply called a K-out-of-N system, is good if there are at least K out of its N components are good. In other words, the system fails when the number of good components is less than K. The number of primary components in a K-out-of-N system is always K. The number of backup components is initially N-K and will be reduced due to failures. For this structure, the backup components are operated the same way as the primary components; hence, there is no clear distinction between backup and primary components.

There are two types of K-out-of-N systems, namely, K-out-of-$N:G$ systems and K-out-of-$N:F$ systems. K-out-of-$N:G$ systems are systems that are good as long as the number of good components is at least K. K-out-of-$N:F$ systems are systems that fail whenever the number of failed components in the systems is at least K. A K-out-of-$N:G$ system is equivalent to a (N-K+1)-out-of-$N:F$ system since K components must be good for the system to be good, which is equivalent to (N-K+1)

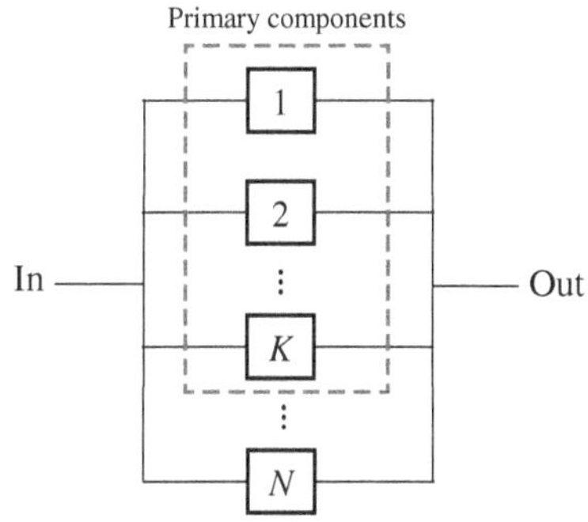

Figure 5.1 Reliability block diagram of a K-out-of-N structure.

components must fail for the system to fail. Thus, there is no need to develop two different methods for the two types of K-out-of-N systems. Unless otherwise specified, K-out-of-N systems in general refer to K-out-of-$N : G$ systems as this book follows.

The reliability block diagram of K-out-of-N system is the same as that of an N-parallel system with the additional information that the number of primary components is K as shown in Figure 5.1.

Example 5.1 A 3-generator system comprises generators 1, 2, and 3 with capacity of 100, 90, and 80 kVA, respectively. This system is used to supply power in the amount of 150 kVA. With this amount of power requirement, at least two of the three available generators operated simultaneously can meet this power requirement. Therefore, this 3-Generator system has a 2-out-of-3 structure to meet the 150 kVA power requirement. The reliability block diagram of this system is given in Figure 5.2. ■

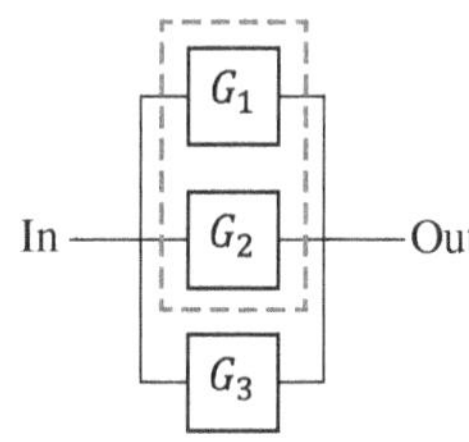

Figure 5.2 Reliability block diagram of the system in Example 5.1.

Example 5.2 The 3-generator system in the 2-out-of-3 structure in Example 5.1 cannot meet the power requirement of 180 kVA. Not all combinations of two generators can supply the 180 kVA requirement. This requirement can only be fulfilled by G_1 and G_2 or G_1 and G_3, but cannot be fulfilled by G_2 and G_3. Therefore, the 3-generator system to meet the 180 kVA power requirement must have a structure with G_1 is arranged in series with G_2 and G_3 in parallel or in standby, not a 2-out-of-3 structure. ■

Exercise 5.1 Determine the system structure in Example 5.2 and its reliability block diagram. Find the lower and upper limits of the power requirement that this structure still holds.

Next, a method to derive the survival matrix of a K-out-of-N system is developed.

5.1.1 Survival Matrix of K-Out-of-N System

The survival matrix of a K-out-of-N system is generated from the survival matrix of an N-parallel system by identifying and removing the system states with less than K good components.

The reason is that states with less than K good components are failed states; therefore, they do not belong to the survival matrix of a K-out-of-N system. An illustration explaining the method to obtain survival matrices of K-out-of-3 systems, for $K = 1, 2, 3$, from the survival matrix of a 3-parallel system is as follows.

The infinitesimal generator matrix for a 3-parallel system has been derived in Section 4.4.1 to produce Equation (4.38), which is

$$\boldsymbol{Q}_S^{(3)} = \begin{bmatrix} \boldsymbol{A}_S^{(3)} & \boldsymbol{b}_S^{(3)} \\ \boldsymbol{0} & 0 \end{bmatrix}$$

$$= \begin{bmatrix} \boldsymbol{A}_1 \oplus \boldsymbol{A}_2 \oplus \boldsymbol{A}_3 & \boldsymbol{I}_1 \otimes \boldsymbol{b}_2 \otimes \boldsymbol{I}_3 & \boldsymbol{b}_1 \otimes \boldsymbol{I}_2 \otimes \boldsymbol{I}_3 & \boldsymbol{I}_1 \otimes \boldsymbol{I}_2 \otimes \boldsymbol{b}_3 & \boldsymbol{0} & \boldsymbol{0} & \boldsymbol{0} & \boldsymbol{0} \\ \boldsymbol{0} & \boldsymbol{A}_1 \oplus \boldsymbol{A}_3 & \boldsymbol{0} & \boldsymbol{0} & \boldsymbol{I}_1 \otimes \boldsymbol{b}_3 & \boldsymbol{0} & \boldsymbol{b}_1 \otimes \boldsymbol{I}_3 & \boldsymbol{0} \\ \boldsymbol{0} & \boldsymbol{0} & \boldsymbol{A}_2 \oplus \boldsymbol{A}_3 & \boldsymbol{0} & \boldsymbol{0} & \boldsymbol{I}_2 \otimes \boldsymbol{b}_3 & \boldsymbol{b}_2 \otimes \boldsymbol{I}_3 & \boldsymbol{0} \\ \boldsymbol{0} & \boldsymbol{0} & \boldsymbol{0} & \boldsymbol{A}_1 \oplus \boldsymbol{A}_2 & \boldsymbol{I}_1 \otimes \boldsymbol{b}_2 & \boldsymbol{b}_1 \otimes \boldsymbol{I}_2 & \boldsymbol{0} & \boldsymbol{0} \\ \boldsymbol{0} & \boldsymbol{0} & \boldsymbol{0} & \boldsymbol{0} & \boldsymbol{A}_1 & \boldsymbol{0} & \boldsymbol{0} & \boldsymbol{b}_1 \\ \boldsymbol{0} & \boldsymbol{0} & \boldsymbol{0} & \boldsymbol{0} & \boldsymbol{0} & \boldsymbol{A}_2 & \boldsymbol{0} & \boldsymbol{b}_2 \\ \boldsymbol{0} & \boldsymbol{0} & \boldsymbol{0} & \boldsymbol{0} & \boldsymbol{0} & \boldsymbol{0} & \boldsymbol{A}_3 & \boldsymbol{b}_3 \\ \boldsymbol{0} & \boldsymbol{0} & \boldsymbol{0} & \boldsymbol{0} & \boldsymbol{0} & \boldsymbol{0} & \boldsymbol{0} & 0 \end{bmatrix}$$

where $\boldsymbol{I}_i$ is an identity matrix with the same dimension as $\boldsymbol{A}_i$, $i = 1, 2, 3$.

The state transition diagram of a 3-parallel system constructed using $\boldsymbol{Q}_S^{(3)}$ has been presented in Figure 4.15. That figure is redrawn here as Figure 5.3 by adding information about possible structures that can be derived from the 3-parallel structure.

The state transition diagram in Figure 5.3 contains state transition diagrams of K-out-of-3 systems for $K = 1, 2, 3$. For $K = 3$, the system has a series structure with one transient state (g, g, g),and other

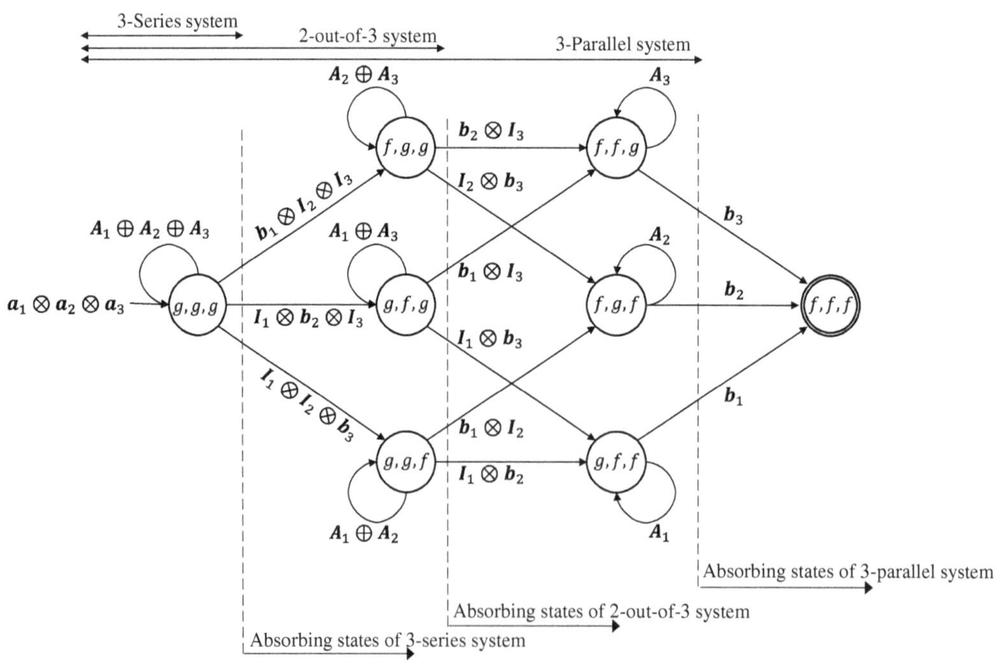

Figure 5.3 State transition diagram of a 3-parallel system.

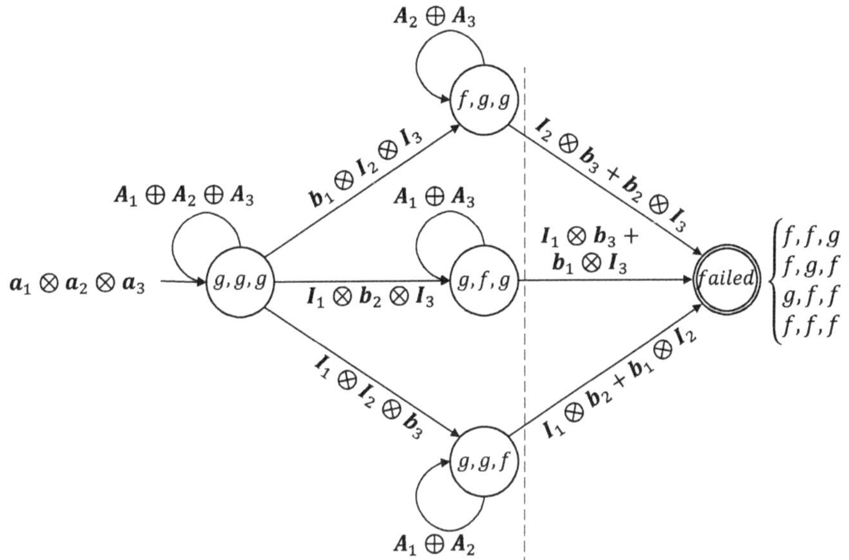

Survival matrix of 2-out-of-3 system | Failure vector of 2-out-of-3 system

$$Q_S = \left[\begin{array}{cccc|c} A_1 \oplus A_2 \oplus A_3 & I_1 \otimes b_2 \otimes I_3 & b_1 \otimes I_2 \otimes I_3 & I_1 \otimes I_2 \otimes b_3 & 0 \\ 0 & A_1 \oplus A_3 & 0 & 0 & I_1 \otimes b_3 + b_1 \otimes I_3 \\ 0 & 0 & A_2 \oplus A_3 & 0 & I_2 \otimes b_3 + b_2 \otimes I_3 \\ 0 & 0 & 0 & A_1 \oplus A_2 & I_1 \otimes b_2 + b_1 \otimes I_2 \\ \hline 0 & 0 & 0 & 0 & 0 \end{array}\right] \quad A_S \quad b_S$$

Figure 5.4 The survival matrix of a 2-out-of-3 system obtained from the infinitesimal generator of a 3-parallel system.

states are absorbing states. For $K = 1$, the system has a parallel structure with one absorbing state (f, f, f), and other states are transient states. For $K = 2$, the system has a 2-out-of-3 structure with states with less than two good components are absorbing states; those are states (g, f, f), (f, g, f), (f, f, g), and (f, f, f). This illustration shows that the survival matrix of a 2-out-of-3 system can be derived using the survival matrix of a 3-parallel system by discarding all states that indicate the system is in failed states. The infinitesimal generator matrix of a 2-out-of-3 system obtained using the state transition diagram in Figure 5.3 is given in Figure 5.4, excluding the transitions within failed states since they are not needed in constructing the survival matrix and failure vector of the 2-out-of-3 system.

In this figure, the survival matrix of a 2-out-of-3 system contains the transient states with two or more good components, namely, states (g, g, g), (f, g, g), (g, f, g), and (g, g, f). These states correspond to block matrices in the survival matrix of 2-out-of-3 system, one block matrix for the three-component survival matrix representing the transient state (g, g, g), and three block matrices for the survival matrices of combinations of two of the three components representing the transient states (f, g, g), (g, f, g), and (g, g, f). Block matrices correspond to states with less than two good components are not in the survival matrix of the 2-out-of-3 system; they are in the failure probability vector. However, calculating the failure probability vector using the transition rates from the transient states to the absorbing states is too complicated compared to using $\boldsymbol{b}_S = -\boldsymbol{A}_S\boldsymbol{u}$ given in Equation (2.12). Furthermore, it is more efficient to generate $\boldsymbol{A}_S$ and calculate $\boldsymbol{b}_S$ from $\boldsymbol{A}_S$, instead of generating both $\boldsymbol{A}_S$ and $\boldsymbol{b}_S$. With this consideration, it is sufficient to generate only the survival matrix of any K-out-of-N systems.

This result shows that the lifetime of a 2-out-of-3 system with components have lifetimes following PH distributions also has a PH distribution. Furthermore, its PH representation can be generated from the survival matrix of a 3-parallel system by removing rows and columns of states with less than K good components. The rows of the initial probability vector are also removed accordingly.

This method can be extended to K-out-of-N systems for arbitrary values of K and N given as Lemma 5.1.

Lemma 5.1 [8] If all components of a K-out-of-N system have lifetimes following independent and not necessarily identical PH distribution, then the system lifetime follows a PH distribution whose representation is obtained by discarding the rows/columns of the infinitesimal generator matrix and initial probability vector of an N-parallel system associated with less than K good components.

Block matrices in the survival matrix can be classified into block diagonal matrices and block off-diagonal matrices. Each block diagonal matrix represents a transition matrix within each transient state. Based on Figure 5.4, its value is the Kronecker sum of survival matrices of good components in that state. Each block off-diagonal matrix represents a transition matrix from one transient state to another transient state due to a component failure. Based on Figure 5.4, its value is the Kronecker products of identity matrices of surviving components and the failure vector of the failing component. (There is only one failing component in each such transition.) Since Kronecker operations are not commutative, the operation ordering must strictly follow component ordering.

In general, the transition matrices in a survival matrix can be depicted in a generic state transition diagram shown in Figure 5.5.

States V and W in Figure 5.5 are transient states; each is a vector whose elements are the state of each component, with V_i and W_i being the ith element of the vector system states V and W, respectively. $\boldsymbol{A}_{VV}$ and $\boldsymbol{A}_{WW}$ are transition matrices within transient states V and W, respectively, which are due to the survival of each element of vector states V and W. $\boldsymbol{A}_{VV}$ and $\boldsymbol{A}_{WW}$ become block diagonal elements of the survival matrix. $\boldsymbol{A}_{VW}$ is the transition matrix from transient state vector V to transient state vector W, which is due to a failure of one good elements in system state vector V. $\boldsymbol{A}_{VW}$ becomes a block off-diagonal element of the survival matrix. From Figure 5.4 and its explanation, $\boldsymbol{A}_{VV}$ and $\boldsymbol{A}_{WW}$ are the Kronecker sums of the survival matrices of the good components at the system states V and W, respectively. Likewise, $\boldsymbol{A}_{VW}$ is the Kronecker products of the identity matrices representing the surviving components and the failure vector of the failed component that causes the transition from V to W. Thus,

$$\boldsymbol{A}_{VV} = \oplus \boldsymbol{A}_i, \quad i|V_i = g \tag{5.1}$$

$$\boldsymbol{A}_{VW} = \otimes \boldsymbol{X}_i, \quad i|V_i = g \tag{5.2}$$

where

$$\begin{aligned} \boldsymbol{X}_i &= \boldsymbol{I}_i \quad \text{if } W_i = g, \\ &= \boldsymbol{b}_i \quad \text{if } W_i = f. \end{aligned}$$

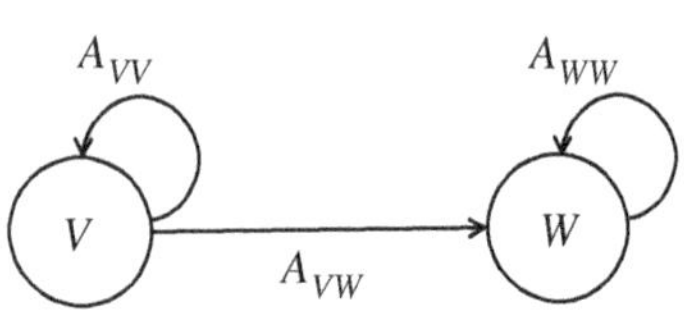

Figure 5.5 States and matrix transitions in a survival matrix.

Since no component has failed before being used, the initial probability vector is

$$\boldsymbol{a}_S = \left[\otimes_1^N \boldsymbol{a}_i \;\; \boldsymbol{0} \;\ldots\; \boldsymbol{0}\right], \tag{5.3}$$

where $\otimes_1^N \boldsymbol{a}_i = \boldsymbol{a}_1 \otimes \boldsymbol{a}_2 \ldots \otimes \boldsymbol{a}_N$, and the dimensions of $\boldsymbol{0}$'s follow the dimensions of the next system states in sequence.

Equations (5.1) and (5.2) do not show the ordering and locations of transition matrices in the survival matrix. The ordering used in Figure 5.4 is based on the result of generating survival matrix of 3-parallel system using Algorithm 4.2. Constructing the survival matrix using Equations (5.1) and (5.2) required a simple algorithm for generating good states of the system, which are combinations of states of each component that lead to the system being good.

Exercise 5.2 Using Equations (5.1) and (5.2) generates the survival matrices of a 2-out-of-3 and 3-parallel systems. Compare the results with Figure 5.4.

Equipped with these understandings, the construction of an algorithm for generating the survival matrix of PH distribution of the lifetime of K-out-of-N system is under way.

5.1.2 Algorithm for Finding the Survival Matrix of *K*-Out-of-*N* System

Equations (5.1) and (5.2) can be utilized to design an algorithm for generating the block matrices of the survival matrix of a K-out-of-N system. Such an algorithm must be supported by another algorithm to generate combinations of good and failed component states that lead to good system states. A good system state is a system state vector whose elements contain at least K g's; otherwise, it is a failed system state. The block diagonal elements of the survival matrix are generated using Equation (5.1) for each good system state, while the block off-diagonal elements are generated using Equation (5.2) for each pair of good system states. Alkaff and Qomarudin [8] proposed an algorithm for generating the survival matrix of a K-out-of-N system using the survival matrix of an N-parallel system by identifying the row/column positions of states in which the system failed. Their positions must be identified based on the values of K and N and the order of PH distribution of each component lifetime. These positions are scattered in the survival matrix of N-parallel system that makes the algorithm to identify and discard those elements complicated. Even though their algorithm is faster, for easier understanding, this book uses the algorithm based on Equations (5.1) and (5.2).

Example 5.3 The survival matrix of a 2-out-of-4 system is generated as follows. The block diagonal matrices, each representing the transition matrix within a system state, are generated using Equation (5.1). For example, the transition matrix within state (g, g, g, g) is $\boldsymbol{A}_1 \oplus \boldsymbol{A}_2 \oplus \boldsymbol{A}_3 \oplus \boldsymbol{A}_4$, within state (g, g, g, f) is $\boldsymbol{A}_1 \oplus \boldsymbol{A}_2 \oplus \boldsymbol{A}_3$, and within state (g, g, f, f) is $\boldsymbol{A}_1 \oplus \boldsymbol{A}_2$. The transition matrix of component 4 is not included in the transition matrix within state (g, g, g, f) since in this state component, 4 has failed. Likewise for state (g, g, f, f). A block off-diagonal matrix representing a transition matrix from a current system state to the next system state is generated using Equation (5.2). For example, the transition matrix from state (g, g, g, g) to state (g, g, g, f) is $\boldsymbol{I}_1 \otimes \boldsymbol{I}_2 \otimes \boldsymbol{I}_3 \otimes \boldsymbol{b}_4$ since the transition is due to the failure of component 4. Likewise, from state (g, g, g, f) to state (g, g, f, f) is $\boldsymbol{I}_1 \otimes \boldsymbol{I}_2 \otimes \boldsymbol{b}_3$ since the transition is due to the failure of component 3, while component 4 has failed when in state (g, g, g, f) and components 1 and 2 survive in that transition. There is no transition from state (g, g, g, g) to (g, g, f, f) because such a transition is due to simultaneous failures of components 3 and 4, which is not possible in continuous-time systems; hence, the transition matrix from state (g, g, g, g) to (g, g, f, f) is $\boldsymbol{0}$. Transitions within failed states and to failed states are ignored since they are not part of the survival matrix. The resulting block diagonal and off-diagonal matrices of a 2-out-of-4 system are given in Table 5.1.

Table 5.1 Block matrices in the survival matrix of a 2-out-of-4 system.

Current system state	Block diagonal matrix	Next system state	Block off-diagonal matrix
		(g, g, g, f)	$\boldsymbol{I}_1 \otimes \boldsymbol{I}_2 \otimes \boldsymbol{I}_3 \otimes \boldsymbol{b}_4$
		(g, g, f, g)	$\boldsymbol{I}_1 \otimes \boldsymbol{I}_2 \otimes \boldsymbol{b}_3 \otimes \boldsymbol{I}_4$
(g, g, g, g)	$\boldsymbol{A}_1 \oplus \boldsymbol{A}_2 \oplus \boldsymbol{A}_3 \oplus \boldsymbol{A}_4$	(g, f, g, g)	$\boldsymbol{I}_1 \otimes \boldsymbol{b}_2 \otimes \boldsymbol{I}_3 \otimes \boldsymbol{I}_4$
		(f, g, g, g)	$\boldsymbol{b}_1 \otimes \boldsymbol{I}_2 \otimes \boldsymbol{I}_3 \otimes \boldsymbol{I}_4$
		Others	$\boldsymbol{0}$
		(g, g, f, f)	$\boldsymbol{I}_1 \otimes \boldsymbol{I}_2 \otimes \boldsymbol{b}_3$
(g, g, g, f)	$\boldsymbol{A}_1 \oplus \boldsymbol{A}_2 \oplus \boldsymbol{A}_3$	(g, f, g, f)	$\boldsymbol{I}_1 \otimes \boldsymbol{b}_2 \otimes \boldsymbol{I}_3$
		(f, g, g, f)	$\boldsymbol{b}_1 \otimes \boldsymbol{I}_2 \otimes \boldsymbol{I}_3$
		Others	$\boldsymbol{0}$
		(g, g, f, f)	$\boldsymbol{I}_1 \otimes \boldsymbol{I}_2 \otimes \boldsymbol{b}_4$
(g, g, f, g)	$\boldsymbol{A}_1 \oplus \boldsymbol{A}_2 \oplus \boldsymbol{A}_4$	(g, f, f, g)	$\boldsymbol{I}_1 \otimes \boldsymbol{b}_2 \otimes \boldsymbol{I}_4$
		(f, g, f, g)	$\boldsymbol{b}_1 \otimes \boldsymbol{I}_2 \otimes \boldsymbol{I}_4$
		Others	$\boldsymbol{0}$
		(g, f, g, f)	$\boldsymbol{I}_1 \otimes \boldsymbol{I}_3 \otimes \boldsymbol{b}_4$
(g, f, g, g)	$\boldsymbol{A}_1 \oplus \boldsymbol{A}_3 \oplus \boldsymbol{A}_4$	(g, f, f, g)	$\boldsymbol{I}_1 \otimes \boldsymbol{b}_3 \otimes \boldsymbol{I}_4$
		(f, f, g, g)	$\boldsymbol{b}_1 \otimes \boldsymbol{I}_3 \otimes \boldsymbol{I}_4$
		Others	$\boldsymbol{0}$
		(f, g, g, f)	$\boldsymbol{I}_2 \otimes \boldsymbol{I}_3 \otimes \boldsymbol{b}_4$
(f, g, g, g)	$\boldsymbol{A}_2 \oplus \boldsymbol{A}_3 \oplus \boldsymbol{A}_4$	(f, g, f, g)	$\boldsymbol{I}_2 \otimes \boldsymbol{b}_3 \otimes \boldsymbol{I}_4$
		(f, f, g, g)	$\boldsymbol{b}_2 \otimes \boldsymbol{I}_3 \otimes \boldsymbol{I}_4$
		Others	$\boldsymbol{0}$
(g, g, f, f)	$\boldsymbol{A}_1 \oplus \boldsymbol{A}_2$		
(g, f, g, f)	$\boldsymbol{A}_1 \oplus \boldsymbol{A}_3$		
(g, f, f, g)	$\boldsymbol{A}_1 \oplus \boldsymbol{A}_4$	Absorbing states	Not required (not part of the survival matrix)
(f, g, g, f)	$\boldsymbol{A}_2 \oplus \boldsymbol{A}_3$		
(f, g, f, g)	$\boldsymbol{A}_2 \oplus \boldsymbol{A}_4$		
(f, f, g, g)	$\boldsymbol{A}_3 \oplus \boldsymbol{A}_4$		

Note that the sequence of operands in Kronecker operators must strictly follow the component sequence number since Kronecker operators are not commutative. ■

The algorithm for generating the survival matrix of a K-out-of-N system is summarized as Algorithm 5.1 as given below

---

Algorithm 5.1

---

1) Generate the vector system states as N-tupple of all combinations of component states, with the ith element of a vector system state is the state of component i, which are "g" or "f."
2) Discard system states with less than K "g" elements.
3) Determine the block diagonal matrix for each system state vector not discarded in step 2 using Equation (5.1).
4) Determine the next states for each current state due to exactly one good component in current state becomes a failed component.
5) Determine the block off-diagonal matrix using Equation (5.2) for each possible pair of current and next states obtained from step 4.
6) Set a value **0** for other block off-diagonal matrices.

Algorithm 5.1 is programmed in MATLAB as **phkofnsys** function. Its MATLAB code is available in the software package **dysrel-am**. The function receives N, K, and $(n_i, \boldsymbol{a}_i, \boldsymbol{A}_i)$, $i = 1, 2, \ldots, N$, as inputs and returns the PH representation of the system lifetime distribution $(n_S, \boldsymbol{a}_S, \boldsymbol{A}_S)$ as the output.

The order of PH representation output is equal to the order of PH distribution of the lifetime of N-parallel system minus the dimensions of the removed block matrices. For identical components, the order is $\sum_{j=K}^{N} C_N^j n^j$, where $C_N^j = N!/(j!(N-j)!)$. The K-out-of-N system is modeled as

$$T_S = \mathrm{kofn}(T_1, T_2, \ldots, T_N; K). \tag{5.4}$$

Exercise 5.3 Prepare a similar table as Table 5.1 for 3-out-of-5 system.

Exercise 5.4 Calculate the order of the PH distribution of the lifetime of a 2-out-of-5 system with identical components having 2-Erlang distributed lifetimes.

Exercise 5.5 Devise an algorithm for calculating the order of the PH distribution of the lifetime of a K-out-of-N system with components having PH-distributed lifetimes of order n_i, $i = 1, 2, \ldots, N$. Validate your algorithm by generating the order and compare the result with the result of Exercise 5.4. ■

Next, the algorithm is illustrated with the simplest case of exponential distribution.

5.1.3 K-Out-of-N System with CFR Components

Manual calculation for generating PH representations of K-out-of-N systems is illustrated with 2-out-of-3 systems having CFR components.

Example 5.4 A 2-out-of-3 system is constructed using components with lifetimes following exponential distribution with $MTTF$ equal to 1, 2, and 5 days. Its survival matrix can be derived using Figure 5.4 as

$$\boldsymbol{A}_s = \begin{array}{c} \begin{array}{cccc} ggg & \qquad gfg & \qquad fgg & \qquad ggf \end{array} \\ \left[\begin{array}{cccc} \boldsymbol{A}_1 \oplus \boldsymbol{A}_2 \oplus \boldsymbol{A}_3 & \boldsymbol{I}_1 \otimes \boldsymbol{b}_2 \otimes \boldsymbol{I}_3 & \boldsymbol{b}_1 \otimes \boldsymbol{I}_2 \otimes \boldsymbol{I}_3 & \boldsymbol{I}_1 \otimes \boldsymbol{I}_2 \otimes \boldsymbol{b}_3 \\ 0 & \boldsymbol{A}_1 \oplus \boldsymbol{A}_3 & 0 & 0 \\ 0 & 0 & \boldsymbol{A}_2 \oplus \boldsymbol{A}_3 & 0 \\ 0 & 0 & 0 & \boldsymbol{A}_1 \oplus \boldsymbol{A}_2 \end{array}\right] \end{array} \begin{array}{l} \\ ggg \\ gfg \\ fgg \\ ggf \end{array} \tag{5.5}$$

The PH representations for the component lifetime distributions are

$$n_1 = n_2 = n_3 = 1, \quad \boldsymbol{a}_1 = \boldsymbol{a}_2 = \boldsymbol{a}_3 = [1],$$

$$\boldsymbol{A}_1 = [-1], \quad \boldsymbol{A}_2 = [-0.5], \quad \boldsymbol{A}_3 = [-0.2].$$

The failure probability vector for component i is $\boldsymbol{b}_i = -\boldsymbol{A}_i\boldsymbol{u}$,

$$\boldsymbol{b}_1 = [1], \quad \boldsymbol{b}_2 = [0.5], \quad \boldsymbol{b}_3 = [0.2].$$

Following the dimensions of the matrix $\boldsymbol{A}_i$, results in $\boldsymbol{I}_i = [1]$, $i = 1, 2, 3$. Substituting these values into Equation (5.5) yields

$$\boldsymbol{A}_S = \begin{bmatrix} [-1] \oplus [-0.5] \oplus [-0.2] & [1] \otimes [0.5] \otimes [1] & [1] \otimes [1] \otimes \{1] & [1] \otimes [1] \otimes [0.2] \\ 0 & [-1] \oplus [-0.2] & 0 & 0 \\ 0 & 0 & [-0.5] \oplus [-0.2] & 0 \\ 0 & 0 & 0 & [-1] \oplus [-0.5] \end{bmatrix}$$

$$= \begin{bmatrix} -1.7 & 0.5 & 1 & 0.2 \\ 0 & -1.2 & 0 & 0 \\ 0 & 0 & -0.7 & 0 \\ 0 & 0 & 0 & -1.5 \end{bmatrix},$$

$$\boldsymbol{a}_S = \left[[1] \otimes [1] \otimes [1] \ 0 \ 0 \ 0 \right] = \left[1 \ 0 \ 0 \ 0 \right].$$

It can be verified that $\boldsymbol{A}_S$ fulfills all requirements for being a valid survival matrix. ∎

Example 5.5 The survival matrix of system in Example 5.4 can be generated using Equations (5.1) and (5.2) as follows. The transient states of this system are (g, g, g), (f, g, g), (g, f, g), and (g, g, f), while (f, f, g), (f, g, f), (g, f, f), and (f, f, f) are absorbing states.

(a) Transition within each transient system state is generated using Equation (5.1):

a) State (g, g, g): $V_1 = V_2 = V_3 = g \rightarrow$

$$A_S((g,g,g),(g,g,g)) = \boldsymbol{A}_1 \oplus \boldsymbol{A}_2 \oplus \boldsymbol{A}_3 = [-1] + [-0.5] + [-0.2] = [-1.7].$$

b) State (g, f, g): $V_1 = V_3 = g$, $V_2 = f \rightarrow$

$$A_S((g,f,g),(g,f,g)) = \boldsymbol{A}_1 \oplus \boldsymbol{A}_3 = [-1] + [-0.2] = [-1.2].$$

c) State (f, g, g): $V_1 = f$, $V_2 = V_3 = g \rightarrow$

$$A_S((f,g,g),(f,g,g)) = \boldsymbol{A}_2 \oplus \boldsymbol{A}_3 = [-0.5] + [-0.2] = [-0.7].$$

d) State (g, g, f): $V_1 = V_2 = g$, $V_3 = f \rightarrow$

$$A_S((g,g,f),(g,g,f)) = \boldsymbol{A}_1 \oplus \boldsymbol{A}_2 = [-1] + [-0.5] = [-1.5].$$

(b) Transition between each possible pair of transient system states is generated using Equation (5.2):

a) From state (g, g, g) to state (g, f, g): $V_1 = W_1 = g \rightarrow \boldsymbol{X}_1 = \boldsymbol{I}_1$; $V_2 = g$, $W_2 = f \rightarrow \boldsymbol{X}_2 = \boldsymbol{b}_2$; $V_3 = W_3 = g \rightarrow \boldsymbol{X}_3 = \boldsymbol{I}_3$. Thus,

$$A_S((g,g,g),(g,f,g)) = \boldsymbol{I}_1 \otimes \boldsymbol{b}_2 \otimes \boldsymbol{I}_3 = [1] \times [0.5] \times [1] = [0.5].$$

b) From state (g, g, g) to state (f, g, g): $V_1 = g$, $W_1 = f \to \boldsymbol{X}_1 = \boldsymbol{b}_1$; $V_2 = W_2 = g \to \boldsymbol{X}_2 = \boldsymbol{I}_2$; $V_3 = W_3 = g \to \boldsymbol{X}_3 = \boldsymbol{I}_3$. Thus,

$$A_S((g, g, g), (f, g, g)) = \boldsymbol{b}_1 \otimes \boldsymbol{I}_2 \otimes \boldsymbol{I}_3 = [1] \times [1] \times [1] = [1].$$

c) From state (g, g, g) to state (g, g, f): $V_1 = W_1 = g \to \boldsymbol{X}_1 = \boldsymbol{I}_1$; $V_2 = W_2 = g \to \boldsymbol{X}_2 = \boldsymbol{I}_2$; $V_3 = g$, $W_3 = f \to \boldsymbol{X}_3 = \boldsymbol{b}_3$. Thus,

$$A_S((g, g, g), (g, g, f)) = \boldsymbol{I}_1 \otimes \boldsymbol{I}_2 \otimes \boldsymbol{b}_3 = [1] \times [1] \times [0.2] = [0.2].$$

d) From state (f, g, g) to states (g, g, g), (g, f, g), and (g, g, f) are impossible. Likewise, from state (g, f, g) to states (g, g, g), (f, g, g), and (g, g, f), as well as from state (g, g, f) to states (g, g, g), (g, f, g), and (f, g, g). Thus, those transitions have transition matrices equal to **0**. Transitions from transient states to absorbing states are ignored since they are not part of the survival matrix.

(c) Putting all the results into a matrix yields

$$\boldsymbol{A}_s = \begin{array}{c} \begin{matrix} ggg & gfg & fgg & ggf \end{matrix} \\ \begin{bmatrix} -1.7 & 0.5 & -1 & -0.2 \\ 0 & -1.2 & 0 & 0 \\ 0 & 0 & -0.7 & 0 \\ 0 & 0 & 0 & -1.5 \end{bmatrix} \end{array} \begin{matrix} ggg \\ gfg \\ fgg \\ ggf \end{matrix}$$

As seen, Examples 5.4 and 5.5 yield the same survival matrix. ■

Exercise 5.6 Derive the survival matrix of a 3-out-of-4 system using the two methods as in Examples 5.4 and 5.5.

Examples 5.4 and 5.5 show that, even though all the components are CFR, a K-out-of-N system is not a CFR. Therefore, as for parallel systems, there is no simple expression for the PH distribution of the lifetime of a K-out-of-N system with arbitrary CFR components. A simple expression is still possible for a K-out-of-N system with identical CFR components. A K-out-of-N system with identical CFR components can be represented in a state transition diagram with the numbers of good components as the system states. The initial state is N, and the absorbing state is $(K-1)$ since the system fails in that state. If the failure rate of each component is λ, then the duration of stay in state k is $k\lambda$, $k = N, N-1, \ldots, K$. The state transition diagram of a K-out-of-N system with CFR components whose lifetimes are identically distributed is presented in Figure 5.6, which is a diagram representation of a hypoexponential distribution $(N\lambda, (N-1)\lambda, \ldots, (K+1)\lambda, K\lambda)$ given in Table 2.1. It is like the N-parallel system, but the absorbing state is state $K-1$, not state 0.

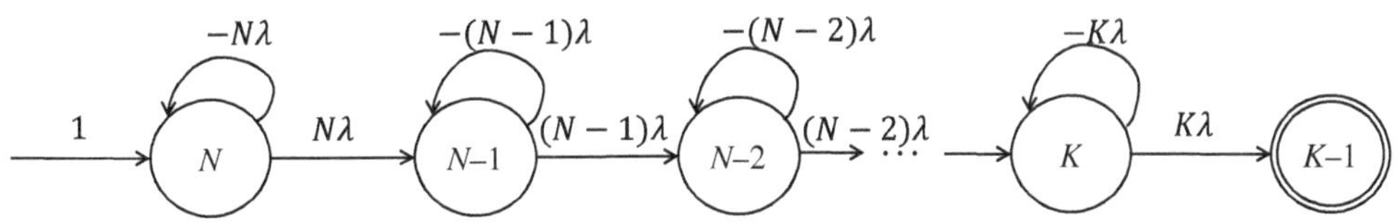

Figure 5.6 State transition diagram of a K-out-of-N system with identical CFR components.

Referring to Table 2.1, this hypoexponential distribution has the following representations:

- Matrix representation:

$$n = N - K + 1, \quad \boldsymbol{a} = \begin{bmatrix}1 & 0 \dots 0\end{bmatrix}, \quad \boldsymbol{A} = \begin{bmatrix} -N\lambda & N\lambda & \dots & 0 & 0 \\ 0 & -(N-1)\lambda & \dots & 0 & 0 \\ \vdots & \vdots & \ddots & \vdots & \vdots \\ 0 & 0 & \dots & -(K+1)\lambda & (K+1)\lambda \\ 0 & 0 & \dots & 0 & -K\lambda \end{bmatrix}. \tag{5.6}$$

- Scalar representation with $\lambda_i = i\lambda, i = K, K+1, \dots, N$:

$$R_S(t) = \sum_{i=K}^{N} \frac{e^{-\lambda_i t}}{P_i} \quad \text{with} \quad P_i = \prod_{j=K, j\neq i}^{N} \left(1 - \frac{\lambda_i}{\lambda_j}\right). \tag{5.7}$$

Example 5.6 The lifetime distribution of a 2-out-of-3 system having identical CFR components with failure rate = 1 has a PH representation generated using Equation (5.6) as

$$n = 3 - 2 + 1 = 2, \quad \boldsymbol{a} = \begin{bmatrix}1 & 0\end{bmatrix}, \quad \boldsymbol{A} = \begin{bmatrix} -3\lambda & 3\lambda \\ 0 & -2\lambda \end{bmatrix} = \begin{bmatrix} -3 & 3 \\ 0 & -2 \end{bmatrix}.$$

Using methods in Examples 5.4 and 5.5 for nonidentical components yields

$$n_S = 4, \quad \boldsymbol{a}_S = [1 \;\; 0 \;\; 0 \;\; 0], \quad \boldsymbol{A}_S = \begin{bmatrix} -3 & 1 & 1 & 1 \\ 0 & -2 & 0 & 0 \\ 0 & 0 & -2 & 0 \\ 0 & 0 & 0 & -2 \end{bmatrix}.$$

Seemingly two different results are equivalent by lumping three identical states (g, g, f), (g, f, g), and (f, g, g) into one super-state 2 representing two good components. The transition from state 1, for three good components (g, g, g), to super-state 2 becomes $1 + 1 + 1 = 3$. Thus, the two methods yield identical results, with the method for identical components producing a smaller order of PH distribution than the method for not necessarily identical components. Such order reduction has significant effect on computation time. ■

Exercise 5.7 Generate the survival matrix of a 3-out-of-5 system with identical CFR components.

Figure 5.6 also shows that the system lifetime is equal to the sum of the duration of stay in each state:

$$T_S = V_N + V_{N-1} + \cdots + V_{K+1} + V_K, \tag{5.8}$$

where V_i is the duration of stay in state i, $i = N, N-1, \dots, K$, and T_S is the system lifetime. Based on the properties of the exponential distribution, V_i, $i = N, N-1, \dots, 1$, have independent exponential distributions with rate $i\lambda$, or with mean $1/i\lambda$. Using Equation (5.8) obtains

$$E[T_S] = E[V_N] + E[V_{N-1}] + \cdots + E[V_K],$$
$$\text{Var}[T_S] = \text{Var}[V_N] + \text{Var}[V_{N-1}] + \cdots + \text{Var}[V_K].$$

Thus, the system *MTTF* and *VTTF* can be calculated as

$$MTTF_S = \frac{1}{N\lambda} + \frac{1}{(N-1)\lambda} + \cdots + \frac{1}{K\lambda} = \frac{1}{\lambda}\sum_{i=K}^{N}\frac{1}{i}, \tag{5.9}$$

$$VTTF_S = \frac{1}{(N\lambda)^2} + \frac{1}{((N-1)\lambda)^2} + \cdots + \frac{1}{(K\lambda)^2} = \frac{1}{\lambda^2}\sum_{i=K}^{N}\frac{1}{i^2}. \tag{5.10}$$

Since $MTTF_{comp} = 1/\lambda$ and $VTTF_{comp} = 1/\lambda^2 = (MTTF_{comp})^2$, Equations (5.9) and (5.10) can be rewritten as

$$MTTF_S = MTTF_{comp} \sum_{i=K}^{N} \frac{1}{i}, \tag{5.11}$$

$$VTTF_S = (MTTF_{comp})^2 \sum_{i=K}^{N} \frac{1}{i^2}. \tag{5.12}$$

Exercise 5.8 Calculate manually the *MTTF*, *VTTF*, and the reliability for a mission of 5 months for a 3-out-of-5 system with identical CFR components having $MTTF = 10$ months. Calculate the percentages of improvements if the backup components are increased by one, or if the primary components are reduced by one, or if the backup and the primary are each increased by one.

Dynamic reliability analyses for K-out-of-N systems, even though they have identical CFR components, are difficult to conduct manually. Using **phkofnsys** is the only reasonable way for such analyses.

Example 5.7 The function **phkofnsys** is utilized to calculate *MTTF* and *SDTTF* of a 2-out-of-5 system with CFR components having $MTTF = 10$ months.

The main program to call the **phkofnsys** function and to calculate *MTTF* and *VTTF* for this example is available as **example5_7.m** file in the package **dysrel-am**. The code is as follows:

MATLAB code for example5_7.m

```
clear;close all
addpath('..\dysrel-am','..\dysrel-am\expokit')
% PH components lifetime (CFR)                          input
T = phnErlang(1,10);
% Number of components
N = 5;
% Number of active components
K = 2;
% PH system
Tc = [];
for n = 1:N
  Tc = [Tc T];
end
Ts  = phkofnsys(Tc,K);                                  model
% Display results                                       output
mttf = phmttf(Ts)
sdttf = phsdttf(Ts)
```

Executing this main program produces the following output:

```
>> example5_7
mttf =
   12.8333
sdttf =
    6.8089
```

■

The main programs for each example and application in this chapter are given as **example5_x.m** and **application5_x.m** files in the software package **dysrel-am**, where x is the example or application number.

Example 5.8 A system has N identical CFR components, each with $MTTF = 10$ months. This system is good if at least three of its components are good (namely, 3-out-of-N system). The PH representations of these systems are generated using **phkofnsys**. Its reliability measures are calculatedr using the resulting PH representation. The results are exhibited in Figure 5.7 for several N. Note that for $N = 3$, the system is 3-series.

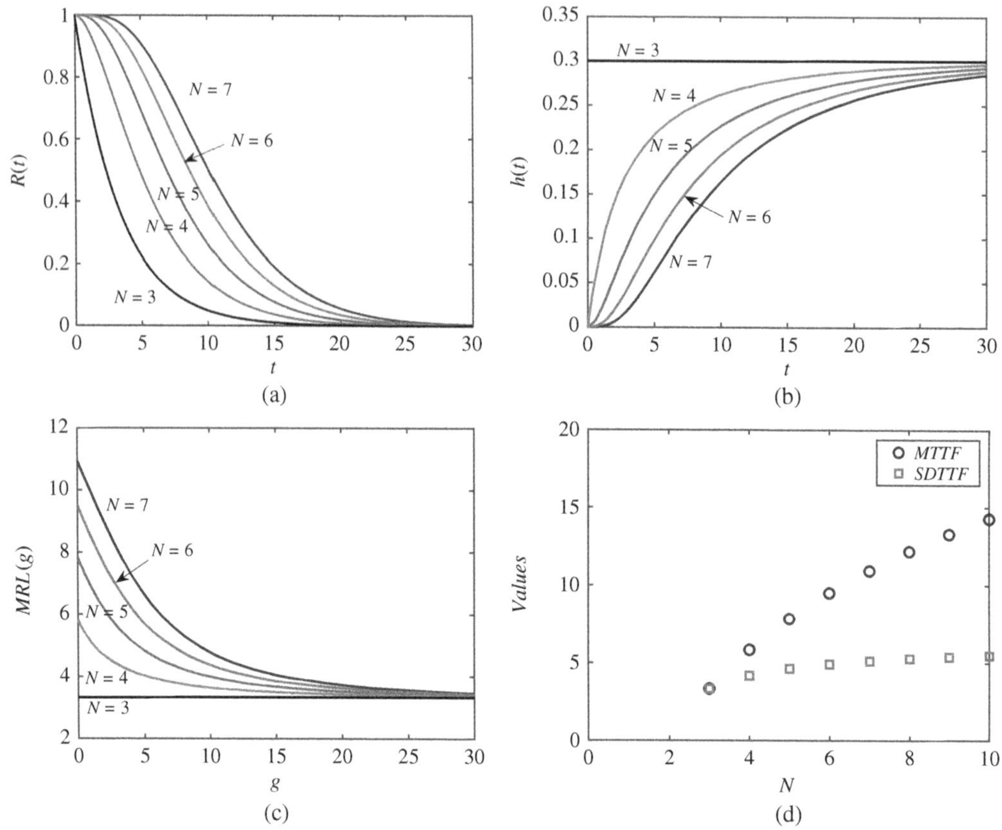

Figure 5.7 Reliability measures of 3-out-of-N systems in Example 5.8. (a) System reliability functions, (b) system hazard functions, (c) system MRL functions, and (d) system MTTF and SDTTF.

Figure 5.7 can be used as a guide for a 3-out-of-N system design. For example, to design a 3-out-of-N system with $MTTF$ longer than 10 months, then from Figure 5.7d, the system must have at least $N = 7$ units or have at least four backup components. To design the same system but have a reliability higher than 0.8 for a mission duration of 5 months, the number of backup components must be at least $6 - 3 = 3$. Likewise, if the requirement is having a remaining life expectancy after being used for 5 months, MRL (5), is greater than 5 months, then the number of backup components must be at least $6 - 3 = 3$ units. If the system must satisfy all the above requirements, then the number of backup components must be four, the highest of all requirements. ■

The results for CFR components are extended for components with lifetimes following higher orders PH distributions.

5.1.4 *K*-Out-of-*N* Systems with Non-CFR Components

The state transition diagram of a K-out-of-N system with nonidentical components is too complicated to draw. It is still possible to produce the state transition diagrams of K-out-of-N systems with

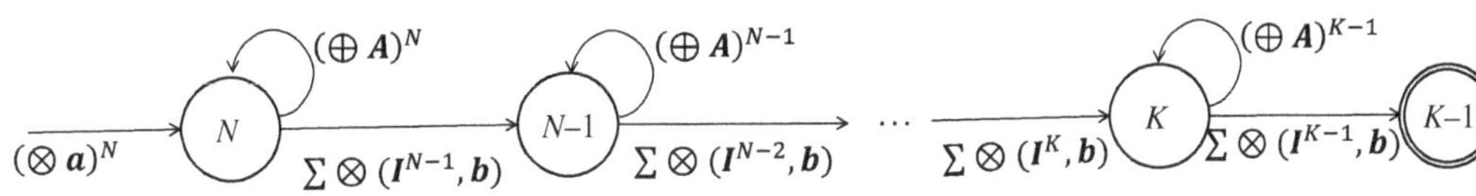

Figure 5.8 State transition diagram of a K-out-of-N system with identical components having PH-distributed lifetimes.

components having identical PH-distributed lifetimes with representation $(n, \boldsymbol{a}, \boldsymbol{A})$ from the state transition diagram of the N-parallel system with the same distribution given in Section 4.4.3 by acknowledging state $K-1$ as the absorbing state. The result is presented in Figure 5.8.

The PH representation $(n_S, \boldsymbol{a}_S, \boldsymbol{A}_S)$ of the distribution of the system lifetime can be derived from Figure 5.8 as

$$n_S = n^N + n^{N-1} + \cdots + n^K, \quad \boldsymbol{a}_S = [(\otimes\boldsymbol{a})^N, \boldsymbol{0}, \ldots, \boldsymbol{0}],$$

$$\boldsymbol{A}_S = \begin{bmatrix} (\oplus\boldsymbol{A})^N & \sum\otimes(\boldsymbol{I}^{N-1}, \boldsymbol{b}) & \ldots & \boldsymbol{0} & \boldsymbol{0} \\ \boldsymbol{0} & (\oplus\boldsymbol{A})^{N-1} & \ldots & \boldsymbol{0} & \boldsymbol{0} \\ \vdots & \vdots & \ddots & \vdots & \vdots \\ \boldsymbol{0} & \boldsymbol{0} & \ldots & (\oplus\boldsymbol{A})^{K+1} & \sum\otimes(\boldsymbol{I}^K, \boldsymbol{b}) \\ \boldsymbol{0} & \boldsymbol{0} & \ldots & \boldsymbol{0} & (\oplus\boldsymbol{A})^K \end{bmatrix}. \tag{5.13}$$

For example, the survival matrix of a 2-out-of-4 system can be obtained using Equation (5.13) for $N = 4$ and $K = 2$ as

$$\boldsymbol{A} = \begin{bmatrix} (\oplus\boldsymbol{A})^4 & \sum\otimes(\boldsymbol{I}^3, \boldsymbol{b}) & \boldsymbol{0} \\ \boldsymbol{0} & (\oplus\boldsymbol{A})^3 & \sum\otimes(\boldsymbol{I}^2, \boldsymbol{b}) \\ \boldsymbol{0} & \boldsymbol{0} & (\oplus\boldsymbol{A})^2 \end{bmatrix},$$

where, as given in Section 4.4.3,

$$(\oplus\boldsymbol{A})^4 = \boldsymbol{A} \oplus \boldsymbol{A} \oplus \boldsymbol{A} \oplus \boldsymbol{A}, (\oplus\boldsymbol{A})^3 = \boldsymbol{A} \oplus \boldsymbol{A} \oplus \boldsymbol{A}, (\oplus\boldsymbol{A})^2 = \boldsymbol{A} \oplus \boldsymbol{A},$$

$$\sum\otimes(\boldsymbol{I}^3, \boldsymbol{b}) = \boldsymbol{b} \otimes \boldsymbol{I} \otimes \boldsymbol{I} \otimes \boldsymbol{I} + \boldsymbol{I} \otimes \boldsymbol{b} \otimes \boldsymbol{I} \otimes \boldsymbol{I} + \boldsymbol{I} \otimes \boldsymbol{I} \otimes \boldsymbol{b} \otimes \boldsymbol{I} +$$

$$\boldsymbol{I} \otimes \boldsymbol{I} \otimes \boldsymbol{I} \otimes \boldsymbol{b}, \quad \sum\otimes(\boldsymbol{I}^2, \boldsymbol{b}) = \boldsymbol{b} \otimes \boldsymbol{I} \otimes \boldsymbol{I} + \boldsymbol{I} \otimes \boldsymbol{b} \otimes \boldsymbol{I} + \boldsymbol{I} \otimes \boldsymbol{I} \otimes \boldsymbol{b}.$$

The order of the resulting PH distribution is $n^4 + n^3 + n^2$, which is smaller than the order of the PH distribution obtained using Algorithm 5.1 for nonidentical components, which is $n^4 + 4n^3 + 6n^2$. For K-out-of-N system, the order of the PH distribution of the system lifetime for nonidentical components is $C_N^N n^N + C_N^{N-1} n^{N-1} + C_N^{N-2} n^{N-2} \cdots + C_N^K n^K$, while the order for the same system with identical components is $n^N + n^{N-1} + \cdots + n^K$. Thus, using a special function for identical components is better than using the general **phkofnsys** function whenever possible.

The algorithm for generating the PH representation of the distribution of lifetime of K-out-of-N system with identical components can be derived from Algorithm 4.3 started with $k = K$ to produce the following recursive equations:

$$n_S = n_S n + n_S, \tag{5.14}$$

$$\boldsymbol{a}_S = [\boldsymbol{a}_S \otimes \boldsymbol{a}, \boldsymbol{0}], \tag{5.15}$$

$$\boldsymbol{I}_S = \boldsymbol{I}_S \otimes \boldsymbol{I}, \tag{5.16}$$

$$\boldsymbol{A}_S(N-k, N-k) = \boldsymbol{A}_S(N-k+1, N-k+1) \oplus \boldsymbol{A}, \tag{5.17}$$

$$\boldsymbol{A}_S(N-k-1, N-k) = \boldsymbol{A}_S(N-k, N-k+1) \otimes \boldsymbol{I} + \boldsymbol{I}_S \otimes \boldsymbol{b}, \tag{5.18}$$

for $k = K, 2, \ldots, N-1$, except Equation (5.18) is until $k = N-2$, with initial conditions $n_S = n^K$, $\boldsymbol{a}_S = (\otimes \boldsymbol{a})^K$, $\boldsymbol{A}_S = \boldsymbol{0}$, $\boldsymbol{A}_S(N-K+1,\ N-K+1) = (\oplus \boldsymbol{A})^K$, $\boldsymbol{A}_S(N-K,\ N-K+1) = \sum \otimes(\boldsymbol{I}^K,\ \boldsymbol{b})$, $\boldsymbol{I}_S = (\otimes \boldsymbol{I})^K$.

These equations are utilized to construct Algorithm 5.2. Its pseudocode is as follows:

Algorithm 5.2

---

input $N, K, (n, \boldsymbol{a}, \boldsymbol{A})$,
$n_S = n^K$
$\boldsymbol{a}_S = (\otimes \boldsymbol{a})^K$
$\boldsymbol{b} = -\boldsymbol{Au}$
$\boldsymbol{A}_S = [\boldsymbol{0}_{n \times n}]_{(N-K+1) \times (N-K+1)}$
$\boldsymbol{A}_S(N-K+1, N-K+1) = (\oplus \boldsymbol{A})^K$
$\boldsymbol{A}_S(N-K, N-K+1) = \sum \otimes(\boldsymbol{I}^K, \boldsymbol{b})$
$\boldsymbol{I}_S = (\otimes \boldsymbol{I})^K$
for $k = K$ to $N-1$
  $n_S = n_S n + n_S$
  $\boldsymbol{a}_S = [\boldsymbol{a}_S \otimes \boldsymbol{a}\ \ \boldsymbol{0}]$
  $\boldsymbol{I}_S = \boldsymbol{I}_S \otimes \boldsymbol{I}$
    $\boldsymbol{A}_S(N-k, N-k) = \boldsymbol{A}_S(N-k+1, N-k+1) \oplus \boldsymbol{A}$
  if $k < N-1$ then
    $\boldsymbol{A}_S(N-k-1, N-k) = \boldsymbol{A}_S(N-k, N-k+1) \otimes \boldsymbol{I} + \boldsymbol{I}_S \otimes \boldsymbol{b}$
  end
  $k = k+1$
end
end

Algorithm 5.2 is programmed as MATLAB function **phkofnidnsys** available in the software package **dysrel-am**.

Exercise 5.9 Derive the survival matrix of a 3-out-of-5 system with identical components using Algorithm 5.2.

Manual calculations are difficult for K-out-of-N systems with components having higher-order PH distributed lifetimes, even for systems with identical components. Using **phkofnsys** or **phkofnidnsys** is the only viable way to analyze such systems.

Example 5.9 To show that **phkofnidnsys** and **phkofnsys** functions produce the same results for K-out-of-N systems with identical components, both are used to generate reliability functions of 2-out-of-N and 3-out-of-N systems with identical components having 2-Erlang distributed lifetimes with $MTTF = 2$. The resulting reliability functions as presented in Figure 5.9 exhibits their identicality. ∎

Example 5.10 Function **phkofnidnsys** is utilized to calculate PH representations of lifetime distributions of K-out-of-7 systems constructed using identical IFR components in Example 3.2. The resulting PH representations are utilized to produce Figure 5.10.

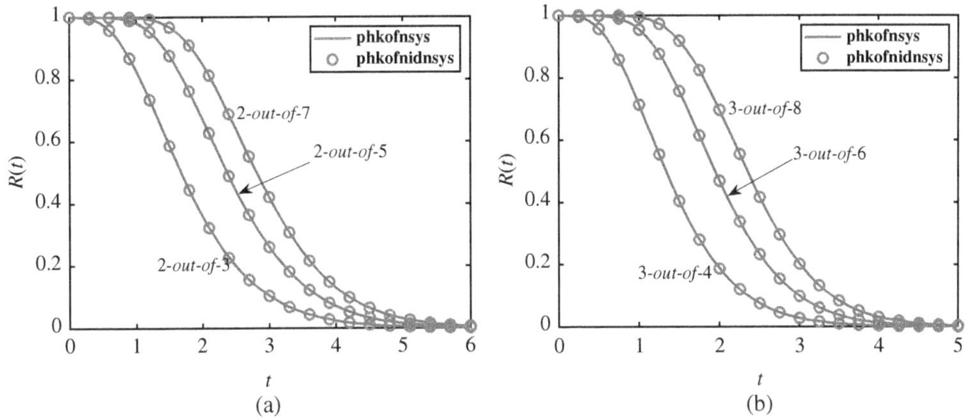

Figure 5.9 Reliability functions of K-out-of-N systems with identical components calculated using phkofnsys and phkofnidnsys. (a) 2-out-of-N systems and (b) 3-out-of-N systems.

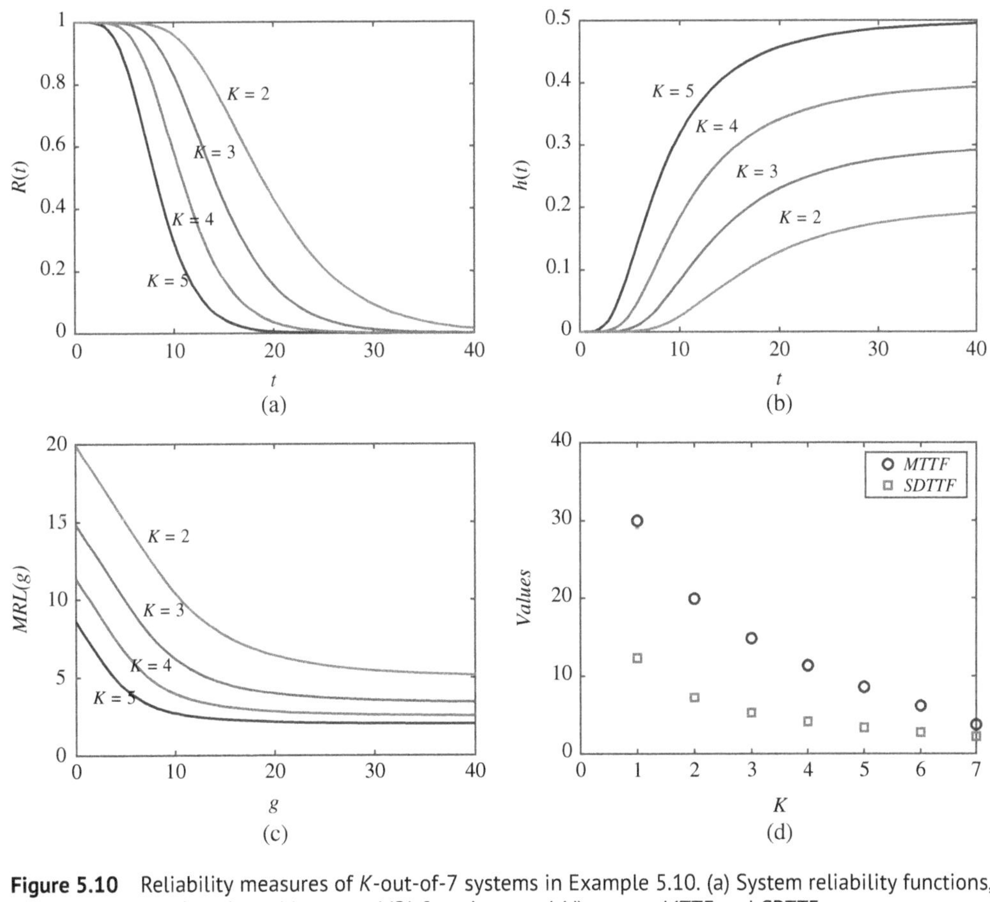

Figure 5.10 Reliability measures of K-out-of-7 systems in Example 5.10. (a) System reliability functions, (b) system hazard functions, (c) system *MRL* functions, and (d) system *MTTF* and *SDTTF*.

Figure 5.10 shows that system reliability and *MRL* functions are higher for lower K, but the hazard function is higher for higher K. If the K-out-of-7 system must have an expected residual lifetime of not less than 5 days after being used for 10 days, then Figure 5.10c suggests the system must have not more than three primary components. ■

Example 5.11 Reliability measures of systems in Example 5.10 for $K = 3$, and several $N's$ are presented in Figure 5.11.

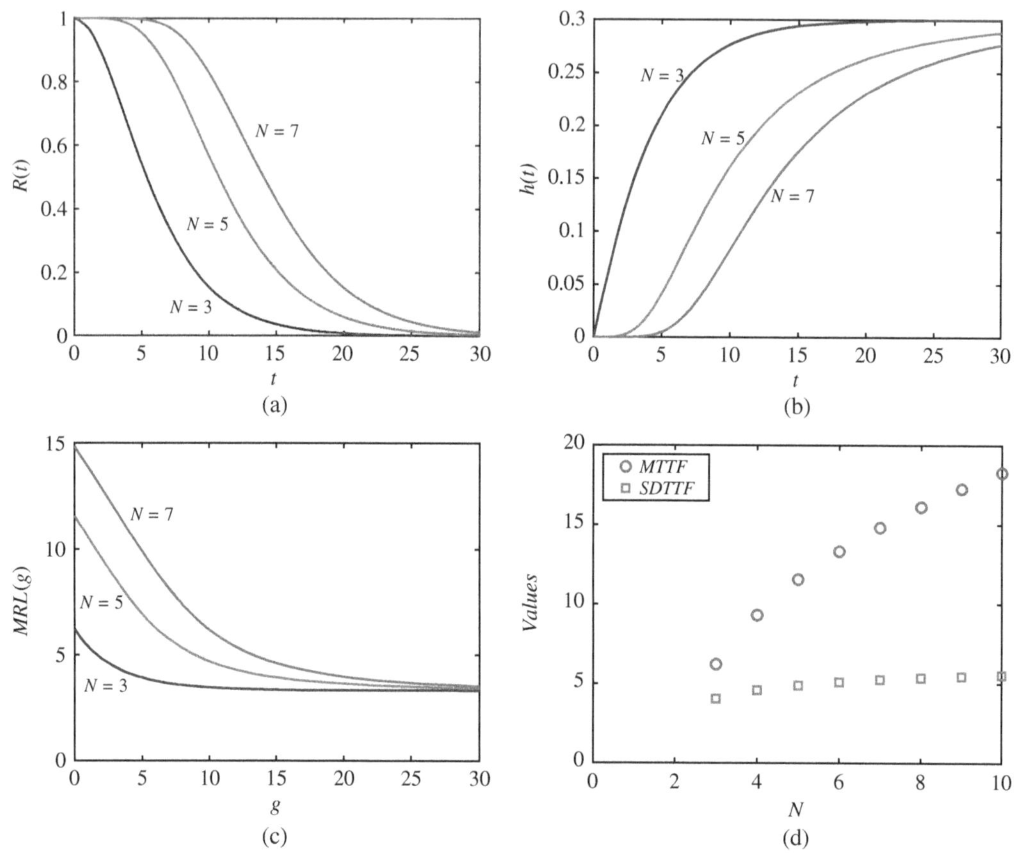

Figure 5.11 Reliability measures of systems in Example 5.11. (a) System reliability function, (b) system hazard function, (c) system *MRL* function, and (d) system *MTTF* and *SDTTF*.

Figure 5.11 shows that the system reliability and *MRL* functions are higher for higher N, but the system hazard function is higher for lower N. If the system is required to have reliability higher than 0.7 for a mission duration of 10 days, then Figure 5.11a suggests that the system must have seven components even though at any time only three components are required to function. ■

Example 5.12 The experiment in Example 5.11 is repeated using DFR components in Example 4.10. The resulting PH representation is used to plot the graphs of system reliability measures, as presented in Figure 5.12.

Figure 5.12b and c shows that the system is DFR for $N = 3$ and becomes IFR for $N = 5$. However, for $N = 3$ the system has a series structure which is the reason it has the same type as its components. To investigate the effect of K and N on the system *CVTTF* as a measure to determine

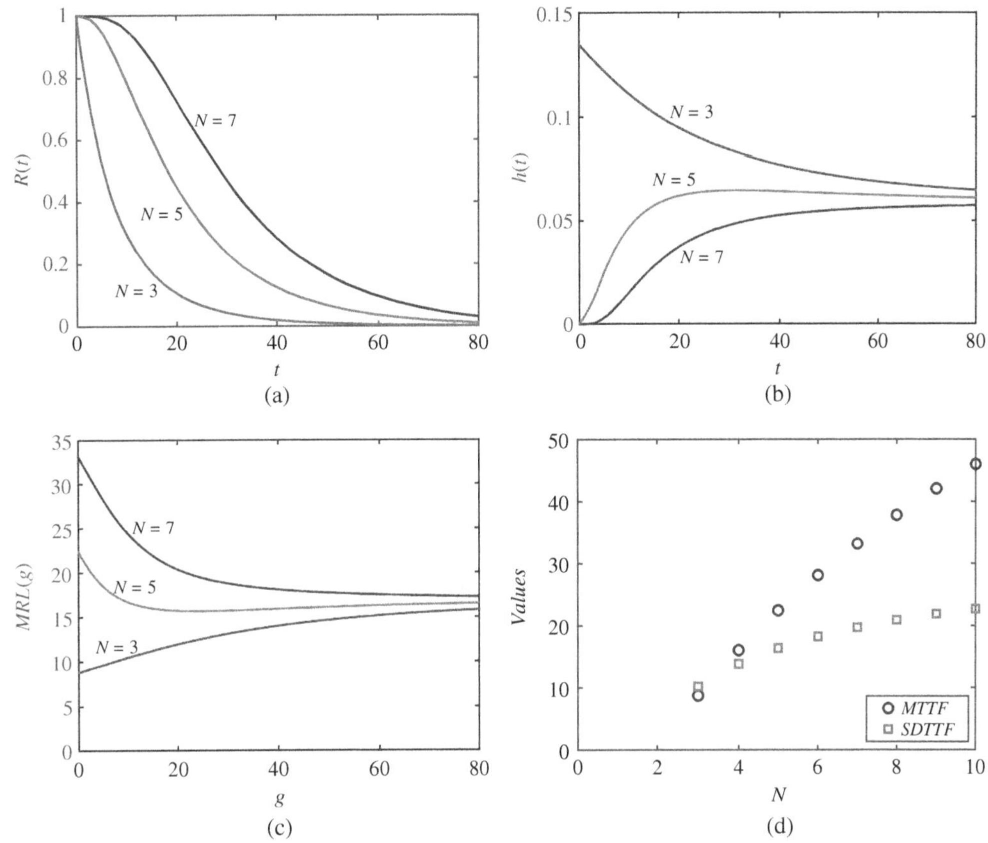

Figure 5.12 Reliability measures of systems in Example 5.12. (a) System reliability function, (b) system hazard function, (c) system *MRL* function, and (d) system *MTTF* and *SDTTF*.

Table 5.2 *CVTTF* of *K*-out-of-*N* systems in Example 5.12.

	N						
K	**1**	**2**	**3**	**4**	**5**	**6**	**7**
1	1.25	0.94	0.80	0.72	0.67	0.62	0.59
2		1.22	0.90	0.76	0.68	0.62	0.58
3			1.17	0.86	0.73	0.65	0.59
4				1.13	0.83	0.70	0.63
5					1.11	0.81	0.68
6						1.09	0.80
7							1.08

the system type, the *CVTTF* of these systems are calculated for several K and N. The results are presented in Table 5.2, which indicates that K-out-of-N systems with identical DFR components having $CVTTF = 1.25$ have $CVTTF < 1$ for any $K < N$. Therefore, the systems become IFR if they have at least one backup component. ■

Examples 5.10–5.12 show that the properties of K-out-of-N systems are like N-parallel systems discussed in Section 4.3.4. Even if all components are DFR, the system can be IFR, while if all the components are IFR or CFR, then the systems are always IFR.

Exercise 5.10 Use **phkofnsys** and **phkofnidnsys** to calculate and plot the system reliability measures of 3-out-of-5 systems with identical components having 4-Erlang distributed lifetimes. Show that the results are identical.

As for parallel and standby systems that have been discussed in Chapter 4, it is possible to determine the optimum number of backup components in K-out-of-N systems that minimize a certain cost function.

Application 5.1 A system with a lifetime following 4-Erlang distribution with $MTTF = 20$ days is employed to carry a mission for s days. The mission may use several identical systems arranged in a K-out-of N structure to increase mission reliability. The interest is to find the optimum value of N that minimizes the expected mission cost for several values of K. The costs associated with the mission comprise the procurement cost $P = 10$ per system and the penalty cost due to mission failure $F = 200$ per mission. The cost of operating the systems is negligible compared to the procurement and penalty costs.

The penalty cost is a random variable with a probability equal to the probability of system failure during the mission of s days, which is $1 - R(s)$. The expected mission total cost TC can be calculated as the sum of procurement cost and expected failure cost, which can be formulated as

$$TC = NP + F\,(1 - R(s)) = 10N + 200(1 - R(s)).$$

The results are presented in Table 5.3.

Table 5.3 The optimum number of backup components in K-out-of-N systems.

	K					
	Mission duration = 15 days			Mission duration = 18 days		
N	1	2	3	1	2	3
1	80.55	–	–	106.96	–	–
2	**44.89**	136.22	–	67.00	166.91	–
3	38.78	87.11	175.77	52.79	125.44	202.65
4	43.10	65.83	128.39	**51.05**	98.01	172.87
5	51.09	**61.12**	97.90	55.36	83.81	144.30
6	60.39	64.63	84.09	62.60	**79.15**	123.13
7	70.14	71.88	**81.49**	71.26	80.62	110.47
8	80.05	80.75	85.27	80.61	85.80	**105.09**
9	90.02	90.30	92.35	90.30	93.12	105.15

Optimum values are given in bold.

Table 5.3 shows that a system with two primary components to carry a mission for duration of 15 days needs to be supported by $N - K = 5 - 2 = 3$ backup components to minimize the total cost of operation and failure. The optimum number of backup components increases to $N - K = 7 - 3 = 4$

if the primary components is increased to $K = 3$ and increases to $N - K = 6 - 2 = 5$ if the mission duration is lengthened to 18 days. The numbers in Table 5.3 show that the optimum number of backup components is increasing with the number of primary components and the mission duration. ∎

Exercise 5.11 Investigate the effects of procurement and penalty costs to the optimum number of backup components.

The previous examples are for identical components. Example 5.13 is for K-out-of-N systems with nonidentical components and a comparison with the method based on scalar distributions.

Example 5.13 Several 3-out-of-N systems are constructed, each with the kth component having k-Erlang distributed lifetimes with $MTTF = 10$ months. The system reliability measures for several N are presented in Figure 5.13. A comparison with the method based on scalar distributions given in many books on system reliability is also given to show that they produce the same results.

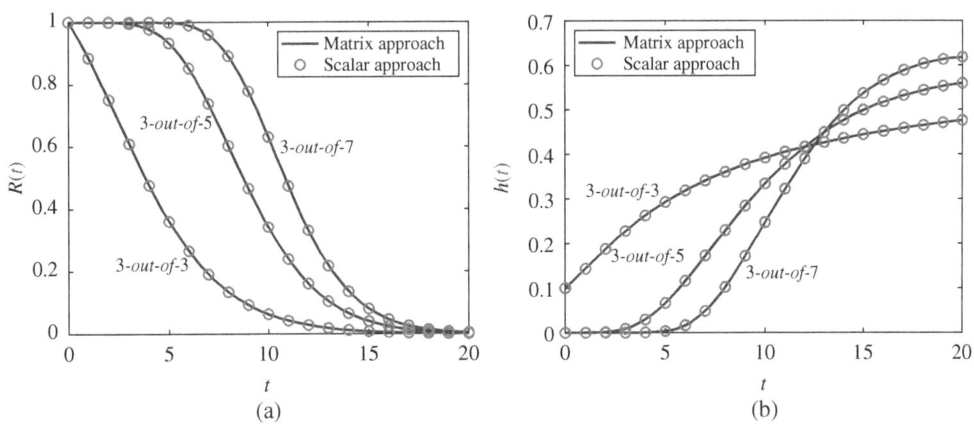

Figure 5.13 Reliability and hazard functions of systems in Example 5.13 calculated using matrix and scalar distributions. (a) System reliability function and (b) system hazard function.

∎

Next, the methods and algorithms for PH distribution are extended to ME distribution.

5.1.5 *K*-Out-of-*N* System in ME Distribution

Algorithms 5.1 and 5.2 are applicable for ME distributions with an additional step for calculating the closing vector $\boldsymbol{v}$ in the ME representation. Equation (5.1) is applicable for calculating the closing vector for system state V, that is,

$$\boldsymbol{v}_V = \otimes \boldsymbol{v}_i, \quad i|V_i = 1 \tag{5.19}$$

The **mekofnsys** function can be constructed from **phkofnsys** function by adding lines for calculating $\boldsymbol{v}_V$ using Equation (5.19). Likewise, **mekofnidnsys** from **phkofnidnsys**. It is left as an exercise for the interested readers to modify these functions.

Example 5.14 A 2-out-of-6 and 3-out-of-7 systems are constructed using identical components in Example 3.3. Their reliability and hazard functions are produced using the PH and ME representations generated using **phkofnsys** and **mekofnsys**. ME distribution is the Jordan canonical form of the PH distribution of the component lifetime. The results are presented in Figure 5.14, which shows that they produced identical functions.

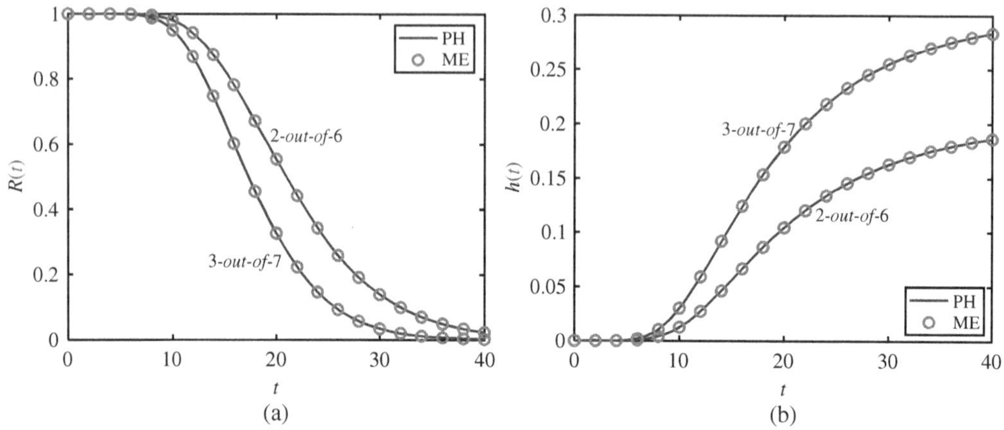

Figure 5.14 Reliability and hazard functions of systems in Example 5.14 calculated using PH and ME distributions. (a) System reliability function and (b) system hazard function. ■

Exercise 5.12 Compare the computation times for calculating reliability measures using PH and ME distributions for the system in Example 5.14.

Real-world systems may have subsystems in combinations of basic structures discussed in Chapter 4, and K-out-of-N structure will be discussed next.

5.1.6 Simple Compound Structures Involving K-Out-of-N Subsystems

Simple compound structures involving K-out-of-N subsystems can be modeled in two levels of structures, one of which is a K-out-of-N structure. In practice, it is common to find several K-out-of-N subsystems arranged in a series structure to form a series-(K-out-of-N) system. There are also several series subsystems arranged in a K-out-of-N structure to form a (K-out-of-N)-series system. Parallel-(K-out-of-N), standby-(K-out-of-N), (K-out-of-N)-parallel, and (K-out-of-N)-standby structures can be expressed in general standby structures, which will be discussed in the next section. Series-(K-out-of-N) and (K-out-of-N)-series systems are modeled in two levels: one level is modeled using **phseriessys** and the other using **phkofnsys**. It is necessary to write a supplementary main program to call the functions in the system model to generate the PH distribution of the system lifetime.

Example 5.15 An electronic system comprises four devices. At least two of the four devices are required at any given time for the system to function (to be good). Each device has three modules: a memory unit, a processor unit, and an I/O unit. All modules are required for the device to function; failure of one module causes the device to fail. Modules in one device cannot be used by other devices.

Since a device fails when one of its three modules fails, each device has a 3-series structure. The system structure with four devices is a 2-out-of-4 structure because the system needs two of the four available devices to function. Considering the modules in one device cannot be used by other devices, the system is a (2-out-of-4)-series system with reliability block diagram as presented in Figure 5.15.

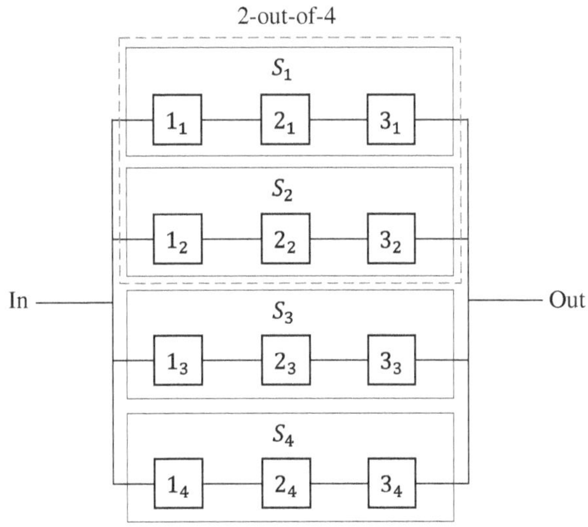

Figure 5.15 Reliability block diagram of the system in Example 5.15.

The system reliability model using the provided MATLAB functions is

$$T_{S_1} = \text{phseriessys}\,(T_{1_1}, T_{2_1}, T_{3_1}),$$

$$T_{S_2} = \text{phseriessys}\,(T_{1_2}, T_{2_2}, T_{3_2}),$$

$$T_{S_3} = \text{phseriessys}\,(T_{1_3}, T_{2_3}, T_{3_3}),$$

$$T_{S_4} = \text{phseriessys}\,(T_{1_4}, T_{2_4}, T_{3_4}),$$

$$T_S = \text{phkofnsys}\,(T_{S_1}, T_{S_2}, T_{S_3}, T_{S_4}; 2).$$

Without ambiguity, the model can be simplified by combining the model for all series subsystems as one model, considering they all have the same number of components:

$$T_{S_i} = \text{phseriessys}\,(T_{1_i}, T_{2_i}, T_{3_i}) \;\; i = 1, 2, 3, 4,$$

$$T_S = \text{phkofnsys}\,(T_{S_1}, T_{S_2}, T_{S_3}, T_{S_4}; 2).$$

However, the model in the main program must be written completely one by one unless they are identical. ■

Example 5.16 Each module in all devices in Example 5.15 is identical. All modules in each device have n-Erlang distributed lifetimes. Module 1 has $n = 1$ and $MTTF = 4$ months, module 2 has $n = 2$ and $MTTF = 6$ months, and module 3 has $n = 3$ and $MTTF = 10$ months. The PH representation of the system lifetime distribution is calculated, and the result is utilized to produce graphs of the system reliability and hazard functions in Figure 5.16.

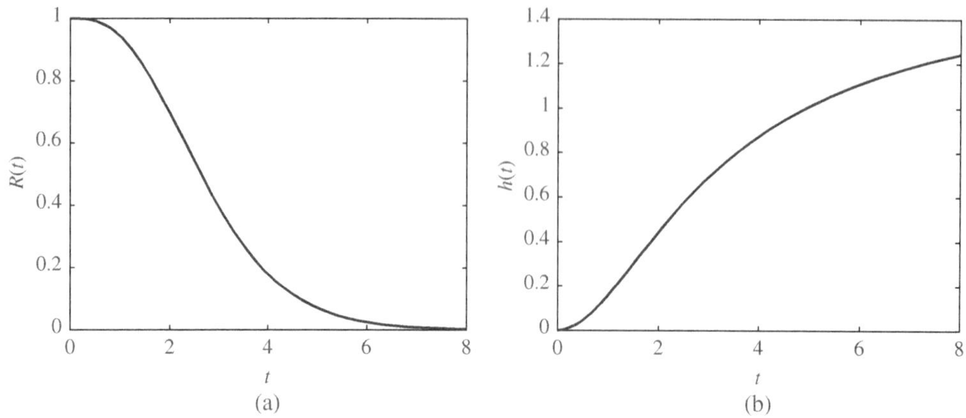

Figure 5.16 Reliability and hazard functions of the system in Example 5.16. (a) Reliability function and (b) hazard function.

The main program to call the required functions to produce the PH representation of the distribution of system lifetime and to generate the system reliability and hazard functions is given in file **example5_16.m**. Its source code is as follows:

MATLAB code for example5_16.m

```
clear;close all
addpath('..\dysrel-am','..\dysrel-am\expokit')
% PH components
% T(modulID,subsystemID)
T(1,1) = phnErlang(1,4);                                    input
T(2,1) = phnErlang(2,6);
T(3,1) = phnErlang(3,10);
T(1,2) = phnErlang(1,4);
T(2,2) = phnErlang(2,6);
T(3,2) = phnErlang(3,10);
T(1,3) = phnErlang(1,4);
T(2,3) = phnErlang(2,6);
T(3,3) = phnErlang(3,10);
T(1,4) = phnErlang(1,4);
T(2,4) = phnErlang(2,6);
T(3,4) = phnErlang(3,10);
% K-of-N of series system                                   model
Tss(1) = phseriessys([T(1,1) T(2,1) T(3,1)]);
Tss(2) = phseriessys([T(1,2) T(2,2) T(3,2)]);
Tss(3) = phseriessys([T(1,3) T(2,3) T(3,3)]);
Tss(4) = phseriessys([T(1,4) T(2,4) T(3,4)]);
Ts = phkofnsys(Tss,2);
ns     = Ts.n;
MTTFs  = phmttf(Ts);                                        statistical output
CVTTFs = phcvttf(Ts);
% Time vector
tmax = 8;
t  = linspace(0,tmax,21);
% Plots                                                     graphical output
[x,Rs] = phrelplot(t,Ts);
[x,hs] = phhazplot(t,Ts);
```

■

Example 5.17 Suppose the modules in one device are compatible with the same modules in other devices. In that case, the system in Example 5.15 becomes a series-(2-out-of-4) system with reliability block diagram in Figure 5.17.

Figure 5.17 Reliability block diagram of the system in Example 5.17.

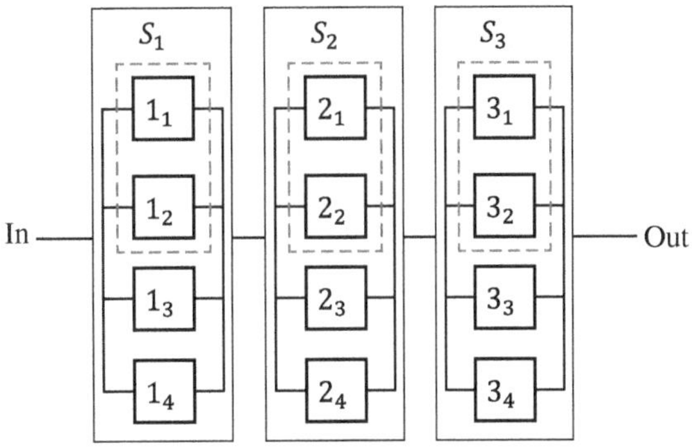

The simplified system reliability model using the provided MATLAB functions is

$$T_{S_i} = \text{phkofnsys}(T_{i_1}, T_{i_2}, T_{i_3}, T_{i_4}; 2)\ \ i = 1, 2, 3,$$

$$T_S = \text{phseriessys}(T_{S_1}, T_{S_2}, T_{S_3}).$$

■

Example 5.18 A comparison of the reliability and hazard functions of the systems in Examples 5.15 and 5.17 using the same distributions as used for Example 5.16 is presented in Figure 5.18.

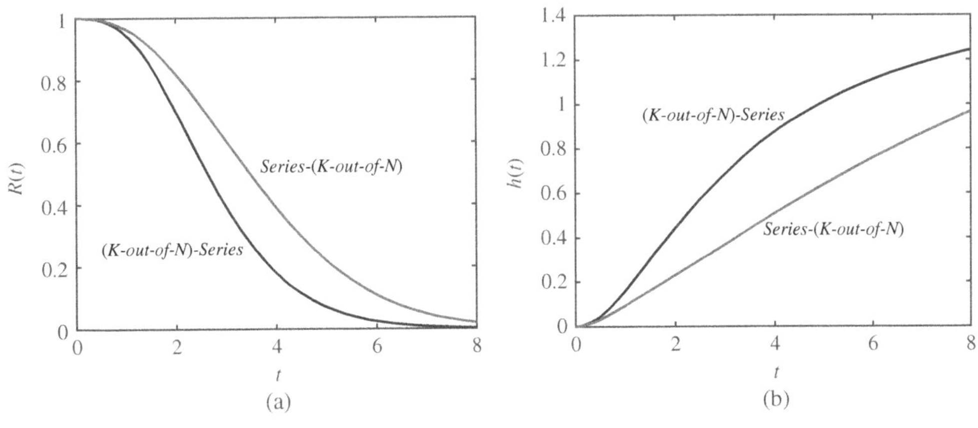

Figure 5.18 Comparison of reliability and hazard functions of the systems in Example 5.15 and 5.17. (a) Reliability function and (b) hazard function.

Figure 5.18 shows that the series-(K-out-of-N) system has higher reliability function and lower hazard function than the (K-out-of-N)-series system. This means that the lifetime of the series-(K-out-of-N) system is stochastically longer than the lifetime of the (K-out-of-N) system. Thus,

$$R_{kn-s}(t) \geq R_{s-kn}(t), \tag{5.20}$$

$$h_{kn-s}(t) \le h_{s-kn}(t), \tag{5.21}$$

$$T_{kn-s} \succcurlyeq T_{s-kn}, \tag{5.22}$$

where *kn-s* and *s-kn* are (*K*-out-of-*N*)-series and series-(*K*-out-of-*N*), respectively.

This conclusion is the same as for the series–parallel system against the parallel–series system and the series–standby system against the standby–series system in Section 4.6. ■

Exercise 5.13 Modules 1 and 2 in Examples 5.15 and 5.17 in all devices are modular compatible, while modules 3 and 4 are not. Draw the system reliability block diagram, construct the model using the available MATLAB functions, and use the model to compare its reliability measures with reliability measures of systems in Examples 5.15 and 5.17 using the distributions in Example 5.16.

Application 5.2 A production line comprises four workstations arranged in series. For line balancing, each workstation requires a different number of machines. Workstations 1, 2, 3, and 4 require two, one, three, and two primary machines, respectively. In addition, each workstation is equipped with several backup machines. The backup machines at workstations 1, 2, and 3 are two units each, while at workstation 4 is one unit. All machines, primary and backup, in each workstation, are identical. All backup machines are operated as if they were primary machines (namely, they are in hot standby) so that they are instantly ready to replace failing primary machines.

The suitable structure for this production line is a series-(*K*-out-of-*N*) structure with the machines at each workstation arranged in a *K*-out-of-*N* structure, and the four workstations are arranged in a series structure. The reliability block diagram of this production line is in Figure 5.19.

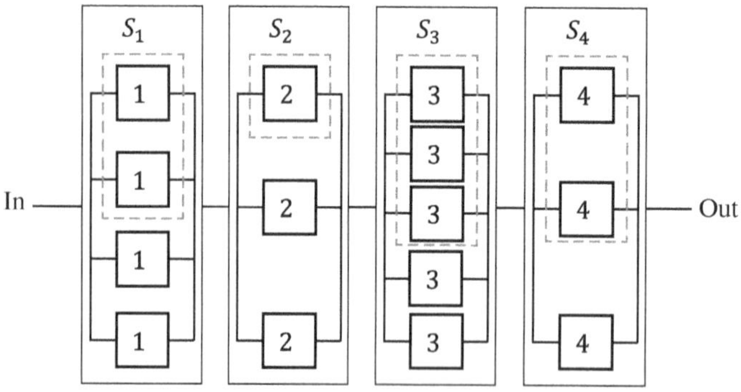

Figure 5.19 Production line in Application 5.2.

The production line can be modeled using the provided MATLAB functions as

$$T_{S_1} = \text{phkofnsys}(T_1, T_1, T_1, T_1; 2),$$

$$T_{S_2} = \text{phkofnsys}(T_2, T_2, T_2; 1),$$

$$T_{S_3} = \text{phkofnsys}(T_3, T_3, T_3, T_3, T_3; 3),$$

$$T_{S_4} = \text{phkofnsys}(T_4, T_4, T_4; 2),$$

$$T_S = \text{phseriessys}(T_{S_1}, T_{S_2}, T_{S_3}, T_{S_4}).$$

Since all *K*-out-of-*N* subsystems use identical components, all **phkofnsys** in the model can be replaced by **phkofnidnsys**. In addition, T_{S_2} can also be modeled as phparallelsys(T_2, T_2, T_2) because *K*-out-of-*N* system with $K = 1$ is equivalent to an *N*-parallel system. Likewise, T_S can also be modeled as phkofnsys(T_{SS}, 4) because a series structure is equivalent to an N-out-of-N structure.

Lifetimes of all machines in each workstation have identical n-Erlang distribution. Machines in different workstations have different *MTTF*, with machines at workstations 1, 2, 3, and 4 having *MTTF* equal to 20, 10, 30, and 40 days, respectively. The system reliability measures are compared for $n = 1, 2$. The results are presented in Figure 5.20.

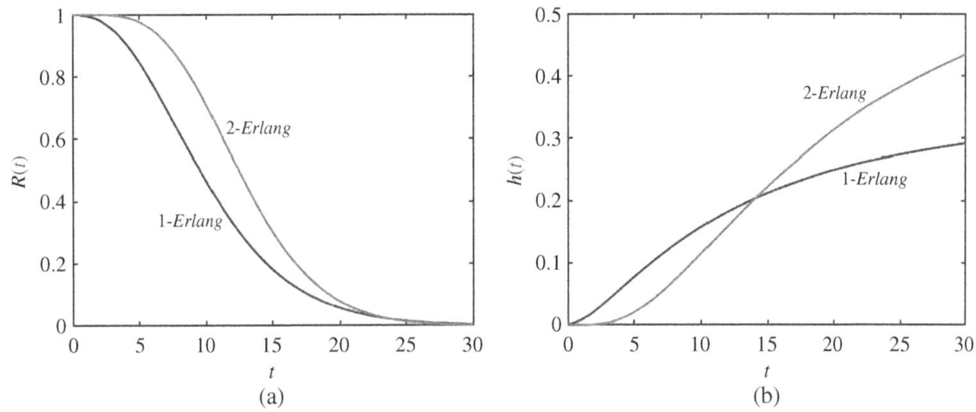

Figure 5.20 Reliability and hazard functions of the system in Application 5.2. (a) System reliability function and (b) system hazard function.

The system *MTTF* and *SDTTF* are also calculated, and the results are 10.36 and 5.46 days for $n = 1$, and 12.91 and 4.73 days for $n = 2$. The numbers show that the system with components having 2-Erlang distributed lifetimes has longer *MTTF* and lower *SDTTF* than the system with components having exponentially distributed lifetimes. The conclusion for *MTTF* is apparent from the graphs of their reliability functions, while for *SDTTF* is obvious from the fact that 2-Erlang distribution has a lower *SDTTF* than exponential distribution with the same *MTTF*. ■

Exercise 5.14 Compare the results and computation times of using **phkofnidnsys** and **phkofnsys** as models for the system in Application 5.2.

The next section extends the results of K-out-of-N systems to K-out-of-N general standby systems by allowing the backup components to be in cold, hot, and warm standby.

5.2 Modeling General Standby Structured Systems

As explained in the introduction to this chapter, general standby structured systems, which will be referred to simply as general standby systems, have several primary components supported by several backup components. The backup components may be operated at full power or partial power or may not be operated at all, depending on the needs. A backup component that is in full power, like the primary components, is in hot standby so that it can immediately replace a failed primary component. Unfortunately, a component in hot standby requires high operating costs and wears out quickly (even fails) as a backup. In contrast, a component in cold standby is a backup in a passive mode that is not operated, so there is no operating cost and it does not wear out or fail while as a backup. Although economical, a component in cold standby requires a heating-up process to

Table 5.4 Comparison of backup components in cold, hot, and warm standby.

Characteristics	Cold standby	Hot standby	Warm standby
Operation mode	Not connected, not operated	Connected, operated in full power, like the primary components	Sleep, operated but not at full power
Failure	Cannot fail	Fails at the same rate as the primary components	Fail at a lower rate than the primary components.
Operating cost	None	Same as the primary components	Less than the primary components
Activation time	Takes a long time	None, ready to use	Takes less time than cold standbys
Other costs	The cost of waiting until ready	None	The cost of waiting is smaller than cold standbys

make it ready to replace a failed primary component. Waiting for a component in cold standby to be ready can disrupt the system operation. Another possibility is to operate a backup component in warm standby that can be made active quickly to replace a failed primary component. Components in warm standby require lower operating costs and wear out more slowly than in hot standby in exchange for being more responsive than a cold standby. Table 5.4 shows a comparison of components in cold, warm, and hot standby.

Modeling warm standby is the most complicated among the three types of standby because a component in warm standby requires two different reliability functions, one for when it is operating as a primary component and another for when it is operating as a backup component. Therefore, although the warm standby is essential in real-world applications, it is rarely discussed in textbooks and only recently has begun to emerge. Some studies on the utilization of warm standby are Jia et al. [9] and Yen and Wang [7] for power systems, Levitin et al. [10] for chemical reactor controllers, Liu et al. [11] for computer systems, Levitin et al. [5] for iron ore processing, Wu et al. [12] for water treatment, Ma et al. [6] for cooling systems, and Jia et al. [13] for oil processing systems.

One way to model a component in warm standby is by making its clock tick slower than the clock of a component in hot standby. Cha et al. [14] and Li et al. [15] introduced a clock deceleration function as a model for components in warm standby. Using this approach, a component in warm standby has the same reliability function as a primary component and as a backup component, but while as a backup component, its clock is decelerated. Another and simpler approach is using a deceleration factor to represent the condition that a component in warm standby is under less stress when it is not a primary component [16]. Components can have varying degrees of warmness from cold to hot by choosing an appropriate value for the deceleration factor. The resulting model has been used in many studies, such as Levitin et al. [17, 18], Lin et al. [19], Ma et al. [6], and Alkaff [20]. Amari et al. [21] proposed the use of continuous-time Markov chains (CTMC) to model a K-out-of-N warm standby system, assuming all components have identically exponentially distributed lifetimes and with identical warmness levels for all backup components, which is still far from being a general standby model. Yen and Wang [7] and Wu et al. [12] also utilized CTMC assuming all components have exponentially distributed lifetimes. A book on warm standbys is Peng et al. [22]. Alkaff et al. [23] used CTMC to extend previous studies to systems whose components have not necessarily identical PH distributed lifetimes. They proved that the PH distribution is closed in a general standby system with backup components having arbitrary and not necessarily identical warmness levels. The scalar deceleration factor in previous studies was expanded into a matrix.

This section discusses the use of CTMC to generate state transition diagrams of general standby systems with components having arbitrary PH-distributed lifetimes, and backup components may have arbitrary deceleration matrices, which means having arbitrary warmness levels. The survival matrix of a component when it is a backup is the survival matrix of that component when it is a primary corrected by a deceleration matrix so that the component in each of its states can experience different decelerations (e.g., when it is new, it experiences different decelerations than when it is already in a state of high stress). The advantages of using the proposed model are the same as those of using PH distributions mentioned in Chapter 3. In addition, and most importantly, the method and model developed here are applicable to all types of standby in any combination.

The development for general standby systems starts with the simplest standby systems with only two components, one primary and one backup.

5.2.1 Modeling 1-Out-of-2 General Standby System

A 1-out-of-2 general standby system has two components arranged in a standby structure, with the backup component allowing any level of warmness. All types of standby, hot, warm, and cold, can be represented in a single model that is applicable to all three through the inclusion of a decelerating matrix. Each component has a PH-distributed lifetime with a representation $(n_i, \boldsymbol{a}_i, \boldsymbol{A}_i)$, $i = 1, 2$, when it is a primary component. As a backup, its survival matrix is corrected by the deceleration matrix $\boldsymbol{D}_i$.

Define d_{ij} as the deceleration factor for the transition rate of component i in state j when it is a backup. The deceleration matrix of component i can be written as

$$\boldsymbol{D}_i = \text{diag}[d_{i1}, d_{i2}, \dots, d_{in_i}], \tag{5.23}$$

where n_i is the number of states of component i. As a deceleration matrix, all elements of $\boldsymbol{D}_i$ is nonnegative less than one. While component 2 is a backup, its survival matrix is adjusted by $\boldsymbol{D}_2$ to become $\boldsymbol{D}_2\boldsymbol{A}_2$. The deceleration matrix $\boldsymbol{D}_2$ reduces all transition rates in $\boldsymbol{A}_2$ by factors equal to the elements of $\boldsymbol{D}_2$. When component 2 becomes the primary replacing component 1 that has failed, the survival matrix of component 2 changes to $\boldsymbol{A}_2$.

The lifetime of a general standby system cannot be expressed in terms of the lifetime of a standby system explained in Chapter 4, because the standby system in Chapter 4 assumes the backup never fails. In a general standby system, the backup component can fail if its warmness level is not $\mathbf{0}$, so it is more like a parallel system. Therefore, the lifetime of a 1-out-of-2 general standby system can be expressed as the maximum of the lifetimes of the two components because the system fails when both have failed. With slight modifications, Theorem 2.4 can be used to generate the PH distribution of a 1-out-of-2 general standby system. While component 1 is still functioning, component 2 is a backup with a survival matrix $\boldsymbol{D}_2\boldsymbol{A}_2$. If component 2 has become the primary component replacing the failed component 1, its survival matrix is $\boldsymbol{A}_2$. Component 2 can fail while still a backup with a failure vector $= -\boldsymbol{D}_2\boldsymbol{A}_2\boldsymbol{u} = \boldsymbol{D}_2\boldsymbol{b}_2$. By using Theorem 2.4 and incorporating the deceleration matrix to component 2 while component 1 is still good, the PH representation $(n_S, \boldsymbol{a}_S, \boldsymbol{A}_S)$ for the distribution of the lifetime of a 1-out-of-2 general standby system can be determined using the equations for maximum operation as

$$n_S = n_1 n_2 + n_1 + n_2, \tag{5.24}$$

$$\boldsymbol{a}_S = \left[\boldsymbol{a}_1 \otimes \boldsymbol{a}_2 \;\; \mathbf{0} \;\; \mathbf{0}\right], \tag{5.25}$$

$$\boldsymbol{A}_S = \begin{bmatrix} \boldsymbol{A}_1 \oplus \boldsymbol{D}_2\boldsymbol{A}_2 & \boldsymbol{I}_1 \otimes \boldsymbol{D}_2\boldsymbol{b}_2 & \boldsymbol{b}_1 \otimes \boldsymbol{I}_2 \\ \mathbf{0} & \boldsymbol{A}_1 & \mathbf{0} \\ \mathbf{0} & \mathbf{0} & \boldsymbol{A}_2 \end{bmatrix}. \tag{5.26}$$

Since there are three possible component states, primary, backup, and failed, the system state in a general standby system is expressed as a combination of the letters p for primary, b for backup, and f for failed. The system states of a 1-out-of-2 general standby system are the following:

- (p, b) for component 1 is the primary and component 2 is the backup,
- (p, f) for component 1 is the primary and component 2 is failed,
- (f, p) for component 1 is failed and component 2 is the primary, and
- (f, f) for components 1 and 2 are failed.

Using these system states, Equations (5.25) and (5.26) produce a state transition diagram in Figure 5.21. This diagram shows that component 2 can fail while acting as a backup. This failure causes the system to transition from state (p, b) representing components 1 and 2 are good to state (p, f) representing component 1 is good and component 2 is failed as a backup. The failure vector of component 2 at that time is $\boldsymbol{D}_2\boldsymbol{b}_2$, which is the failure vector $\boldsymbol{b}_2$ corrected by the deceleration matrix $\boldsymbol{D}_2$ because component 2 is still a backup. The survival matrix of component 2 in state (p, b) is also corrected by the same factor to become $\boldsymbol{D}_2\boldsymbol{A}_2$. The transition vector from state (f, p) to (f, f) is $\boldsymbol{b}_2$, no longer corrected by the deceleration matrix because in state (f, p) component 2 is already the primary component, replacing component 1 that has failed. The matrix $\boldsymbol{I}_i$ in Equation (5.26) and Figure 5.21 represents the identity matrix with a dimension equal to the order of the PH distribution of the lifetime of component i.

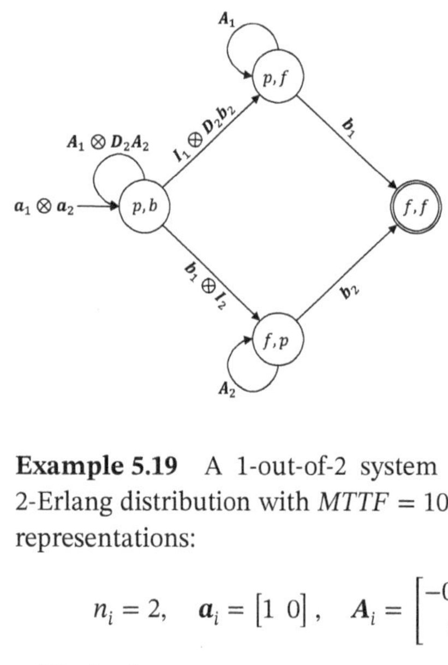

Figure 5.21 State transition diagram of a 1-out-of-2 general standby system.

Example 5.19 A 1-out-of-2 system has two identical components with lifetimes following 2-Erlang distribution with $MTTF = 100$. The PH distribution of the component lifetimes has PH representations:

$$n_i = 2, \quad \boldsymbol{a}_i = \begin{bmatrix} 1 & 0 \end{bmatrix}, \quad \boldsymbol{A}_i = \begin{bmatrix} -0.02 & 0.02 \\ 0 & -0.02 \end{bmatrix}, \quad i = 1, 2.$$

The backup components are operated at half power or with a deceleration factor 0.5 for all component states. The PH representation of the system lifetime distribution is calculated using Equations (5.24)–(5.26) as

$$n_S = n_1 n_2 + n_1 + n_2 = 8.$$

$$\boldsymbol{a}_S = \begin{bmatrix} \boldsymbol{a}_1 \otimes \boldsymbol{a}_2 & \boldsymbol{0} & \boldsymbol{0} \end{bmatrix} = \begin{bmatrix} \begin{bmatrix} 1 & 0 \end{bmatrix} \otimes \begin{bmatrix} 1 & 0 \end{bmatrix} & \begin{bmatrix} 0 & 0 \end{bmatrix} & \begin{bmatrix} 0 & 0 \end{bmatrix} \end{bmatrix} = \begin{bmatrix} 1 & 0 & 0 & 0 & 0 & 0 & 0 & 0 \end{bmatrix}.$$

$$\boldsymbol{b}_1 = \boldsymbol{b}_2 = -\boldsymbol{A}_1\boldsymbol{u} = -\begin{bmatrix} -0.02 & 0.02 \\ 0 & -0.02 \end{bmatrix} \begin{bmatrix} 1 \\ 1 \end{bmatrix} = \begin{bmatrix} 0 \\ 0.02 \end{bmatrix}.$$

$$\boldsymbol{D}_2 = \begin{bmatrix} 0.5 & 0 \\ 0 & 0.5 \end{bmatrix} = 0.5\boldsymbol{I}_2.$$

$$\boldsymbol{D}_2\boldsymbol{A}_2 = 0.5\boldsymbol{I}_2 \begin{bmatrix} -0.02 & 0.02 \\ 0 & -0.02 \end{bmatrix} = \begin{bmatrix} -0.01 & 0.01 \\ 0 & -0.01 \end{bmatrix}.$$

$$\boldsymbol{D}_2\boldsymbol{b}_2 = 0.5\boldsymbol{I}_2 \begin{bmatrix} 0 \\ 0.02 \end{bmatrix} = \begin{bmatrix} 0 \\ 0.01 \end{bmatrix}$$

$$\boldsymbol{A}_S = \begin{bmatrix} \boldsymbol{A}_1 \oplus \boldsymbol{D}_2\boldsymbol{A}_2 & \boldsymbol{I}_1 \otimes \boldsymbol{D}_2\boldsymbol{b}_2 & \boldsymbol{b}_1 \otimes \boldsymbol{I}_2 \\ \boldsymbol{0} & \boldsymbol{A}_1 & \boldsymbol{0} \\ \boldsymbol{0} & \boldsymbol{0} & \boldsymbol{A}_2 \end{bmatrix}$$

$$= \begin{bmatrix} \begin{bmatrix} -0.02 & 0.02 \\ 0 & -0.02 \end{bmatrix} \oplus \begin{bmatrix} -0.01 & 0.01 \\ 0 & -0.01 \end{bmatrix} & \boldsymbol{I}_1 \otimes \begin{bmatrix} 0 \\ 0.01 \end{bmatrix} & \begin{bmatrix} 0 \\ 0.02 \end{bmatrix} \otimes \boldsymbol{I}_2 \\ \boldsymbol{0} & \begin{bmatrix} -0.02 & 0.02 \\ 0 & -0.02 \end{bmatrix} & \boldsymbol{0} \\ \boldsymbol{0} & \boldsymbol{0} & \begin{bmatrix} -0.02 & 0.02 \\ 0 & -0.02 \end{bmatrix} \end{bmatrix}$$

$$= \begin{bmatrix} -0.03 & 0.01 & 0.02 & 0 & 0 & 0 & 0 & 0 \\ 0 & -0.03 & 0 & 0.02 & 0.01 & 0 & 0 & 0 \\ 0 & 0 & -0.03 & 0.01 & 0 & 0 & 0.02 & 0 \\ 0 & 0 & 0 & -0.03 & 0 & 0.01 & 0 & 0.02 \\ 0 & 0 & 0 & 0 & -0.02 & 0.02 & 0 & 0 \\ 0 & 0 & 0 & 0 & 0 & -0.02 & 0 & 0 \\ 0 & 0 & 0 & 0 & 0 & 0 & -0.02 & 0.02 \\ 0 & 0 & 0 & 0 & 0 & 0 & 0 & 0.02 \end{bmatrix}.$$

■

Exercise 5.15 Determine the survival matrix of a 1-out-of-2 general standby system in Example 5.19 with the backup component has deceleration matrix $\begin{bmatrix} 0.3 & 0 \\ 0 & 0.7 \end{bmatrix}$.

Exercise 5.16 Redo Exercise 5.15 for deceleration matrix $=\boldsymbol{0}$ and simplify the resulting survival matrix. In the case of the component, lifetimes have exponential distributions, showing that the result is equivalent to the survival matrix of a 2-standby system.

Example 5.20 The PH representation of the lifetime distribution of 1-out-of-2 system in Example 5.19 is utilized to plot the graphs of the system reliability and hazard functions for several values of $\boldsymbol{D}_2$. The **phparallelsys** program provided in Section 4.4 can be used for this purpose with slight modifications. The results are shown in Figure 5.22. The system reliability and hazard functions for the cold standby are also calculated using **phstandsys**, which produce identical graphs. Thus, the cold and hot standbys, which were expressed in two different models in Chapter 4, can be unified in the model given by Equations (5.25) and (5.26) by choosing suitable values for $\boldsymbol{D}_2$. ■

To simplify the writing, the triplet $(n_i, \boldsymbol{a}_i, \boldsymbol{A}_i)$ is the representation of the PH distribution of the lifetime of component i as a primary component, and $(n_i, \boldsymbol{a}_i, \boldsymbol{A}_i')$ is its representation as a backup component, with $\boldsymbol{A}_i' = \boldsymbol{D}_i\boldsymbol{A}_i$. Using $\boldsymbol{A}_i'$ can also make the model more general than using $\boldsymbol{D}_i\boldsymbol{A}_i$ as

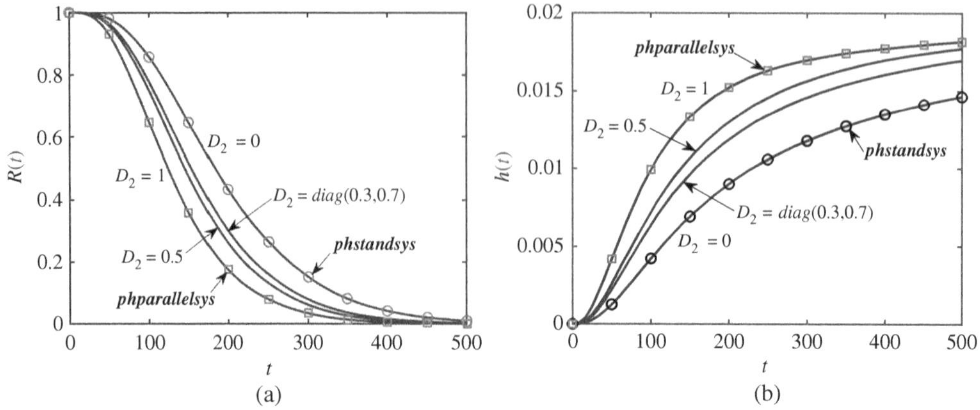

Figure 5.22 Reliability and hazard functions of 1-out-of-2 general standby systems in Example 5.20. (a) Reliability function and (b) hazard function.

long as all elements of $\boldsymbol{A}_i$ are not smaller in magnitude than the same elements of $\boldsymbol{A}_i'$. If all diagonal elements in the deceleration matrix have the same value, then $\boldsymbol{D}_i$ in Equation (5.23) can be written as $d_i\boldsymbol{I}_i$, which can be simplified to d_i because $\boldsymbol{D}_i\boldsymbol{A}_i = d_i\boldsymbol{I}_i\boldsymbol{A}_i = d_i\boldsymbol{A}_i$. Thus, the deceleration matrix reduces to a scalar deceleration factor. The backup component i with a deceleration factor d_i is said to have a warmness level d_i or operated at capacity d_i, where $0 \leq d_i \leq 1$.

Next, the results of Example 5.20 is mathematically formulated and proved.

5.2.2 Closure Properties of PH Distribution in General Standby Systems

For Equations (5.24)–(5.26) to be valid as a model for a 1-out-of-2 general standby system, they must be applicable to all types of standbys depending on the values of $\boldsymbol{D}_2$, that is,

a) $\boldsymbol{D}_2 = \boldsymbol{I}$ for the hot standby,
b) $\boldsymbol{D}_2 \neq \boldsymbol{0}, \boldsymbol{I}$, or $\boldsymbol{0} < \boldsymbol{D}_2 < \boldsymbol{I}$, for the warm standby, and
c) $\boldsymbol{D}_2 = \boldsymbol{0}$ for the cold standby.

If propositions (a)–(c) together with Equations (5.24)–(5.26) are satisfied, then PH distribution is closed in 1-out-of-2 general standby systems. Proposition (a) is obvious from substituting $\boldsymbol{D}_2 = \boldsymbol{I}$ into Equations (5.24)–(5.26) will result in Theorem 2.4 for the maximum of two PH-distributed variables as a model for a 2-parallel system, which is a hot standby system. Proposition (b) is the definition of the warm standby by reducing transition rates among component states by a factor $\boldsymbol{D}_2$ which has nonnegative elements less than 1. Proposition (c) requires a mathematical proof that if $\boldsymbol{D}_2 = \boldsymbol{0}$, then Equations (5.25)–(5.26) are equivalent to equations in Theorem 2.3 for the cold standby system.

Lemma 5.2 [23] If $D_2 = 0$, then Equations (5.25) and (5.26) become the initial probability vector and survival matrix of the cold standby structure, equivalent to those given by Theorem 2.3.

The proof of Lemma 5.2 is given in Alkaff et al. [23] using Laplace–Stieltjes transform. This book provides simpler proof using properties of Kronecker product.[1]

---

1 Advanced material, can be omitted in lecture.

Equations (5.25) and (5.26) for $\boldsymbol{D}=\boldsymbol{0}$ are

$$\boldsymbol{a} = \left[\boldsymbol{a}_1 \otimes \boldsymbol{a}_2 \ \boldsymbol{0} \ \boldsymbol{0}\right], \tag{5.27}$$

$$\boldsymbol{A} = \begin{bmatrix} \boldsymbol{A}_1 \oplus \boldsymbol{0} & \boldsymbol{I}_1 \otimes \boldsymbol{0} & \boldsymbol{b}_1 \otimes \boldsymbol{I}_2 \\ \boldsymbol{0} & \boldsymbol{A}_1 & \boldsymbol{0} \\ \boldsymbol{0} & \boldsymbol{0} & \boldsymbol{A}_2 \end{bmatrix}. \tag{5.28}$$

From the properties of the Kronecker operation, $\boldsymbol{A}_1 \oplus \boldsymbol{0} = \boldsymbol{A}_1 \otimes \boldsymbol{I}_2 + \boldsymbol{I}_1 \otimes \boldsymbol{0} = \boldsymbol{A}_1 \otimes \boldsymbol{I}_2$, and Equation (5.28) becomes

$$\boldsymbol{A} = \begin{bmatrix} \boldsymbol{A}_1 \otimes \boldsymbol{I}_2 & \boldsymbol{0} & \boldsymbol{b}_1 \otimes \boldsymbol{I}_2 \\ \boldsymbol{0} & \boldsymbol{A}_1 & \boldsymbol{0} \\ \boldsymbol{0} & \boldsymbol{0} & \boldsymbol{A}_2 \end{bmatrix}. \tag{5.29}$$

The initial state cannot reach the block (2, 2) in Equation (5.29), so the second row and column of Equations (5.27) and (5.29) can be discarded to yield

$$\boldsymbol{a} = \left[\boldsymbol{a}_1 \otimes \boldsymbol{a}_2 \ \boldsymbol{0}\right], \tag{5.30}$$

$$\boldsymbol{A} = \begin{bmatrix} \boldsymbol{A}_1 \otimes \boldsymbol{I}_2 & \boldsymbol{b}_1 \otimes \boldsymbol{I}_2 \\ \boldsymbol{0} & \boldsymbol{A}_2 \end{bmatrix}. \tag{5.31}$$

Equations (5.30) and (5.31) should be shown equivalent to the equations given by Theorem 2.3, namely, PH representations $\left(\left[\boldsymbol{a}_1 \otimes \boldsymbol{a}_2 \ \boldsymbol{0}\right], \begin{bmatrix} \boldsymbol{A}_1 \otimes \boldsymbol{I}_2 & -\boldsymbol{A}_1\boldsymbol{u}_1 \otimes \boldsymbol{I}_2 \\ \boldsymbol{0} & \boldsymbol{A}_2 \end{bmatrix}\right)$ and $\left(\left[\boldsymbol{a}_1 \ \boldsymbol{0}\right], \begin{bmatrix} \boldsymbol{A}_1 & -\boldsymbol{A}_1\boldsymbol{u}_1\boldsymbol{a}_2 \\ \boldsymbol{0} & \boldsymbol{A}_2 \end{bmatrix}\right)$ are equivalent. They can be proved equivalent by showing that their reliability functions are identical, that is,

$$\left[\boldsymbol{a}_1 \otimes \boldsymbol{a}_2 \ \boldsymbol{0}\right] e^{\begin{bmatrix} \boldsymbol{A}_1 \otimes \boldsymbol{I}_2 & -\boldsymbol{A}_1\boldsymbol{u}_1 \otimes \boldsymbol{I}_2 \\ \boldsymbol{0} & \boldsymbol{A}_2 \end{bmatrix} t} \begin{bmatrix} \boldsymbol{u}_1 \otimes \boldsymbol{u}_2 \\ \boldsymbol{u}_2 \end{bmatrix} = \left[\boldsymbol{a}_1 \ \boldsymbol{0}\right] e^{\begin{bmatrix} \boldsymbol{A}_1 & -\boldsymbol{A}_1\boldsymbol{u}_1\boldsymbol{a}_2 \\ \boldsymbol{0} & \boldsymbol{A}_2 \end{bmatrix} t} \begin{bmatrix} \boldsymbol{u}_1 \\ \boldsymbol{u}_2 \end{bmatrix}, \tag{5.32}$$

for $\forall t \geq 0$.

Use Taylor series expansion for exponential function

$$e^{\boldsymbol{A}} = \boldsymbol{I} + \frac{\boldsymbol{A}}{1!} + \frac{\boldsymbol{A}^2}{2!} + \cdots = \sum_{k=0}^{n} \frac{\boldsymbol{A}^n}{n!},$$

to transform identity in Equation (5.32) into

$$\left[\boldsymbol{a}_1 \otimes \boldsymbol{a}_2 \ \boldsymbol{0}\right] \begin{bmatrix} \boldsymbol{A}_1 \otimes \boldsymbol{I}_2 & -\boldsymbol{A}_1\boldsymbol{u}_1 \otimes \boldsymbol{I}_2 \\ \boldsymbol{0} & \boldsymbol{A}_2 \end{bmatrix}^n \begin{bmatrix} \boldsymbol{u}_1 \otimes \boldsymbol{u}_2 \\ \boldsymbol{u}_2 \end{bmatrix} = \left[\boldsymbol{a}_1 \ \boldsymbol{0}\right] \begin{bmatrix} \boldsymbol{A}_1 & -\boldsymbol{A}_1\boldsymbol{u}_1\boldsymbol{a}_2 \\ \boldsymbol{0} & \boldsymbol{A}_2 \end{bmatrix}^n \begin{bmatrix} \boldsymbol{u}_1 \\ \boldsymbol{u}_2 \end{bmatrix},$$

for $n = 0, 1, 2, \ldots$.

The proof for $n = 0$ is straightforward as

$$(\boldsymbol{a}_1 \otimes \boldsymbol{a}_2)\boldsymbol{I}(\boldsymbol{u}_1 \otimes \boldsymbol{u}_2) = (\boldsymbol{a}_1\boldsymbol{u}_1 \otimes \boldsymbol{a}_2\boldsymbol{u}_2) = 1 = \boldsymbol{a}_1\boldsymbol{I}\boldsymbol{u}_1 = 1.$$

For other n, it is sufficient to prove for $n = 1, 2$, and the rest will follow by recursion,

$$\left[\boldsymbol{a}_1 \otimes \boldsymbol{a}_2 \ \boldsymbol{0}\right] \begin{bmatrix} \boldsymbol{A}_1 \otimes \boldsymbol{I}_2 & -\boldsymbol{A}_1\boldsymbol{u}_1 \otimes \boldsymbol{I}_2 \\ \boldsymbol{0} & \boldsymbol{A}_2 \end{bmatrix} \begin{bmatrix} \boldsymbol{u}_1 \otimes \boldsymbol{u}_2 \\ \boldsymbol{u}_2 \end{bmatrix} = \left[\boldsymbol{a}_1 \ \boldsymbol{0}\right] \begin{bmatrix} \boldsymbol{A}_1 & -\boldsymbol{A}_1\boldsymbol{u}_1\boldsymbol{a}_2 \\ \boldsymbol{0} & \boldsymbol{A}_2 \end{bmatrix} \begin{bmatrix} \boldsymbol{u}_1 \\ \boldsymbol{u}_2 \end{bmatrix}. \tag{5.33}$$

$$\left[\boldsymbol{a}_1 \otimes \boldsymbol{a}_2 \ \boldsymbol{0}\right] \begin{bmatrix} \boldsymbol{A}_1 \otimes \boldsymbol{I}_2 & -\boldsymbol{A}_1\boldsymbol{u}_1 \otimes \boldsymbol{I}_2 \\ \boldsymbol{0} & \boldsymbol{A}_2 \end{bmatrix}^2 \begin{bmatrix} \boldsymbol{u}_1 \otimes \boldsymbol{u}_2 \\ \boldsymbol{u}_2 \end{bmatrix} = \left[\boldsymbol{a}_1 \ \boldsymbol{0}\right] \begin{bmatrix} \boldsymbol{A}_1 & -\boldsymbol{A}_1\boldsymbol{u}_1\boldsymbol{a}_2 \\ \boldsymbol{0} & \boldsymbol{A}_2 \end{bmatrix}^2 \begin{bmatrix} \boldsymbol{u}_1 \\ \boldsymbol{u}_2 \end{bmatrix}. \tag{5.34}$$

Proof of Equation (5.33) starts with

$$\begin{aligned}[\boldsymbol{a}_1 \otimes \boldsymbol{a}_2\ \ \boldsymbol{0}] \begin{bmatrix} \boldsymbol{A}_1 \otimes \boldsymbol{I}_2 & -\boldsymbol{A}_1\boldsymbol{u}_1 \otimes \boldsymbol{I}_2 \\ \boldsymbol{0} & \boldsymbol{A}_2 \end{bmatrix} &= [(\boldsymbol{a}_1 \otimes \boldsymbol{a}_2)(\boldsymbol{A}_1 \otimes \boldsymbol{I}_2)\ \ (\boldsymbol{a}_1 \otimes \boldsymbol{a}_2)(-\boldsymbol{A}_1\boldsymbol{u}_1 \otimes \boldsymbol{I}_2)] \\ &= [(\boldsymbol{a}_1\boldsymbol{A}_1 \otimes \boldsymbol{a}_2\boldsymbol{I}_2)\ \ (\boldsymbol{a}_1(-\boldsymbol{A}_1\boldsymbol{u}_1) \otimes \boldsymbol{a}_2\boldsymbol{I}_2)] \\ &= [(\boldsymbol{a}_1\boldsymbol{A}_1 \otimes \boldsymbol{a}_2)\ \ (\boldsymbol{a}_1(-\boldsymbol{A}_1\boldsymbol{u}_1) \otimes \boldsymbol{a}_2)]\,. \end{aligned} \tag{5.35}$$

Substituting Equation (5.35) into the left-hand side of Equation (5.33) yields

$$\begin{aligned}[(\boldsymbol{a}_1\boldsymbol{A}_1 \otimes \boldsymbol{a}_2)\ \ (\boldsymbol{a}_1(-\boldsymbol{A}_1\boldsymbol{u}_1) \otimes \boldsymbol{a}_2)] \begin{bmatrix} \boldsymbol{u}_1 \otimes \boldsymbol{u}_2 \\ \boldsymbol{u}_2 \end{bmatrix} &= [(\boldsymbol{a}_1\boldsymbol{A}_1 \otimes \boldsymbol{a}_2)(\boldsymbol{u}_1 \otimes \boldsymbol{u}_2)\ \ (\boldsymbol{a}_1(-\boldsymbol{A}_1\boldsymbol{u}_1) \otimes \boldsymbol{a}_2\boldsymbol{u}_2)] \\ &= (\boldsymbol{a}_1\boldsymbol{A}_1\boldsymbol{u}_1) \otimes (\boldsymbol{a}_2\boldsymbol{u}_2) + (\boldsymbol{a}_1(-\boldsymbol{A}_1\boldsymbol{u}_1))(\boldsymbol{a}_2\boldsymbol{u}_2) \\ &= (\boldsymbol{a}_1\boldsymbol{A}_1\boldsymbol{u}_1) + (\boldsymbol{a}_1(-\boldsymbol{A}_1\boldsymbol{u}_1)\boldsymbol{a}_2)\boldsymbol{u}_2 \\ &= \boldsymbol{a}_1 \begin{bmatrix} \boldsymbol{A}_1 & -\boldsymbol{A}_1\boldsymbol{u}_1\boldsymbol{a}_2 \end{bmatrix} \begin{bmatrix} \boldsymbol{u}_1 \\ \boldsymbol{u}_2 \end{bmatrix} \\ &= [\boldsymbol{a}_1\ \ \boldsymbol{0}] \begin{bmatrix} \boldsymbol{A}_1 & -\boldsymbol{A}_1\boldsymbol{u}_1\boldsymbol{a}_2 \\ \boldsymbol{0} & \boldsymbol{A}_2 \end{bmatrix} \begin{bmatrix} \boldsymbol{u}_1 \\ \boldsymbol{u}_2 \end{bmatrix}, \end{aligned}$$

which is the right-hand side of Equation (5.33), and the proof is complete.

To prove Equation (5.34) starts with

$$\begin{aligned}\begin{bmatrix} \boldsymbol{A}_1 \otimes \boldsymbol{I}_2 & -\boldsymbol{A}_1\boldsymbol{u}_1 \otimes \boldsymbol{I}_2 \\ \boldsymbol{0} & \boldsymbol{A}_2 \end{bmatrix} \begin{bmatrix} \boldsymbol{u}_1 \otimes \boldsymbol{u}_2 \\ \boldsymbol{u}_2 \end{bmatrix} &= \begin{bmatrix} (\boldsymbol{A}_1 \otimes \boldsymbol{I}_2)(\boldsymbol{u}_1 \otimes \boldsymbol{u}_2) - \boldsymbol{A}_1\boldsymbol{u}_1 \otimes \boldsymbol{I}_2\boldsymbol{u}_2 \\ \boldsymbol{A}_2\boldsymbol{u}_2 \end{bmatrix} \\ &= \begin{bmatrix} (\boldsymbol{A}_1\boldsymbol{u}_1 \otimes \boldsymbol{I}_2\boldsymbol{u}_2) - \boldsymbol{A}_1\boldsymbol{u}_1 \otimes \boldsymbol{I}_2\boldsymbol{u}_2 \\ \boldsymbol{A}_2\boldsymbol{u}_2 \end{bmatrix} = \begin{bmatrix} \boldsymbol{0} \\ \boldsymbol{A}_2\boldsymbol{u}_2 \end{bmatrix}, \end{aligned} \tag{5.36}$$

Writing the left-hand side of Equation (5.34) as

$$\begin{aligned}&[\boldsymbol{a}_1 \otimes \boldsymbol{a}_2\ \ \boldsymbol{0}] \begin{bmatrix} \boldsymbol{A}_1 \otimes \boldsymbol{I}_2 & -\boldsymbol{A}_1\boldsymbol{u}_1 \otimes \boldsymbol{I}_2 \\ \boldsymbol{0} & \boldsymbol{A}_2 \end{bmatrix}^2 \begin{bmatrix} \boldsymbol{u}_1 \otimes \boldsymbol{u}_2 \\ \boldsymbol{u}_2 \end{bmatrix} \\ &= \left([\boldsymbol{a}_1 \otimes \boldsymbol{a}_2\ \ \boldsymbol{0}] \begin{bmatrix} \boldsymbol{A}_1 \otimes \boldsymbol{I}_2 & -\boldsymbol{A}_1\boldsymbol{u}_1 \otimes \boldsymbol{I}_2 \\ \boldsymbol{0} & \boldsymbol{A}_2 \end{bmatrix}\right) \left(\begin{bmatrix} \boldsymbol{A}_1 \otimes \boldsymbol{I}_2 & -\boldsymbol{A}_1\boldsymbol{u}_1 \otimes \boldsymbol{I}_2 \\ \boldsymbol{0} & \boldsymbol{A}_2 \end{bmatrix} \begin{bmatrix} \boldsymbol{u}_1 \otimes \boldsymbol{u}_2 \\ \boldsymbol{u}_2 \end{bmatrix}\right). \end{aligned} \tag{5.37}$$

Substituting Equations (5.35) and (5.36) into Equation (5.37) yields

$$\begin{aligned}[\boldsymbol{a}_1 \otimes \boldsymbol{a}_2\ \ \boldsymbol{0}] \begin{bmatrix} \boldsymbol{A}_1 \otimes \boldsymbol{I}_2 & -\boldsymbol{A}_1\boldsymbol{u}_1 \otimes \boldsymbol{I}_2 \\ \boldsymbol{0} & \boldsymbol{A}_2 \end{bmatrix}^2 \begin{bmatrix} \boldsymbol{u}_1 \otimes \boldsymbol{u}_2 \\ \boldsymbol{u}_2 \end{bmatrix} &= [(\boldsymbol{a}_1\boldsymbol{A}_1 \otimes \boldsymbol{a}_2)\ \ (\boldsymbol{a}_1(-\boldsymbol{A}_1\boldsymbol{u}_1) \otimes \boldsymbol{a}_2)] \begin{bmatrix} \boldsymbol{0} \\ \boldsymbol{A}_2\boldsymbol{u}_2 \end{bmatrix} \\ &= -(\boldsymbol{a}_1\boldsymbol{A}_1\boldsymbol{u}_1) \otimes (\boldsymbol{a}_2\boldsymbol{A}_2\boldsymbol{u}_2) \\ &= -(\boldsymbol{a}_1\boldsymbol{A}_1\boldsymbol{u}_1)(\boldsymbol{a}_2\boldsymbol{A}_2\boldsymbol{u}_2). \end{aligned} \tag{5.38}$$

Simplifying the right-hand side of Equation (5.34) yields

$$\begin{aligned}[\boldsymbol{a}_1\ \ \boldsymbol{0}] \begin{bmatrix} \boldsymbol{A}_1 & -\boldsymbol{A}_1\boldsymbol{u}_1\boldsymbol{a}_2 \\ \boldsymbol{0} & \boldsymbol{A}_2 \end{bmatrix}^2 \begin{bmatrix} \boldsymbol{u}_1 \\ \boldsymbol{u}_2 \end{bmatrix} &= \left([\boldsymbol{a}_1\ \ \boldsymbol{0}] \begin{bmatrix} \boldsymbol{A}_1 & -\boldsymbol{A}_1\boldsymbol{u}_1\boldsymbol{a}_2 \\ \boldsymbol{0} & \boldsymbol{A}_2 \end{bmatrix}\right) \left(\begin{bmatrix} \boldsymbol{A}_1 & -\boldsymbol{A}_1\boldsymbol{u}_1\boldsymbol{a}_2 \\ \boldsymbol{0} & \boldsymbol{A}_2 \end{bmatrix} \begin{bmatrix} \boldsymbol{u}_1 \\ \boldsymbol{u}_2 \end{bmatrix}\right) \\ &= [\boldsymbol{a}_1\boldsymbol{A}_1\ \ \boldsymbol{a}_1(-\boldsymbol{A}_1\boldsymbol{u}_1\boldsymbol{a}_2)] \begin{bmatrix} \boldsymbol{A}_1\boldsymbol{u}_1 - \boldsymbol{A}_1\boldsymbol{u}_1\boldsymbol{a}_2\boldsymbol{u}_2 \\ \boldsymbol{A}_2\boldsymbol{u}_2 \end{bmatrix} \\ &= [\boldsymbol{a}_1\boldsymbol{A}_1\ \ \boldsymbol{a}_1(-\boldsymbol{A}_1\boldsymbol{u}_1\boldsymbol{a}_2)] \begin{bmatrix} \boldsymbol{0} \\ \boldsymbol{A}_2\boldsymbol{u}_2 \end{bmatrix} \\ &= -(\boldsymbol{a}_1\boldsymbol{A}_1\boldsymbol{u}_1)(\boldsymbol{a}_2\boldsymbol{A}_2\boldsymbol{u}_2), \end{aligned} \tag{5.39}$$

which is the same as Equation (5.38) derived from the left-hand side of Equation (5.34).

Thus, Equation (5.34) is proved.

Generalization of Lemma 5.2 yields the following theorem.

Theorem 5.1 If $T_1 \backsim PH(n_1, \boldsymbol{a}_1, \boldsymbol{A}_1)$ and $T_2 \backsim PH(n_2, \boldsymbol{a}_2, \boldsymbol{A}_2)$ are the lifetimes of components 1 and 2, respectively, and they are independent and not necessarily identical, then the lifetime of a 1-out-of-2 general standby system constructed using these components has a PH distribution with representation $(n_S, \boldsymbol{a}_S, \boldsymbol{A}_S)$ given by Equations (5.24)–(5.26).

This theorem can be applied to the 1-out-of-N general standby system by using recursion.

Lemma 5.3 If all components in a 1-out-of-N general standby system have independent PH distributions, then the system lifetime has a PH distribution with a representation that can be generated recursively using Equations (5.24)–(5.26).

The proof is carried out recursively. If the lifetime of subsystem S_j, consisting of components $i = 1, \dots, j$, has a PH distribution, and the lifetime of the backup component $(j+1)$ has a PH distribution, then according to Theorem 5.1, the lifetime of subsystem S_{j+1} consisting of components $i = 1, \dots, j+1$, has a PH distribution. Repeat for $j = 2, 3, \dots, N$ to complete the proof.

As explained in Section 5.1, the survival matrix of the distribution of the lifetime of a K-out-of-N system can be obtained from the survival matrix of the distribution of the lifetime of a 1-out-of-N system by eliminating the system states that have less than K good components. The following lemma is obvious.

Lemma 5.4 The PH representation of the distribution of the lifetime of a K-out-of-N general standby system can be obtained from the PH representation of the distribution of the lifetime of a 1-out-of-N general standby system by setting the deceleration matrices of the first K good components at any system state equal to identity matrices and removing the rows and columns of the survival matrix and the columns of the initial probability vector corresponding to system states with less than K good components.

The state transition diagram and infinitesimal generator matrix of a 3-general standby system can be derived using Lemmas 5.3 and 5.4. Since components 2 and 3 in this system are backup components, their survival matrices and failure vectors must be corrected by their deceleration matrices. The resulting survival matrix for a 3-general standby system is similar to a 3-parallel system with an adjustment for the survival matrices and failure vectors of its backup components, that is,

$$\boldsymbol{A}_S = \begin{bmatrix} \boldsymbol{A}_1 \oplus \boldsymbol{A}_2' \oplus \boldsymbol{A}_3' & \boldsymbol{I}_1 \otimes \boldsymbol{b}_2' \otimes \boldsymbol{I}_3 & \boldsymbol{b}_1 \otimes \boldsymbol{I}_2 \otimes \boldsymbol{I}_3 & \boldsymbol{I}_1 \otimes \boldsymbol{I}_2 \otimes \boldsymbol{b}_3' & \boldsymbol{0} & \boldsymbol{0} & \boldsymbol{0} \\ \boldsymbol{0} & \boldsymbol{A}_1 \oplus \boldsymbol{A}_3' & \boldsymbol{0} & \boldsymbol{0} & \boldsymbol{I}_1 \otimes \boldsymbol{b}_3' & \boldsymbol{0} & \boldsymbol{b}_1 \otimes \boldsymbol{I}_3 \\ \boldsymbol{0} & \boldsymbol{0} & \boldsymbol{A}_2 \oplus \boldsymbol{A}_3' & \boldsymbol{0} & \boldsymbol{0} & \boldsymbol{I}_2 \otimes \boldsymbol{b}_3' & \boldsymbol{b}_2 \otimes \boldsymbol{I}_3 \\ \boldsymbol{0} & \boldsymbol{0} & \boldsymbol{0} & \boldsymbol{A}_1 \oplus \boldsymbol{A}_2' & \boldsymbol{I}_1 \otimes \boldsymbol{b}_2' & \boldsymbol{b}_1 \otimes \boldsymbol{I}_2 & \boldsymbol{0} \\ \boldsymbol{0} & \boldsymbol{0} & \boldsymbol{0} & \boldsymbol{0} & \boldsymbol{A}_1 & \boldsymbol{0} & \boldsymbol{0} \\ \boldsymbol{0} & \boldsymbol{0} & \boldsymbol{0} & \boldsymbol{0} & \boldsymbol{0} & \boldsymbol{A}_2 & \boldsymbol{0} \\ \boldsymbol{0} & \boldsymbol{0} & \boldsymbol{0} & \boldsymbol{0} & \boldsymbol{0} & \boldsymbol{0} & \boldsymbol{A}_3 \end{bmatrix}. \tag{5.40}$$

The survival matrices and failure vectors of components 2 and 3 in Equation (5.40) are adjusted by deceleration matrices to become $\boldsymbol{A}_2', \boldsymbol{A}_3'$ and $\boldsymbol{b}_2', \boldsymbol{b}_3'$, respectively. The sequence of the states to produce Equation (5.40) is (p, b, b), (p, f, b), (f, p, b), (p, b, f), (p, f, f), (f, p, f), and (f, f, p), following the sequence of the states generated using the algorithm for a 3-parallel system.

The survival matrix of a 2-out-of-3 general standby system is derived using Equation (5.40) by removing the rows and columns corresponding to the system states with one good component, namely, rows 5–7 and columns 5–7. In this system, the survival matrix of component 2 is not corrected by the deceleration matrix because component 2 is a primary component. When there are only two good components left, both are primaries with their survival matrices not corrected by the deceleration matrix. The resulting survival matrix for a 2-out-3 general standby system is

$$\boldsymbol{A}_S = \begin{bmatrix} \boldsymbol{A}_1 \oplus \boldsymbol{A}_2 \oplus \boldsymbol{A}_3' & \boldsymbol{I}_1 \otimes \boldsymbol{b}_2 \otimes \boldsymbol{I}_3 & \boldsymbol{b}_1 \otimes \boldsymbol{I}_2 \otimes \boldsymbol{I}_3 & \boldsymbol{I}_1 \otimes \boldsymbol{I}_2 \otimes \boldsymbol{b}_3' \\ \boldsymbol{0} & \boldsymbol{A}_1 \oplus \boldsymbol{A}_3 & \boldsymbol{0} & \boldsymbol{0} \\ \boldsymbol{0} & \boldsymbol{0} & \boldsymbol{A}_2 \oplus \boldsymbol{A}_3 & \boldsymbol{0} \\ \boldsymbol{0} & \boldsymbol{0} & \boldsymbol{0} & \boldsymbol{A}_1 \oplus \boldsymbol{A}_2 \end{bmatrix}. \tag{5.41}$$

The state transition diagrams of a 3-general standby system and a 2-out-of-3 general standby system can be constructed using Equations (5.40) and (5.41), respectively. The results are presented as Figures 5.23 and 5.24, respectively.

Note that the system starts from state (p, b, b) in the 3-general standby system and from state (p, p, b) in the 2-out-of-3 general standby system. These state descriptions are related to the 3-general standby system only needs one primary component, while the 2-out-of-3 general standby system needs two primary components. Consequently, the survival matrix and failure vector of component 2 at the initial state in the 3-general standby system are corrected by the deceleration matrix but not in the 2-out-of-3 general standby system. From this illustration, it is apparent that the state transition diagram of the K-out-of-N general standby system (which also means its survival matrix) cannot be directly obtained by simply eliminating some rows and columns of the survival matrix of the N-general standby system associated with system states that have less than K good components. It is still necessary to adjust the warmness levels of the primary

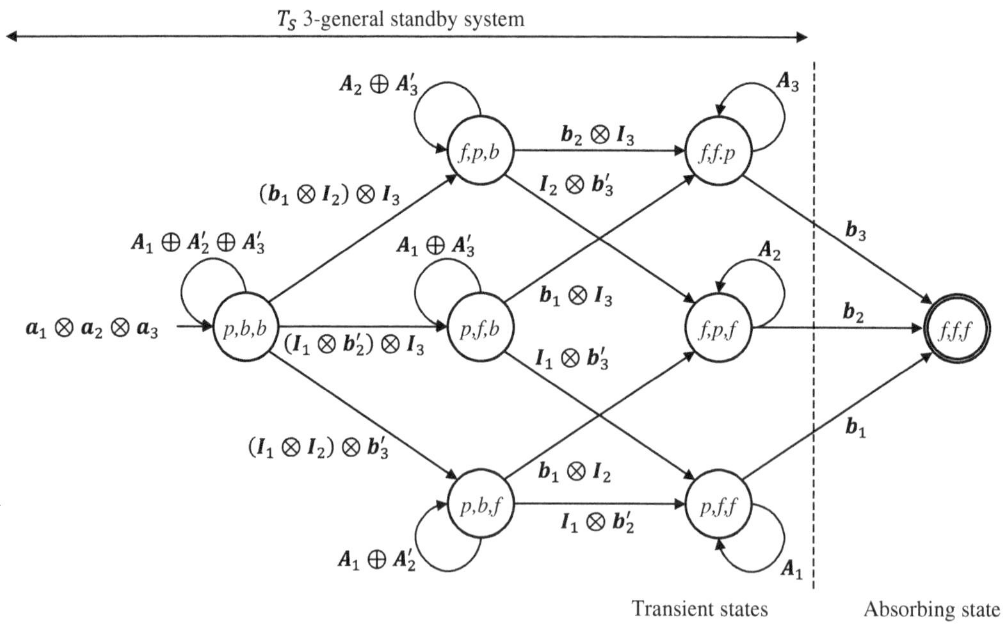

Figure 5.23 State transition diagram of a 3-general standby system.

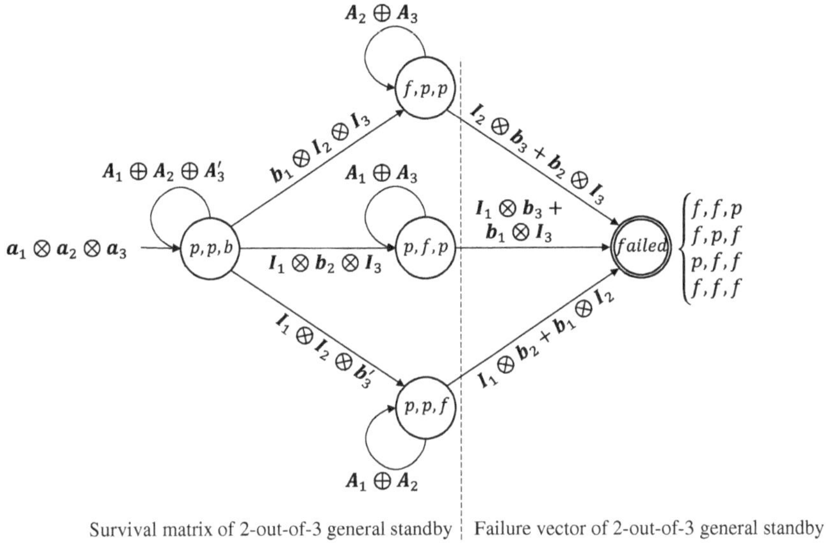

Figure 5.24 State transition diagram of a 2-out-of-3 general standby system.

components in accordance with Lemma 5.4. This requirement causes the algorithm for finding the PH representation of the distribution of the lifetime of a K-out-of-N general standby system to be more complicated than a K-out-of-N system.

The next section extends the model of K-out-of-N system into K-out-of-N general standby system by incorporating deceleration matrices for backup components.

5.2.3 Modeling K-Out-of-N General Standby System

A K-out-of-N general standby system has K primary and $(N - K)$ backup components in any combination of warmness levels, from cold to hot. If all backups are in hot standby, then the deceleration matrix is $\boldsymbol{I}$, and the system is a K-out-of-N system, as discussed in Section 5.1. The replacement for a failed primary is always the backup at the top of the list of good backup components. The survival matrix of a K-out-of-N general standby system is derived from the survival matrix of a K-out-of-N system by identifying the primary and backup components in each system state. The block matrix of each system state in the survival matrix is determined by the role of each component in that system state: as a primary (p), a backup (b), or failed (f). This process is shown for a 2-out-of-4 general standby system.

The survival matrix for a 2-out-of-4 general standby system has 11 block diagonal matrices denoted by $\boldsymbol{A}_{j,j}$, $j = 1, 2, \ldots, 11$. As for a K-out-of-N system, each block diagonal matrix is the Kronecker sum of the survival matrices of good primary and backup components in that state, with survival matrices of backup components adjusted by their deceleration matrices. The block diagonal matrices of the survival matrix of a 2-out-of-4 system are as shown in Table 5.5. The state numbers can be generated arbitrarily.

Example 5.21 Block $\boldsymbol{A}_{1,1}$ is the transition matrix within the system state (p, p, b, b). In this state, components 1 and 2 are primary, and 3 and 4 are backups. The survival matrices of primary components 1 and 2 are $\boldsymbol{A}_1$ and $\boldsymbol{A}_2$, respectively, and the survival matrices of backup components 3 and 4 are $\boldsymbol{A}_3'$ and $\boldsymbol{A}_4'$, respectively. Therefore, block $\boldsymbol{A}_{1,1}$ is $\boldsymbol{A}_1 \oplus \boldsymbol{A}_2 \oplus \boldsymbol{A}_3' \oplus \boldsymbol{A}_4'$.

Table 5.5 Block diagonal matrices in the survival matrix of a 2-out-of-4 general standby system.

Number	System state	Primary	Backup	Block	Block diagonal matrix
1	(p, p, b, b)	$\{1, 2\}$	$\{3, 4\}$	$\boldsymbol{A}_{1,1}$	$\boldsymbol{A}_1 \oplus \boldsymbol{A}_2 \oplus \boldsymbol{A}'_3 \oplus \boldsymbol{A}'_4$
2	(f, p, p, b)	$\{2, 3\}$	$\{4\}$	$\boldsymbol{A}_{2,2}$	$\boldsymbol{A}_2 \oplus \boldsymbol{A}_3 \oplus \boldsymbol{A}'_4$
3	(p, f, p, b)	$\{1, 3\}$	$\{4\}$	$\boldsymbol{A}_{3,3}$	$\boldsymbol{A}_1 \oplus \boldsymbol{A}_3 \oplus \boldsymbol{A}'_4$
4	(p, p, f, b)	$\{1, 2\}$	$\{4\}$	$\boldsymbol{A}_{4,4}$	$\boldsymbol{A}_1 \oplus \boldsymbol{A}_2 \oplus \boldsymbol{A}'_4$
5	(p, p, b, f)	$\{1, 2\}$	$\{3\}$	$\boldsymbol{A}_{5,5}$	$\boldsymbol{A}_1 \oplus \boldsymbol{A}_2 \oplus \boldsymbol{A}'_3$
6	(f, p, f, p)	$\{2, 4\}$	–	$\boldsymbol{A}_{6,6}$	$\boldsymbol{A}_2 \oplus \boldsymbol{A}_4$
7	(f, f, p, p)	$\{3, 4\}$	–	$\boldsymbol{A}_{7,7}$	$\boldsymbol{A}_3 \oplus \boldsymbol{A}_4$
8	(p, f, f, p)	$\{1, 4\}$	–	$\boldsymbol{A}_{8,8}$	$\boldsymbol{A}_1 \oplus \boldsymbol{A}_4$
9	(p, f, p, f)	$\{1, 3\}$	–	$\boldsymbol{A}_{9,9}$	$\boldsymbol{A}_1 \oplus \boldsymbol{A}_3$
10	(f, p, p, f)	$\{2, 3\}$	–	$\boldsymbol{A}_{10,10}$	$\boldsymbol{A}_2 \oplus \boldsymbol{A}_3$
11	(p, p, f, f)	$\{1, 2\}$	–	$\boldsymbol{A}_{11,11}$	$\boldsymbol{A}_1 \oplus \boldsymbol{A}_2$

Block $\boldsymbol{A}_{2,2}$ is the transition matrix within the system state (p, f, p, b). In this state, component 2 has failed, components 1 and 3 are primary, and component 4 is backup (component 3 replaces the role of the failed component 2 as a primary). Therefore, block $\boldsymbol{A}_{2,2}$ is $\boldsymbol{A}_1 \oplus \boldsymbol{A}_3 \oplus \boldsymbol{A}'_4$. The survival matrix of component 2 is not included in the block matrix $\boldsymbol{A}_{2,2}$ because component 2 in that state is failed. ■

Exercise 5.17 Determine the first six block diagonal matrices of a 3-out-of-5 general standby system.

The block off-diagonal matrix $\boldsymbol{A}_{j,k}$, $j, k = 1, 2, \ldots, 11$, of the survival matrix can be derived by identifying component failures that cause transitions from system state j to system state k. As in a K-out-of-N system, each block off-diagonal matrix is the Kronecker product of the failure vector of the failing component that triggers the transition and the identity matrices representing surviving components in that transition. The failing component could be primary or backup component; hence, its failure vector must be adjusted as necessary. Some block off-diagonal matrices in the survival matrix of a 2-out-of-4 general standby system are given in Table 5.6.

Example 5.22 The transition from system state 1 (p, p, b, b) to system state 3 (p, f, p, b) is due to the failure of component 2, which is a primary component. The failure vector of component 2 as a primary is $\boldsymbol{b}_2$. Surviving components are represented by identity matrices with dimensions equal to the dimensions of their survival matrices. Thus, block off-diagonal matrix $\boldsymbol{A}_{1,3}$ is $\boldsymbol{I}_1 \otimes \boldsymbol{b}_2 \otimes \boldsymbol{I}_3 \otimes \boldsymbol{I}_4$.

The transition from system state 1 (p, p, b, b) to system state 4 (p, p, f, b) is caused by the failure of backup component 3. The failure vector of component 3 as a backup is $\boldsymbol{b}'_3 = \boldsymbol{D}_3\boldsymbol{b}_3$, not $\boldsymbol{b}_3$. Thus, block off-diagonal matrix $\boldsymbol{A}_{1,4}$ is $\boldsymbol{I}_1 \otimes \boldsymbol{I}_2 \otimes \boldsymbol{b}'_3 \otimes \boldsymbol{I}_4$.

There is no transition from system state 1 (p, p, b, b) to system state 8 (p, f, f, p) because such a transition occurs when components 2 and 3 fail simultaneously, which is impossible in the continuous-time model. Thus, the block off-diagonal matrix $\boldsymbol{A}_{1,8}$ is $\boldsymbol{0}$.

Table 5.6 Some block off-diagonal matrices in the survival matrix of a 2-out-of-4 general standby system.

From		To				
State j	**System state**	**State k**	**System state**	**Failure event**	**Block**	**Block off-diagonal matrix**
1	(p, p, b, b)	2	(f, p, p, b)	1 (primary) fails	$\boldsymbol{A}_{1,2}$	$\boldsymbol{b}_1 \otimes \boldsymbol{I}_2 \otimes \boldsymbol{I}_3 \otimes \boldsymbol{I}_4$
		3	(p, f, p, b)	2 (primary) fails	$\boldsymbol{A}_{1,3}$	$\boldsymbol{I}_1 \otimes \boldsymbol{b}_2 \otimes \boldsymbol{I}_3 \otimes \boldsymbol{I}_4$
		4	(p, p, f, b)	3 (backup) fails	$\boldsymbol{A}_{1,4}$	$\boldsymbol{I}_1 \otimes \boldsymbol{I}_2 \otimes \boldsymbol{b}'_3 \otimes \boldsymbol{I}_4$
		5	(p, p, b, f)	4 (backup) fails	$\boldsymbol{A}_{1,5}$	$\boldsymbol{I}_1 \otimes \boldsymbol{I}_2 \otimes \boldsymbol{I}_3 \otimes \boldsymbol{b}'_4$
		Others	Impossible		$\boldsymbol{A}_{1,k}$	$\boldsymbol{0}$
2	(f, p, p, b) 1 has failed	6	(f, p, f, p)	3 (primary) fails	$\boldsymbol{A}_{2,6}$	$\boldsymbol{I}_2 \otimes \boldsymbol{b}_3 \otimes \boldsymbol{I}_4$
		7	(f, f, p, p)	2 (primary) fails	$\boldsymbol{A}_{2,7}$	$\boldsymbol{b}_2 \otimes \boldsymbol{I}_3 \otimes \boldsymbol{I}_4$
		10	(f, p, p, f)	4 (backup) fails	$\boldsymbol{A}_{2,10}$	$\boldsymbol{I}_2 \otimes \boldsymbol{I}_3 \otimes \boldsymbol{b}'_4$
		Others	Impossible		$\boldsymbol{A}_{2,k}$	$\boldsymbol{0}$
3	(p, f, p, b) 2 has failed	7	(f, f, p, p)	1 (primary) fails	$\boldsymbol{A}_{3,7}$	$\boldsymbol{b}_1 \otimes \boldsymbol{I}_3 \otimes \boldsymbol{I}_4$
		8	(p, f, f, p)	3 (primary) fails	$\boldsymbol{A}_{3,8}$	$\boldsymbol{I}_1 \otimes \boldsymbol{b}_3 \otimes \boldsymbol{I}_4$
		9	(p, f, p, f)	4 (backup) fails	$\boldsymbol{A}_{3,9}$	$\boldsymbol{I}_1 \otimes \boldsymbol{I}_3 \otimes \boldsymbol{b}'_4$
		Others	Impossible		$\boldsymbol{A}_{3,k}$	$\boldsymbol{0}$

The transition from system state 3 (p, f, p, b) to system state 8 (p, f, f, p) is caused by the failure of primary component 3. The failure vector of component 3 as a primary is $\boldsymbol{b}_3$. Thus, block off-diagonal matrix $\boldsymbol{A}_{3,8}$ is $\boldsymbol{I}_1 \otimes \boldsymbol{b}_3 \otimes \boldsymbol{I}_4$. Component 2 is not represented in $\boldsymbol{A}_{3,8}$ because it has failed in state 3. ■

Exercise 5.18 Determine the block off-diagonal matrices in the survival matrix of a 3-out-of-5 general standby system representing transition matrices from state (p, f, p, p, b) to all other states.

Note that the terms in all block matrices strictly follow the sequence of the components since Kronecker operations are not commutative.

Figure 5.5 can be utilized to generate the block matrices of the survival matrix of a K-out-of-N general standby system. The definitions of $\boldsymbol{A}_{VV}$, $\boldsymbol{A}_{WW}$, and $\boldsymbol{A}_{VW}$ in Equations (5.1) and (5.2) must be extended to include primary and backup components. Using Figures 5.23 and 5.24, Tables 5.5 and 5.6 obtains

$$\boldsymbol{A}_{VV} = \oplus \boldsymbol{X}_i, \quad i \mid V_i \neq f \tag{5.42}$$

$$\boldsymbol{A}_{VW} = \otimes \boldsymbol{Y}_i, \quad i \mid V_i \neq f \tag{5.43}$$

where

$$\begin{aligned}
\boldsymbol{X}_i &= \boldsymbol{A}_i \quad \text{if } V_i = p, \\
&= \boldsymbol{A}'_i \quad \text{if } V_i = b, \\
\boldsymbol{Y}_i &= \boldsymbol{I}_i \quad \text{if } W_i \neq f, \\
&= \boldsymbol{b}_i \quad \text{if } V_i = p \quad \text{and} \quad W_i = f, \\
&= \boldsymbol{b}'_i \quad \text{if } V_i = b \quad \text{and} \quad W_i = f.
\end{aligned}$$

Equations (5.42) and (5.43) can be utilized to generate the survival matrix of K-out-of-N general standby systems for any values of K and N.

Example 5.23 The block diagonal and off-diagonal matrices in the survival matrix of a 2-out-of-4 general standby system corresponding to states (p, f, p, b) and (p, f, p, f) can be generated using Equations (5.42) and (5.43) as follows:

Within state $V = (p, f, p, b)$:

$$(V_1 = p, V_2 = f, V_3 = p, V_4 = b) \rightarrow (\boldsymbol{X}_1 = \boldsymbol{A}_1, \boldsymbol{X}_3 = \boldsymbol{A}_3, \boldsymbol{X}_4 = \boldsymbol{A}_4')$$
$$\rightarrow \boldsymbol{A}_{VV} = \boldsymbol{A}_1 \oplus \boldsymbol{A}_3 \oplus \boldsymbol{A}_4'.$$

From $V = (p, f, p, b)$ to $W = (p, f, p, f)$ (component 2 has failed before transition):

$$(V_1 = p, W_1 \neq f \rightarrow \boldsymbol{Y}_1 = \boldsymbol{I}_1; V_3 = p, W_3 \neq f \rightarrow \boldsymbol{Y}_3 = \boldsymbol{I}_3, V_4 = b, W_4 = f \rightarrow \boldsymbol{Y}_4 = \boldsymbol{b}_4')$$
$$\rightarrow \boldsymbol{A}_{VW} = \boldsymbol{I}_1 \otimes \boldsymbol{I}_3 \otimes \boldsymbol{b}_4'.$$

■

Exercise 5.19 Using Equations (5.42) and (5.43) generates the transition matrices within state (p, f, p, p, b) and from state (p, f, p, p, b) to all other possible states in the survival matrix of a 3-out-of-5 general standby system.

The algorithm for generating the PH distribution of the lifetime of a K-out-of-N general standby system is formalized by creating a set P to store the index of the primary components. The number of elements in P at any time is K. If i is an element of P, then component i is a primary, and its PH representation is $(n_i, \boldsymbol{a}_i, \boldsymbol{A}_i)$. If i is greater than the largest element of P, then component i is a backup, and its representation is $(n_i, \boldsymbol{a}_i, \boldsymbol{A}_i')$ where $\boldsymbol{A}_i' = \boldsymbol{D}_i\boldsymbol{A}_i$. The block diagonal matrix of each system state is generated using Equation (5.42), and each block off-diagonal matrix is generated using Equation (5.43). The algorithm is referred to as Algorithm 5.3 as described below.

Algorithm 5.3

---

1) Generate the system states as N-tuples of component states that contain at least K good components.
2) The first K good components are given values p, the remaining good components are given values b, and failed components are given values f.
3) Determine the block diagonal matrix for each state obtained from step 2 using Equation (5.42).
4) Determine the next states for each current states due. The possible next states are states with exactly one good component in the current state became failed.
5) Determine the block off-diagonal matrix using Equation (5.43) for each pair of a current state and its next states obtained from step 4.
6) Set a value **0** for other block off-diagonal matrices.

Algorithm 5.3 is programmed in MATLAB as **phgensys** function. The function receives K, N, $(n_i, \boldsymbol{a}_i, \boldsymbol{A}_i)$, $i = 1, 2, \ldots, N$, and $\boldsymbol{D}_i$, $i = K+1, K+2, \ldots, N$, as inputs.

The next section discusses the use of Algorithm 5.3 and **phgensys** function for analyzing the reliability of general standby systems.

5.3 Reliability Analysis of General Standby Systems

The method in Section 5.2 is utilized to calculate reliability measures of general standby systems with various types of components. The development starts with the simplest case of general standby systems with components having exponentially distributed lifetimes.

5.3.1 General Standby Systems with CFR Components

The method derived in Section 5.2 is illustrated with its application for general standby systems having CFR components. Manual calculation is reasonable for such systems.

Example 5.24 Components 1, 2, and 3 have exponentially distributed lifetimes with *MTTF* equal to 1, 2, and 5 months, respectively. A 3-general standby system is constructed using these components with components 2 and 3 as backup components with warmness levels of 0.8 and 0.5, respectively. The same components are also used to construct a 2-out-of-3 general standby system with the same warmness level for component 3.

The survival matrix for the 3-general standby system can be derived using Equation (5.40). The survival matrices and failure vectors of the primary and backups components in this system can be determined from the failure rate (=1/*MTTF*) of each component and its warmness level. The results are

$$\boldsymbol{A}_1 = [-1], \boldsymbol{b}_1 = [1],$$

$$\boldsymbol{A}_2 = [-0.5], \boldsymbol{A}_2' = 0.8[-0.5] = [-0.4], \boldsymbol{b}_2 = [0.5], \boldsymbol{b}_2' = 0.8[0.5] = [0.4],$$

$$\boldsymbol{A}_3 = [-0.2], \boldsymbol{A}_3' = 0.5[-0.2] = [-0.1], \boldsymbol{b}_3 = [0.2], \boldsymbol{b}_3' = 0.5[0.2] = [0.1].$$

Since they are scalars, the bracket representing the matrix can be ignored.

Substituting them into Equation (5.40) yields

$$\boldsymbol{A}_S = \begin{bmatrix} -1-0.4-0.1 & (1)(0.4)(1) & (1)(1)(1) & (1)(1)(0.1) & 0 & 0 & 0 \\ 0 & -1-0.1 & 0 & 0 & (1)(0.1) & 0 & (1)(1) \\ 0 & 0 & -0.5-0.1 & 0 & 0 & (1)(0.1) & (0.5)(1) \\ 0 & 0 & 0 & -1-0.4 & (1)(0.4) & (1)(1) & 0 \\ 0 & 0 & 0 & 0 & -1 & 0 & 0 \\ 0 & 0 & 0 & 0 & 0 & -0.5 & 0 \\ 0 & 0 & 0 & 0 & 0 & 0 & -0.2 \end{bmatrix}$$

$$= \begin{bmatrix} -1.5 & 0.4 & 1 & 0.1 & 0 & 0 & 0 \\ 0 & -1.1 & 0 & 0 & 0.1 & 0 & 1 \\ 0 & 0 & -0.6 & 0 & 0 & 0.1 & 0.5 \\ 0 & 0 & 0 & -1.4 & 0.4 & 1 & 0 \\ 0 & 0 & 0 & 0 & -1 & 0 & 0 \\ 0 & 0 & 0 & 0 & 0 & -0.5 & 0 \\ 0 & 0 & 0 & 0 & 0 & 0 & -0.2 \end{bmatrix}.$$

The survival matrix of a 2-out-of-3 general standby system can be calculated using Equation (5.41). In this system, component 2 is a primary. Its survival matrix $\boldsymbol{A}_2$ and failure vector $\boldsymbol{b}_2$ are not adjusted by a warmness level. When there are only two good components, both are primary. Substituting them into Equation (5.41) yields

$$\boldsymbol{A}_S = \begin{bmatrix} -1-0.5-0.1 & (1)(0.5)(1) & (1)(1)(1) & (1)(1)(0.1) \\ 0 & -1-0.2 & 0 & 0 \\ 0 & 0 & -0.5-0.2 & 0 \\ 0 & 0 & 0 & -1-0.5 \end{bmatrix} = \begin{bmatrix} -1.6 & 0.5 & 1 & 0.1 \\ 0 & -1.2 & 0 & 0 \\ 0 & 0 & -0.7 & 0 \\ 0 & 0 & 0 & -1.5 \end{bmatrix}.$$

■

Example 5.25 The survival matrix of the 2-out-of-3 general standby system in Example 5.24 is generated using Equations (5.42) and (5.43).

The transient states of this system are (p, p, b), (p, f, p), and (f, p, p). The transition matrices within the system states and between pairs of system states are calculated using $\boldsymbol{A}_1, \boldsymbol{b}_1, \boldsymbol{A}_2, \boldsymbol{b}_2, \boldsymbol{A}_3, \boldsymbol{A}_3', \boldsymbol{b}_3, \boldsymbol{b}_3'$ as given in Example 5.24.

(a) Transitions within each transient system state:

1) State (p, p, b): $V_1 = V_2 = p, V_3 = b \rightarrow \boldsymbol{X}_1 = \boldsymbol{A}_1, \boldsymbol{X}_2 = \boldsymbol{A}_2, \boldsymbol{X}_3 = \boldsymbol{A}_3'$.

$$A_S((p,p,b),(p,p,b)) = \boldsymbol{A}_1 \oplus \boldsymbol{A}_2 \oplus \boldsymbol{A}_3' = [-1] + [-0.5] + [-0.1] = [-1.6].$$

2) State (p, f, p): $V_1 = V_3 = p, V_2 = f \rightarrow \boldsymbol{X}_1 = \boldsymbol{A}_1, \boldsymbol{X}_3 = \boldsymbol{A}_3$.

$$A_S((p,f,p),(p,f,p)) = \boldsymbol{A}_1 \oplus \boldsymbol{A}_3 = [-1] + [-0.2] = [-1.2].$$

3) State (f, p, p): $V_1 = f; V_2 = V_3 = p, \rightarrow \boldsymbol{X}_2 = \boldsymbol{A}_2, \boldsymbol{X}_3 = \boldsymbol{A}_3$.

$$A_S((f,p,p)(f,p,p)) = \boldsymbol{A}_2 \oplus \boldsymbol{A}_3 = [-0.5] + [-0.2] = [-0.7].$$

4) State (p, p, f): $V_1 = V_2 = p, V_3 = f \rightarrow \boldsymbol{X}_1 = \boldsymbol{A}_1, \boldsymbol{X}_2 = \boldsymbol{A}_2$.

$$A_S((p,p,f),(p,p,f)) = \boldsymbol{A}_1 \oplus \boldsymbol{A}_2 = [-1] + [-0.5] = [-1.5].$$

(b) Transitions between possible pairs of transient system states:

1) From state (p, p, b) to (p, f, p): $V_1, W_1 \neq f \rightarrow \boldsymbol{Y}_1 = \boldsymbol{I}_1$; $V_2 = p, W_2 = f \rightarrow \boldsymbol{Y}_2 = \boldsymbol{b}_2$; $V_3, W_3 \neq f \rightarrow \boldsymbol{Y}_3 = \boldsymbol{I}_3$. Thus,

$$A_S((p,p,b),(p,f,p)) = \boldsymbol{I}_1 \otimes \boldsymbol{b}_2 \otimes \boldsymbol{I}_3 = [0.5].$$

2) From state (p, p, b) to (f, p, p): $V_1 = p, W_1 = f \rightarrow \boldsymbol{Y}_1 = \boldsymbol{b}_1$; $V_2, W_2 \neq f \rightarrow \boldsymbol{Y}_2 = \boldsymbol{I}_2$; $V_3, W_3 \neq f \rightarrow \boldsymbol{Y}_3 = \boldsymbol{I}_3$. Thus,

$$A_S((p,p,b),(f,p,p)) = \boldsymbol{b}_1 \otimes \boldsymbol{I}_2 \otimes \boldsymbol{I}_3 = [1].$$

3) From state (p, p, b) to (p, p, f): $V_1, W_1 \neq f \rightarrow \boldsymbol{Y}_1 = \boldsymbol{I}_1; V_2, W_2 \neq f \rightarrow \boldsymbol{Y}_2 = \boldsymbol{I}_2; V_3 = b, W_3 = f \rightarrow \boldsymbol{Y}_3 = \boldsymbol{b}_3'$. Thus,

$$A_S((p,p,b),(p,p,f)) = \boldsymbol{I}_1 \otimes \boldsymbol{I}_2 \otimes \boldsymbol{b}_3' = [0.1].$$

4) Other transitions among transient states are **0**.

The survival matrix obtained from the derived transition matrices following the same sequence of states as in Example 5.24 is

$$\boldsymbol{A}_S = \begin{bmatrix} -1.6 & 0.5 & 1 & 0.1 \\ 0 & -1.2 & 0 & 0 \\ 0 & 0 & -0.7 & 0 \\ 0 & 0 & 0 & -1.5 \end{bmatrix},$$

the same as that produced in Example 5.24. ■

Exercise 5.20 Generate the survival matrix of 1-out-of-3 and 2-out-of-3 general standby systems with CFR components having failure rate 2, 5, and 10 per month. The warmness levels for component 2 and 3 while as backup components are 1 and 0, respectively.

Examples 5.24 and 5.25 show that the PH distributions of the system lifetimes are not simple. A simple PH distribution is still possible when the components and warmness levels each are identical, which sometimes is referred to as a homogeneous general standby system. As in K-out-of-N systems with identical components, the system states can be represented by the number of good components whose values are between K and N. At system state i, $i = K, K+1, \ldots, N$, the number of primary components is K, and the number of backup components is $i-K$. The transition rate from state i to state $i-1$ is the sum of the failure rates of all good components in state i, which is

$$\lambda_i = (K + (i-K)d)\lambda, \quad i = K, K+1, \ldots, N, \tag{5.44}$$

where λ is the component failure rate, and d is the warmness level of all backup components.

The state transition diagram of a homogeneous K-out-of-N general standby system with CFR components (having identical components and warmness levels) constructed using Equation (5.44) is given in Figure 5.25. This state transition diagram is not applicable for general standby systems with non-identical CFR components or warmness levels, which sometimes is referred to as heterogeneous K-out-of-N general standby systems.

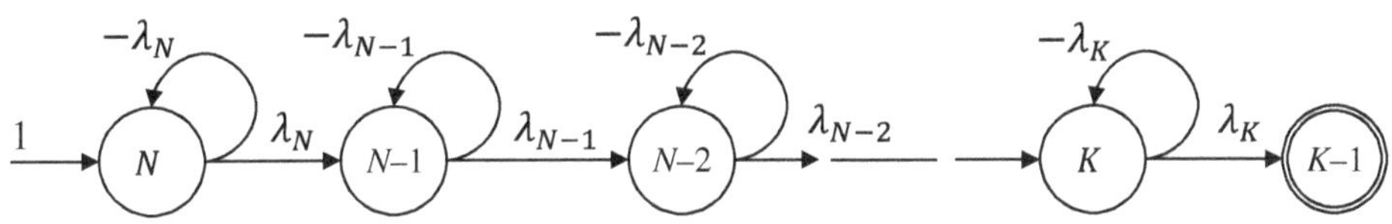

Figure 5.25 State transition diagram of a homogeneous K-out-of-N general standby system with CFR components.

Figure 5.25 indicates that the lifetime of a homogeneous K-out-of-N general standby system has a hypoexponential distribution. Matrix and scalar representations of the system lifetime distribution can be obtained from Table 2.1, with λ_i is given by Equation (5.44).

- Matrix representation:

$$n = N - K + 1, \quad \boldsymbol{a} = \begin{bmatrix} 1 & 0 & \ldots & 0 \end{bmatrix},$$

$$\boldsymbol{A} = \begin{bmatrix} -(K+(N-K)d)\lambda & (K+(N-K)d)\lambda & \ldots & 0 \\ 0 & -(K+(N-1-K)d)\lambda & \ldots & 0 \\ \vdots & \vdots & \ddots & \vdots \\ 0 & 0 & \ldots & (K+d)\lambda \\ 0 & 0 & \ldots & -K\lambda \end{bmatrix}. \tag{5.45}$$

- Scalar representation:

$$R_S(t) = \sum_{i=K}^{N} \frac{e^{-\lambda_i t}}{P_i} \quad \text{with} \quad P_i = \prod_{j=K, j\neq i}^{N} \left(1 - \frac{\lambda_i}{\lambda_j}\right). \tag{5.46}$$

Using the same method as in the derivation for the K-out-of-N system with identical CFR components in Subsection 5.1.3 yields

$$MTTF_S = \frac{1}{(K+(N-K)d)\lambda} + \frac{1}{(K+(N-1-K)d)\lambda} + \cdots + \frac{1}{K\lambda} = \frac{1}{\lambda}\sum_{i=K}^{N}\frac{1}{(K+(i-K)d)}, \tag{5.47}$$

$$\begin{aligned} VTTF_S &= \frac{1}{((K+(N-K)d)\lambda)^2} + \frac{1}{((K+(N-1-K)d)\lambda)^2} + \cdots + \frac{1}{(K\lambda)^2} \\ &= \frac{1}{\lambda^2}\sum_{i=K}^{N}\frac{1}{(K+(i-K)d)^2}. \end{aligned} \tag{5.48}$$

Equations (5.47) and (5.48) can be rewritten as

$$MTTF_S = MTTF_{comp}\sum_{i=K}^{N}\frac{1}{(K+(i-K)d)}, \tag{5.49}$$

$$VTTF_S = (MTTF_{comp})^2\sum_{i=K}^{N}\frac{1}{(K+(i-K)d)^2}. \tag{5.50}$$

Example 5.26 Example 5.24 is repeated using identical CFR components with $MTTF = 1$ and identical warmness level $= 0.5$. The survival matrix of 3-general standby system generated using Equation (5.45) is

$$\boldsymbol{A} = \begin{bmatrix} -(1+(3-1)(0.5))(1) & (1+(3-1)(0.5))(1) & 0 \\ 0 & -(1+(2-1)(0.5))(1) & (1+(2-1)(0.5))(1) \\ 0 & 0 & -(1)(1) \end{bmatrix} = \begin{bmatrix} -2 & 2 & 0 \\ 0 & -1.5 & 1.5 \\ 0 & 0 & -1 \end{bmatrix}.$$

The survival matrix of 3-general standby system generated using Equation (5.40) is

$$\boldsymbol{A}_S = \begin{bmatrix} -1-0.5-0.5 & (1)(0.5)(1) & (1)(1)(1) & (1)(1)(0.5) & 0 & 0 & 0 \\ 0 & -1-0.5 & 0 & 0 & (1)(0.5) & 0 & (1)(1) \\ 0 & 0 & -1-0.5 & 0 & 0 & (1)(0.5) & (1)(1) \\ 0 & 0 & 0 & -1-0.5 & (1)(0.5) & (1)(1) & 0 \\ 0 & 0 & 0 & 0 & -1 & 0 & 0 \\ 0 & 0 & 0 & 0 & 0 & -1 & 0 \\ 0 & 0 & 0 & 0 & 0 & 0 & -1 \end{bmatrix}$$

$$= \left[\begin{array}{c|ccc|ccc} -2 & 0.5 & 1 & 0.5 & 0 & 0 & 0 \\ \hline 0 & -1.5 & 0 & 0 & 0.5 & 0 & 1 \\ 0 & 0 & -1.5 & 0 & 0 & 0.5 & 1 \\ 0 & 0 & 0 & -1.5 & 0.5 & 1 & 0 \\ \hline 0 & 0 & 0 & 0 & -1 & 0 & 0 \\ 0 & 0 & 0 & 0 & 0 & -1 & 0 \\ 0 & 0 & 0 & 0 & 0 & 0 & -1 \end{array}\right].$$

Again, the two results obtained using the two methods are identical. If states with two good components, (p, b, f), (f, p, b), and (p, f, b), and states with one good component, (p, f, f), (f, p, f), and (f, f, p), in the method based on Equation (5.40) are lumped into state 2 and state 1 in the method based on Equation (5.45), respectively, then the survival matrices obtained using the two methods are identical. ■

Example 5.27 A general standby system has five identical CFR components with $MTTF$ = 10 days. The system requires at least three good components to function properly. All backup components have an identical warmness level of 0.5.

Using Equations (5.49) and (5.50) yields

$$MTTF_S = 10\sum_{i=3}^{5}\frac{1}{(K+(i-K)0.5)} = 10\left(\frac{1}{3}+\frac{1}{3+0.5}+\frac{1}{3+1}\right) = 8.69 \text{ days.}$$

$$VTTF_S = 100\sum_{i=3}^{5}\frac{1}{(K+(i-K)0.5)^2} = 100\left(\frac{1}{3^2}+\frac{1}{(3+0.5)^2}+\frac{1}{(3+1)^2}\right) = 25.52 \text{ days}^2$$

$$SDTTF_S = 5.05 \text{ days.}$$

The effect of the backup component warmness level on the system $MTTF$ and $SDTTF$ is shown in Table 5.7.

Table 5.7 Effect of warmness levels on *MTTF* and *SDTTF* of the system in Example 5.27.

Warmness level	0	0.2	0.4	0.6	0.8	1.0
System *MTTF*	10.00	9.40	8.91	8.49	8.14	7.83
System *SDTTF*	5.77	5.43	5.17	4.95	4.77	4.62

Table 5.7 shows that the system $MTTF$ and $SDTTF$ decrease with the increasing warmness level of the backup components. The system with all cold standbys has the highest expected lifetime, and the system with all hot standbys has the lowest expected lifetime. It is because the higher the warmness level, the higher the possibility of the backup components failing while serving as backup. ■

Exercise 5.21 A 2-out-of-6 general standby system has identical CFR components and identical warmness levels for the backup components as given in Example 5.27. Draw its state transition diagram, generate its survival matrix, and calculate its $MTTF$ and $SDTTF$.

The development is continued for general standby systems with components having lifetimes following higher-order PH distributions.

5.3.2 General Standby System with Non-CFR Components

Reliability analysis of general standby systems with components having higher-order PH-distributed lifetimes requires using **phgensys** function.

Example 5.28 Several 2-out-of-5 general standby systems are constructed using nonidentical components with component k having lifetimes following k-Erlang (λ_k), where $\lambda_k = 5 + k$, $k = 1, 2, \ldots, 5$. The warmness levels for the jth backup components, $j = 1, 2, 3$, are $d_j = 1.0, 0.5, 0.0$. The PH distribution for the system lifetime is generated using **phgensys**, and its representation is used to generate the reliability measures of the system in Figure 5.26.

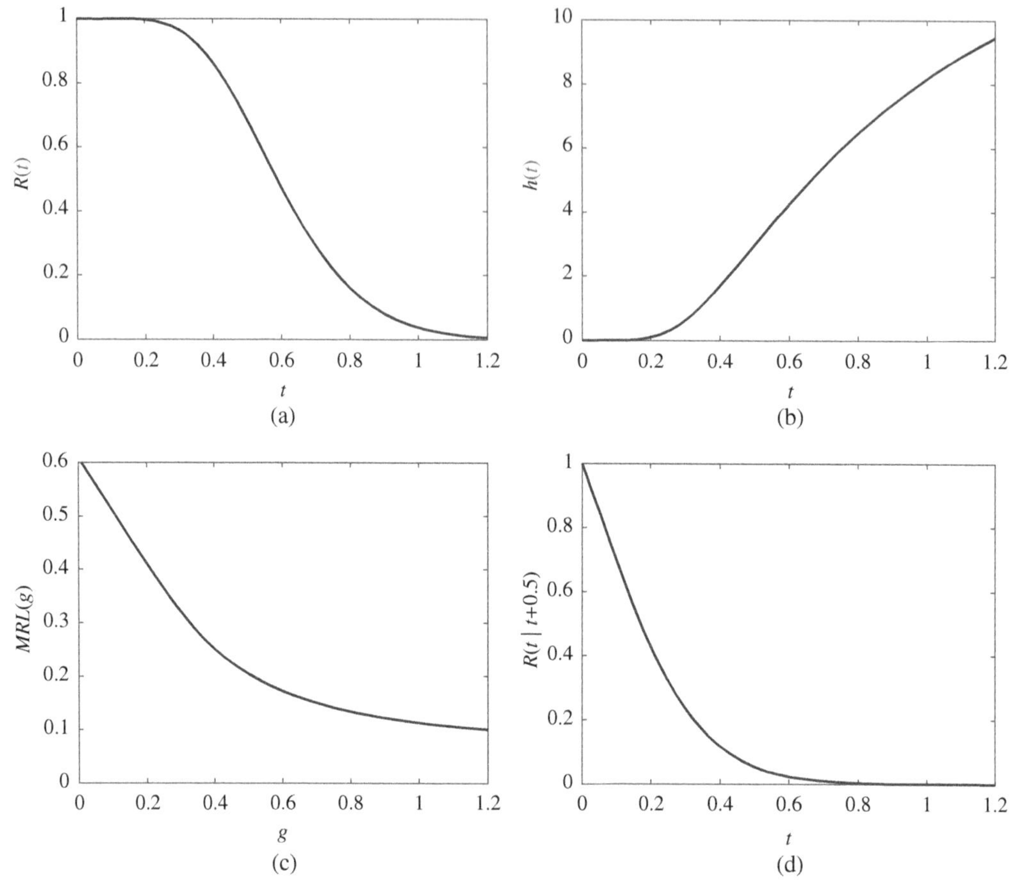

Figure 5.26 Reliability measures of the system in Example 5.28. (a) System reliability function, (b) system hazard function, (c) system *MRL* function, and (d) system residual reliability function (age 0.5).

The main program to call **phgensys** to produce Figure 5.26 is available in **example5.28.m** file in the software package **dysrel-am**. Its source code is as follows:

MATLAB code for example5_28.m

```
clear;close all
addpath('..\dysrel-am','..\dysrel-am\expokit')
K = 2;                                          input
N = 5;
% PH components (n-Erlang)
Tc = [];
for n = 1:N
  mttf = n/(5+n);
  T = phnErlang(n,mttf);
  Tc = [Tc T];
end
% Warmness level for backup components
D = [1 0.5 0];
% PH representation of system
Ts  = phgensys(Tc,K,D);                         model
% Time vector
tmax = 1.2;
t  = linspace(0,tmax,21);
```

```
% Reliability plots                                output
[x,Rs]  = phrelplot(t,Ts);
[x,hs]  = phhazplot(t,Ts);
[x,RRs] = phresrelplot(t,Ts,0.5);
```

■

Example 5.29 A 3-out-of-7 general standby system is constructed using the IFR components of Example 3.2, with all backup components assigned identical warmness levels. The PH representation obtained using **phgensys** is used to examine the effect of the backup component warmness level d on system reliability measures. The results are shown in Figure 5.27.

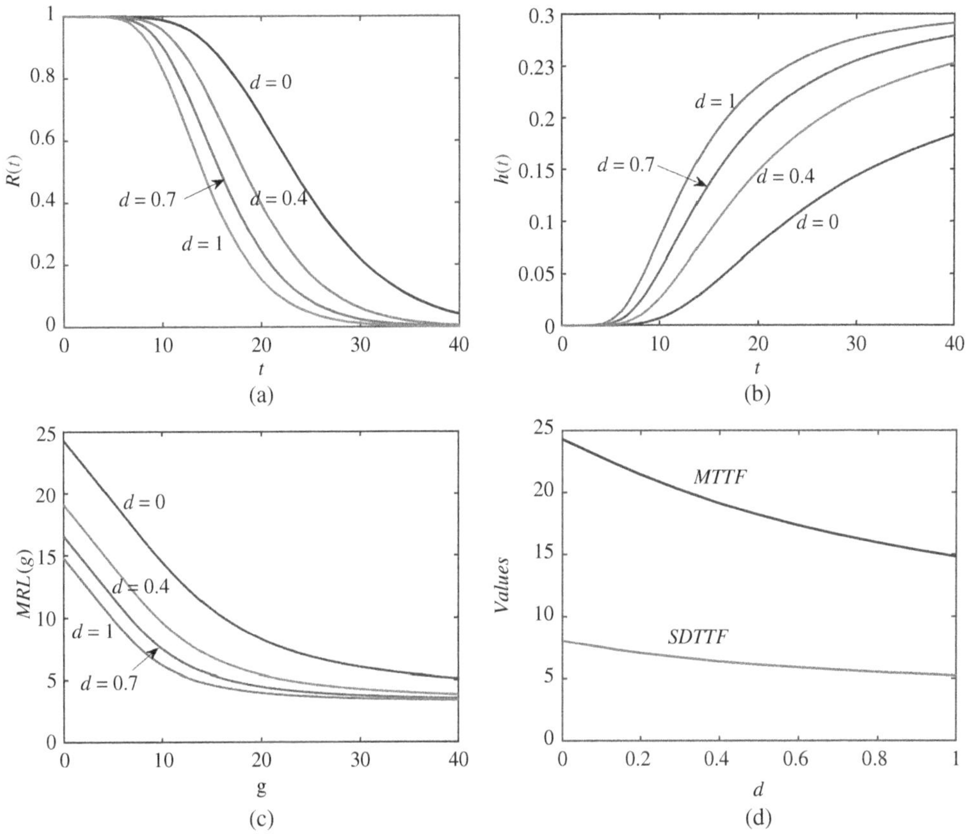

Figure 5.27 Reliability measures of the system in Example 5.29. (a) System reliability function, (b) system hazard function, (c) system *MRL* function, (d) system *MTTF* and *SDTTF*. ■

The graphs in Figure 5.27 show that the system reliability measures are better for the lower warmness level. For $d_1 > d_2$ holds

$$
\begin{aligned}
R_{d_1}(t) &\leq R_{d_2}(t),\\
h_{d_1}(t) &\geq h_{d_2}(t)\\
MRL_{d_1}(t) &\leq MRL_{d_2}(t),\\
T_{d_1} &\preccurlyeq T_{d_2},\\
MTTF_{d_1} &\leq MTTF_{d_2},\\
SDTTF_{d_1}(t) &\leq SDTTF_{d_2}(t).
\end{aligned}
$$

Example 5.30 Example 5.29 is repeated using the DFR components of Example 4.10. The results are presented in Figure 5.28.

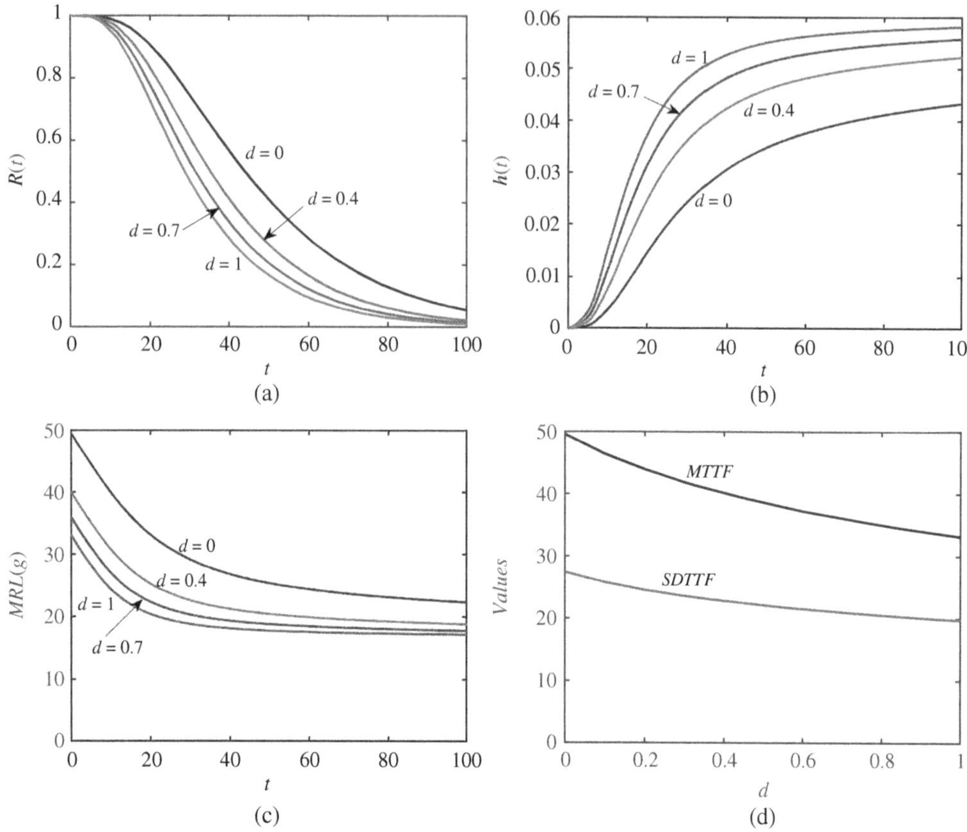

Figure 5.28 Reliability measures of the system in Example 5.30. (a) System reliability function, (b) system hazard function, (c) system *MRL* function, and (d) system *MTTF* and *SDTTF*.

The graphs in Figure 5.28 show that a system with DFR components behaves similarly to systems with IFR components. That is, system reliability measures improve as the warmness level decreases. ■

It is also possible to generate a simpler survival matrix of homogeneous K-out-of-N general standby systems. For such a system, the survival matrices and failure vectors of the last $N-K$ components are adjusted by the same deceleration matrix, while the first K components are not. The notation to express the survival matrix of an arbitrary homogeneous K-out-of-N general standby system is similar to K-out-of-N system with adjustments. The adjustments are the first K components in any state are primary that are without adjustment, and the remaining components are backups that need adjustment. The following example illustrates a method to derive the survival matrix of a homogeneous K-out-of-N general standby system from the survival matrix of a K-out-of-N system with identical components.

Example 5.31 The survival matrix of a homogeneous 2-out-of-4 general standby system can be derived using Equation (5.13) for a 2-out-of-4 system with identical components by adjusting the survival matrices and failures vectors of the backup components at each state. The result is

$$A = \begin{bmatrix} A_{1,1} & A_{1,2} & 0 \\ 0 & A_{2,2} & A_{2,3} \\ 0 & 0 & A_{3,3} \end{bmatrix},$$

where

$$A_{1,1} = A \oplus A \oplus A' \oplus A', A_{2,2} = A \oplus A \oplus A', A_{3,3} = A \oplus A,$$

$$A_{1,2} = b \otimes I \otimes I \otimes I + I \otimes b \otimes I \otimes I + I \otimes I \otimes b' \otimes I + I \otimes I \otimes I \otimes b',$$

$$A_{2,3} = b \otimes I \otimes I + I \otimes b \otimes I + I \otimes I \otimes b'.$$

Using notations in Section 5.1.4 expresses the block matrices recursively as follows:

$$A_{2,2} = A \oplus A \oplus A' = A_{3,3}A', A_{1,1} = A \oplus A \oplus A' \oplus A' = A_{2,2} \oplus A'.$$

$$A_{1,2} = (b \otimes I \otimes I + I \otimes b \otimes I + I \otimes I \otimes b') \otimes I + I \otimes I \otimes I \otimes b' = A_{2,3} \otimes I + (\otimes I)^3 b'$$

For simplicity, $A_{2,3}$ is written as $\sum \otimes (I^2, b, b')$, with b's are for the first K good components and b''s are for the remaining good components. ■

The algorithm for homogeneous general standby systems can be derived from Algorithm 5.2 by adjusting the survival matrices and failure vectors of backup components using their warmness levels to produce Algorithm 5.4. Using the result of Example 5.31, Algorithm 5.4 is given by the following recursion equations:

$$n_S = n_S n + n_S, \tag{5.51}$$

$$a_S = \left[a_S \otimes a \;\; 0\right], \tag{5.52}$$

$$I_S = I_S \otimes I, \tag{5.53}$$

$$A_S(N-k, N-k) = A_S(N-k+1, N-k+1) \oplus A', \tag{5.54}$$

$$A_S(N-k-1, N-k) = A_S(N-k, N-k+1) \otimes I + I_S \otimes b', \tag{5.55}$$

for $k = K, K+1, \ldots, N-1$, except Equation (5.55) is until $k = N-2$, with initial conditions $n_S = n^K$, $a_S = (\otimes a)^K$, $A_S = 0$,$A_S(N-K+1, N-K+1) = (\oplus A)^K$, $A_S(N-K, N-K+1) = \sum \otimes (I^K, b, b')$, $I_S = (\otimes I)^K$.

A pseudocode for Algorithm 5.4 is as follows:

Algorithm 5.4

---

input $N, K, D, (n, a, A)$
$n_S = n^K$
$a_S = (\otimes a)^K$
$b = -Au$
$I_S = (\otimes I)^K$
$A' = DA$
$b' = Db$
$A_S = [0_{n \times n}]_{(N-K+1)\times(N-K+1)}$
$A_S(N-K+1, N-K+1) = (\oplus A)^K$
$A_S(N-K, N-K+1) = \sum \otimes (I^K, b, b')$

for $k = K$ to $N - 1$
 $n_S = n_S n + n_S$
 $\boldsymbol{a}_S = [\boldsymbol{a}_S \otimes \boldsymbol{a}\ \ \boldsymbol{0}]$
 $\boldsymbol{I}_S = \boldsymbol{I}_S \otimes \boldsymbol{I}$
  $\boldsymbol{A}_S(N-k, N-k) = \boldsymbol{A}_S(N-k+1, N-k+1) \oplus \boldsymbol{A}'$
 if $k < N - 1$ then
  $\boldsymbol{A}_S(N-k-1, N-k) = \boldsymbol{A}_S(N-k, N-k+1) \otimes \boldsymbol{I} + \boldsymbol{I}_S \otimes \boldsymbol{b}'$
 end
 $k = k + 1$
end
end

Algorithm 5.4 is programmed in MATLAB as **phgenidnsys** function.

Exercise 5.22 Generate the survival matrix of a 3-out-of-5 general standby system with identical components and warmness levels using Algorithm 5.4.

Example 5.32 To show that **phgenidnsys** and **phgensys** functions produce identical PH representations, both are applied to 2-out-of-N and 3-out-of-N homogeneous general standby systems having 2-Erlang distributed lifetimes with $MTTF = 2$ and warmness levels of the backup components are 0.5. The resulting reliability functions are given as Figure 5.29, which shows they are identical.

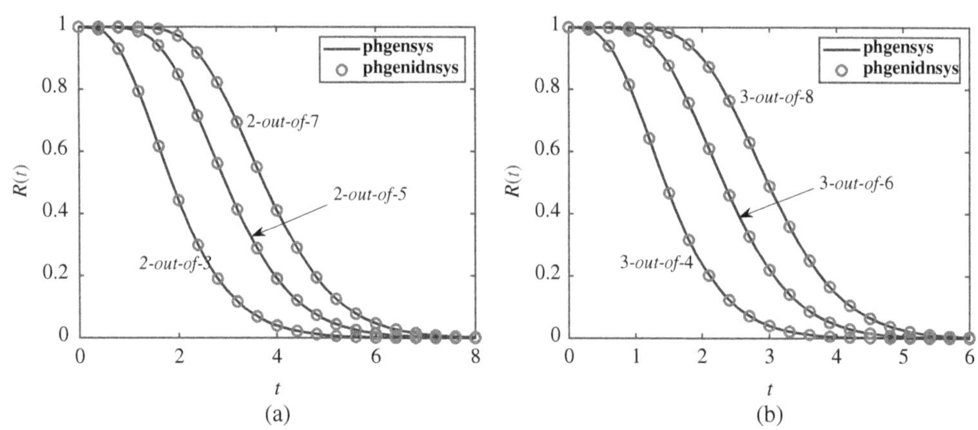

Figure 5.29 Reliability functions of homogeneous K-out-of-N general standby systems calculated using phgensys and phgenidnsys. (a) 2-out-of-N systems and (b) 3-out-of-N systems. ■

As for K-out-of-N systems, the PH distribution of the lifetime of a general standby system with identical components and warmness levels obtained using **phgenidnsys** has a smaller order than obtained using **phgensys**. Therefore, it is better to use **phgenidnsys** than **phgensys** whenever possible.

Exercise 5.23 For a homogeneous 2-out-of-6 general standby system having lifetimes following 4-Erlang distribution, compare the order of the PH distribution of the system lifetime generated using **phgensys** and **phgenidnsys**, and the computation time for producing system reliability function and calculating $MTTF$.

The optimum number of backup components that minimize the total costs of operation and failure in general standby systems can be found in a similar manner as for K-out-of-N systems.

Application 5.3 Optimum numbers of backup systems in homogeneous general standby systems will be determined using the same cost parameters as in Application 5.1 for $K = 1$, 2, and $d = 0$, 0.5, 1.0. The results are given in Table 5.8.

Table 5.8 Optimum numbers of components in general standby systems.

	Mission duration = 20 days					
	$d = 0$		$d = 0.5$		$d = 1$	
N	$K = 1$	$K = 2$	$K = 1$	$K = 2$	$K = 1$	$K = 2$
1	123.31	–	123.31	–	123.31	–
2	30.23	182.42	52.37	182.42	84.19	182.42
3	**30.18**	103.06	**37.10**	122.25	66.37	149.84
4	40	51.22	41.38	77.59	**60.60**	123.66
5	50	**50.71**	50.25	**62.18**	61.67	106.32
6	60	60.02	60.04	63.38	66.61	96.97
7	70	70	70.01	70.84	73.75	**93.81**
8	80	80	80	80.19	82.12	95.11

Optimum values are given in bold.

Table 5.8 shows that the general standby system with two primary components and warmness level = 0.5 needs to be supported by $N - K = 5 - 2 = 3$ backup systems to minimize the total cost of operation and failure for carrying a mission of 20 days. As seen from the table, the optimum number of backup systems increases with the warmness level of the backup systems. ■

Exercise 5.24 Investigate the effect of component *MTTF* to the optimum number of backup components.

Next, the method for general standby systems in PH distribution is generalized to ME distribution.

5.3.3 General Standby System in ME Distribution

Algorithm 5.3 is applicable for ME distribution with an additional step for calculating the closing vector $\boldsymbol{v}$ in the ME representation, the same as for **mekofnsys**. Likewise, the function is referred to as **megensys**.

Example 5.33 A 1-out-of-3 general standby system is constructed using components in Examples 3.1, 3.2, and 3.3 as components 1, 2, and 3, respectively. The backup components are given several warmness levels: all 0, all 0.5, and [0.5,0]. The ME distributions for the component lifetimes are obtained from their PH distributions using their respective matrix similarity transformation. The PH and ME distributions for the system lifetimes are generated using **phgensys** and **megensys**. The resulting PH and ME representations are utilized to generate the reliability and hazard functions in Figure 5.30, which shows they are identical.

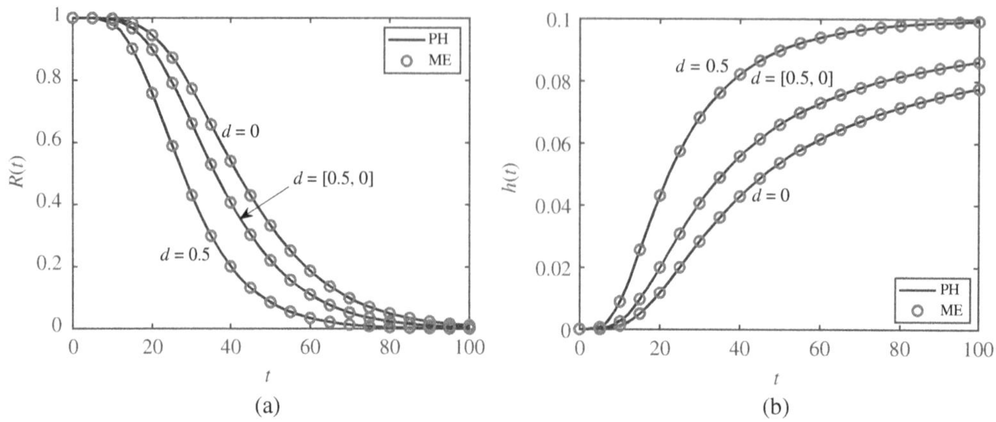

Figure 5.30 Comparison of using PH and ME distribution for general standby systems in Example 5.33. (a) System reliability function and (b) system hazard function. ■

Exercise 5.25 Compare the computation times of using PH and ME distributions in calculating reliability measures of K-out-of-N general standby systems for several values of K and N.

A method to analyze reliability of systems comprises subsystems with different structures involving general standby structures is developed in the next sections. The development starts with simple compound structures.

5.3.4 Simple Compound General Standby Structured Systems

A simple compound general structured system can be modeled in two-level structures, one of which is a general standby structure.

Example 5.34 The system in Example 5.15 is made into a general standby system by providing different warmness levels to backup devices.

The simplified system reliability model in the provided MATLAB functions is

$$T_{S_i} = \text{phseriessys}(T_{1_i}, T_{2_i}, T_{3_i}) \quad i = 1, 2, 3, 4,$$
$$T_S = \text{phgensys}(T_{S_1}, T_{S_2}, T_{S_3}, T_{S_4}; 2; d_1, d_2).$$

where d_1 and d_2 are the warmness levels of backup devices 1 and 2, respectively.

Likewise, the system in Example 5.17 can also be made into a general standby system by providing different warmness levels to backup modules.

The simplified system reliability model in the provided MATLAB functions is

$$T_{S_i} = \text{phgensys}\,(T_{i_1}, T_{i_2}, T_{i_3}, T_{i_4}; 2; d_{1_i}, d_{2_i}), \quad i = 1, 2, 3,$$
$$T_S = \text{phseriessys}\,(T_{S_1}, T_{S_2}, T_{S_3}).$$

where d_{1_i} and d_{2_i} are the warmness levels of backup modules 1 and 2, respectively, in the ith device. ■

Example 5.35 The reliability and hazard functions of the systems in Example 5.34 with the same distribution as in Example 5.16 are compared for the warmness level $d_{1_i} = d_1 = 0.8$ and $d_{2_i} = d_2 = 0.4$ for i = 1, 2, 3. A comparison of the resulting reliability and hazard functions is depicted in Figure 5.31.

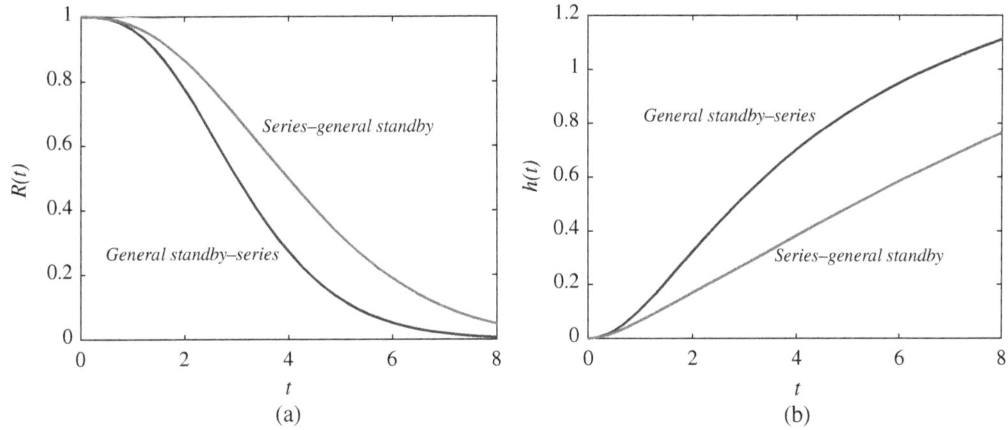

Figure 5.31 Reliability and hazard functions of series-(general standby) and (general standby)-series systems. ■

The graphs in Figure 5.31 show that (general standby)–series system has a lower reliability function and a higher hazard function than the series–(general standby) system, the same behavior as series-(K-out-of-N) and (K-out-of-N)-series systems. It means that the lifetime of the (general standby)–series system is stochastically shorter than the lifetime of the series–(general standby) system.

$$R_{g-s}(t) \leq R_{s-g}(t), \tag{5.56}$$

$$h_{g-s}(t) \geq h_{s-g}(t), \tag{5.57}$$

$$T_{g-s} \preccurlyeq T_{s-g}, \tag{5.58}$$

where g-s and s-g denote the (general standby)–series and series–(general standby), respectively. The residual reliability function, the *MRL* function, and *MTTF* have the same behavior as the reliability function.

Exercise 5.26 Show that Equations (5.56)–(5.58) apply for all types of components (DFR, CFR, and IFR).

Exercise 5.27 Derive the relationships for other reliability measures.

Example 5.36 The production line in Application 5.2 is made into a series–(general standby) system by providing warmness levels to backup machines.

The system reliability model of the production line in the provided MATLAB function is

$$T_{S_1} = \text{phgensys}(T_1, T_1, T_1, T_1; 2; d_{11}, d_{21}),$$

$$T_{S_2} = \text{phgensys}(T_2, T_2, T_2; 1; d_{12}, d_{22}),$$

$$T_{S_3} = \text{phgensys}(T_3, T_3, T_3, T_3, T_3; 3; d_{13}, d_{23}),$$

$$T_{S_4} = \text{phgensys}(T_4, T_4, T_4; 2; d_{14}),$$

$$T_S = \text{phseriessys}(T_{S_1}, T_{S_2}, T_{S_3}, T_{S_4}).$$

where d_{ij} is the warmness level of backup machine i at workstation j. ■

Example 5.37 Machines in workstations 1, 2, 3, and 4 have the same lifetime distributions as in Application 5.2. Each backup component is given a warmness level of 0.8 for the first backup and 0.4 for the second backup. Thus, $d_{11} = d_{12} = d_{13} = d_{14} = 0.8$, and $d_{21} = d_{22} = d_{23} = 0.4$.

The PH representation of the distribution of system lifetime is generated using **phgensys**. The result is used to calculate and plot the system reliability and hazard functions in Figure 5.32. The system *MTTF* and *SDTTF* are also calculated to be 13.00 and 6.61 days for $n = 1$ and 15.97 and 5.65 days for $n = 2$.

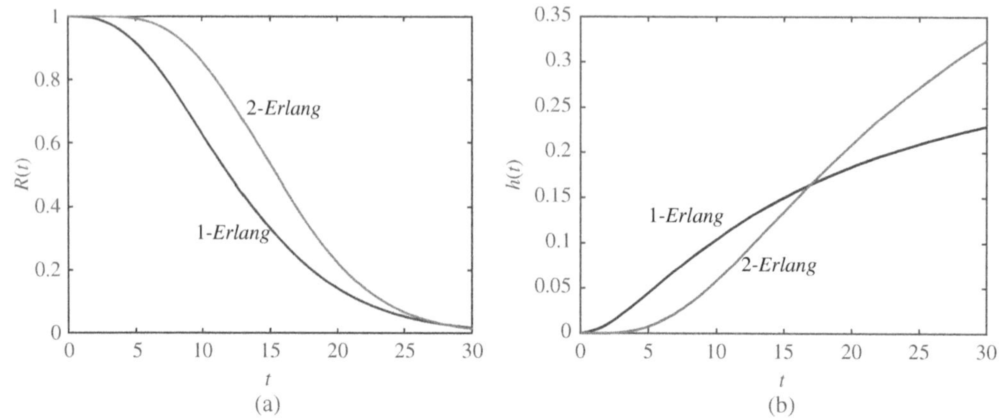

Figure 5.32 Reliability and hazard functions of the system in Example 5.37. (a) System reliability function and (b) system hazard function. ■

Exercise 5.28 Investigate the effect of reversing the warmness levels of the first and second backup components to the system reliability and hazard functions. Draw a conclusion from the result.

The modeling process of compound structure systems can be simplified by using only general standby function as a unifying model for all structures as described in the next section.

5.3.5 General Standby System as a Unifying Model

All models discussed in Chapter 4 and Section 5.1 can be considered as special forms of the general standby model. The series structure is an N-out-of-N general standby structure, and the parallel structure is a 1-out-of-N general standby structure with a warmness level of 1. Likewise, the K-out-of-N system is a K-out-of-N general standby system with a warmness level of 1. The use of general standby system as unifying models for systems in previous examples is presented.

A combination of general standby structures as a model for the (K-out-of-N)-series system in Example 5.34 is

$$T_{S_i} = \text{phgensys}(T_{1_i}, T_{2_i}, T_{3_i}; 3) \quad i = 1, 2, 3, 4,$$

$$T_S = \text{phgensys}(T_{S_1}, T_{S_2}, T_{S_3}, T_{S_4}; 2; d_1, d_2).$$

The series-(K-out-of-N) system in Example 5.34 modeled in a combination of only general standby system structures results in

$$T_{S_i} = \text{phgensys}(T_{i_1}, T_{i_2}, T_{i_3}, T_{i_4}; 2; d_1, d_2) \quad i = 1, 2, 3,$$

$$T_S = \text{phgensys}(T_{S_1}, T_{S_2}, T_{S_3}; 3).$$

The model of production line in Example 5.36 in a combination of several general standby structures is

$$T_{S_1} = \text{phgensys}(T_1, T_1, T_1, T_1; 2; d_{11}, d_{21}),$$

$$T_{S_2} = \text{phgensys}(T_2, T_2, T_2; 1; d_{12}, d_{22}),$$

$$T_{S_3} = \text{phgensys}(T_3, T_3, T_3, T_3, T_3; 3; d_{13}, d_{23}),$$

$$T_{S_4} = \text{phgensys}(T_4, T_4, T_4; 2; d_{14}),$$

$$T_S = \text{phgensys}(T_{S_1}, T_{S_2}, T_{S_3}, T_{S_4}; 4).$$

Example 5.38 A manufacturing company uses a production line with four workstations. Each workstation has several primary and backup machines. All machines at each workstation are identical. Their data are presented in Table 5.9. The reliability function of workstation 1 is $R(t) = e^{-(0.004555t)^{1.45}}$, which is a Weibull distribution with $MTTF \approx 200$ and $CVTTF \approx 0.7$.

Table 5.9 Workstation data for production line in Example 5.38.

Work station	Distribution of machine lifetime	Machine *MTTF*	Machine *CVTTF*	Number of primary machines	Number of backup machines	Backup warmness level
1	Weibull	200	0.7	3	1	0.8
2	2-Erlang	150	$1/\sqrt{2}$	1	2	1.0, 0.5
3	Exponential	250	1	3	2	0.7, 0.0
4	2-Coxian	120	1.2	2	1	0.5

A PH distribution approximation for the Weibull distribution of the lifetime of workstation 1 is given by (see Section 2.5)

$$n = 3, \quad \boldsymbol{a} = \begin{bmatrix} 1 & 0 & 0 \end{bmatrix}, \quad \boldsymbol{A} = \begin{bmatrix} -0.0123 & 0.0068 & 0.0045 \\ 0 & -0.0123 & 0.0123 \\ 0 & 0 & -0.0123 \end{bmatrix}.$$

A system reliability model for this production line in **phgensys** functions is

$$T_{S_1} = \text{phgensys}(T_{11}, T_{12}, T_{13}, T_{14}; 3; 0.8)$$

$$T_{S_2} = \text{phgensys}(T_{21}, T_{22}, T_{23}; 1; 1.0,0.5)$$

$$T_{S_3} = \text{phgensys}(T_{31}, T_{32}, T_{33}, T_{34}, T_{35}; 3; 0.7,0.0)$$

$$T_{S_4} = \text{phgensys}(T_{41}, T_{42}, T_{43}; 2; 0.5)$$

$$T_S = \text{phgensys}(T_{S_1}, T_{S_2}, T_{S_3}, T_{S_4}; 4)$$

The model is employed to calculate the PH representation of the system lifetime distribution, which is then used to generate graphs of the reliability and hazard functions of each workstation and the production line in Figure 5.33.

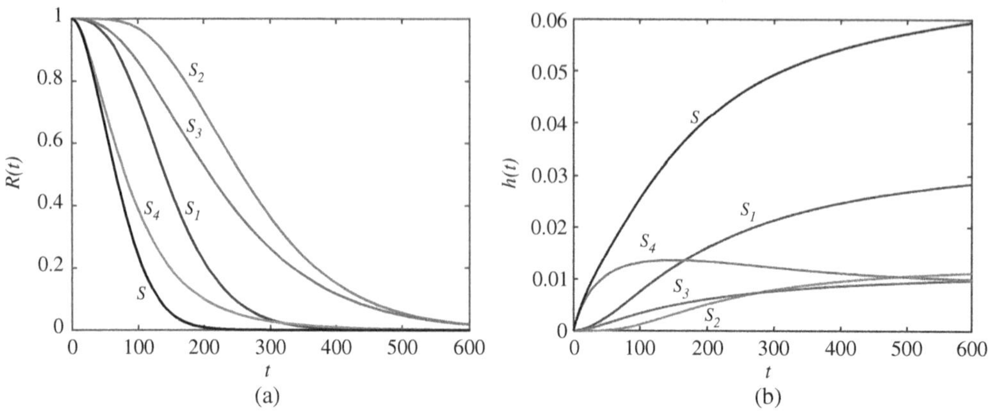

Figure 5.33 Reliability and hazard functions of the workstations and system in Example 5.38. (a) Reliability function and (b) hazard function.

The main program for this example is given in **example5_38.m** file on the software package **dysrel-am**. Its source code is as follows:

MATLAB code for example5_38.m

```
clear;close all
addpath('../dysrel-am','../dysrel-am/expokit')

% PH Components                                                  input
% Weibull R(t) = exp(-0.1t^1.5) (Trivedi Bobbio)
T(1).a = [1 0 0];
T(1).A = 9.0947/200*[-0.27 0.147 0.099;0 -0.27 0.27;0 0 -0.27];
T(1).n = 3;
T(2) = phnErlang(2,150);
T(3) = phnErlang(1,250);
T(4) = ph2Coxian(1.2,120);
% System PH representation                                       model
S(1) = phgensys([T(1),T(1),T(1),T(1)],3,0.8);
S(2) = phgensys([T(2),T(2),T(2)],1,[1 0.5]);
S(3) = phgensys([T(3),T(3),T(3),T(3),T(3)],3,[0.7 0]);
S(4) = phgensys([T(4),T(4),T(4)],2,0.5);
S(5) = phgensys(S(1:4),4);
% Define time span                                               output
tmax = 600;
t = [0:4:24 linspace(28,tmax,21)];
U = []; V = [];
for j = 1:5
[Rt,ht] = phrelhaz(t,S(j));
[x,u] = smoothing(t,Rt);
[x,v] = smoothing(t,ht);
U = [U u']; V = [V v'];
end
% Plots
lgnd = {'S_1';'S_2';'S_3';'S_4';'S'};
multiplot(5,x,U,lgnd,'t','R(t)');
multiplot(5,x,V,lgnd,'t','h(t)');
```

■

Example 5.39 Even though all machines in each workstation are identical, this production line cannot be modeled only using **phgenidnsys** because the warmness levels for backup components in workstations 2 and 3 are not identical. This production line can be modeled in a combination of **phgenidnsys** for workstations 1 and 4 and **phgensys** for workstations 2 and 3. Comparisons of PH orders and computation times between using mixed **phgenidnsys-phgensys** and using only **phgensys** for all workstations are presented in Table 5.10. As seen, the computation time using mixed **phgenidnsys-phgensys** is three times faster.

Table 5.10 PH orders and computation times for Example 5.39 using phgensys and phgenidnsys.

Model	PH order	Computation time (seconds)
phgensys	1,572,480	230.99
phgenidnsys-phgensys	539,136	73.03

■

Exercise 5.29 Make the same comparison as Table 5.10 for the case all backup components are in cold standby. Draw a conclusion from the result.

The simple compound structure is made complex in the next section.

5.3.6 Complex Compound Structured Systems

Systems with non-simple compound structures require more hierarchies of subsystems and sub-subsystems in different structures. The sub-subsystems and subsystems may have series, parallel, K-out-of-N, and general standby structures. As explained in Section 4.4, the PH representation of lifetime distribution of a complex structured system can be calculated recursively in a hierarchical manner using a combination of the provided MATLAB functions. Modeling a complex structured system is illustrated with the following applications.

Application 5.4 A power supply-demand system in a factory consists of three identical generators operated simultaneously, even though the power supplied from two generators is sufficient to meet the demand of three production lines. The power from the generators is distributed to three identical production lines using one of two available distribution channels. The two channels are operated in a 2-standby structure with an imperfect switch having reliability of 0.9. Each production line has three compact machines in series. The factory needs one production line to meet the production target. The second and third production lines are backups in warm standby with warmness levels d_1 and d_2, respectively. The system reliability block diagram is as presented in Figure 5.34. Identical components are denoted using the same numbers.

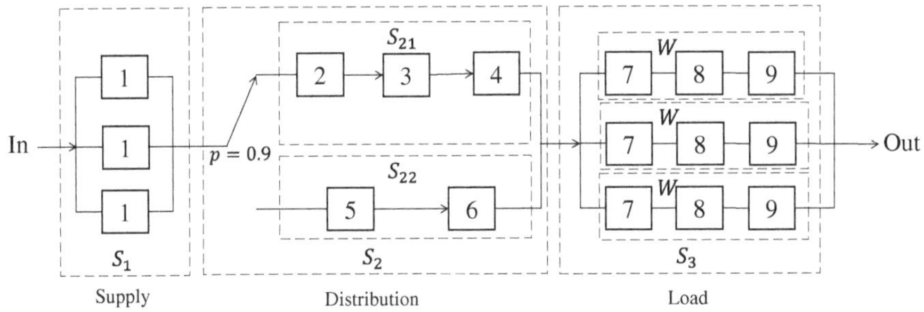

Figure 5.34 Reliability block diagram of power supply-demand in Application 5.4.

The system reliability model using the provided MATLAB functions is

$$T_{S_1} = \text{phkofnsys}(T_1, T_1, T_1; 2)$$
$$S_{21} = \text{phseriessys}(T_2, T_3, T_4)$$
$$S_{22} = \text{phseriessys}(T_5, T_6)$$
$$T_{S_2} = \text{phstandsys}(S_{21}, S_{22}; 0.9)$$
$$T_W = \text{phseriessys}(T_7, T_8, T_9)$$
$$T_{S_3} = \text{phgensys}(T_W, T_W, T_W; 1; d_1, d_2)$$
$$T_S = \text{phseriessys}(T_{S_1}, T_{S_2}, T_{S_3})$$

The model for S_1 may use **phkofnidnsys**, but not for S_3 since the backup components have different warmness levels.

All components in this system have PH-distributed lifetimes. All generators have lifetimesfollowing an identical 2-Erlang distributions with MTTF = 100 days. The lifetimes of components 2, 3, 4, 5, and 6 in the distribution subsystem follow exponential distributions with $MTTF$ = 200, 300, 400, 150, and 250 days, respectively. Machines 7, 8, and 9 have 2-Coxian distributed lifetimes with $MTTF$ = 25 days and $CVTTF$ = 1.1, 1.2, and 1.3, respectively. The values of d_1 and d_2 are set at 1 and 0, respectively.

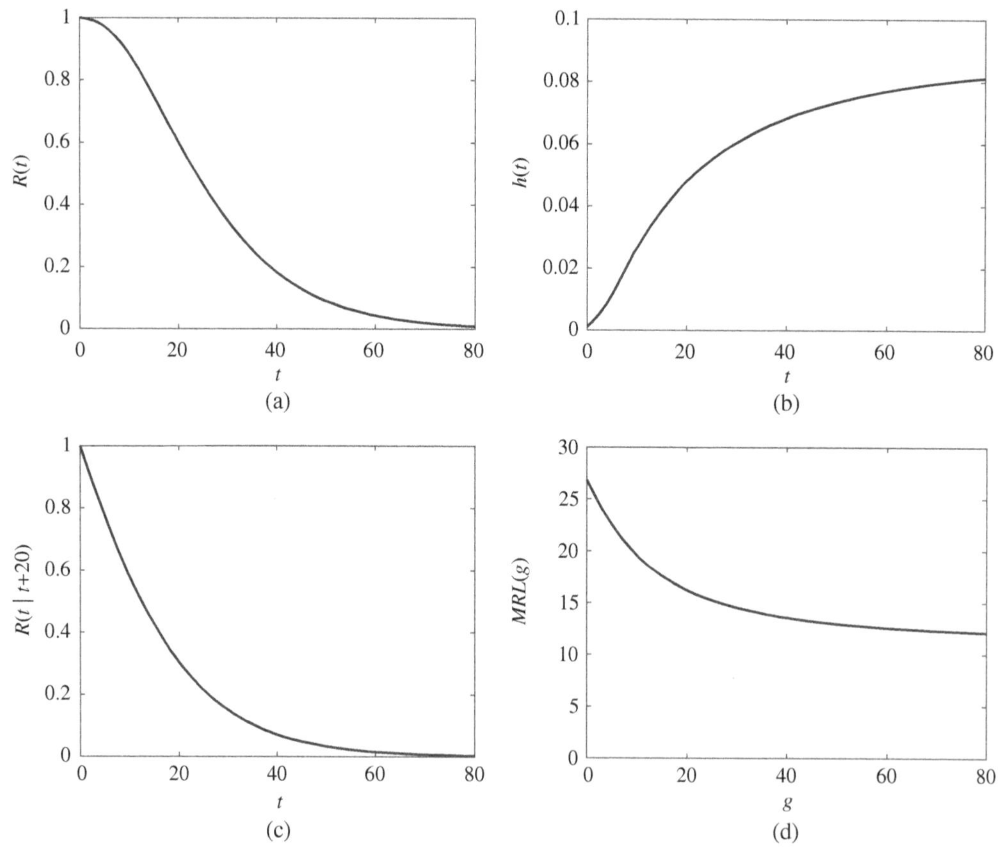

Figure 5.35 Reliability measures of the system in Application 5.4. (a) System reliability function, (b) system hazard function, (c) residual system reliability function age 20 days, and (d) system *MRL* function.

The model is employed to calculate the PH representation of the system lifetime distribution. The resulting PH representation is used to produce graphs of some system reliability measures as presented in Figure 5.35. Calculations of the system *MTTF* and *SDTTF* yield 61.93 and 32.15 days, respectively.

The effect of warmness levels of backup production lines on the system reliability measures is investigated by varying their warmness levels.

The energy consumption of the production lines depends on the warmness levels of the backup components. Assume that the remaining power after being consumed by the primary components is sufficient to meet the needs of a total warmness level equal to 1, or $d_1 + d_2 = 1$. The model is used to calculate the system reliability measures for several values of d_1 and d_2. The results are shown in Figure 5.36.

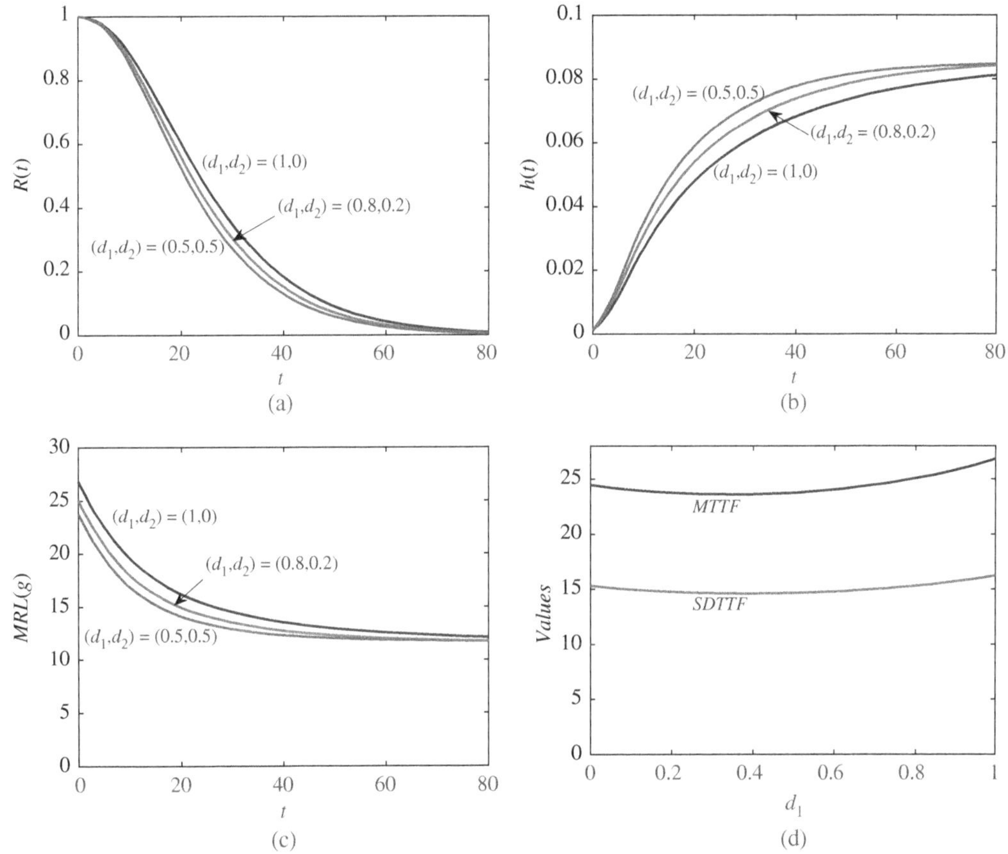

Figure 5.36 Reliability measures of system in Application 5.4 for several warmness levels. (a) System reliability function, (b) system hazard function, (c) system *MRL* function, (d) system *MTTF* and *SDTTF*.

The graphs in Figure 5.36 show that the system reliability measures are better by giving a higher warmness level for the first backup production line. Interestingly, Figure 5.36d shows that the system is worst not at $d_1 = 0$, but at the value of d_1 that makes the *MTTF* smallest, which is $d_1 = 0.35$, and therefore $d_2 = 0.65$; this means the second backup is warmer than the first backup. As will be shown later, the optimum warmness level ordering for identical backup components is a decreasing order; backup components earlier in the sequence should have higher warmness levels. Figure 5.36d shows that system $MTTF$ is higher for $d_1 > 0.5$, and reaches a maximum value at $d_1 = 1$, thus $d_2 = 0$.

This system can also be modeled using only the **phgensys** function as the unifying model to simplify the modeling process. However, **phgensys** can handle imperfect switches only for the cold standby structure in Section 4.4. Its extension to a general standby system is left as future research for interested readers.

The reliability model for this system in **phgensys** functions for all structures is

$$T_{S_1} = \text{phgensys}(T_1, T_1, T_1; 2; 1)$$

$$S_{21} = \text{phgensys}(T_2, T_3, T_4; 3)$$

$$S_{22} = \text{phgensys}(T_5, T_6; 2)$$

$$T_{S_2} = \text{phgensys}(S_{21}, S_{22}; 1; 0; 0.9)$$

$$T_W = \text{phgensys}(T_7, T_8; 2)$$

$$T_{S_3} = \text{phgensys}(T_W, T_W, T_W; 1; d_1, d_2)$$

$$T_S = \text{phgensys}(T_{S_1}, T_{S_2}, T_{S_3}; 3)$$

The system reliability and hazard functions calculated using the first and second models for Application 5.4 with $d_1 = 1$, $d_2 = 0$ produce identical graphs, as presented in Figure 5.37.

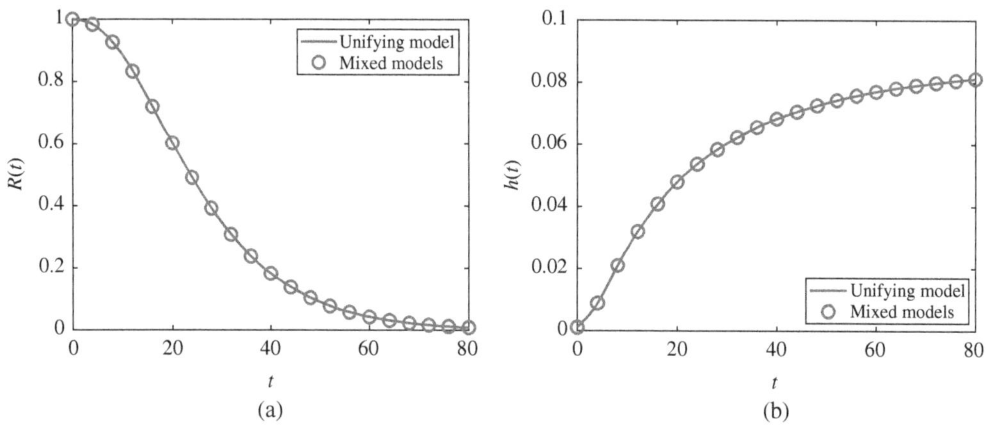

Figure 5.37 Comparison of reliability and hazard functions of Application 5.4 calculated using only phgensys function and combination of several functions. (a) Reliability function and (b) hazard function.

A comparison of complexities between using a specific model for each structure and the unifying model for all structures to produce Figure 5.37 is presented in Table 5.11.

Table 5.11 PH orders and computation times of Application 5.4.

Model	PH order	Computation time (seconds)
Specific models for each structure	4,960	0.3469
The unifying model for all structures	7,440	0.4081

Table 5.11 shows the trade-off between the convenience and efficiency of using the general standby system as a unifying model. The convenience of the general standby system as a model for the entire system comes at the expense of increased PH order and computation time. The PH order increases by around 50%, and the computation time increases by almost 20%. This increase will be more pronounced as the system becomes more complex and the PH orders of the component lifetime distributions are higher. Thus, using a general standby system as a unifying model is suitable for systems that are not too complex or the orders of PH distributions of the component lifetimes are not too high. ■

Exercise 5.30 Reproduce Figure 5.36a–c for $(d_1, d_2) = (0, 1), (0.1,0.9)$, and $(0.2,0.8)$, respectively. Draw a conclusion from the results.

Application 5.5 A transmission system used to transmit power from three identical power plants to supply one region has a reliability block diagram as given in Figure 5.38. Assume each power plant has a capacity of 100 MVA and lifetime following 2-Erlang distribution with $MTTF = 200$ weeks. All transformers are identical with lifetimes follow 2-Erlang distributions with $MTTF = 200$ weeks and all stabilizers are identical with lifetimes follow exponential distribution with $MTTF = 300$ weeks. Likewise, assume the two protections are identical with lifetimes follow 2-Erlang distribution with $MTTF = 400$ weeks and the two connectors are also identical with lifetimes follow 3-Erlang distribution with $MTTF = 300$ weeks. Use G, PC, and L to denote the generator, protection-connection, and load subsystems, respectively.

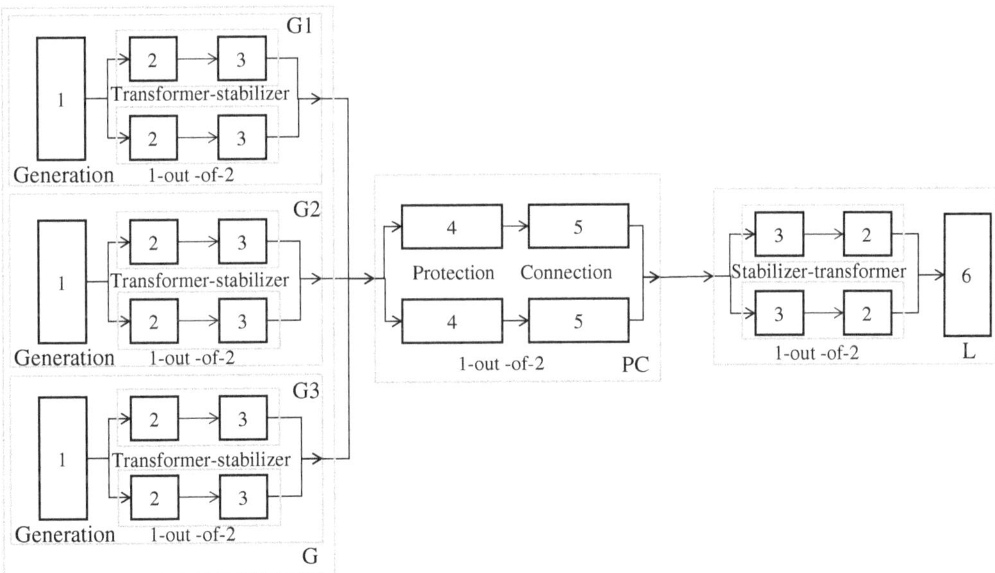

Figure 5.38 HV transmission line from three power plants to supply a region.

The three generators are in a K-out-of-3 general standby structure, with the value of K depends on the load. The system model using only **phgensys** functions is

$$S_{TS1} = \text{phgensys}([T_2, T_3], 2)$$

$$S_{TS2} = \text{phgensys}([T_2, T_3], 2)$$

$$S_{TS} = \text{phgensys}([S_{TS1}, S_{TS2}], 1{,}1.0)$$
$$T_{G1} = \text{phgensys}([T_1, S_{TS}], 2)$$
$$T_{G2} = \text{phgensys}([T_1, S_{TS}], 2)$$
$$T_{G3} = \text{phgensys}([T_1, S_{TS}], 2)$$
$$T_G = \text{phgensys}([T_{G1}, T_{G2}, T_{G3}], K, [D_1, D_2])$$
$$S_{PC1} = \text{phgensys}([T_4, T_5], 2)$$
$$S_{PC2} = \text{phgensys}([T_4, T_5], 2)$$
$$T_{PC} = \text{phgensys}([S_{PC1}, S_{PC2}], 1{,}1.0)$$
$$S_{ST1} = \text{phgensys}([T_2, T_3], 2)$$
$$S_{ST2} = \text{phgensys}([T_2, T_3], 2)$$
$$T_{ST} = \text{phgensys}([S_{ST1}, S_{ST2}], 1{,}1.0)$$
$$T_S = \text{phgensys}([T_G, T_{PC}, T_{ST}], 3)$$

Note that the value 1.0 in the models for T_{TS}, T_{PC}, and T_{ST} is the switch reliability.

The system reliability model can be simplified by using the same model for identical components and subsystems, independent from the component sequence; hence, subsystem S_{TS1}, S_{TS2}, S_{ST1}, and S_{ST2} are also identics. The simplified model is

$$S_{TS1} = \text{phgensys}([T_2, T_3], 2)$$
$$S_{TS} = \text{phgensys}([S_{TS1}, S_{TS1}], 1{,}1.0)$$
$$T_{G1} = \text{phgensys}([T_1, S_{TS}], 2)$$
$$T_G = \text{phgensys}([T_{G1}, T_{G1}, T_{G1}], K, [D_1, D_2])$$
$$T_{PC1} = \text{phgensys}([T_4, T_5], 2)$$
$$T_{PC} = \text{phgensys}([T_{PC1}, T_{PC1}], 1{,}1.0)$$
$$T_S = \text{phgensys}([T_G, T_{PC}, S_{TS}], 3)$$

The software package can accept the simplified model as well.

Suppose that the load is constant at all times during the system operation. Several possible scenarios are examined.

Scenario 1 Assume that all generators are operated in full capacity, which means $D_1 = D_2 = 1$. The values of K are determined as follows:

- Load ≤ 100: $K = 1$. The generator subsystem model is

$$T_G = \text{phgensys}([T_{G1}, T_{G1}, T_{G1}], 1, [1, 1])$$

- $100 <$ Load ≤ 200: $K = 2$. The generator subsystem model is

$$T_G = \text{phgensys}([T_{G1}, T_{G1}, T_{G1}], 2, 1)$$

- $200 <$ Load ≤ 300: $K = 3$. The generator subsystem model is

$$T_G = \text{phgensys}([T_{G1}, T_{G1}, T_{G1}], 3)$$

Executing the model using the given data will produce the PH representation of the system lifetime distribution, from which the system reliability and hazard functions to fulfill the load was calculated and plot as presented in Figure 5.39.

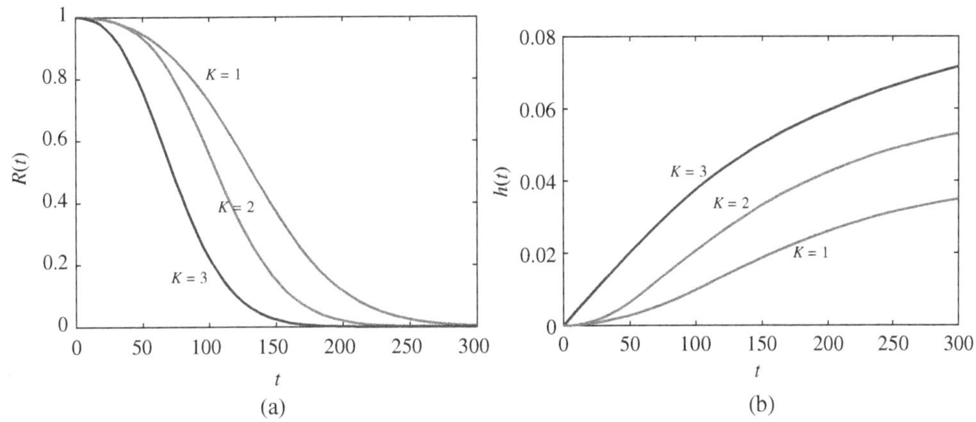

Figure 5.39 System reliability and hazard functions with hot standby for different loads. (a) Reliability function and (b) hazard function.

Scenario 2 Suppose the generators are operated only when necessary, they are in cold standby when not needed and are operated at full capacity when needed. With primary components always having a deceleration factor equal to 1, the values of K, D_1, and D_2 will be as follows:

- Load ≤ 100: $K = 1$, $D_1 = 0$, $D_2 = 0$. The generator subsystem model is

$$T_G = \text{phgensys}([T_{G1}, T_{G1}, T_{G1}], 1, [0, 0])$$

- $100 < \text{Load} \leq 200$: $K = 2$, $D_2 = 0$ (with one backup generator). The generator subsystem model is

$$T_G = \text{phgensys}([T_{G1}, T_{G1}, T_{G1}], 2, 0)$$

- $200 < \text{Load} \leq 300$: $K = 3$ (without backup generator). The generator subsystem model is

$$T_G = \text{phgensys}([T_{G1}, T_{G1}, T_{G1}], 3)$$

The system reliability and hazard functions for this scenario are given in Figure 5.40.

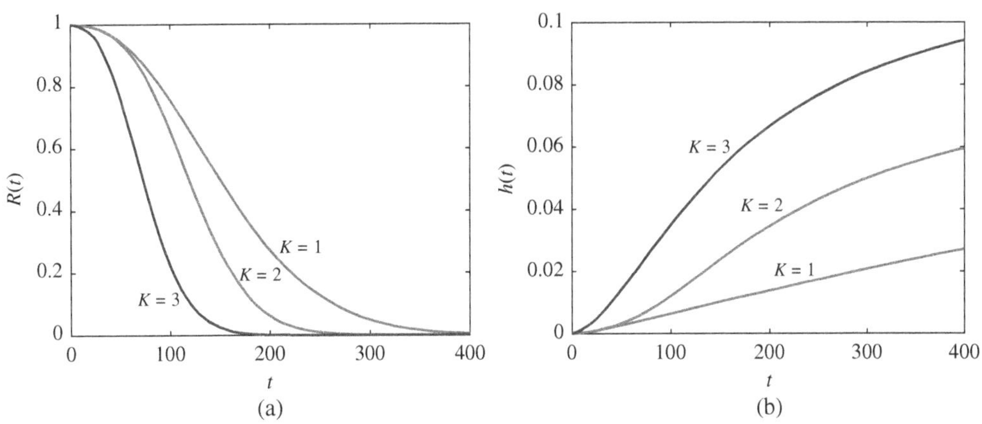

Figure 5.40 System reliability and hazard functions with cold standby for different load. (a) Reliability function and (b) hazard function.

Scenario 3 Suppose the backup generators can be operated at half power whenever sufficient. Assume that operating at half power reduces the deterioration process of the system condition by the same amount as the reduction in the operational capacity. Thus, the values of K, D_1, and D_2 are as follows:

1) Load ≤ 100: $K = 1, D_1 = 0, D_2 = 0$.
2) $100 <$ Load ≤ 150: $K = 2, D_1 = 0.5, D_2 = 0$.
3) $150 <$ Load ≤ 200: $K = 2, D_1 = 1, D_2 = 0$.
4) $200 <$ Load ≤ 250: $K = 3, D_1 = 1, D_2 = 0.5$.
5) $200 <$ Load ≤ 300: $K = 3, D_1 = D_2 = 1$.

Cases 2 and 4 cannot be modeled directly using **phgensys** since the model for $K = 2$ will automatically assign $D_1 = 1$ when $K = 2$ and $D_1 = D_2 = 1$ when $K = 3$. For these two cases the values of D_1 and D_2 are used to adjust the survival matrices in lifetime distributions of the second and third generators, respectively. Thus, case 2 must be modeled using **phgensys** with $K = 2, \boldsymbol{A}_2' = D_1\boldsymbol{A}, D_1 = 0.5$, and $D_2 = 0$. For this case, the lifetime distribution for Generator 2 is not T_{G1}, it is denoted by T_{G2}. Likewise, case 4 must be modeled using **phgensys** with $K = 3$ and $\boldsymbol{A}_3' = D_2\boldsymbol{A}, D_2 = 0.5$ and denotes its lifetime distribution as T_{G3}. Thus, the generator subsystem models are

- Case 1: Load ≤ 100: T_G=phgensys([T_{G1}, T_{G1}, T_{G1}], 1, [0, 0]).
- Case 2: $100 <$ Load ≤ 150: T_G=phgensys([T_{G1}, T_{G2}, T_{G1}], 2, 0) with $\boldsymbol{A}_2' = 0.5\boldsymbol{A}$.
- Case 3: $150 <$ Load ≤ 200: T_G=phgensys([T_{G1}, T_{G1}, T_{G1}], 2, 0).
- Case 4: $200 <$ Load ≤ 250: T_G=phgensys([T_{G1}, T_{G1}, T_{G3}], 3) with $\boldsymbol{A}_3' = 0.5\boldsymbol{A}$.
- Case 5: $250 <$ Load ≤ 300: T_G=phgensys([T_{G1}, T_{G1}, T_{G1}], 3).

The system reliability and hazard functions calculated using this model is presented in Figure 5.41.

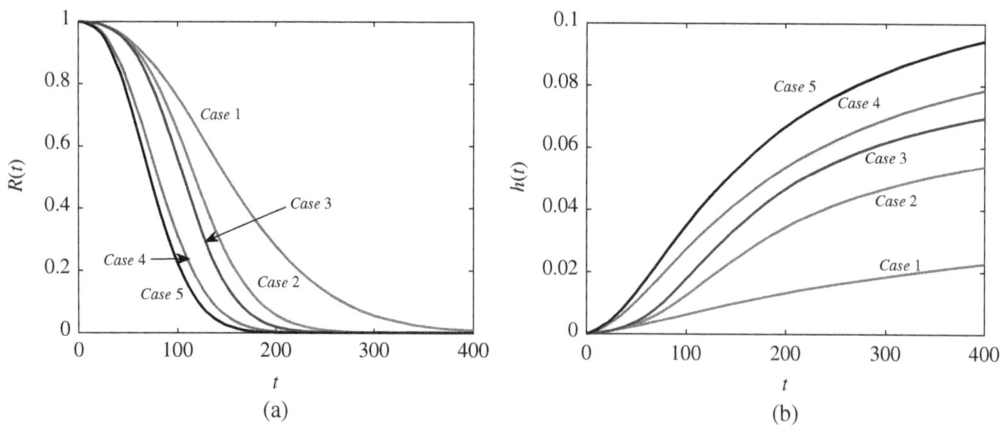

Figure 5.41 System reliability and hazard functions with generators can be operated in half power. (a) Reliability function and (b) hazard function.

Scenario 4 The system is operated following scenario 5.1, but the load may exceed 300 MVA which makes the generator must be operated in maximum capacity above their nominal capacity. Suppose that each generator can provide up to 10% higher than its nominal capacity of 100 MVA.

However, because of this over capacity, the generators will deteriorate 20% faster, which can be translated as having warmness level equal to 1.2. in this case, D becomes an accelerator factor, not a deceleration factor. The survival matrices for their lifetimes becomes $\boldsymbol{A}_i' = 1.2\boldsymbol{A}_i, i = 1, 2, 3$. The generator subsystem models are

- Case 1: Load ≤ 100: $K = 1$. The generator subsystem model is

$$T_G = \text{phgensys}([T_{G1}, T_{G1}, T_{G1}], 1, [1, 1]).$$

- Case 2: $100 <$ Load ≤ 200: $K = 2$. The generator subsystem model is

$$T_G = \text{phgensys}([T_{G1}, T_{G1}, T_{G1}], 2, 1).$$

- Case 3: $200 <$ Load ≤ 300: $K = 3$. The generator subsystem model is

$$T_G = \text{phgensys}([T_{G1}, T_{G1}, T_{G1}], 3).$$

- Case 4: $300 <$ Load ≤ 330: $K = 3$. The generator subsystem model is

$$T_G = \text{phgensys}\left(\left[T_{G1}', T_{G1}', T_{G1}'\right], 3\right) \text{ with } \boldsymbol{A}' = 1.2\boldsymbol{A}.$$

The system reliability and hazard functions calculated using this model and data is presented in Figure 5.42.

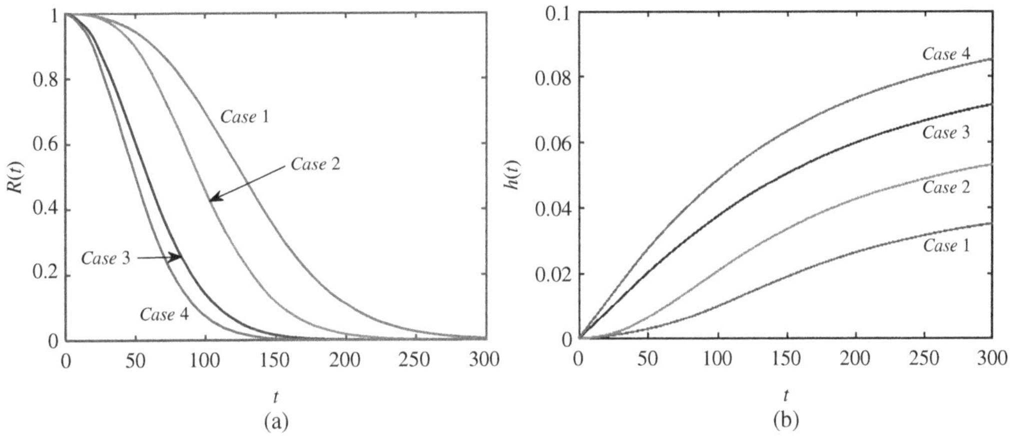

Figure 5.42 System reliability and hazard function with maximum capacity. (a) Reliability function and (b) hazard function.

The next scenario is different from all previous scenarios that assume the load is constant all the time; instead, the load is a function time.

Scenario 5 Suppose the load is dynamic, increasing over time following

$$L(t) = \min(80 + 2t, 300),$$

where t is time in week, and $L(t)$ is load at time t.

The generators are operated as in scenario 5.2. In the beginning of operation, the system has 1-out-of-3 general standby structure (3-standby) since one generator 1 is sufficient to fulfill a load of 80 MVA. The load becomes 100 MVA at week 10, which means at $t > 10$ the second generator is activated, and the system is in a 2-out-of-3 general standby structure with the third generator in cold standby. Again, the load becomes 200 at week 60, so at $t > 60$ all generators are operated in full power and the system has 3-out-of-3 general standby structure (3-series). Thus, the generator subsystem model is different for each time interval and is given by the following models in **phgensys**:

- $0 \le t < 10$: only one generator is needed. The generator subsistem model is

$$T_G = \text{phgensys}([T_{G1}, T_{G1}, T_{G1}], 1, [0, 0]).$$

- $10 \le t < 60$: two generators are needed. The generator subsistem model is

$$T_G = \text{phgensys}([RT_{G1}(10), T_{G1}, T_{G1}], 2, [0]).$$

- $60 \le t$: all generators are needed. The generator subsistem model is

$$T_G = \text{phgensys}([RT_{G1}(60), RT_{G1}(50), T_{G1}], 3).$$

where $RT(g)$ is the residual lifetime at age g whose PH representation is $(n, \mathbf{c}(g), \boldsymbol{A})$.

In the model for $10 \le t < 60$, the first generator has been utilized for 10 weeks (its age is 10 weeks), while the second and third generator have not been utilized yet (their ages are 0). Therefore, the first generator is using residual lifetime. Likewise, the first and second generators in the model for $t \ge 60$ are using their residual lifetimes after being used for 60 and 50 weeks, respectively.

With this adjustment in the main program, the system reliability and hazard function are calculated and plotted, and the result is shown Figure 5.43. The jumps in the hazard function occur when the structure changes from parallel to 2-out-of-3 at $t = 10$ which has a higher failure rate, and again from 2-out-of-3 to series at $t = 60$ also which has a higher failure rate.

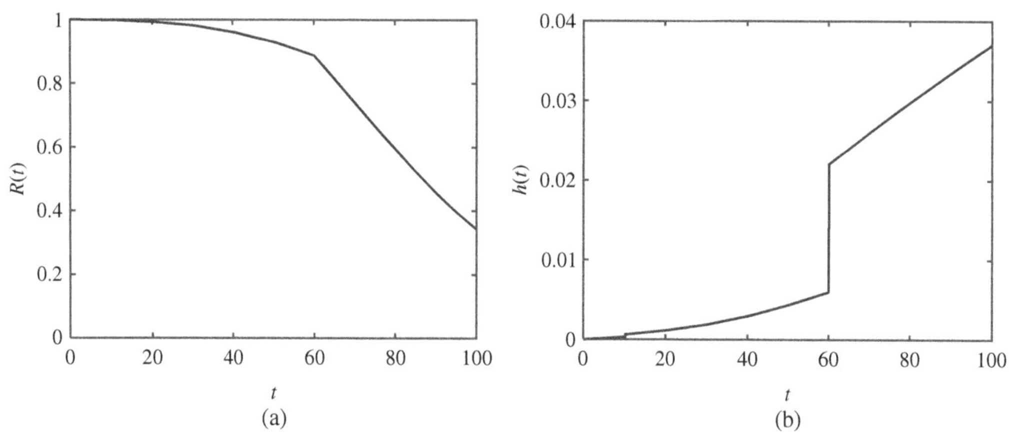

Figure 5.43 System reliability and hazard functions with cold standby for dynamic load. (a) Reliability function and (b) hazard function.

There are other scenarios in operating the system. Some of them can be modeled using the provided functions; others need special models that must be derived specifically for each operational scenario. ■

As in the case of imperfect switch standby systems discussed in Section 4.6 that backup component affects system reliability; the reliability of general standby systems is also affected by backup component ordering even though the systems use a perfect switch.

5.4 Backup Component Ordering

The order of nonidentical backup components in hot standby has no effect on the reliability of the K-out-of-N system. However, Shaked and Shanthikumar [24] showed the influence of backup component order on system reliability in some K-out-of-N general standby systems. Zhai et al. [25] proved that a 1-out-of-N warm standby system will be most reliable if its components are sequenced in descending order of their hazard functions. Roy and Gupta [4] proved that the K-out-of-N system with $K > 1$ support by two backup components in cold standby will have the highest reliability if the two backup components are sequenced according to the ascending order of their hazard functions. Alkaff [26] shows that K-out-of-N general standby systems with identical backup components having different warmness levels will have the highest reliability when the backup components are sequenced in descending order of their warmness levels. This section investigates the effects of ordering backup components based on their hazard functions or warmness levels on reliability measures of general standby systems. There is no specific computer program or function prepared for finding the effects of component orderings on system reliability measures. Instead, investigating such effects is conducted by writing a main program to call the **phgensys** function several times in accordance with the number of orderings compared.

First, orderings based on backup component hazard functions are investigated.

5.4.1 Hazard Functions Ordering

The effect of ordering backup components based on their hazard functions is investigated for different component types and numbers of primary components.

Example 5.40 The lifetimes of components in a 1-out-of-5 general standby system have nonidentical 4-Erlang distributions (IFR). The primary component has $MTTF = 1$,and the ith backup component, $i = 1, 2, 3, 4$, has $MTTF = 1, 5, 20$, and 50, respectively.

As shown in Chapter 3, the hazard function of k-stage Erlang for the same k is lower for higher $MTTF$. Thus, the hazard functions of the backup components of this system follow the ordering

$$h_4(t) < h_3(t) < h_2(t) < h_1(t),$$

where $h_i(t)$ is the hazard function of the ith backup component.

Several 1-out-of-5 general standby systems are constructed using these components in various sequences of backup components. All backup components have the same warmness level; either all of them are 0 (cold), 0.2 (warm, slightly cold), 0.6 (warm, slightly hot), or 1 (hot). The backup component orders compared are given in Table 5.12. Comparisons of system reliability functions in these three backup component orderings for several warmness levels are shown in Figure 5.44.

Table 5.12 Backup component orderings for Example 5.40.

System number	Type	Backup component order
S_1	Ascending	4-3-2-1
S_2	Descending	1-2-3-4
S_3	Arbitrary	1-3-4-2

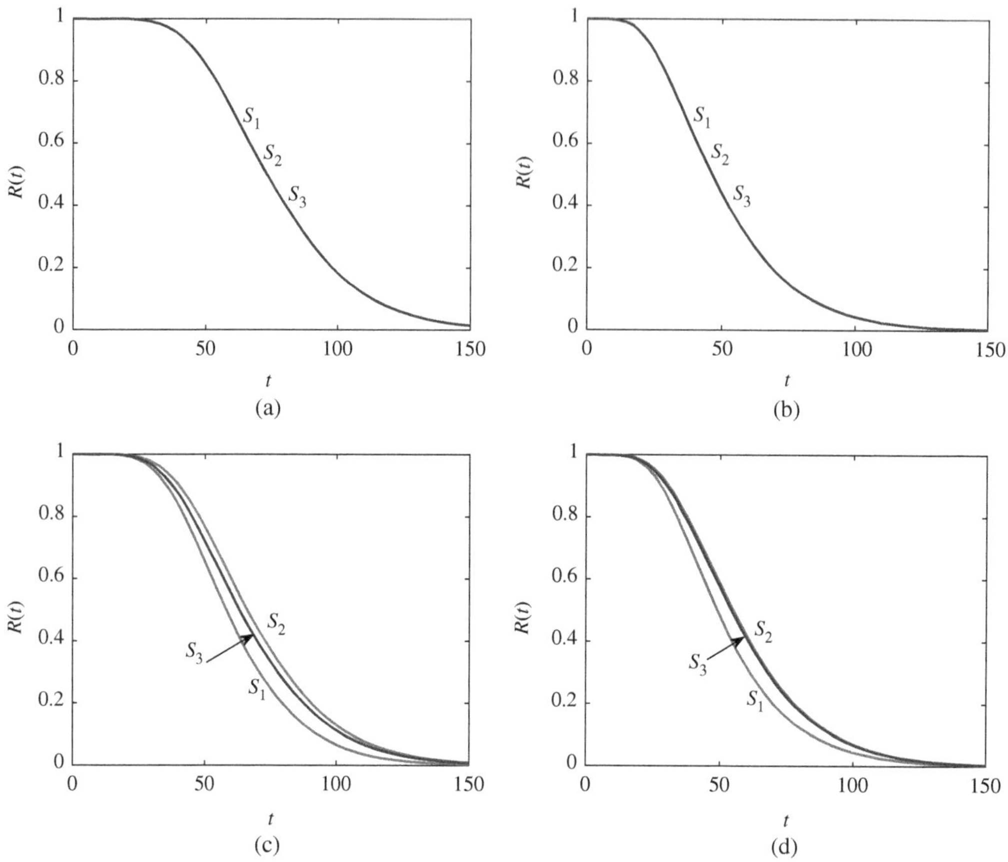

Figure 5.44 Reliability functions of systems in Example 5.40. (a) Warmness level = 0, (b) warmness level = 1, (c) warmness level = 0.2, and (d) warmness level = 0.6.

The main program for Example 5.40 is as follows:

MATLAB code for example5_40.m

```
clear;close all
addpath('../dysrel-am','../dysrel-am/expokit')

% Warmness level                                          input
D = [0 0.2 0.6 1];
% Components PH (Primary)
n = 4;
Tp = phnErlang(n,1);
% Components PH (Backup)
Tb(1) = phnErlang(n,1);
Tb(2) = phnErlang(n,5);
Tb(3) = phnErlang(n,20);
Tb(4) = phnErlang(n,50);
% Time span
tmax = 150;
t = linspace(0,tmax,21);
% Backup ordering
CID = [4 3 2 1;
```

```
    1 2 3 4;
    1 3 4 2];
for i = 1:length(D)
  U = [];
  for j = 1:3
    disp(['S = ',num2str(j)])
    Tc = Tp;
    for k = 1:4
        Tc = [Tc Tb(CID(j,k))];
    end
    Ts = phgensys(Tc,1,D(i)*ones(1,4));              model
    Rt = phrel(t,Ts);
    [x,u] = smoothing(t,Rt);
    U = [U u'];
  end
  % Reliability plots                                output
  lgnd = {'S_1';'S_2';'S_3'};
  multiplot(3,x,U,lgnd,'t','R(t)');
end
```

The graphs in Figure 5.44 show that the order of backup components, with all having the same warmness levels of 0 or 1, has no effect on the system reliability, as shown in Figure 5.44a and b. These results are in accordance with the system definitions. The lifetime of a system with all backup components in hot standby is the maximum of the component lifetimes, and the maximum of several variables produces the same value regardless of their order. Likewise, the lifetime of a system with all backup components in cold standby is the sum of the component lifetimes, and the sum of several variables is independent of their order. For warmness levels equal to 0.2 and 0.6, the order of components following the decreasing order of hazard functions gives higher reliability functions. This result is in accordance with the findings of Zhai et al. [25] that a 1-out-of-N system with warm standby components should be arranged in the decreasing order of their hazard functions in order to maximize its reliability function. The increase in reliability value at $t = 50$ for the 1-out-of-5 system with warmness level = 0.2 using the best compared to the worst ordering is approximately 10%. ■

Example 5.41 A 2-out-of-6 general standby system has two identical primary components, and the distributions of the lifetimes of all its components and their ordering are as in Example 5.40. The resulting reliability functions for several backup component ordering are presented in Figure 5.45.

The graphs in Figure 5.45 show that the 2-out-of-6 general standby system with warmness level $\neq 1$ behaves differently from the 1-out-of-5 general standby system. There are two significant differences between the K-out-of–$(4 + K)$ general standby systems for $K = 1$ and $K = 2$. First, the optimum order of hazard functions of the backup components in cold standby for $K = 2$ as shown in Figure 5.45a is increasing, whereas for $K = 1$ is not affected by the order. Second, the optimum order of hazard functions of the backup components in non-hot standby for $K = 2$, as shown in Figure 5.45c and d, are increasing, whereas for $K = 1$ is decreasing. This result is in accordance with the finding of Roy and Gupta [4]. The increase in reliability value at $t = 20$ for the 2-out-of-6 system with warmness level = 0.2 using the best compared to the worst ordering is approximately 5%.

The effect of component warmness levels on system reliability can be seen more clearly through the system $MTTF$. Figure 5.46 shows the effects of backup component hazard functions ordering

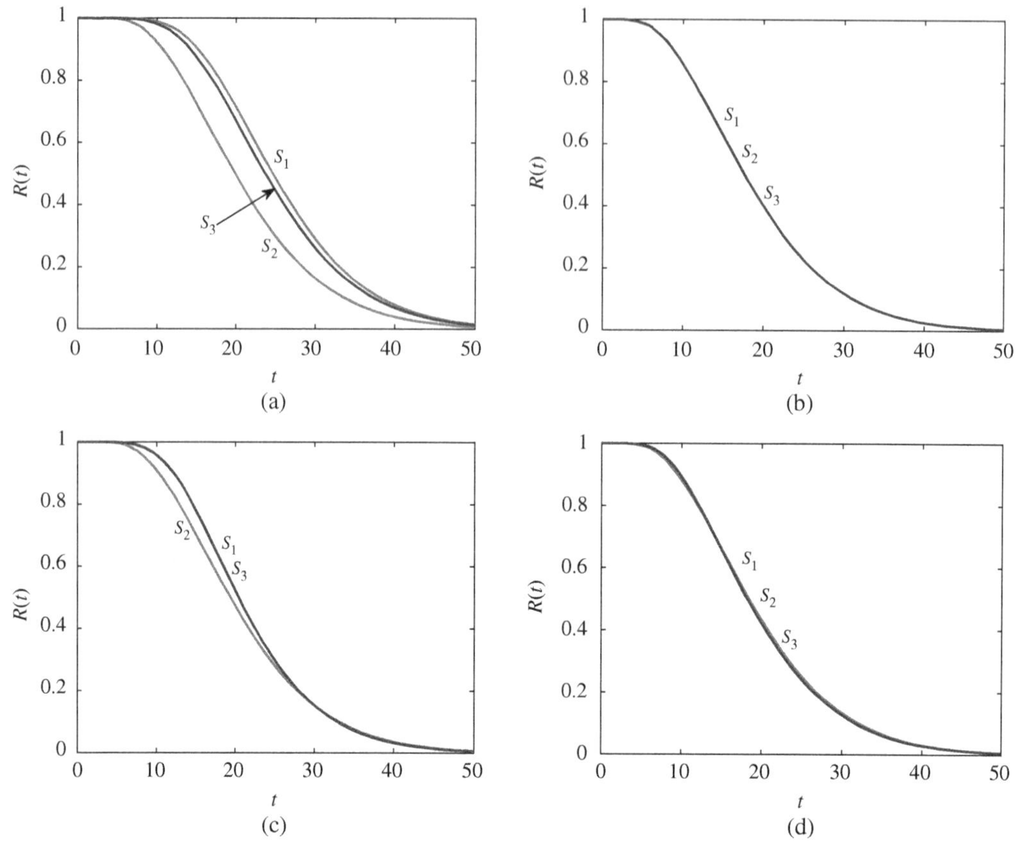

Figure 5.45 Reliability functions of systems in Example 5.41. (a) Warmness level = 0, (b) warmness level = 1, (c) warmness level = 0.2, (d) warmness level = 0.6.

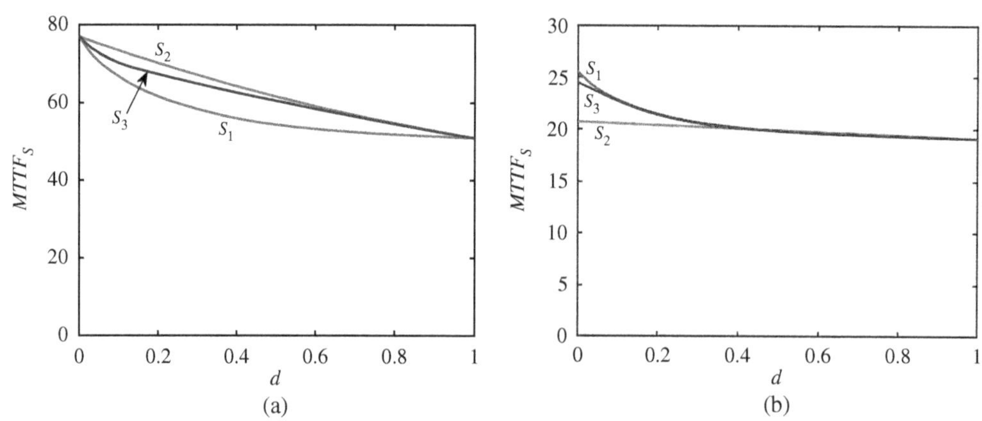

Figure 5.46 *MTTF* of systems in Examples 5.40 and 5.41 as a function of warmness level. (a) $K = 1$ and (b) $K = 2$.

to $MTTF$ of K-out-of-($K+4$) system when warmness levels of the backup components are varied from 0 to 1. The decreasing order has the highest $MTTF$ for $K = 1$ and has the lowest $MTTF$ for $K = 2$. The increase in $MTTF$ of the 1-out-of-5 system with warmness level = 0.2 using the best compared to the worst orderings is approximately 20%, while for the 2-out-of-6 system with the same warmness level is approximately 10%. ■

Example 5.42 Example 5.41 is repeated using DFR components having 2-Coxian distributed lifetimes with the same $MTTF$ as Example 5.40 and $CVTTF = 1.2$. The result is presented in Figure 5.47.

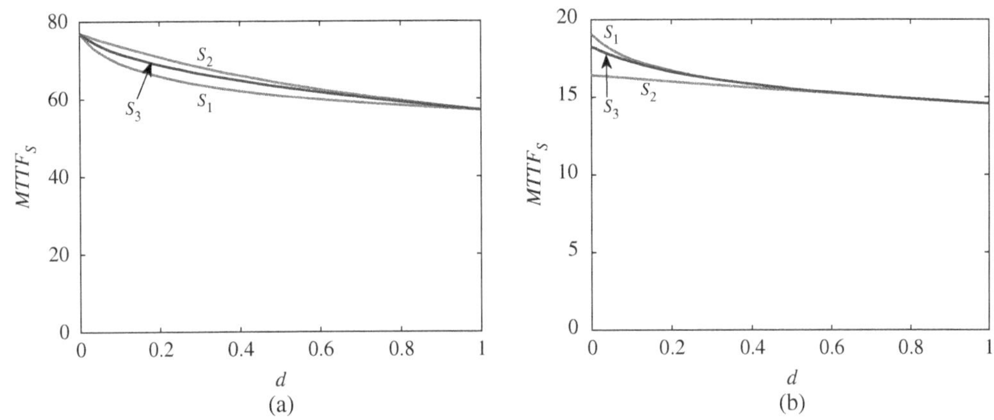

Figure 5.47 $MTTF$ of system with DFR components. (a) $K = 1$ and (b) $K = 2$.

Example 5.42 shows general standby systems with DFR components have the same optimum backup component ordering as the IFR components. This conclusion also applies to CFR components. ■

Exercise 5.31 Investigate the effects of backup component orderings in K-out-of-(K+4) systems for $K = 1$ and 2, where the lifetimes of kth backup components follow k-Erlang distribution, $k = 1$, 2, 3, 4, with $MTTF = 10$ and the primary component(s) have an exponential distribution with the same $MTTF$. Draw conclusions from the results.

Another way of ordering backup components is based on their warmness levels as given in the next section.

5.4.2 Warmness Levels Ordering

The ordering of backup components can also be based on warmness levels. The effect of ordering warmness levels on the reliability measures of general standby systems is investigated using backup components with an identical lifetime distribution but nonidentical warmness levels to ensure that the effect is solely due to warmness levels, not hazard functions.

Example 5.43 General standby systems are constructed using IFR components whose lifetimes follow an identical 4-Erlang distribution with $MTTF = 10$. The backup components have different warmness levels, varying from hot to cold. Several warmness level orderings are given in

Table 5.13. The resulting system reliability functions are depicted in Figure 5.48, which shows that the descending order is better than the ascending order for the cases of $K = 1$ and $K = 2$. The increase in reliability value at $t = 20$ for 1-out-of-5 and 2-out-of-6 systems using the best compared to the worst orderings is approximately 25% and 50%, respectively.

Table 5.13 Warmness level ordering.

System number	Ordering type	Warmness order
S_1	Ascending	0, 0.3, 0.7, 1
S_2	Descending	1, 0.7, 0.3, 0
S_3	Arbitrary	1, 0.3, 0, 0.7

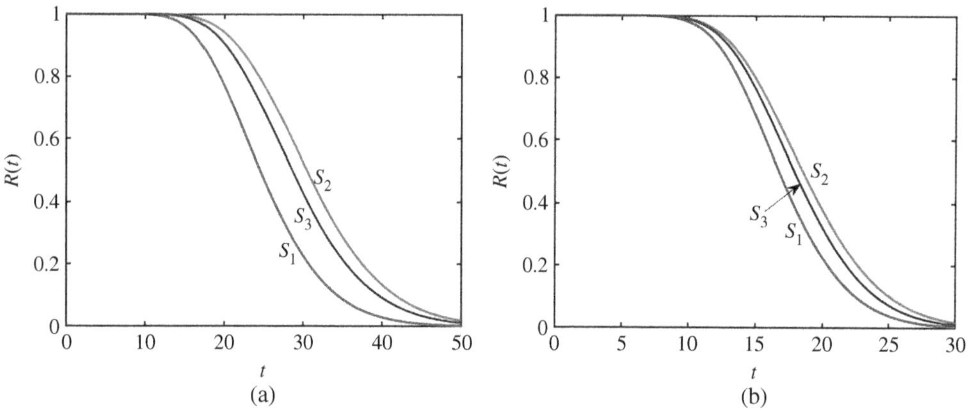

Figure 5.48 The effect of warmness level ordering in general standby systems with identical IFR components. (a) 1-out-of-5 and (b) 2-out-of-6.

■

Example 5.44 Example 5.43 is repeated using identical DFR components having 2-Coxian distributed lifetimes with $MTTF = 10$ and $CVTTF = 1.2$. Comparisons of the resulting reliability functions are presented in Figure 5.49.

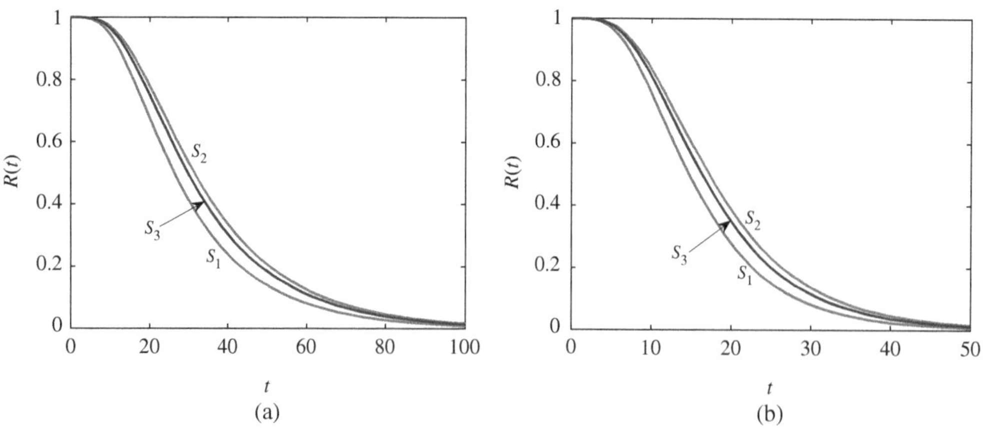

Figure 5.49 The effect of warmness level ordering in general standby systems with identical DFR components. (a) 1-out-of-5 and (b) 2-out-of-6.

■

Examples 5.43 and 5.44 indicate that the optimum ordering of the backup components follows decreasing warmness levels. This descending optimum ordering holds for all types of components. Unlike hazard function ordering, optimal warmness level ordering is always descending regardless of the number of primary components in the system.

Exercise 5.32 Calculate and compare the system *MTTF* under several backup component orderings based on warmness levels given in Table 5.13. Draw a conclusion from this result.

Further investigations to understand the behavior of general standby systems to component orderings are given as problems. The next section develops a method for finding the optimum warmness levels in general standby systems that minimizes a cost function.

5.5 Optimum Warmness Level[2]

A component in hot standby requires a significant operating cost in exchange for immediately replacing a failed primary component. In contrast, a component in cold standby does not require any operating cost, but it does require a preparation time to be ready to substitute a failed primary component. This waiting period disrupts the overall system operation and can even disrupt the operation of other systems. This operational loss is also a cost; thus, the two terms are interchangeable without confusion. Opportunity cost measures the opportunity loss due to the system being unable to operate while waiting for a backup component to substitute a failed primary component. Both costs must be considered when determining the appropriate type of standby.

Hot and cold standbys are two extremes of the warmness level with opposite characteristics. In between are warm standbys that require less operating and opportunity costs. Each of these costs depends on the warmness level of the component; the higher the warmness level, the higher the operating cost and the lower the opportunity cost (see Table 5.4). The warmness level that minimizes the total cost can be found by considering the sum of these two costs as a criterion. These costs are related to component lifetime, making them random; therefore, the objective is to minimize the expected total cost. From Section 5.2, it is known that the warmness level is related to *MTTF*, the higher the warmness level, the shorter the *MTTF*. To make a fair comparison, the objective to be minimized should be the expected total cost per unit time to operate a general standby system until it fails. Currently, modeling this problem assumes that the component lifetimes are exponentially distributed. This problem is first proposed in Alkaff [26].

The formulation of this problem starts with modeling the cost functions related to operating cost and opportunity loss.

5.5.1 Cost Models in General Standby Systems

The operating and opportunity costs associated with a general standby system are functions of the warmness levels. These costs also depend on the state of the system and the transitions between states. The system state before and after an event of failure are denoted by S' and S, respectively. These costs can be formulated as follows.

The operating cost of a general standby system is the sum of the operating costs of all system components. The unit operating cost, or the cost to operate per unit time, of components i as a primary and a backup are denoted by OC_i and OC'_i, respectively. The unit operating cost at state S,

2 Advanced material, can be omitted from lecture.

$UOC(S)$, is the sum of the unit operating costs of the good components at state S,

$$UOC(S) = \sum_{i|S_i=p} OC_i + \sum_{i|S_i=b} OC'_i. \tag{5.59}$$

The expected operating costs while in state S, $EOC(S)$, can be calculated as

$$EOC(S) = \left(\sum_{i|S_i=p} OC_i + \sum_{i|S_i=b} OC'_i \right) U_S. \tag{5.60}$$

where U_S is the duration of stay in state S.

The value of OC'_i is a portion of OC_i because a backup component is not always operated at full power or in hot standby. Therefore, the unit operating cost as a backup component is the unit operating cost as a primary multiplied by a function of its warmness level,

$$OC'_i = OC_i g_i(d_j). \tag{5.61}$$

The function $g_i(d_j)$ is a monotonic increasing function with respect to d_j, the warmness level of component i as the jth backup, with a value between 0 and 1, having characteristics

$$g_i(0) = 0, \tag{5.62}$$

$$g_i(1) = 1. \tag{5.63}$$

Equations (5.62) and (5.63) imply that components in hot and cold standby require maximum and no operating cost, respectively. A simple function that satisfies these characteristics is

$$g_i(d_j) = d_j^{b_i}, \quad b_i > 0, \quad j = 1, 2, \ldots, N - K, \quad i = j + K. \tag{5.64}$$

Exercise 5.33 Calculate the unit operating cost of a 2-out-of-4 general standby system with a warmness level $d_1 = 1$ and $d_2 = 0$ in the beginning of operation if the unit operating cost for each component operating at full power is 10.

Opportunity loss occurs during a transition from system state S' to system state S due to a failure of a primary component. There is no opportunity loss for the transition caused by failure of a backup component because the system operation is not disrupted. If the failing primary component is replaced by component i, then the unit opportunity loss, denoted by OL_i, is the opportunity loss due to waiting for the backup component i to be ready. This opportunity loss occurs only if the failing component is a primary. Hence,

$$\begin{aligned} OL(S' \to S) &= OL_i \quad \text{if } B \neq \emptyset, \\ &= 0 \quad \text{if } B = \emptyset, \end{aligned} \tag{5.65}$$

where $B = \{i \mid S'_i = b\} \cap \{i \mid S_i = p\}$ denotes the set of indices of the backup component in state S' that becomes a primary in state S. If B is empty, then the transition is caused by a failure of a backup component, so no opportunity loss occurs. B has at most one element because, for a continuous-time system, at any time instant only one failure is possible.

Since the waiting time for a backup component to be ready depends on its warmness level, Equation (5.65) must be corrected by a function of the warmness level of component i as the jth backup. If the failed primary component is replaced by component i as the jth backup with a warmness level d_j, then the opportunity loss is

$$\begin{aligned} OL(S' \to S) &= OL_i h_i(d_j) \quad \text{if } B \neq \emptyset, \\ &= 0 \quad \text{if } B = \emptyset, \end{aligned} \tag{5.66}$$

where $h_i(d_j)$ is a monotonic decreasing function with respect to d_j, having values between 0 and 1, with characteristics

$$h_i(1) = 0, \tag{5.67}$$

$$h_i(0) = 1. \tag{5.68}$$

Equation (5.67) means there is no opportunity loss for components in hot standby. Equation (5.68) states that components in cold standby cause the maximum opportunity loss, for they take the longest time to be ready. A simple function that satisfies these characteristics is

$$h_i(d_j) = (1 - d_j)^{c_i}, \quad c_i > 1, \quad j = 1, 2, \ldots, N - K, \quad i = j + K. \tag{5.69}$$

A realistic value for c_i is greater than or equal to 1 to indicate the heating time decrease faster than the increase in the warmness level. For example, the time to increase a warmness level from 0.4 to 1.0 (an increase of 0.6) is more than twice the time to increase a warmness level from 0.7 to 1.0 (an increase of 0.3).

Exercise 5.34 Calculate the opportunity loss for the system in Exercise 5.33 when the primary components failed for the first and second times if the unit opportunity loss is 10 and $c_2 = 2$.

Operating cost and opportunity loss are summed up to calculate the total cost for a given system state. The total cost for all states is the sum of the product of probability of being in each state and the total cost at that state. The probability of being in state S is the multiplication of the probability of being in state $S'$and the transition probability from state S' to state S summed over all state S',

$$P[S] = \sum_{S'} P[S']P[S' \to \mathrm{S}]. \tag{5.70}$$

Starting from the probability of being at the initial state, which is equal to 1, the probability of being at all subsequent states can be calculated using Equation (5.70) recursively by knowing the transition probabilities $P[S' \to \mathrm{S}]$.

The expected operating cost $EOC(S)$ occurs with probability $P[S]$. Thus, the expected total operating cost until the system fails is

$$TOC(\mathbf{d}) = \sum_{S} P[S]EOC(S), \tag{5.71}$$

where $\boldsymbol{d}$ is a vector of the warmness levels of all backup components.

The opportunity loss $OL[S' \to S]$ occurs when the system makes a transition from S' to S. This transition occurs with probability $P[S']P[S' \to S]$. Thus, the expected opportunity loss for each transition is

$$EOL(S', S) = P[S']P[S' \to S]OL[S' \to S], \tag{5.72}$$

and the expected total opportunity loss until the system fails is

$$TOL(\mathbf{d}) = \sum_{S',S} EOL(S', S). \tag{5.73}$$

The expected total cost is the sum of the expected total operating cost and opportunity loss, that is,

$$TC(\mathbf{d}) = TOC(\mathbf{d}) + TOL(\mathbf{d}). \tag{5.74}$$

The average total cost AC is the total cost per unit time until the system fails. The time until the system fails is the system expected lifetime, $MTTF_S$, which is also a function of $\boldsymbol{d}$. Thus,

$$AC(\mathbf{d}) = \frac{TC(\mathbf{d})}{MTTF_S(\mathbf{d})}. \tag{5.75}$$

where $MTTF_S$ is calculated as the average duration of stay across all states,

$$MTTF_S(\mathbf{d}) = \sum_S P[S]U_S. \tag{5.76}$$

The average cost can be optimized by choosing the appropriate values of warmness levels. The optimum warmness levels for all backup components are the value of $\mathbf{d}$ that minimizes $AC(\mathbf{d})$. If there are no functional constraints on the warmness levels of backup components, then the cost optimization model is

$$\begin{aligned} &\min AC(\mathbf{d}), \\ &\text{subject to } \mathbf{0} \leq \mathbf{d} \leq \mathbf{1}. \end{aligned} \tag{5.77}$$

Given that component operation consumes some amount of power, and the power availability is often limited, in practice it is sometimes necessary to add power availability constraints. If a_j is the power requirement of the jth backup when fully operated (in hot standby or when it has become a primary component), then its power requirement for operation with a warm level d_j is $a_j d_j$. The cost optimization model with power limitation constraints is

$$\begin{aligned} &\min AC(\mathbf{d}) \\ &\text{subject to } \mathbf{a}^T\mathbf{d} \leq A \\ &\qquad \mathbf{0} \leq \mathbf{d} \leq \mathbf{1} \end{aligned} \tag{5.78}$$

where A is the remaining power capacity for the backup components after fulfilling the power requirements of all primary components, and $\boldsymbol{a}$ is the vector of power requirements of the backup components.

Since the lifetimes of all components are assumed to follow exponential distribution, the properties of exponential distributions useful for formulating this problem are reminded.

5.5.2 Some Properties of Exponential Distribution

The general standby system discussed in this section uses components with independently and exponentially distributed lifetimes that need not be identical. The probabilistic quantities required to calculate the cost components can be determined using several properties of the exponential distribution.

For X_i, $i = 1, \dots, N$, having independent exponential distributions with rates λ_i, or $X_i \backsim \exp(\lambda_i)$, it is known that $P[X_i > t] = e^{-\lambda_i t}$, $E[X_i] = \frac{1}{\lambda_i}$, and

$$\begin{aligned} P[\min(X_1, \dots, X_N) > t] &= P[X_1 > t]P[X_2 > t] \dots P[X_N > t] \\ &= e^{-(\lambda_1 + \lambda_2 + \dots + \lambda_N)t} \end{aligned}$$

Hence,

$$\min(X_1, \dots, X_N) \backsim \exp\left(\sum_{i=1}^{N} \lambda_i\right) \tag{5.79}$$

$$E[\min(X_1, \dots, X_N)] = \frac{1}{\sum_{i=1}^{N} \lambda_i} \tag{5.80}$$

$$\begin{aligned} P[X_i < X_j] &= \int_0^\infty P[X_j > X_i \mid X_i = u] f_{X_i}(u) du = \int_0^\infty e^{-\lambda_j u} \left(\lambda_i e^{-\lambda_i u}\right) du \\ &= \frac{\lambda_i}{\lambda_i + \lambda_j}, i \neq j; \quad i,j = 1, 2, \dots, N \end{aligned} \tag{5.81}$$

Because of the memoryless property of the exponential distribution, a good component in all system states has the same failure rate, regardless of its age.

The development of the problem formulation starts with the simpler case of general standby systems with identical components and warmness levels.

5.5.3 Homogeneous General Standby Systems

A homogeneous general standby system is a general standby system whose components are identical, and all backup components have identical warmness level. Let λ be the failure rate of all components; the failure rate of backup components is corrected by its warmness level d to become $d\lambda$.

The states of a homogeneous system can be expressed by the number of good components in each state. All system states have K good primary components and k good backup components, $k = N - K, N - K - 1, \ldots, 0$. Transitions from one state to the next are due to a component failure. Transitions due to failures of primary components incur opportunity loss, while transitions due to failures of backup components do not. Therefore, there are two transitions from state $K + k$ to $K + k - 1$. A transition due to failure of a primary component occurs when one of K primary components fail before one of k backup components. The time until one of the primary components fails, W_K, is the minimum of the lifetimes of all K primary components. According to Equation (5.79), W_K is exponentially distributed with rate $K\lambda$. Likewise, the time until one of the backup components fails, V_k, is the minimum of the lifetimes of all k backup components. Also, according to Equation (5.79), V_k is exponentially distributed with rate $kd\lambda$. The probability that a transition occurs due to failure of a primary component is when $W_K < V_k$. From Equation (5.81), this probability is $\frac{K\lambda}{(K+kd)\lambda} = \frac{K}{K+kd}$, and the probability of a transition occurring due to failure of a backup component is $1 - \frac{K}{K+kd} = \frac{kd}{K+kd}$.

A probability transition diagram can be constructed using these probability values. But it is necessary to add the opportunity losses arising from the transitions in the diagram, which is $Lh(d)$, when a transition occurs due to the failure of a primary and 0 when a transition occurs due to the failure of a backup. Given the components and warmness levels are identical, all transitions have the same opportunity losses. The operating cost occurs in each system state, and its value depends on the system state. Since they are identical, all components also have identical unit operating costs. The unit operating cost at state $K + k$ is $KOC + kOCg(d)$, which is the sum of the unit operating cost of K primary and k backup components. The results are presented in Figure 5.50 without including the failed system state. Note that the total probability out of a state is equal to 1, as it is supposed to be. Since the probability at the initial state is equal to 1, the probability at each state is also equal to 1.

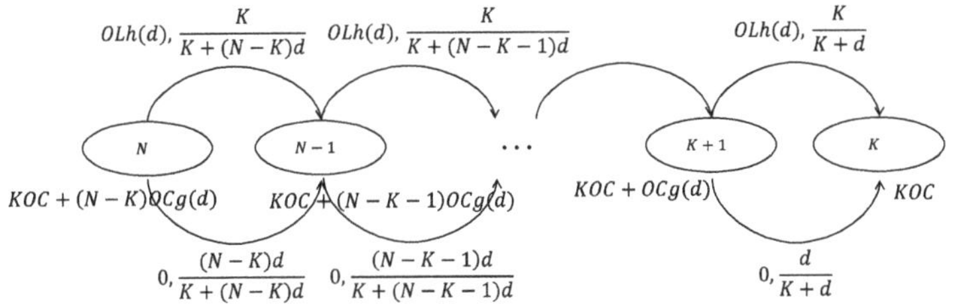

Figure 5.50 Cost-probability transition diagram for a homogeneous general standby system.

According to Equation (5.79), duration of stay in state $K+k$, $LOS(K+k)$ has exponential distribution with rate $K\lambda + kd\lambda$, which is the sum of the failure rates of good components at that state,

$$LOS(K+k) \backsim \exp(K\lambda + kd\lambda). \tag{5.82}$$

Using Equations (5.80) and (5.82) obtains the expected duration of stay in state $K+k$,

$$U_{K+k} = \frac{1}{K\lambda + kd\lambda} = \frac{1}{(K+kd)\lambda}. \tag{5.83}$$

Equation (5.83) also shows that there is only one state at each stage. Therefore, $P[S'] = P[S] = 1$, and $P[K+k] = 1$, $k = 0, 1, \dots, N-K$. Using Equation (5.76) yields

$$MTTF(d) = \sum_{k=0}^{N-K} U_{K+k} P[K+k] = \sum_{k=0}^{N-K} U_{K+k} = \frac{1}{\lambda} \sum_{k=0}^{N-K} \frac{1}{K+kd}. \tag{5.84}$$

Since all components are identical, Equation (5.59) becomes

$$UOC(K+k) = (K + kg(d))OC. \tag{5.85}$$

Using Equations (5.60), (5.83), and (5.85) obtains

$$EOC(K+k) = ((K + kg(d))OC)U_{K+k} = \frac{OC}{\lambda} \frac{(K + kg(d))}{(K+kd)}. \tag{5.86}$$

Substituting Equation (5.86) into Equation (5.71) with $P[K+k] = 1$ yields

$$TOC(d) = \frac{OC}{\lambda} \sum_{k=0}^{N-K} \frac{K + kg(d)}{K+kd}. \tag{5.87}$$

The opportunity loss $OLh(d)$ occurs with probability $\frac{K}{K+kd}$. From Equation (5.72),

$$EOL((K+k), (K+k-1)) = OLh(d)\frac{K}{K+kd}. \tag{5.88}$$

Substituting Equation (5.88) into Equation (5.73) yields

$$TOL(d) = OLh(d) \sum_{k=1}^{N-K} \frac{K}{K+kd}. \tag{5.89}$$

Note that the sum in Equation (5.89) does not include $k = 0$.

Using Equations (5.87) and (5.89) obtains the total cost until the system fails,

$$TC(d) = \frac{OC}{\lambda} \sum_{k=0}^{N-K} \frac{K + kg(d)}{K+kd} + OL \sum_{k=1}^{N-K} \frac{h(d)K}{K+kd}. \tag{5.90}$$

The average total cost per unit time until the system failure, $AC(d)$, is calculated by dividing $TC(d)$ in Equation (5.90) by $MTTF$ in Equation (5.84).

Example 5.45 The optimum warmness level for the backup component of a 1-out-of-2 general standby system is to be calculated for $\lambda = 1$, $OC = 1$, $OL = 1$, $b = 1$, $c = 2$.

Substituting these costs and system parameters into Equation (5.90) yields

$$TC(d) = \frac{1}{1} \sum_{k=0}^{2-1} \frac{1+kd}{1+kd} + 1 \sum_{k=1}^{2-1} \frac{(1-d)^2 1}{1+kd}$$

$$= 1(1+1) + \frac{(1-d)^2}{1+d} = \frac{2 + 2d + (1-d)^2}{1+d},$$

and into Equation (5.84) yields

$$MTTF(d) = \frac{1}{1}\sum_{k=0}^{2-1}\frac{1}{1+kd} = \frac{1}{1+0d} + \frac{1}{1+1d} = 1 + \frac{1}{1+d} = \frac{2+d}{1+d}.$$

The average total cost until system failure calculated using Equation (5.75) is

$$AC(d) = \frac{2+2d+(1-d)^2}{1+d} : \left(\frac{2+d}{1+d}\right) = \frac{2+2d+(1-d)^2}{2+d}.$$

Equating the derivative of $AC(d)$ with respect to d to 0 yields

$$AC(d)' = \frac{(2+2(1-d)(-1))(2+d) - 1(2+2d+(1-d)^2)}{(2+d)^2} = 0.$$

Simplifying it yields

$$4d + 2d^2 - 2 - 2d - 1 + 2d - d^2 = 0,$$

and solving it produces an optimum warmness level = 0.65. ■

Example 5.46 A general standby system has system parameters and costs as in Example 5.45. The optimum warmness levels of the backup components for several K and N are calculated and the results are presented in Table 5.14. The number of backup components is $N - K$. For example, the 2-out-of-5 general standby system has two primary components (row 2) and three backup components (column 3). The optimum warmness level for the three backup components in this system is 0.62. The optimum warmness level for a 1-out-of-2 general standby system is given in the table as 0.65 from one primary component and one backup component, consistent with the results of Example 5.45.

Table 5.14 Optimum warmness level of backup components in Example 5.46.

Number of backup components ($N-K$)	Number of primary components (K)					
	1	2	3	4	5	6
1	0.65	0.80	0.86	0.89	0.91	0.92
2	0.49	0.71	0.79	0.84	0.87	0.89
3	0.31	0.62	0.73	0.79	0.83	0.86
4	0	0.54	0.68	0.75	0.79	0.82
5	0	0.45	0.62	0.70	0.76	0.79
6	0	0.36	0.56	0.66	0.72	0.76

Table 5.14 shows that the optimum warmness level of the backup components increases as the number of primary components increases and decreases as the number of backup components increases. ■

Exercise 5.35 Use Table 5.14 to determine the optimum warmness level for a 3-out-of-6 general standby system as in Example 5.46.

Exercise 5.36 Investigate the effects of increasing the component: (1) unit operating cost, (2) failure rate, and (3) unit opportunity loss, each to the optimum warmness level. Draw conclusions from the results.

The discussion continues with general standby systems having nonidentical components or warmness levels in the next section.

5.5.4 Heterogeneous General Standby Systems

Heterogeneous general standby systems may have nonidentical components or nonidentical warmness levels, or both. General standby system usually is heterogeneous; hence, it is sufficient to say general standby systems without mentioning its heterogenicity. The equations derived in Section 5.5.1 are applicable to all cases. The system states for heterogeneous systems are defined as combinations of the states of each component.

The state of each component in a general standby system is either primary (p), backup (b), or fail (f). The duration of stay in each system state is the minimum of the lifetimes of good components in that state. Since all component lifetimes are exponentially distributed, the duration of stay in state S, $LOS(S)$, and its expected value, U_S, can be determined from Equations (5.79) and (5.80) as

$$LOS(S) \backsim \exp\left(\sum_{i|S_i=p} \lambda_i + \sum_{i|S_i=b} \lambda_i'\right), \tag{5.91}$$

$$U_S = \left(\sum_{i|S_i=p} \lambda_i + \sum_{i|S_i=b} \lambda_i'\right)^{-1}, \tag{5.92}$$

where $\lambda_i' = d_j\lambda_i$.

The transition probability from state S' to state S, $P[S' \to S]$, is the probability that the lifetime of the component causing failure is less than or equal to the minimum lifetime of all components in state S'. Using Equations (5.79)–(5.81) obtains

$$P[S' \to S] = \frac{\theta}{\left(\sum_{i|S_i'=p} \lambda_i + \sum_{i|S_i'=b} \lambda_i'\right)}, \tag{5.93}$$

where θ is the failure rate of the component that is causing the transition, which is the component that is good in state S' and fail in the state, S. Thus,

$$\theta = \lambda_i \mid i \in \{m \mid S_m' \neq f\} \cap \{m \mid S_m = f\}. \tag{5.94}$$

Only the index of the component that causes this transition satisfies Equation (5.94), which is in any event of failure always only one component. Substituting Equation (5.92) into Equation (5.93) obtains

$$P[S' \to S] = \theta U_{S'}. \tag{5.95}$$

The cost-probability transition diagram for a 1-out-of-3 general standby system is used as an illustration. The failure rates of components 1, 2, and 3 as the primary, first backup, and second backup are λ_1, $\lambda_2' = d_1\lambda_2$ and $\lambda_3' = d_2\lambda_3$, respectively. The initial state of the system is (p, b, b); therefore, $P(p, b, b) = 1$, and $OC(p, b, b) = OC_1 + OC_2 g_2(d_1) + OC_3 g_3(d_2)$. There are three possible transitions and next states due to the failure of one of the three components.

- *Transitions $pbb \to fpb$ and state fpb*

$$S' = pbb, S = fpb$$

$\{i \mid S_i' \neq f\} \cap \{i \mid S_i = f\} = \{1, 2, 3\} \cap \{1\} = \{1\} \to \theta = \lambda_1$(component 1 failed).

$\{i \mid S_i' = b\} \cap \{i \mid S_i = p\} = \{2, 3\} \cap \{2\} = \{2\}$ (component 2 becomes primary).

$OL(pbb \rightarrow fpb) = OL_2 h_2(d_1)$ (opportunity loss of waiting for component 2 as first backup to become primary).

$$U_{pbb} = \left(\sum_{i|S_i=p} \lambda_i + \sum_{i|S_i=b} \lambda'_i \right)^{-1} = \frac{1}{\lambda_1 + \lambda'_2 + \lambda'_3},$$

$$P[pbb \rightarrow fpb] = \lambda_1 U_{pbb} = \frac{\lambda_1}{\lambda_1 + \lambda'_2 + \lambda'_3},$$

$$P[fpb] = P[pbb \rightarrow fpb]P[pbb] = \frac{\lambda_1}{\lambda_1 + \lambda'_2 + \lambda'_3},$$

$$OC[fpb] = OC_2 + OC_3 g_3(d_2).$$

- *Transition pbb → pfb and state pfb*

$$S' = pbb, \quad S = pfb.$$

$$\{i \mid S'_i \neq f\} \cap \{i \mid S_i = f\} = \{1, 2, 3\} \cap \{2\} = \{2\} \rightarrow \theta = \lambda'_2;$$

$$\{i \mid S'_i = b\} \cap \{i \mid S_i = p\} = \{2, 3\} \cap \{1\} = \{\}; L(S' \rightarrow S) = 0;$$

$$P[pbb \rightarrow pfb] = \lambda'_2 U_{pbb} = \frac{\lambda'_2}{\lambda_1 + \lambda'_2 + \lambda'_3}.$$

$$P[pfb] = P[pbb \rightarrow pfb]P[pbb] = \frac{\lambda'_2}{\lambda_1 + \lambda'_2 + \lambda'_3},$$

$$OC[pfb] = OC_1 + OC_3 g_3(d_2).$$

There is no opportunity loss since the failing component is a backup component.

- *Transition pbb → pbf and state pbf*

$$S' = pbb, \quad S = pbf.$$

$$\{i \mid S'_i \neq f\} \cap \{i \mid S_i = f\} = \{1, 2, 3\} \cap \{3\} = \{3\} \rightarrow \theta = \lambda'_3.$$

$$\{i \mid S'_i = b\} \cap \{i \mid S_i = p\} = \{2, 3\} \cap \{1\} = \{\}; \quad L(S' \rightarrow S) = 0.$$

$$P[pbb \rightarrow pbf] = \lambda'_3 U_{pbb} = \frac{\lambda'_3}{\lambda_1 + \lambda'_2 + \lambda'_3},$$

$$P[pbf] = P[pbb \rightarrow pbf]P[pbb] = \frac{\lambda'_3}{\lambda_1 + \lambda'_2 + \lambda'_3},$$

$$OC[pbf] = OC_1 + OC_2 g_2(d_1).$$

There is no opportunity loss since the failing component is a backup component.

The process continues by finding the costs and probabilities of the transitions *fpb → ffp*, *fpb → fpf*, *pfb → ffp*, *pfb → pff*, *pbf → fpf*, *pbf → pff*, and the probabilities of being in states *ffp*, *fpf*, and *pff*.

This information is utilized to construct the cost-probability transition diagram of this 3-general standby system. The result is presented in Figure 5.51 without the absorbing states. In this diagram $OL'_2 = OL_2 h_2(d_1)$ and $OL'_3 = OL_3 h_3(d_2)$.

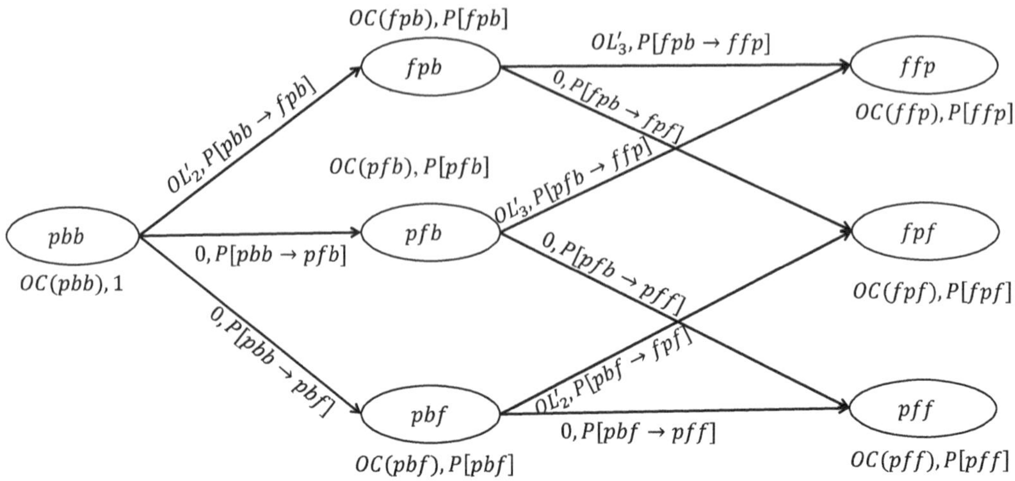

Figure 5.51 Cost-probability transition diagram of a 1-out-of-3 general standby system.

The following state probabilities can be derived using Figure 5.51.

$$P[ffp] = P[fpb \rightarrow ffp]P[fpb] + P[pfb \rightarrow ffp]P[pfb]$$
$$= \left(\frac{\lambda_2}{\lambda_2 + \lambda'_3}\right)\left(\frac{\lambda_1}{\lambda_1 + \lambda'_2 + \lambda'_3}\right) + \left(\frac{\lambda_1}{\lambda_1 + \lambda'_3}\right)\left(\frac{\lambda'_2}{\lambda_1 + \lambda'_2 + \lambda'_3}\right).$$
$$P[fpf] = P[fpb \rightarrow fpf]P[fpb] + P[pbf \rightarrow fpf]P[pbf]$$
$$= \left(\frac{\lambda'_3}{\lambda_2 + \lambda'_3}\right)\left(\frac{\lambda_1}{\lambda_1 + \lambda'_2 + \lambda'_3}\right) + \left(\frac{\lambda_1}{\lambda_1 + \lambda'_2}\right)\left(\frac{\lambda'_3}{\lambda_1 + \lambda'_2 + \lambda'_3}\right).$$
$$P[pff] = P[pfb \rightarrow pff]P[pfb] + P[pbf \rightarrow pff]P[pbf]$$
$$= \left(\frac{\lambda'_3}{\lambda_1 + \lambda'_3}\right)\left(\frac{\lambda'_2}{\lambda_1 + \lambda'_2 + \lambda'_3}\right) + \left(\frac{\lambda'_2}{\lambda_1 + \lambda'_2}\right)\left(\frac{\lambda'_3}{\lambda_1 + \lambda'_2 + \lambda'_3}\right).$$

Example 5.47 The optimum warmness level of the backup component in a 1-out-of-2 general standby system is to be calculated for $\lambda_1 = 1$, $\lambda_2 = 2$, $OC_1 = OC_2 = 1$, $OL_2 = 1$, $b_2 = 1$, $c_2 = 2$.

Since this system has nonidentical components, the good system states are (pb), (pf),and (fp). The unit operating costs at all states are

$$OC(pb) = OC_1 + OC_2 g_2(d) = 1 + 1(d)^1 = 1 + d,$$
$$OC(pf) = OC_1 = 1, \quad OC(fp) = OC_2 = 1.$$

The probabilities at each state and between each pair of states are

$$P[pb] = 1, \quad P[pb \rightarrow pf] = \frac{\lambda'_2}{\lambda_1 + \lambda'_2} = \frac{2d}{1 + 2d},$$
$$P[pf] = P[pb]P[pb \rightarrow pf] = \frac{2d}{1 + 2d},$$
$$P[pb \rightarrow fp] = \frac{\lambda_1}{\lambda_1 + \lambda'_2} = \frac{1}{1 + 2d},$$
$$P[fp] = P[pb]P[pb \rightarrow fp] = \frac{1}{1 + 2d}.$$

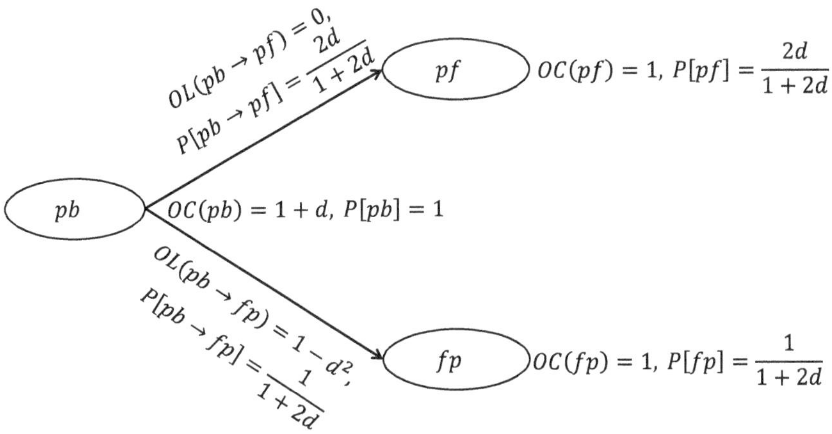

Figure 5.52 Cost-probability transition diagram of Example 5.47.

The opportunity losses at all possible transitions are

$$OL[pb \to pf] = 0, \quad OL[pb \to fp] = OL_2 h_2(d) = 1(1-d)^2 = (1-d)^2.$$

The cost-probability transition diagram of this system is given in Figure 5.52.
The expected duration of stays at each state are

$$U_{pb} = \frac{1}{\lambda_1 + \lambda_2'} = \frac{1}{1+2d}, \quad U_{pf} = \frac{1}{\lambda_1} = 1, \quad U_{fp} = \frac{1}{\lambda_2} = \frac{1}{2}.$$

The expected operating costs at each state are

$$EOC(pb) = OC(pb)U_{pb} = (1+d)\frac{1}{1+2d} = \frac{1+d}{1+2d},$$

$$EOC(pf) = OC(pf)U_{pf} = (1)(1) = 1,$$

$$EOC(fp) = OC(fp)U_{fp} = (1)\left(\frac{1}{2}\right) = \frac{1}{2}.$$

The expected lifetime of the system is

$$MTTF(d) = \sum_S P[S]U_S = P_{pb}U_{pb} + P_{pf}U_{pf} + P_{fp}U_{fp}$$

$$= (1)\left(\frac{1}{1+2d}\right) + \left(\frac{2d}{1+2d}\right)(1) + \left(\frac{1}{1+2d}\right)\left(\frac{1}{2}\right) = \frac{1+2d+\frac{1}{2}}{1+2d} = \frac{\frac{3}{2}+2d}{1+2d}$$

The total cost until the system fails is the sum of all costs at each state and each state transition, that is,

$$TC(d) = TOC(d) + TOL(d)$$

$$= P[pb]EO(pb) + P[pf]EO(pf) + P[fp]EO(fp)$$

$$+ P[pb]P[pb \to fp]OL[pb \to fp] + P[pb]P[pb \to pf]OL[pb \to pf]$$

$$= (1)\left(\frac{1+d}{1+2d}\right) + \left(\frac{2d}{1+2d}\right)(1) + \left(\frac{1}{1+2d}\right)\left(\frac{1}{2}\right) + (1)\left(\frac{1}{1+2d}\right)(1)(1-d)^2 + 0$$

$$= \frac{\frac{5}{2}+d+d^2}{1+2d}.$$

The average total cost per unit time is calculated as

$$AC(d) = \frac{TC(d)}{MTTF(d)} = \frac{\frac{5}{2} + d + d^2}{\frac{3}{2} + 2d}.$$

Set the derivative of $AC(d)$ with respect to d equal to 0 to obtain

$$d^{opt} = 0.77.$$

■

Example 5.48 The order of the components in Example 5.47 is reversed. So, the failure rates become $\lambda_1 = 2$, $\lambda_2 = 1$, and the other parameters are unchanged.

$$U_{pb} = \frac{1}{\lambda_1 + \lambda_2'} = \frac{1}{2+d}, \quad U_{pf} = \frac{1}{\lambda_1} = \frac{1}{2}, \quad U_{fp} = \frac{1}{\lambda_2} = 1,$$

$$EOC(pb) = OC(pb)U_{pb} = (1+d)\frac{1}{2+d} = \frac{1+d}{2+d},$$

$$EOC(pf) = OC(pf)U_{pf} = (1)\left(\frac{1}{2}\right) = \frac{1}{2}, \quad EO(fp) = O(fp)U_{fp} = (1)(1) = 1.$$

$$P[pb] = 1, \quad P[pb \to pf] = \frac{\lambda_2'}{\lambda_1 + \lambda_2'} = \frac{d}{2+d}, \quad P[pf] = P[pb]P[pb \to pf] = \frac{d}{2+d},$$

$$P[pb \to fp] = \frac{\lambda_1}{\lambda_1 + \lambda_2'} = \frac{2}{2+d}, \quad P[fp] = P[pb]P[pb \to fp] = \frac{2}{2+d}.$$

Following the same procedure as in Example 5.47 produces $d_{opt} = 0.78$. ■

Examples 5.47 and 5.48 show that the optimum warmness level depends on the components ordering, it changes with a change in components ordering.

Exercise 5.37 Compare the results of Examples 5.47 and 5.48. What conclusion can be made from this comparison? To strengthen the conclusion, calculate the optimum warmness level for another value of λ_2.

Exercise 5.38 Investigate the effects of increasing: (1) unit operating cost and (2) unit opportunity loss each to the optimum warmness level for the system in Example 5.47.

Finding optimum warmness levels for arbitrary K-out-of-N general standby systems is a difficult task to do manually. For this purpose, a MATLAB program for calculating the optimum warmness levels of backup components in general standby systems is prepared as an application package **owl-am**, separate from **dysrel-am**. This program takes λ_i, OC_i, OL_i $i = 1, 2, \ldots N$ and b_j, $c_j, j = K+1, \ldots, N$ as inputs. The following examples are calculated using **owl-am**.

Example 5.49 A K-out-of-N general standby system has identical components with $\lambda_i = 1$, $OC_i = 1$, $OL_i = 1.5$ for $i = 1, 2, \ldots, N$, and $b_j = 1$, $c_j = 2$ for $j = K+1, \ldots, N$. The system has two backup components, which may have different warmness levels. The effect of the number of primary components on the average total cost function is expressed in the contours of the average total cost function with respect to the warmness levels of the two backup components, denoted by d_1 and d_2. The resulting contours are shown in Figure 5.53. The diagonal lines represent identical warmness levels. The lower left corner represents both backup components are in cold standby,

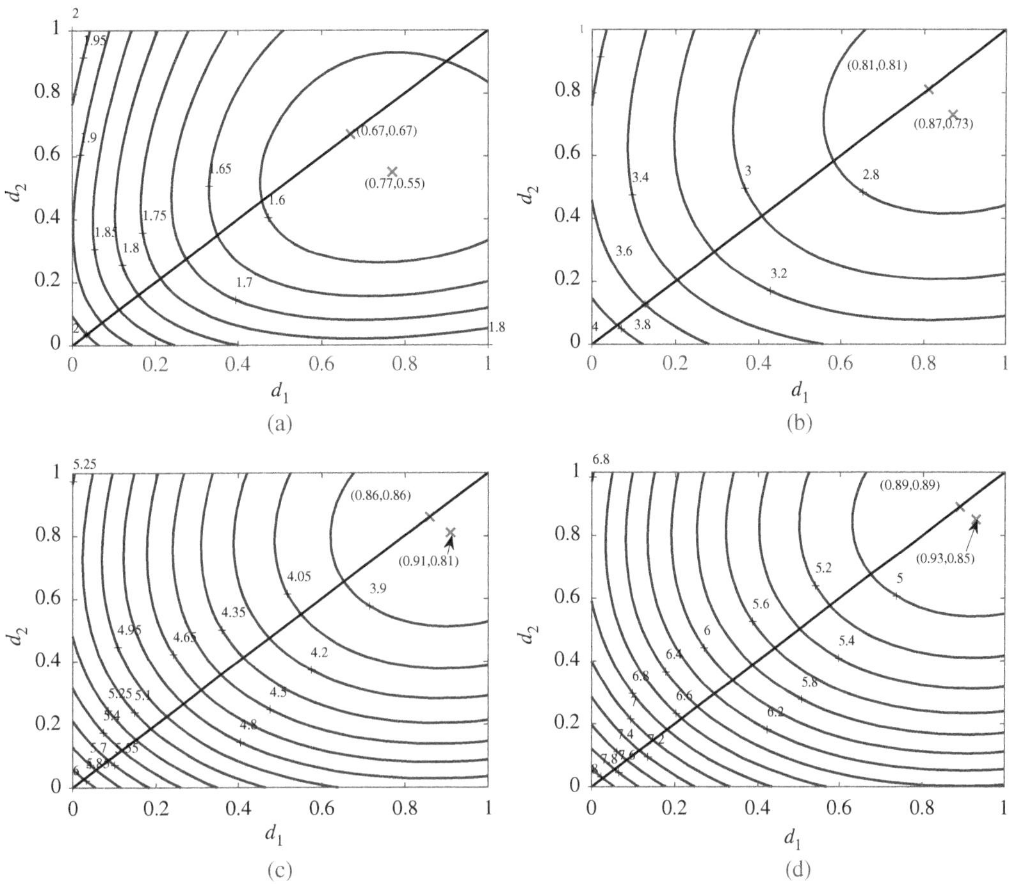

Figure 5.53 Contours of the average total cost of the system in Example 5.49. (a) $K = 1$ (1-out-of-3), (b) $K = 2$ (2-out-of-4), (c) $K = 3$ (3-out-of-5), and (d) $K = 4$ (4-out-of-6).

and the upper right corner represents both are in hot standby. Figure 5.53 shows that the optimum warmness levels are not on the diagonal line, meaning they are not identical.

A comparison of average total costs for several configurations is given in Table 5.15. The table shows a very small difference between identical and nonidentical warmness levels for systems with identical components. Thus, for practical reasons, systems with identical components should

Table 5.15 Comparison of average total costs of several system configurations.

K	Configuration	d_1	d_2	*AC*	% increase
2 (2-out-of-4 system)	Nonidentical warmness	0.87	0.73	2.70	–
	Identical warmness	0.81	0.81	2.71	0.30
	All hot	1.00	1.00	2.77	2.48
	All cold	0.00	0.00	4.00	48.04
4 (4-out-of-6 system)	Nonidentical warmness	0.93	0.85	4.82	–
	Identical warmness	0.89	0.89	4.83	0.12
	All hot	1.00	1.00	4.86	0.95
	All cold	0.00	0.00	8.00	66.00

have identical warmness levels. However, using backups only in hot or cold standby significantly increases the average cost. In addition, the more primary components, the greater the percentage increase in cost due to using all backup components in cold standby. It is due to the more frequent waiting for the readiness of the backup components to replace the failed primary components. Conversely, the more primary components, the lower the percentage increase in cost due to using all backup components in hot standby.

The main program to call **owl-am** for this example is given in **example5_49.m**. Its source code is as follows:

MATLAB code for example5_49.m

```
clear;close all
addpath('../owl-am')
% Number of primary components
K = 4;
% Number of backup components
B = 2;
% Contour step in plot
contourstep = 0.2;
% Constant
b = 1;
c = 2;
% Operation cost
Oi = 1;
% Opportunity cost
Li = 1.5;
% Vector of rate (lambda) for all components (choose one)
lambda  = ones(K+B,1); % Identic
lambda2 = lambda; % Non-identic
for i = 1:B
    lambda2(K+i) = lambda2(K+i-1)+0.5;
end
% Vector of primary components
d = ones(K,1);
g = d.^b;
h = (1-d).^c;
L = zeros(K,1);
O = ones(K,1);
% d vectors
deltad = 0.01;
d1 = 0:deltad:1;
d2 = 0:deltad:1;
% Calculate MTTF and Accumulated Cost for all possible d
dpos = [];
for i = 1:length(d1)
    for j = 1:length(d2)
        disp(['i = ',num2str(i),', j = ',num2str(j)])
        % Vector of backup components
        for k = 1:B
            if k == 1, x = d1(i); elseif k == 2, x = d2(j); end
            d(K+k) = x; h(K+k) = (1-x)^c; g(K+k) = d(K+k)^b;
            L(K+k) = Li; O(K+k) = Oi;
        end
        % Calculate MTTF and Accumulated Cost
        [ACval(i,j),TCval(i,j),MTTFval(i,j)] = ...
          calculateAC_markov(K,B,O,L,lambda,d,g,h,1);
    end
end
```

```
[AC_heterogen,idx] = min(ACval(:));
[row,col]    = ind2sub(size(ACval),idx);
% Find optimum warmness level (homogeneous)
% Calculate MTTF and Accumulated Cost for all possible d
for i = 1:length(d1)
    disp(['i = ',num2str(i)])
    % Vector of backup components
    for k = 1:B
        if k == 1, x = d1(i); elseif k == 2, x = d1(i); end
        d(K+k) = x; h(K+k) = (1-x)^c; g(K+k) = d(K+k)^b; L(K+k) = Li;
        O(K+k) = Oi;
    end
    % Calculate MTTF and Accumulated Cost
    [ACval2(i),TCval2(i),MTTFval2(i)] = calculateAC_markov(K,B,O,L,lambda,d,g,h,1);
end
[AC_homogen,idx] = min(ACval2);
% Show results in command window
disp('Results:')
d_homogen = d1(idx)
AC_homogen
d_heterogen = [d1(row) d2(col)]
AC_heterogen
AC_hot = ACval(end,end)
AC_cold = ACval(1,1)
%% Plot 2-D Contour
figure
[D2,D1] = meshgrid(d2,d1);
hct = contour(D1,D2,ACval,'b','LineWidth',2,'ShowText','off','LevelStep',...
  contourstep); hold on
clabel(hct,'LabelSpacing',1000,'Color','k','FontWeight','bold',...
  'FontName','Times New Roman','FontSize',15);
plot(D1(row,1),D2(1,col),'rx','MarkerSize',10,'LineWidth',2)
text(D1(row,1),D2(1,col),['  (',num2str(D1(row,1)),',',...
  num2str(D2(1,col)),')'],'FontName','Times New Roman','FontSize',15,...
  'FontWeight','bold');
plot(d_homogen,d_homogen,'rx','MarkerSize',10,'LineWidth',2)
text(d_homogen,d_homogen,['  (',num2str(d_homogen),',',...
  num2str(d_homogen),')'],'FontName','Times New Roman','FontSize',15,...
  'FontWeight','bold')
plot([0 1],[0 1]','k','LineWidth',2)
set(gca,'FontName','Times New Roman','FontSize',16,'LineWidth',1.5,...
  'FontWeight','bold');xlabel('\itd_1');ylabel('\itd_2')
```

■

Exercise 5.39 Complete Table 5.15 with $K = 1$ and 3, and check whether their behavior are consistent.

Example 5.50 A general standby system has identical components with

$$\lambda_i = 1, \quad OC_i = 1, \quad OL_i = 1.5, \quad i = 1, 2, \dots N, \quad \text{and} \quad b_j = 1, \quad c_j = 2, \quad j = K + 1, \dots, N.$$

All components have the same power requirements, or $a_j = a$. The power limitation constraint in Equation (5.78) can be simplified to

$$\sum_{j=1}^{K} d_j \leq d_{limit} \tag{5.96}$$

where $d_{limit} = \frac{A}{a}$ is the sum of the warmness levels of all backup components. If $d_{limit} \geq N\text{-}K$, then any warmness level can still be served by the available power. Thus, the power availability constraint is not binding, and the problem becomes unconstrained. The contour of the average total cost function for $d_{limit} = 1$ is presented in Figure 5.54 with the diagonal line and shading indicating the feasible region.

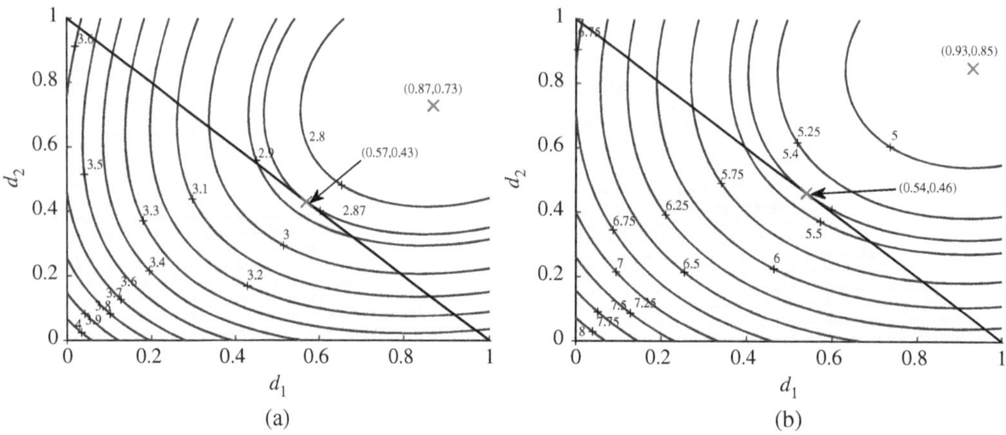

Figure 5.54 Contours of average total cost with power availability constraints. (a) 2-out-of-4 and (b) 4-out-of-6.

The presence of this constraint causes the optimum warmness level to shift from (0.87,0.73) to (0.57,0.43) for the 2-out-of-4 system and from (0.93,0.85) to (0.54,0.46) for the 4-out-of-6 system. It is due to the power availability not being sufficient for the unconstrained warmness levels, as can be seen that the unconstrained warmness levels are outside the feasible region. A $d_{limit} = 1.6$ makes the unconstrained optimum warmness levels for the 2-out-of-4 general standby system is in the feasible region, thus still its optimum warmness levels. In contrast, the unconstrained optimum warmness levels for the 4-out-of-6 general standby system for $d_{limit} = 1.6$ is still outside the feasible region; hence, it is still not the optimum warmness levels. ■

Exercise 5.40 Determine the value of d_{limit} to make the energy constraint becomes unbinding for both general standby systems in Example 5.50.

Exercise 5.41 What can be said about the values of d_{limit} that make the energy constraint becomes unbinding for 1-out-of-3 and 3-out-of-5 general standby systems for the same cost and component parameters as in Example 5.50?

Example 5.51 The average total costs as functions of d_{limit} for the systems in Example 5.50 are shown in Figure 5.55 for several values of K. The vertical axis is the percentage increase in the average total cost with the power constraint over the average total cost without the power constraint.

Figure 5.55 shows that the average total cost increases with decreasing d_{limit}. At $d_{limit} = 0$, all the backup components are in cold standby. The existence of opportunity loss makes the cold standbys not the optimal choice even though its operating cost is minimum. If all the backup components are in hot standby, then $d_{limit} = 2$ for Figure 5.55a and $d_{limit} = 4$ for Figure 5.55b, which are the numbers of backup components in those two systems. After reaching a certain value less than

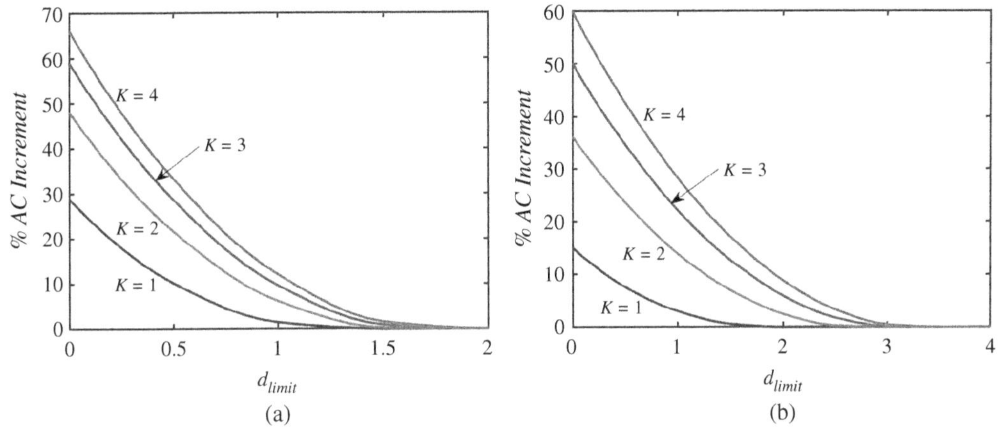

Figure 5.55 The effect of d_{limit} on average total cost. (a) K-out-of-($K+2$) and (b) K-out-of-($K+4$).

the number of backup components, increasing d_{limit} can no longer reduce the average total cost. It means using all backup only in hot standby is also not optimal even though its opportunity loss is zero. When the average total cost curve can no longer decrease, it shows that d_{limit} is too large and is no longer a constraint. The effective power requirement can be calculated from this d_{limit} value. The power supply equals to the effective power requirement is sufficient to produce an average total cost equal to the average total cost without the power limitation constraint. The availability of power exceeding the effective requirement is wasteful.

Table 5.16 shows the average total cost of a general standby system with four primary components. This table can be used to justify an increase in power supply. For example, increasing d_{limit} from 0.5 to 1.0 for a 4-out-of-6 system (using two backup components) would reduce the average total cost from 6.42 to 5.40, a savings of 1.02 times the unit operating cost of the component (since the unit operating cost is chosen equal to 1). If the cost of increasing the power capacity by 100% (from $d_{limit} = 0.5$ to $d_{limit} = 1.0$) is less than 1.02 times the unit operating cost of all backup components, then increasing power availability can be economically justified. The reduction in average total cost when d_{limit} is increased from 1.0 to 1.5 for a 4-out-of-6 system is 5.40 – 4.89 = 0.51 times the unit operating cost, not as great as the savings from increasing d_{limit} from 0.5 to 1.0. This means the law of diminishing returns of limited resources applies.

Table 5.16 Effect of d_{limit} on the average total cost for $K = 4$.

Number of backup components	Average total cost		
	$d_{limit} = 0.5$	$d_{limit} = 1$	$d_{limit} = 1.5$
2	6.42	5.40	4.89
4	7.84	7.06	6.44

■

Exercise 5.42 Complete Table 5.16 with the number of backup components equal to 1 and 3.

All the examples discussed so far are for general standby systems with identical components. Example 5.52 is for nonidentical components.

Example 5.52 A general standby system has several nonidentical components with parameters $\lambda_i = 0.5i$, $OC_i = 1$, $OL_i = 0.5$ for $i = 1, 2, \ldots, N$ and $b_j = 1$, $c_j = 2$, for $j = K+1, \ldots, N$. The optimum warmness level of each backup component and the optimum average total cost until failure are presented in Table 5.17.

Table 5.17 The effect of the number of primary and backup components on the optimum warmness level.

Number of primary components	Optimum warmness level								
	Backup components = 3				Backup components = 4				
	AC	d_1	d_2	d_3	AC	d_1	d_2	d_3	d_4
1	1.36	0	0	0	1.44	0	0	0	0
2	2.74	0.49	0.28	0.23	2.90	0.46	0.25	0.20	0.20
3	3.93	0.74	0.57	0.47	4.16	0.74	0.56	0.46	0.41
4	5.05	0.84	0.72	0.62	5.33	0.84	0.72	0.62	0.56
5	6.12	0.89	0.80	0.72	6.44	0.89	0.80	0.73	0.66

Although Table 5.17 shows the optimum warmness level follows the descending order rule, in certain cases with backup components having large differences in the values of OC and λ, the optimum warmness levels may not follow the descending order rule. Investigation of such cases is given in problems. ■

Exercise 5.43 Complete Table 5.17 with the number of backups = 1 and 2 and draw the graphs and contours of average total cost for all cases.

Real-world systems such as production lines operate in series–standby structures. The development of a method to determine optimum warmness levels should covers such systems as given in the next section.

5.5.5 Series–(General Standby) Systems

The method to find the optimum warmness levels of backup components is extended to series–(general standby) systems. Let the system consist of M general standby subsystems connected in series and $MTTF_m$ is the expected lifetime of the general standby subsystem m, $m = 1, 2, \ldots, M$. The average total cost of the entire system per unit time, AC_S, calculated using Equation (5.75) is

$$AC_S = \frac{TC_S}{MTTF_S}, \tag{5.97}$$

where TC_S and MTT_S are the total cost and expected lifetime of the system, respectively. Since general standby subsystems are connected in series, the system fails when one of the subsystems fails. Thus,

$$MTTF_S = \min(MTTF_1, MTTF_2, \ldots, MTTF_M). \tag{5.98}$$

TC_S is the sum of the expected total cost of each subsystem until the system fails, which is expected to be at $MTTF_S$. The expected total cost of the subsystem m, TC_m, is calculated using Equation (5.74) until this subsystem fails, which is expected to occur at $MTTF_m$. Thus, the expected total cost of the subsystem m until the system fails, $TC_{m,S}$, is

$$TC_{m,S} = \left(\frac{TC_m}{MTTF_m}\right) MTTF_S = (AC_m)(MTTF_S), \tag{5.99}$$

where AC_m is the average total cost of the subsystem m.

The total cost until the system fails is

$$TC_S = \sum_{m=1}^{M} TC_{m,S} \tag{5.100}$$

Substituting Equations (5.99) and (5.100) into Equation (5.97) obtains

$$AC_S = \sum_{m=1}^{M} AC_m \tag{5.101}$$

Equation (5.101) implies that minimizing AC_S without constraint is the same as minimizing AC_m, $m = 1, 2, \ldots, M$. Thus, the problem can be decomposed into problems of finding the optimum warmness levels for each general standby subsystem, and the global optimum warmness levels are obtained by aggregating the local optimum warmness levels of each subsystem. This conclusion is stated in the following Theorem 5.2.

Theorem 5.2 The optimum warmness levels for all backup components of a series–(general standby) system without constraints is the aggregate of the optimum warmness levels of the backup components of each subsystem.

The cost optimization formulation with power limitation constraints is

$$\begin{aligned} & min\, AC_S(\boldsymbol{d}_1, \boldsymbol{d}_2 \ldots, \boldsymbol{d}_M) \\ & \text{subject to} \sum_{m=1}^{M} \boldsymbol{a}_m^T \boldsymbol{d}_m \leq A \\ & \boldsymbol{0} \leq \boldsymbol{d}_m \leq \boldsymbol{1}, \quad m = 1, 2, \ldots, M \end{aligned} \tag{5.102}$$

where $\boldsymbol{a}_m$ and $\boldsymbol{d}_m$ are vectors of the full power requirements and the warmness levels of backup components in subsystem m, respectively. If $A \geq \sum_{m=1}^{M} \boldsymbol{a}_m \boldsymbol{d}_m^u$, where $\boldsymbol{d}_m^u$ is the unconstrained optimum warmness levels, then this constraint becomes unbinding and by Theorem 5.2 can decompose the problem into an unconstrained cost optimization for each subsystem. Providing a power supply to backup components more than A is wasteful because it will not reduce the average total cost. This result is stated as the following lemma.

Lemma 5.5 If the power availability for all backup components is greater than or equal to the power requirement to achieve the optimum warmness levels of backup components in each subsystem, then the problem of determining the optimum warmness levels of the series–(general standby) system can be decomposed into the unconstrained problem of determining the optimum warmness levels of each subsystem.

According to Lemma 5.5, if the power availability for all backup components is greater than or equal to the power requirement to achieve the optimum warmness level of each subsystem, then the global optimum warmness levels of the entire system are the aggregate of the local optimum warmness levels of each subsystem.

The software package **owl-am** can be utilized to find the optimum warmness levels of series–(general standby) systems.

Application 5.6 The method is applied to an eight-station serial production line for making casing and assembling component, with a workstation for each production stage. Table 5.18 presents the list of workstation numbers, the production process, and the number of primary machines to balance the line. Each station has three backups.

Table 5.18 Process description and the number of primary at each station.

Workstation number	Production process	Number of primary
1	Preparing sheet	1
2	Outer cutting	1
3	Inner cutting	2
4	Drilling	1
5	Coating and printing	2
6	Mounting modules	3
7	Assembling and wiring	2
8	Mounting accessories	1

Table 5.19 displays data for each workstation. All machines at each workstation, primaries and backups, are identical. The opportunity loss for each backup machine is equal to its operating cost, so $OL_m = OC_m$, and the parameter $b_m = 1$, $c_m = 2$, is the same for all backup machines.

Table 5.19 Workstation data for Application 5.2.

Station m	1	2	3	4	5	6	7	8
The failure rate of each machine (λ_m)	1	1.5	2	1.2	1.6	2.5	1	1.5
Number of primary machines (K_m)	1	1	2	1	2	3	2	1
Number of backup machines ($N_m - K_m$)	1	2	3	1	2	3	3	2
Power requirement of each machine (a_m)	1	1.5	1.5	1.5	1	2	1.5	1
Full operating cost of each machine (OC_m)	1	1.5	2	1	1	1	1.5	1.5

The optimum warmness levels without and with power constraints for this series–(general standby) system are given in Table 5.20.

This table shows that if the power supply is sufficient to meet the power requirement of the optimum warmness levels without constraints, then the optimum warmness levels with constraints are equal to the unconstrained warmness levels. The table shows a slight difference between the optimum value without constraints and the optimum value with power constraints equal to the power requirement of the optimum warmness level without constraints. This slight difference occurs because the optimization method used is a meta-heuristic method and there is a possibility that the solution is not unique. ■

Exercise 5.44 Complete Table 5.20 with a power constraint equal to 75% and 125% of the total power requirement for unconstrained optimum warmness levels.

The sequence of the subsystems in a series–(general standby) system does not affect the optimum warmness levels. For example, if subsystem 1 is switched with subsystem 3, the optimum warmness level of subsystem 1 after the sequence is switched is the same as before it is switched with subsystem 3. Any subsystem sequencing will produce the same optimum warmness level for each subsystem. The proof of this conclusion is given in Problems.

Table 5.20 Optimum warmness levels without and with power supply constraints.

Station m	1	2	3	4	5	6	7	8	AC_S
Backup					**Without power constraint**				
1	0.64	0.77	0.91	0.71	0.88	0.95	0.80	0.77	24.51
2		0.56	0.80		0.75	0.89	0.61	0.55	
3			0.71			0.83	0.44		
Backup					**With power constraint** $A = \frac{1}{2}\sum_{m=1}^{M}\sum_{j=1}^{N_m-K_m} a_{jm}d_{jm}^u$ **(binding)**				
1	0.17	0.21	0.70	0.12	0.61	0.66	0.23	0.37	26.77
2		0.12	0.57		0.48	0.57	0.10	0.24	
3			0.48			0.51	0.04		
Backup					**With power constraint** $A = \sum_{m=1}^{M}\sum_{j=1}^{N_m-K_m} a_{jm}d_{jm}^u$ **(unbinding)**				
1	0.64	0.77	0.90	0.71	0.89	0.95	0.79	0.78	24.51
2		0.55	0.81		0.75	0.89	0.61	0.55	
3			0.70			0.83	0.44		

5.5.6 Discrete Warmness Levels

Warmness levels, in practice, are generally given as ordinal discrete values because they are easier to set. For example, the warmness levels are expressed in ordinal values 0 (cold, C) to 5 (hot, H), with the values in between indicating the levels of warmness such as slightly warm (SW), warm (W), too warm (TW), and almost hot (AH). The correspondence of these qualitative measures to the quantitative warmness levels can be made in identical or nonidentical intervals. Table 5.21 is an example of a discrete warmness level spectrum with identical intervals.

The formulation of the discrete warmness level optimization problem can be derived from Equation (5.102) by replacing continuous variables d_{jm} with values between 0 and 1 into discrete

Table 5.21 Examples of discrete warmness levels and their descriptions.

Discrete warmness levels	0	1	2	3	4	5
Description of warmness	C	SW	W	TW	AH	H
Warmness level value	0.0	0.2	0.4	0.6	0.8	1.0

variables with certain possible values. The cost optimization formulation with power constraints for discrete warmness levels is

$$\min\ AC_S(\boldsymbol{d}_1, \boldsymbol{d}_2 \ldots, \boldsymbol{d}_M)$$

$$\text{subject to} \sum_{m=1}^{M} \boldsymbol{a}_m \boldsymbol{d}_m \le A$$

$$\boldsymbol{d}_m = \left\{ d^i_{jm}, i = 1, 2, \ldots, I_{jm}; j = 1, 2, \ldots, N_m - K_m \right\} \tag{5.103}$$

where $\left\{ d^i_{jm}, i = 1, 2, \ldots, I_{jm}; j = 1, 2, \ldots, N_m - K_m \right\}$ is the set of possible discrete values of $\boldsymbol{d}_m$, $m = 1, 2, \ldots, M$.

Application 5.6 (continued). The optimum discrete warmness levels based on Table 5.21 for the system in Application 5.6 calculated using Equation (5.103) are given in Table 5.22. The calculation also utilized **owl-am**.

Table 5.22 Optimum discrete warmness levels for Application 5.5.

Station m	1	2	3	4	5	6	7	8	AC_S
Backup				Without power constraints					
1	0.6	0.8	1	0.8	0.8	1	0.8	0.8	24.58
2		0.6	0.8		0.8	0.8	0.6	0.6	
3			0.8			0.8	0.4		
Backup				With power constraint $A = \frac{1}{2}\sum_{m=1}^{M}\sum_{j=1}^{K_m} a_{jm} d^u_{jm}$					
1	0	0.2	0.8	0.2	0.6	0.6	0.2	0.4	26.88
2		0	0.6		0.6	0.6	0.2	0.2	
3			0.4			0.6	0		

Comparing Tables 5.20 and 5.22 shows that the optimum discrete warmness levels cannot be obtained directly by rounding the optimum continuous warmness levels to the nearest discrete values. The average total cost using discrete warmness levels is slightly higher than using continuous one. This is realistic considering that the search space for discrete values is smaller than continuous ones.

Exercise 5.45 Calculate the discrete optimum warmness levels for Tables 5.15 and 5.17.

The development so far assumes that warmness levels are fixed for the entire operation. With the availability of new technology, it is plausible to have warmness levels adjusted during operation to make warmness levels become dynamic.

5.6 Dynamic Warmness Level

One of the advantages of modeling general standby systems using state transition diagrams is that the warmness level of each backup component can be made dynamic following its sequence to become a primary component. In a general standby system with dynamic warmness levels, the warmness level of a backup component must be adjusted each time the backup list changes. Current technology makes it possible to adjust the warmness levels of backup components online whenever needed. When a primary component fails, the list of the backup components changes, shifting one order up: the first backup component becomes the primary, the second backup component becomes the first backup, and so on. Given that the optimum warmness level of the backup components in most cases is in descending order, adjusting the warmness level usually increases the warmness level of the backup components.

Dynamic warmness levels are very useful in reducing the time required to prepare the first backup component to become the primary component and at the same time reduce operating costs for the next backups. For example, if the first backup component is in hot standby, then there is no need to increase its warmness level when it replaces the failed primary component. At that time, the warmness level of the second backup component is increased to hot because it is now becoming the first backup component. When the primary component fails again, the second backup component can be operated immediately. In this case, there is a saving in waiting for the second backup component to replace the next failed primary component, and, at the same time, it saves operating costs because the second backup component is operated as a non-hot standby until the first failure on the primary component occurs. Backup components that are not in cold standby may fail. If this happens, only the warmness levels of the backup components in the sequence below the failed backup component are adjusted.

One form of realization of the dynamic warmness level of the backup components in a 2-out-of-6 general standby system is illustrated in Table 5.23.

Table 5.23 2-out-of-6 general standby system with dynamic warmness levels.

Event	System state	Backup components	Deceleration matrix					
			1	2	3	4	5	6
Beginning of operation	(p, p, b, b, b, b)	{3, 4, 5, 6}	$\boldsymbol{I}$	$\boldsymbol{I}$	$\boldsymbol{D}_1$	$\boldsymbol{D}_2$	$\boldsymbol{D}_3$	$\boldsymbol{D}_4$
Primary 1 fails, backup 1 becomes primary, warmness levels of backups 2, 3, and 4 are adjusted from $\boldsymbol{D}_2, \boldsymbol{D}_3, \boldsymbol{D}_4$ to $\boldsymbol{D}_1, \boldsymbol{D}_2, \boldsymbol{D}_3$.	(f, p, p, b, b, b)	{4, 5, 6}	–	$\boldsymbol{I}$	$\boldsymbol{I}$	$\boldsymbol{D}_1$	$\boldsymbol{D}_2$	$\boldsymbol{D}_3$
Backup 2 fails, warmness level of backup 3 is adjusted from $\boldsymbol{D}_3$ to $\boldsymbol{D}_2$.	(f, p, p, b, f, b)	{4, 6}	–	$\boldsymbol{I}$	$\boldsymbol{I}$	$\boldsymbol{D}_1$	–	$\boldsymbol{D}_2$
Primary 2 fails, backup 1 becomes primary, warmness level of backup 2 is adjusted from $\boldsymbol{D}_2$ to $\boldsymbol{D}_1$.	(f, p, f, p, f, b)	{6}	–	$\boldsymbol{I}$	–	$\boldsymbol{I}$	–	$\boldsymbol{D}_1$
Backup 1 fails, no adjustments.	(f, p, f, p, f, f)	{}		$\boldsymbol{I}$	–	$\boldsymbol{I}$	–	–

Example 5.53 A 2-out-of-6 general standby system with dynamic warmness levels has six identical components whose lifetimes follow an independent 4-Erlang distribution with $MTTF = 10$ days. The warmness level of the jth backup component, $j = 1, 2, 3, 4$, is

$$d_j = \frac{(N-K-j)}{(N-K-1)}, \quad j = 1, \ldots, 4.$$

Thus, the warmness levels of the first, second, third, and fourth backup components in the beginning of operations are $d_1 = 1$, $d_2 = 2/3$, $d_3 = 1/3$, $d_4 = 0$, respectively, following the optimum sequence from the highest to the lowest warmness level.

A comparison between the reliability and hazard functions with dynamic and static warmness levels is presented in Figure 5.56. Their *MTTF* and *SDTTF* are 16.083 and 3.726, respectively, for the dynamic warmness level and 18.849 and 4.613, respectively, for the static warmness level.

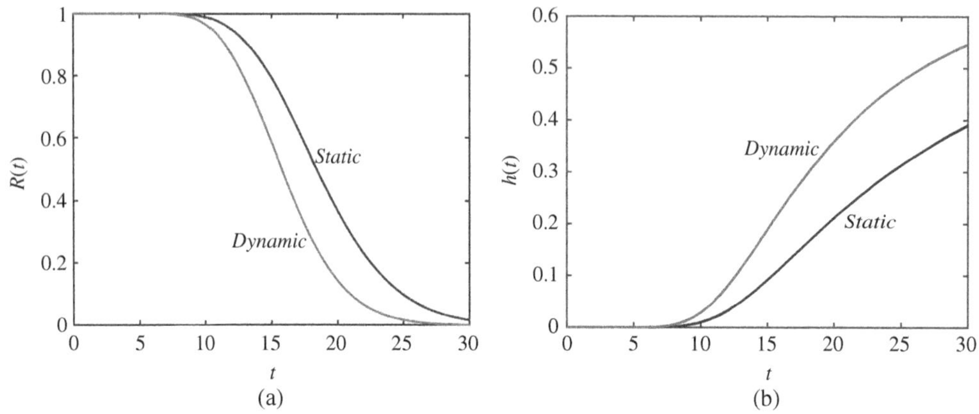

Figure 5.56 Reliability and hazard functions of a general backup system with dynamic and static warmness levels.

The resulting numbers and figures show that the dynamic warmness level has worse reliability than the static warmness level. It makes sense because the static warmness level does not increase the warmness level of the backup component when a component fails. This deficiency is compensated for by the ability of the system with a dynamic warmness level to prepare for the takeover of the failed primary component more quickly. This speed will reduce the opportunity loss of waiting for the readiness of the backup component while maintaining its operation cost low. ■

Exercise 5.46 Illustrate a possible realization of a 2-out-of-7 general standby system with dynamic warmness levels similar to Table 5.23.

Exercise 5.47 Repeat Example 5.53 for a 2-out-of-7 general standby system.

Dynamic warmness levels can be directly applied to complex systems that have been discussed in the previous sections by selecting the appropriate MATLAB function module. The use of the concept of dynamic warmness levels in determining the optimum warmness level requires different cost models from the static warmness level cost model. This problem is attractive to future research. Likewise, extending the method to model and to find optimum warmness levels in general standby systems with components having PH distributed lifetimes is still an open challenge for future research.

5.7 Problems

5.1 Determine the orders of PH distributions of lifetimes of systems in Examples 5.15 and 5.17, assuming all modules have PH-distributed lifetimes of the same order.

5.2 Compare the orders of PH distributions of lifetimes of (K-out-of-N)-series and series-(K-out-of-N) systems having the same components. What is the conclusion of this comparison? Is this conclusion related to the fact that a series-(K-out-of-N) system has a higher reliability function than (K-out-of-N)-series systems with the same components?

5.3 Explain why a series-(K-out-of-N) system has higher reliability than a (K-out-of-N)-series system with the same components. Can this proposition be proven mathematically?

5.4 Determine the PH representation of the lifetime distribution of a 1-out-of-2 system having two identical components whose lifetimes follow a 2-Erlang distribution with $MTTF = 200$, if the backup component is operated as:
a) Hot standby.
b) Cold standby.
c) Warm standby with a deceleration factor of 0.5.

5.5 Determine the diagonal block matrix that represents:
a) State $(1, 0, 1, 0, 1, 1)$ in 2-out-of-6 and 3-out-of-6 systems
b) State $(0, 1, 1, 1, 0, 1)$ in 2-out-of-6 and 3-out-of-6 systems.
c) Repeat questions a. and b. for general standby systems.

5.6 Determine the off-diagonal block matrices representing the transition matrices from state $(1, 0, 1, 0, 1, 1)$ to all possible states in
a) A 2-out-of-6 system.
b) A 2-out-of-6 general standby system

5.7 Devise an algorithm for **megenidnsys**.

5.8 Create a state transition diagram of 2-out-of-4 cold standby system that is more efficient (fewer transitions) than using the state transition diagram of a general standby system. Use the result to explain a method to reduce the state transition diagram of a general standby system when all or some backup components are cold standby.

5.9 Based on the result of Problem 8, devise an algorithm to generate the survival matrix of a K-out-of-N cold standby system. Compare the order of PH distribution of the system lifetime derived using general standby structure and using cold standby structure and computation time for producing system reliability measures for several K and N. Draw a conclusion.

5.10 Investigate the effect of component ordering based on hazard function on a general standby system when the primary component is also included. Thus, when using descending order, the primary components are those with the highest failure rates. Demonstrate through experiments and logical considerations that ordering all components according to the optimum ordering rules will produce better reliability measures than ordering only the backup components.

5.11 Prove the order of the backup components or the order of all components has no effect on the reliability measures of the N-standby and N-parallel systems.

5.12 Modify the reliability model and the main program for Application 5.2 if the same machines in different workstations are compatible.
a) Compare the order and computation between using compatible and incompatible machines.
b) Compare also the reliability and hazard functions.

5.13 All backup machines at each workstation in the production line in Application 5.2 have the same warmness level. The sum of the warmness level values is limited to no more than half the number of backup machines.
a) Which is better, giving a higher warmness value to workstations with fewer backup machines or to workstations with more backup machines?
b) Repeat question a. from the point of view of lower or higher *MTTF*.
c) Repeat questions a. and b. by considering both the number of backup machines and the *MTTF*s.

5.14 Modify the main program of Application 5.4 if the company has three identical production lines and only two of them are needed, if
a) The backup line is only operated if one of the production lines fails.
b) The backup line is operated simultaneously with the primary lines.
c) The backup line is ready to operate with a readiness level of 0.5.

5.15 Write as general standby systems the models of
a) parallel-(K-out-of-N) systems and (K-out-of-N)-parallel systems.
b) parallel–standby systems and standby–parallel systems.

5.16 Write the reliability model of the system in Application 5.5 operated using scenario 3 for dynamic load.

5.17 Design another scenario for Application 5.5 and construct its model.

5.18 Investigate the effect of ordering the backup components in a K-out-of-N general standby system based on the hazard function for $K = 3$.

5.19 Investigate the optimal order of backup components if the ith backup component has non-identical hazard function $h_i(t)$ and warmness level d_i.

5.20 Manually calculate the optimum warmness level of a 2-out-of-3 general standby system with the same parameter values as Example 5.45. Verify the results using the package software **owl-am**.

5.21 Too small of a unit opportunity cost causes the cold standby to be the optimum. Determine the largest opportunity cost value so that the cold backup remains the optimum for several values of K and N.

5.22 Show that the average total cost for a nonidentical general standby system depends on the value of $L/(O/\lambda)$. Thus, it is possible for a backup component lower in the backup list to have a higher warmness level than the backup component higher in the backup list.
a) Conduct an experiment to show such a possibility.
b) Prove that optimum warmness levels depend on $L/(O/\lambda)$.

5.23 Extend Theorem 5.2 to systems with backup components having different power requirements.

5.24 Define $d_{limit}(min)$ as the smallest value of d_{limit} at which the power availability constraint is still non-binding. Plot the graph of $d_{limit}(min)$ as a function of the number of primary components with the number of backup components as the legend.

5.25 Show mathematically and experimentally that the sequence of workstations in a series–(general standby) system has no effect on the optimum warmness level of the backup machines at each workstation.

5.26 Show that optimum warmness level ordering for static warmness levels holds for dynamic warmness levels.

References

1 Wells, C.E., 2014. Reliability analysis of a single warm-standby system subject to repairable and nonrepairable failures. European Journal of Operational Research 235, 180–186. https://doi.org/10.1016/j.ejor.2013.12.027.

2 Eryilmaz, S., 2012. On the mean residual life of a k-out-of-n: G system with a single cold standby component. European Journal of Operational Research 222, 273–277. https://doi.org/10.1016/j.ejor.2012.05.012.

3 Eryilmaz, S., 2013. Reliability of a K-out-of-n system equipped with a single warm standby component. IEEE Transactions on Reliability 62, 499–503. https://doi.org/10.1109/TR.2013.2259202.

4 Roy, A., Gupta, N., 2020. Reliability function of k-out-of-n system equipped with two cold standby components. Communications in Statistics - Theory and Methods 50(24), 5759–5778. https://doi.org/10.1080/03610926.2020.1737122.

5 Levitin, G., Xing, L., Xiang, Y., 2020. Series phased-mission systems with heterogeneous warm standby components. Computers and Industrial Engineering 145, 106552. https://doi.org/10.1016/j.cie.2020.106552.

6 Ma, X., Liu, B., Yang, L., Peng, R., Zhang, X., 2020. Reliability analysis and condition-based maintenance optimization for a warm standby cooling system. Reliability Engineering and System Safety 193, 106588. https://doi.org/10.1016/j.ress.2019.106588.

7 Yen, T.-C., Wang, K.-H., 2020. Cost benefit analysis of four retrial systems with warm standby units and imperfect coverage. Reliability Engineering and System Safety 202, 107006. https://doi.org/10.1016/j.ress.2020.107006.

8 Alkaff, A., Qomarudin, M.N., 2020. Modeling and analysis of system reliability using phase-type distribution closure properties. Applied Stochastic Models in Business and Industry 36, 548–569. https://doi.org/10.1002/asmb.2509.

9 Jia, H., Liu, D., Li, Y., Ding, Y., Liu, M., Peng, R., 2020. Reliability evaluation of power systems with multi-state warm standby and multi-state performance sharing mechanism. Reliability Engineering and System Safety 204, 107139. https://doi.org/10.1016/j.ress.2020.107139.

10 Levitin, G., Xing, L., Luo, L., 2019. Influence of failure propagation on mission abort policy in heterogeneous warm standby systems. Reliability Engineering and System Safety 183, 29–38. https://doi.org/10.1016/j.ress.2018.11.006.

11 Liu, B., Wen, Y., Kang, S., Qiu, Q., 2020. A multiple warm standby repairable system under N-policy with multiple vacations following Markovian arrival process. Communications in Statistics - Theory and Methods 49, 3609–3634. https://doi.org/10.1080/03610926.2019.1710758.

12 Wu, H., Li, Y.-F., Bérenguer, C., 2020. Optimal inspection and maintenance for a repairable k-out-of-n: G warm standby system. Reliability Engineering and System Safety 193, 106669. https://doi.org/10.1016/j.ress.2019.106669.

13 Jia, H., Levitin, G., Ding, Y., Song, Y., 2019. Reliability analysis of standby systems with multi-state elements subject to constant transition rates. Quality and Reliability Engineering International 35, 318–328. https://doi.org/10.1002/qre.2401.

14 Cha, J.H., Mi, J., Yun, W.Y., 2008. Modelling a general standby system and evaluation of its performance. Applied Stochastic Models in Business and Industry 24, 159–169. https://doi.org/10.1002/asmb.704.

15 Li, X., Zhang, Z., Wu, Y., 2009. Some new results involving general standby systems. Applied Stochastic Models in Business and Industry 25, 632–642. https://doi.org/10.1002/asmb.770.

16 Alhadeed, A.A., Yang, S.-S., 2005. Optimal simple step-stress plan for cumulative exposure model using log-normal distribution. IEEE Transactions on Reliability 54, 64–68. https://doi.org/10.1109/TR.2004.841704.

17 Levitin, G., Xing, L., Dai, Y., 2017. Reliability versus expected mission cost and uncompleted work in heterogeneous warm standby multiphase systems. IEEE Transactions on Systems, Man, and Cybernetics: Systems 47, 462–473. https://doi.org/10.1109/TSMC.2015.2505643.

18 Levitin, G., Xing, L., Dai, Y., 2013. Optimal sequencing of warm standby elements. Computers and Industrial Engineering 65, 570–576. https://doi.org/10.1016/j.cie.2013.05.001.

19 Lin, Y.-H., Li, X.-Y., Kang, R., 2017. A state transfer scheduling optimization framework for standby systems. IEEE Transactions on Reliability 66, 1101–1109. https://doi.org/10.1109/TR.2017.2760918.

20 Alkaff, A., 2021. Discrete time dynamic reliability modeling for systems with multistate components. Reliability Engineering and System Safety 209, 107462. https://doi.org/10.1016/j.ress.2021.107462.

21 Amari, S.V., Pham, H., Misra, R.B., 2012. Reliability characteristics of k-out-of-n warm standby systems. IEEE Transactions on Reliability 61, 1007–1018. https://doi.org/10.1109/TR.2012.2220891.

22 Peng, R., Zhai, Q., Yang, J., 2021. Reliability Modelling and Optimization of Warm Standby Systems. Springer-Verlag, Singapore.

23 Alkaff, A., Qomarudin, M.N., Purwantini, E., Wiratno, S.E., 2021. Dynamic reliability modeling for general standby systems. Computers and Industrial Engineering 161, 107615. https://doi.org/10.1016/j.cie.2021.107615.

24 Shaked, M., Shanthikumar, J.G., 2007. Stochastic Orders. Springer.

25 Zhai, Q., Yang, J., Peng, R., Zhao, Y., 2015. A study of optimal component order in a general 1-out-of-n warm standby system. IEEE Transactions on Reliability 64, 349–358. https://doi.org/10.1109/TR.2014.2337073.

26 Alkaff, A., 2023. Optimum warmness levels in general standby systems. Reliability Engineering and System Safety 237, 109366. https://doi.org/10.1016/j.ress.2023.109366.

6

Network Structured Systems

Learning objectives:

1) Understand the derivations and proofs of all equations for network reliability modeling.
2) Able to construct reliability models of network structured systems.
3) Able to evaluate the reliability of general networks including capacitated networks.
4) Able to utilize the software for reliability analyses of complex networks.

Abbreviation

CTMC	continuous-time Markov chain
IE	inclusion–exclusion
ME	matrix-exponential
PH	phase-type
SDP	sum of disjoint products
CVTTF	coefficient of variations of time to failure
MC	minimum cut
MP	minimum path
MRL	mean residual life
MTTF	mean time to failure
SDTTF	standard deviation of time to failure
VTTF	variance of time to failure

Notations

T	time to failure, lifetime
t	time, mission duration
g	age
n	order, dimension
$\boldsymbol{a}$	initial probability vector
$\boldsymbol{A}$	survival matrix
$\boldsymbol{I}$	identity matrix
$\boldsymbol{u}$	unitary vector

$\boldsymbol{b}$	failure probability vector
K	number of disjoint products
N	number of components in a system
$(n_i, \boldsymbol{a}_i, \boldsymbol{A}_i)$	PH representation of lifetime distribution of component i
$(n_i, \boldsymbol{a}_i, \boldsymbol{A}_i, \boldsymbol{v}_i)$	ME representation of lifetime distribution of component i
$(n_S, \boldsymbol{a}_S, \boldsymbol{A}_S)$	PH representation of lifetime distribution of network
$(n_S, \boldsymbol{a}_S, \boldsymbol{A}_S, \boldsymbol{v}_S)$	ME representation of lifetime distribution of network
$R(t)$	reliability function
$RR(t \mid g)$	residual reliability function of system with age g
$h(t)$	hazard function
$MRL(g)$	mean residual life function
$\otimes$	Kronecker product
$\oplus$	Kronecker sum
$\vee$	disjunction
$\wedge$	conjunction
X_S	system state
X_i	state of component i
$\overline{X}$	negation or inverse of X
Y	disjoint product term
DP	disjoint product
D	number of variables in a disjoint product term
M_P	number of minimum paths
M_C	number of minimum cuts

Networks are frequently encountered and used in daily life, such as electricity networks, telephone networks, and transportation networks. Activities can come to a halt or be severely disrupted when a network malfunctions. Given the critical role of networks, one important consideration in network design is their reliability. Subsequently, the term network reliability emerged, which measures the probability that a network will be able to perform its function properly. Systems with a network structure are more complex than systems with structures discussed in previous chapters. This complexity causes network reliability analysis to require special methods. Such methods have subsequently evolved to also be used for systems that are physically not networks, such as electromechanical systems, whose reliability studies are represented as networks [1].

The uniqueness of a network structure, which differs from other structures, is that the network has one or more components that function as bridges, namely, components that connect one subsystem to another. The presence of these bridges makes the system structure cannot be expressed in terms of basic structures such as series, parallel, standby, or any combination of them. Consequently, the methods discussed in Chapters 4 and 5 cannot be used directly to analyze network reliability. Special methods that consider the uniqueness of the network need to be developed for network reliability analysis. Many analytical modeling are available for such purpose, as discussed in Chaturvedi [2]. Basically, there are two popular analytical models in network reliability analysis. The first model is the classical method based on the principle of inclusion and exclusion (IE) in probability theory, which is then known as the IE model. The second model uses the concept of sum of disjoint products (SDP) in probability theory, which was introduced later that is more structured and efficient in many ways, especially from a programming perspective. It is referred to as the SDP model.

SDP models currently are available for static reliability analysis using numerical values of component reliability to determine the numerical value of network reliability. For practical purpose, however, dynamic analysis of network reliability using reliability functions is highly desirable because it can provide more detailed information that cannot be obtained from static analysis [3, 4]. The problem is that dynamic reliability analysis processes component reliability functions as input to produce network reliability functions as output is not easy. As explained in previous chapters, this difficulty can be overcome if the component reliability functions are expressed in a matrix-form distribution, because such a distribution has closure properties that are useful in simplifying dynamic network reliability analysis.

This chapter discusses the use of matrix-form distributions for dynamic network reliability analysis as proposed in Alkaff [5] and Alkaff et al. [6]. The existing static SDP model for static network reliability analysis is developed further to be applicable to dynamic network reliability analysis where the lifetimes of all components follow matrix-form distributions. First, it is necessary to derive the closure properties of the matrix-form distribution in network structures in SDP representation. These closure properties differ from the closure properties of the matrix-form distribution in CTMC employed in Chapters 4 and 5. For simplicity and flexibility, network reliability analysis needs to use ME distribution. Using the ME distribution will simplify the proof of its closure properties in an SDP representation. Obviously, the method developed can still be used for PH distribution since it is a special case of ME distribution. Practically, the distributions of component lifetimes are still expressed in PH distributions, while the output is the network lifetime distribution in an ME representation. All the advantages of using the PH distribution for dynamic system reliability analysis still prevail.

The main difference between network reliability analysis methods and the system reliability analysis methods discussed previously lies in the models used. Chapter 4 explained two models to system reliability analysis: the state transition model and the combinatorial model. While both models can be used for all types of structures, including network structures, in practice, the combinatorial model is more efficient for systems with network structures. The functioning (or failure) of a network in the combinatorial model is expressed as a combination of the functioning (or failure) of its components. This model does not require a state transition diagram of CTMC. The IE and SDP models mentioned above are derivatives of the combinatorial model. For completeness, this book also develops the state transition model for network reliability analysis where the network is presented in state transition diagram, similar to that have been used in pervious chapters. Comparison between the two models is discussed.

This chapter begins with the derivation of network structure functions using a combinatorial model. Network structure functions can be derived from the relationship between the network being good and its components being good. Network structure functions can also be derived from the relationship between the network being failed and its components being failed. These relationships are expressed in Boolean operations on Boolean variables representing the states of network components that results in a Boolean function. Since the reliability function is a real-valued function calculated using arithmetic operations, the Boolean function must be converted into an arithmetic function. A method for converting a Boolean structure function into an SDP representation will be discussed. That the ME distribution is closed under arithmetic operations in the SDP representation is proved. These closure properties are utilized to design an algorithm to find an ME representation of the network reliability function from the ME or PH representations of the reliability functions of its components. The algorithm is then combined with a capacitated network analysis algorithm to determine the capacitated network reliability and its optimization. In the final section, network modeling using the state transition model is presented, along with comparisons between using the state transition and combinatorial models in terms of their complexities and computation time.

One thing to keep in mind is that network reliability analysis is more complex than non-network system reliability analysis. Therefore, as long as it is still possible to use the methods described in Chapters 4 and 5, the dynamic network reliability analysis methods presented in this chapter should be avoided.

6.1 Network Structure

A network is generally represented as a graph with a flow. A graph is a set of nodes connected by a set of arcs. Thus, the components of a network are nodes or arcs or both depending on the purpose of the analysis. In network reliability analysis, there is a source node (or nodes) as the origin of the flow and a target node (or nodes) as the destination of the flow so that the flow is from the source node(s) to the target node(s). The arcs connecting the nodes can have a certain direction from one starting (tail) node to one ending (head) node. Arcs can also be undirected, in which case the flow can occur in both directions (back and forth). A network is said to be functioning or in good status if the network is able to deliver flows from the source node(s) to the target node(s). Conversely, a network is said to fail if the connection from the source node(s) to the target node(s) is broken due to failures of one or more network components, so that the source node cannot deliver flows to the target node, or the amount of flow does not meet a requirement. This book defines network structure as a structure in which there is at least one component that functions as a bridge to connect one subsystem with one or more other subsystems so that the subsystems cannot be said to have series or parallel structures.

Example 6.1 The simplest network structure is as shown in Figure 6.1. This network serves to send flow from source node A to target node D using arcs 1, 2, 3, 4, and 5. To reach the target node, intermediate nodes B and C are utilized. For the purpose of network reliability analysis, network arcs are generally directional, with the overall direction being from the source node to the target node. There are also arcs that allow two reversible directions if both directions can be used to send flow from the source node to the target node. The arcs represented by components 1, 2, 4, and 5 have one direction (unidirectional) required to send flow from source A to target D. The arc represented by component 3 has two directions (bidirectional) that can be used to send flow from source node A to target node D. If the source node or target node changes, then the direction of flow in the arcs can also change following the flow direction. As a convention, this book uses alphabet and numeric as notations for nodes and arcs, respectively. ■

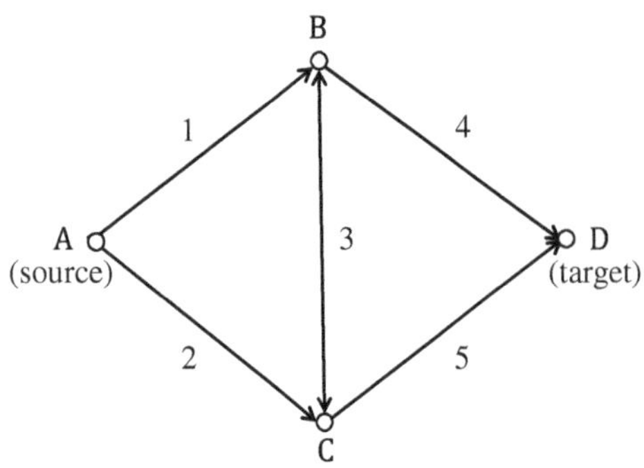

Figure 6.1 Simple network.

In network reliability analysis, network components are considered to be parts of the network that can fail or assumed can fail, while parts of the network that cannot fail or are assumed to be never fail are not network components. For the purpose of modeling network structures, network components need to be assigned different identities. To make network reliability diagram consistent with system reliability block diagram, this book represents network components in rectangular block. Such a consistency is necessary since complex systems may composed of basic and network structures.

Example 6.2 If it is assumed that nodes cannot fail, then the network components are its arcs. With this assumption, the network in Figure 6.1 for reliability analysis is presented as Figure 6.2 with network components are represented in rectangle and given identity 1, 2, 3, 4, and 5. Component 3 in this network functions as a bridge that connects the upper part of the network to the lower part. Although, in this example, component 3 has two directions, this does not mean that a bridge must have two directions. Component 3 can have one direction, for example, from top to bottom, and still functions as a bridge. The presence of component 3 makes the network cannot be analyzed as series or parallel structures. Because of component 3, components 1 and 4, as well as components 2 and 5, are not connected in a series structure. For the same reason, components 1 and 2, as well as components 4 and 5, are not connected in a parallel structure. ∎

Figure 6.2 Simple network with arcs as components.

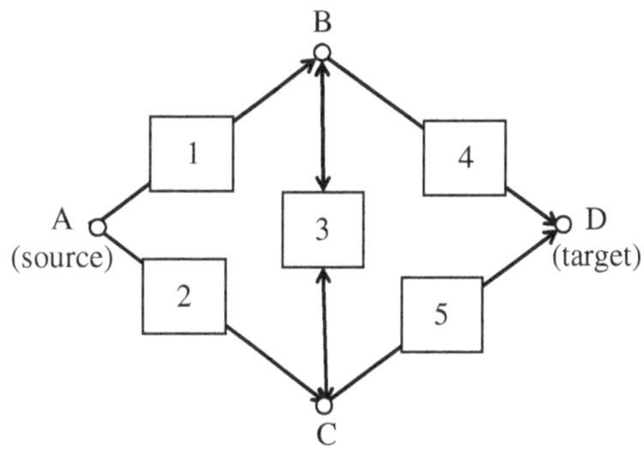

There are several other variations related to the network reliability analysis for the network in Figure 6.1 depending on the assumed failures of its components.

Example 6.3 If the nodes of a network can fail and the arcs cannot, then the components of the network are the nodes. A graph representation of the network in Figure 6.1 with nodes as its components is given in Figure 6.3a.

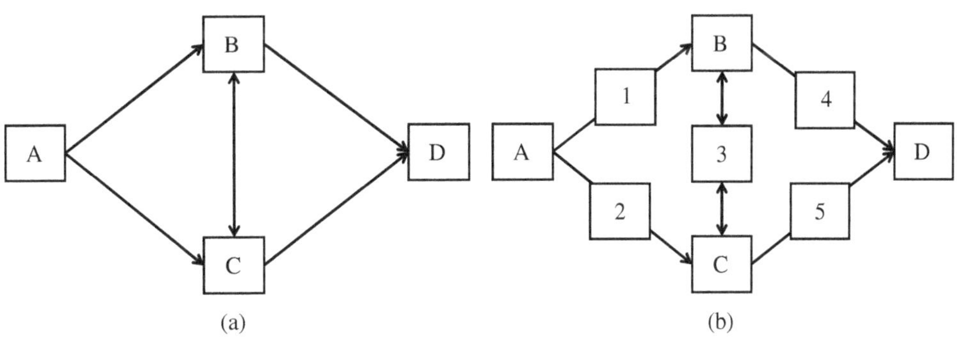

Figure 6.3 Simple network with different component definitions. (a) Nodes are components; (b) nodes and arcs are components.

If nodes and arcs can fail, then the components of the network are the nodes and arcs. A graph representation of the network in Figure 6.1 with arcs and nodes as its components is given in Figure 6.3b. ■

It is possible that not all nodes or arcs in a network can fail; only some of them can fail. Network diagrams can be adapted to the nodes and arcs that are components.

Henceforth, systems with network structures are referred to as networks, and the reliability of systems with network structures is termed network reliability.

Exercise 6.1 Create a reliability block diagram of Example 6.1 to represent a network with nodes B and C, and arcs 1, 2, 4, and 5, as its components.

6.2 Network Reliability Modeling

Reliability analysis of networks is generally simpler using a combinatorial model, which represents the network state as a combination of the states of its components. As explained in Section 4.1, the combinatorial model uses the states of the network and its components as state variables, namely:

- X_S is a variable representing the network state, and
- X_i is a variable representing the state of component i, $i = 1, 2, \ldots, N$.

The relationship between the network state and the states of its components can be expressed as

$$X_S = \varphi(X_1, X_2, \ldots, X_N), \tag{6.1}$$

or

$$\overline{X}_S = \psi(\overline{X}_1, \overline{X}_2, \ldots, \overline{X}_N), \tag{6.2}$$

with φ and ψ represent network structure functions whose equations depend on the network structure being modeled. Equation (6.1) expresses the relationship between the good states of the components (X_i, $i = 1, = 2, \ldots, N$) and the good state of the network (X_S), while Equation (6.2) expresses the relationship between the failed states of the components ($\overline{X}_i, i = 1, 2, \ldots, N$) and the failed state of the network ($\overline{X}_S$). Analysts have the choice of using either one according to their needs and ease of application. Ease is often determined by the number of terms in the resulting structure function. Since φ and ψ are two different functions that express the relationship between two different variables, they are certain to have different terms, one function may be more concise than the other. However, they are equivalent since both represent the same network.

Variables in Equation (6.1) is Boolean variables, namely,

$X_i = 1$ if component i, $i = 1, 2, \ldots, N$, is good, and
$X_i = 0$ if component i, $i = 1, 2, \ldots, N$, is failed.

Also,

$X_S = 1$ if the network is good, and
$X_S = 0$ if the network is failed.

Thus, X_S is a Boolean function of X_i, $i = 1, 2, \ldots, N$, as given by Equation (6.1). All algebraic operations used in structural functions φ and ψ are Boolean operations, including the operations "conjunction" ($\wedge$), "disjunction" ($\vee$), and "inverse" ($\overline{X}_i$). The Boolean operation of conjunction in Boolean algebra is identical to the operation of arithmetic multiplication. For simplicity, conjunction is expressed by multiplication. The same rule applies to Equation (6.2), but using the inverse of variables representing the states of components and system,

$\overline{X}_i = 1$ if component i, $i = 1, 2, \dots, N$, is failed, and
$\overline{X}_i = 0$ if component i, $i = 1, 2, \dots, N$, is good.

Also,

$\overline{X}_S = 1$ if the network is failed, and
$\overline{X}_S = 0$ if the network is good.

In dynamic reliability analysis, the state variable is a function time to show that its value changes over time. To show that X's are functions of time, they are presented as

$X_i(t) = 1$ if component i, $i = 1, 2, \dots, N$, is good at time t, and
$X_i(t) = 0$ if component i, $i = 1, 2, \dots, N$, is failed at time t.

Also,

$X_S(t) = 1$ if the network is good at time t, and
$X_S(t) = 0$ if the network is failed at time t.

The state variable is presented as $X(t)$ is only when time plays a crucial role. For simplicity, the process is using X_i and at the last step to determine the network reliability function then X_i is replaced with $R_i(t)$.

6.2.1 Minimum Path and Cut

A network structure function can be derived from the minimum paths or minimum cuts of the network. A path is defined as a set of components that, if all of them are good, then the network is guaranteed to be good. A minimum path is a path with as few elements as possible to ensure the network is good. If any of the elements of the minimum path is failed, then the path does not guarantee the network to be good. Conversely, a cut is defined as a set of components that, if all of them are failed, then the network is guaranteed to be failed. A minimum cut is a cut with as few elements as possible to ensure the network is failed. If any of the elements of the minimum cut are good, then the cut cannot guarantee the network to be failed. Minimum paths and cuts can be found from the network structure. There are many algorithms that can be used to find the minimum paths or cuts of a graph.

Example 6.4 The network in Figure 6.2 has many paths; one of these paths is {1,3,4}. If all the components in this set are good, then the network is guaranteed to be good. However, this is not a minimum path because the network remains good even if component 3 is failed. The minimum path of these paths is {1,4} because if either of the two elements in this set is failed, then the remaining components in the set cannot guarantee the network to be good. The minimum paths in this network can be identified from the graph as {1,4}, {2,5}, {1,3,5}, and {2,3,4}. All these sets can be verified as minimum paths.

Likewise, the network in Figure 6.3a has minimum paths {A,B,D} and {A,C,D}, while the network in Figure 6.3b has minimum paths {A, 1, B, 4, D}, {A, 2, C, 5, D}, {A, 1, B, 3, C, 5, D}, and {A, 2, C, 3, B, 4, D}.

The network in Figure 6.2 has many cuts; one of these cuts is {1,2,3}. If all components in this set fail, then the network is guaranteed to fail. But, this is not a minimum cut, because the network remains failed even if component 3 is good. The minimum cut is {1,2}; if any one of the two elements of this set are good, then the remaining components cannot guarantee the network is failed. This network can be verified to have minimum cuts {1,2}, {4,5}, {1,3,5}, and {2,3,4}.

The network in Figure 6.3a has minimum cuts {A}, {B, C}, {D}, and the network in Figure 6.3b has minimum cuts {A}, {1,2}, {1,3,5}, {2,3,4}, {4,5}, {D}, {2,B}, {5,B}, {1,C}, {4,C}, and {B,C}. ■

Exercise 6.2 Find the minimum paths and minimum cuts of the network in Exercise 6.1.

Unless otherwise specified, this book assumes that nodes are perfect (never fail) and arcs are imperfect (can fail). This assumption means that network components are arcs of the network, while nodes are not network components. Thus, the network graph representation is as shown in Figure 6.2. However, the methods described in this chapter can still be used for networks with network components consisting of nodes, arcs, or both.

The development of mathematical expressions for network structure functions is presented next.

6.2.2 Network Structure Function

One way to express the structure function of a network is by using minimum paths or minimum cuts. Let MP_j, $j = 1, 2, \ldots, M_P$, be the minimum path j, where M_P is the number of minimum paths of the network. The source node sends flows to the target node using the minimum paths. A minimum path can be used to send flows from the source node to the target node if the minimum path is good, which means all components in the minimum path are good. Therefore, the state of a minimum path in network reliability analysis is the product (conjunction operation) of the state variables of components in the minimum path. The network is good if flow from the source node to the target node can occur, and this flow occurs when one of the minimum paths (disjunction operation) is good. Thus, the relationship between X_S and X_i, $i = 1, 2, \ldots, N$, using minimal paths is

$$X_S = \bigvee_{j=1}^{M_P} \left(\prod_{i \in MP_j} X_i \right). \tag{6.3}$$

Each minimum path in Equation (6.3) is expressed as the multiplication (conjunction operation) of all the variables that represent the components of the minimum path ($\prod_{i \in MP_j} X_i$). Because the network is good if one of its minimum paths is good, the network state is the result of the disjunction operation ($\vee$) of the states of M_P minimum paths of the network.

Following the same argument, the network structure function using minimum cuts is

$$\overline{X}_S = \bigvee_{j=1}^{M_C} \left(\prod_{i \in MC_j} \overline{X}_i \right), \tag{6.4}$$

where MC_j is the minimum cut j, $j = 1, 2, \ldots, M_C$, and M_C is the number of minimum cuts of the network. The network structure function in minimum cuts uses the inverse variable because it represents the relationship between component failures and network failure, in contrast to using minimum paths that express the relationship between components being good and network being good. Each term in Equation (6.4) is a multiplication (conjunction operation) of the state variables in the minimum cut represented by that term. Because the network fails if one of its minimum cuts fails, the terms representing the minimum cuts are combined with disjunction operations.

Example 6.5 The structure function of network in Figure 6.2 can be derived from all its minimum paths, namely, {1,4}, {2,5}, {1,3,5}, and {2,3,4}. The minimum path {1,4} is expressed in component state variables as X_1X_4, {2,5} is expressed as X_2X_5, {1,3,5} is expressed as $X_1X_3X_5$, and {2,3,4} is expressed as $X_2X_3X_4$. Thus, the network structure function expressed in its minimum paths is

$$X_S = X_1X_4 \vee X_2X_5 \vee X_1X_3X_5 \vee X_2X_3X_4.$$

The structure function of this network can also be derived from all its minimum cuts, namely, {1,2}, {4,5}, {1,3,5}, and {2,3,4}. The minimum cut {1,2} is expressed as $\overline{X}_1\overline{X}_2$, {4,5} is expressed as $\overline{X}_4\overline{X}_5$, {1,3,5} is expressed as $\overline{X}_1\overline{X}_3\overline{X}_5$, and {2,3,4} is expressed as $\overline{X}_2\overline{X}_3\overline{X}_4$. Thus, the network structure function in its minimum cuts is

$$\overline{X}_S = \overline{X}_1\overline{X}_2 \vee \overline{X}_4\overline{X}_5 \vee \overline{X}_1\overline{X}_3\overline{X}_5 \vee \overline{X}_2\overline{X}_3\overline{X}_4.$$

■

Example 6.6 The structure function of network in Figure 6.3a expressed using the minimum paths can be derived from the minimum paths obtained in Example 6.4,

$$X_S = X_AX_BX_D \vee X_AX_CX_D.$$

The network structure function expressed in minimum cuts obtained in Example 6.4 is

$$\overline{X}_S = \overline{X}_A \vee \overline{X}_B\overline{X}_C \vee \overline{X}_D.$$

These two representations are equivalent since they are expressing the same network. Their equivalency can be shown using De Moivre's inverse law,

$$\overline{X_A \vee X_B} = \overline{X}_A\overline{X}_B. \tag{6.5}$$

Taking the inverse of the network structure function in minimum paths yields

$$\begin{aligned}\overline{X}_S &= \overline{X_AX_BX_D \vee X_AX_CX_D} = (\overline{X}_A \vee \overline{X}_B \vee \overline{X}_D)(\overline{X}_A \vee \overline{X}_C \vee \overline{X}_D) = \overline{X}_A \vee (\overline{X}_B\overline{X}_C) \vee \overline{X}_D\\ &= \overline{X}_A \vee \overline{X}_B\overline{X}_C \vee \overline{X}_D,\end{aligned}$$

the same as that produced using the minimum cuts. ■

The minimum paths of a network can be found using the **minpaths** function from MATLAB. It is included as part of **dysrel-am**. The nodes and arcs of the network are input to the **minpaths** function to produce the minimum paths of the network. The arc numbers can be automatically generated by the **minpaths** function. Therefore, it is not necessary to have arc numbers in the network diagram. However, in case users want to define the arc number for their analysis, **minpaths** function in MATLAB has been modified to accept the numbers (note that MATLAB refers to arc as edge). As will be clear later, the minimum paths produced by the **minpaths** function are ordered from the minimum path with the least numbers of arcs to the most numbers of arcs.

Example 6.7 The minimum paths of the network in Figure 6.1 are generated using the MATLAB function **minpaths** without specifying the arc numbers. The MATLAB code is as follows:

MATLAB code for example6_7.m (using autogenerated arc numbers)

```
nodes = {'A','B','C','D'};
startNode = {'A','A','B','C','B','C'};
endNode = {'B','C','D','D','C','B'};
weights = [1 1 1 1 1 1];
```

```
% ==== Source & Target ====
src = 'A';
tgt = 'D';
% Create digraph
G = digraph(startNode,endNode,weights,nodes);
% Find minimum paths
mp = minpaths(G,startNode,endNode,src,tgt);
```

The minimum paths of the network in Figure 6.2 with specific arc numbers are generated using the MATLAB function **minpaths**. The MATLAB code is as follows:

MATLAB code for example6_7.m (using user-defined arc numbers)

```
nodes = {'A','B','C','D'};
startNode = {'A','A','B','C','B','C'};
endNode = {'B','C','D','D','C','B'};
weights = [1 1 1 1 1 1];
edgeLabels = [1 2 4 5 3 3];
% ==== Source & Target ====
src = 'A';
tgt = 'D';
% Create digraph
G = digraph(startNode,endNode,weights,nodes);
% Find minimum paths
mp = minpaths(G,startNode,endNode,src,tgt,edgeLabels);
```

Running both MATLAB codes produces the following output:

```
All possible paths from A to D (arc-number sequences):
Path 1: 1 -> 4
Path 2: 2 -> 5
Path 3: 1 -> 3 -> 5
Path 4: 2 -> 3 -> 4
```

The output is expressed in the structure function format as

$$X_1X_4 \vee X_2X_5 \vee X_1X_3X_5 \vee X_2X_3X_4,$$

the same expression as produced using manual derivation in Example 6.5. ■

Exercise 6.3 Manually derive the network structure functions in Exercise 6.1 using minimum paths. Compare the results with those obtained using **minpaths** function with autogenerated path numbers and user-defined arc numbers.

Using minimum paths to construct a network structure function yields the network reliability from the reliability of its components. Conversely, using minimum cuts yields the network unreliability (vulnerability) from the unreliability (vulnerability) of its components. Both yield equivalent results; that is, the result of using minimum paths is the inverse of the result of using minimum cuts. Since the inverse of the reliability function is the CDF of network lifetime, using minimum cuts yields the CDF of network lifetime from the CDF of its component lifetimes. The choice of using minimum paths or cuts depends on the number of terms in the structure function; the fewer the terms, the easier it is. The network in Figure 6.2 can be represented using minimum paths or minimum cuts with the same complexity since the network has four minimum paths and also four minimum cuts.

Next, a method to convert structure function from Boolean to arithmetic is developed.

6.2.3 Converting Boolean Functions to Arithmetic Function

Network reliability, like system reliability in general, is the result of arithmetic operations on the reliability of its components. Not all Boolean operations are equivalent to arithmetic operations. Conjunction is identical with arithmetic multiplication, but disjunction is not equivalent to arithmetic addition. Therefore, disjunction operation in a network structure function must first be converted into appropriate arithmetic operations. In addition, Boolean variables must satisfy

$$X^n = X, \tag{6.6}$$

and

$$\overline{X} = 1 - X. \tag{6.7}$$

The easiest way to convert a Boolean structure function to an arithmetic structure function is to convert the disjunction operation into an arithmetic operation and applies identities given by Equations (6.6) and (6.7) to the resulting expression. The Boolean disjunction operation can be converted into an arithmetic operation by using Equations (6.5) and (6.7),

$$X_1 \vee X_2 = 1 - \overline{\overline{X_1} \vee \overline{X_2}} = 1 - \overline{X}_1\overline{X}_2 = 1 - (1 - X_1)(1 - X_2). \tag{6.8}$$

Applying Equations (6.6–6.8) to $X_1 \vee X_1X_2$ yields

$$\begin{aligned} X_1 \vee X_1X_2 &= 1 - (1 - X_1)(1 - X_1X_2) \\ &= 1 - 1 + X_1 + X_1X_2 - X_1^2X_2 = X_1 + X_1X_2 - X_1X_2 = X_1. \end{aligned}$$

Hence, the disjunction of two terms with the second term being a subset of the first term results in the first term only, which is the largest term (since X is either 0 or 1, it holds true $X_1 \geq X_1X_2$). Likewise, $X_1X_2 \vee X_1X_2 = X_1X_2$. These uniqueness of operations with Boolean variables must be maintained when converting Boolean operations to arithmetic operations: some eliminations are required.

Equation (6.8) can be employed to transform the structure function with Boolean operations in Equation (6.3) into a structure function with arithmetic operations,

$$X_S = 1 - \prod_{j=1}^{M_P} (1 - Y_j), \tag{6.9}$$

where

$$Y_j = \prod_{i \in MP_j} X_i. \tag{6.10}$$

Likewise, Equation (6.4) is transformed into

$$\overline{X}_S = 1 - \prod_{j=1}^{M_C} (1 - Y_j), \tag{6.11}$$

where

$$Y_j = \prod_{i \in MC_j} \overline{X}_i. \tag{6.12}$$

Equations (6.9) and (6.11) do not directly produce structure functions with valid arithmetic operations; the results must still be verified to ensure that there are no exponentiation operations in each term and that no term is subset of another term. These requirements make the use of Equations (6.9) and (6.11) cumbersome. Finding another way to transform a Boolean function into an arithmetic function is necessary.

Example 6.8 A Boolean structure function $X_S = X_1X_2 \vee X_2X_3 \vee X_1X_2X_3$ is converted into an arithmetic structure function with the following steps:

Because $X_1X_2X_3 \leq X_2X_3$, $X_2X_3 \vee X_1X_2X_3 = X_2X_3$. Hence,

$$\begin{aligned} X_S &= X_1X_2 \vee X_2X_3 \vee X_1X_2X_3 = X_1X_2 \vee X_2X_3 \\ &= 1 - (1 - X_1X_2)(1 - X_2X_3) = X_1X_2 + X_2X_3 - X_1X_2^2X_3. \end{aligned}$$

Using Equation (6.6) to remove exponentiation, that is $X_2^2 = X_2$, yields

$$X_S = X_1X_2 + X_2X_3 - X_1X_2X_3.$$

■

This example shows that the arithmetic structure function is the sum of each term minus the conjunction of the two terms. This concept, known as the IE principle, can be extended to many terms. By the IE principle, the arithmetic structure function of a network is:

> The sum of all minimum paths – the sum of the conjunctions of every two minimum paths + the sum of the conjunctions of every three minimum paths - + · · ·

Its mathematical expression is

$$X_S = \sum_{j=1}^{M_P} Y_j - \sum_{j=1}^{M_P} \sum_{\substack{k=1 \\ k \neq j}}^{M_P} Y_jY_k + \sum_{j=1}^{M_P} \sum_{\substack{k=1 \\ k \neq j}}^{M_P} \sum_{\substack{l=1 \\ l \neq k}}^{M_P} Y_jY_kY_l \mp \dots \tag{6.13}$$

Example 6.9 A 2-out-of-3 system is good if at least two of its three components are good. The minimum paths of this system consist of two of the three components, namely, {1,2}, {1,3} and {2,3}. The structure function of the 2-out-of-3 system using minimum paths in Boolean operations is

$$X_S = X_1X_2 \vee X_1X_3 \vee X_2X_3.$$

The network structure function with arithmetic operations can be found using Equation (6.13) in the following steps:

$$\begin{aligned} &\sum Y_j = X_1X_2 + X_1X_3 + X_2X_3, \\ &\sum\sum Y_jY_k = (X_1X_2)(X_1X_3) + (X_1X_2)(X_2X_3) + (X_1X_3)(X_2X_3) \\ &\qquad\qquad = X_1^2X_2X_3 + X_1X_2^2X_3 + X_1X_2X_3^2 = 3X_1X_2X_3, \\ &\sum\sum\sum Y_jY_kY_l = (X_1X_2)(X_1X_3)(X_2X_3) = X_1^2X_2^2X_3^2 = X_1X_2X_3. \end{aligned}$$

Hence,

$$X_S = (X_1X_2 + X_1X_3 + X_2X_3) - (3X_1X_2X_3) + (X_1X_2X_3) = X_1X_2 + X_1X_3 + X_2X_3 - 2X_1X_2X_3.$$

With the resulting structure function in arithmetic operations, the network reliability value R_S can be found by substituting the component reliability value for each component state variable; that is, X_i is replaced with R_i and X_S with R_S:

$$R_S = R_1R_2 + R_1R_3 + R_2R_3 - 2R_1R_2R_3.$$

In static reliability analysis for $R_1 = 0.9$, $R_2 = 0.8$, and $R_3 = 0.7$ yields

$$R_S = 0.9 \times 0.8 + 0.9 \times 0.7 + 0.8 \times 0.7 - 2 \times 0.9 \times 0.8 \times 0.7 = 0.902.$$

In dynamic reliability analysis, X_i represents $X_i(t)$, and replace $X_i(t)$ by reliability function $R_i(t)$. The obtained network reliability function is

$$R_S(t) = R_1(t)R_2(t) + R_1(t)R_3(t) + R_2(t)R_3(t) - 2R_1(t)R_2(t)R_3(t).$$

If the lifetime of all components is exponentially distributed with $MTTF = 1$, then $R_i(t) = e^{-t}$, $i = 1, 2, 3$. Substituting them into $R_S(t)$ yields

$$R_S(t) = e^{-t}e^{-t} + e^{-t}e^{-t} + e^{-t}e^{-t} + 2e^{-t}e^{-t}e^{-t} = 3e^{-2t} - 2e^{-3t}.$$

■

The result of using the IE principle in Equation (6.13) is called the IE representation of the network structure function in minimum path, referred to as the network IE representation. The resulting arithmetic function contains many terms that can still be simplified through elimination and combination using the properties of Boolean variables. This method, although easy to do manually, generally produces functions with many multiplicative terms. There are many methods to simplify an IE representation, one of which is proposed in Schäfer et al. [7].

Another more popular and efficient method is to convert the disjunction operations into SDP terms. The method is called SDP method, and the result is called the network SDP representation. If the structure function is derived using minimum paths, then the terms in the network SDP representation are the minimum paths that have been made disjoint to each other. The meaning of being disjoint is that under certain conditions, only one term can have a value of 1, while the other terms under the same conditions must have a value of 0. This disjoint process must also ensure that the properties of Boolean variables, such as $X_1X_1 = X_1$, $X_1 \vee X_1X_2 = X_1$, and $X_1\overline{X}_1 = 0$ are maintained.

There are many algorithms for generating the simplest SDP representation of a Boolean structure function. An efficient one is the algorithm based on multivariable inversion as proposed in Heidtmann [8], which was later modified in Soh and Rai [9], Luo and Trivedi [10], and Chaturvedi and Misra [11]. The process is based on identity

$$\begin{aligned} X_1 \vee X_2 &= 1 - (1 - X_1)(1 - X_2) = X_1 + X_2 - X_1X_2 \\ &= X_1 + (1 - X_1)X_2 = X_1 + \overline{X}_1X_2. \end{aligned}$$

Hence,

$$X_1 \vee X_2 = X_1 + \overline{X}_1X_2. \tag{6.14}$$

Equation (6.14) suggests that to make the second term disjoint from the first term, multiply the second term with the inverse of the first term.

Equation (6.14) can be used for disjunction operations involving terms that are multiplications of several Boolean variables. For example,

$$X_1X_2 \vee X_1X_3 = X_1X_2 + \overline{X_1X_2}X_1X_3.$$

Following de Moivre's Theorem, $\overline{X_1X_2} = \overline{X}_1 \vee \overline{X}_2$. Using Equation (6.14) by multiplying the second term $(\overline{X}_2)$ by the inverse of the first term $\overline{X}_1$, namely, X_1, yields

$$\overline{X_1X_2} = \overline{X}_1 + X_1\overline{X}_2.$$

Therefore,

$$X_1X_2 \vee X_1X_3 = X_1X_2 + (\overline{X}_1 + X_1\overline{X}_2)X_1X_3 = X_1X_2 + \overline{X}_1X_1X_3 + X_1\overline{X}_2X_1X_3.$$

Use Boolean variable identity $\overline{X}_1 X_1 = 0$ and $X_1 X_1 = X_1$ to obtain

$$X_1X_2 \vee X_1X_3 = X_1X_2 + X_1\overline{X}_2X_3. \tag{6.15}$$

Equation (6.15) shows that to convert a Boolean disjunction operation into an arithmetic addition operation, the second term is simply multiplied by the inverse of the element of the first term that is not present in the second term. This result simplifies the process of converting Boolean functions into SDP representations.

The general form of an SDP representation of a network is the SDP terms produced through the process of disjointing the Boolean terms. If DY_k, $k = 1, 2, \dots, K$, are the disjoint products obtained from the network structure function using the minimum paths, then

$$X_S = \sum_{k=1}^{K} \left(\prod_{i \in DY_k} Z_i \right), \tag{6.16}$$

where Z_i can be X_i, inverse $\overline{X}_i$, or multivariable inverse $\overline{X_iX_j}$, depending on the resulting DY_k. Equation (6.16) is called the SDP representation of the network. There are many equivalent SDP representations for a single Boolean structure function. Equation (6.16) can be used for the structure function in minimum cuts by replacing the variables used with their inverse.

Example 6.10 The SDP representation for the Boolean structure function of a 2-out-of-3 system can be found in the following steps:

$$X_S = X_1X_2 \vee X_1X_3 \vee X_2X_3.$$

1) Make the second term disjoints from the first term; that is, multiply the second term by the inverse of the element of the first term that is not present in the second term, namely, X_2. The second term becomes $X_1\overline{X}_2X_3$.
2) Make the third term disjoints from the first term; that is, multiply the third term by the inverse of the element of the first term that is not present in the third term, namely, X_1. The third term becomes $\overline{X}_1X_2X_3$.
3) Make the transformed third term disjoints from the second term. However, the transformed third term already disjoints with the second term (the product of the transformed third term and the second term is $(\overline{X}_1X_2X_3)(X_1X_3) = 0$), so there is no need for disjointing.

The resulting SDP representation for a 2-out-of-3 system is

$$X_S = X_1X_2 + X_1\overline{X}_2X_3 + \overline{X}_1X_2X_3. \tag{6.17}$$

All terms in Equation (6.17) are disjoint to each other; that is, only one of them can have a value 1. If the first term is 1, then $X_1 = X_2 = 1$, and $\overline{X}_1 = \overline{X}_2 = 0$. Thus, the second and third terms are 0. It can be verified that if the second term is 1, then the first and third terms must be 0, and if the third term is 1, then the first and second terms are 0.

For dynamic reliability analysis, X_i and $\overline{X}_i$ in the resulting SDP representation represent $X_i(t)$ and $\overline{X}_i(t)$, respectively. To find the network reliability function, replace them with $R_i(t)$ and $(1 - R_i(t))$, respectively,

$$R_S(t) = R_1(t)R_2(t) + R_1(t)(1 - R_2(t))R_3(t) + (1 - R_1(t))R_2(t)R_3(t). \tag{6.18}$$

If the lifetime of all components is exponentially distributed with $MTTF = 1$, then $R_i(t) = e^{-t}$, $i = 1, 2, 3$. Substitute them into Equation (6.18) to produce the network reliability function

$$R_S(t) = e^{-t}e^{-t} + e^{-t}(1 - e^{-t})e^{-t} + (1 - e^{-t})e^{-t}e^{-t} = 3e^{-2t} - 2e^{-3t},$$

the same as obtained using the IE method.

In static reliability analysis for $R_1 = 0.9$, $R_2 = 0.8$, and $R_3 = 0.7$, yields

$$R_S = 0.9 \times 0.8 + 0.9 \times (1 - 0.8) \times 0.7 + (1 - 0.9) \times 0.8 \times 0.7 = 0.902,$$

the same as obtained using the IE method. ■

The SDP representation resulting from Example 6.10 is equivalent to the IE representation resulting from Example 6.9. To verify their equivalency, use Boolean variable identity $\overline{X}_i = 1 - X_i$,

$$\begin{aligned} X_1X_2 + X_1\overline{X}_2X_3 + \overline{X}_1X_2X_3 &= X_1X_2 + X_1(1 - X_2)X_3 + (1 - X_1)X_2X_3 \\ &= X_1X_2 + X_1X_3 + X_2X_3 - 2X_1X_2X_3, \end{aligned}$$

which is the IE representation for the 2-out-of-3 system generated in Example 6.9.

This comparison shows that although both are equivalent, the SDP representation obtained is more concise than the IE representation. Moreover, the SDP representation is also easier to interpret physically. The terms in Equation (6.17) can be interpreted as:

1) The first term states that the network is good if components 1 and 2 are good.
2) The second term states that the network is still good even if component 2 failed as long as components 1 and 3 are good.
3) The third term states that the network is still good even if component 1 failed as long as components 2 and 3 are good.

The steps to convert Boolean structure function into arithmetic structure function in SDP representation can be used to construct an algorithm for this purpose as given next.

6.2.4 Algorithm for Converting Boolean Function into SDP Representation

The process of converting Boolean structure functions into SDP representation can be performed recursively starting from the first term to the last term.

Example 6.11 The structure function of network in Figure 6.2 using the minimum paths generated in Example 6.5 is

$$X_S = X_1X_4 \vee X_2X_5 \vee X_1X_3X_5 \vee X_2X_3X_4.$$

The SDP representation can be found in the following steps:

1) The first term X_1X_4 remains as it is.
2) The second term X_2X_5 is made disjoint from the first term → multiply the second term by the inverse of variable(s) in the first term that is not in the second term, namely, $\overline{X_1X_4}$.
   The second term becomes $\overline{X_1X_4}X_2X_5$.
3) The third term $X_1X_3X_5$ is made disjoint from the first term → multiply by the inverse of variable in the first term that is not in third term, namely, $\overline{X}_4$. The third term becomes $\overline{X}_4X_1X_3X_5$.
   Make the new third term disjoint from the second term → multiply by the inverse of variable in the second term that is not in the new third term, namely, $\overline{X}_2$. The third term becomes $\overline{X}_2\overline{X}_4X_1X_3X_5$.
4) The fourth term $X_2X_3X_4$ is made disjoint from the first term → multiply by the inverse of variable in the first term that is not in the fourth term, namely, $\overline{X}_1$ to become $\overline{X}_1X_2X_3X_4$.
   Make the new fourth term disjoint from the second term → multiply by the inverse of variable in the second term that is not in the fourth term, namely, $\overline{X}_5$. The fourth term becomes $\overline{X}_5\overline{X}_1X_2X_3X_4$.

Make the new fourth term disjoint from the third term. But they are already disjoint, since $(X_1X_3X_5)(\overline{X}_5\overline{X}_1X_2X_3X_4) = 0$ (there is multiplication $X_1\overline{X}_1 = 0$) → no changes.
The fourth term remains as $\overline{X}_5\overline{X}_1X_2X_3X_4$.

The SDP representation is the arithmetic sum of the resulting terms of each step, namely,

$$X_S = X_1X_4 + \overline{X_1X_4}X_2X_5 + \overline{X}_2\overline{X}_4X_1X_3X_5 + \overline{X}_5\overline{X}_1X_2X_3X_4.$$

Writing the function following the order of its components results in

$$X_S = X_1X_4 + \overline{X_1X_4}X_2X_5 + X_1\overline{X}_2X_3\overline{X}_4X_5 + \overline{X}_1X_2X_3X_4\overline{X}_5.$$

To find the network reliability function from the SDP representation, let X_i represents $X_i(t)$ and replaces $X_i(t)$ with $R_i(t)$. The obtained network reliability function is

$$\begin{aligned}R_S(t) &= R_1(t)R_4(t) + (1 - R_1(t)R_4(t))R_2(t)R_5(t)\\ &\quad + R_1(t)(1 - R_2(t))R_3(t)(1 - R_4(t))R_5(t)\\ &\quad + (1 - R_1(t))R_2(t)R_3(t)R_4(t)(1 - R_5(t)).\end{aligned} \tag{6.19}$$

For static reliability analysis with the reliability of arcs 1–5, $R_1 = 0.8$, $R_2 = 0.7$, $R_3 = 0.6$, $R_4 = 0.5$, $R_5 = 0.9$, the network reliability is

$$\begin{aligned}R_S &= 0.8 \times 0.5 + (1 - 0.8 \times 0.5) \times 0.7 \times 0.9 + 0.8 \times (1 - 0.7) \times 0.6 \times (1 - 0.5) \times 0.9\\ &\quad + (1 - 0.8) \times 0.7 \times 0.6 \times 0.5 \times (1 - 0.9) = 0.847.\end{aligned}$$

■

Each term in Equation (6.19) is not necessarily a valid reliability function. For example, $R_1(t)R_4(t)R_5(t)(1 - R_3(t))$ is not a reliability function of a coherent system as assumed in this book. However, the complete SDP representation will produce a valid reliability function.

Note that $\overline{X_1X_3} = 1 - X_1X_3$ is not the same as $\overline{X}_1\overline{X}_3 = (1 - X_1)(1 - X_3)$. $\overline{X_1X_3}$ is called a multivariable inverse, and $\overline{X}_1$ is called a univariable inverse.

Example 6.12 The structure function of network in Figure 6.2 using minimum cuts is

$$\overline{X}_S = \overline{X}_1\overline{X}_2 \vee \overline{X}_4\overline{X}_5 \vee \overline{X}_1\overline{X}_3\overline{X}_5 \vee \overline{X}_2\overline{X}_3\overline{X}_4.$$

The SDP representation can be found using the following steps:

1) The first term $\overline{X}_1\overline{X}_2$ remains as it is.
2) The second term $\overline{X}_4\overline{X}_5$ is made disjoint from the first term → multiply by the inverse of $\overline{X}_1\overline{X}_2 = \overline{\overline{X}_1\overline{X}_2}$. The second term becomes $\overline{\overline{X}_1\overline{X}_2}\overline{X}_4\overline{X}_5$.
3) The third term $\overline{X}_1\overline{X}_3\overline{X}_5$ is made disjoint from the first term → multiply by the inverse of $\overline{X}_2 = X_2$ to become $\overline{X}_1X_2\overline{X}_3\overline{X}_5$. Make the new third term disjoint from the second term → multiply by the inverse of $\overline{X}_4 = X_4$. The third term becomes $\overline{X}_1X_2\overline{X}_3X_4\overline{X}_5$.
4) The fourth term $\overline{X}_2\overline{X}_3\overline{X}_4$ is made disjoint from the first term → multiply by the inverse of $\overline{X}_1 = X_1$ to become $X_1\overline{X}_2\overline{X}_3\overline{X}_4$. Make the new fourth term disjoint from the second term → multiply by the inverse of $\overline{X}_5 = X_5$ to get $X_1\overline{X}_2\overline{X}_3\overline{X}_4X_5$. Make the new fourth term disjoint from the third term. But, it is already disjoint: $(\overline{X}_1\overline{X}_3\overline{X}_5)(X_1\overline{X}_2\overline{X}_3\overline{X}_4X_5) = 0$ (there is $X_1\overline{X}_1 = 0$) → The fourth term remains as $X_1\overline{X}_2\overline{X}_3\overline{X}_4X_5$.

The SDP representation is the arithmetic sum of the terms resulting from each step, namely,

$$\overline{X}_S = \overline{X}_1\overline{X}_2 + \overline{\overline{X}_1\overline{X}_2}\overline{X}_4\overline{X}_5 + \overline{X}_1X_2\overline{X}_3X_4\overline{X}_5 + X_1\overline{X}_2\overline{X}_3\overline{X}_4X_5.$$

The unreliability function, or CDF of the network lifetime, is obtained by replacing $\overline{X}_i$ with $F_i(t)$, and X_i with $R_i(t) = 1 - F_i(t)$,

$$F_S(t) = F_1(t)F_2(t) + (1 - F_1(t)F_2(t))F_4(t)F_5(t) + F_1(t)(1 - F_2(t))F_3(t)(1 - F_4(t))F_5(t)$$
$$+ (1 - F_1(t))F_2(t)F_3(t)F_4(t)(1 - F_5(t)).$$

For static unreliability analysis with the same data as in Example 6.11, unreliability, or vulnerability of arcs 1–5, are calculated as $F_i = 1 - R_i$. Hence, $F_1 = 0.2$, $F_2 = 0.3$, $F_3 = 0.4$, $F_4 = 0.5$, $F_5 = 0.1$, and the network unreliability (vulnerability) is

$$F_S = 0.2 \times 0.3 + (1 - 0.2 \times 0.3)0.5 \times 0.1 + 0.2(1 - 0.3)0.4(1 - 0.5) \times 0.1$$
$$+ (1 - 0.2)0.3 \times 0.4 \times 0.5(1 - 0.1) = 0.153 = 1 - R_S.$$

Thus, calculations in Examples 6.11 and 6.12 produced the same results. ■

As with Example 6.6, the two SDP representations resulting from Examples 6.11 and 6.12 are equivalent, and the resulting reliability functions are identical.

Exercise 6.4 Determine the SDP representation for the network in Exercise 6.1.

The SDP representation generated by a structure function is not unique. A structure function can have multiple SDP representations. In general, each change in the order of terms in a structure function will produce a different SDP representation. Furthermore, the same structure function can produce different SDP representations if different algorithms are used.

Example 6.13 A structure function $X_S = X_1X_2X_3 \vee X_2X_4$ processed in the manner described in Example 6.11 produces an SDP representation

$$X_S = X_1X_2X_3 + \overline{X_1X_3}X_2X_4.$$

If the order of the terms in the structure function is reversed to $X_S = X_2X_4 \vee X_1X_2X_3$, then the same method will produce

$$X_S = X_2X_4 + \overline{X}_4X_1X_2X_3.$$

Since they originate from the same structure function, the two SDP representations are equivalent, that is

$$X_1X_2X_3 + \overline{X_1X_3}X_2X_4 = X_1X_2X_3 + (1 - X_1X_3)X_2X_4 = X_2X_4 + X_1X_2X_3 - X_1X_2X_3X_4$$
$$= X_2X_4 + X_1X_2X_3(1 - X_4) = X_2X_4 + X_1X_2X_3\overline{X}_4.$$

■

The process of finding an SDP representation is more complex than shown in Examples 6.11 and 6.12. The steps in both examples need to be supplemented by checking for any terms that can be absorbed by other terms. If so, they must be removed from the SDP representation. This relates to the previously mentioned property of the Boolean disjunction operation, namely, $X_1 \vee X_1 = X_1$ and $X_1 \vee X_1X_2 = X_1$. Essentially, one term is a subset of another term if all variables in the other term are present in the term being examined in exactly the same format, that is, both are inverses or both are not inverses.

The algorithm for generating SDP representation from a structure function is finding a complete representation of each disjoint multiplication term, absorbs terms that are part of other terms, removes exponentiation, and removes terms containing one variable with its inverse. Such an algorithm that proposed by Heidtmann [8] is programmed in MATLAB to produce the **sdprep** function. This function accepts minimum paths as inputs.

```
function SDPterms = sdprep(minpaths)
% Generate Sum of Disjoint Products (SDP) terms of a network
% using KDH88 algorithm
%   minpaths : minimum paths of a network (cell format)
%              Ex. minpaths = {[1 2];[3 4];[2 4]}
%   SDPterms : SDP terms (array format)

% Convert minimum paths to array of binary representation
numofpath = length(minpaths);
% Find the number of components
N = 1;
for i = 1:numofpath
    q = minpaths{i};
    if N < max(q)
        N = max(q);
    end
end
binarypaths = -ones(numofpath,N);
for i = 1:numofpath
    pathlength = length(minpaths{i});
    for j = 1:pathlength
        binarypaths(i,minpaths{i}(j)) = 0;
    end
end
% KDH88 algorithm
x = size(binarypaths);
n = x(2);
m = x(1);
SDPterms = [];
for k = 1:m
    product = binarypaths(k,:);
    SDPterms = nextstep(n,1,k,binarypaths,product);
end
clear nextstep
end
function dp = nextstep(n,j,k,minproduct,product)
% Function to generate Disjoint Product from minproduct
persistent result;
if j<k
    [mask,r,disjoint] = masking(n,j,product,minproduct);
    if disjoint
        nextstep(n,j+1,k,minproduct,product);
    else
        dproduct = product;
        for s = 1:r
            for i = 1:n
                if (product(i) == mask(s)) && (minproduct(j,i)<0)
                    dproduct(i) = -1;
                end
            end
            nextstep(n,j+1,k,minproduct,dproduct);
            for i = 1:n
                if product(i) == mask(s)
```

```
                    if minproduct(j,i) == 0
                        dproduct(i) = 0;
                    else
                        dproduct(i) = mask(s);
                    end
                end
            end
        end
        for i = 1:n
            if (product(i)<0) && (minproduct(j,i) == 0)
                dproduct(i) = j;
                disjoint = 1;
            end
        end
        if disjoint
            nextstep(n,j+1,k,minproduct,dproduct);
        end
    end
    dp = result;
else
    result = [result;product];
    dp = result;
end
end
function [mask,r,disjoint] = masking(n,j,product,minproduct)
disjoint = 0;
r = 0;
s = 1;
mask = 0;
while not(disjoint) && (s<j)
    b = 0;
    c = 0;
    for i = 1:n
        if product(i)  ==  s
            if minproduct(j,i)  ==  0
                b = 1;
            else
                c = 1;
            end
        end
    end
    if b
        if c
            r = r+1;
            mask(r) = s;
        else
            disjoint = 1;
        end
    end
    s = s+1;
end
end
```

Example 6.14 The **sdprep** function is employed to find the SDP representation of the network in Figure 6.2 from its Boolean structure function derived in Example 6.5,

$$X_S = X_1X_4 \vee X_2X_5 \vee X_1X_3X_5 \vee X_2X_3X_4.$$

This input to **sdprep** function is

```
≫ minpaths = {[1 4];[2 5];[1 3 5];[2 3 4]};
```

The output of **sdprep** function is

```
≫ sdp = sdprep(minpaths)
sdp =
     0    -1    -1     0    -1
     1     0    -1     1     0
     0     2     0     1     0
     1     0     0     0     2
```

This output is interpreted as 0 for X_i, -1 for void, single positive number as univariable inversion $\overline{X}_i$, and several same positive numbers as multivariable inversion $\overline{X_i X_j}$

$$[0,-1,-1,0,-1] = X_1X_4$$

$$[1,0,-1,1,0] = \overline{X_1X_4}X_2X_5$$

$$[0,2,0,1,0] = X_1\overline{X}_2X_3\overline{X}_4X_5$$

$$[1,0,0,0,2] = \overline{X}_1X_2X_3X_4\overline{X}_5$$

Therefore, the SDP representation is

$$X_S = X_1X_4 + \overline{X_1X_4}X_2X_5 + X_1\overline{X}_2X_3\overline{X}_4X_5 + \overline{X}_1X_2X_3X_4\overline{X}_5.$$

If the order of the structure function terms is changed to

$$X_S = X_1X_4 \vee X_1X_3X_5 \vee X_2X_3X_4 \vee X_2X_5,$$

The input to **sdprep** is

```
≫ minpaths={[1 4];[1 3 5];[2 3 4];[2 5]};
```

The output of **sdprep** is

```
≫ sdp = sdprep(minpaths)
sdp =
     0    -1    -1     0    -1
     0    -1     0     1     0
     1     0     0     0    -1
     1     0     3     3     0
     0     0     2     1     0
```

Its interpretation is

$$[0,-1,-1,0,-1] = X_1X_4$$

$$[0,-1,0,1,0] = X_1X_3\overline{X}_4X_5$$

$$[1,0,0,0,-1] = \overline{X}_1X_2X_3X_4$$

$$[1,0,3,3,0] = \overline{X}_1X_2\overline{X_3X_4}X_5$$

$$[0,0,2,1,0] = X_1X_2\overline{X}_3\overline{X}_4X_5$$

The resulting SDP representation is

$$X_S = X_1X_4 + X_1X_3\overline{X}_4X_5 + \overline{X}_1X_2X_3X_4 + \overline{X}_1X_2\overline{X_3X_4}X_5 + X_1X_2\overline{X}_3\overline{X}_4X_5.$$

These two results are equivalent. The proof is given in Problem. However, the first SDP representation contains less terms than the second SDP representation; thus, it is more efficient since it requires less computation. This example shows that it is advisable to arrange the minimum paths following the first sequence, which is the minimum paths with less components that are placed earlier in the sequence. The output of **minpaths** function is already ordered following this rule.

■

Example 6.15 The network in Figure 6.4 is inputted to **minpaths** function to generate its minimum paths. They are used as input to **sdprep** function to generate an SDP representation of the network.

Figure 6.4 Network for Example 6.15.

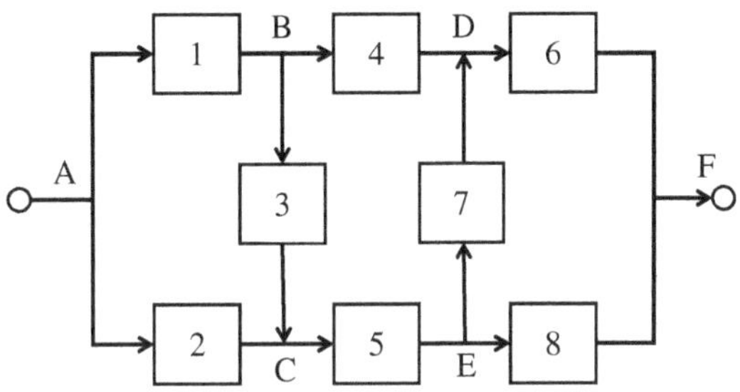

The main program for generating the minimum paths and an SDP representation of the network is available in **example6_15.m** file in the **dysrel-am** software package. This program uses autogenerated arc numbers that happen to produce the same arc numbers as in Figure 6.4.

MATLAB code for example6_15.m

```
clear; close all
addpath('..\dysrel-am')
% ==== Define directed graph ====
nodes = {'A','B','C','D','E','F'};                          input
startNode = {'A','B','D','A','C','E','B','E'};
endNode = {'B','D','F','C','E','F','C','D'};
weights = [1 1 1 1 1 1 1 1];
% ==== Source & Target ====
src = 'A';
tgt = 'F';
% Create digraph                                            generate minpaths
G = digraph(startNode,endNode,weights,nodes);
% Find minimum paths
mp = minpaths(G,startNode,endNode,src,tgt);
% Generate SDP                                              generate sdp
sdp = sdprep(mp)
```

Executing this program produces output as shown in Figure 6.5.

The minimum paths of the network derived from the output of **minpaths** are

$$\{1,4,6\};\{2,5,8\};\{1,3,5,8\};\{2,5,7,6\};\{1,3,5,7,6\}.$$

Hence, the network structure function in minimum paths is

$$X_S = X_1X_4X_6 \vee X_2X_5X_8 \vee X_1X_3X_5X_8 \vee X_2X_5X_6X_7 \vee X_1X_3X_5X_6X_7.$$

```
>> example6_15_net8comp
All possible paths from A to F (branch number sequences):
Path 1: 1 -> 4 -> 6
Path 2: 2 -> 5 -> 8
Path 3: 1 -> 3 -> 5 -> 8
Path 4: 2 -> 5 -> 7 -> 6
Path 5: 1 -> 3 -> 5 -> 7 -> 6

sdp =

     0    -1    -1     0    -1     0    -1    -1
     1     0    -1     1     0     1    -1     0
     0     2     0     1     0     1    -1     0
     1     0    -1     1     0     0     0     2
     0     4     0     1     0     0     0     2
```

Figure 6.5 Output of minpaths and sdprep for network in Figure 6.4.

The SDP representation of the network derived from the output of **sdprep** is

$$X_S = X_1X_4X_6 + \overline{X_1X_4X_6}X_2X_5X_8 + X_1\overline{X}_2X_3\overline{X_4X_6}X_5X_8 + \overline{X_1X_4}X_2X_5X_6X_7\overline{X}_8 + X_1\overline{X}_2X_3\overline{X}_4X_5X_6X_7\overline{X}_8.$$

■

Exercise 6.5 Reverse the direction of arcs 7 and 8 in Figure 6.4 and determine the new network structure function using the **minpaths** function. Use the results to derive an SDP representation of the network using the **sdprep** function.

SDP representation is used to generate the network reliability function from the reliability functions of its components that are assumed to follow ME distribution. This process is made simple by employing the closure properties of ME distribution. The next section shows that ME distribution is closed in SDP representation of a network.

6.3 Closure Properties of ME Distribution in SDP Representation

Each network component is assumed to be multistate with a lifetime following ME or PH distribution. The network reliability function is derived from the ME or PH representations of the component reliability functions and the SDP representation of the network. As will be seen later, the resulting network reliability function has an ME representation, even though the component reliability functions have PH distributions. However, as PH distribution is closed in coherent systems [12], the result actually is an ME representation of a PH distribution. To generalize the results, the component reliability functions are assumed to have ME distribution, which is a generalization of the PH distribution (see Chapter 2). In this way, the ME distribution is closed under operations in SDP representation: inputting ME representations of the component reliability functions into an SDP representation yields an ME representation of the network reliability function. The following development uses structure functions in minimum paths. Analogies for structure functions in minimum cuts can be developed in a similar manner.

Equation (6.16) can be expressed as

$$X_S = \sum_{k=1}^{K} Y_k, \tag{6.20}$$

where $Y_k = \prod_{i \in DP_k} Z_i$.

Each disjoint product term Y_k contains multiplications of the variables representing the states of the components, and possibly their inverses. Note that each term in the SDP representation contains at least one variable without an inverse. Therefore, an SDP representation contains three basic operations:

a) Multiplication of variables in the form X_iX_j.
b) Multiplication by univariable or multivariable inverses in the form $X_i\overline{X_j}$ or $X_i\overline{X_jX_k}$.
c) The SDP terms in the form $Y_i + Y_j$.

Each basic operation can be extended to many variables recursively.

The network reliability function is obtained by substituting the ME representations of the component reliability functions into the SDP representation of the network. The resulting function from each term is a matrix exponential function, but not necessarily a reliability function. Defined $G_{Y_k}(t)$ as the function resulted from disjoint product terms Y_k when the variables X_i, $\overline{X}_i$, and $\overline{X_iX_j}$ in Y_k are replaced by $R_i(t)$, $1 - R_i(t)$, and $1 - R_i(t)R_j(t)$, respectively. In this way, Equation (6.20) becomes

$$R_S(t) = \sum_{k=1}^{K} G_{Y_k}(t). \tag{6.21}$$

The mathematical expression for $R_S(t)$ when $R_i(t)$, $i = 1, 2, \ldots, N$, have ME distributions can be derived using the properties of the ME distribution for the basic operations in the SDP representation. Let the reliability function of component i, $i = 1, 2, \ldots, N$, have ME representations $(\boldsymbol{a}_i, \boldsymbol{A}_i, \boldsymbol{v}_i)$, or

$$R_i(t) = \boldsymbol{a}_i e^{\boldsymbol{A}_i t} \boldsymbol{v}_i, \quad i = 1, 2, \ldots, N, \quad t \geq 0. \tag{6.22}$$

It is necessary to design an algorithm that can generate an ME representation of the network reliability function from the ME representations of its component reliability functions and the SDP representation of the network. The process begins by finding an ME representation for $G_{Y_k}(t)$ based on algebraic operations in Y_k and using the results to find an ME representation for $R_S(t)$. The process is simplified using the closure properties of ME distribution in SDP representation derived in Alkaff et al. [6] as described next.

6.3.1 Multiplication Operation Without Inverse

The simplest form of multiplication without an inverse is

$$Y = X_1X_2. \tag{6.23}$$

Substituting $R_i(t)$ in Equation (6.22) into Equation (6.23) yields

$$G_Y(t) = R_1(t)R_2(t) = \left(\boldsymbol{a}_1 e^{\boldsymbol{A}_1 t} \boldsymbol{v}_1\right)\left(\boldsymbol{a}_2 e^{\boldsymbol{A}_2 t} \boldsymbol{v}_2\right). \tag{6.24}$$

Using the Kronecker multiplication property as used for the minimum operation yields

$$G_Y(t) = \left(\boldsymbol{a}_1 e^{\boldsymbol{A}_1 t} \boldsymbol{v}_1\right) \otimes \left(\boldsymbol{a}_2 e^{\boldsymbol{A}_2 t} \boldsymbol{v}_2\right) = (\boldsymbol{a}_1 \otimes \boldsymbol{a}_2)\left(e^{\boldsymbol{A}_1 t} \otimes e^{\boldsymbol{A}_2 t}\right)(\boldsymbol{v}_1 \otimes \boldsymbol{v}_2).$$

Hence,

$$G_Y(t) = (\boldsymbol{a}_1 \otimes \boldsymbol{a}_2)\left(e^{(\boldsymbol{A}_1 \oplus \boldsymbol{A}_2)t}\right)(\boldsymbol{v}_1 \otimes \boldsymbol{v}_2) = \boldsymbol{a}_Y e^{\boldsymbol{A}_Y t} \boldsymbol{v}_Y. \tag{6.25}$$

This result is formalized as the following theorem.

Theorem 6.1 If $X_1 \sim ME(n_1, \boldsymbol{a}_1, \boldsymbol{A}_1, \boldsymbol{v}_1)$ and $X_2 \sim ME(n_2, \boldsymbol{a}_2, \boldsymbol{A}_2, \boldsymbol{v}_2)$ are independent, then $Y = X_1X_2$ has an ME representation $(n_Y, \boldsymbol{a}_Y, \boldsymbol{A}_Y, \boldsymbol{v}_Y)$, where

$$n_Y = n_1n_2, \quad \boldsymbol{a}_Y = \boldsymbol{a}_1 \otimes \boldsymbol{a}_2, \quad \boldsymbol{A}_Y = \boldsymbol{A}_1 \oplus \boldsymbol{A}_2, \quad \boldsymbol{v}_Y = \boldsymbol{v}_1 \otimes \boldsymbol{v}_2. \tag{6.26}$$

Example 6.16 X_1 and X_2 are state variables of components having lifetimes following exponential distribution with $MTTF = 10$ and 2-Erlang distribution with $MTTF = 20$. The PH representations, which are also the ME representations, for the lifetimes distributions of X_1 and X_2 are

$$n_1 = 1, \quad n_2 = 2, \quad \boldsymbol{a}_1 = [1], \quad \boldsymbol{a}_2 = \begin{bmatrix} 1 & 0 \end{bmatrix}, \quad \boldsymbol{v}_1 = \begin{bmatrix} 1 \end{bmatrix}, \quad \boldsymbol{v}_2 = \begin{bmatrix} 1 \\ 1 \end{bmatrix},$$

$$\boldsymbol{A}_1 = \begin{bmatrix} -0.1 \end{bmatrix}, \quad \boldsymbol{A}_2 = \begin{bmatrix} -0.1 & 0.1 \\ 0 & -0.1 \end{bmatrix}.$$

ME representation $(n_Y, \boldsymbol{a}_Y, \boldsymbol{A}_Y, \boldsymbol{v}_Y)$ for $Y = X_1X_2$ calculated using Equation (6.26) is

$$n_Y = 1 \times 2 = 2, \quad \boldsymbol{a}_Y = \begin{bmatrix} 1 \end{bmatrix} \otimes \begin{bmatrix} 1 & 0 \end{bmatrix} = \begin{bmatrix} 1 & 0 \end{bmatrix}, \quad \boldsymbol{v}_Y = \begin{bmatrix} 1 \end{bmatrix} \otimes \begin{bmatrix} 1 \\ 1 \end{bmatrix} = \begin{bmatrix} 1 \\ 1 \end{bmatrix},$$

$$\boldsymbol{A}_Y = \begin{bmatrix} -0.1 \end{bmatrix} \oplus \begin{bmatrix} -0.1 & 0.1 \\ 0 & -0.1 \end{bmatrix} = \begin{bmatrix} -0.2 & 0.1 \\ 0 & -0.2 \end{bmatrix}.$$

Substitute these values into Equation (6.25) to produce $G_Y(t)$. ∎

6.3.2 Multiplication Operation with Inverse

The simplest form of multiplication with an inverse variable is

$$Y = X_1\overline{X}_2. \tag{6.27}$$

Use $\overline{X}_2 = 1 - X_2$ to change Equation (6.27) to

$$Y = X_1(1 - X_2) = X_1 - X_1X_2. \tag{6.28}$$

Substituting $R_i(t)$ in Equation (6.22) into Equation (6.28) produces

$$G_Y(t) = R_1(t) - R_1(t)R_2(t) = \boldsymbol{a}_1e^{\boldsymbol{A}_1t}\boldsymbol{v}_1 - \left(\boldsymbol{a}_1e^{\boldsymbol{A}_1t}\boldsymbol{v}_1\right)\left(\boldsymbol{a}_2e^{\boldsymbol{A}_2t}\boldsymbol{v}_2\right).$$

The second term on the right side is the same as Equation (6.24) that can be used to produce

$$\begin{aligned} G_Y(t) &= \boldsymbol{a}_1e^{\boldsymbol{A}_1t}\boldsymbol{v}_1 - (\boldsymbol{a}_1 \otimes \boldsymbol{a}_2)\left(e^{\boldsymbol{A}_1t \oplus \boldsymbol{A}_2t}\right)(\boldsymbol{v}_1 \otimes \boldsymbol{v}_2) \\ &= [\boldsymbol{a}_1 \quad -\boldsymbol{a}_1 \otimes \boldsymbol{a}_2]\begin{bmatrix} e^{\boldsymbol{A}_1t} & \boldsymbol{0} \\ \boldsymbol{0} & e^{(\boldsymbol{A}_1 \oplus \boldsymbol{A}_2)t} \end{bmatrix}\begin{bmatrix} \boldsymbol{v}_1 \\ \boldsymbol{v}_1 \otimes \boldsymbol{v}_2 \end{bmatrix} = \boldsymbol{a}_Ye^{\boldsymbol{A}_Yt}\boldsymbol{v}_Y. \end{aligned} \tag{6.29}$$

This result is formalized as the following theorem.

Theorem 6.2 If $X_1 \sim ME(n_1, \boldsymbol{a}_1, \boldsymbol{A}_1, \boldsymbol{v}_1)$ and $X_2 \sim ME(n_2, \boldsymbol{a}_2, \boldsymbol{A}_2, \boldsymbol{v}_2)$ are independent, then $Y = X_1\overline{X}_2$ produce $G_Y(t)$ that can be expressed in a matrix exponential function $\boldsymbol{a}_Ye^{\boldsymbol{A}_Yt}\boldsymbol{v}_Y$ with

$$n_Y = n_1 + n_1n_2, \quad \boldsymbol{a}_Y = [\boldsymbol{a}_1 \quad -\boldsymbol{a}_1 \otimes \boldsymbol{a}_2],$$

$$\boldsymbol{A}_Y = \begin{bmatrix} \boldsymbol{A}_1 & \boldsymbol{0} \\ \boldsymbol{0} & \boldsymbol{A}_1 \oplus \boldsymbol{A}_2 \end{bmatrix}, \quad \boldsymbol{v}_Y = \begin{bmatrix} \boldsymbol{v}_1 \\ \boldsymbol{v}_1 \otimes \boldsymbol{v}_2 \end{bmatrix}. \tag{6.30}$$

Equation (6.30) cannot be called an ME representation because $G_Y(t)$ is not a valid reliability function, in contrast to the result of the operation without inverse that produces a valid reliability function.

Multiplication by multivariable inverses such as $X_1\overline{X_2X_3}$ is carried out by defining $Z_2 = X_2X_3$ so that $X_1\overline{Z}_2$. Use Theorem 6.1 to obtain the ME representation for Z_2, and continue with using Theorem 6.2 to obtain the matrix exponential expression for $X_1\overline{Z}_2$.

Example 6.17 With X_1 and X_2 as in Example 6.16, the matrix exponential expression $\boldsymbol{a}_Y e^{\boldsymbol{A}_Y t}\boldsymbol{v}_Y$ for $Y = X_1\overline{X}_2$ is

$$n_Y = 1 + 1 \times 2 = 3, \quad \boldsymbol{a}_Y = \begin{bmatrix}1 - 1 \otimes \begin{bmatrix}1 & 0\end{bmatrix}\end{bmatrix} = \begin{bmatrix}1 & -1 & 0\end{bmatrix},$$

$$\boldsymbol{A}_Y = \begin{bmatrix} -0.1 & 0 \\ 0 & -0.1 \oplus \begin{bmatrix} -0.1 & 0.1 \\ 0 & -0.1 \end{bmatrix} \end{bmatrix} = \begin{bmatrix} -0.1 & 0 & 0 \\ 0 & -0.2 & 0.1 \\ 0 & 0 & -0.2 \end{bmatrix}, \quad \boldsymbol{v}_Y = \begin{bmatrix} 1 \\ 1 \otimes \begin{bmatrix} 1 \\ 1 \end{bmatrix} \end{bmatrix} = \begin{bmatrix} 1 \\ 1 \\ 1 \end{bmatrix}.$$

■

6.3.3 Addition Operation

The simplest form of addition is

$$X_S = Y_1 + Y_2. \tag{6.31}$$

where Y_1 and Y_2 are disjoint product terms resulting from the multiplication operations of state variables in minimum paths. According to Theorems 6.1 and 6.2, $G_{Y1}(t)$ and $G_{Y2}(t)$ may have ME representations or matrix exponential functions with parameters $(n_{Y_1}, \boldsymbol{a}_{Y_1}, \boldsymbol{A}_{Y_1}, \boldsymbol{v}_{Y1})$ and $(n_{Y_2}, \boldsymbol{a}_{Y_2}, \boldsymbol{A}_{Y_2}, \boldsymbol{v}_{Y2})$, that is,

$$G_{Y1}(t) = \boldsymbol{a}_{Y_1} e^{\boldsymbol{A}_{Y_1} t}\boldsymbol{v}_{Y1} \quad \text{and} \quad G_{Y_2}(t) = \boldsymbol{a}_{Y_2} e^{\boldsymbol{A}_{Y_2} t}\boldsymbol{v}_{Y2}.$$

Substituting them into Equation (6.31) yields

$$\begin{aligned} R_S(t) &= \boldsymbol{a}_{Y_1} e^{\boldsymbol{A}_{Y_1} t}\boldsymbol{v}_{Y1} + \boldsymbol{a}_{Y_2} e^{\boldsymbol{A}_{Y_2} t}\boldsymbol{v}_{Y2} \\ &= [\boldsymbol{a}_{Y_1} \ \ \boldsymbol{a}_{Y_2}] \begin{bmatrix} e^{\boldsymbol{A}_{Y_1} t} & \boldsymbol{0} \\ \boldsymbol{0} & e^{\boldsymbol{A}_{Y_2} t} \end{bmatrix} \begin{bmatrix} \boldsymbol{v}_{Y_1} \\ \boldsymbol{v}_{Y_2} \end{bmatrix}, \end{aligned} \tag{6.32}$$

which is a matrix exponential function with an ME representation because it is a valid ME distribution. This result is formalized as the following theorem.

Theorem 6.3 If $G_{Y1}(t) = \boldsymbol{a}_{Y_1} e^{\boldsymbol{A}_{Y_1} t}\boldsymbol{v}_{Y1}$ and $G_{Y_2}(t) = \boldsymbol{a}_{Y_2} e^{\boldsymbol{A}_{Y_2} t}\boldsymbol{v}_{Y2}$ are matrix exponential functions obtained from two disjoint product terms, then $R_S(t) = G_{Y1}(t) + G_{Y_2}(t)$ has an ME representation $(n_S, \boldsymbol{a}_S, \boldsymbol{A}_S, \boldsymbol{v}_S)$, where

$$n_S = n_1 + n_2, \quad \boldsymbol{a}_S = [\boldsymbol{a}_{Y_1} \ \ \boldsymbol{a}_{Y_2}]. \quad \boldsymbol{A}_S = \begin{bmatrix} \boldsymbol{A}_{Y_1} & \boldsymbol{0} \\ \boldsymbol{0} & \boldsymbol{A}_{Y_2} \end{bmatrix}, \quad \boldsymbol{v}_S = \begin{bmatrix} \boldsymbol{v}_{Y_1} \\ \boldsymbol{v}_{Y_2} \end{bmatrix}. \tag{6.33}$$

Example 6.18 An ME representation for the reliability function of a system with structure function $X_S = X_1X_2 \vee X_1X_3$ is calculated using Theorems 6.1–6.3. Assume the reliability functions of components 1 and 2 are the same as in Example 6.16, and component 3 has lifetime following exponential distribution with $MTTF = 5$.

The PH representation of the reliability function of component 3 is

$$n_3 = 1, \quad \boldsymbol{a}_3 = \begin{bmatrix}1\end{bmatrix}, \quad \boldsymbol{A}_3 = \begin{bmatrix}-0.2\end{bmatrix}, \quad \boldsymbol{v}_3 = \begin{bmatrix}1\end{bmatrix}.$$

The SDP representation of this system is

$$X_S = X_1X_2 + X_1X_3\overline{X}_2.$$

It was obtained in Example 6.16 that the ME representation for $Y_1 = X_1X_2$ is

$$n_{Y_1} = 2, \quad \boldsymbol{a}_{Y_1} = \begin{bmatrix}1 & 0\end{bmatrix}, \quad \boldsymbol{A}_{Y_1} = \begin{bmatrix}-0.2 & 0.1\\ 0 & -0.2\end{bmatrix}, \quad \boldsymbol{v}_{Y_1} = \begin{bmatrix}1\\1\end{bmatrix}.$$

An ME representation of $Z = X_1X_3$ is derived using Theorem 6.1 as

$$n_Z = 1 \times 1 = 1, \quad \boldsymbol{a}_Z = \begin{bmatrix}1\end{bmatrix} \otimes \begin{bmatrix}1\end{bmatrix} = \begin{bmatrix}1\end{bmatrix}, \quad \boldsymbol{v}_Z = \begin{bmatrix}1\end{bmatrix} \otimes \begin{bmatrix}1\end{bmatrix} = \begin{bmatrix}1\end{bmatrix},$$
$$\boldsymbol{A}_Z = \begin{bmatrix}-0.1\end{bmatrix} \oplus \begin{bmatrix}-0.2\end{bmatrix} = \begin{bmatrix}-0.3\end{bmatrix}.$$

A matrix exponential representation of $Y_2 = Z\overline{X}_2$ is derived using Theorem 6.2 as

$$n_{Y_2} = 1 + 1 \times 2 = 3, \quad \boldsymbol{a}_{Y_2} = \begin{bmatrix}1 - 1 \otimes \begin{bmatrix}1 & 0\end{bmatrix}\end{bmatrix} = \begin{bmatrix}1 & -1 & 0\end{bmatrix}, \quad \boldsymbol{v}_{Y_2} = \begin{bmatrix}1\\ 1 \otimes \begin{bmatrix}1\\1\end{bmatrix}\end{bmatrix} = \begin{bmatrix}1\\1\\1\end{bmatrix},$$

$$\boldsymbol{A}_{Y_2} = \begin{bmatrix}-0.3 & 0\\ 0 & -0.1 \oplus \begin{bmatrix}-0.1 & 0.1\\ 0 & -0.1\end{bmatrix}\end{bmatrix} = \begin{bmatrix}-0.3 & 0 & 0\\ 0 & -0.2 & 0.1\\ 0 & 0 & -0.2\end{bmatrix}.$$

An ME representation for $R_S(t)$ is obtained using Theorem 6.3 as

$$n_S = 2 + 3 = 5, \quad \boldsymbol{a}_S = \begin{bmatrix}\begin{bmatrix}1 & 0\end{bmatrix} & \begin{bmatrix}1 & -1 & 0\end{bmatrix}\end{bmatrix} = \begin{bmatrix}1 & 0 & 1 & -1 & 0\end{bmatrix}, \quad \boldsymbol{v}_Y = \begin{bmatrix}\begin{bmatrix}1\\1\end{bmatrix}\\ \begin{bmatrix}1\\1\\1\end{bmatrix}\end{bmatrix} = \begin{bmatrix}1\\1\\1\\1\\1\end{bmatrix},$$

$$\boldsymbol{A}_Y = \begin{bmatrix}\begin{bmatrix}-0.2 & 0.1\\ 0 & -0.2\end{bmatrix} & \boldsymbol{0}\\ \boldsymbol{0} & \begin{bmatrix}-0.3 & 0 & 0\\ 0 & -0.2 & 0.1\\ 0 & 0 & -0.2\end{bmatrix}\end{bmatrix} = \begin{bmatrix}-0.2 & 0.1 & 0 & 0 & 0\\ 0 & -0.2 & 0 & 0 & 0\\ 0 & 0 & -0.3 & 0 & 0\\ 0 & 0 & 0 & -0.2 & 0.1\\ 0 & 0 & 0 & 0 & -0.2\end{bmatrix}.$$

■

Exercise 6.6 Determine an ME representation for the lifetime distribution of a network having structure function $X_S = X_1 \vee X_2X_3 \vee X_2X_4$ if each component has an exponentially distributed lifetime with $MTTF = 10$ hours.

Closure properties of ME distribution in SDP representation of a network are utilized in the next section to construct an algorithm for deriving ME representation of network reliability function from ME representations of its component reliability functions.

6.4 Algorithm for Generating ME Representation of Network Reliability Function

The two multiplication rules given in Theorems 6.1 and 6.2 can be combined into one and extended to many variables. Since there is at least one variable without an inverse in each disjoint product term, the sequence of variables in each term can be arranged so that the first variable is without an inverse. Let D_k be the number of variables in the kth disjoint product term, Y_k, and i, $i = 1, 2, \ldots, D_k$, be the order of variables in the disjoint product term DP_k. Theorems 6.1 and 6.2 can be utilized recursively to prove the following lemma.

Lemma 6.1 If component i has a lifetime distribution with ME representation $(n_i, \boldsymbol{a}_i, \boldsymbol{A}_i, \boldsymbol{v}_i)$, $i = 1, 2, \ldots, N$, then each disjoint product Y_k in the SDP representation has a matrix exponential function with representation $\left(n_Y^{(D_k)}, \boldsymbol{a}_Y^{(D_k)}, \boldsymbol{A}_Y^{(D_k)}, \boldsymbol{v}_Y^{(D_k)}\right)$ computed recursively as follows:

For $i = 1, \ldots, N$,

Read Z_i

Repeat until $Z_i = X_i$

$$n_Y^{(i)} = n_i, \quad \boldsymbol{a}_Y^{(i)} = \boldsymbol{a}_i, \quad \boldsymbol{A}_Y^{(i)} = \boldsymbol{A}_i, \quad \boldsymbol{v}_Y^{(i)} = \boldsymbol{v}_i. \tag{6.34}$$

For $j = i + 1, \ldots, N$,

Repeat until $Z_j = X_j$ or $\overline{X}_j$

If $Z_j = X_j$, then calculate

$$n_Y^{(j)} = n_Y^{(j-1)} \times n_j, \tag{6.35}$$

$$\boldsymbol{a}_Y^{(j)} = \boldsymbol{a}_Y^{(j-1)} \otimes \boldsymbol{a}_j, \tag{6.36}$$

$$\boldsymbol{A}_Y^{(j)} = \boldsymbol{A}_Y^{(j-1)} \oplus \boldsymbol{A}_j, \tag{6.37}$$

$$\boldsymbol{v}_Y^{(j)} = \boldsymbol{v}_Y^{(j-1)} \otimes \boldsymbol{v}_j, \tag{6.38}$$

Else

$$n_Y^{(j)} = n_Y^{(j-1)} + n_Y^{(j-1)} \times n_j, \tag{6.39}$$

$$\boldsymbol{a}_Y^{(j)} = \left[\boldsymbol{a}_Y^{(j-1)} \ \ -\boldsymbol{a}_Y^{(j-1)} \otimes \boldsymbol{a}_j\right], \tag{6.40}$$

$$\boldsymbol{A}_Y^{(j)} = \begin{bmatrix} \boldsymbol{A}_Y^{(j-1)} & \boldsymbol{0} \\ \boldsymbol{0} & \boldsymbol{A}_Y^{(j-1)} \oplus \boldsymbol{A}_j \end{bmatrix}, \tag{6.41}$$

$$\boldsymbol{v}_Y^{(j)} = \begin{bmatrix} \boldsymbol{v}_Y^{(j-1)} \\ \boldsymbol{v}_Y^{(j-1)} \otimes \boldsymbol{v}_j \end{bmatrix}. \tag{6.42}$$

SDP representation involves addition operations. The ME representation for $R_S(t)$ can be found using Theorem 6.3 recursively. Let K denote the number of disjoint product terms in the SDP representation with $k = 1, 2, \ldots, K$ specifying their sequence. The following lemma can be proved recursively using Theorem 6.3.

Lemma 6.2 The sum of all disjoint product terms $\sum_{k=1}^{K} G_{Y_k}(t)$ in an SDP representation has an ME representation $(n_S, \boldsymbol{a}_S, \boldsymbol{A}_S, \boldsymbol{v}_S)$ calculated recursively as follows:

For $k = 2, \ldots, K$, calculate

$$n_S^{(k)} = n_S^{(k-1)} + n_{Y_k}, \tag{6.43}$$

$$\boldsymbol{a}_S^{(k)} = \left[\boldsymbol{a}_S^{(k-1)} \ \ \boldsymbol{a}_Y^{(k)}\right], \tag{6.44}$$

$$\boldsymbol{A}_S^{(k)} = \begin{bmatrix} \boldsymbol{A}_S^{(k-1)} & \boldsymbol{0} \\ \boldsymbol{0} & \boldsymbol{A}_Y^{(k)} \end{bmatrix}, \tag{6.45}$$

$$\boldsymbol{v}_S^{(k)} = \begin{bmatrix} \boldsymbol{v}_S^{(k-1)} \\ \boldsymbol{a}_Y^{(k)} \end{bmatrix}, \tag{6.46}$$

with initial conditions

$$n_S^{(1)} = n_Y^{(1)}, \quad \boldsymbol{a}_S^{(1)} = \boldsymbol{a}_Y^{(1)}, \quad \boldsymbol{A}_S^{(1)} = \boldsymbol{A}_Y^{(1)}, \quad \boldsymbol{v}_S^{(1)} = \boldsymbol{v}_Y^{(1)}. \tag{6.47}$$

The recursive Equations (6.34–6.47) is utilized to construct Algorithm 6.1 which can be utilized to generate an ME representation of the network reliability function from the output of **sdprep** and the PH or ME representations of the component reliability functions. Its pseudo code is as follows.

Algorithm 6.1

Input SDP, $n_i, \boldsymbol{a}_i, \boldsymbol{A}_i, i = 1, 2, \ldots, N$
Z_i: the ith term of SDP
For $i = 1, \ldots, N,$
  Read Z_i
  Repeat until $Z_i = X_i$
  $n_{Y_i} = n_i, \boldsymbol{a}_{Y_i} = \boldsymbol{a}_i, \boldsymbol{A}_{Y_i} = \boldsymbol{A}_i, \boldsymbol{v}_{Y_i} = \boldsymbol{v}_i$
  For $j = i + 1, \ldots, N,$
    Repeat until $Z_j = X_j$ *or* $\overline{X}_j$
    If $Z_j = X_j$, then calculate
      $n_{Y_j} = n_{Y_i} n_j$
      $\boldsymbol{a}_Y^{(j)} = \boldsymbol{a}_{Y_i} \otimes \boldsymbol{a}_j$
      $\boldsymbol{A}_{Y_j} = \boldsymbol{A}_{Y_i} \oplus \boldsymbol{A}_j$
      $\boldsymbol{v}_{Y_j} = \boldsymbol{v}_{Y_i} \otimes \boldsymbol{v}_j$
    Else
      $n_{Y_j} = n_{Y_i} + n_{Y_i} n_j$
      $\boldsymbol{a}_{Y_i} = \begin{bmatrix} \boldsymbol{a}_{Y_i} - \boldsymbol{a}_{Y_i} \otimes \boldsymbol{a}_j \end{bmatrix}$
      $\boldsymbol{A}_{Y_i} = \begin{bmatrix} \boldsymbol{A}_{Y_i} & \boldsymbol{0} \\ \boldsymbol{0} & \boldsymbol{A}_{Y_i} \oplus \boldsymbol{A}_j \end{bmatrix}$
      $\boldsymbol{v}_{Y_i} = \begin{bmatrix} \boldsymbol{v}_{Y_i} \\ \boldsymbol{v}_{Y_i} \otimes \boldsymbol{v}_j \end{bmatrix}$
  $j = j + 1$
  end
  $n_S = n_{Y_1}, \boldsymbol{a}_S = \boldsymbol{a}_{Y_1}, \boldsymbol{A}_S = \boldsymbol{A}_{Y_1}, \boldsymbol{v}_S = \boldsymbol{v}_{Y_1}$
  For $k = 2, \ldots, K$, calculate
    $n_S = n_S + n_{Y_k}$
    $\boldsymbol{a}_S = \begin{bmatrix} \boldsymbol{a}_S & \boldsymbol{a}_{Y_k} \end{bmatrix}$
    $\boldsymbol{A}_S = \begin{bmatrix} \boldsymbol{A}_S & \boldsymbol{0} \\ \boldsymbol{0} & \boldsymbol{A}_{Y_k} \end{bmatrix}$
    $\boldsymbol{v}_S = \begin{bmatrix} \boldsymbol{v}_S \\ \boldsymbol{v}_{Y_k} \end{bmatrix}$
  $k = k + 1$
  end
end

However, the output of **sdprep** is based on component ordering, not necessarily the first variable in each disjoint term is without inversion. The complete algorithm based on arbitrary SDP representation is programmed in MATLAB as **menetcomb** function, available in the software package **dysrel-am**.

Since it is known that $\boldsymbol{A}_i$, $i = 1, 2, \ldots, N$, is nonsingular, its Kronecker sum is also nonsingular. Therefore, matrix $\boldsymbol{A}_S$ produced is nonsingular as a condition for a valid ME distribution.

Example 6.19 Equations (6.34–6.47) is used to find an ME representation of the reliability function of a 2-out-of-3 system with components having lifetimes following exponential distribution with $MTTF = 1$.

The SDP representation of a 2-out-of-3 system is $X_S = X_1X_2 + X_1X_3\overline{X}_2 + X_2X_3\overline{X}_1$.

The ME representation for the lifetime distribution of all components is

$$n_i = 1, \quad \boldsymbol{a}_i = [1], \quad \boldsymbol{A}_i = [-1], \quad \boldsymbol{v}_i = [1], \quad i = 1, 2, 3.$$

The process of calculating the ME representation of the network reliability function is carried out with the following steps:

1) The ME representation of the first term, $Y_1 = X_1X_2$, is calculated using Equations (6.35–6.38). The result is

$$n_{Y_1} = 1 \times 1 = 1, \quad \boldsymbol{a}_{Y_1} = [1] \otimes [1] = [1],$$
$$\boldsymbol{A}_{Y_1} = [-1] \oplus [-1] = [-2], \quad \boldsymbol{v}_{Y_1} = [1] \otimes [1] = [1].$$

2) The representation of the second term $Y_2 = X_1X_3\overline{X}_2$ is calculated recursively using Equations (6.35–6.42). First, calculate the representation for $Z = X_1X_3$ using Equations (6.35–6.38) to yield

$$n_Z = 1 \times 1 = 1, \quad \boldsymbol{a}_Z = [1] \otimes [1] = [1],$$
$$\boldsymbol{A}_Z = [-1] \oplus [-1] = [-2], \quad \boldsymbol{v}_Z = [1] \otimes [1] = [1].$$

Second, calculate the representation for $Y_2 = Z\overline{X}_2$ using Equations (6.39–6.42)

$$n_{Y_2} = 1 + 1 \times 1 = 2, \quad \boldsymbol{a}_{Y_2} = [1 - 1 \otimes 1] = [1 - 1],$$
$$\boldsymbol{A}_{Y_2} = \begin{bmatrix} -2 & 0 \\ 0 & -1 \oplus -2 \end{bmatrix} = \begin{bmatrix} -2 & 0 \\ 0 & -3 \end{bmatrix}, \quad \boldsymbol{v}_{Y_2} = \begin{bmatrix} 1 \\ 1 \otimes 1 \end{bmatrix} = \begin{bmatrix} 1 \\ 1 \end{bmatrix}.$$

3) The representation of the third term $X_2X_3\overline{X}_1$ is calculated in the same way as the second term to produce

$$n_{Y_3} = 1 + 1 \times 1 = 2, \quad \boldsymbol{a}_{Y_3} = [1 - 1 \otimes 1] = [1 - 1],$$
$$\boldsymbol{A}_{Y_3} = \begin{bmatrix} -2 & 0 \\ 0 & -1 \oplus -2 \end{bmatrix} = \begin{bmatrix} -2 & 0 \\ 0 & -3 \end{bmatrix}, \quad \boldsymbol{v}_{Y_3} = \begin{bmatrix} 1 \\ 1 \otimes 1 \end{bmatrix} = \begin{bmatrix} 1 \\ 1 \end{bmatrix}.$$

4) The ME representation for the system reliability function is calculated recursively using Equations (6.43–6.46). The result is

$$n_S = 1 + 2 + 2 = 5, \quad \boldsymbol{a}_S = \begin{bmatrix} 1 & [1 - 1] & [1 - 1] \end{bmatrix} = \begin{bmatrix} 1 & 1 & -1 & 1 & -1 \end{bmatrix},$$

$$\boldsymbol{A}_S = \begin{bmatrix} -2 & 0 & 0 \\ 0 & \begin{bmatrix} -2 & 0 \\ 0 & -3 \end{bmatrix} & 0 \\ 0 & 0 & \begin{bmatrix} -2 & 0 \\ 0 & -3 \end{bmatrix} \end{bmatrix} = \begin{bmatrix} -2 & 0 & 0 & 0 & 0 \\ 0 & -2 & 0 & 0 & 0 \\ 0 & 0 & -3 & 0 & 0 \\ 0 & 0 & 0 & -2 & 0 \\ 0 & 0 & 0 & 0 & -3 \end{bmatrix}, \quad \boldsymbol{v}_S = \begin{bmatrix} 1 \\ \begin{bmatrix} 1 \\ 1 \end{bmatrix} \\ \begin{bmatrix} 1 \\ 1 \end{bmatrix} \end{bmatrix} = \begin{bmatrix} 1 \\ 1 \\ 1 \\ 1 \\ 1 \end{bmatrix}.$$

The presence of negative numbers in elements of $\boldsymbol{a}_S$ indicates that the result is an ME distribution, not PH, even though the inputs are PH distributions.

For comparison, the PH representation of the lifetime distribution of a 2-out-of-3 system using the state transition model for non-identical K-out-of-N systems given in Section 5.1 is

$$\boldsymbol{a}_S = \begin{bmatrix} 1 & 0 & 0 & 0 \end{bmatrix}.$$

$$\boldsymbol{A}_S = \begin{bmatrix} -1-1-1 & (1)(1)(1) & (1)(1)(1) & (1)(1)(1) \\ 0 & -1-1 & 0 & 0 \\ 0 & 0 & -1-1 & 0 \\ 0 & 0 & 0 & -1-1 \end{bmatrix} = \begin{bmatrix} -3 & 1 & 1 & 1 \\ 0 & -2 & 0 & 0 \\ 0 & 0 & -2 & 0 \\ 0 & 0 & 0 & -2 \end{bmatrix}.$$

The system reliability functions resulting from these two different representations are presented in Figure 6.6, which shows that they are indeed identical.

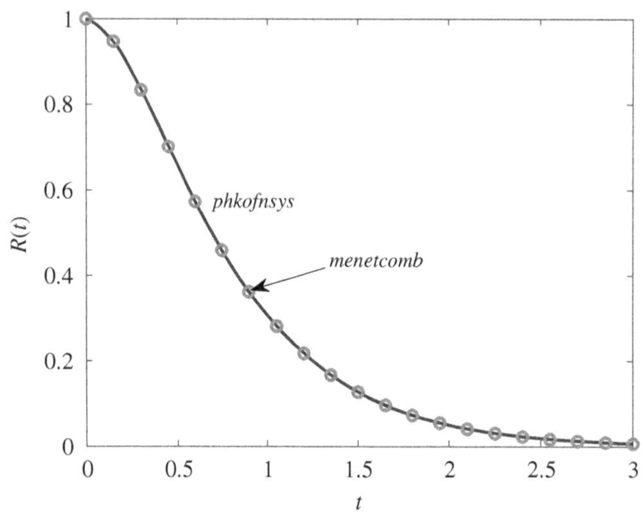

Figure 6.6 The system reliability function of Example 6.19 calculated using combinatorial and state transition models.

This equivalency means the result of using combinatorial model is an ME representation of PH distribution obtained by using state transition model. ■

This example shows that using the state transition model results in a simpler PH representation than the ME representation resulting from the combinatorial model. However, the survival matrix of the PH representation is an upper diagonal matrix, while the survival matrix of the ME representation is diagonal. Computing-derived quantities of the upper diagonal matrix are slower than the diagonal matrix. The increase in dimensionality of the ME representation is compensated by its survival matrix being a diagonal matrix, while for PH representation obtained using state transition model is an upper diagonal matrix. A more reasonable comparison of computing time using combinatorial and state transition models will be discussed later.

Exercise 6.7 Redo Exercise 6.6 using Algorithm 6.1.

Example 6.20 The **menetcomb** function is used to generate ME representations of the network in Figure 6.2 using its minimum paths. All components have identical lifetime distribution, with the same parameter as in Example 6.19. The Boolean structure function for this network can be written as

$$X_S = X_1X_4 \vee X_2X_5 \vee X_1X_3X_5 \vee X_2X_3X_4.$$

The SDP representation generated using **sdprep** is

$$X_S = X_1X_4 + \overline{X_1X_4}X_2X_5 + X_1\overline{X}_2X_3\overline{X}_4X_5 + \overline{X}_1X_2X_3X_4\overline{X}_5.$$

The ME representation of the network reliability function generated using **menetcomb** is

$$n_S = 11, \quad \boldsymbol{a}_S = \begin{bmatrix} 1 & 1 & -1 & 1 & -1 & -1 & 1 & 1 & -1 & -1 & 1 \end{bmatrix},$$

$$\boldsymbol{A}_S = \begin{bmatrix} -2 & 0 & 0 & 0 & 0 & 0 & 0 & 0 & 0 & 0 & 0 \\ 0 & -2 & 0 & 0 & 0 & 0 & 0 & 0 & 0 & 0 & 0 \\ 0 & 0 & -4 & 0 & 0 & 0 & 0 & 0 & 0 & 0 & 0 \\ 0 & 0 & 0 & -3 & 0 & 0 & 0 & 0 & 0 & 0 & 0 \\ 0 & 0 & 0 & 0 & -4 & 0 & 0 & 0 & 0 & 0 & 0 \\ 0 & 0 & 0 & 0 & 0 & -4 & 0 & 0 & 0 & 0 & 0 \\ 0 & 0 & 0 & 0 & 0 & 0 & -5 & 0 & 0 & 0 & 0 \\ 0 & 0 & 0 & 0 & 0 & 0 & 0 & -3 & 0 & 0 & 0 \\ 0 & 0 & 0 & 0 & 0 & 0 & 0 & 0 & -4 & 0 & 0 \\ 0 & 0 & 0 & 0 & 0 & 0 & 0 & 0 & 0 & -4 & 0 \\ 0 & 0 & 0 & 0 & 0 & 0 & 0 & 0 & 0 & 0 & -5 \end{bmatrix},$$

$$\boldsymbol{v}_S = \begin{bmatrix} 1 & 1 & 1 & 1 & 1 & 1 & 1 & 1 & 1 & 1 & 1 \end{bmatrix}^T.$$

■

Example 6.21 The structure function for this network also can be written as

$$X_1X_4 \vee X_1X_3X_5 \vee X_2X_3X_4 \vee X_2X_5.$$

The SDP representation generated using **sdprep** is

$$X_1X_4 + X_1X_3\overline{X}_4X_5 + \overline{X}_1X_2X_3X_4 + \overline{X}_1X_2\overline{X_3X_4}X_5 + X_1X_2\overline{X}_3\overline{X}_4X_5.$$

The ME representation of the network reliability function obtained using **menetcomb** is

$$n_S = 13, \quad \boldsymbol{a}_S = \begin{bmatrix} 1 & 1 & -1 & 1 & -1 & 1 & -1 & -1 & 1 & 1 & -1 & -1 & 1 \end{bmatrix},$$

$$\boldsymbol{A}_S = \begin{bmatrix} -2 & 0 & 0 & 0 & 0 & 0 & 0 & 0 & 0 & 0 & 0 & 0 & 0 \\ 0 & -3 & 0 & 0 & 0 & 0 & 0 & 0 & 0 & 0 & 0 & 0 & 0 \\ 0 & 0 & -4 & 0 & 0 & 0 & 0 & 0 & 0 & 0 & 0 & 0 & 0 \\ 0 & 0 & 0 & -3 & 0 & 0 & 0 & 0 & 0 & 0 & 0 & 0 & 0 \\ 0 & 0 & 0 & 0 & -4 & 0 & 0 & 0 & 0 & 0 & 0 & 0 & 0 \\ 0 & 0 & 0 & 0 & 0 & -2 & 0 & 0 & 0 & 0 & 0 & 0 & 0 \\ 0 & 0 & 0 & 0 & 0 & 0 & -3 & 0 & 0 & 0 & 0 & 0 & 0 \\ 0 & 0 & 0 & 0 & 0 & 0 & 0 & -4 & 0 & 0 & 0 & 0 & 0 \\ 0 & 0 & 0 & 0 & 0 & 0 & 0 & 0 & -5 & 0 & 0 & 0 & 0 \\ 0 & 0 & 0 & 0 & 0 & 0 & 0 & 0 & 0 & -3 & 0 & 0 & 0 \\ 0 & 0 & 0 & 0 & 0 & 0 & 0 & 0 & 0 & 0 & -4 & 0 & 0 \\ 0 & 0 & 0 & 0 & 0 & 0 & 0 & 0 & 0 & 0 & 0 & -4 & 0 \\ 0 & 0 & 0 & 0 & 0 & 0 & 0 & 0 & 0 & 0 & 0 & 0 & -5 \end{bmatrix},$$

$$\boldsymbol{v}_S = \begin{bmatrix} 1 & 1 & 1 & 1 & 1 & 1 & 1 & 1 & 1 & 1 & 1 & 1 & 1 \end{bmatrix}^T.$$

■

It is apparent that arranging the terms of the structure function with fewer variables placed earlier in the sequence results in a simpler (smaller orders) ME representation compared to using a structure function with unordered terms. It is advisable to first sort the resulting structure function by the number of elements in each term before using it to find the SDP representation of the network reliability function. The **minpaths** function provided in **dysrel-am** is equipped with sorting algorithm so that the minimum paths generated using this function are already sorted. The original minpaths function from MATLAB does not produce sorted minimum paths. Example 6.7 was using **minpaths** function from MATLAB that produced unordered minimum paths, while Example 6.15 was using the modified **minpaths** function from **dysrel-am** that produced ordered minimum paths.

Example 6.22 ME representations of the network lifetime distribution resulting from Examples 6.20 and 6.21 are used to graph the network reliability and hazard functions. The results are presented in Figure 6.7. As seen, both representations produce the same graphs indicating that those ME representations are equivalent.

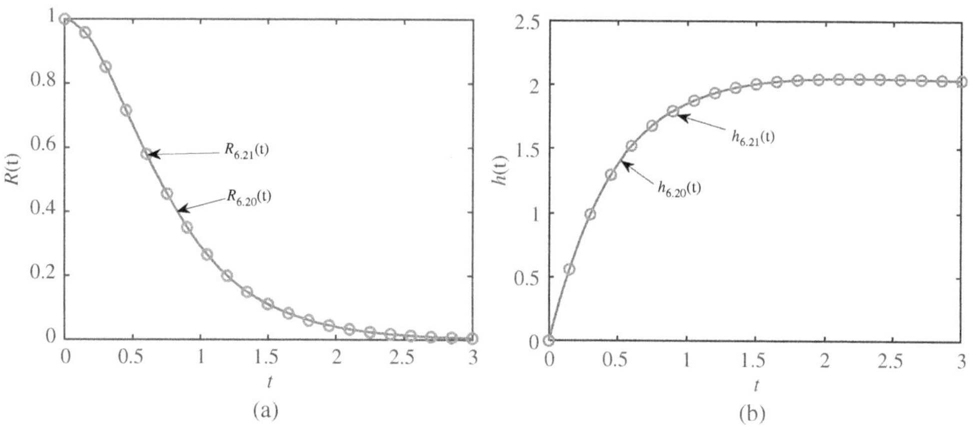

Figure 6.7 Reliability and hazard function of Examples 6.20 and 6.21.

■

Example 6.23 The network in Figure 6.4 is analyzed for its reliability using the **menetcomb** function.

The SDP representation generated in Example 6.15 is input to **menetcomb**. The ME representation generated is used to calculate network reliability measures. The component lifetimes are assumed to have identical distributions with $MTTF = 100$. Three cases are considered: the exponential, 2-Erlang, and 2-Coxian distributions. The results are presented in Table 6.1 and Figure 6.8.

Table 6.1 Comparison of *MTTF*, *SDTTF*, order, and computation time.

Distribution of component lifetime	Network lifetime			Computation time (s)
	MTTF	***SDTTF***	**Order**	
Exponential	55.60	38.06	23	0.02
2-Erlang	69.90	31.40	1616	0.39
2-Coxian ($CVTTF = 1.2$)	49.70	36.42	1616	0.25

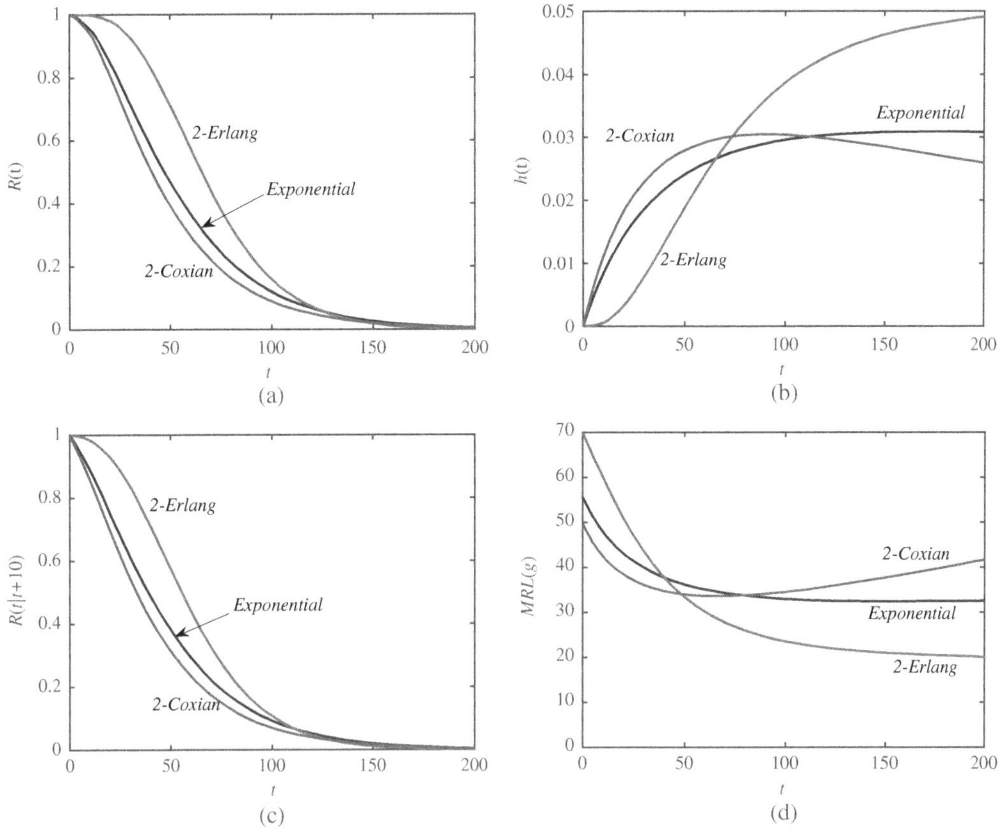

Figure 6.8 Graphs of functional reliability measures of network in Figure 6.4. (a) Reliability function, (b) hazard function, (c) residual reliability function for age 10, and (d) *MRL* function. ■

Exercise 6.8 Compare the reliability functions of the networks in Figure 6.4 and Exercise 6.5 for exponential and 2-Erlang distributions.

As with system reliability calculations, transforming the PH distribution of component lifetimes into an ME distribution in a Jordan canonical form can speed up the computation times for calculating network reliability measures.

Example 6.24 The network in Figure 6.4 is constructed using components with identical lifetime distributions. In this experiment, an APH distribution and a CPH distribution with the following representations are used:

$$\text{APH}: n_1 = 4, \quad \boldsymbol{\alpha}_1 = \begin{bmatrix}1 & 0 & 0 & 0\end{bmatrix}, \quad \boldsymbol{A}_1 = \begin{bmatrix} -5 & 4 & 1 & 0 \\ 0 & -4 & 3 & 0 \\ 0 & 0 & -3 & 1 \\ 0 & 0 & 0 & -2 \end{bmatrix}.$$

$$\text{CPH}: n_2 = 4, \quad \boldsymbol{\alpha}_2 = \begin{bmatrix}1 & 0 & 0 & 0\end{bmatrix}, \quad \boldsymbol{A}_2 = \begin{bmatrix} -5 & 4 & 1 & 0 \\ 0 & -4 & 3 & 0 \\ 0 & 1 & -3 & 1 \\ 0 & 1 & 0 & -2 \end{bmatrix}.$$

The results of similarity transformations to the two PH distributions are ME distributions with the following representations:

ME 1:

$$n_1 = 4, \quad \boldsymbol{a}_1 = [1 \quad 4 \quad 6.5 \quad 2.3333], \quad \boldsymbol{A}_1 = \begin{bmatrix} -5 & 0 & 0 & 0 \\ 0 & -4 & 0 & 0 \\ 0 & 0 & -3 & 0 \\ 0 & 0 & 0 & -2 \end{bmatrix}.$$

$$\boldsymbol{v}_1 = [0.6667 - 0.5 \quad 0 \quad 1]^T.$$

ME 2:

$$n_2 = 4, \quad \boldsymbol{a}_2 = [-1.625 - 0.071 \quad 0.625 \quad 0.2415], \quad \boldsymbol{A}_2 = \begin{bmatrix} -3 & 0 & 0 & 0 \\ 0 & -1 & 0 & 0 \\ 0 & 0 & -5 & 1 \\ 0 & 0 & 0 & -5 \end{bmatrix},$$

$$\boldsymbol{v}_2 = [0 - 17.6 - 0.4 \quad 0]^T.$$

The results of using two types of distributions are given in Figure 6.9 and Table 6.2.

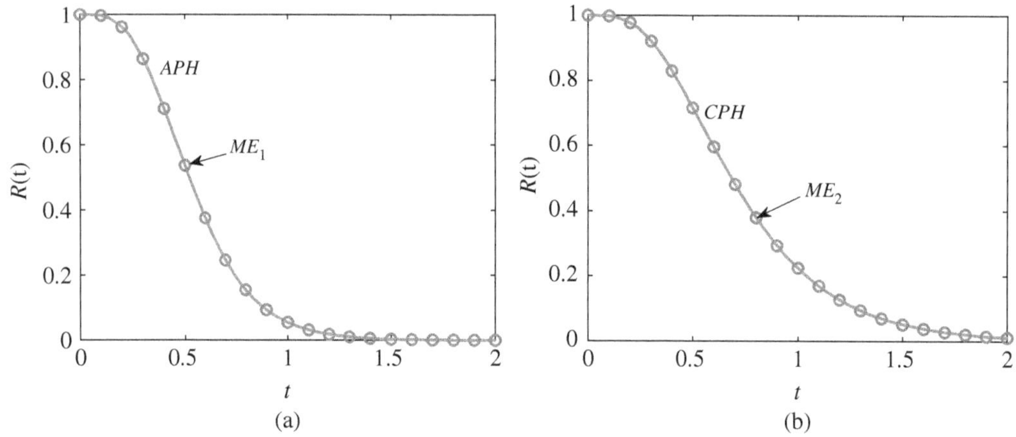

Figure 6.9 Network reliability function in Example 6.24. (a) APH distribution and (b) CPH distribution.

Table 6.2 *MTTF* and *SDTTF* of the network in Example 6.24 with their computation time including to produce Figure 6.9.

Distribution of component lifetime	Network lifetime			Computation time (s)
	MTTF	*SDTTF*	Order	
APH	0.5564	0.2490	185,984	21.64
ME1	0.5564	0.2490	185,984	18.81
CPH	0.7571	0.3986	185,984	hours
ME2	0.7571	0.3986	185,984	24.10

Figure 6.9 and Table 6.2 show that the PH distributions and their similar ME distributions produce identical reliability measures. However, the computation times for both PH distributions are slower than that for their similar ME distributions. The difference in computation times between using PH distributions and ME distributions is more pronounced for the CPH distribution. Therefore, it is advisable to transform CPH distribution to ME distribution. These results confirm the conclusion in Section 4.7. ■

Examples in previous sections show that SDP representation grows rapidly with the number of components in the network. The next section presents a method to avoid dimensionality explosion in networks with many components.

6.5 Complex Network Structured Systems

Networks typically connect multiple subsystems. These connected subsystems have various structures, including network structures. Analysis of such complex networks requires a step-by-step approach, from the component level to the system level, progressing through several subsystem levels depending on the complexity of the network. As with the complex systems discussed in Chapters 4 and 5, analysis of complex networks combines the methods has been discussed in this chapter with the methods described in Chapters 4 and 5. This is achieved by modeling the network system in a hierarchical manner. Initially, several components in a basic structure are combined into a single subsystem. A PH/ME representation for subsystem reliability function with that structure is calculated using the methods described in Chapter 4. If a general standby system exists, the methods described in Chapter 5 are used. If the resulting combination still yields multiple subsystems in known structures, they are considered sub-subsystems and combined into a single subsystem using the methods described in Chapter 4 or 5. This process continues until only subsystems connected in a simpler network structure. If all component reliability functions follow PH/ME distributions, then the reliability function of each subsystem has PH/ME representation. The resulting network structure is represented in SDP, and the ME distribution of the entire network lifetime is calculated using the method in this chapter. It is possible to have one or more subsystems in the form of a simple network such as Figure 6.2. By recursively using Theorems 6.1–6.3, it can be proven that the network reliability function has an ME distribution.

Example 6.25 Figure 6.10 is the reliability block diagram of a complex system involving network and basic structures. If all components have PH or ME distributions, then the methods of Chapters 4–6 can be used to model and analyze the reliability of the entire network.

The subsystem consisting of components 2–6 can be represented by a single block with an ME representation that can be calculated using the method of this chapter. The subsystem consisting of components 7–9 can be represented by a single block with an ME representation that can be calculated as a 3-standby (if all backups are in cold standby) or as a 3-general standby (if the backups are in warm standby). Thus, Figure 6.10 reduces to Figure 6.11.

Next, component 1 and subsystem SS1 in Figure 6.11 are combined into a single subsystem S1 whose ME representation can be derived using **meseriessys**. Subsystem SS2 and component 10 are also combined into S2 in a similar manner. Finally, what remains is a simple network as in Figure 6.2 whose ME representation of its lifetime distribution can be found from the SDP representation and the ME representations of lifetime distributions of all its subsystems.

The networks has two identical bridge networks, namely, bridge network {2,3,4,5,6} in Figure 6.10 and {S1,S2,11,12,13} in Figure 6.11. To simplify the system reliability model for the overall network,

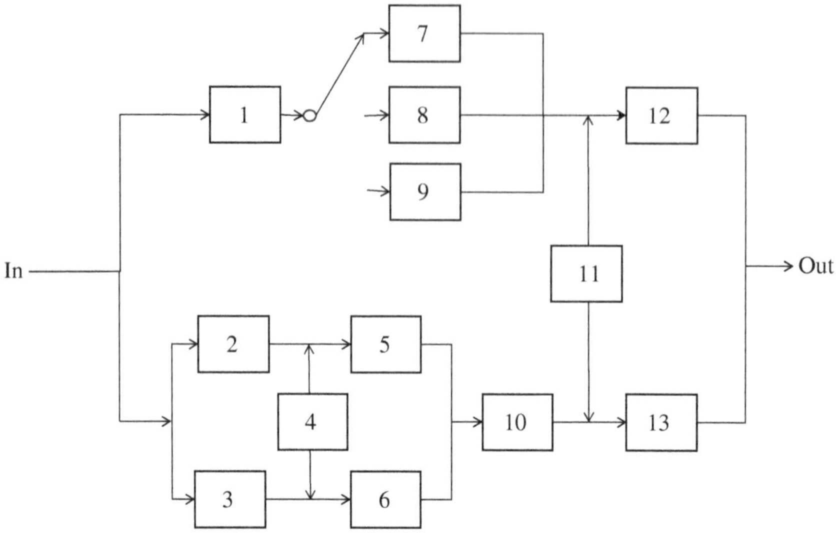

Figure 6.10 General form of complex network system.

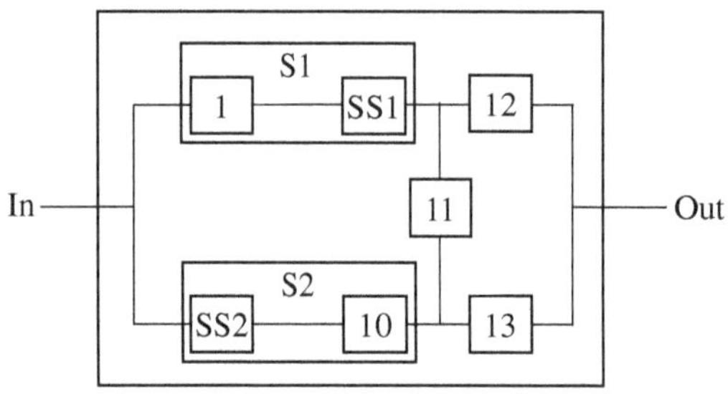

Figure 6.11 Simplified block diagram of Figure 6.10.

the two identical bridge networks are modeled using one **sdprep** function. The components in these two networks must follow the same sequence to make them identical and can be represented by the same SDP representation. Note that {S1,S2,11,12,13} follows the same sequence as {2,3,4,5,6}. Therefore, both can be represented by a generic minimum paths {1,4}, {2,5}, {1,3,5}, and {2,3,4}, and the SDP representation is generated using this generic minimum path.

The system reliability model for the network in Figure 6.10 with the help of Figure 6.11 using the prepared MATLAB functions is

$$T_{SS1} = \text{mestandsys}(T_7, T_8, T_9; 1, 1)$$

$$T_{S1} = \text{meseriessys}(T_1, T_{SS1})$$

$$SDP = \text{sdprep}(\{1,4\}, \{2,5\}, \{1,3,5\}, \{2,3,4\})$$

$$T_{SS2} = \text{menetcomb}(SDP; [T_2 - T_6])$$

$$T_{S2} = \text{meseriessys}(T_{SS2}, T_{10})$$

$$T_S = \text{menetcomb}(SDP, [T_{S1}, T_{S2}, T_{11}, T_{12}, T_{13}])$$

In this model, components 7, 8, 9 are assumed to be arranged in cold standby with perfect switch; hence, it is represented with **mestandsys** function with values of $p_1 = p_2 = 1$. Network reliability measures are calculated for the network that has identical components. Three different lifetime distributions in Example 6.23.are used, and the results are compared. The main program to generate this output is given in the **example6_25.m** file in the **dysrel-am** software package. The source code is as follows:

MATLAB code for example6_25.m

```
clear;close all
addpath('..\dysrel-am','..\dysrel-am\expokit')
% ME distributions                                          input
ME(1) = menErlang(1,100);
ME(2) = menErlang(2,100);
ME(3) = me2Coxian(1.2,100);
U = []; V = [];
tmax = 200;
t = linspace(0,tmax,21);
for j = 1:3
  % ME Components lifetime
  for i = 1:13
    T(i) = ME(j);
  end
  % Network representation                                  model
  Tss1 = mestandsys(T(7:9),[1 1]);
  Ts1 = meseriessys([T(1) Tss1]);
  mpath = {[1 4];[2 5];[1 3 5];[2 3 4]};
  sdp = sdprep(mpath);
  Tss2 = menetcomb(sdp,T(2:6));
  Ts2 = meseriessys([T(10) Tss2]);
  Ts = menetcomb(sdp,[Ts1 Ts2 T(11) T(12) T(13)]);
  % Compute reliability and hazard                          output
  [Rs,hs] = merelhaz(t,Ts);
  [x,y] = smoothing(t,Rs);
  [x,v] = smoothing(t,hs);
  U = [U y']; V = [V v'];
end
% Plot                                                      graphical output
lgnd = {'Exponential','2-Erlang','2-Coxian'};
multiplot(3,x,U,lgnd,'t','R(t)');
multiplot(3,x,V,lgnd,'t','h(t)');
```

Executing this program produces the output in Figure 6.12.

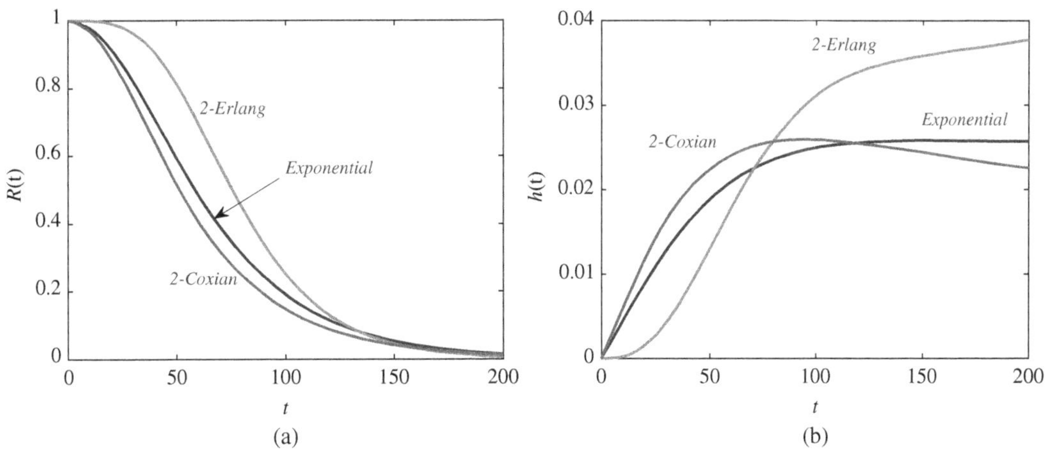

Figure 6.12 Outputs of example6_25.m. (a) Reliability function and (b) hazard function.

■

Exercise 6.9 Compare the network reliability functions with and without component 11 for the case of exponential distribution.

Application 6.1 A power system in a manufacturing company has three main subsystems: power supply, power distribution, and power consumption. The reliability block diagram of this system is shown in Figure 6.13, with identical components represented by the same numbers. The supply subsystem employs four identical generators, each with a 2-Erlang-distributed lifetime having $MTTF = 100$. The power supply requirement is load dependent, with the idle generator positioned as a warm standby with a warmness level of 0.5. The capacity of each generator is 75 kVA. Components 2 and 3 in the distribution subsystem have a 2-Erlang-distributed lifetime with $MTTF = 500$, the two bridges are identical with an exponentially distributed lifetime with $MTTF = 200$, and the switch is perfect. The load subsystem consists of three identical motors. All motors have a 2-Coxian-distributed lifetime with $MTTF = 200$ and $CVTTF = 1.2$. At least one motor must be operating, with the remainder operating in cold standby connected through a perfect switch. The power requirement for each motor is 100 kVA.

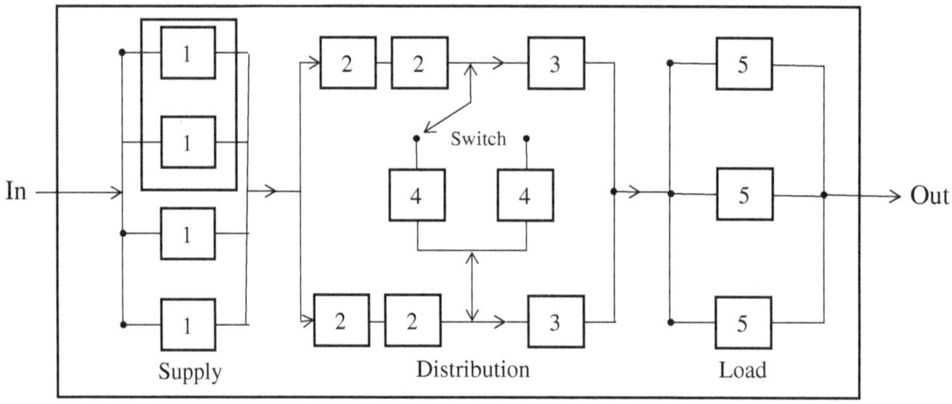

Figure 6.13 Reliability block diagram of an electrical network.

By grouping some components arranged in known structures into subsystems, the reliability block diagram of this system can be simplified to the one shown in Figure 6.14. This system is modeled through a hierarchy as in Example 6.25, namely, by first finding the PH or ME representation for subsystems S1, S21, S22, S3, and S4. The two component 3 are distinguished by presenting them as components 31 and 32.

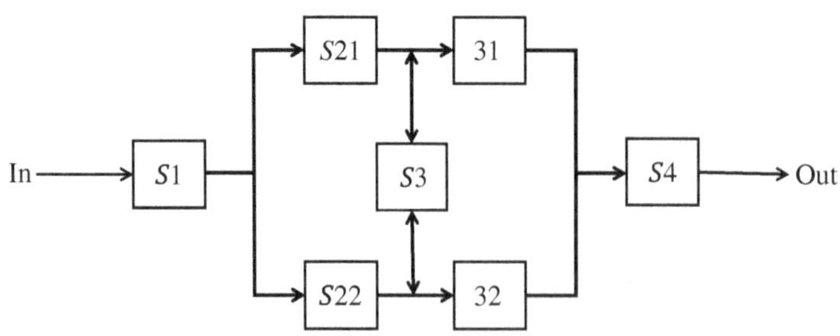

Figure 6.14 Simplified block diagram of Figure 6.13.

Based on their individual capacities, the number of generators required is one more than the number of motors being operated. For example, if one motor is operated, two other motors will serve as backups in cold standby; to meet this requirement, two generators must be turned on, and two others are backups in warm standby. This is because a 100-kVA power requirement of one motor cannot be met by a single 75-kVA generator.

The system reliability model using the provided MATLAB function is as follows:

$$T_{S1} = \text{megensys}(T_1, T_1, T_1, T_1; k+1; 0.5, 0.5)$$

$$T_{S21} = \text{meseriessys}(T_2, T_2)$$

$$T_{S22} = \text{meseriessys}(T_2, T_2)$$

$$T_{S3} = \text{mestandsys}(T_4, T_4)$$

$$T_{S4} = \text{megensys}(T_5, T_5, T_5; k; 0; 0)$$

$$SDP = \text{sdprep}(\{S21, 31\}, \{S22, 32\}, \{S21, S3, 32\}, \{S22, S3, 31\})$$

$$T_{SS} = \text{menetcomb}(SDP; [T_{S21}, T_{S22}, T_{S3}, T_{31}, T_{31}])$$

$$T_S = \text{meseriessys}(T_{S1}, T_{SS}, T_{S4})$$

In this model, there is a variable k that states the number of motors operated, which also means the number of generators that must be operated to meet the motor power requirement, namely, $k + 1$. The value of k can be 1, 2, or 3.

The system reliability model and the lifetime distribution data for each component are used to call the necessary MATLAB functions to generate an ME representation of the system lifetime distribution. Using the resulting ME representation, the system reliability and hazard function for $k = 1, 2, 3$ are presented in Figure 6.15.

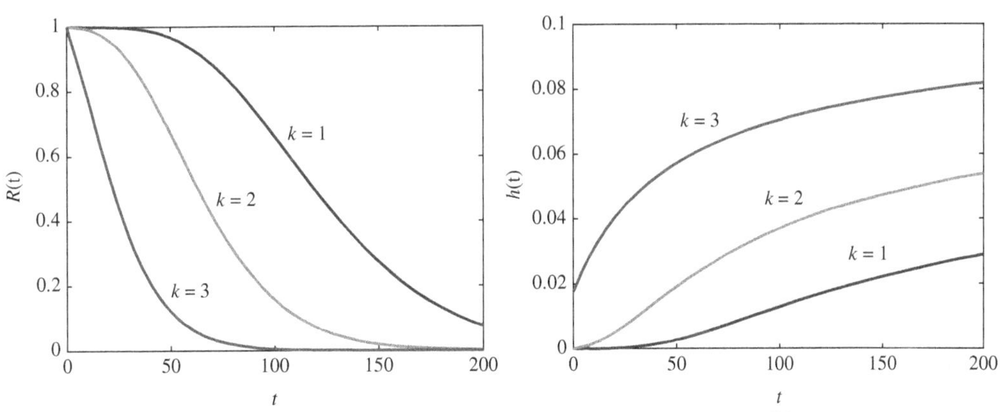

Figure 6.15 Reliability and hazard functions of the system in Application 6.1. (a) Reliability function and (b) hazard function.

■

Exercise 6.10 Compare the network reliability and hazard functions for certain k and several warmness levels of backup generators.

Application 6.2 A communication network is used to connect multiple hubs. The connection between two distant hubs is facilitated by several relay stations. The connection between two hubs can use two incompatible communication links, thus creating parallel–series structures. All communication links are bidirectional. This communication network is shown in Figure 6.16, with black nodes representing hubs or sources and white nodes representing relays. Not all relays support two links; for example, relays *c* and *e* use only one link. This network is analyzed for its reliability in terms of its ability to facilitate information flow from one hub to another.

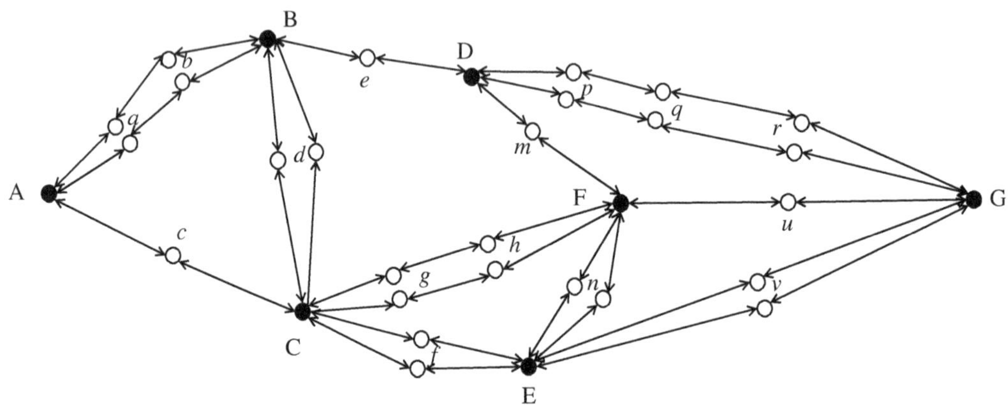

Figure 6.16 Communication network with backup link.

This network can be simplified by combining the series-connected arcs into a single arc. This simplification produces a simplified communication network as shown in Figure 6.17. The PH or ME representation of each arc after merging series links is the result of the minimum operation on the PH or ME representation of each arc before merging, which can be found using **phseriessys** or **meseriessys**.

This diagram can be further simplified by combining two backup arcs into one arc. The standby structure in each arc is a general standby, which can be a cold, hot, or warm standby. The PH

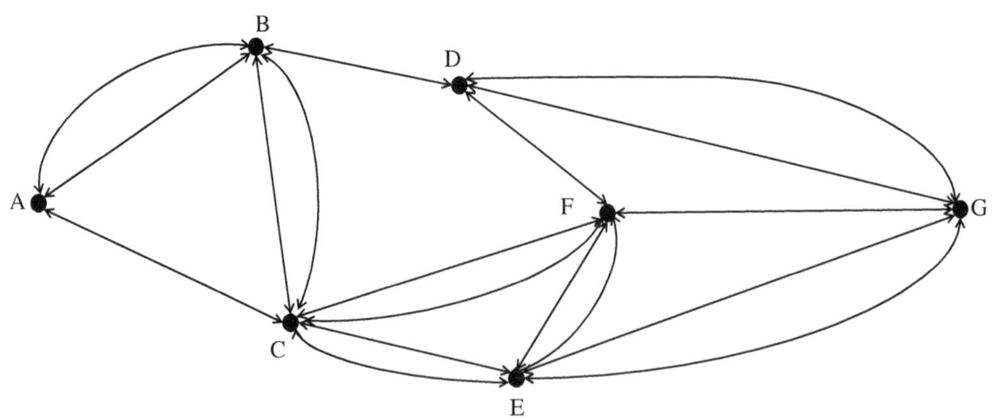

Figure 6.17 Communication network Figure 6.16 with combined series connection.

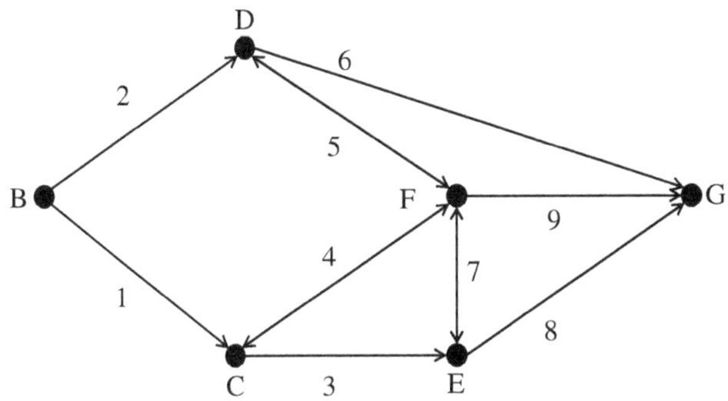

Figure 6.18 A simplified diagram of communication network in Figure 6.17.

or ME representation of each arc resulting from combining the primary and backup arcs can be found using **phgensys** or **megensys**. The simplified network is analyzed for its reliability using its minimum paths, which are first converted to SDP representation. The minimum paths depend on the nodes used as source and target nodes. The arc directions are also adjusted to the direction of communication, so some arcs can be considered unidirectional. For example, if the network reliability analysis is for sending information from node B to G, then paths BD and BC are unidirectional and path BC is parallel to path BAC. The network diagram can be simplified further as shown in Figure 6.18, with network components given sequential numbers generated by MATLAB (but they are not used in the model).

All components are assumed to have identically exponentially distributed lifetimes, which is a PH distribution of order 1, with $MTTF = 100$. The PH representation of the lifetime distribution of the subsystems resulting from the combination of serial arcs is calculated first using the **phseriessys** function. After that, the PH representations for the lifetime distributions of the subsystems formed by combining primary and backup paths are calculated using the **phgensys** function.

In this application, arcs are denoted by their start and end notes. For example, the arc 6 is represented as DG. The lifetime of arc DG, T_{DG}, in Figures 6.16 and 6.17 is

$$T_{DG} = \text{phgensys}[\text{phseriessys}(T_{Dp}, T_{pq}, T_{qr}, T_{rG}), \text{phseriessys}(T_{Dp}, T_{pq}, T_{qr}, T_{rG})],$$

where T_{xy} is the lifetime of arc x–y. Finally, the ME representation for the lifetime distribution of the entire network is calculated from the PH representation of each subsystem resulting from the cascade combination and the simplified network SDP representation. The resulting ME representation is used to calculate reliability measures for each network configuration. It is possible to use all functions in ME distributions. For example, **phseriessys** is replaced with **meseriessys** and **phgensys** is replaced with **megensys**.

Referring to Figures 6.16–6.18, the system reliability model for this network in the provided MATLAB functions is

$$T_{BC1} = \text{phseriessys}(T_{BA}, T_{\text{AC}}),$$

$$T_{AC} = \text{phseriessys}(T_{Ac}, T_{cC}),$$

$$T_{BA1} = \text{phseriessys}(T_{Bb}, T_{ba}, T_{aA}),$$

$$T_{BA} = \text{phgensys}(T_{BA1}, T_{BA1}; 1; [d]),$$

$$T_{BC21} = \text{phseriessys}(T_{Bd}, T_{dC}),$$

$$T_{BC2} = \text{phgensys}(T_{BC21}, T_{BC21}; 1; [d]),$$

$$T_{BC} = \text{phparallelsys}(T_{BC1}, T_{BC2}),$$

$$T_{BD} = \text{phseriessys}(T_{Be}, T_{eD}),$$

$$T_{DF} = \text{phseriessys}(T_{Dm}, T_{mF}),$$

$$T_{FD} = T_{DF},$$

$$T_{CE1} = \text{phseriessys}(T_{Cf}, T_{fE}),$$

$$T_{CE} = \text{phgensys}(T_{CE1}, T_{CE1}; 1; [d]),$$

$$T_{CF1} = \text{phseriessys}(T_{Cg}, T_{gh}, T_{hF}),$$

$$T_{CF} = \text{phgensys}(T_{CF1}, T_{CF1}; 1; [d]),$$

$$T_{FC} = T_{CF},$$

$$T_{EF1} = \text{phseriessys}(T_{En}, T_{nF}),$$

$$T_{EF} = \text{phgensys}(T_{CE1}, T_{EF1}; 1; [d]),$$

$$T_{DG1} = \text{phseriessys}(T_{Dp}, T_{pq}, T_{qr}, T_{rG}),$$

$$T_{DG} = \text{phgensys}(T_{DG1}, T_{DG1}; 1; [d]),$$

$$T_{EG1} = \text{phseriessys}(T_{Ev}, T_{vG}),$$

$$T_{EG} = \text{phgensys}(T_{EG1}, T_{CG1}; 1; [d]),$$

$$T_{FG} = \text{phseriessys}(T_{Fu}, T_{uG}),$$

$$Graph = (BC, BD, CE, CF, FC, DF, FD, DG, EF, FE, EG, FG)$$

$$mpaths = \text{minpaths}(Graph),$$

$$SDP = \text{sdprep}(mpaths),$$

$$T_S = \text{menetcomb}(SDP, [T_{BC}, T_{BD}, T_{CE}, T_{CF}, T_{DF}, T_{DG}, T_{EF}, T_{EG}, T_{FG}]).$$

A comparison of the reliability, hazard, residual reliability, and *MRL* functions generated for several standby configurations is presented in Figure 6.19. The value of $\boldsymbol{D}$ in this figure represents warmness levels of the backup component as described in Section 5.2. A comparison of their *MTTF*and *SDTTF* is presented in Table 6.3. To show the importance of minimum path ordering, the order of ME distributions of network lifetime distribution and computations times are compared between without and with minimum path ordering. It is apparent that ordering minimum paths can reduce the computation times by 20%, even though reduction in the order of ME distribution is not significant.

Reliability modeling using the original (unsimplified) network graph representation in Figure 6.17 is given as in Problems. ■

Exercise 6.11 Compare the network reliability functions for arcs CF and DF are uni- and bidirectional (there are four combinations!). Show that the network with both arcs is bidirectional and has the highest reliability.

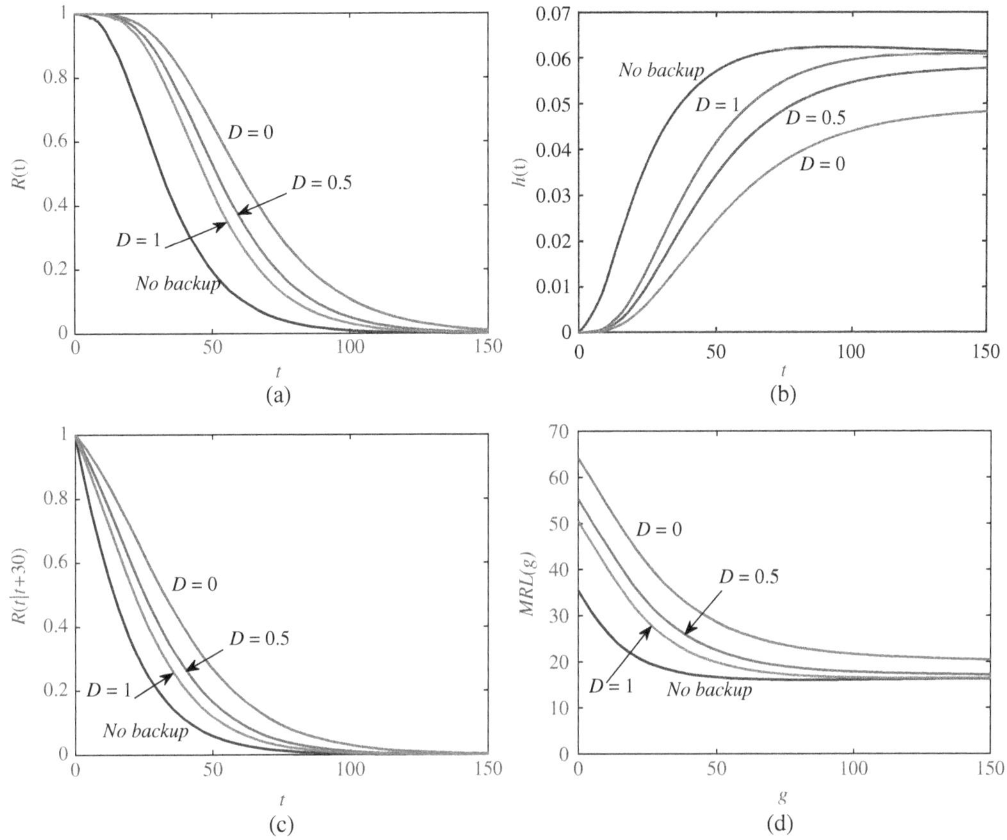

Figure 6.19 Functional reliability measures of network in Application 6.2. (a) Reliability function, (b) hazard function, (c) residual reliability function (age 100), and (d) *MRL* function.

Table 6.3 *MTTF* and *SDTTF* of network in Application 6.2.

			ME order/computation time (s)	
Configuration	***MTTF***	***SDTTF***	**Without minpath ordering**	**With minpath ordering**
One line, no backup	35.42	19.66	497/0.51	363/0.42
One main line and one cold backup (2-standby)	64.20	28.33	144,481/67.60	141,823/59.53
One main line and one warm backup (2-general standby, $D = 0, 5$)	55.27	24.12	144,481/79.69	141,823/63.18
One main line and one hot backup (2-parallel)	50.47	22.24	144,481/85.92	141,823/66.29

6.6 General Network Reliability Analysis

Network in general does not have specific source or target nodes. As in Application 6.2, any node can be either a source or a target node, unless certain operational requirements must be met. But the source and target nodes define the minimum paths of the networks, and minimum paths determine the network reliability. Thus, network reliability is different for each pair of source-target nodes. A general network can also have multiple sources and multiple targets. Flow can occur from any source node to any target node, and network reliability analysis needs to consider all possible desired flows. Furthermore, network can also have flow capacity constraints on its arcs. These constraints affect network reliability, which makes the need for additional analysis. The development in previous sections is extended for general network reliability analysis.

6.6.1 Network with Arbitrary Source and Target Nodes

To calculate network reliability, it is necessary first to define the source and target nodes. The resulting network reliability is the probability of the flow being delivered from the selected source node to the selected target node. For example, the reliability of the communication network in Application 6.2 is a measure of the probability that the network is able to deliver information from node B to node G. If the source node or target node is changed, then the network reliability also changes. The network reliability from B to G differs from the network reliability from B to F. If not all arcs are bidirectional, then the network reliability from G to B and from G to B are different.

A general network may contain bidirectional arcs. For network reliability analysis, the directions of the arcs must be associated with the desired source and target nodes. After selecting B and G, respectively, as the source and target nodes in Application 6.2, then the arcs BA, BC, AC are simply considered unidirectional. Other arcs also may have their directions adjusted to reach the target node. The role of reliability analysts is still needed to determine the realistic direction of the arcs from a network reliability perspective.

Example 6.26 The general network with all bidirectional arcs in Figure 6.20a has a graph representation in Figure 6.20b when used to determine its reliability to carry out a mission of

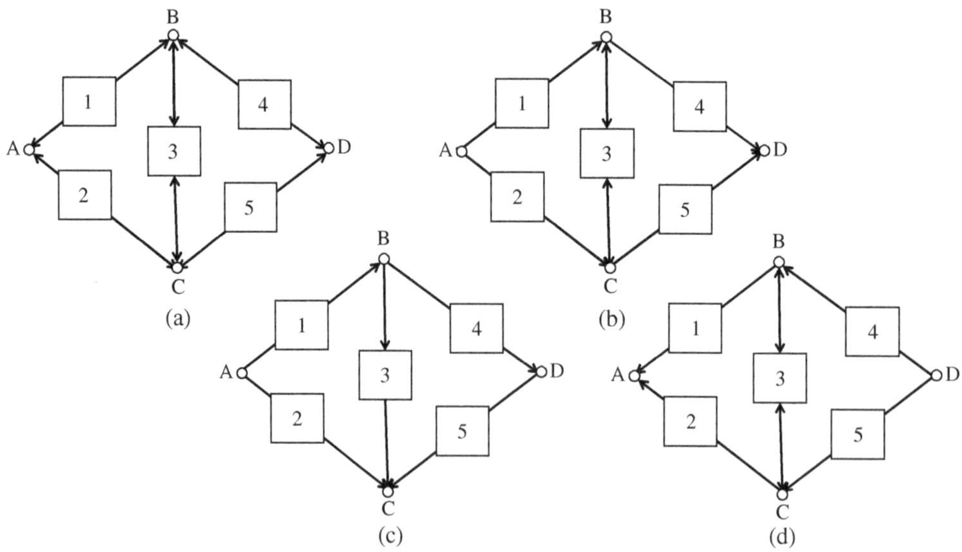

Figure 6.20 Graph representations for the network reliability analysis in Example 6.26. (a) Simple general network, (b) A–D network, (c) A–C network, and (d) D–A network.

delivering flows from node A to node D (refer to as A–D network). In this graph representation, arcs 1, 2, 4, and 5, are unidirectional, while arc 3 is still bidirectional because both directions are required to send flows from A to D. Likewise, Figure 6.20c is a graph representation for reliability analysis of A–C network, while Figure 6.20d is a graph representation for reliability analysis of D–A network. ■

Example 6.27 All arcs in the network of Figure 6.20a have 2-Erlang-distributed lifetimes with *MTTFs* equal to 10, 15, 20, 25, and 30, respectively. A comparison of the reliability and hazard functions of the network with three pairs of sources and targets described in Example 6.26 is presented in Figure 6.21.

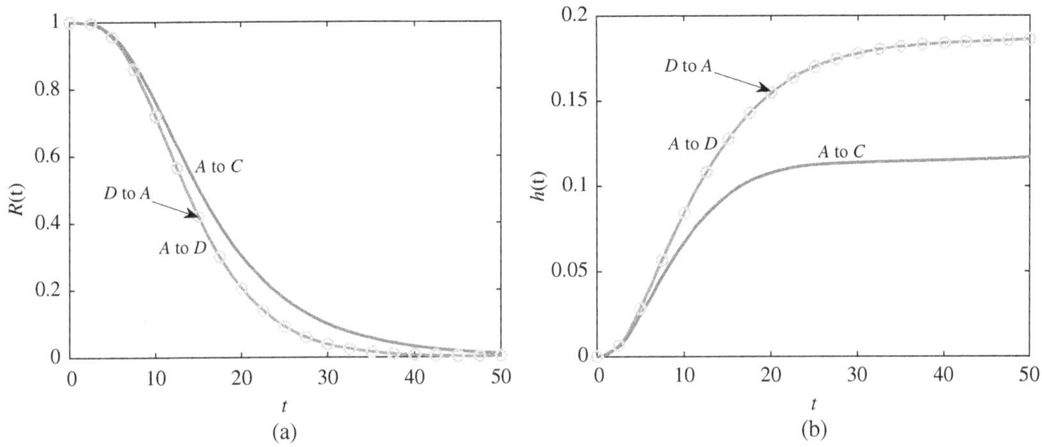

Figure 6.21 Reliability and hazard functions of the networks in Example 6.27. (a) Reliability function and (b) hazard function.

Figure 6.21 shows that the A–C network has higher reliability function than the A–D network. This is because the A–C network has one path with one component while the A–D network has paths with at least two components, and the reliability of a path is the product of the reliability of its components. It is only a coincidence that the reliability functions of the A–D network and D–A network are identical. It is because both have the same paths, although in different directions. However, if not all arcs are bidirectional, then the network reliability function will be affected when the source and target are reversed. ■

As a summary, once the source and target of the network are defined, the network reliability measures can be determined in the same way as in the previous sections:

1) Define the source node and the target node of the network.
2) Generate and order minimum paths of the network using **minpaths** function.
3) Find SDP representation of the network from the resulting minimum paths using **sdprep** function.
4) Calculate ME representation of the network lifetime distribution from the SDP representation of the network using **menetcomb** function
5) Use **merelplot, mehazplot, mestat,** and other functions as needed to calculate network reliability measures from the resulting ME representation.

Next, the discussion of general networks continues with networks having multi sources and targets.

6.6.2 Multisource–Multitarget Network

The reliability of a multisource–multitarget network depends on the desired definition of reliability. This definition of reliability depends on the mission it is intended to fulfill. The mission of a multisource–multitarget network can be defined, among other things, as being able to deliver flows from all sources to all targets, from at least one source to all targets, or from at least one source to at least one target. Each network mission definition results in a network reliability formulation, which may differ from the network reliability formulation for other missions. Different missions may also result in different graph representations.

Example 6.28 The network in Figure 6.22 has a mission to deliver flows from one of the sources to one of the targets. There are four ways to deliver flows from the sources to the targets in this network, namely, from A to C using the A–C network, from A to D using the A–D network, from B to C using the B–C network, and from B to D using the B–D network. The mission is fulfilled if one of these four networks is good.

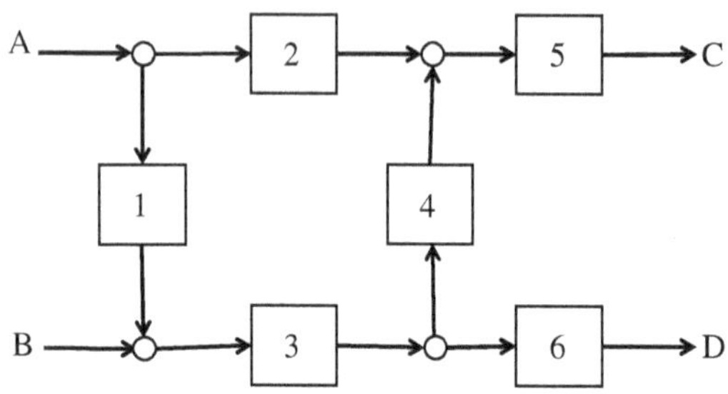

Figure 6.22 Multisource–multitarget network in Example 6.28.

The A–C network has minimum paths {2,5} and {1,3,4,5} that produce a network structure function

$$X_{A-C} = X_2X_5 \vee X_1X_3X_4X_5,$$

where X_{A-C} is the state variable for A–C network.

Use the same method to determine the structure functions of A–D, B–C, and B–D networks. The results are

$$X_{A-D} = X_1X_3X_6, \quad X_{B-C} = X_3X_4X_5, \quad X_{B-D} = X_3X_6.$$

Since the mission of the network is to deliver flows from any source to any target, the network structure function is

$$X_S = X_{A-C} \vee X_{A-D} \vee X_{B-C} \vee X_{B-D} = (X_2X_5 \vee X_1X_3X_4X_5) \vee (X_1X_3X_6) \vee (X_3X_4X_5) \vee (X_3X_6).$$

Simplifying $X_1X_3X_6 \vee X_3X_6 = X_3X_6$ and $X_1X_3X_4X_5 \vee X_3X_4X_5 = X_3X_4X_5$ yields

$$X_S = X_2X_5 \vee X_3X_6 \vee X_3X_4X_5.$$

Since the resulting structure function does not contain X_1, it means that component 1 does not affect the reliability of the network with a mission of delivering flows from one of the sources to one of the targets. This is because component 1 functions only to deliver flow from A to D, and this mission can be fulfilled by flows from B to D without using component 1. The ME representation for the network lifetime distribution can be derived by first converting X_S to SDP representation. ■

Example 6.29 If the network mission is to deliver flows from all sources to all targets, then the network structure function is the result of the operation of multiplying (conjunction) the structure functions of all possible networks from any source to any target,

$$X_S = (X_{A-C})(X_{A-D})(X_{B-C})(X_{B-D}) = (X_2X_5 \vee X_1X_3X_4X_5)(X_1X_3X_6)(X_3X_4X_5)(X_3X_6).$$

After some simplifications,

$$X_S = X_1X_2X_3X_5X_6 \vee X_1X_3X_4X_5X_6 = X_1X_3X_5X_6(X_2 \vee X_4).$$

■

Example 6.30 If the network mission is to deliver flows to all targets regardless of their sources, then tracing all desired flows yields

$$\begin{aligned} X_S &= ((X_{A-C}) \vee (X_{B-C}))((X_{A-D}) \vee (X_{B-D})) \\ &= ((X_2X_5 \vee X_1X_3X_4X_5) \vee (X_3X_4X_5))((X_1X_3X_6) \vee (X_3X_6)) \\ &= (X_2X_5 \vee X_3X_4X_5)(X_3X_6) = X_3X_5X_6(X_2 \vee X_4). \end{aligned}$$

The resulting structure function indicates that the network can fulfill its mission if components 3, 5, and 6 are all good and one of component 2 or 4 is good. Component 1 has no effect on mission reliability.

■

Based on their mission description, the graph representations for the networks in Examples 6.29 and 6.30 can be simplified to Figure 6.23a,b, respectively. Essentially, source or target nodes that are not required to send or receive flows can be merged into one. It can be verified that the structure functions generated using the simplified graph representations are the same as the structure functions generated using the original graph representations.

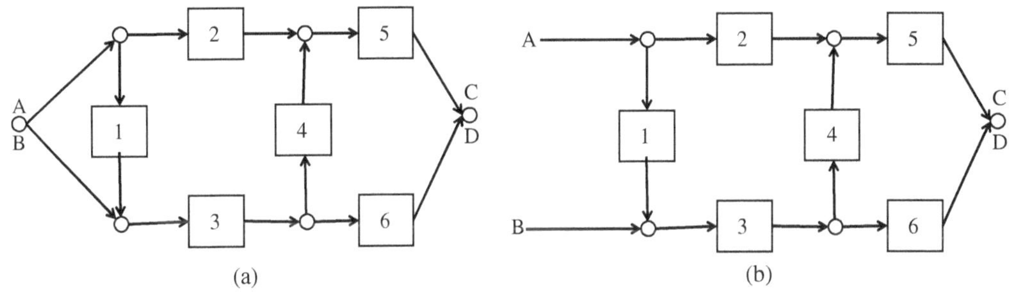

Figure 6.23 Simplification of the graph representations of the networks in Examples 6.29 and 6.30. (a) Graph representation of Example 6.29 and (b) graph representation of Example 6.30.

Example 6.31 A comparison of the reliability and hazard functions of a network with several missions given in Examples 6.28–6.30 is shown in Figure 6.24. The graphs are produced using all component lifetimes having an identical 2-Erlang distribution with $MTTF = 100$. According to the description, the network in Example 6.29 carries the most demanding mission, so its reliability function is the lowest and its hazard function is the highest. Conversely, the network in Example 6.28 carries the least demanding mission, which makes its reliability function the highest and its hazard function is the lowest. Denote $R_{6.28}(t)$, $R_{6.29}(t)$, $R_{6.30}(t)$, $h_{6.28}(t)$, $h_{6.29}(t)$, and $h_{6.30}(t)$ as the reliability and hazard functions in Examples 6.28–6.30, respectively. Then,

$$R_{6.28}(t) > R_{6.30}(t) > R_{6.29}(t),$$

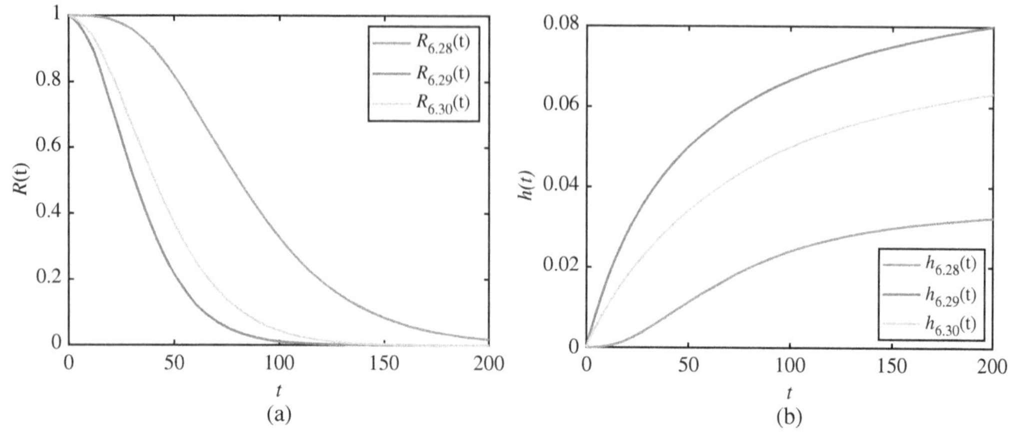

Figure 6.24 Reliability and hazard functions of Examples 6.28–6.30.

and

$$h_{6.28}(t) < h_{6.30}(t) < h_{6.29}(t).$$

■

Exercise 6.12 Show that the results of Example 6.31 should be obvious from the results of Examples 6.28–6.30, which means the results of Example 6.31 is distribution free.

Multisource–multitarget networks can also be analyzed in the same way as networks with an arbitrary source and target with an additional step to define all required combinations of sources and target nodes:

1) Define all required combinations of source node(s) and target node(s).
2) Generate and order minimum paths of the network using **minpaths** function for each pair of source–target nodes.
3) Determine the structure function of each pair of source–target nodes.
4) Determine the overall network structure function from the structure function of each pair of source–target nodes based on the required combinations.
5) Find SDP representation of the network from the resulting network structure function using the **sdprep** function.
6) Calculate reliability measures from SDP representation of the network using the **menetcomb** function.
7) Use **merelplot, mehazplot, mestat**, and other functions as needed to calculate network reliability measures from the ME representation.

6.6.3 Capacitated Network

General network has arcs with limited capacity. The flow on each arc is restricted to not exceed the capacity of the arc it passes through. This limitation requires calculating the flow on each minimum path and checking that the flow demand is met. The amount of flow delivered using a minimum path must be subtracted from the capacity of that minimum path to yield the residual capacity of each arc in the minimum path. The flow demand fulfillment is the sum of the flow through the minimum paths used to deliver the flow. This problem has been discussed in

Lin et al. [14] for static reliability analysis of a capacitated network, which then follows by others. They introduced the term *d*-minpath, which is a collection of minimum paths that can be used to deliver *d* amount of flow. The available problem description and method are extended to dynamic network reliability analysis with an objective of finding the network reliability function and other network reliability measures for networks with multistate components. The following are some provisions related to flow in a capacitated network.

Denote the arc capacity *i* by b_i. If arc *i* has a flow of x_i, then it applies

$$x_i \leq b_i, \tag{6.48}$$

and

$$d_i = b_i - x_i, \tag{6.49}$$

where d_i is the residual capacity of arc *i* after being used to send a flow of x_i.

The minimum path capacity is the minimum of the residual capacities of all arcs that form the minimum path. Let c_k be the residual capacity of minimum path *k*. It follows from the definition of capacity that

$$c_k = \min_{i \in MP_k} d_i, \tag{6.50}$$

The flow received by a target node is the sum of the flows on each minimum path leading to the target,

$$A_{target} = \sum_{k \in MP_{target}} c_k. \tag{6.51}$$

where MP_{target} is the set of minimum paths to the target node and A_{target} is the amount of flow received by the target node.

Example 6.32 Figure 6.25 is a capacitated network whose arcs have weights represent (capacities, *MTTF*) given in parentheses at each arc. The mission is to deliver maximum flow from A to B.

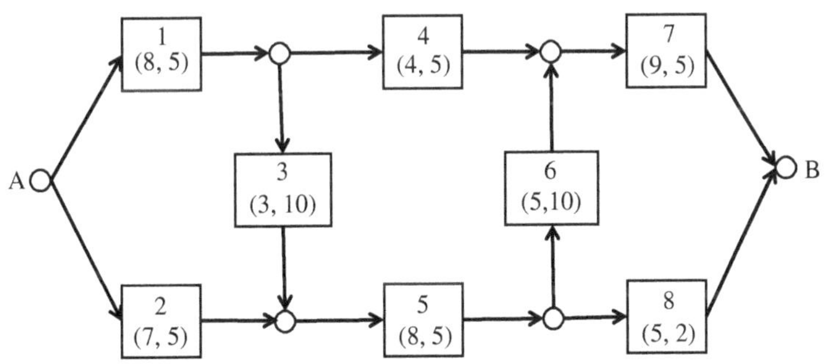

Figure 6.25 Capacitated network for Example 6.32.

This network has minimum paths with the following respective capacities:

1) The first minimum path chosen is {1,4,7}, which none of its arcs have been used. This path has a capacity calculated using Equation (6.50) as $c_1 = \min(d_1, d_4, d_7) = \min(b_1, b_4, b_7) = \min(8, 4, 9) = 4$. Following Equation (6.50), delivering flow of 4 units from A to B causes the residual capacities of arc 1 to be $d_1 = b_1 - x_1 = 8 - 4 = 4$, arc 4 to be $d_4 = b_4 - x_4 = 4 - 4 = 0$

(exhausted), and arc 7 to be $d_7 = b_7 - x_7 = 9 - 4 = 5$. The residual capacity of path {1,4,7} is $\min(d_1, d_4, d_7) = 0$, which means minimum path {1,4,7} is saturated. Find another minimum path to increase the flow.

2) The second minimum path chosen is {2,5,8}, which is unused yet. It has a capacity of $c_2 = \min(d_2, d_5, d_8) = \min(b_2, b_5, b_8) = \min(7, 8, 5) = 5$. Send flow of 5 units from A to B. This flow causes the residual capacities of arc 2 to be $d_2 = b_2 - x_2 = 7 - 5 = 2$, arc 5 to be $d_5 = b_5 - x_5 = 8 - 5 = 3$, and arc 8 to be $d_8 = b_8 - x_8 = 0$ (exhausted). The {2,5,8} path is saturated. Find another minimum path to increase the flow.
3) The third minimum path chosen is {1,3,5,6,7}, which has a capacity of $c_3 = \min(d_1, b_3, d_5, b_6, d_7) = \min(4, 3, 3, 5, 5) = 3$. Send flow of 3 units from A to B. The residual capacities of arc 1 to be $d_1 = d_1 - x_1 = 4 - 3 = 1$, arc 3 to be exhausted, arc 5 to be exhausted, arc 6 to be $d_6 = b_6 - x_6 = 5 - 3 = 2$, and arc 7 to be $d_7 = d_7 - x_7 = 5 - 3 = 2$. The path {1,3,5,6,7} is saturated. Find another minimum path to increase the flow.
4) There is no other minimum path that can be used to increase the flow from A to B. The iteration is stopped.

Following Equation (6.51), the maximum flow from A to B is 12 units, and it is achievable using minimum paths {1,4,7} for 4 units, {2,5,8} for 5 units, and {1,3,5,7} for 3 units. Since all these minimum paths must be utilized to deliver maximum flow, the network structure function for this mission is conjunction of the structure functions of these minimum paths,

$$X_S = (X_1X_4X_7)(X_2X_5X_8)(X_1X_3X_5X_6X_7) = X_1X_2X_3X_4X_5X_6X_7X_8,$$

and the network reliability function

$$R_S(t) = R_1(t)R_2(t)R_3(t)R_4(t)R_5(t)R_6(t)R_7(t)R_8(t).$$

■

There are other alternatives for delivering 12 units from A to B. One is to use the minimum paths {1,4,7} for 4 units, {2,5,8} for 5 units, {2,5,6,7} for 2 units, and {1,3,5,6,7} for 1 unit. It is a coincidence that all alternatives in this example produce the same network reliability for carrying out the mission of delivering of 12 units from A to B because the minimum paths of other alternatives also utilize all arcs. If the mission is delivering less than the maximum flow, then some alternatives may have different reliability. Likewise, if delivering maximum flow does not utilize all arcs, then there are also alternatives for delivering maximum flow.

Example 6.33 The network in Example 6.32 is used for a mission of sending flow of at least 8 units from A to B.

To fulfill the mission requirement of sending a flow of at least 8 units, not all of the minimum paths are required, some of them are sufficient. From the results of Example 6.32, this mission can be served by using the following combinations of minimum paths:

1) {1,4,7} for 4 units and {2,5,8} for 5 units, with a total flow of 9 units.
2) {1,4,7} for 4 units and {2,5,6,7} for 5 units, with a total flow of 9 units.
3) {1,3,5,8} for 3 units and {2,5,6,7} for 5 units, with a total flow of 8 units.
4) {1,3,5,6,7} for 3 units and {2,5,8} for 5 units, with a total flow of 8 units.

Other combinations require more than two minimum paths, such as {1,4,7} for 4 units, {1,3,5,8} for 3 units, and {2,5,8} for 2 units with a total of 10 units. However, this combination is the superset of the combination in point 2, which will be absorbed in the network structure function.

The resulting network structure function is

$$
\begin{aligned}
X_S &= (X_{1,4,7})(X_{2,5,8}) \vee (X_{1,4,7})(X_{2,5,6,7}) \vee (X_{1,3,5,8})(X_{2,5,6,7}) \vee (X_{1,3,5,6,7})(X_{2,5,8}) \\
&= X_1X_2X_4X_5X_7X_8 \vee X_1X_2X_4X_5X_6X_7 \vee X_1X_2X_3X_5X_6X_7X_8 \vee X_1X_2X_3X_5X_6X_7X_8 \\
&= X_1X_2X_4X_5X_7X_8 \vee X_1X_2X_4X_5X_6X_7 \vee X_1X_2X_3X_5X_6X_7X_8.
\end{aligned}
$$

As seen, combinations in points 3 and 4 contain the same arcs; therefore, the disjunction operation in the network structure function eliminates one of them.

An SDP representation of the resulting network structure function is

$$X_S = X_1X_2X_4X_5X_7X_8 + X_1X_2X_4X_5X_6X_7\overline{X}_8 + X_1X_2X_3\overline{X}_4X_5X_6X_7X_8.$$

The network reliability for this mission can be determined as

$$
\begin{aligned}
R_S(t) = {} & R_1(t)R_2(t)R_4(t)R_5(t)R_7(t)R_8(t) + R_1(t)R_2(t)R_4(t)R_5(t)R_6(t)R_7(t)(1 - R_8(t)) \\
& + R_1(t)R_2(t)R_3(t)(1 - R_4(t))R_5(t)R_6(t)R_7(t)R_8(t).
\end{aligned}
$$

Examining the network reliability functions derived in Examples 6.32 and 6.33, the network reliability for delivering 12 units is lower than that for delivering 8 units. ■

Example 6.34 Structure functions of the network in Figure 6.25 for minimum amounts of flows to be delivered from A to B are determined as given in Table 6.4.

Table 6.4 shows that the network structure functions contain more disjunction operations for lower delivery requirements. Furthermore, delivering a minimum flow of 10, 11, or 12 units is identical, even though delivering 10 units has more feasible routes than delivering 12 units. However, they are using all arcs, which means all arcs must be good to deliver these amounts.

Assume that all arcs have lifetime following 2-Erlang distribution. The reliability and hazard functions of the network in Figure 6.25 for delivering 6, 8, 10, and 12 units of flow from A to B are shown in Figure 6.26. ■

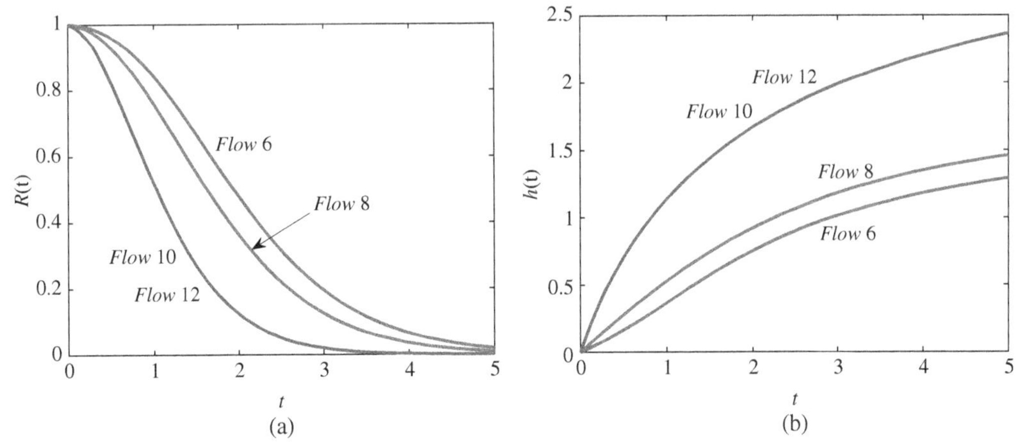

Figure 6.26 Network reliability and hazard functions to meet the flow requirements. (a) Reliability function and (b) hazard function.

Table 6.4 Combinations of minimum paths for delivering minimum amounts of flows.

Minimum flow	Combination of minimum paths
3	{1,4,7} or {1,3,5,8} or {1,3,5,6,7} or {2,5,8} or {2,5,6,7} $X_S = X_1X_4X_7 \vee X_1X_3X_5X_8 \vee X_1X_3X_5X_6X_7 \vee X_2X_5X_8 \vee X_2X_5X_6X_7$
4	{1,4,7} or {2,5,8} or {2,5,6,7} $X_S = X_1X_4X_7 \vee X_2X_5X_8 \vee X_2X_5X_6X_7$
5	{2,5,8} or {2,5,6,7} or [{1,4,7} and {1,3,5,8}] or [{1,4,7} and {1,3,5,6,7}] $X_S = X_2X_5X_8 \vee X_2X_5X_6X_7 \vee X_1X_3X_4X_5X_7X_8 \vee X_1X_3X_4X_5X_6X_7$
6	[{1,4,7} and {1,3,5,8}] or [{1,4,7} and {1,3,5,6,7}] or [{1,4,7} and {2,5,8}] or [{1,4,7} and [2,5,6,7}] or [{2,5,8} and {2,5,6,7}] or [{1,3,5,8} and {2,5,6,7}] or [{1,3,5,6,7} and {2,5,8}] (The last two combinations can be eliminated.) $X_S = X_1X_3X_4X_5X_7X_8 \vee X_1X_3X_4X_5X_6X_7 \vee X_1X_2X_4X_5X_7X_8 \vee X_1X_2X_4X_5X_6X_7 \vee X_2X_5X_6X_7X_8$
7	Same as 6 $X_S = X_1X_3X_4X_5X_7X_8 \vee X_1X_3X_4X_5X_6X_7 \vee X_1X_2X_4X_5X_7X_8 \vee X_1X_2X_4X_5X_6X_7 \vee X_2X_5X_6X_7X_8$
8	[{1,4,7} and {2,5,8}] or [{1,4,7} and {2,5,6,7}] or [{1,3,5,8} and {2,5,6,7}] or [{1,3,5,6,7} and {2,5,8}] (The last two routes contain the same elements.) $X_S = X_1X_2X_4X_5X_7X_8 \vee X_1X_2X_4X_5X_6X_7 \vee X_1X_2X_3X_5X_6X_7X_8$
9	[{1,4,7} and {2,5,8}] or [{1,4,7} and [2,5,6,7}] or [{1,4,7} and {1,3,5,8} and {2,5,6,7}] or [{1,4,7} and {1,3,5,6,7} and {2,5,8}] (The last two combinations are eliminated.) $X_S = X_1X_2X_4X_5X_7X_8 \vee X_1X_2X_4X_5X_6X_7$
10	[{1,4,7} and {1,3,5,8} and {2,5,6,7}] or [{1,4,7} and {1,3,5,6,7} and {2,5,8}] or [{2,5,8} and {2,5,6,7} and {1,4,7}] or [{2,5,8} and {2,5,6,7} and {1,3,5,6,7}] (All combinations contains all arcs.) $X_S = X_1X_2X_3X_4X_5X_6X_7X_8$
11	[{1,4,7} and {1,3,5,8} and {2,5,6,7}] or [{1,4,7} and {1,3,5,6,7} and {2,5,8}] or [{2,5,8} and {2,5,6,7} and {1,4,7}] (All combinations contain all arcs.) $X_S = X_1X_2X_3X_4X_5X_6X_7X_8$
12	[{1,4,7} and {1,3,5,8} and {2,5,6,7}] or [{1,4,7} and {1,3,5,6,7} and {2,5,8}] (All combinations contain all arcs.) $X_S = X_1X_2X_3X_4X_5X_6X_7X_8$

Exercise 6.13 Show that the orderings of reliability and hazard functions in Figure 6.26 are distribution free.

The algorithm for calculating network reliability measures for delivering flow to meet the demand in the target node can be summarized as follows:

1) Generate and order minimum paths of the network using **minpaths** function for the given pair of source–target nodes.
2) Find all combinations of minimum paths that can deliver flow greater than or equal to demand.
3) Determine the structure function of each combination of minimum path.

4) Determine the network structure function as disjunction of structure functions of all combinations of minimum paths.
5) Find the SDP representation of the network from the network structure function using **sdprep** function.
6) Calculate ME representation of the network lifetime distribution from SDP representation of the network using **menetcomb** function.
7) Use **merelplot, mehazplot,** and **mestat** functions as needed to calculate network reliability measures from the ME representation.

This algorithm can be slightly modified in the case the source node has limited capacity, which could be less than the demand or less than the maximum flow. The feasible flow must be the minimum of source supply, target demand, and maximum flow.

6.6.4 Optimizing Capacitated Network Reliability[1]

Previous examples show that there are alternatives for delivering some amount of flow from the source node to the target node, and different delivery methods have different structure functions. In this case, there are several combinations of minimum paths that can meet the flow demand. It is possible that not all minimum paths are needed to meet the flow demand. If the objective is to find network reliability for delivering a certain amount of flows irrespective of the routes taken, then all those alternatives can be utilized, and the network structure functions are disjunctions of those alternatives. In addition, each combination of minimum paths produces different network reliability measures, so comparisons need to be made to select the minimum path combination that provides the best reliability measure. If the objective is to find the most reliable route, then the reliability measures of all available alternatives are ranked to determine the most reliable routes.

Example 6.35 Table 6.4 shows there are three routes for delivering 8 units from A to B, namely, route 1 is {1,4,7} and {2,5,8}, route 2 is {1,4,7} and {2,5,6,7}, and route 3 is {1,3,5,8} and {2,5,6,7}. Let all arcs have lifetime following exponential distribution. The reliability functions for delivering 8 units using the three route alternatives are presented in Figure 6.27.

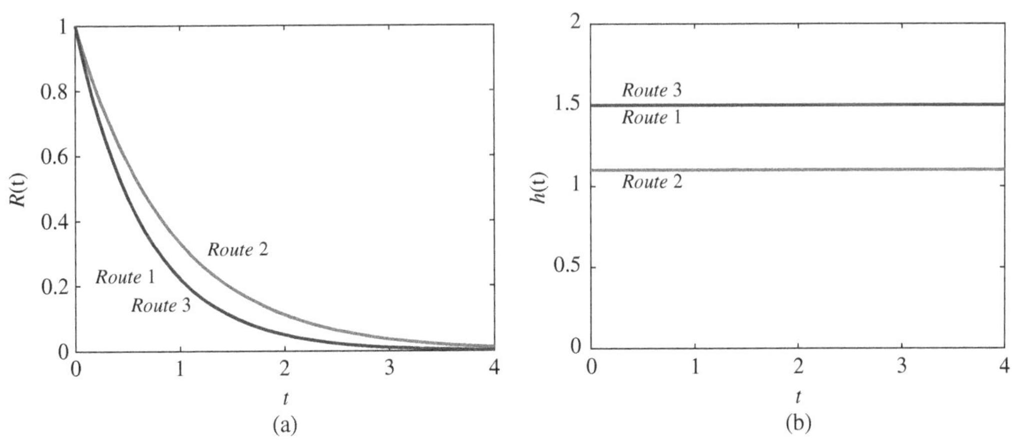

Figure 6.27 Reliability and hazard functions of alternatives in Example 6.35. (a) Reliability function and (b) hazard function.

Figure 6.27 shows that the most reliable route to deliver 8 units is route 2. ■

1 Advanced material, can be omitted from lecture.

The illustration shows that there is an optimum alternative for delivering flow from a source node to a target node, namely, the alternative with the highest reliability function. It is necessary to design a systematic method to find the optimum delivery method that has the highest reliability to meet the needs of a large number of flows. The available methods for static network reliability analysis are extended to the dynamic reliability analysis of a capacitated network with the goal of finding the optimum network reliability function and other reliability measures for a mission of delivering a certain amount of flow from the source to the target nodes.

Assume that arc i has an exponentially distributed lifetime with failure rate λ_i. Since a minimum path is constructed using arcs in a series structure, the lifetime of minimum path MP_j is the minimum of the lifetimes of all its arcs. Therefore, the lifetime of minimum path MP_j is also exponentially distributed with a failure rate equal to the sum of the failure rates of all arcs in MP_j,

$$\lambda_{MP_j} = \sum_{i \in MP_j} \lambda_i. \tag{6.52}$$

The reliability of MP_j increases as λ_{MP_j} decreases. Thus, maximizing the reliability of MP_j is equivalent to minimizing λ_{MP_j}, and finding a minimum path with the highest reliability is equivalent to finding a minimum path with the lowest failure rate, which is the sum of the failure rate of arcs in the minimum path. Since fulfilling demand is using a route that may require several minimum paths, the best route is the route with the lowest total failure rate among all possible routes.

Example 6.36 Assume that all arcs in the network in Figure 6.25 have exponential distributions. The best route in Example 6.33 can be determined as follows:

- Route 1 comprises of minimum paths {1,4,7} and {2,5,8}.

$$X_{Route_1} = (X_1X_4X_7)(X_2X_5X_8) = X_1X_2X_4X_5X_7X_8.$$

$$\lambda_{route_1} = \lambda_1 + \lambda_2 + \lambda_4 + \lambda_5 + \lambda_7 + \lambda_8 = \frac{1}{5} + \frac{1}{5} + \frac{1}{5} + \frac{1}{5} + \frac{1}{5} + \frac{1}{2} = 1.5.$$

- Route 2 comprises of minimum paths {1,4,7} and {2,5,6,7}

$$X_{Route_2} = X_{MP_1}X_{MP_3} = (X_1X_4X_7)(X_2X_5X_6X_7) = X_1X_2X_4X_5X_6X_7.$$

$$\lambda_{route_2} = \lambda_1 + \lambda_2 + \lambda_4 + \lambda_5 + \lambda_6 + \lambda_7 = \frac{1}{5} + \frac{1}{5} + \frac{1}{5} + \frac{1}{5} + \frac{1}{5} + \frac{1}{10} = 1.1.$$

- Route 3 comprises of minimum paths {1,3,5,8} and {2,5,6,7}

$$X_{Route_3} = X_{MP_4}X_{MP_3} = (X_1X_3X_5X_8)(X_2X_5X_6X_7) = X_1X_2X_3X_5X_6X_7X_8.$$

$$\lambda_{route_3} = \lambda_1 + \lambda_2 + \lambda_3 + \lambda_5 + \lambda_6 + \lambda_7 + \lambda_8 = \frac{1}{5} + \frac{1}{5} + \frac{1}{10} + \frac{1}{5} + \frac{1}{10} + \frac{1}{5} + \frac{1}{2} = 1.5.$$

The route with the smallest total failure rate is route 2. Thus, delivering 8 units using route 2 will get the maximum reliability. This result is the same as obtained in Example 6.35. ■

The arc failure rate can be used as an analogy to the flow cost; hence, finding the maximum reliability for a flow delivery mission on a capacitated network is equivalent to finding the minimum cost flow on that network. Using this analogy, the problem of maximum reliability network flow problem can be formulated as a minimum cost network flow problem, and algorithms for solving the minimum cost flow problem can be applied to solve the problem of maximizing the reliability of flow delivery mission. One of them is the negative cycle canceling algorithm [15]. This algorithm is slightly modified to be applicable for maximizing network reliability instead of minimizing cost. The algorithm is referred to as Algorithm 6.2, which can be described as follows:

Algorithm 6.2

1) Determine the demand.
2) Find a minimum path and its capacity. Deliver along this minimum path a flow in the amount of its capacity. Find the residual demand and network because of this flow.
3) Repeat step 2 using residual network until there is no demand or minimum path.
4) Find a negative cycle in the residual network.
   a) If there is no negative cycle, then the route chosen is optimum. Determine the network structure function as the product of the structure functions of the minimum paths in the route chosen. Stop.
   b) Else, find the capacity of the negative cycle. Return the flow along this cycle in the amount of its capacity. Update the residual network.
5) Repeat step 4.

Example 6.37 Find the best route for Example 6.35 using negative cycle canceling algorithm.

Demand $D = 8$.

$$MP_1 = \{1, 4, 7\},\quad c_1 = \min(8, 4, 9) = 4,\quad x = \min(D, c_1) = 4,\quad D = D - x = 4;$$
$$d_1 = 8 - 4 = 4,\quad d_4 = 4 - 4 = 0,\quad d_7 = 9 - 4 = 5.$$

$$MP_2 = \{2, 5, 8\},\quad c_2 = \min(7, 8, 5) = 5,\quad x = \min(D, c_2) = 4,\quad D = D - x = 0;$$
$$d_1 = 8 - 4 = 4,\quad d_2 = 7 - 4 = 3,\quad d_8 = 5 - 4 = 1$$

The resulting residual network is presented in Figure 6.28. For an easier information to follow, the arc $MTTF$ is converted to failure rate $\lambda = 1/MTTF$.

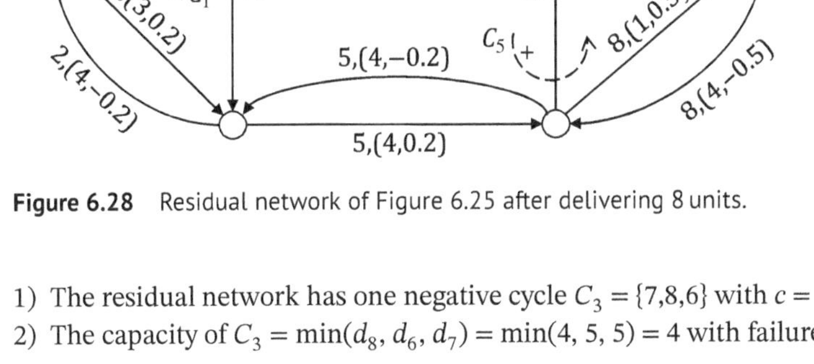

Figure 6.28 Residual network of Figure 6.25 after delivering 8 units.

1) The residual network has one negative cycle $C_3 = \{7,8,6\}$ with $c = 0.2–0.5 + 0.1 = –0.2$.
2) The capacity of $C_3 = \min(d_8, d_6, d_7) = \min(4, 5, 5) = 4$ with failure rate (cost) $= –0.2$.

Reroute the delivery of 4 units along C_3. The residual network after rerouting is presented in Figure 6.29.

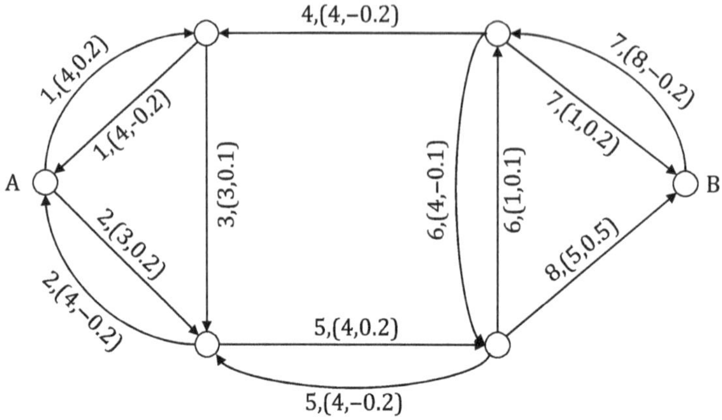

Figure 6.29 Residual network of Figure 6.28 after rerouting 4 units.

3) There is no negative cycle in residual network in Figure 6.29. The current delivery route is optimum. For example, cycle {4,3,5,6} has $c = -0.2 + 0.1 + 0.2 + 0.1 = 0.2 > 0$.

The optimum route can be obtained from Figure 6.29 by following paths with reverse directions from the original network. Those paths are {1,4} with capacity 4 units, paths {2,5,6} with capacity 4 units, and arc 7 for 8 units. It means the delivery is using minimum paths {1,4,7} for 4 units and {2,5,6,7} for 4 units, which is route 2. The 8 units flow through arc 7 is from path {1,4,7} for the amount of 4 units and from path {2,5,6,7} for the amount of 4 units (arc 7 is used by both paths). Therefore, the result of using negative cycle canceling algorithm is the same as using comparison of all possible route alternatives as given in Example 6.36. ■

Exercise 6.14 Show that the current solution is not optimum if *MTTF* of arc 1 is changed from 5 to 2. Find the new optimum routes to send the flow so that the network reliability is maximized.

The development of methods for network reliability modeling and analysis so far is based on combinatorial models. The next section uses state transition model as a tool for network reliability modeling and analysis. Comparison between using combinatorial and state transition models is also presented.

6.7 Network Reliability Analysis Using State Transition Model

The network reliability function can be derived by representing the network in a state transition diagram. A network state is a good (transient) state if the good components in that network state form a path from source to target nodes. Otherwise, the network state is a failed (absorbing) state. There are transitions from one network state to other network states due to a failure in one of the good components in the previous state. Since it is in continuous time, only one good component can fail from one state to the next state. The transition matrix from one network state to another network state can be derived using the same rule as for general standby systems.

The network reliability analysis algorithm using the state transition model is referred to as Algorithm 6.3

Algorithm 6.3

---

1) Generate network states as combinations of component states starting from a network state with all components that are good.
2) Generate minimum paths of the network using the **minpaths** function.
3) Classify the network states into transient states and absorbing states by checking whether good components in a state form at least one minimum path from the source to the target. If it is, then it is a transient state; otherwise, it is an absorbing state.
4) For each transient state, generate the next state for each failure of one good component in that transient state.
5) Repeat step 4 until there is no transient state.
6) Determine the transition matrix within each transient state using Equation (5.1).
7) Determine the transition matrices between each transient state and its next states using Equation (5.2).
8) Construct a survival matrix using the transition matrices generated in steps 6 and 7.
9) Determine the initial probability vector from the network initial state.
10) Use the results of steps 8 and 9 to calculate network reliability measures.

Example 6.38 The network survival matrix for the network in Figure 6.3a is generated following Algorithm 6.3. The minimum paths for this network are A–B–D and A–C–D. Assume component i has PH-distributed lifetime with representation $(n_i, \boldsymbol{a}_i, \boldsymbol{A}_i)$, $i = 1, 2, 3, 4$, where 1, 2, 3, 4 represent A, B, C, D, respectively.

1) The network states are (g, g, g, g), (g, g, g, f), (g, g, f, g), (g, f, g, g), (f, g, g, g), and other states with more than one failed component, where (g, g, g, g) is the initial state.
2) The initial state is a transient state. The states (g, g, f, g) and (g, f, g, g) are transient states because the good components in these states form a path from source to target, namely, the path {A,B,D} for the state (g, g, f, g) and the path {A,C,D} for the state (g, f, g, g). On the contrary, the states (g, g, g, f), (f, g, g, g), and other states with more than one failed component are absorbing states because the good components in these states do not form a path from source to target.
3) Find the next states for each transient state:
   - The next states of (g, g, f, g) are (g, g, f, f), (g, f, f, g), and (f, g, f, g). All of them are absorbing states, so they do not have next states.
   - The next states of (g, f, g, g) are (g, f, g, f), (g, f, f, g), and (f, f, g, g). All of them are absorbing states, so they do not have next states.
4) This network has three transient states (g, g, g, g), (g, g, f, g), and (g, f, g, g). The transition matrices in these transient states are determined using Equation (5.1):
   - Transition matrix within state (g, g, g, g) is $\boldsymbol{A}_1 \oplus \boldsymbol{A}_2 \oplus \boldsymbol{A}_3 \oplus \boldsymbol{A}_4$.
   - Transition matrix within state (g, g, f, g) is $\boldsymbol{A}_1 \oplus \boldsymbol{A}_2 \oplus \boldsymbol{A}_4$.
   - Transition matrix within state (g, f, g, g) is $\boldsymbol{A}_1 \oplus \boldsymbol{A}_3 \oplus \boldsymbol{A}_4$.
5) The transition matrices from transient states to their corresponding next states are determined using Equation (5.2):
   - Transition matrix from (g, g, g, g), to (g, g, f, g) is $\boldsymbol{I}_1 \otimes \boldsymbol{I}_2 \otimes \boldsymbol{b}_3 \otimes \boldsymbol{I}_4$.

- Transition matrix from (g, g, g, g) to (g, f, g, g) is $\boldsymbol{I}_1 \otimes \boldsymbol{b}_2 \otimes \boldsymbol{I}_3 \otimes \boldsymbol{I}_4$.
- There is no transition from (g, g, f, g) to (g, f, g, g) or vice versa.

6) The network survival matrix is

$$\boldsymbol{A}_S = \begin{bmatrix} \boldsymbol{A}_1 \oplus \boldsymbol{A}_2 \oplus \boldsymbol{A}_3 \oplus \boldsymbol{A}_4 & \boldsymbol{I}_1 \otimes \boldsymbol{I}_2 \otimes \boldsymbol{b}_3 \otimes \boldsymbol{I}_4 & \boldsymbol{I}_1 \otimes \boldsymbol{b}_2 \otimes \boldsymbol{I}_3 \otimes \boldsymbol{I}_4 \\ \boldsymbol{0} & \boldsymbol{A}_1 \oplus \boldsymbol{A}_2 \oplus \boldsymbol{A}_4 & \boldsymbol{0} \\ \boldsymbol{0} & \boldsymbol{0} & \boldsymbol{A}_1 \oplus \boldsymbol{A}_3 \oplus \boldsymbol{A}_4 \end{bmatrix}.$$

7) The initial probability vector is

$$\boldsymbol{a}_S = [\boldsymbol{a}_1 \otimes \boldsymbol{a}_2 \otimes \boldsymbol{a}_3 \otimes \boldsymbol{a}_4 \quad \boldsymbol{0} \quad \boldsymbol{0}]$$

The network states, the transition matrix within each transient state, and the transition matrices among transient states are used to construct the state transition diagram for the network in Figure 6.3a. The result is presented as Figure 6.30.

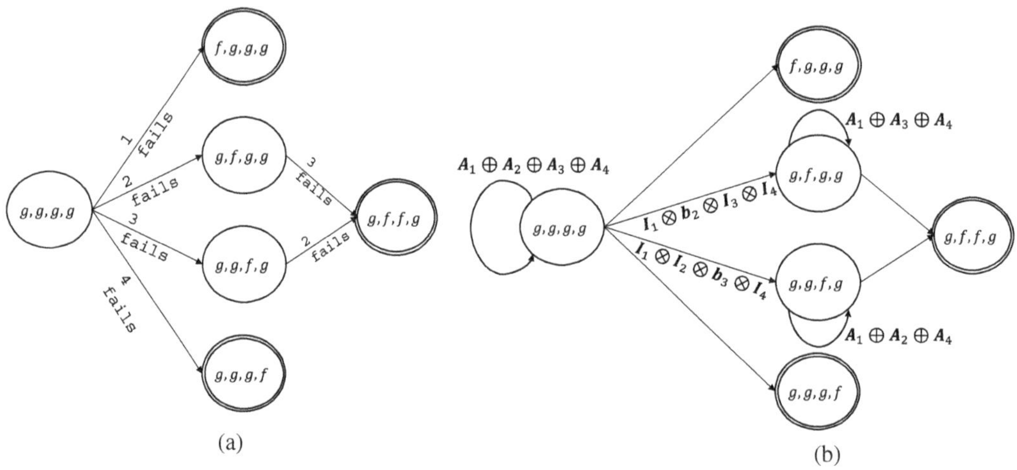

Figure 6.30 State transition diagram of network in Figure 6.3a. ■

Example 6.39 The minimum paths of the network in Figure 6.3a are {A,B,D} and {A,C,D}. The structure function is $X_S = X_A X_B X_D \vee X_A X_C X_D = X_A X_B X_D + X_A \overline{X}_B X_C X_D$. The survival matrix and the initial and closing probability vectors of the network lifetime distribution derived using Lemmas 6.1 and 6.2 from the SDP representation are

$$\boldsymbol{A}_S = \begin{bmatrix} \boldsymbol{A}_1 \oplus \boldsymbol{A}_2 \oplus \boldsymbol{A}_4 & \boldsymbol{0} & \boldsymbol{0} \\ \boldsymbol{0} & \boldsymbol{A}_1 \oplus \boldsymbol{A}_3 \oplus \boldsymbol{A}_4 & \boldsymbol{0} \\ \boldsymbol{0} & \boldsymbol{0} & \boldsymbol{A}_1 \oplus \boldsymbol{A}_2 \oplus \boldsymbol{A}_3 \oplus \boldsymbol{A}_4 \end{bmatrix},$$

$$\boldsymbol{a}_S = [\boldsymbol{a}_1 \otimes \boldsymbol{a}_2 \otimes \boldsymbol{a}_4 \quad \boldsymbol{a}_1 \otimes \boldsymbol{a}_3 \otimes \boldsymbol{a}_4 \quad -\boldsymbol{a}_1 \otimes \boldsymbol{a}_2 \otimes \boldsymbol{a}_3 \otimes \boldsymbol{a}_4],$$

$$\boldsymbol{v}_S = [\boldsymbol{v}_1 \otimes \boldsymbol{v}_2 \otimes \boldsymbol{v}_4 \quad \boldsymbol{v}_1 \otimes \boldsymbol{v}_3 \otimes \boldsymbol{v}_4 \quad -\boldsymbol{v}_1 \otimes \boldsymbol{v}_2 \otimes \boldsymbol{v}_3 \otimes \boldsymbol{v}_4],$$

where $\boldsymbol{v}_i = \boldsymbol{u}_i$ since the lifetime of component i, $i = 1, 2, 3, 4$, has PH distribution. ■

The resulting representations in Examples 6.38 and 6.39 have the same order $n_1n_2n_3n_4 + n_1n_2n_4 + n_1n_3n_4$, but their survival matrices and initial probability vectors are different. Due to the presence of negative elements in the initial probability vector, the representation produced in Example 6.39 is an ME representation, while the one produced using the state transition diagram in Example 6.38

is a PH representation. It is the main difference between the combinatorial and state transition models. The combinatorial model produces an ME distribution even if the component lifetimes have PH distributions. On the contrary, the state transition model always generates the network lifetime distribution in accordance with the distribution of its component lifetimes: if all components have PH-distributed lifetimes, then the network lifetime has PH distribution, but if any component has ME-distributed lifetime, then the network lifetime has ME distribution. However, the results are equivalent but in different representations.

Example 6.40 The PH/ME representations obtained from the combinatorial and state transition models in Examples 6.38 and 6.39 are utilized to generate the network reliability and hazard functions. The components are assumed to have lifetimes following identical 2-Erlang distribution with $MTTF = 100$. The results are presented in Figure 6.31, which shows that both models produce identical functions. This means that the PH representation in Example 6.38 is equivalent to the ME representation in Example 6.39.

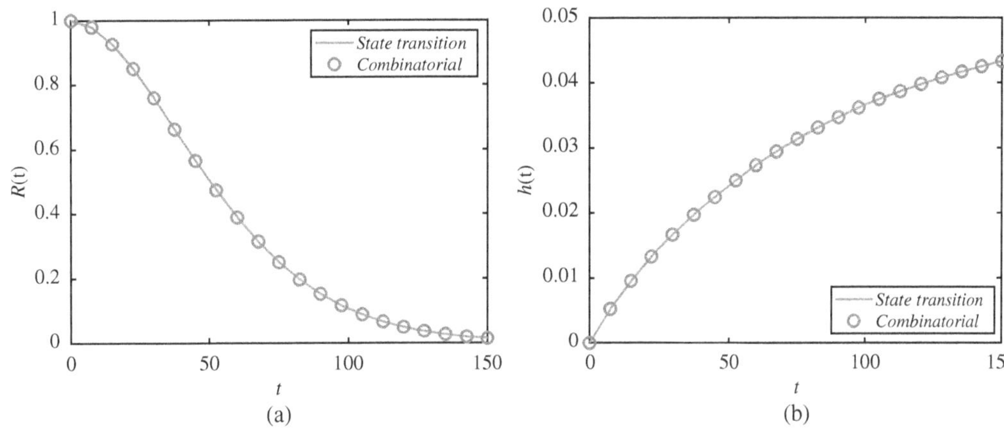

Figure 6.31 Reliability and hazard functions of the network in Figure 6.3a calculated using combinatorial and state transition models.

■

Exercise 6.15 List the states of the network in Figure 6.2 if it is represented in state transition model. Use the states to determine the order of the PH distribution of the network lifetime. Then, use Algorithm 6.1 to determine the order of ME distribution of the network lifetime generated using combinatorial model. Draw a conclusion from the result.

Large networks require computer assistance to generate and classify network states, to determine the transition matrix within each transient state, between each possible pair of transient states, and finally to produce a PH/ME representation of the network lifetime distribution. Algorithm 6.3 is programmed in MATLAB as **menettrans** function.

Example 6.41 Function **menettrans** is used to generate PH representations of the network reliability functions in Application 6.2 using the simplified network diagram. The lifetime distribution of each arc is the same as in Application 6.2. The results are used to generate network reliability and hazard functions depicted in Figure 6.32. Comparison with the results of using the combinatorial model used in Application 6.2 as well shows that the state transition and combinatorial models produce identical graphs.

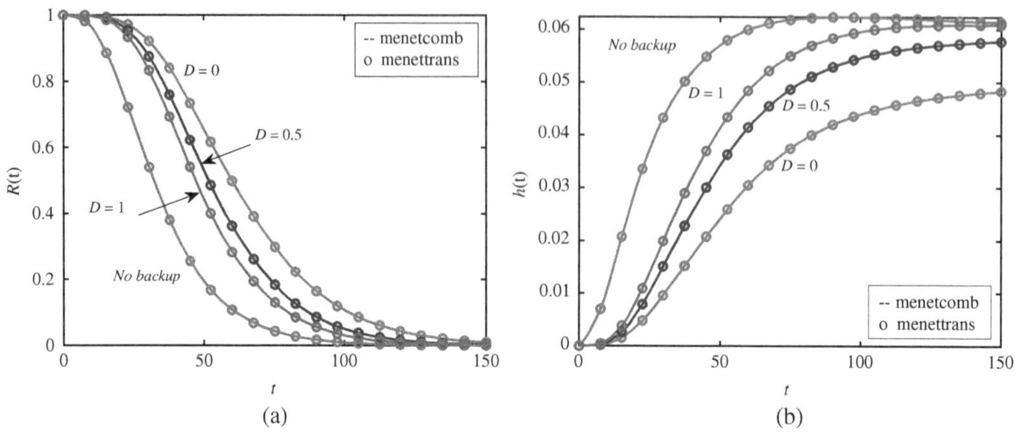

Figure 6.32 Reliability and hazard functions of network in Application 6.2 generated using combinatorial and state transition models. (a) Reliability function and (b) hazard function.

Complexities and computation times using the two models are presented in Table 6.5. The combinatorial model in this experiment is using ordered minimum paths.

Table 6.5 Comparison of orders of PH/ME distributions and computation times to produce Figure 6.32.

	Order of distribution		Computation time (s)	
Configuration	**menetcomb**	**menettrans**	**menetcomb**	**menettrans**
One line, no backup	323	631	0.04	0.63
One main line and one backup line (2-standby)	95 023	118 075	4.21	14.14
One main line and one warm backup ($D = 0, 5$)	95 023	118 075	4.54	15.06
One main line and one hot backup line (2-parallel)	95 023	118 075	5.28	14.86

Table 6.5 shows that using combinatorial model produced network reliability function with lower order than using state transition model. The computation time for producing Figure 6.32 using combinatorial model is also faster than using state transition model. Even though the difference in distribution orders is not significant, the difference in computation times is significant. It is because the survival matrix of the network reliability function obtained using the combinatorial model is diagonal, while using the state transition model is upper diagonal, and computation of diagonal matrices is faster than upper diagonal matrices with the same dimension. ■

6.8 Problems

6.1 Create a graph representation for the network in Example 6.1 with the network components being nodes A and D and arcs 1, 3, and 5. Determine the network structure function by using minimum path and minimum cut.

6.2 Create a table of the properties of Boolean operations applicable to network reliability analysis using combinatorial model.

6.3 Find an SDP representation for the structure function $X_1X_5 \vee X_1X_3X_4$. Interchange the sequence of the terms and find its SDP representation. Prove that the two resulting SDP representations are equivalent.

6.4 Show that the Boolean operation "∨" on structure function $X_1X_5 \vee \overline{X}_1X_2X_3X_4$ can be directly replaced by the arithmetic operation "+."

6.5 Prove that the two SDP representations produced in Example 6.14 are equivalent.

6.6 Derive the structure function for the 2-out-of-4 system using combinatorial model and transform the result into SDP representation.

6.7 Use the result of Problem 6.6 to derive the survival matrix of a 2-out-of-4 system and compare it with its survival matrix derived using state transition model.

6.8 Formulate series–parallel and parallel–series systems in combinatorial models. Use the models to show that series–parallel system always has higher reliability than parallel–series system with the same components.

6.9 The standby structure cannot be expressed using the combinatorial model described in this book. Suggest a method to transform a network with arcs in standby structure so that combinatorial model can still be used.

6.10 Write the structure function of the network in Figure P.6.1 using minimum path. Manually derived its SDP representation and the network reliability function assuming all components have the same reliability value.

6.11 Generate the SDP representation of the network in Figure P.6.1 using **sdprep** function and compare it with the result of Problem 6.10.

6.12 Find a simplified graph for the network in Application 6.2 if it is utilized to send information from G to B, instead of from B to G. Based on the simplified network, compare the network reliability function and *MTTF* for delivering information from node B to node G with from node G to node B.

6.13 Prove that the longer the distance between the source and the target, the lower the reliability of the network. Distance between two nodes in a network is the minimum number of arcs connecting the two nodes.

6.14 Derive a network reliability model for the network in Application 6.2 without combining B–A–C to B–C. Use the model and the necessary functions (**minpath, sdprep, menetomb**) to generate network reliability measures for the same distributions used in Application 6.2. Compare the order of the network reliability function obtain in this problem with the result of Application 6.2.

6.15 Use the model in Problem 6.14 to produce the same network reliability measure as in Application 6.2. Compare your results and computation time with the results and computation time of Application 6.2. Make conclusions from the comparison.

6.16 In multisource–multitarget network, which one has higher reliability: from all sources to all targets, from any source to any target, from any source to all targets, and from all sources to any target.

6.17 Show that in a capacitated network, the higher the flow demand, the lower the network reliability. Use this result to show that the network lifetime is shorter stochastically for higher flow demand.

6.18 Find a combination of routes that give maximum reliability for sending 6 units of flow in the network in Figure 6.25 using cycle canceling algorithm. Compare the network reliability functions for all possible routes to send 6 units of flow and show that the result of cycle canceling algorithm indeed has the maximum reliability.

6.19 Suppose the direction of arc 6 in Figure 6.25 is reversed. Can it be proved that the network reliability always becomes lower because of reversion? If not, determine the condition that the new network has lower reliability than the original network.

6.20 Prepare a table like Table 6.4 for the network in Problem 6.18.

6.21 Use the software package to compare the dimensions of the survival matrix of a 2-out-of-5 general standby system derived using combinatorial and state transition models for all components having identical 2-Erlang-distributed lifetimes. Compare also the computation times to produce all system reliability measures using the two models in the case of all backups have warmness levels = 0.0, 0.5, and 1. Does warmness level affect computation time?

6.22 Can combinatorial model be simplified for identical components? Compare the dimension and computation times for generating reliability measures of homogeneous and heterogeneous K-out-of-N systems using state transition and combinatorial models. Draw a conclusion from the results. Although not required, for simplicity, assume all components have 2-Erlang-distributed lifetimes.

6.23 Compare **menetcomb** and **menettrans** applied to the network in Figure P.6.1 in terms of their computation times for generating the network reliability functions. Assume all components have lifetimes following identical distributions:

a) Exponential.
b) 2-Erlang.

Provide an explanation and conclusion from the results of this experiment. Use any value for component $MTTF$.

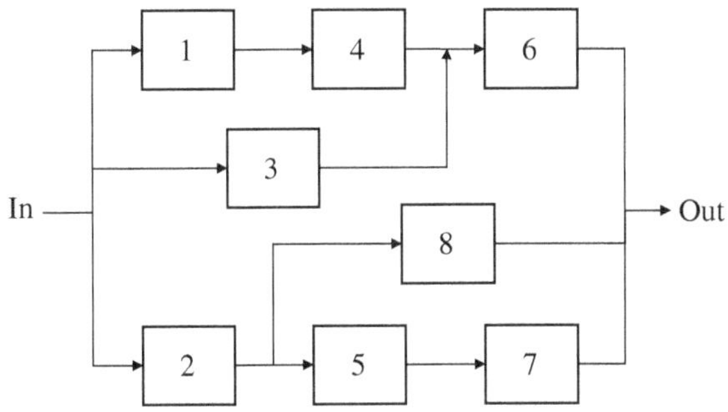

Figure P.6.1 Network for problems in this chapter.

6.24 Add a bridge component 9 to connect the output of component 3 to the input of component 8 in the network in Figure P.6.1. Repeat Problem 6.22 for this network.

6.25 Determine the effect of this additional component to the network reliability function and *MTTF*. Use the distributions in Problem 6.22.

References

1 Lin, S., Wang, Y., Jia, L., Zhang, H., 2018. Reliability assessment of complex electromechanical systems: A network perspective. Qual Reliab Eng Int 34, 772–790 https://doi.org/10.1002/qre.2289.

2 Chaturvedi, S.K., 2016. Network Reliability: Measures and Evaluation. Scrivener Publishing, Beverly, MA.

3 Ardakan, M.A., Mirzaei, Z., Hamadani, A.Z., Elsayed, E.A., 2017. Reliability optimization by considering time-dependent reliability for components. Qual Reliab Eng Int 33, 1641–1654 https://doi.org/10.1002/qre.2132.

4 Rebello, S., Yu, H., Ma, L., 2018. An integrated approach for system functional reliability assessment using Dynamic Bayesian Network and Hidden Markov Model. Reliab Eng Syst Saf 180, 124–135 https://doi.org/10.1016/j.ress.2018.07.002.

5 Alkaff, A., 2021. Discrete time dynamic reliability modeling for systems with multistate components. Reliab Eng Syst Saf 209, 107462 https://doi.org/10.1016/j.ress.2021.107462.

6 Alkaff, A., Qomarudin, M.N., Bilfaqih, Y., 2021a. Network reliability analysis: Matrix-exponential approach. Reliab Eng Syst Saf 212, 107591 https://doi.org/10.1016/j.ress.2021.107591.

7 Schäfer, L., García, S., Srithammavanh, V., 2018. Simplification of inclusion–exclusion on intersections of unions with application to network systems reliability. Reliab Eng Syst Saf 173, 23–33 https://doi.org/10.1016/j.ress.2018.01.003.

8 Heidtmann, K.D., 1989. Smaller sums of disjoint products by subproduct inversion. IEEE Trans Reliab 38(3), 305–311 https://doi.org/10.1109/24.44172.

9 Soh, S., Rai, S., 1991. CAREL: Computer aided reliability evaluator for distributed computing networks. IEEE Trans Parallel Distrib Syst 2, 199–213 https://doi.org/10.1109/71.89065.

10 Luo, T., Trivedi, K.S., 1998. An improved algorithm for coherent-system reliability. IEEE Trans Reliab 47, 73–78 https://doi.org/10.1109/24.690911.

11 Chaturvedi, S.K., Misra, K.B., 2002. An efficient multi-variable inversion algorithm for reliability evaluation of complex systems using path sets. Int J Reliab Qual Saf Eng 09, 237–259 https://doi.org/10.1142/S0218539302000809.

12 Assaf, D., Levikson, B., 1982. Closure of phase type distributions under operations arising in reliability theory. Ann Probab 10(1), 265–269.

13 Alkaff, A., Qomarudin, M.N., Purwantini, E., Wiratno, S.E., 2021b. Dynamic reliability modeling for general standby systems. Comput Ind Eng 161, 107615 https://doi.org/10.1016/j.cie.2021.107615.

14 Lin, J.S., Jane, C.C., Yuan, J., 1995. On reliability evaluation of capacitated-flow networks in terms of minimal pathsets. Networks 25(3), 131–138 https://doi.org/10.1002/net.3230250306.

15 Ahuja, R.K., Magnanti, T.L., Orlin, J.B., 1993. Network Flows: Theory, Algorithms, and Applications. Prentice Hall, Upper Saddle River.

7

Discrete-Time Systems

Learning objectives:
1) Understand the derivations of all reliability measures in DPH and MG distributions.
2) Understand important properties of DPH and MG distributions in system reliability.
3) Able to model complex structured discrete-time systems.
4) Able to utilize the software to calculate reliability measures of discrete-time systems.

Abbreviations

CFR	constant failure rate
DFR	decreasing failure rate
DPH	discrete phase-type
DTMC	discrete-time Markov chain
IE	inclusion-exclusion
IFR	increasing failure rate
MG	matrix-geometric
NB	negative binomial
SDP	sum of disjoint products
CVTTF	coefficient of variations of time to failure
MRL	mean residual life
MTTF	mean time to failure
SDTTF	standard deviation of time to failure
VTTF	variance of time to failure

Notations

T	time to failure
k	time, mission duration in discrete values
m	age
n	order, dimension
$\boldsymbol{Q}$	infinitesimal generator matrix
$\boldsymbol{a}$	initial probability vector

$\boldsymbol{A}$	survival matrix
$\boldsymbol{I}$	identity matrix
$\boldsymbol{u}$	unitary vector
$\mathbf{b}$	failure probability vector
$\boldsymbol{D}$	deceleration matrix
d	deceleration factor, warmness level
K	number of primary components in a system
N	number of components in a system
$(n_i, \boldsymbol{a}_i, \boldsymbol{A}_i)$	DPH representation of lifetime distribution of component i
$(n_i, \boldsymbol{a}_i, \boldsymbol{A}_i, \boldsymbol{v}_i)$	MG representation of lifetime distribution of component i
$(n_S, \boldsymbol{a}_S, \boldsymbol{A}_S)$	DPH representation of system lifetime distribution
$(n_S, \boldsymbol{a}_S, \boldsymbol{A}_S, \boldsymbol{v}_S)$	MG representation of system lifetime distribution
$p(k)$	probability mass function
$F(k)$	cumulative distribution function
$R(k)$	discrete-time reliability function
$RR(k \mid g)$	discrete-time residual reliability function of system with age g
$h(k)$	discrete-time hazard function
$MRL(g)$	mean residual life function
$\otimes$	Kronecker product
$\Phi_{\{.\}}(z)$	probability generating function
X	component or system state
$\overline{X}$	negation or inverse of X
Y	disjoint product term
MP	minimum path
MC	minimum cut

System reliability is not always modeled in continuous time where the system and its component conditions are assumed to change at any time. Systems in some applications are used periodically, and thus, the conditions change periodically, such as the landing gears and tires of aircraft. The lifetime of batteries is strongly influenced by the number of times they are charged, not the duration of their use. In many cases, systems are inspected periodically; despite their conditions deteriorating continuously, the systems need to be modeled in discrete time to be in line with the results of inspection. Furthermore, discretization of continuous-time system can make system reliability analysis more efficient as shown in Roy and Dasgupta [1]. For this reason, several distributions in continuous-time reliability analysis are discretized because they are needed in practice; these are discrete-time Weibull distributions [2] and discrete-time gamma distributions [3]. In addition, there are several lifetime models specifically intended for discrete-time reliability analysis as discussed in discrete-time reliability book by Nair, Sankaran, and Balakrishnan [4], including their categorization into several types of distributions. The aforementioned lifetime models are for systems whose conditions are either good or failed.

Reliability analysis of discrete-time systems with multistate components are discussed in Chryssaphinou, Limnios, and Malefaki [5]; Yu et al. [6]; Jia, Xing, and Song [7], among others. However, these publications used scalar transitions for modeling multistate systems. On the contrary, this book uses matrix transitions to model multistate system reliability by representing the system lifetime in discrete phase-type (DPH) distribution. The use of DPH distribution in system reliability analysis is not yet as popular as PH distribution in the reliability analysis of systems with many components, despite both having the same complexity. Steady-state analysis of discrete-time systems is studied in Ruiz-Castro [8]. Alkaff [9, 10] describes methods for modeling system reliability in discrete time using DPH distribution and uses the model for dynamic system reliability analysis in discrete time. Bilfaqih, Qomarudin, and Sahal [11] described methods for using MG distribution for dynamic system reliability analysis in discrete time for computation time efficiency. The book specifically for the use of DPH distribution in system reliability analysis is Eryilmaz [12]; however, it is for research and does not cover systems in multiple, let alone complex, structures.

The development of methods for reliability modeling and analysis of discrete-time systems follows the development of continuous-time systems closely. State space and combinatorial models, as well as hybrid models, are employed as the basis for development. The properties of DPH distribution are utilized to design algorithms for calculating DPH representation of system reliability function from the DPH representation of its component reliability functions. The systems may have various structures as discussed in Chapters 4–6 with the same assumption as in those chapters that the component lifetimes are independent but not necessarily identical. The development starts with single structured systems, which include series, parallel, and standby with perfect and imperfect switches, K-out-of-N, and general standby structures, and then it continues for complex structured systems containing combinations of those single structures. In the presence of a bridge, the DPH distribution must be generalized into a matrix-geometric (MG) distribution [9]. The system reliability measures are then calculated using the resulting DPH or MG representations of the system lifetime distribution. They include functional system reliability measures, such as reliability, hazard, and *MRL* functions, as well as statistical reliability measures, such as *MTTF*, *SDTTF*, and *CVTTF*. All reliability measures are calculated using only matrix algebra.

This chapter starts with descriptions of discrete-time Markov chain (DTMC) and its relationship with DPH distribution. Reliability measures in DPH distribution are then derived. MG distribution is introduced as a generalization of DPH distribution, and a method to transform DPH distribution into MG distribution to reduce computation time is presented. Some properties of DPH and MG distributions useful for system reliability modeling and analysis are presented. The properties are utilized for devising algorithms for calculating DPH representations of reliability function of systems in various single structures and in complex structures. The algorithms are prepared as MATLAB functions that can be used for system reliability modeling. Numerical results of using the functions for system reliability analysis are presented in graphs and tables, including their computation times.

This chapter is a translation of material in previous chapters for continuous time into discrete time. Many examples and applications are also repetitions from examples and applications in continuous-time system discussed in previous chapters. Therefore, this chapter is presented at a faster pace than the previous chapters.

7.1 Discrete-Time Markov Chain (DTMC)

DTMC is a discrete-state random process in discrete time that obeys the Markov property. The discrete states can be numbered as state i, $i = 1, 2, \dots, n$. The process is inspected periodically and, as a result of such inspection, the process may still be in the same state as in the previous inspection or out of that state. Let $X(k)$ be the state of the process at period k, which may take any value i, $i = 1, 2, \dots, n$. In the next inspection period, the process will be in state $X(k+1)$ that also may take any value j, $j = 1, 2, \dots, n$. Because of the Markov property in discrete time, the duration of stay of DTMC in each state has a geometry distribution with a parameter equal to the probability that the process stays at the same state in the next inspection period [13]. The process is assumed to be stationary, which makes this probability does not change over time.

Define

$$q_{ij} = P[X(k+1) = j|X(k) = i], \tag{7.1}$$

constant for all k due to the stationary assumption.

The q_{ij} is referred to as the transition probability from state i to state j. If $j = i$, then the process stays in state i, and q_{ii} is the probability that the process stays in state i. If T_i denotes the duration of stay in state i, then

$$T_i \sim \text{geo}(q_{ii}), \tag{7.2}$$

where q_{ii} is the probability that the process remains in state i at the beginning and end of an inspection period.

The probability mass function (PMF) of T_i in Equation (7.2) is

$$P[T_i = k] = P_{T_i}(k) = 0 \quad \text{for} \quad k = 0 = q_{ii}^{k-1}(1-q_{ii}) \quad \text{for} \quad k = 1, 2, \dots. \tag{7.3}$$

Equation (7.3) implies that the system stays in any state for at least one period.

The CDF of T_i can be derived as

$$\begin{aligned} P[T_i \le k] &= \sum_{j=1}^{k} P(T_i = j) = \sum_{j=1}^{k} q_{ii}^{j-1}(1-q_{ii}) \\ &= (1-q_{ii})\left(1+q_{ii}+q_{ii}^2+\cdots q_{ii}^{k-1}\right) = \frac{(1-q_{ii})\left(1-q_{ii}^k\right)}{1-q_{ii}}, \end{aligned}$$

which can be simplified to

$$F_{T_i}(k) = P[T_i \le k] = 1 - q_{ii}^k. \tag{7.4}$$

The expected duration of stay in state i is

$$\begin{aligned} E[T_i] &= \sum_{k=1}^{\infty} kP[T_i = k] = \sum_{k=1}^{\infty} kq_{ii}^{k-1}(1-q_{ii}) \\ &= (1-q_{ii})\left(1+2q_{ii}+3q_{ii}^2+\cdots\right) \\ &= (1-q_{ii})\frac{d}{dq_{ii}}\left(q_{ii}+q_{ii}^2+q_{ii}^3+\cdots\right) \\ &= (1-q_{ii})\frac{d}{dq_{ii}}\left(\left(1+q_{ii}+q_{ii}^2+q_{ii}^3+\cdots\right)-1\right) \\ &= (1-q_{ii})\frac{d}{dq_{ii}}\left(\frac{1}{1-q_{ii}}-1\right) = (1-q_{ii})\frac{1}{(1-q_{ii})^2}. \end{aligned}$$

Thus,

$$E[T_i] = \frac{1}{1-q_{ii}}, \tag{7.5}$$

and parameter q_{ii} can be detemined from the expected duration of stay in state i as

$$q_{ii} = 1 - \frac{1}{E[T_i]}. \tag{7.6}$$

As in CTMC, the states of DTMC are classified into transient and absorbing states. State i is a transient state if $0 \le q_{ii} < 1$, and an absorbing state if $q_{ii} = 1$. It follows from Equation (7.5) that the expected duration of stay in an absorbing state is $1/0 \to \infty$, as it should be staying in an absorbing state forever. On the contrary, the expected duration of stay in any transient state is finite as it should stay in any transient state temporarily.

Being the distribution of the duration of stay in states of DTMC, geometry distribution must possess memoryless property, that is,

$$P[T \le k + x | T > x] = \frac{P[x < T \le k + x]}{P[T > x]} = \frac{P[T \le k + x] - P[T \le x]}{P[T > x]}$$
$$= \frac{1 - q_{ii}^{k+x} - \left(1 - q_{ii}^{x}\right)}{1 - \left(1 - q_{ii}^{k}\right)} = 1 - q_{ii}^{k} = P[T \le k].$$

With this property, the probability that the process remains in state i in any inspection period is constant equal q_{ii}, irrespective of how long the process has been in that state.

The transition probability is used to model DTMC as described next.

7.1.1 Modeling DTMC

DTMC mechanism is expressed in an infinitesimal generator matrix $\boldsymbol{Q}$ whose elements are the transition probabilities from each state to all other states,

$$\boldsymbol{Q} = [q_{ij}] = \begin{bmatrix} q_{11} & q_{12} & \cdots & q_{1n} \\ q_{21} & q_{22} & \cdots & q_{2n} \\ \cdots & \cdots & \cdots & \cdots \\ q_{n1} & q_{n2} & \cdots & q_{nn} \end{bmatrix}. \tag{7.7}$$

The sum of all transition probabilities from state i must be equal to 1, or

$$\sum_j q_{ij} = 1. \tag{7.8}$$

Since the probability of staying in an absorbing state following Equation (7.8) is equal to 1, the transition probabilities into other states from the absorbing state must be 0.

The description of DTMC implies characteristics of the infinitesimal generator matrix $\boldsymbol{Q}$ for discrete-time systems can be summarized as:

- All elements are nonnegative.
- The row sum equals one.
- All off-diagonal elements in the row representing each absorbing state are zero, while the diagonal element is one.

The requirement that the row sum of $\boldsymbol{Q}$ equals one can be expressed as

$$\boldsymbol{Q}\boldsymbol{u} = \boldsymbol{u}, \tag{7.9}$$

where $\boldsymbol{u}$ is a unitary column vector, that is a vector with all elements 1.

Let $X(k)$ be the state of the process at period k, and $\pi_i(k)$ is the probability that the process at period k is in state i,

$$\pi_i(k) = P[X(k) = i]. \tag{7.10}$$

The relationship between unconditional probability $\pi_j(k+1)$ and $\pi_i(k)$ can be derived using the law of total probability,

$$\pi_j(k+1) = P[X(k+1) = j] = \sum_{i=1}^{n} P[X(k+1) = j | X(k) = i] P[X(k) = i] = \sum_{i=1}^{n} \pi_i(k) q_{ij}. \tag{7.11}$$

The state probability vector at period k, denoted by $\boldsymbol{\pi}(k)$, is the probability that the process is at any state at period k,

$$\boldsymbol{\pi}(k) = [\pi_1(k) \;\; \pi_2(k) \;\; \cdots \;\; \pi_n(k)]. \tag{7.12}$$

Using vector notation, Equation (7.11) can be written as

$$\pi_j(k+1) = \boldsymbol{\pi}(k)\begin{bmatrix} q_{1j} \\ q_{2j} \\ \cdots \\ q_{nj} \end{bmatrix},$$

and

$$\boldsymbol{\pi}(k+1) = \boldsymbol{\pi}(k)\boldsymbol{Q}. \tag{7.13}$$

Equation (7.13) means the probability of being in any state in the future can be determined from the matrix $\boldsymbol{Q}$ and initial probability vector $\boldsymbol{\pi}(0)$; that is,

$$\boldsymbol{\pi}(k) = \boldsymbol{\pi}(0)\boldsymbol{Q}^k. \tag{7.14}$$

Example 7.1 A four-state DTMC has the following description. States 1, 2, and 3 are transient states, and state 4 is an absorbing state. The duration of stays in states 1, 2, and 3 follow geometry distributions with mean 10, 5, and 2 days, respectively. When the process exits state 1, the probabilities of making transitions to states 2 and 3 are 0.8 and 0.2, respectively. The probability of making transitions to states 3 and 4 when the process is out of state 2 are 0.6 and 0.4, respectively. When the process exits state 3, the probabilities of making transitions to states 2 and 4 are 0.4 and 0.6, respectively.

Note that the sum of the probabilities of making transitions from each transient state to all other states is equal to one, and no information is required about the duration of staying in the absorbing state or the transition probabilities out of the absorbing state.

The DTMC description provides the following information:

- The probability of remaining in each state can be calculated using Equation (7.6) as

$$q_{11} = 1 - 1/E[T_1] = 1 - 1/0.1 = 0.9,$$

$$q_{22} = 1 - 1/E[T_2] = 0.8,$$

$$q_{33} = 1 - 1/E[T_3] = 0.5.$$

- The transition probability from state 1 to states 2 and 3 can be calculated as the transition probability of out of state 1, $(1 - q_{11})$, multiply by the probability of making transition to states 2 and 3 when the process being out of state 1,

$$q_{12} = (1 - q_{11}) \times 0.8 = 0.08,$$

$$q_{13} = (1 - q_{11}) \times 0.2 = 0.02.$$

Other transition probabilities q_{ij} can be calculated accordingly.

- The infinitesimal generator matrix of this DTMC following Equation (7.7) and the results of calculating q_{ij}, $i, j = 1, 2, 3$, is

$$\boldsymbol{Q} = [q_{ij}] = \begin{bmatrix} 0.9 & 0.08 & 0.02 & 0 \\ 0 & 0.8 & 0.12 & 0.08 \\ 0 & 0.2 & 0.5 & 0.3 \\ 0 & 0 & 0 & 1 \end{bmatrix}.$$

Note that all off-diagonal elements in the row corresponding to the absorbing state (state 4) are zero and the diagonal element in that row is one. The sum of all elements of each row equal to one can be verified using Equation (7.9) as

$$\begin{bmatrix} 0.9 & 0.08 & 0.02 & 0 \\ 0 & 0.8 & 0.12 & 0.08 \\ 0 & 0.2 & 0.5 & 0.3 \\ 0 & 0 & 0 & 1 \end{bmatrix} \begin{bmatrix} 1 \\ 1 \\ 1 \\ 1 \end{bmatrix} = \begin{bmatrix} 1 \\ 1 \\ 1 \\ 1 \end{bmatrix}.$$

Other information that can be derived from the process description is:

- The duration of stay in state 1 can be determined using Equation (7.2) as

$$T_1 \sim \text{geo}(0.9).$$

- When in state 1, the probability that the process is there for more than two days can be calculated using Equation (7.4) as

$$P[T_1 > 2] = 1 - P[T_1 \leq 2] = 1 - \left(1 - q_{11}^2\right) = q_{11}^2 = 0.81.$$

■

Exercise 7.1 Calculate other q_{ij}'s and show the results match the elements of $\boldsymbol{Q}$.

Exercise 7.2 Determine the probability of the process being in any state in the next two periods if the process now is known to be in state 1.

Next, a diagram for DTMC description is presented.

7.1.2 State Transition Diagram

Like CTMC, DTMC is often presented in a state transition diagram, with the weight of each arc representing the transition probability from a start (source) node to an end (target) node.

Example 7.2 The state transition diagram of DTMC in Example 7.1 is exhibited in Figure 7.1.

■

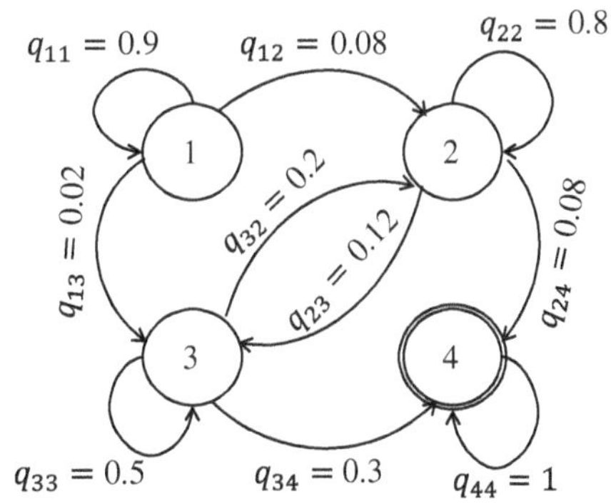

Figure 7.1 State transition diagrams of DTMC in Example 7.1.

Example 7.3 The transition probability out from state 2 can be calculated from Figure 7.1 to be $q_{23} + q_{24} = 0.12 + 0.08 = 0.2$. Another way to calculate this transition probability is using the complement of transition probability staying in state 2, namely, $1 - q_{22} = 1 - 0.8 = 0.2$, or $q_{22} = 1 - (q_{21} + q_{23} + q_{24}) = 1 - 0 - 0.12 - 0.08 = 0.2$. ■

Exercise 7.3 Prove that in any DTMC, $q_{ii} = 1 - \sum_{j \neq i} q_{ij}$.

The next section discusses the relationship between DTMC and DPH distribution.

7.2 Discrete Phase-Type Distribution

As PH distribution is related to CTMC, DPH distribution is related to DTMC, with one absorbing state such as DTMC in Example 7.1. To use DPH distribution for a process that has more than one absorbing state, all absorbing states must be lumped into one superstate.

7.2.1 DTMC with One Absorbing State

By arranging the state such that the absorbing state is the last and referring to it as the $(n+1)^{th}$ state, the infinitesimal generator matrix of a DTMC has a structure

$$\boldsymbol{Q} = \left[\begin{array}{cccc|c} q_{11} & q_{12} & \cdots & q_{1n} & q_{1,n+1} \\ q_{21} & q_{22} & \cdots & q_{2n} & q_{2,n+1} \\ \cdots & \cdots & \cdots & \cdots & - \\ q_{n1} & q_{n2} & \cdots & q_{nn} & q_{n,n+1} \\ \hline 0 & 0 & 0 & 0 & 1 \end{array}\right],$$

that can be represented in general as

$$\boldsymbol{Q} = \begin{bmatrix} \boldsymbol{A} & \boldsymbol{b} \\ \boldsymbol{0} & 1 \end{bmatrix}, \tag{7.15}$$

where $\boldsymbol{A}$ is the transient matrix, which is a substochastic matrix such that $\boldsymbol{I} - \boldsymbol{A}$ is nonsingular, representing the transition probabilities among transient states, and $\boldsymbol{b}$ is the absorbing vector representing the transition probabilities from all transient states to the absorbing state.

Using Equation (7.9)

$$\boldsymbol{Qu} = \begin{bmatrix} \boldsymbol{A} & \boldsymbol{b} \\ \boldsymbol{0} & 1 \end{bmatrix} \boldsymbol{u} = \begin{bmatrix} \boldsymbol{A} & \boldsymbol{b} \\ \boldsymbol{0} & 1 \end{bmatrix} \begin{bmatrix} \boldsymbol{u} \\ 1 \end{bmatrix} = \boldsymbol{u},$$

yields

$$\boldsymbol{Au} + \boldsymbol{b} = \boldsymbol{u}, \tag{7.16}$$

or

$$\boldsymbol{b} = (\boldsymbol{I} - \boldsymbol{A})\boldsymbol{u}. \tag{7.17}$$

Example 7.4 The infinitesimal generator of DTMC in Example 7.1 can be written as

$$\boldsymbol{Q} = \left[\begin{array}{ccc|c} 0.9 & 0.08 & 0.02 & 0 \\ 0 & 0.8 & 0.12 & 0.08 \\ 0 & 0.2 & 0.5 & 0.3 \\ \hline 0 & 0 & 0 & 1 \end{array}\right].$$

Therefore, the transient matrix and absorbing vector of this DTMC are

$$A = \begin{bmatrix} 0.9 & 0.08 & 0.02 \\ 0 & 0.8 & 0.12 \\ 0 & 0.2 & 0.5 \end{bmatrix}, \quad b = \begin{bmatrix} 0 \\ 0.08 \\ 0.3 \end{bmatrix}.$$

For Equation (7.17) to be satisfied, $(I - A)u$ must be equal to b.

$$(I - A)u = \left(\begin{bmatrix} 1 & 0 & 0 \\ 0 & 1 & 0 \\ 0 & 0 & 1 \end{bmatrix} - \begin{bmatrix} 0.9 & 0.08 & 0.02 \\ 0 & 0.8 & 0.12 \\ 0 & 0.2 & 0.5 \end{bmatrix} \right) \begin{bmatrix} 1 \\ 1 \\ 1 \end{bmatrix} = \begin{bmatrix} 0.1 & -0.08 & -0.02 \\ 0 & 0.2 & -0.12 \\ 0 & -0.2 & 0.5 \end{bmatrix} \begin{bmatrix} 1 \\ 1 \\ 1 \end{bmatrix} = \begin{bmatrix} 0 \\ 0.08 \\ 0.3 \end{bmatrix} = b,$$

as should be.

It can be verified that $(I - A)$ is nonsingular. ■

Based on the properties of matrix Q, matrix A is an $n \times n$ matrix having nonnegative elements with row sum less than or equals 1 representing the transitions from n transient states to n transient states. Transient matrices with these characteristics are always nonsingular (have an inverse) [14]. The absorbing vector b is $n \times 1$ vector having nonnegative elements less than or equal to one representing the transition probabilities from n transient states to one absorbing state. The row sum of matrix A is less than or equal to one because elements of vector b that are always nonnegative are not included as elements of A.

7.2.2 Distribution of Absorption Time

DPH distribution is defined as the distribution of absorption time in a DTMC with one absorbing state. It is generally assumed that the process does not start from the absorbing state. If T denotes the absorption time for a DTMC having infinitesimal generator Q as given by Equation (7.15), then the PMF of T is [14]

$$\begin{aligned} P_T(k) = P[T = k] &= 0 \qquad k = 0 \\ &= aA^{k-1}b \qquad k = 1, 2, \ldots, \end{aligned} \tag{7.18}$$

where a is the initial probability vector with each element representing the probability that the process starts from each transient state. It is the same as $\pi(0)$ in Equation (7.12) but excluding the probability starting at the absorbing state. Given $P[T = k] = 0$ for $k = 0$, Equation (7.18) implies that the process does not start from the absorbing state. Therefore, the elements of a sum to 1,

$$au = 1. \tag{7.19}$$

The CDF of T can be derived as:

$$\begin{aligned} F_T(k) &= P[T \le k] = 1 - P[T > k] = 1 - (P[T = k+1] + P[T = k+2] + \cdots) \\ &= 1 - a(A^k + A^{k+1} + \cdots)b = 1 - aA^k(I - A)^{-1}b. \end{aligned}$$

Equation (7.17) can be rewritten as $(I - A)^{-1}b = u$. Hence,

$$F_T(k) = 1 - aA^k u \qquad k = 0, 1, 2, \ldots. \tag{7.20}$$

The random variable T having the CDF in Equation (7.20) is said to have a DPH distribution (or be DPH-distributed) with representation (n, a, A) and written in short as $T \backsim DPH(n, a, A)$ or $T \backsim DPH_n(a, A)$.

Example 7.5 For DPH distribution of order 1, the initial probability vector $\boldsymbol{a}$ is a scalar with a value equal to 1. The transient matrix $\boldsymbol{A}$ and the absorbing vector $\boldsymbol{b}$ are also scalars. Let state 1 be the transient state. The transient matrix has only one element, which is the transition probability from state 1 to state 1; hence, $\boldsymbol{A} = [q_{11}]$. The absorbing vector also has only one element, which is the transition probability out of state 1; hence, $\boldsymbol{b} = [1 - q_{11}]$. The PMF for the absorption time is given by Equation (7.18) with $\boldsymbol{a} = [1]$, $\boldsymbol{A} = [q_{11}]$, $\boldsymbol{b} = [1 - q_{11}]$; hence,

$$P_T(k) = P[T = k] = \boldsymbol{a}\boldsymbol{A}^{k-1}\boldsymbol{b} = (1)(q_{11})^{k-1}(1 - q_{11}) \qquad k = 1, 2, \ldots,$$

which is the PMF of geometry distribution as given in Equation (7.3). ■

Example 7.5 shows that the geometry distribution is a DPH distribution of order 1. In other words, DPH distribution is a generalization of geometry distribution by replacing the scalar transition with a matrix transition.

DPH distribution can be expressed in a state transition diagram with two states, namely, a transient state to represent all transient states and an absorbing state to represent one absorbing state, like the state transition diagram of PH distribution, with an exception that in DPH distribution, there is a transition from the absorbing state to itself. Following its infinitesimal generator matrix $\boldsymbol{Q}$ given by Equation (7.15), the state transition diagram of DPH distribution is as exhibited in Figure 7.2.

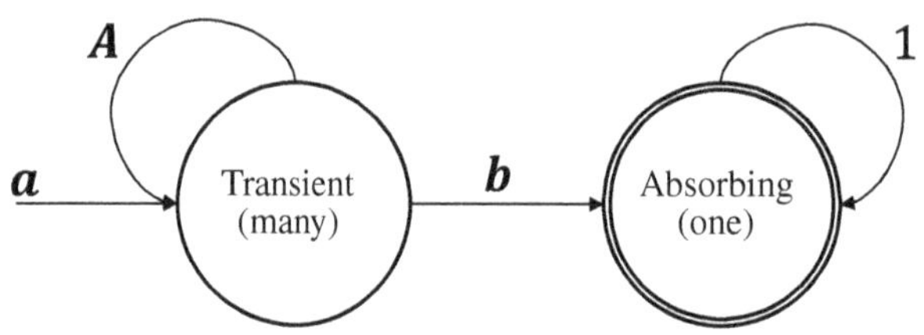

Figure 7.2 State transition diagram of DPH distribution.

Example 7.6 Suppose that the DTMC in Example 7.1 is started from state 1 with probability 0.6 and from state 2 with probability 0.4, which means the probability starting from state 3 is 0. The DPH distribution of the absorption time of DTMC in Example 7.1 has a DPH representation

$$n = 3, \quad \boldsymbol{a} = \begin{bmatrix} 0.6 & 0.4 & 0 \end{bmatrix}, \quad \boldsymbol{A} = \begin{bmatrix} 0.9 & 0.08 & 0.02 \\ 0 & 0.8 & 0.12 \\ 0 & 0.2 & 0.5 \end{bmatrix}.$$

Its state transition diagram using transition matrices is exhibited in Figure 7.3. ■

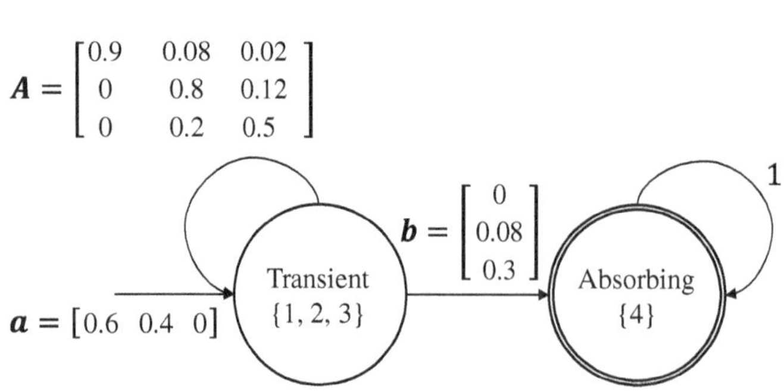

Figure 7.3 DPH distribution for the absorption time of DTMC in Example 7.4.

As for PH distribution, there are two types of DPH distribution: acyclic DPH and cyclic DPH with the same definition as for PH distribution.

Application 7.1 The condition of a radar sensor is inspected at the beginning of each month. The inspection results show that its condition can be classified into four states: Excellent, Decent, Fair, and Broken. The equipment is initially in the Excellent state. Equipment in the "Excellent" state at the beginning of a month will be in the "Decent" state at the next inspection with a probability of 0.04. Equipment in the "Decent" state at the beginning of a month will be in the "Fair" state with a probability of 0.01 and at the Broken state with a probability of 0.01 in the next inspection. Equipment in the "Fair" condition at the beginning of a month will be at "Broken" state at the next inspection with a probability of 0.02. Broken equipment will remain at the "Broken" state forever (until it is repaired). This description is summarized in Table 7.1.

Table 7.1 State and transition probability of Application 7.1.

		Transition probabilities			
State number	**State description**	**1**	**2**	**3**	**4**
1	Excellent	–	0.04	–	–
2	Decent	–	–	0.01	0.01
3	Fair	–	–	–	0.02
4	Broken	–	–	–	1.00

The process underlying the changes in the equipment states as described is a DTMC with three transient states (Excellent, Decent, and Fair) and one absorbing state (Broken). The DTMC can be represented in a state transition diagram in Figure 7.4. The probability of returning to each state i, $i = 1, 2, 3, 4$, is

$$q_{ii} = 1 - \sum_{j \neq i} q_{ij}.$$

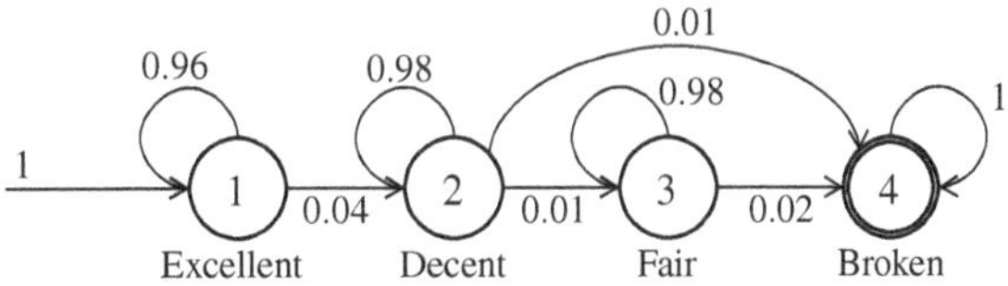

Figure 7.4 State transition diagram of the equipment states in Application 7.1.

Thus, $q_{11} = 1 - 0.04 = 0.96$, $q_{22} = 1 - (0.01 + 0.01) = 0.98$ and $q_{33} = 1 - 0.02 = 0.98$. Since the equipment is initially known to have excellent condition, the initial probability of starting from the Excellent state is 1; thus, the initial probability of starting from other states must be 0.

The infinitesimal generator matrix and initial probability vector of the DTMC underlying this equipment states can be obtained from Figure 7.4 as

$$\boldsymbol{Q} = \left[\begin{array}{ccc|c} 0.96 & 0.04 & 0 & 0 \\ 0 & 0.98 & 0.01 & 0.01 \\ 0 & 0 & 0.98 & 0.02 \\ \hline 0 & 0 & 0 & 1 \end{array}\right], \quad \boldsymbol{a} = [[1\ \ 0\ \ 0]\ \ 0].$$

The duration of use until the equipment fails is the same as the absorption time in the DTMC underlying the change in the equipment states. Thus, the duration of use until failure for this equipment has a DPH distribution with a representation

$$n = 3, \quad \boldsymbol{a} = \begin{bmatrix} 1 & 0 & 0 \end{bmatrix}, \quad \boldsymbol{A} = \begin{bmatrix} 0.96 & 0.04 & 0 \\ 0 & 0.98 & 0.01 \\ 0 & 0 & 0.98 \end{bmatrix}.$$

The state transition diagram or transient matrix provides other information as well, such as:

- The duration of stay at the "Excellent" state is geo(0.96).
- The expected duration of stay at the "Decent" state is $1/(1 - 0.98) = 50$ months.

The probability that the equipment does not last more than 20 months can be calculated using Equation (7.20) as

$$\begin{aligned} P[T \leq 20] &= F_T(20) = 1 - \boldsymbol{a}\boldsymbol{A}^k\boldsymbol{u}\,|_{k=20} \\ &= 1 - \begin{bmatrix} 1 & 0 & 0 \end{bmatrix} \begin{bmatrix} 0.96 & 0.04 & 0 \\ 0 & 0.98 & 0.01 \\ 0 & 0 & 0.98 \end{bmatrix}^{20} \begin{bmatrix} 1 \\ 1 \\ 1 \end{bmatrix} = 0.0599. \end{aligned}$$

Suppose for safety reasons management considers the "Fair" state as bad as the "Broken" state and must be repaired. This provision causes the DTMC underlying the equipment states to have two absorbing states, the "Fair" and "Broken" states. The state transition diagram for this equipment becomes as shown in Figure 7.5a. Combining the two absorbing states into one produces a

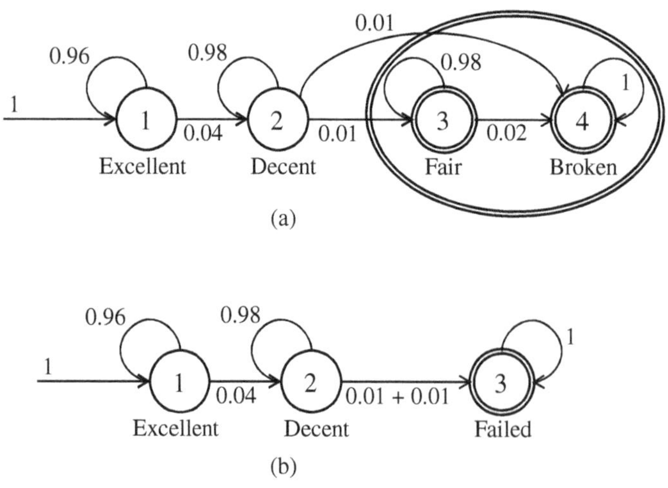

Figure 7.5 State transition diagram of the equipment in Application 7.1. (a) State transition diagram of the equipment in Application 7.1 with two absorbing states and (b) state transition diagram of the equipment in Application 7.1 with one absorbing state.

new state transition diagram given in Figure 7.5b. The transition probability from state 2 to state 3 in Figure 7.5b is the sum of the transition probabilities from state 2 to state 3 and to state 4 in Figure 7.5a.

The infinitesimal generator matrix of the DTMC underlying the equipment states becomes

$$Q = \left[\begin{array}{cc|c} 0.96 & 0.04 & 0 \\ 0 & 0.98 & 0.02 \\ \hline 0 & 0 & 1 \end{array}\right].$$

The distribution for duration of use until the equipment fails has a DPH representation

$$n = 2, \quad \boldsymbol{a} = \begin{bmatrix} 1 & 0 \end{bmatrix}, \quad \boldsymbol{A} = \begin{bmatrix} 0.96 & 0.04 \\ 0 & 0.98 \end{bmatrix}.$$

■

Exercise 7.4 Calculate the probability that the equipment is in "Excellent" state more than 10 months before being in the "Decent" state.

Methods for estimating the parameters of the DPH distribution and finding its minimum representation are given in Bobbio et al. [15].

Next, some popular discrete distributions that are special forms of DPH distribution are presented.

7.2.3 Some Popular DPH Distributions

DPH distribution is a class of distributions. Several popular discrete distributions are members of the DPH distribution class. Basically, all distributions formed from the geometry distribution are in the DPH distribution class. Table 7.2 provides widely used probability models as members of the DPH distribution class in various representations. In this table, the scalar representations are PMFs in scalar functions.

Table 7.2 shows similarity with Table 2.1 for PH distribution with the role of exponential distribution in PH distribution being replaced by geometry distribution in DPH distribution.

A generalization of DPH distribution is presented next.

7.2.4 Matrix-Geometric Distribution

MG distribution is a generalization of DPH distribution. A discrete random variable T has MG distribution if its PMF can be represented as [16]

$$\begin{aligned} P_T(k) = P[T = k] &= 0 && k = 0 \\ &= \boldsymbol{s}\boldsymbol{A}^{k-1}\boldsymbol{c} && k = 1, 2, \ldots, \end{aligned} \tag{7.21}$$

where $\boldsymbol{s}$ is a starting row vector, $\boldsymbol{c}$ is a closing column vector, and $\boldsymbol{A}$ is a square matrix. Unlike DPH distribution that has strict restrictions on the parameters $\boldsymbol{a}$, $\boldsymbol{A}$, and $\boldsymbol{u}$, there is no restriction on the parameters $\boldsymbol{s}$, $\boldsymbol{A}$, and $\boldsymbol{c}$ in MG distribution. The only requirements are Equation (7.21) must be a valid PMF and the moments must be finite.

The CDF of MG distribution can be derived as

$$\begin{aligned} F_T(k) &= P[T \le k] = 1 - P[T > k] = 1 - (P[T = k+1] + P[T = k+2] + \cdots) \\ &= 1 - \boldsymbol{s}(\boldsymbol{A}^k + \boldsymbol{A}^{k+1} + \cdots)\boldsymbol{c} = 1 - \boldsymbol{s}\boldsymbol{A}^k(\boldsymbol{I} - \boldsymbol{A})^{-1}\boldsymbol{c}. \end{aligned}$$

Table 7.2 Some popular members of DPH distribution class.

No.	Distribution	Representation	
1.	Geometry (p)	Diagram	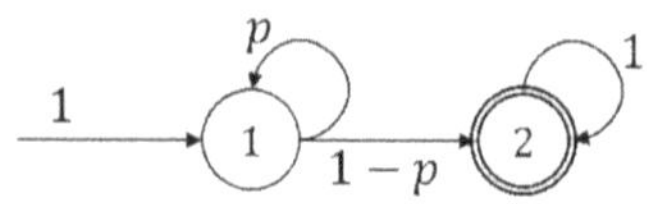
		Scalar	$P[T=k]=p^{k-1}(1-p)$
		Matrix	$n=1,\quad \boldsymbol{a}=[1],\quad \boldsymbol{A}=[p]$
		Mean	$1/(1-p)$
		Variance	$p/(1-p)^2$
2.	n-Mixed-geometry (a_i, p_i) $0<a_i<1$ $\sum_{i=1}^{n} a_i = 1$ (mixture of geometry)	Diagram	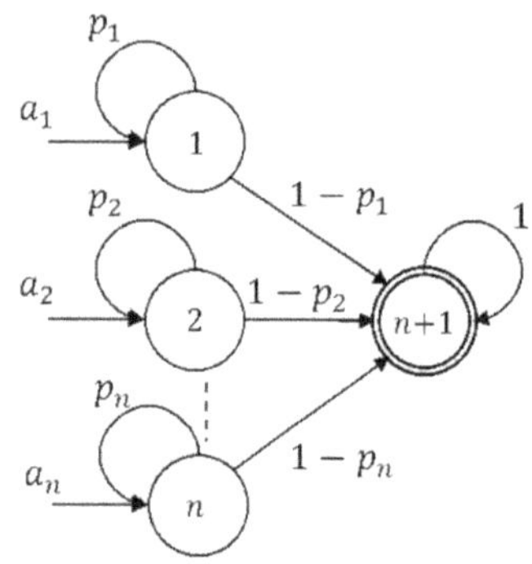
		Scalar	$P[T=k]=\sum_{i=1}^{n} a_i p_i^k (1-p_i)$
		Matrix	$n=n \quad \boldsymbol{a}=\begin{bmatrix} a_1 & a_2 \dots a_n \end{bmatrix} \quad \boldsymbol{A}=\begin{bmatrix} p_1 & 0 & \dots & 0 \\ 0 & p_2 & \dots & 0 \\ \vdots & \vdots & \ddots & \vdots \\ 0 & 0 & \dots & p_n \end{bmatrix}$
		Mean	$a_1/(1-p_1)+a_2/(1-p_2)+\cdots+a_n/(1-p_n)$
		Variance	Complicated
3.	n-Negative binomial (p) (sum of identical geometry)	Diagram	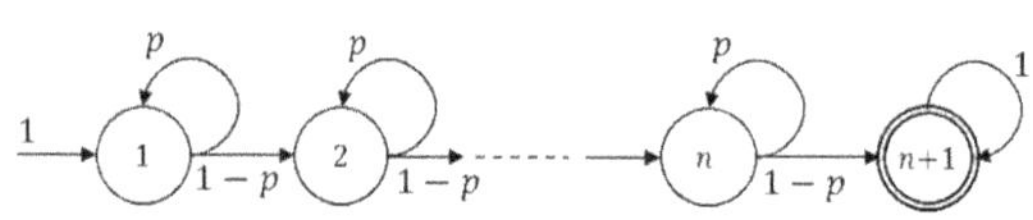
		Scalar	$P[T=k]=\sum_{r}^{n} \binom{k+r-1}{k} p^r (1-p)^{k-r}$
		Matrix	$n=n \quad \boldsymbol{a}=\begin{bmatrix} 1 & 0 \dots 0 \end{bmatrix} \quad \boldsymbol{A}=\begin{bmatrix} p & 1-p & \dots & 0 \\ 0 & p & \dots & 0 \\ \vdots & \vdots & \ddots & \vdots \\ 0 & 0 & \dots & p \end{bmatrix}$
		Mean	$n/(1-p)$
		Variance	$np/(1-p)^2$

Table 7.2 (Continued)

No.	Distribution	Representation	
4.	n-Hypogeometry (p_i) (sum of nonidentical geometry)	Diagram	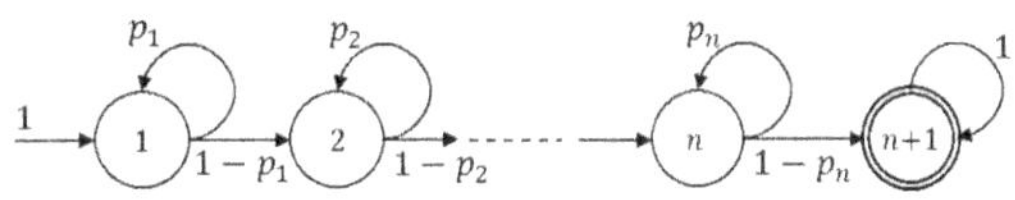
		Scalar	Complicated
		Matrix	$n = n, \quad \boldsymbol{a} = \begin{bmatrix} 1 & 0 \ldots 0 \end{bmatrix}$, $\boldsymbol{A} = \begin{bmatrix} p_1 & p_1 & \cdots & 0 & 0 \\ 0 & p_2 & \cdots & 0 & 0 \\ \vdots & \vdots & \ddots & \vdots & \vdots \\ 0 & 0 & \cdots & p_{n-1} & p_{n-1} \\ 0 & 0 & \cdots & 0 & p_n \end{bmatrix}$
		Mean	$1/(1-p_1)+1/(1-p_2)+\cdots+1/(1-p_n)$
		Variance	$p_1/(1-p_1)^2+p_2/(1-p_2)^2+\cdots+p_n/(1-p_n)^2$
5.	n-Deterministic	Diagram	
		Scalar	$P[T=k]=1$ for $k=n$, and 0 for other k
		Matrix	$n = n \cdot \boldsymbol{a} = \begin{bmatrix} 1 & 0 \ldots 0 \end{bmatrix}, \quad \boldsymbol{A} = \begin{bmatrix} 0 & 1 & \ldots & 0 & 0 \\ 0 & 0 & \ldots & 0 & 0 \\ \vdots & \vdots & \ddots & \vdots & \vdots \\ 0 & 0 & \ldots & 0 & 1 \\ 0 & 0 & \ldots & 0 & 0 \end{bmatrix}$
		Mean	n
		Variance	0

It can be written as

$$F_T(k) = 1 - \boldsymbol{a}\boldsymbol{A}^k\boldsymbol{v}, \qquad k = 0, 1, 2, \ldots, \tag{7.22}$$

where $\boldsymbol{a} = \boldsymbol{s}$ and $\boldsymbol{v} = (\boldsymbol{I} - \boldsymbol{A})^{-1}\boldsymbol{c}$.

Equation (7.22) shows that DPH and MG distributions have the same form of CDF, but DPH distribution has only two parameters $\boldsymbol{a}$ and $\boldsymbol{A}$, while MG distribution has three parameters, $\boldsymbol{a}$,$\boldsymbol{A}$, and $\boldsymbol{v}$. With this similarity, all equations for DPH distribution are applicable for MG distribution by replacing the unitary column vector $\boldsymbol{u}$ with a column vector $\boldsymbol{v}$.

A variable T with PMF as given by Equation (7.22) is said to have MG distribution, or MG-distributed, with representation (n,$\boldsymbol{a}$,$\boldsymbol{A}$,$\boldsymbol{v}$) and is denoted by $T \sim MG(n, \boldsymbol{a}, \boldsymbol{A}, \boldsymbol{v})$ or $T \sim MG_n(\boldsymbol{a}, \boldsymbol{A}, \boldsymbol{v})$.

As for PH distribution, DPH distribution can be transformed into MG distribution in a Jordan canonical form by using matrix similarity transformation.

$$F_T(t) = 1 - \boldsymbol{a}\boldsymbol{A}^k\boldsymbol{v} = 1 - \boldsymbol{a}\boldsymbol{P}\boldsymbol{P}^{-1}\boldsymbol{A}^k\boldsymbol{P}\boldsymbol{P}^{-1}\boldsymbol{v}$$
$$= 1 - \boldsymbol{a}\boldsymbol{P}(\boldsymbol{P}^{-1}\boldsymbol{A}\boldsymbol{P})^k\boldsymbol{P}^{-1}\boldsymbol{v}.$$

The resulting CDF can be written as

$$F_T(t) = 1 - \boldsymbol{a}'(\boldsymbol{A}')^k\boldsymbol{v}', \tag{7.23}$$

where

$$\boldsymbol{a}' = \boldsymbol{a}\boldsymbol{P}, \quad \boldsymbol{A}' = \boldsymbol{P}^{-1}\boldsymbol{A}\boldsymbol{P}, \quad \boldsymbol{v}' = \boldsymbol{P}^{-1}\boldsymbol{v}. \tag{7.24}$$

Example 7.7 The transient matrix of DPH distribution in Example 7.1 has eigenvalues

$$\lambda_1 = 0.9, \quad \lambda_2 = 0.4344, \quad \lambda_3 = 0.8656,$$

and eigenvectors

$$e_1 = \begin{bmatrix} 1 \\ 0 \\ 0 \end{bmatrix}, \quad e_2 = \begin{bmatrix} 0.0134 \\ -0.3282 \\ 1 \end{bmatrix}, \quad e_3 = \begin{bmatrix} -4.8384 \\ 1.8282 \\ 1 \end{bmatrix}.$$

Note that all eigenvalues are negatives less than 1 guaranteeing the CDF is convergent. The similarity transformation matrix constructed using the eigenvectors is

$$P = \begin{bmatrix} 1 & 0.0134 & -4.8384 \\ 0 & -0.3282 & 1.8282 \\ 0 & 1 & 1 \end{bmatrix}.$$

Use the transformation matrix to obtain

$$\boldsymbol{a}' = \boldsymbol{a}\boldsymbol{P} = [1 \;\; 0.0134 - 4.8384],$$

$$\boldsymbol{A}' = \boldsymbol{P}^{-1}\boldsymbol{A}\boldsymbol{P} = \begin{bmatrix} 0.9 & 0 & 0 \\ 0 & 0.4344 & 0 \\ 0 & 0 & 0.8656 \end{bmatrix}, \quad \boldsymbol{v}' = \boldsymbol{P}^{-1}\boldsymbol{v} = \begin{bmatrix} 3.975 \\ 0.3841 \\ 0.6159 \end{bmatrix}.$$

The resulting MG distribution in a Jordan canonical form has representation $(3, \boldsymbol{a}', \boldsymbol{A}', \boldsymbol{v}')$ as given above. ■

A method for finding a minimum representation and estimating the parameters of an MG distribution from measurement data is given in Akar [17].

The next section derives reliability measures for systems with DPH-distributed lifetimes. The results are also applicable to MG distribution by replacing the DPH parameters with MG parameters.

7.3 Reliability Measures in DPH Distribution

DTMC is closely related to a reliability model of a system in discrete time. Transient states can be used to represent good conditions that the system is considered still functioning, while the absorbing state can be used to represent failed conditions that the system is considered to not function as expected. With this analogy, the absorption time is equivalent to the time until the system fails, or the time to failure of the system. Consequently, the system that its conditions can be represented in DTMC has time to failure that follows DPH distribution. Acknowledging time to failure has DPH distribution opens the doors to derive all system reliability measures.

Reliability function in DPH distribution can be derived from Equation (7.20) as

$$R(k) = P[T > k] = 1 - P[T \leq k] = \boldsymbol{a}\boldsymbol{A}^k\boldsymbol{u} \quad k = 0, 1, 2, \dots. \tag{7.25}$$

There are many definitions of hazard function in discrete time; one widely used is [18]

$$h(k) = \frac{P_T(k)}{P[T \geq k]} = \frac{R(k-1) - R(k)}{R(k-1)}. \tag{7.26}$$

Substituting Equation (7.25) into Equation (7.26) yields

$$h(k) = 1 - (\boldsymbol{a}\boldsymbol{A}^k\boldsymbol{u})(\boldsymbol{a}\boldsymbol{A}^{k-1}\boldsymbol{u})^{-1} \qquad k = 1, 2, \dots.$$

Another equation for hazard function is [19]

$$h(k) = \ln\left(\frac{R(k-1)}{R(k)}\right), \quad k = 1, 2, \dots. \tag{7.27}$$

If the discrete-time model is a result of discretization of continuous-time model, hazard function calculated using Equation (7.27) produces a better approximation to the original continuous hazard function [9].

The residual reliability function of a system of age m (still good after being used for m periods) is

$$R(k+m|T > m) = \frac{P[(T > k+m) \wedge (T > m)]}{P[T > m]} = \frac{P[T > k+m]}{P[T > m]}.$$

Substituting Equation (7.25) obtains

$$R(k+m|T > m) = \frac{\boldsymbol{a}\boldsymbol{A}^{k+m}\boldsymbol{u}}{\boldsymbol{a}\boldsymbol{A}^m\boldsymbol{u}} = \left(\frac{\boldsymbol{a}\boldsymbol{A}^m}{\boldsymbol{a}\boldsymbol{A}^m\boldsymbol{u}}\right)\boldsymbol{A}^k\boldsymbol{u}.$$

Thus,

$$R(k+m|T > m) = \boldsymbol{c}(m)\boldsymbol{A}^k\boldsymbol{u}, \tag{7.28}$$

where $\boldsymbol{c}(m) = \boldsymbol{a}\boldsymbol{A}^m(\boldsymbol{a}\boldsymbol{A}^m\boldsymbol{u})^{-1}$.

Equation (7.28) implies that at any time the remaining lifetime of a system has a DPH distribution with the same survival matrix as when it was new (does not change over time), but with initial probability vector corrected by an age-dependent factor $\boldsymbol{A}^m(\boldsymbol{a}\boldsymbol{A}^m\boldsymbol{u})^{-1}$ from $\boldsymbol{a}$ to $\boldsymbol{c}(m) = \boldsymbol{a}\boldsymbol{A}^m(\boldsymbol{a}\boldsymbol{A}^m\boldsymbol{u})^{-1}$.

The $MTTF$ is calculated as the expected value of lifetime,

$$E[T] = \sum_{k=1}^{\infty} kP[T = k] = \sum_{k=1}^{\infty} k\boldsymbol{a}\boldsymbol{A}^{k-1}\boldsymbol{b} = \boldsymbol{a}\left(\sum_{k=1}^{\infty} k\boldsymbol{A}^{k-1}\right)\boldsymbol{b} = \boldsymbol{a}\frac{d}{d\boldsymbol{A}}\left(\sum_{k=1}^{\infty} \boldsymbol{A}^k\right)\boldsymbol{b} = \boldsymbol{a}(\boldsymbol{I} - \boldsymbol{A})^{-2}\boldsymbol{b}.$$

Using Equation (7.17) obtains

$$MTTF = \boldsymbol{a}(\boldsymbol{I} - \boldsymbol{A})^{-1}\boldsymbol{u}. \tag{7.29}$$

Equation (7.29) can also be derived using the relationship between mean value and reliability function as for PH distribution with integration is replaced with summation,

$$E[T] = \sum_{k=0}^{\infty} R(k) = \sum_{k=0}^{\infty} \boldsymbol{a}\boldsymbol{A}^k\boldsymbol{u} = \boldsymbol{a}(\boldsymbol{I} + \boldsymbol{A} + \boldsymbol{A}^2 + \dots)\boldsymbol{u} = \boldsymbol{a}(\boldsymbol{I} - \boldsymbol{A})^{-1}\boldsymbol{u}.$$

Equations for moments are easier to find using probability generating function (PGF) of PMF, defined as

$$\Phi_T(z) = E[z^T] = \sum_{k=1}^{\infty} z^k P[T = k] = \boldsymbol{a} z \left(\sum_{k=1}^{\infty} (z\boldsymbol{A})^{k-1} \right) \boldsymbol{b}.$$

For $T \backsim DPH(n, \boldsymbol{a}, \boldsymbol{A})$, the PGF is

$$\Phi_T(z) = \boldsymbol{a} z \left(\sum_{k=1}^{\infty} (z\boldsymbol{A})^{k-1} \right) \boldsymbol{b}.$$

Using identity $\sum_{k=1}^{\infty} (z\boldsymbol{A})^{k-1} = (\boldsymbol{I} - z\boldsymbol{A})^{-1}$ results in

$$\Phi_T(z) = \boldsymbol{a}(z^{-1}\boldsymbol{I} - \boldsymbol{A})^{-1}\boldsymbol{b}. \tag{7.30}$$

The first moment can be derived using PGF as

$$E[T] = \sum_{k=1}^{\infty} kP[T = k] = \sum_{k=1}^{\infty} kz^{k-1}P[T = k] |_{z=1} = E[Tz^{T-1}] |_{z=1} = \frac{d}{dz}\Phi_T(z) |_{z=1}.$$

Using Equation (7.30) and (7.17) obtains

$$E[T] = \frac{d}{dz}\boldsymbol{a}(z^{-1}\boldsymbol{I} - \boldsymbol{A})^{-1}\boldsymbol{b} |_{z=1} = \boldsymbol{a}(-(z^{-1}\boldsymbol{I} - \boldsymbol{A})^{-2}(-z^{-2})\boldsymbol{b} |_{z=1} = \boldsymbol{a}(\boldsymbol{I} - \boldsymbol{A})^{-1}\boldsymbol{u},$$

as given in Equation (7.29).

The kth factorial moment of T can be found as the kth derivative of $\Phi_T(z)$,

$$E[T(T-1)\dots(T-k+1)] = \frac{d^k}{dz^k}\Phi_T(z) |_{z=1} = k!\boldsymbol{a}(\boldsymbol{I} - \boldsymbol{A})^{-k} \boldsymbol{A}^{k-1}\boldsymbol{u}, \quad k = 1, 2, \dots. \tag{7.31}$$

For $k = 2$ results in

$$E[T(T-1)] = E[T^2] - E[T] = 2!\boldsymbol{a}(\boldsymbol{I} - \boldsymbol{A})^{-2} \boldsymbol{A}\boldsymbol{u}.$$

VTTF can be derived as

$$VTTF = E[T^2] - E[T]^2 = E[T(T-1)] + E[T] - E[T]^2 = 2!\boldsymbol{a}(\boldsymbol{I} - \boldsymbol{A})^{-2}\boldsymbol{A}\boldsymbol{u} + MTTF - MTTF^2.$$

Substituting Equation (7.29) yields

$$VTTF = 2\boldsymbol{a}(\boldsymbol{I} - \boldsymbol{A})^{-2} \boldsymbol{A}\boldsymbol{u} + \boldsymbol{a}(\boldsymbol{I} - \boldsymbol{A})^{-1}\boldsymbol{u} - (\boldsymbol{a}(\boldsymbol{I} - \boldsymbol{A})^{-1}\boldsymbol{u})^2.$$

For numerical efficiency, matrix inversions in calculating *MTTF* and *VTTF* are avoided by defining $\boldsymbol{c} = \boldsymbol{a}(\boldsymbol{I} - \boldsymbol{A})^{-1}$ and $\boldsymbol{d} = (\boldsymbol{I} - \boldsymbol{A})^{-1} \boldsymbol{A}\boldsymbol{u}$ so that $\boldsymbol{c}(\boldsymbol{I} - \boldsymbol{A}) = \boldsymbol{a}$ and $(\boldsymbol{I} - \boldsymbol{A})\boldsymbol{d} = \boldsymbol{A}\boldsymbol{u}$. Thus,

$$MTTF = \boldsymbol{cu}, VTTF = \boldsymbol{2cd} + \boldsymbol{cu} + (\boldsymbol{cu})^2. \tag{7.32}$$

where $\boldsymbol{c}$ and $\boldsymbol{d}$ are the solutions of systems of linear equations $\boldsymbol{c}(\boldsymbol{I} - \boldsymbol{A}) = \boldsymbol{a}$ and $(\boldsymbol{I} - \boldsymbol{A})\boldsymbol{d} = \boldsymbol{A}\boldsymbol{u}$, respectively.

SDTTF and *CVTTF* are calculated as $\sqrt{(VTTF)}$ and $SDTTF/MTTF$, respectively.

MRL function can be derived by using Equation (7.28) and the relationship between mean value and reliability function. Follow the steps for deriving mean value using reliability function to result in

$$MRL(m) = \sum_{k=0}^{\infty} R(k + m|T > m) = \boldsymbol{c}(m)(\boldsymbol{I} - \boldsymbol{A})^{-1}\boldsymbol{u}, \tag{7.33}$$

where $\boldsymbol{c}(m) = \boldsymbol{a}\boldsymbol{A}^m(\boldsymbol{a}\boldsymbol{A}^m\boldsymbol{u})^{-1}$.

Example 7.8 Reliability measures of a system with lifetime following geometry distribution with $MTTF = 5$ days can be calculated as follows.

Geometry distribution is a DPH distribution of order 1, which means $\boldsymbol{a} = \boldsymbol{u} = [1]$. Substituting them and the value of $MTTF$ into Equation (7.29) yields

$$[1](\boldsymbol{I} - \boldsymbol{A})^{-1}[1] = 5.$$

The survival matrix $\boldsymbol{A}$ of this system can be calculated as

$$\boldsymbol{A} = 1 - 1/5 = 0.8.$$

Thus, the system lifetime has DPH distribution with representation $(n, \boldsymbol{a}, \boldsymbol{A}) = (1, [1], [0.8])$. The system reliability measures can be derived as follows:

- The system reliability function is given by Equation (7.25) as

$$R(k) = \boldsymbol{a}\boldsymbol{A}^k\boldsymbol{u} = [1]0.8^k[1] = 0.8^k, \quad k = 0, 1, 2, \ldots.$$

- The system hazard function is given by Equation (7.26) as

$$h(k) = 1 - (\boldsymbol{a}\boldsymbol{A}^k\boldsymbol{u})(\boldsymbol{a}\boldsymbol{A}^{k-1}\boldsymbol{u})^{-1} = 1 - (0.8^k)(0.8^{k-1})^{-1} = 0.2, \quad k = 1, 2, \ldots.$$

The constant value hazard function shows that the system is CFR.

- The residual reliability function of age = 5 days is given by Equation (7.28) as

$$R(k + 5|T > 5) = \boldsymbol{a}\boldsymbol{A}^m(\boldsymbol{a}\boldsymbol{A}^m\boldsymbol{u})^{-1}\boldsymbol{A}^k\boldsymbol{u}\,|_{m=5}$$
$$= (1)(0.8^5(0.8^5)^{-1})0.8^k(1) = 0.8^k, \quad k = 1, 2, \ldots.$$

The residual reliability function at any age is equal to the reliability function due to the system is CFR.

- The MRL function is given by Equation (7.33) as

$$MRL(m) = \boldsymbol{a}\boldsymbol{A}^m(\boldsymbol{a}\boldsymbol{A}^m\boldsymbol{u})^{-1}(\boldsymbol{I} - \boldsymbol{A})^{-1}\boldsymbol{u} = (1)0.8^m(0.8^m)^{-1}(1 - 0.8)^{-1}(1) = 5 \text{ days}.$$

The residual life expectancy at any age is constant and equal to the expected lifetime is due to the fact that the system is CFR.

- The $VTTF$ is given by Equations (7.32) and (7.29) as

$$VTTF = 2\boldsymbol{a}(\boldsymbol{I} - \boldsymbol{A})^{-2}\,\boldsymbol{A}\boldsymbol{u} + MTTF - MTTF^2 = 2[1][0.2]^{-2}[0.8][1] + 5 - 25 = 20 \text{ days}^2,$$

the same as calculated using equation for variance of geometry distribution in Table 7.2. ■

As for PH distribution, calculating reliability measures in DPH distribution needs computer support. The MATLAB functions to generate reliability measures in DPH distribution are available in the software package **dysrel-am**. They are **dphrelplot**, **dphhazplot**, **dphresrelplot**, and **dphmrlplot** for plotting reliability, hazard, residual reliability, and MRL functions, respectively, and **dphmttf**, **dphvttf**, and **dphsdttf** for calculating $MTTF$, $VTTF$, and $SDTTF$, respectively.

Example 7.9 Reliability measures of systems in Application 7.1 are calculated using the provided MATLAB functions. The main program to receive the input and call the necessary functions is available in **example7_9.m** file in the software package **dysrel-am**. Its source code is as follows:

MATLAB code for example7_9.m

```
clear; close all; addpath('..\dysrel-am')
% DPH representation                                    input
T.a = [1 0 0];
T.A = [0.96 0.04 0;
    0 0.98 0.01;
    0 0 0.98];
T.n = 3;
% Plots                                                 graphical output
age = 5;
k = 0:30:300;
dphrelplot(k,T)
dphhazplot(k,T)
dphresrelplot(k,T,age)
dphmrlplot(k,T)
% Statistic                                             statistical output
mttf = dphmttf(T)
vttf = dphvttf(T)
sdttf = dphsdttf(T)
```

Executing this program results in Figure 7.6. ∎

Exercise 7.5 Compare the reliability and residual reliability functions of ages 10 and 20 months in one figure for the system in Application 7.1 and use the graphs to conclude about the system reliability behavior.

Example 7.10 Reliability measures of DPH distributions in Table 7.2 are calculated and graphed using the provided MATLAB functions. All distributions are chosen to have $MTTF = 10$. They are compared in Figure 7.7 and Table 7.3.

- Geometry distribution: $n = 1, \boldsymbol{a} = \begin{bmatrix} 1 \end{bmatrix}, \boldsymbol{A} = \begin{bmatrix} 0.9 \end{bmatrix}$.
- 2-NB distribution: $n = 2, \boldsymbol{a} = \begin{bmatrix} 1 & 0 \end{bmatrix}, \boldsymbol{A} = \begin{bmatrix} 0.8 & 0.2 \\ 0 & 0.8 \end{bmatrix}$.
- 2-Mixed-geometry distribution: $n = 2, \boldsymbol{a} = \begin{bmatrix} 0.5 & 0.5 \end{bmatrix}, \boldsymbol{A} = \begin{bmatrix} 0.75 & 0 \\ 0 & 15/16 \end{bmatrix}$.
- 2-Hypogeometry distribution: $n = 2, \boldsymbol{a} = \begin{bmatrix} 1 & 0 \end{bmatrix}, \boldsymbol{A} = \begin{bmatrix} 0.5 & 0.5 \\ 0 & 0.875 \end{bmatrix}$.
- 10-Deterministic: $n = 10, \boldsymbol{a} = \begin{bmatrix} 1 & 0 & \dots & 0 \end{bmatrix}, \boldsymbol{A} = \begin{bmatrix} 0 & 1 & 0 & \dots & 0 \\ 0 & 0 & 1 & \dots & 0 \\ 0 & 0 & 0 & \dots & 0 \\ \dots & \dots & \dots & \dots & 1 \\ 0 & 0 & 0 & \dots & 0 \end{bmatrix}$.

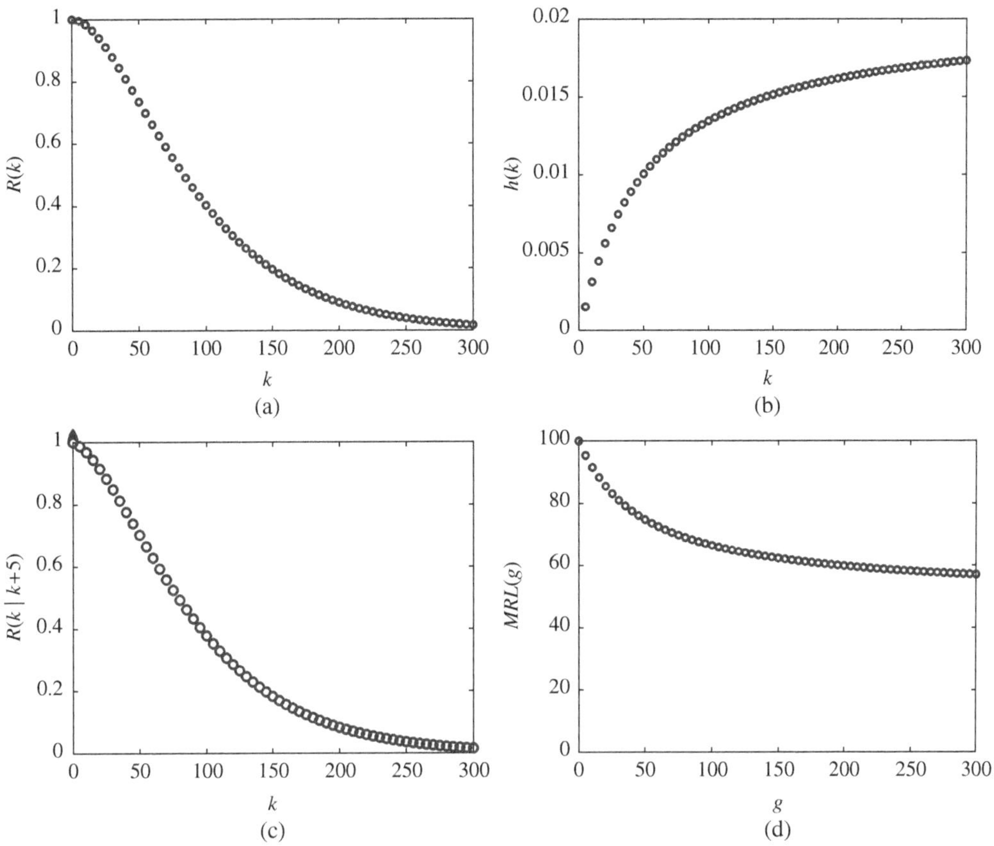


```
≫ example7_9
mttf =
  100.0000
vttf =
   4.9000e+03
sdttf =
   70.0000
```

Figure 7.6 Outputs of **example7_9.m**. (a) Reliability function, (b) hazard function, (c) residual reliability function (age = 5), and (d) *MRL* function.

Table 7.3 Statistical reliability measures of distributions in Example 7.10.

Distribution	*MTTF*	*SDTTF*
Geometry	10	9.49
2-NB	10	6.32
2-Mixed-geometry	10	12.73
2-Hypogeometry	10	7.62
10-Deterministic	10	0

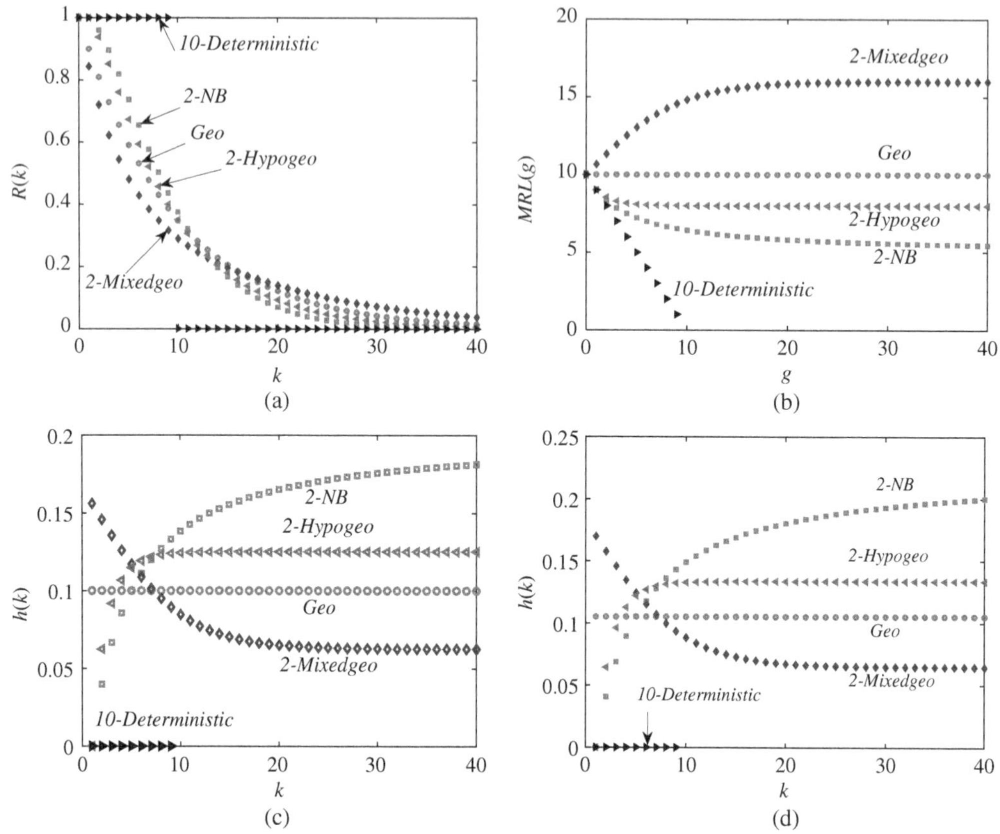

Figure 7.7 Functional reliability measures of distributions in Example 7.10. (a) Reliability function, (b) *MRL* function, (c) hazard function using Equation (7.26), (d) hazard function using Equation (7.27).

The graphs in Figure 7.7 show that geometry distribution has constant hazard function, 2-NB distribution has increasing hazard function, and mixed-geometry distribution has decreasing hazard function. Thus, the geometry distribution represents the lifetime distribution of CFR systems, negative binomial (NB) distributions represent IFR systems, and mixed-geometry distributions represent DFR systems. Unlike PH distribution, CFR distribution does not imply $CVTTF = 1$, and vice versa. ■

Reliability measures in MG distribution can utilize the equations in DPH distribution by replacing DPH representation with MG representation. Therefore, the functions for MG distribution are similar to DPH distribution, but with additional input the value of closing vector $\boldsymbol{v}$ that is not needed in DPH distribution. They are **mgrelplot, mghazplot, mgresrelplot, mgmrlplot, mgmttf, mgvttf,** and **mgsdttf**.

Example 7.11 Reliability measures for system in Application 7.1 are calculated using its similar MG distribution to the original DPH distribution derived in Example 7.7. The results are shown in Figure 7.8. Comparison with Figure 7.6 shows that they are indeed identical as they should.

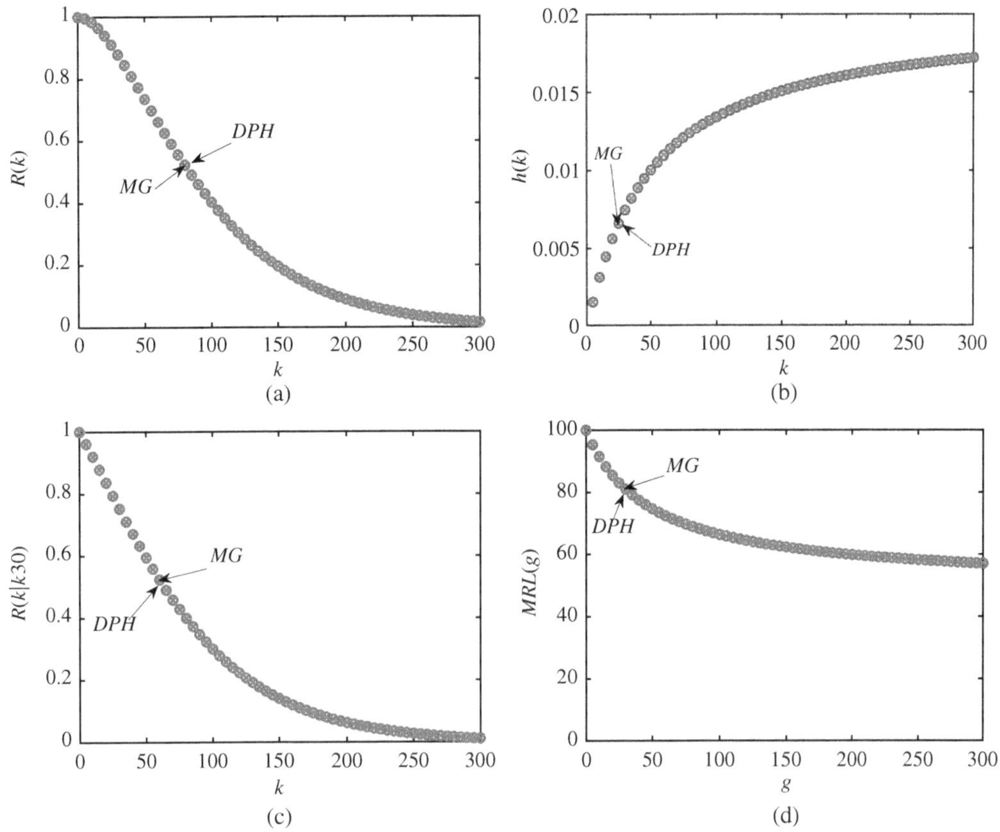

Figure 7.8 Reliability measures of the system in Application 7.1 using DPH and its similar MG distribution. (a) Reliability function, (b) hazard function, (c) residual reliability function (age = 30), and (d) *MRL* function.

```
≫ example7_11
mttf_dph =
  100.0000
vttf_dph =
   4.9000e+03
sdttf_dph =
   70.0000
mttf_mg =
  100.0000
vttf_mg =
   4.9000e+03
sdttf_mg =
   70.0000
```

■

The use of DPH or MG distributions in discrete-time system reliability modeling and analysis relies on the important properties of these distributions as presented next.

7.4 Closure Properties of DPH and MG Distributions

Sections 2.6 and 2.7 describe the closure properties of PH and ME distributions in the state space model of continuous-time systems. Section 6.4 describes the closure properties of ME distribution in the combinatorial model of continuous-time systems. The same properties hold for DPH and MG distributions for discrete-time systems. While the closure properties of PH and ME distributions rely on Kronecker product and Kronecker sum, the closure properties of DPH and MG distributions rely only on Kronecker product. Their properties in state transition and combinatorial models are discussed next.

7.4.1 Closure Properties in State Transition Model

Random variables T_1 and T_2 having independent and not necessarily identical DPH distributions with representations $(n_1, \boldsymbol{a}_1, \boldsymbol{A}_1)$ and $(n_2, \boldsymbol{a}_2, \boldsymbol{A}_2)$ satisfy properties given by the following theorems [9, 20]. The proofs for these theorems follow the same logic as proving the similar theorems in continuous time.

Theorem 7.1 $T_S = \min(T_1, T_2) \sim DPH(n_S, \boldsymbol{a}_S, \boldsymbol{A}_S)$ with

$$n_S = n_1 n_2, \qquad \boldsymbol{a}_S = \boldsymbol{a}_1 \otimes \boldsymbol{a}_2, \qquad \boldsymbol{A}_S = \boldsymbol{A}_1 \otimes \boldsymbol{A}_2. \tag{7.34}$$

The closure property for minimum operation in DPH distribution is different than in PH distribution, namely, in DPH distribution $\boldsymbol{A}_S = \boldsymbol{A}_1 \otimes \boldsymbol{A}_2$, while in PH distribution is $\boldsymbol{A}_S = \boldsymbol{A}_1 \oplus \boldsymbol{A}_2$.

Theorem 7.2 $T_S = \text{mix}(T_1, T_2) \sim DPH(n_S, \boldsymbol{a}_S, \boldsymbol{A}_S)$ with

$$n_S = n_1 + n_2, \quad \boldsymbol{a}_S = \begin{bmatrix} p_1\boldsymbol{a}_1 & p_2\boldsymbol{a}_2 \end{bmatrix}, \quad \boldsymbol{A}_S = \begin{bmatrix} \boldsymbol{A}_1 & \boldsymbol{0} \\ \boldsymbol{0} & \boldsymbol{A}_2 \end{bmatrix}, \tag{7.35}$$

the same as for PH distribution.

Theorem 7.3 $T_S = \text{add}(T_1, T_2) = T_1 + T_2 \sim DPH(n_S, \boldsymbol{a}_S, \boldsymbol{A}_S, \boldsymbol{v}_S)$ with

$$n_S = n_1 + n_2, \quad \boldsymbol{a}_S = \begin{bmatrix} \boldsymbol{a}_1 & \boldsymbol{0} \end{bmatrix}, \quad \boldsymbol{A}_S = \begin{bmatrix} \boldsymbol{A}_1 & \boldsymbol{b}_1\boldsymbol{a}_2 \\ \boldsymbol{0} & \boldsymbol{A}_2 \end{bmatrix}, \tag{7.36}$$

the same as for PH distribution.

Theorem 7.4 $T_S = \max(T_1, T_2) \sim DPH(n_S, \boldsymbol{a}_S, \boldsymbol{A}_S)$ with

$$n_S = n_1 n_2 + n_1 + n_2, \qquad \boldsymbol{a}_S = \begin{bmatrix} \boldsymbol{a}_1 \otimes \boldsymbol{a}_2 & \boldsymbol{0} & \boldsymbol{0} \end{bmatrix},$$

$$\boldsymbol{A}_S = \begin{bmatrix} \boldsymbol{A}_1 \otimes \boldsymbol{A}_2 & \boldsymbol{A}_1 \otimes \boldsymbol{b}_2 & \boldsymbol{b}_1 \otimes \boldsymbol{A}_2 \\ \boldsymbol{0} & \boldsymbol{A}_1 & \boldsymbol{0} \\ \boldsymbol{0} & \boldsymbol{0} & \boldsymbol{A}_2 \end{bmatrix}. \tag{7.37}$$

The closure property for maximum operation in DPH distribution is also different than in PH distribution, namely, block (1,1) of survival matrix in DPH distribution is $\boldsymbol{A}_1 \otimes \boldsymbol{A}_2$ while in PH distribution it is $\boldsymbol{A}_1 \oplus \boldsymbol{A}_2$, and the transition matrices for surviving components in block (1,2) and (1,3) are the survival matrices of component 1 and component 2, $\boldsymbol{A}_1$ and $\boldsymbol{A}_2$, respectively, while in PH distribution, they are identity matrices, $\boldsymbol{I}_1$ and $\boldsymbol{I}_2$, respectively.

Example 7.12 Random variables T_1 and T_2 have independent and identical geometry distribution with parameter p.

The DPH representations of T_1 and T_2 are

$$n_1 = n_2 = 1, \quad \boldsymbol{a}_1 = \boldsymbol{a}_2 = \left[1\right], \quad \boldsymbol{A}_1 = \boldsymbol{A}_2 = \left[p\right].$$

The DPH representations $(n_X, \boldsymbol{a}_X, \boldsymbol{A}_X)$ of $X = \min(T_1, T_2)$ can be obtained using Theorem 7.1 as follows:

$$n_X = n_1 n_2 = 1, \quad \boldsymbol{a}_X = \boldsymbol{a}_1 \otimes \boldsymbol{a}_2 = \left[1\right] \otimes \left[1\right] = \left[1\right],$$

$$\boldsymbol{A}_X = \boldsymbol{A}_1 \otimes \boldsymbol{A}_2 = \left[p\right] \otimes [p] = [p^2].$$

Thus, $X = \min(T_1, T_2)$ has geometry distribution with DPH representation $\left(1, [1], [p^2]\right)$. This result can be extended to more than two variables, and to random variables with nonidentical geometry distributions.

The DPH representations $(n_Y, \boldsymbol{a}_Y, \boldsymbol{A}_Y)$ of $Y = \text{add}(T_1, T_2)$ can be obtained using Theorem 7.3 as follows:

$$n_Y = n_1 + n_2 = 2, \quad \boldsymbol{a}_S = \left[\boldsymbol{a}_1 \ \ \boldsymbol{0}\right] = \left[1 \ \ 0\right],$$

$$\boldsymbol{b}_1 = (\boldsymbol{I} - \boldsymbol{A}_1)\boldsymbol{u} = (1 - p)(1) = 1 - p,$$

$$\boldsymbol{A}_Y = \begin{bmatrix} \boldsymbol{A}_1 & \boldsymbol{b}_1\boldsymbol{a}_2 \\ \boldsymbol{0} & \boldsymbol{A}_2 \end{bmatrix} = \begin{bmatrix} p & (1-p)(1) \\ 0 & p \end{bmatrix} = \begin{bmatrix} p & 1-p \\ 0 & p \end{bmatrix}.$$

Thus, $Y = \text{add}(T_1, T_2)$ has 2-NB distribution with DPH representation $\left(2, \left[1 \ \ 0\right], \begin{bmatrix} p & 1-p \\ 0 & p \end{bmatrix}\right)$. This result can also be extended to any number of random variables with identical and independent geometry distribution. ■

Exercise 7.6 Redo Example 7.12 for nonidentical geometry random variables and determine the resulting distributions based on Table 7.2.

Example 7.13 Random variables T_1 and T_2 have independent and nonidentical DPH distributions with representations $(n_1, \boldsymbol{a}_1, \boldsymbol{A}_1)$ and $(n_2, \boldsymbol{a}_2, \boldsymbol{A}_2)$, where

$$n_1 = n_2 = 2, \quad \boldsymbol{a}_1 = \boldsymbol{a}_2 = \left[1 \ \ 0\right], \quad \boldsymbol{A}_1 = \begin{bmatrix} 0.2 & 0.8 \\ 0 & 0.4 \end{bmatrix}, \quad \boldsymbol{A}_2 = \begin{bmatrix} 0.5 & 0.5 \\ 0 & 0.8 \end{bmatrix}.$$

The types of distributions for T_1 and T_2 can be deduced from their DPH representations and using Table 7.2 as 2-hypogeometry (0.2, 0.4) and 2-hypogeometry (0.5,0.8), respectively.

DPH representation of $T_3 = \min(T_1, T_2)$ calculated using Equation (7.34) is

$$n_3 = n_1 n_2 = 4, \quad \boldsymbol{a}_3 = \boldsymbol{a}_1 \otimes \boldsymbol{a}_2 = \left[1 \ \ 0\right] \otimes \left[1 \ \ 0\right] = \left[1 \ \ 0 \ \ 0 \ \ 0\right],$$

$$\boldsymbol{A}_3 = \boldsymbol{A}_1 \otimes \boldsymbol{A}_2 = \begin{bmatrix} 0.2 & 0.8 \\ 0 & 0.4 \end{bmatrix} \otimes \begin{bmatrix} 0.5 & 0.5 \\ 0 & 0.8 \end{bmatrix} = \begin{bmatrix} 0.1 & 0.1 & 0.4 & 0.4 \\ 0 & 0.16 & 0 & 0.64 \\ 0 & 0 & 0.2 & 0.2 \\ 0 & 0 & 0 & 0.32 \end{bmatrix}.$$

DPH representation of $T_4 = \text{mix}(T_1, T_2)$ with mixing probabilities 0.4 and 0.6 calculated using Equation (7.35) is

$$n_4 = n_1 + n_2 = 4,$$

$$\boldsymbol{a}_4 = [p_1\boldsymbol{a}_1\ p_2\boldsymbol{a}_2] = [0.4\,[1\ 0]\ 0.6\,[1\ 0]] = [0.4\ 0\ 0.6\ 0],$$

$$\boldsymbol{A}_4 = \begin{bmatrix} \begin{bmatrix} 0.2 & 0.8 \\ 0 & 0.4 \end{bmatrix} & \boldsymbol{0} \\ \boldsymbol{0} & \begin{bmatrix} 0.5 & 0.5 \\ 0 & 0.8 \end{bmatrix} \end{bmatrix} = \begin{bmatrix} 0.2 & 0.8 & 0 & 0 \\ 0 & 0.4 & 0 & 0 \\ 0 & 0 & 0.5 & 0.5 \\ 0 & 0 & 0 & 0.8 \end{bmatrix}.$$

DPH representation of $T_5 = \text{add}(T_1, T_2)$ calculated using Equation (7.36) is

$$n_5 = n_1 + n_2 = 4, \quad \boldsymbol{a}_5 = [\boldsymbol{a}_1\ \boldsymbol{0}] = [[1\ 0]\ [0\ 0]] = [1\ 0\ 0\ 0],$$

$$\boldsymbol{b}_1 = (\boldsymbol{I} - \boldsymbol{A}_1)\boldsymbol{u} = \left(\begin{bmatrix} 1 & 0 \\ 0 & 1 \end{bmatrix} - \begin{bmatrix} 0.2 & 0.8 \\ 0 & 0.4 \end{bmatrix} \right) \boldsymbol{u} = \begin{bmatrix} -0.8 & 0.8 \\ 0 & 0.6 \end{bmatrix} \begin{bmatrix} 1 \\ 1 \end{bmatrix} = \begin{bmatrix} 0 \\ 0.6 \end{bmatrix},$$

$$\boldsymbol{A}_5 = \begin{bmatrix} \boldsymbol{A}_1 & \boldsymbol{b}_1\boldsymbol{a}_2 \\ \boldsymbol{0} & \boldsymbol{A}_2 \end{bmatrix} = \begin{bmatrix} \begin{bmatrix} 0.2 & 0.8 \\ 0 & 0.4 \end{bmatrix} & \begin{bmatrix} 0 \\ 0.6 \end{bmatrix} [1\ 0] \\ \boldsymbol{0} & \begin{bmatrix} 0.5 & 0.5 \\ 0 & 0.8 \end{bmatrix} \end{bmatrix} = \begin{bmatrix} 0.2 & 0.8 & 0 & 0 \\ 0 & 0.4 & 0.6 & 0 \\ 0 & 0 & 0.5 & 0.5 \\ 0 & 0 & 0 & 0.8 \end{bmatrix}.$$

DPH representation of $T_6 = \max(T_1, T_2)$ calculated using Equation (7.37) is

$$n_6 = n_1 n_2 + n_1 + n_2 = 8,$$

$$\boldsymbol{a}_6 = [[1\ 0] \otimes [1\ 0]\ \boldsymbol{0}\ \boldsymbol{0}] = [1\ 0\ 0\ 0\ 0\ 0\ 0\ 0],$$

$$\boldsymbol{b}_2 = (\boldsymbol{I} - \boldsymbol{A}_2)\boldsymbol{u} = \left(\begin{bmatrix} 1 & 0 \\ 0 & 1 \end{bmatrix} - \begin{bmatrix} 0.5 & 0.5 \\ 0 & 0.8 \end{bmatrix} \right) \boldsymbol{u} = \begin{bmatrix} 0.5 & -0.5 \\ 0 & 0.2 \end{bmatrix} \begin{bmatrix} 1 \\ 1 \end{bmatrix} = \begin{bmatrix} 0 \\ 0.2 \end{bmatrix},$$

$$\boldsymbol{A}_6 = \begin{bmatrix} \begin{bmatrix} 0.2 & 0.8 \\ 0 & 0.4 \end{bmatrix} \otimes \begin{bmatrix} 0.5 & 0.5 \\ 0 & 0.8 \end{bmatrix} & \begin{bmatrix} 0.2 & 0.8 \\ 0 & 0.4 \end{bmatrix} \otimes \begin{bmatrix} 0 \\ 0.2 \end{bmatrix} & \begin{bmatrix} 0 \\ 0.6 \end{bmatrix} \otimes \begin{bmatrix} 0.5 & 0.5 \\ 0 & 0.8 \end{bmatrix} \\ \boldsymbol{0} & \begin{bmatrix} 0.2 & 0.8 \\ 0 & 0.4 \end{bmatrix} & \boldsymbol{0} \\ \boldsymbol{0} & \boldsymbol{0} & \begin{bmatrix} 0.5 & 0.5 \\ 0 & 0.8 \end{bmatrix} \end{bmatrix}$$

$$= \begin{bmatrix} 0.1 & 0.1 & 0.4 & 0.4 & 0 & 0 & 0 & 0 \\ 0 & 0.16 & 0 & 0.64 & 0.04 & 0.16 & 0 & 0 \\ 0 & 0 & 0.2 & 0.2 & 0 & 0 & 0.3 & 0.3 \\ 0 & 0 & 0 & 0.32 & 0 & 0.08 & 0 & 0.48 \\ 0 & 0 & 0 & 0 & 0.2 & 0.8 & 0 & 0 \\ 0 & 0 & 0 & 0 & 0 & 0.4 & 0 & 0 \\ 0 & 0 & 0 & 0 & 0 & 0 & 0.5 & 0.5 \\ 0 & 0 & 0 & 0 & 0 & 0 & 0 & 0.8 \end{bmatrix}.$$

As a comparison, DPH representations of T_j, $j = 3, 4, 5, 6$, are calculated using the software package **dysrel-am** to produce the following results:

```
>> example7_13
a3 =
     1     0     0     0
A3 =
    0.1000    0.1000    0.4000    0.4000
         0    0.1600         0    0.6400
         0         0    0.2000    0.2000
         0         0         0    0.3200

n3 =
     4
a4 =
    0.4000         0    0.6000         0
A4 =
    0.2000    0.8000         0         0
         0    0.4000         0         0
         0         0    0.5000    0.5000
         0         0         0    0.8000
n4 =
     4
a5 =
     1     0     0     0
A5 =

    0.2000    0.8000         0         0
         0    0.4000    0.6000         0
         0         0    0.5000    0.5000
         0         0         0    0.8000
n5 =
     4
a6 =
     1     0     0     0     0     0     0     0
A6 =
    0.1000    0.1000    0.4000    0.4000         0         0         0         0
         0    0.1600         0    0.6400    0.0400    0.1600         0         0
         0         0    0.2000    0.2000         0         0    0.3000    0.3000
         0         0         0    0.3200         0    0.0800         0    0.4800
         0         0         0         0    0.2000    0.8000         0         0
         0         0         0         0         0    0.4000         0         0
         0         0         0         0         0         0    0.5000    0.5000
         0         0         0         0         0         0         0    0.8000
n6 =
     8
```

This is identical to the results of manual calculations. ■

Exercise 7.7 Redo Example 7.13 for T_1 and T_2, respectively, have geometry and 2-NB distributions with the same $MTTF = 10$.

MG distribution has the same closure properties as DPH distribution. For T_1 and T_2 are random variables with independent and not necessarily identical MG distributions with representation $(n_1, \boldsymbol{a}_1, \boldsymbol{A}_1, \boldsymbol{v}_1)$ and $(n_2, \boldsymbol{a}_2, \boldsymbol{A}_2, \boldsymbol{v}_2)$, respectively, the following theorems are applicable [9].

Theorem 7.5 $T_S = \min(T_1, T_2) \sim MG(n_S, \boldsymbol{a}_S, \boldsymbol{A}_S, \boldsymbol{v}_S)$, where

$$n_S = n_1 n_2, \quad \boldsymbol{a}_S = \boldsymbol{a}_1 \otimes \boldsymbol{a}_2, \quad \boldsymbol{A}_S = \boldsymbol{A}_1 \otimes \boldsymbol{A}_2, \quad \boldsymbol{v}_S = \boldsymbol{v}_1 \otimes \boldsymbol{v}_2. \tag{7.38}$$

Theorem 7.6 $T_S = \text{mix}(T_1, T_2) \sim MG(n_S, \boldsymbol{a}_S, \boldsymbol{A}_S, \boldsymbol{v}_S)$, where

$$n_S = n_1 + n_2, \quad \boldsymbol{a}_S = \begin{bmatrix} p_1 \boldsymbol{a}_1 & p_2 \boldsymbol{a}_2 \end{bmatrix}, \quad \boldsymbol{A}_S = \begin{bmatrix} \boldsymbol{A}_1 & \boldsymbol{0} \\ \boldsymbol{0} & \boldsymbol{A}_2 \end{bmatrix}, \quad \boldsymbol{v}_S = \begin{bmatrix} \boldsymbol{v}_1 \\ \boldsymbol{v}_2 \end{bmatrix}. \tag{7.39}$$

Theorem 7.7 $T_S = \text{add}(T_1, T_2) = T_1 + T_2 \sim MG(n_S, \boldsymbol{a}_S, \boldsymbol{A}_S, \boldsymbol{v}_S)$, where

$$n_S = n_1 + n_2, \quad \boldsymbol{a}_S = \begin{bmatrix} \boldsymbol{a}_1 & \boldsymbol{0} \end{bmatrix}, \quad \boldsymbol{A}_S = \begin{bmatrix} \boldsymbol{A}_1 & \boldsymbol{b}_1 \boldsymbol{a}_2 \\ \boldsymbol{0} & \boldsymbol{A}_2 \end{bmatrix}, \quad \boldsymbol{v}_S = \begin{bmatrix} \boldsymbol{v}_1 \\ \boldsymbol{v}_2 \end{bmatrix}. \tag{7.40}$$

Theorem 7.8 $T_S = \max(T_1, T_2) \sim MG(n_S, \boldsymbol{a}_S, \boldsymbol{A}_S, \boldsymbol{v}_S)$ with

$$n_S = n_1 n_2 + n_1 + n_2, \quad \boldsymbol{a}_S = \begin{bmatrix} \boldsymbol{a}_1 \otimes \boldsymbol{a}_2 & \boldsymbol{0} & \boldsymbol{0} \end{bmatrix},$$

$$\boldsymbol{A}_S = \begin{bmatrix} \boldsymbol{A}_1 \otimes \boldsymbol{A}_2 & \boldsymbol{A}_1 \otimes \boldsymbol{b}_2 & \boldsymbol{b}_1 \otimes \boldsymbol{A}_2 \\ \boldsymbol{0} & \boldsymbol{A}_1 & \boldsymbol{0} \\ \boldsymbol{0} & \boldsymbol{0} & \boldsymbol{A}_2 \end{bmatrix}, \quad \boldsymbol{v}_S = \begin{bmatrix} \boldsymbol{v}_1 \otimes \boldsymbol{v}_2 \\ \boldsymbol{v}_1 \\ \boldsymbol{v}_2 \end{bmatrix}. \tag{7.41}$$

The closure properties for MG distribution are slightly different than for ME distribution for minimum and maximum operations, and the same for mixing and addition operations. The differences are the same as between DPH and PH distributions.

Example 7.14 DPH distributions in Example 7.13 are transformed into MG distributions in a Jordan canonical form. MG distributions for $T_j, j = 3, 4, 5, 6$, are calculated using Theorems 7.5–7.8. The resulting MG representations are utilized to generate the reliability functions, *MTTF* and *SDTTF*, and compare them with the results of using the original DPH distributions. The comparisons are presented in Figure 7.9 and Table 7.4. It is apparent using DPH and its similar MG distributions produce identical results.

Table 7.4 Comparison of statistical reliability measures calculated using DPH and MG distributions.

	MTTF		*SDTTF*	
Lifetime	**DPH**	**MG**	**DPH**	**MG**
T_3	2.74	2.74	1.01	1.01
T_4	5.37	5.37	4.22	4.22
T_5	9.92	9.92	4.84	4.84
T_6	7.18	7.18	4.57	4.57

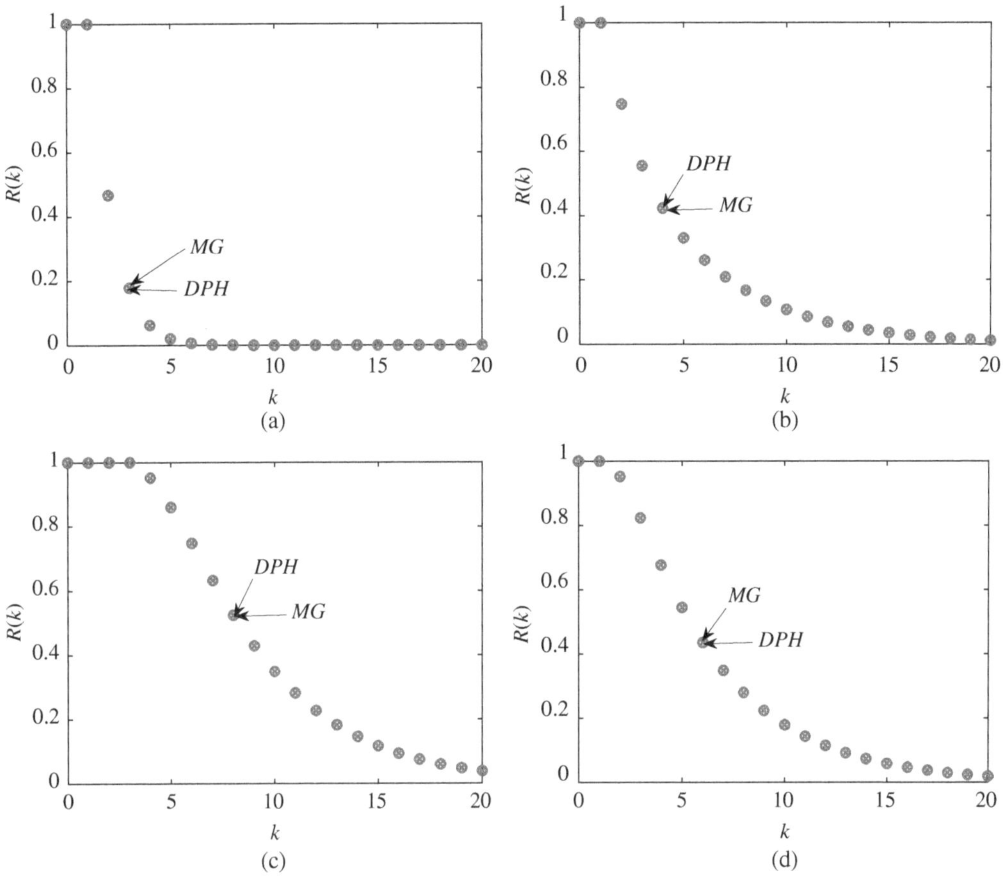

Figure 7.9 Comparison of reliability functions calculated using DPH and MG distributions. (a) Reliability function for T_3, (b) reliability function for T_4, (c) reliability function for T_5, and (d) reliability function for T_6. ■

7.4.2 Closure Properties of MG Distribution in Combinatorial Model

Theorems 7.1–7.8 are based on state transition model. Theorems that are based on combinatorial model as used in Section 6.4 for ME distribution in continuous-time systems are also applicable for MG distribution in discrete-time systems. Let X_1 and X_2 represent system states with duration of stays having independent and not necessarily identical MG distributions with representations $(n_1, \boldsymbol{a}_1, \boldsymbol{A}_1, \boldsymbol{v}_1)$ and $(n_2, \boldsymbol{a}_2, \boldsymbol{A}_2, \boldsymbol{v}_2)$, respectively. The following theorems are analogous to the theorems for ME distribution. The proofs follow the same logic as the proofs for ME distribution [9].

Theorem 7.9 $X_S = X_1X_2$ has MG distribution with representation $(n_S, \boldsymbol{a}_S, \boldsymbol{A}_S, \boldsymbol{v}_S)$, where

$$n_S = n_1n_2, \quad \boldsymbol{a}_S = \boldsymbol{a}_1 \otimes \boldsymbol{a}_2, \quad \boldsymbol{A}_S = \boldsymbol{A}_1 \otimes \boldsymbol{A}_2, \quad \boldsymbol{v}_S = \boldsymbol{v}_1 \otimes \boldsymbol{v}_2. \tag{7.42}$$

It is slightly different than ME distribution that MG distribution produces $\boldsymbol{A}_S = \boldsymbol{A}_1 \otimes \boldsymbol{A}_2$ while ME distribution produces $\boldsymbol{A}_S = \boldsymbol{A}_1 \oplus \boldsymbol{A}_2$.

Theorem 7.10 $X_S = X_1\overline{X}_2$ has MG distribution with representation $(n_S, \boldsymbol{a}_S, \boldsymbol{A}_S, \boldsymbol{v}_S)$, where

$$n_S = n_1 + n_1n_2, \quad \boldsymbol{a}_S = \left[\boldsymbol{a}_1 \ -\boldsymbol{a}_1 \otimes \boldsymbol{a}_2\right],$$

$$A_S = \begin{bmatrix} A_1 & 0 \\ 0 & A_1 \otimes A_2 \end{bmatrix}, \quad v_S = \begin{bmatrix} v_1 \\ v_1 \otimes v_2 \end{bmatrix}. \tag{7.43}$$

As a consequence of Theorem 7.9, Theorem 7.10 for discrete-time system is using $A_1 \otimes A_2$ instead of $A_1 \oplus A_2$.

Theorem 7.11 $X_S = X_1 + X_2$ has MG distribution with representation (n_S, a_S, A_S, v_S), where

$$n_S = n_1 + n_2, \quad a_S = \begin{bmatrix} a_1 & a_2 \end{bmatrix}, \quad A_S = \begin{bmatrix} A_1 & 0 \\ 0 & A_2 \end{bmatrix}, \quad v_S = \begin{bmatrix} v_1 \\ v_2 \end{bmatrix}. \tag{7.44}$$

Example 7.15 The duration of stays in state X_1 and X_2 have MG distribution as given in Example 7.13. The system state is $X_S = X_1 + \overline{X}_1 X_2$.

The DPH distribution in Example 7.13 is an MG distribution with $v_i = u_i$. Hence,

$$n_1 = n_2 = 2, \quad a_1 = a_2 = \begin{bmatrix} 1 & 0 \end{bmatrix},$$

$$A_1 = \begin{bmatrix} 0.2 & 0.8 \\ 0 & 0.4 \end{bmatrix}, \quad A_2 = \begin{bmatrix} 0.5 & 0.5 \\ 0 & 0.8 \end{bmatrix}, \quad v_1 = v_2 = \begin{bmatrix} 1 \\ 1 \end{bmatrix}.$$

Use Theorem 7.10 to determine the MG representation for $Z = \overline{X}_1 X_2$,

$$n_Z = n_2 + n_1 n_2 = 6,$$

$$a_Z = \begin{bmatrix} a_2 & -a_1 \otimes a_2 \end{bmatrix} = \begin{bmatrix} \begin{bmatrix} 1 & 0 \end{bmatrix} & -\begin{bmatrix} 1 & 0 \end{bmatrix} \otimes \begin{bmatrix} 1 & 0 \end{bmatrix} \end{bmatrix} = \begin{bmatrix} 1 & 0 & -1 & 0 & 0 & 0 \end{bmatrix},$$

$$A_Z = \begin{bmatrix} A_2 & 0 \\ 0 & A_1 \otimes A_2 \end{bmatrix} = \begin{bmatrix} \begin{bmatrix} 0.5 & 0.5 \\ 0 & 0.8 \end{bmatrix} & 0 \\ 0 & \begin{bmatrix} 0.2 & 0.8 \\ 0 & 0.4 \end{bmatrix} \otimes \begin{bmatrix} 0.5 & 0.5 \\ 0 & 0.8 \end{bmatrix} \end{bmatrix}$$

$$= \begin{bmatrix} 0.5 & 0.5 & 0 & 0 & 0 & 0 \\ 0 & 0.8 & 0 & 0 & 0 & 0 \\ 0 & 0 & 0.1 & 0.1 & 0.4 & 0.4 \\ 0 & 0 & 0 & 0.16 & 0 & 0.64 \\ 0 & 0 & 0 & 0 & 0.2 & 0.2 \\ 0 & 0 & 0 & 0 & 0 & 0.32 \end{bmatrix},$$

$$v_Z = \begin{bmatrix} v_2 \\ v_1 \otimes v_2 \end{bmatrix} = \begin{bmatrix} 1 & 1 & 1 & 1 & 1 & 1 \end{bmatrix}^T.$$

Use Theorem 7.11 to determine MG representation of $X_S = X_1 + Z$,

$$n_S = n_1 + n_Z = 8,$$

$$a_S = \begin{bmatrix} a_1 & a_z \end{bmatrix} = \begin{bmatrix} \begin{bmatrix} 1 & 0 \end{bmatrix} & \begin{bmatrix} 1 & 0 & -1 & 0 & 0 & 0 \end{bmatrix} \end{bmatrix} = \begin{bmatrix} 1 & 0 & 1 & 0 & -1 & 0 & 0 & 0 \end{bmatrix},$$

$$A_S = \begin{bmatrix} A_1 & 0 \\ 0 & A_Z \end{bmatrix} = \begin{bmatrix} \begin{bmatrix} 0.2 & 0.8 \\ 0 & 0.4 \end{bmatrix} & \mathbf{0} \\ \mathbf{0} & \begin{bmatrix} 0.5 & 0.5 & 0 & 0 & 0 & 0 \\ 0 & 0.8 & 0 & 0 & 0 & 0 \\ 0 & 0 & 0.1 & 0.1 & 0.4 & 0.4 \\ 0 & 0 & 0 & 0.16 & 0 & 0.64 \\ 0 & 0 & 0 & 0 & 0.2 & 0.2 \\ 0 & 0 & 0 & 0 & 0 & 0.32 \end{bmatrix} \end{bmatrix}$$

$$= \begin{bmatrix} 0.2 & 0.8 & 0 & 0 & 0 & 0 & 0 & 0 \\ 0 & 0.4 & 0 & 0 & 0 & 0 & 0 & 0 \\ 0 & 0 & 0.5 & 0.5 & 0 & 0 & 0 & 0 \\ 0 & 0 & 0 & 0.8 & 0 & 0 & 0 & 0 \\ 0 & 0 & 0 & 0 & 0.1 & 0.1 & 0.4 & 0.4 \\ 0 & 0 & 0 & 0 & 0 & 0.16 & 0 & 0.64 \\ 0 & 0 & 0 & 0 & 0 & 0 & 0.2 & 0.2 \\ 0 & 0 & 0 & 0 & 0 & 0 & 0 & 0.32 \end{bmatrix},$$

$$v_S = \begin{bmatrix} v_1 \\ v_2 \end{bmatrix} = \begin{bmatrix} 1 & 1 & 1 & 1 & 1 & 1 & 1 & 1 \end{bmatrix}^T.$$

Note that the result is an MG distribution since $\boldsymbol{a}_S$ is not a probability vector. ■

The closure properties of DPH and MG distributions are employed for reliability modeling and analysis of systems having structures as discussed in Chapters 4–6. First discussion is systems with series structure.

7.5 Series Structured Discrete-Time System

Following the development of continuous-time series systems, the relationship between the lifetime of a series system and the lifetimes of its components in discrete time is also given by a recursion equation

$$T_S^{(i)} = \min(T_S^{(i-1)}, T_i) \qquad i = 2, 3, \dots N, \tag{7.45}$$

where $T_S^{(i-1)} = \min(T_1, T_2, \dots, T_{i-1}), i = 2, 3, \dots N$, and $T_S^{(1)} = T_1$.

The DPH representation (n_S, $\boldsymbol{a}_S$, $\boldsymbol{A}_S$) of the lifetime distribution of discrete-time series systems can be generated by applying Theorem 7.1 to Equation (7.45) recursively to produce the following recursion equations [9]:

$$n_S^{(i)} = n_S^{(i-1)} n_i, \tag{7.46}$$

$$\boldsymbol{a}_S^{(i)} = \boldsymbol{a}_S^{(i-1)} \otimes \boldsymbol{a}_i, \tag{7.47}$$

$$\boldsymbol{A}_S^{(i)} = \boldsymbol{A}_S^{(i-1)} \otimes \boldsymbol{A}_i, \tag{7.48}$$

for $i = 2, 3, \dots, N$, with initial conditions

$$n_S^{(1)} = n_1, \quad \boldsymbol{a}_S^{(1)} = \boldsymbol{a}_1, \quad \boldsymbol{A}_S^{(1)} = \boldsymbol{A}_1. \tag{7.49}$$

Recursion equations (7.46)–(7.49) are utilized to construct Algorithm 7.1. Its pseudo-code is as follows:

Algorithm 7.1

```
input N, (n_i, a_i, A_i), i = 1, 2,…, N
n_S = n_1
a_S = a_1
A_S = A_1
for i = 2 to N
    n_S = n_S n_i
    a_S = a_S ⊗ a_i
    A_S = A_S ⊗ A_i
    i = i + 1
end
end
```

This algorithm is programmed in MATLAB as **dphseriessys** function available in the software package **dysrel-am**.

Algorithm 7.1 produces a DPH distribution of the lifetime of an N-series discrete-time system with representation $(n_S, \boldsymbol{a}_S, \boldsymbol{A}_S)$ given by

$$n_S = n_1 n_2 \dots n_N, \tag{7.50}$$

$$\boldsymbol{a}_S = \boldsymbol{a}_1 \otimes \boldsymbol{a}_2 \otimes \cdots \otimes \boldsymbol{a}_N, \tag{7.51}$$

$$\boldsymbol{A}_S = \boldsymbol{A}_1 \otimes \boldsymbol{A}_2 \otimes \cdots \otimes \boldsymbol{A}_N. \tag{7.52}$$

The state transition diagram of an N-series system in discrete time is presented in Figure 7.10, where $\boldsymbol{b}_S = (I - \boldsymbol{A}_S)\boldsymbol{u}$ as given in Equation (7.17).

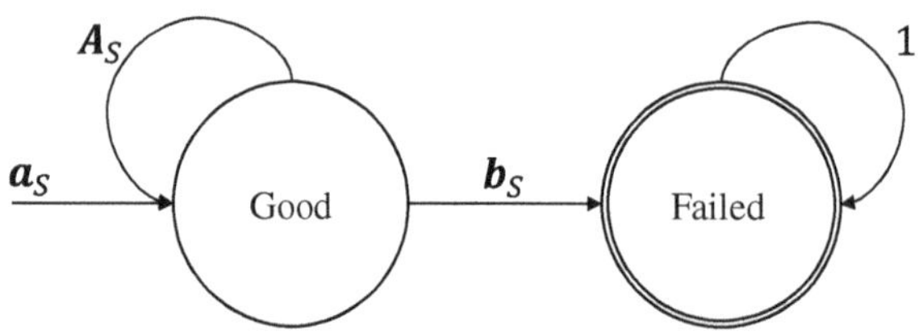

Figure 7.10 State transition diagram of an N-series system in discrete time.

Let g represents a good component and f represents a failed component. The good system state in Figure 7.10 is an N-tuple (g, g, … , g), and the failed system states are all states that contain at least one f element in the N-tuples system states lumped into one failed superstate.

Series systems in continuous and discrete time have the same complexity equal to nN, where n is the order of the DPH distribution of component lifetimes and N is the number of components in the system.

Example 7.16 A 3-series discrete-time system uses three components with lifetimes following geometry distributions with $MTTF = 2$, 5, and 10 months.

The DPH representations of the distributions of component lifetimes can be derived from their geometry distributions and $MTTF$ as

$$n_1 = n_2 = n_3 = 1, \quad \boldsymbol{a}_1 = \boldsymbol{a}_2 = \boldsymbol{a}_3 = [1],$$

$$\boldsymbol{A}_1 = 1 - \frac{1}{2} = [0.5], \quad \boldsymbol{A}_2 = 1 - \frac{1}{5} = [0.8], \quad \boldsymbol{A}_3 = 1 - \frac{1}{10} = [0.9].$$

The calculation of the DPH representation of the lifetime distribution of this series system is carried out recursively using Equations (7.46)–(7.49) as follows:

- The algorithm starts with the initial conditions for $(i) = 1$ given by Equation (7.49):

$$n_S^{(1)} = n_1 = 1, \quad \boldsymbol{a}_S^{(1)} = \boldsymbol{a}_1 = [1], \quad \boldsymbol{A}_S^{(1)} = \boldsymbol{A}_1 = [0.5].$$

- Use Equations (7.46)–(7.48) for $(i) = 2$:

$$n_S^{(2)} = n_S^{(1)} n_2 = 1, \quad \boldsymbol{a}_S^{(2)} = \boldsymbol{a}_S^{(1)} \otimes \boldsymbol{a}_2 = [1] \otimes [1] = [1],$$

$$\boldsymbol{A}_S^{(2)} = \boldsymbol{A}_S^{(1)} \otimes \boldsymbol{A}_2 = [0.5] \otimes [0.8] = [0.4].$$

- Use Equations (7.46)–(7.48) for $(i) = 3$:

$$n_S^{(3)} = n_S^{(2)} n_3 = 1, \quad \boldsymbol{a}_S^{(3)} = \boldsymbol{a}_S^{(2)} \otimes \boldsymbol{a}_3 = [1] \otimes [1] = [1],$$

$$\boldsymbol{A}_S^{(3)} = \boldsymbol{A}_S^{(2)} \otimes \boldsymbol{A}_3 = [0.4] \otimes [0.9] = [0.36].$$

- Because $i = 3 = N$, the algorithm stops.

Thus, the DPH representation for the distribution of lifetime of this 3-series discrete-time system is $(n_S, \boldsymbol{a}_S, \boldsymbol{A}_S) = (1, [1], [0.36])$. Since its order is 1, it is geometry distribution, and the system is CFR. Use Equation (7.25) to obtain the system reliability function

$$R_S(k) = [1][0.36]^k\,[1] = (0.36)^k, \quad k = 0, 1, 2.$$

Other reliability measures can be obtained using their respective equations,

$$h_S(k) = 1 - (\boldsymbol{a}\boldsymbol{A}^k\boldsymbol{u})(\boldsymbol{a}\boldsymbol{A}^{k-1}\boldsymbol{u})^{-1} = 1 - (0.36^k)(0.36^{k-1})^{-1} = 0.64, \quad k = 1, 2, \ldots$$

$$\begin{aligned} RR_S(k + m|T > m) &= \boldsymbol{a}\boldsymbol{A}^m(\boldsymbol{a}\boldsymbol{A}^m\boldsymbol{u})^{-1}\boldsymbol{A}^k\boldsymbol{u} \\ &= (0.36^m(0.36^m)^{-1})0.36^k = (0.36)^k, \quad k = 0, 1, 2, \ldots \\ MRL_S(m) &= \boldsymbol{a}\boldsymbol{A}^m(\boldsymbol{a}\boldsymbol{A}^m\boldsymbol{u})^{-1}(\boldsymbol{I} - \boldsymbol{A})^{-1}\boldsymbol{u} \\ &= (1)0.36^m((1)0.36^m(1))^{-1}(1 - 0.36)^{-1}(1) = 1.56, \quad m = 0, 1, 2, \ldots \end{aligned}$$

In general, if component i has lifetime following a geometry distribution with parameter p_i, $i = 1, 2, \ldots, N$, then the lifetime of N-series discrete time system has a geometry distribution with parameter $p_1 p_2 \cdots p_N$. ■

Exercise 7.8 Determine the DPH distribution of the lifetime of an N-series system in discrete time with identical CFR components having $MTTF = b$ months. Represent the system $MTTF$ in term of N and b.

Systems with components having lifetimes following DPH distributions of order greater than one require longer manual calculations.

Example 7.17 A 2-series discrete-time system is constructed using two components having 2-NB distributed lifetimes with $MTTF = 5$ and 10 months.

The value of p for 2-NB distribution with $MTTF = 5$ can be obtained using information in Table 7.2 as

$$\frac{2}{1-p} = 5 \rightarrow p = \frac{3}{5} = 0.6.$$

Likewise, for $MTTF = 10$ results in

$$\frac{2}{1-p} = 10 \rightarrow p = \frac{8}{10} = 0.8.$$

Therefore, DPH representations of components 1 and 2 are

$$n_1 = n_2 = 2, \quad \boldsymbol{a}_1 = \boldsymbol{a}_2 = \begin{bmatrix} 1 & 0 \end{bmatrix}, \quad \boldsymbol{A}_1 = \begin{bmatrix} 0.6 & 0.4 \\ 0 & 0.6 \end{bmatrix}, \quad \boldsymbol{A}_2 = \begin{bmatrix} 0.8 & 0.2 \\ 0 & 0.8 \end{bmatrix}.$$

Use Algorithm 7.1 to determine the DPH representation for the system lifetime

$$n_S = n_1 n_2 = 4.$$

$$\boldsymbol{a}_S = \boldsymbol{a}_1 \otimes \boldsymbol{a}_2 = \begin{bmatrix} 1 & 0 \end{bmatrix} \otimes \begin{bmatrix} 1 & 0 \end{bmatrix} = \begin{bmatrix} 1 & 0 & 0 & 0 \end{bmatrix}.$$

$$\boldsymbol{A}_S = \boldsymbol{A}_1 \otimes \boldsymbol{A}_2 = \begin{bmatrix} 0.6 & 0.4 \\ 0 & 0.6 \end{bmatrix} \otimes \begin{bmatrix} 0.8 & 0.2 \\ 0 & 0.8 \end{bmatrix} = \begin{bmatrix} 0.48 & 0.12 & 0.32 & 0.08 \\ 0 & 0.48 & 0 & 0.32 \\ 0 & 0 & 0.48 & 0.12 \\ 0 & 0 & 0 & 0.48 \end{bmatrix}.$$

DPH distribution of the lifetime of this series system has representation $(n_S, \boldsymbol{a}_S, \boldsymbol{A}_S)$ as given in the abovementioned calculations.

For comparison, the PH representation obtained using the **dphseriessys** function is

```
ns =
     4
as =
     1     0     0     0
As =
    0.4800    0.1200    0.3200    0.0800
         0    0.4800         0    0.3200
         0         0    0.4800    0.1200
         0         0         0    0.4800
```

which is the same result as manual calculations. ■

The difficulty in calculating DPH representation of the lifetime distribution of series systems with components having lifetime distributions following DPH distributions with high orders can be avoided by using **dphseriessys,** and the results can be utilized to calculate system reliability measures using the provided MATLAB functions.

Example 7.18 The **dphseriessys** function is used to calculate the DPH representation of the lifetime distribution of a 5-series system. The kth component of this system has lifetime following k-NB distributions with $MTTF = 100$, $k = 1, 2, \ldots, 5$. The DPH representation obtained from **dphseriessys** is then used to generate and plot the system reliability and hazard functions and to calculate the system $MTTF$ and $SDTTF$. The main program for this purpose is given in **example7_18.m** file in the software package **dysrel-am**. Its MATLAB code is as follows:

MATLAB code for example7_18.m

```
clear;close all; addpath('..\dysrel-am')
% PH Components                                    input
N = 5;
mttfc = 100;
for k = 1:N
  T(k) = dphnegativebinomial(k,mttfc);
end
% PH Series system                                 model
Ts = dphseriessys(T);
% Time vector
kmax = 100;
k = 0:5:kmax;
% Reliability plot                                 graphical output
dphrelplot(k,Ts);
% Hazard plot
dphhazplot(k,Ts);
% Result in command window                         statistical output
mttf = dphmttf(Ts)
sdttf = dphsdttf(Ts)
```

Executing this file produces Figure 7.11.

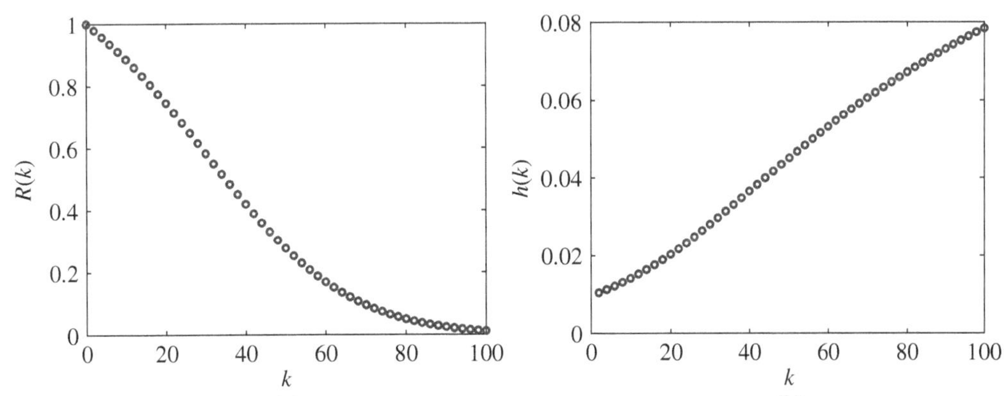

Figure 7.11 Output of **example7_18.m**.

```
>> example7_18
mttf =
   38.1763
sdttf =
   23.3343
```

■

Example 7.19 Three components are arranged in a series structure. Their lifetimes have the following distributions:

- Component 1 has a 3-mixed-geometry distribution with DPH representation:

$$n_1 = 3, \quad \boldsymbol{a}_1 = \begin{bmatrix} 0.5 & 0.3 & 0.2 \end{bmatrix}, \quad \boldsymbol{A}_1 = \begin{bmatrix} 0.9 & 0 & 0 \\ 0 & 0.8 & 0 \\ 0 & 0 & 0.5 \end{bmatrix}.$$

- Component 2 has a geometry distribution with DPH representation:

$$n_2 = 1, \quad \boldsymbol{a}_2 = [1], \quad \boldsymbol{A}_2 = [0.8].$$

- Component 3 has 3-hypogeometry distribution with DPH representation:

$$n_3 = 3, \quad \boldsymbol{a}_3 = \begin{bmatrix} 1 & 0 & 0 \end{bmatrix}, \quad \boldsymbol{A}_3 = \begin{bmatrix} 0.9 & 0.1 & 0 \\ 0 & 0.8 & 0.2 \\ 0 & 0 & 0.5 \end{bmatrix}.$$

The reliability and hazard functions of the system and all its components are shown in Figure 7.12.

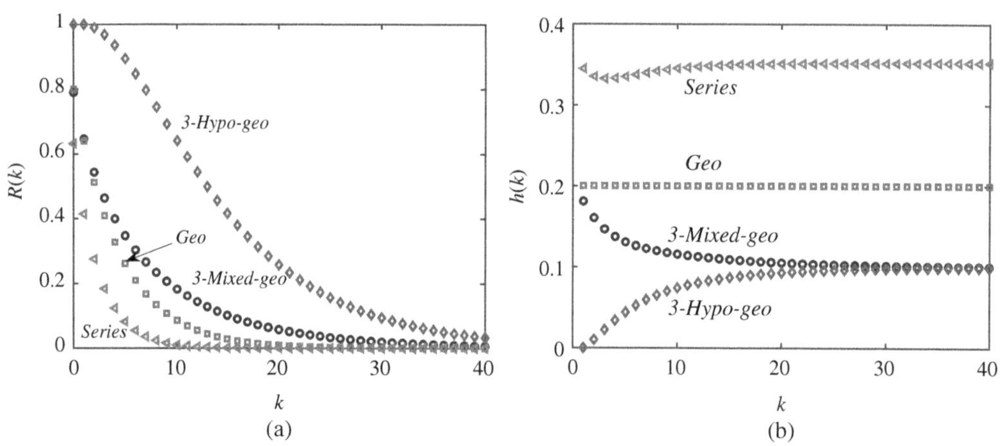

Figure 7.12 Reliability and hazard functions of the system and components in Example 7.19. (a) Reliability functions and (b) hazard functions.

Figure 7.12a shows that the system reliability function at any time is always lower than the lowest reliability function of its components at that time. On the contrary, the system hazard function at any time is higher than the highest hazard function of its components at that time as shown in Figure 7.12b. The expected lifetimes of the system and all components calculated using **dphmttf** are $MTTF_1 = 27.00$, $MTTF_2 = 14.29$, $MTTF_3 = 16.67$, and $MTTF_S = 5.37$, which is smaller than the smallest component $MTTF$. These results show that series systems in discrete time have the same behavior as series systems in continuous time derived in Section 4.3. ■

Exercise 7.9 Conduct experiments to show that all properties of continuous-time series systems hold for discrete-time series systems

An algorithm for discrete-time series systems with components having MG distributions can be constructed using Theorem 7.5 recursively to produce the recursion equations for MG distribution. The recursion equations are the same as for DPH distribution with an additional equation for calculating closing vectors:

$$n_S^{(i)} = n_S^{(i-1)} n_i, \tag{7.53}$$

$$\boldsymbol{a}_S^{(i)} = \boldsymbol{a}_S^{(i-1)} \otimes \boldsymbol{a}_i, \tag{7.54}$$

$$\boldsymbol{A}_S^{(i)} = \boldsymbol{A}_S^{(i-1)} \otimes \boldsymbol{A}_i, \tag{7.55}$$

$$\boldsymbol{v}_S^{(i)} = \boldsymbol{v}_S^{(i-1)} \otimes \boldsymbol{v}_i, \tag{7.56}$$

for $i = 2, 3, \dots, N$, with initial conditions

$$n_S^{(1)} = n_1, \quad \boldsymbol{a}_S^{(1)} = \boldsymbol{a}_1, \quad \boldsymbol{A}_S^{(1)} = \boldsymbol{A}_1, \quad \boldsymbol{v}_S^{(1)} = \boldsymbol{v}_1. \tag{7.57}$$

Algorithm 7.1 can be modified slightly to construct Algorithm 7.2 for discrete-time series systems with components having MG distribution. Its pseudo-code is derived from the pseudo-code of Algorithm 7.1 by adding lines for calculating $\boldsymbol{v}_S$ recursively.

Algorithm 7.2

```
input N, (n_i, a_i, A_i, v_i), i = 1, 2, ..., N
n_S = n_1
a_S = a_1
A_S = A_1
v_S = v_1
for i = 2 to N
    n_S = n_S n_i
    a_S = a_S ⊗ a_i
    A_S = A_S ⊗ A_i
    v_S = v_S ⊗ v_i
    i = i + 1
end
end
```

The algorithm is programmed in MATLAB as **mgseriessys** function, available in the software package **dysrel-am**.

Example 7.20 Four discrete-time series systems are constructed using N identical components with lifetimes following DPH distributions.

$$\text{System 1: } n_1 = 4, \boldsymbol{a}_1 = \begin{bmatrix} 1 & 0 & 0 & 0 \end{bmatrix}, \boldsymbol{A}_1 = \begin{bmatrix} 0.8 & 0.1 & 0.1 & 0 \\ 0 & 0.6 & 0.2 & 0.2 \\ 0 & 0 & 0.5 & 0.3 \\ 0 & 0 & 0 & 0.4 \end{bmatrix}.$$

$$\text{System 2: } n_2 = 4, \boldsymbol{a}_2 = \begin{bmatrix} 1 & 0 & 0 & 0 \end{bmatrix}, \boldsymbol{A}_2 = \begin{bmatrix} 0.8 & 0.1 & 0.1 & 0 \\ 0 & 0.6 & 0.2 & 0.2 \\ 0 & 0 & 0.5 & 0.3 \\ 0 & 0 & 0 & 0.5 \end{bmatrix}.$$

$$\text{System 3: } n_3 = 4, \boldsymbol{a}_3 = \begin{bmatrix} 1 & 0 & 0 & 0 \end{bmatrix}, \boldsymbol{A}_3 = \begin{bmatrix} 0.5 & 0.3 & 0.1 & 0.1 \\ 0 & 0.7 & 0.1 & 0.1 \\ 0 & 0 & 0.8 & 0.1 \\ 0 & 0 & 0 & 0.7 \end{bmatrix}.$$

$$\text{System 4: } n_4 = 3, \boldsymbol{a}_4 = \begin{bmatrix} 1 & 0 & 0 & 0 \end{bmatrix}, \boldsymbol{A}_4 = \begin{bmatrix} 0.8 & 0.1 & 0.1 & 0 \\ 0 & 0.6 & 0.2 & 0.2 \\ 0 & 0.1 & 0.5 & 0.3 \\ 0.2 & 0 & 0 & 0.4 \end{bmatrix}.$$

These DPH distributions are transformed into MG distributions in a Jordan canonical form using matrix similarity transformations. The resulting MG distributions are

$$\text{System 1: } n_1 = 4, \boldsymbol{a}_1 = \begin{bmatrix} -0.5 & 0.3333 & 1 & 0.25 \end{bmatrix},$$

$$\boldsymbol{A}_1 = \begin{bmatrix} 0.6 & 0 & 0 & 0 \\ 0 & 0.5 & 0 & 0 \\ 0 & 0 & 0.8 & 0 \\ 0 & 0 & 0 & 0.4 \end{bmatrix}, \quad \boldsymbol{v}_1 = \begin{bmatrix} 7 \\ 4 \\ 2.9167 \\ 1 \end{bmatrix}.$$

$$\text{System 2: } n_2 = 4, \boldsymbol{a}_2 = \begin{bmatrix} -1.5 & 0.1667 & 0.1 & 1.3333 \end{bmatrix},$$

$$\boldsymbol{A}_2 = \begin{bmatrix} 0.6 & 0 & 0 & 0 \\ 0 & 0.8 & 0 & 0 \\ 0 & 0 & 0.5 & 0.1 \\ 0 & 0 & 0 & 0.5 \end{bmatrix}, \quad \boldsymbol{v}_2 = \begin{bmatrix} 3.6667 \\ 19 \\ 20 \\ 1 \end{bmatrix}.$$

$$\text{System 3: } n_3 = 4, \quad \boldsymbol{a}_3 = \begin{bmatrix} 1 & 1.3333 & 1.5 & 0 \end{bmatrix},$$

$$\boldsymbol{A}_3 = \begin{bmatrix} 0.5 & 0 & 0 & 0 \\ 0 & 0.8 & 0 & 0 \\ 0 & 0 & 0.7 & 0 \\ 0 & 0 & 0 & 0.7 \end{bmatrix}, \quad \boldsymbol{v}_3 = \begin{bmatrix} -0.1667 \\ 2 \\ -1 \\ 1 \end{bmatrix}.$$

System 4: $n_4 = 4, \boldsymbol{a}_4 = [2.5\ 0\ 0.5 - 0.5i\ 0.5 + 0.5i]$,

$$\boldsymbol{A}_4 = \begin{bmatrix} 0.9 & 0 & 0 & 0 \\ 0 & 0.4 & 0 & 0 \\ 0 & 0 & 0.5-0.1i & 0 \\ 0 & 0 & 0 & 0.5+0.1i \end{bmatrix}, \quad \boldsymbol{v}_4 = \begin{bmatrix} 0.5882 \\ -1 \\ 0.2059 - 0.6765i \\ 0.2059 + 0.6765i \end{bmatrix}.$$

Note that components of systems 1 and 2 have acyclic DPH distributions with real and distinct eigenvalues for system 1, and real and repeated eigenvalues for system 2. Components of systems 3 and 4 have cyclic DPH distributions with real and distinct eigenvalues for system 3 and complex eigenvalues for system 4.

System reliability functions for $N = 3$ and 6 calculated using **dphseriessys** and **mgseriessys** are presented in Figure 7.13. Comparisons of computation times for producing these graphs are given in Table 7.5.

Figure 7.13 shows that DPH distributions and their similar MG distributions produced identical graphs. However, Table 7.5 shows that the computation times using DPH distributions are faster than using MG distributions for lower order, but slower for higher orders. Therefore, transforming

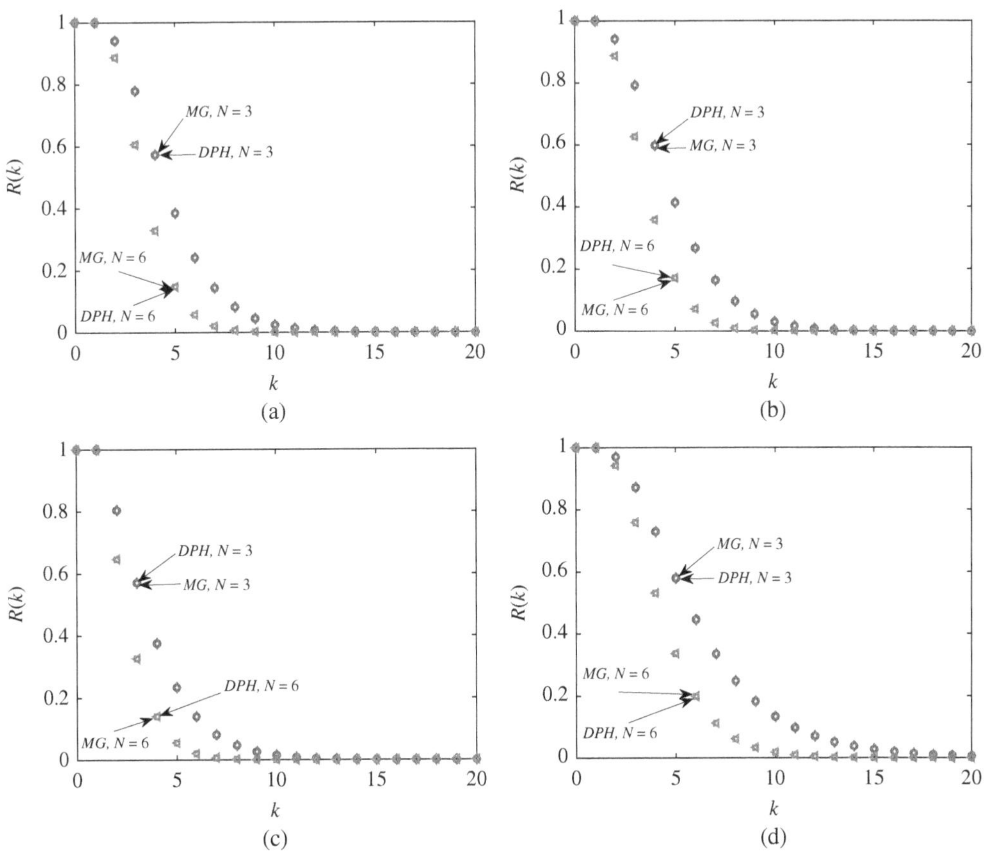

Figure 7.13 Reliability functions of systems in Example 7.20. (a) System 1, (b) system 2, (c) system 3, and (d) system 4.

Table 7.5 Computation times for producing Figure 7.13 (in seconds).

	$N = 3$ (order = 64)		$N = 6$ (order = 4,096)	
System	**DPH**	**MG**	**DPH**	**MG**
1	0.005	0.218	51.260	0.233
2	0.005	0.231	51.649	0.275
3	0.006	0.234	53.804	0.247
4	0.014	0.221	69.978	0.235

DPH into MG distribution in a Jordan canonical form is worthwhile for series systems having high orders of DPH distributions, which could due to the number of components in the system or the orders of DPH distributions of the component lifetimes. ■

Next, the use of one of DPH and MG distribution closure property for systems with parallel structure is presented.

7.6 Parallel Structured Discrete-Time Systems

The relationship between the lifetime of a parallel system, T_S, and the lifetimes of its components T_i, $i = 1, 2, \dots, N$, is given by a recursion equation

$$T_S^{(i)} = \max\left(T_S^{(i-1)}, T_i\right) \qquad i = 2, 3, \dots N, \tag{7.58}$$

where $T_S^{(i-1)} = \max(T_1, T_2, \dots, T_{i-1}), i = 2, 3, \dots N$, and $T_S^{(1)} = T_1$.

Equation (7.58) is applicable for both continuous and discrete time.

The DPH representation (n_S, $\boldsymbol{a}_S$, $\boldsymbol{A}_S$) of the system lifetime distribution in discrete time can be generated by applying Theorem 7.4 to Equation (7.58) recursively to produce recursion equations [9]

$$n_S^{(i)} = n_S^{(i-1)} n_i + n_S^{(i-1)} + n_i, \tag{7.59}$$

$$\boldsymbol{a}_S^{(i)} = \left[\boldsymbol{a}_S^{(i-1)} \otimes \boldsymbol{a}_i \;\; \boldsymbol{0} \;\; \boldsymbol{0}\right], \tag{7.60}$$

$$\boldsymbol{A}_S^{(i)} = \begin{bmatrix} \boldsymbol{A}_S^{(i-1)} \otimes \boldsymbol{A}_i & \boldsymbol{A}_S^{(i-1)} \otimes \boldsymbol{b}_i & \boldsymbol{b}_S^{(i-1)} \otimes \boldsymbol{A}_i \\ \boldsymbol{0} & \boldsymbol{A}_S^{(i-1)} & \boldsymbol{0} \\ \boldsymbol{0} & \boldsymbol{0} & \boldsymbol{A}_i \end{bmatrix}, \tag{7.61}$$

for $i = 2, 3, \dots, N$, with initial conditions given by Equation (7.49), $\boldsymbol{b}_i = (\boldsymbol{I} - \boldsymbol{A}_i)\boldsymbol{u}$, and $\boldsymbol{b}_S^{(i-1)} = \left(\boldsymbol{I} - \boldsymbol{A}_S^{(i-1)}\right)\boldsymbol{u}$.

Equations (7.59)–(7.61) with initial conditions in Equation (7.49) are the core of Algorithm 7.3 for generating a DPH representation of the distribution of the lifetime of a parallel discrete-time system. The algorithm in pseudo-code is as follows:

Algorithm 7.3

```
input N, (n_i, a_i, A_i), i = 1, 2, ..., N
n_S = n_1
a_S = a_1
A_S = A_1
for i = 2 to N
    n_S = n_S n_i + n_S + n_i
    a_S = [a_S ⊗ a_i 0 0]
    b_S = (I − A_S)u
    b_i = (I − A_i)u
          [ A_S ⊗ A_i   A_S ⊗ b_i   b_S ⊗ A_i ]
    A_S = [    0           A_S          0     ]
          [    0            0          A_i    ]
    i = i + 1
end
end
```

This algorithm is programmed in MATLAB as the **dphparallelsys** function, available in the software package **dysrel-am**.

Manual calculation for parallel systems using Algorithm 7.3 is more complicated than for series systems. It is still possible for parallel systems with few CFR components.

Example 7.21 The DPH representation of the lifetime distribution of a 3-parallel system with the same components as given in Example 7.16 is derived as follows.

Use $\boldsymbol{A}_i$, $i = 1, 2, 3$ in Example 7.16 to determine

$$\boldsymbol{b}_1 = (\boldsymbol{I} - \boldsymbol{A}_1)\boldsymbol{u} = ([1] - [0.5])[1] = [0.5],$$

$$\boldsymbol{b}_2 = (\boldsymbol{I} - \boldsymbol{A}_2)\boldsymbol{u} = ([1] - [0.8])[1] = [0.2],$$

$$\boldsymbol{b}_3 = (\boldsymbol{I} - \boldsymbol{A}_3)\boldsymbol{u} = ([1] - [0.9][1]) = [0.1].$$

The iteration using Algorithm 7.3 is as follows:

- Initial conditions are determined using Equation (7.49):

$$n_S^{(1)} = n_1 = 1, \quad \boldsymbol{a}_S^{(1)} = \boldsymbol{a}_1 = [1], \quad \boldsymbol{A}_S^{(1)} = \boldsymbol{A}_1 = [0.5].$$

- Second and third iterations to find n_S using Equation (7.59):

$$n_S^{(2)} = n_S^{(1)} n_2 + n_S^{(1)} + n_2 = 3, \quad n_S^{(3)} = n_S^{(2)} n_3 + n_S^{(2)} + n_3 = 7.$$

- Second and third iterations to find $\boldsymbol{a}_S$ using Equation (7.60):

$$\boldsymbol{a}_S^{(2)} = \left[\boldsymbol{a}_S^{(1)} \otimes \boldsymbol{a}_2 \;\; \boldsymbol{0} \;\; \boldsymbol{0}\right] = \left[1 \otimes 1 \;\; 0 \;\; 0\right] = \left[1 \;\; 0 \;\; 0\right],$$

$$\boldsymbol{a}_S^{(3)} = \left[\boldsymbol{a}_S^{(2)} \otimes \boldsymbol{a}_3 \;\; \boldsymbol{0} \;\; \boldsymbol{0}\right] = \left[\left[1 \;\; 0 \;\; 0\right] \otimes [1] \;\; \left[0 \;\; 0 \;\; 0\right] \;\; 0\right] = \left[1 \;\; 0 \;\; 0 \;\; 0 \;\; 0 \;\; 0 \;\; 0\right].$$

- Second and third iterations to find $\boldsymbol{A}_S$ using Equation (7.61):

$$\boldsymbol{A}_S^{(2)} = \begin{bmatrix} \boldsymbol{A}_S^{(1)} \otimes \boldsymbol{A}_2 & \boldsymbol{A}_S^{(1)} \otimes \boldsymbol{b}_2 & \boldsymbol{b}_S^{(1)} \otimes \boldsymbol{A}_2 \\ \boldsymbol{0} & \boldsymbol{A}_S^{(1)} & \boldsymbol{0} \\ \boldsymbol{0} & \boldsymbol{0} & \boldsymbol{A}_2 \end{bmatrix} = \begin{bmatrix} \boldsymbol{A}_1 \otimes \boldsymbol{A}_2 & \boldsymbol{A}_1 \otimes \boldsymbol{b}_2 & b_1 \otimes \boldsymbol{A}_2 \\ \boldsymbol{0} & \boldsymbol{A}_S^{(1)} & \boldsymbol{0} \\ \boldsymbol{0} & \boldsymbol{0} & \boldsymbol{A}_2 \end{bmatrix}$$

$$= \begin{bmatrix} [0.5] \otimes [0.8] & [0.5] \otimes [0.2] & [0.5] \otimes [0.8] \\ 0 & [0.5] & 0 \\ 0 & 0 & [0.8] \end{bmatrix} = \begin{bmatrix} 0.4 & 0.1 & 0.4 \\ 0 & 0.5 & 0 \\ 0 & 0 & 0.8 \end{bmatrix},$$

$$\boldsymbol{b}_S^{(2)} = \left(\boldsymbol{I} - \begin{bmatrix} 0.4 & 0.1 & 0.4 \\ 0 & 0.5 & 0 \\ 0 & 0 & 0.8 \end{bmatrix} \right) \begin{bmatrix} 1 \\ 1 \\ 1 \end{bmatrix} = \begin{bmatrix} 0.1 \\ 0.5 \\ 0.2 \end{bmatrix},$$

$$\boldsymbol{A}_S^{(3)} = \begin{bmatrix} \boldsymbol{A}_S^{(2)} \otimes \boldsymbol{A}_3 & \boldsymbol{A}_S^{(2)} \otimes \boldsymbol{b}_3 & \boldsymbol{b}_S^{(2)} \otimes \boldsymbol{A}_3 \\ \boldsymbol{0} & \boldsymbol{A}_S^{(2)} & \boldsymbol{0} \\ \boldsymbol{0} & \boldsymbol{0} & \boldsymbol{A}_3 \end{bmatrix}$$

$$= \begin{bmatrix} \begin{bmatrix} 0.4 & 0.1 & 0.4 \\ 0 & 0.5 & 0 \\ 0 & 0 & 0.8 \end{bmatrix} \otimes [0.9] & \begin{bmatrix} 0.4 & 0.1 & 0.4 \\ 0 & 0.5 & 0 \\ 0 & 0 & 0.8 \end{bmatrix} \otimes [0.1] & \begin{bmatrix} 0.1 \\ 0.5 \\ 0.2 \end{bmatrix} \otimes [0.9] \\ \boldsymbol{0} & \begin{bmatrix} 0.4 & 0.1 & 0.4 \\ 0 & 0.5 & 0 \\ 0 & 0 & 0.8 \end{bmatrix} & \boldsymbol{0} \\ \boldsymbol{0} & \boldsymbol{0} & [0.9] \end{bmatrix}$$

$$= \begin{bmatrix} \begin{bmatrix} 0.36 & 0.09 & 0.36 \\ 0 & 0.45 & 0 \\ 0 & 0 & 0.72 \end{bmatrix} & \begin{bmatrix} 0.04 & 0.01 & 0.04 \\ 0 & 0.05 & 0 \\ 0 & 0 & 0.08 \end{bmatrix} & \begin{bmatrix} 0.09 \\ 0.45 \\ 0.18 \end{bmatrix} \\ \begin{bmatrix} 0 & 0 & 0 \\ 0 & 0 & 0 \\ 0 & 0 & 0 \end{bmatrix} & \begin{bmatrix} 0.4 & 0.1 & 0.4 \\ 0 & 0.5 & 0 \\ 0 & 0 & 0.8 \end{bmatrix} & \begin{bmatrix} 0 \\ 0 \\ 0 \end{bmatrix} \\ \begin{bmatrix} 0 & 0 & 0 \end{bmatrix} & \begin{bmatrix} 0 & 0 & 0 \end{bmatrix} & [0.9] \end{bmatrix}$$

$$= \begin{bmatrix} 0.36 & 0.09 & 0.36 & 0.04 & 0.01 & 0.04 & 0.09 \\ 0 & 0.45 & 0 & 0 & 0.05 & 0 & 0.45 \\ 0 & 0 & 0.72 & 0 & 0 & 0.08 & 0.18 \\ 0 & 0 & 0 & 0.4 & 0.1 & 0.4 & 0 \\ 0 & 0 & 0 & 0 & 0.5 & 0 & 0 \\ 0 & 0 & 0 & 0 & 0 & 0.8 & 0 \\ 0 & 0 & 0 & 0 & 0 & 0 & 0.9 \end{bmatrix},$$

$$\boldsymbol{b}_S^{(3)} = \left(\boldsymbol{I} - \begin{bmatrix} 0.36 & 0.09 & 0.36 & 0.04 & 0.01 & 0.04 & 0.09 \\ 0 & 0.45 & 0 & 0 & 0.05 & 0 & 0.45 \\ 0 & 0 & 0.72 & 0 & 0 & 0.08 & 0.18 \\ 0 & 0 & 0 & 0.4 & 0.1 & 0.4 & 0 \\ 0 & 0 & 0 & 0 & 0.5 & 0 & 0 \\ 0 & 0 & 0 & 0 & 0 & 0.8 & 0 \\ 0 & 0 & 0 & 0 & 0 & 0 & 0.9 \end{bmatrix} \right) \begin{bmatrix} 1 \\ 1 \\ 1 \\ 1 \\ 1 \\ 1 \\ 1 \end{bmatrix} = \begin{bmatrix} 0.01 \\ 0.05 \\ 0.02 \\ 0.1 \\ 0.5 \\ 0.2 \\ 0.1 \end{bmatrix}.$$

This example illustrates the difficulties of using manual calculation.
Using **dphparallelsys** produces the following results:

```
N = 2
As =
    0.4000    0.1000    0.4000
         0    0.5000         0
         0         0    0.8000
bs =
    0.1000
    0.5000
    0.2000
N = 3
As =
    0.3600    0.0900    0.3600    0.0400    0.0100    0.0400    0.0900
         0    0.4500         0         0    0.0500         0    0.4500
         0         0    0.7200         0         0    0.0800    0.1800
         0         0         0    0.4000    0.1000    0.4000         0
         0         0         0         0    0.5000         0         0
         0         0         0         0         0    0.8000         0
         0         0         0         0         0         0    0.9000
bs =
    0.0100
    0.0500
    0.0200
    0.1000
    0.5000
    0.2000
    0.1000
```

which is equal to manual calculations. ■

The state transition diagram of discrete-time parallel systems can be drawn only for a small N by recursively using Equations (7.60) and (7.61)

Equation (7.61) for $N = 2$ gives

$$\boldsymbol{A}_S^{(2)} = \begin{bmatrix} \boldsymbol{A}_S^{(1)} \otimes \boldsymbol{A}_2 & \boldsymbol{A}_S^{(1)} \otimes \boldsymbol{b}_2 & \boldsymbol{b}_S^{(1)} \otimes \boldsymbol{A}_2 \\ \boldsymbol{0} & \boldsymbol{A}_S^{(1)} & \boldsymbol{0} \\ \boldsymbol{0} & \boldsymbol{0} & \boldsymbol{A}_2 \end{bmatrix} = \begin{bmatrix} \boldsymbol{A}_1 \otimes \boldsymbol{A}_2 & \boldsymbol{A}_1 \otimes \boldsymbol{b}_2 & \boldsymbol{b}_1 \otimes \boldsymbol{A}_2 \\ \boldsymbol{0} & \boldsymbol{A}_1 & \boldsymbol{0} \\ \boldsymbol{0} & \boldsymbol{0} & \boldsymbol{A}_2 \end{bmatrix}, \tag{7.62}$$

and

$$\boldsymbol{b}_S^{(2)} = \left(\boldsymbol{I}_S^{(2)} - \boldsymbol{A}_S^{(2)}\right)\boldsymbol{u}, \tag{7.63}$$

with

$$\boldsymbol{I}_S^{(2)} = \begin{bmatrix} \boldsymbol{I}_1 \otimes \boldsymbol{I}_2 & \boldsymbol{0} & \boldsymbol{0} \\ \boldsymbol{0} & \boldsymbol{I}_1 & \boldsymbol{0} \\ \boldsymbol{0} & \boldsymbol{0} & \boldsymbol{I}_2 \end{bmatrix}, \tag{7.64}$$

to match the dimensions of $\boldsymbol{A}_S^{(2)}$

Substituting Equations (7.62) and (7.64) into Equation (7.63) yields

$$\boldsymbol{b}_S^{(2)} = \begin{bmatrix} (\boldsymbol{I}_1 \otimes \boldsymbol{I}_2 - \boldsymbol{A}_1 \otimes \boldsymbol{A}_2) & -\boldsymbol{A}_1 \otimes \boldsymbol{b}_2 & -\boldsymbol{b}_1 \otimes \boldsymbol{A}_2 \\ \boldsymbol{0} & (\boldsymbol{I}_1 - \boldsymbol{A}_1) & \boldsymbol{0} \\ \boldsymbol{0} & \boldsymbol{0} & (\boldsymbol{I}_2 - \boldsymbol{A}_2) \end{bmatrix} \begin{bmatrix} \boldsymbol{u}_1 \otimes \boldsymbol{u}_2 \\ \boldsymbol{u}_1 \\ \boldsymbol{u}_2 \end{bmatrix}$$
$$= \begin{bmatrix} (\boldsymbol{I}_1 \otimes \boldsymbol{I}_2 - \boldsymbol{A}_1 \otimes \boldsymbol{A}_2)(\boldsymbol{u}_1 \otimes \boldsymbol{u}_2) - (\boldsymbol{A}_1 \otimes \boldsymbol{b}_2)\boldsymbol{u}_1 - (\boldsymbol{b}_1 \otimes \boldsymbol{A}_2)\boldsymbol{u}_2 \\ (\boldsymbol{I} - \boldsymbol{A}_1)\boldsymbol{u}_1 \\ (\boldsymbol{I} - \boldsymbol{A}_2)\boldsymbol{u}_2 \end{bmatrix}.$$

The second and third block rows of $\boldsymbol{b}_S^{(2)}$ are $\boldsymbol{b}_1$ and $\boldsymbol{b}_2$, respectively. The first term in the first block row of $\boldsymbol{b}_S^{(2)}$ can be simplified as follows:

Defining $\boldsymbol{x}_i = \boldsymbol{A}_i\boldsymbol{u}_i$ results in $\boldsymbol{b}_i = (\boldsymbol{I}_i - \boldsymbol{A}_1)\boldsymbol{u}_1 = \boldsymbol{u}_i - \boldsymbol{x}_1$. The first term of the first block row of $\boldsymbol{b}_S^{(2)}$ becomes

$$(\boldsymbol{I}_1 \otimes \boldsymbol{I}_2 - \boldsymbol{A}_1 \otimes \boldsymbol{A}_2)(\boldsymbol{u}_1 \otimes \boldsymbol{u}_2) = \boldsymbol{u}_1 \otimes \boldsymbol{u}_2 - (\boldsymbol{A}_1\boldsymbol{u}_1) \otimes (\boldsymbol{A}_2\boldsymbol{u}_2) = \boldsymbol{u}_1 \otimes \boldsymbol{u}_2 - \boldsymbol{x}_1 \otimes \boldsymbol{x}_2.$$

The second and third terms of the first block row of $\boldsymbol{b}_S^{(2)}$ become

$$\begin{aligned} -(\boldsymbol{A}_1 \otimes \boldsymbol{b}_2)\boldsymbol{u}_1 - (\boldsymbol{b}_1 \otimes \boldsymbol{A}_2)\boldsymbol{u}_2 &= -(\boldsymbol{A}_1\boldsymbol{u}_1 \otimes \boldsymbol{b}_2) - (\boldsymbol{b}_1 \otimes \boldsymbol{A}_2\boldsymbol{u}_2) \\ &= -(\boldsymbol{x}_1 \otimes \boldsymbol{b}_2) - (\boldsymbol{b}_1 \otimes \boldsymbol{x}_2) \\ &= -\boldsymbol{x}_1 \otimes (\boldsymbol{u}_2 - \boldsymbol{x}_2) - (\boldsymbol{u}_1 - \boldsymbol{x}_1) \otimes \boldsymbol{x}_2 \\ &= -\boldsymbol{x}_1 \otimes \boldsymbol{u}_2 + \boldsymbol{x}_1 \otimes \boldsymbol{x}_2 - \boldsymbol{u}_1 \otimes \boldsymbol{x}_2 + \boldsymbol{x}_1 \otimes \boldsymbol{x}_2. \end{aligned}$$

Combining the three terms gives

$$\begin{aligned} &(\boldsymbol{I}_1 \otimes \boldsymbol{I}_2 - \boldsymbol{A}_1 \otimes \boldsymbol{A}_2)(\boldsymbol{u}_1 \otimes \boldsymbol{u}_2) - (\boldsymbol{A}_1 \otimes \boldsymbol{b}_2)\boldsymbol{u}_1 - (\boldsymbol{b}_1 \otimes \boldsymbol{A}_2)\boldsymbol{u}_2 \\ &= \boldsymbol{u}_1 \otimes \boldsymbol{u}_2 - \boldsymbol{x}_1 \otimes \boldsymbol{x}_2 - \boldsymbol{x}_1 \otimes \boldsymbol{u}_2 + \boldsymbol{x}_1 \otimes \boldsymbol{x}_2 - \boldsymbol{u}_1 \otimes \boldsymbol{x}_2 + \boldsymbol{x}_1 \otimes \boldsymbol{x}_2 \\ &= \boldsymbol{u}_1 \otimes \boldsymbol{u}_2 - \boldsymbol{x}_1 \otimes \boldsymbol{u}_2 - \boldsymbol{u}_1 \otimes \boldsymbol{x}_2 + \boldsymbol{x}_1 \otimes \boldsymbol{x}_2 \\ &= (\boldsymbol{u}_1 - \boldsymbol{x}_1) \otimes (\boldsymbol{u}_2 - \boldsymbol{x}_2) = \boldsymbol{b}_1 \otimes \boldsymbol{b}_2. \end{aligned}$$

Thus,

$$\boldsymbol{b}_S^{(2)} = \begin{bmatrix} \boldsymbol{b}_1 \otimes \boldsymbol{b}_2 \\ \boldsymbol{b}_1 \\ \boldsymbol{b}_2 \end{bmatrix}. \tag{7.65}$$

The first block row of the failure vector $\boldsymbol{b}_S^{(2)}$ in Equation (7.65) represents components 1 and 2 failing at the same period of observation. Unlike continuous-time systems, it is possible in discrete-time systems for two components to fail at the same period.

The state transition diagram for $N = 2$ constructed using Equations (7.62) and (7.65) is given in Figure 7.14.

Note that in discrete-time systems, there is a transition from state (g, g) to (f, f), which does not present in continuous-time systems. It is due to the possibility in discrete time for two components failed in the same inspection period.

For $N = 3$, Equation (7.61) gives

$$\boldsymbol{A}_S^{(3)} = \begin{bmatrix} \boldsymbol{A}_S^{(2)} \otimes \boldsymbol{A}_3 & \boldsymbol{A}_S^{(2)} \otimes \boldsymbol{b}_3 & \boldsymbol{b}_S^{(2)} \otimes \boldsymbol{A}_1 \\ \boldsymbol{0} & \boldsymbol{A}_S^{(2)} & \boldsymbol{0} \\ \boldsymbol{0} & \boldsymbol{0} & \boldsymbol{A}_3 \end{bmatrix}. \tag{7.66}$$

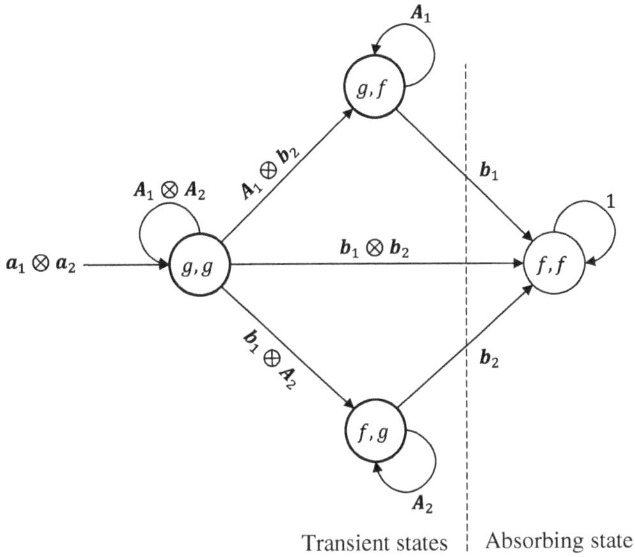

Figure 7.14 State transition diagram of 2-parallel discrete-time systems.

Substituting $\boldsymbol{A}_S^{(2)}$, $\boldsymbol{b}_S^{(2)}$, and $\boldsymbol{I}_S^{(2)}$ in Equations (7.62), (7.65), and (7.64) into Equation (7.66) yields

$$\boldsymbol{A}_S^{(3)} = \begin{bmatrix} \begin{bmatrix} \boldsymbol{A}_1 \otimes \boldsymbol{A}_2 & \boldsymbol{A}_1 \otimes \boldsymbol{b}_2 & \boldsymbol{b}_1 \otimes \boldsymbol{A}_2 \\ \boldsymbol{0} & \boldsymbol{A}_1 & \boldsymbol{0} \\ \boldsymbol{0} & \boldsymbol{0} & \boldsymbol{A}_2 \end{bmatrix} \otimes \boldsymbol{A}_3 & \begin{bmatrix} \boldsymbol{A}_1 \otimes \boldsymbol{A}_2 & \boldsymbol{A}_1 \otimes \boldsymbol{b}_2 & \boldsymbol{b}_1 \otimes \boldsymbol{A}_2 \\ \boldsymbol{0} & \boldsymbol{A}_1 & \boldsymbol{0} \\ \boldsymbol{0} & \boldsymbol{0} & \boldsymbol{A}_2 \end{bmatrix} \otimes \boldsymbol{b}_3 & \begin{bmatrix} \boldsymbol{b}_1 \otimes \boldsymbol{b}_2 \\ \boldsymbol{b}_1 \\ \boldsymbol{b}_2 \end{bmatrix} \otimes \boldsymbol{A}_3 \\ \boldsymbol{0} & \begin{bmatrix} \boldsymbol{A}_1 \otimes \boldsymbol{A}_2 & \boldsymbol{A}_1 \otimes \boldsymbol{b}_2 & \boldsymbol{b}_1 \otimes \boldsymbol{A}_2 \\ \boldsymbol{0} & \boldsymbol{A}_1 & \boldsymbol{0} \\ \boldsymbol{0} & \boldsymbol{0} & \boldsymbol{A}_2 \end{bmatrix} & \boldsymbol{0} \\ \boldsymbol{0} & \boldsymbol{0} & \boldsymbol{A}_3 \end{bmatrix}. \tag{7.67}$$

Simplifying the block matrix in the first block row and column of $\boldsymbol{A}_S^{(3)}$ yields

$$\begin{bmatrix} \boldsymbol{A}_1 \otimes \boldsymbol{A}_2 & \boldsymbol{A}_1 \otimes \boldsymbol{b}_2 & \boldsymbol{b}_1 \otimes \boldsymbol{A}_2 \\ \boldsymbol{0} & \boldsymbol{A}_1 & \boldsymbol{0} \\ \boldsymbol{0} & \boldsymbol{0} & \boldsymbol{A}_2 \end{bmatrix} \otimes \boldsymbol{A}_3 = \begin{bmatrix} \boldsymbol{A}_1 \otimes \boldsymbol{A}_2 \otimes \boldsymbol{A}_3 & \boldsymbol{A}_1 \otimes \boldsymbol{b}_2 \otimes \boldsymbol{A}_3 & \boldsymbol{b}_1 \otimes \boldsymbol{A}_2 \otimes \boldsymbol{A}_3 \\ \boldsymbol{0} & \boldsymbol{A}_1 \otimes \boldsymbol{A}_3 & \boldsymbol{0} \\ \boldsymbol{0} & \boldsymbol{0} & \boldsymbol{A}_2 \otimes \boldsymbol{A}_3 \end{bmatrix}$$

Substituting this result back to $\boldsymbol{A}_S^{(3)}$ in Equation (7.67) produces

$$\boldsymbol{A}_S^{(3)} = \begin{bmatrix} \boldsymbol{A}_1 \otimes \boldsymbol{A}_2 \otimes \boldsymbol{A}_3 & \boldsymbol{A}_1 \otimes \boldsymbol{b}_2 \otimes \boldsymbol{A}_3 & \boldsymbol{b}_1 \otimes \boldsymbol{A}_2 \otimes \boldsymbol{A}_3 & \boldsymbol{A}_1 \otimes \boldsymbol{A}_2 \otimes \boldsymbol{b}_3 & \boldsymbol{A}_1 \otimes \boldsymbol{b}_2 \otimes \boldsymbol{b}_3 & \boldsymbol{b}_1 \otimes \boldsymbol{A}_2 \otimes \boldsymbol{b}_3 & \boldsymbol{b}_1 \otimes \boldsymbol{b}_2 \otimes \boldsymbol{A}_3 \\ \boldsymbol{0} & \boldsymbol{A}_1 \otimes \boldsymbol{A}_3 & \boldsymbol{0} & \boldsymbol{0} & \boldsymbol{A}_1 \otimes \boldsymbol{b}_3 & \boldsymbol{0} & \boldsymbol{b}_1 \otimes \boldsymbol{A}_3 \\ \boldsymbol{0} & \boldsymbol{0} & \boldsymbol{A}_2 \otimes \boldsymbol{A}_3 & \boldsymbol{0} & \boldsymbol{0} & \boldsymbol{A}_2 \otimes \boldsymbol{b}_3 & \boldsymbol{b}_2 \otimes \boldsymbol{A}_3 \\ \boldsymbol{0} & \boldsymbol{0} & \boldsymbol{0} & \boldsymbol{A}_1 \otimes \boldsymbol{A}_2 & \boldsymbol{A}_1 \otimes \boldsymbol{b}_2 & \boldsymbol{b}_1 \otimes \boldsymbol{A}_2 & \boldsymbol{0} \\ \boldsymbol{0} & \boldsymbol{0} & \boldsymbol{0} & \boldsymbol{0} & \boldsymbol{A}_1 & \boldsymbol{0} & \boldsymbol{0} \\ \boldsymbol{0} & \boldsymbol{0} & \boldsymbol{0} & \boldsymbol{0} & \boldsymbol{0} & \boldsymbol{A}_2 & \boldsymbol{0} \\ \boldsymbol{0} & \boldsymbol{0} & \boldsymbol{0} & \boldsymbol{0} & \boldsymbol{0} & \boldsymbol{0} & \boldsymbol{A}_3 \end{bmatrix}. \tag{7.68}$$

Failure vector $\boldsymbol{b}_S^{(3)} = -\boldsymbol{A}_S^{(3)}\boldsymbol{u}$ can be derived the same way as $\boldsymbol{b}_S^{(2)}$, but it requires longer steps. The result should intuitively be

$$\boldsymbol{b}_S^{(3)} = \begin{bmatrix} \boldsymbol{b}_1 \otimes \boldsymbol{b}_2 \otimes \boldsymbol{b}_3 & \boldsymbol{b}_1 \otimes \boldsymbol{b}_3 & \boldsymbol{b}_2 \otimes \boldsymbol{b}_3 & \boldsymbol{b}_1 \otimes \boldsymbol{b}_2 & \boldsymbol{b}_1 & \boldsymbol{b}_2 & \boldsymbol{b}_3 \end{bmatrix}^T, \tag{7.69}$$

where $\boldsymbol{b}_1 \otimes \boldsymbol{b}_2 \otimes \boldsymbol{b}_3$ is the failure vector of components 1–3 failed in the same inspection period, $\boldsymbol{b}_1 \otimes \boldsymbol{b}_3$ is the failure vector of components 1 and 3 failed in the same inspection period, and others can be interpreted accordingly.

The resulting infinitesimal generator matrix of 3-parallel system is

$$\boldsymbol{Q}_S^{(3)} = \begin{bmatrix} \boldsymbol{A}_S^{(3)} & \boldsymbol{b}_S^{(3)} \\ \boldsymbol{0} & 1 \end{bmatrix}.$$

The state transition diagram of a 3-parallel system in discrete time constructed using $\boldsymbol{Q}_S^{(3)}$ yields Figure 7.15. By parallel structure definition, states with at least one good component are transient states, and the state with no good component is the absorbing state. The state transition diagram of N-parallel systems is too complicated for an arbitrary N. Because of the possibility that more than one component fails simultaneously in a period of inspection makes the state transition diagram of a discrete-time parallel system more complicated (contains more transitions) than its continuous-time counterpart which makes its survival matrix denser.

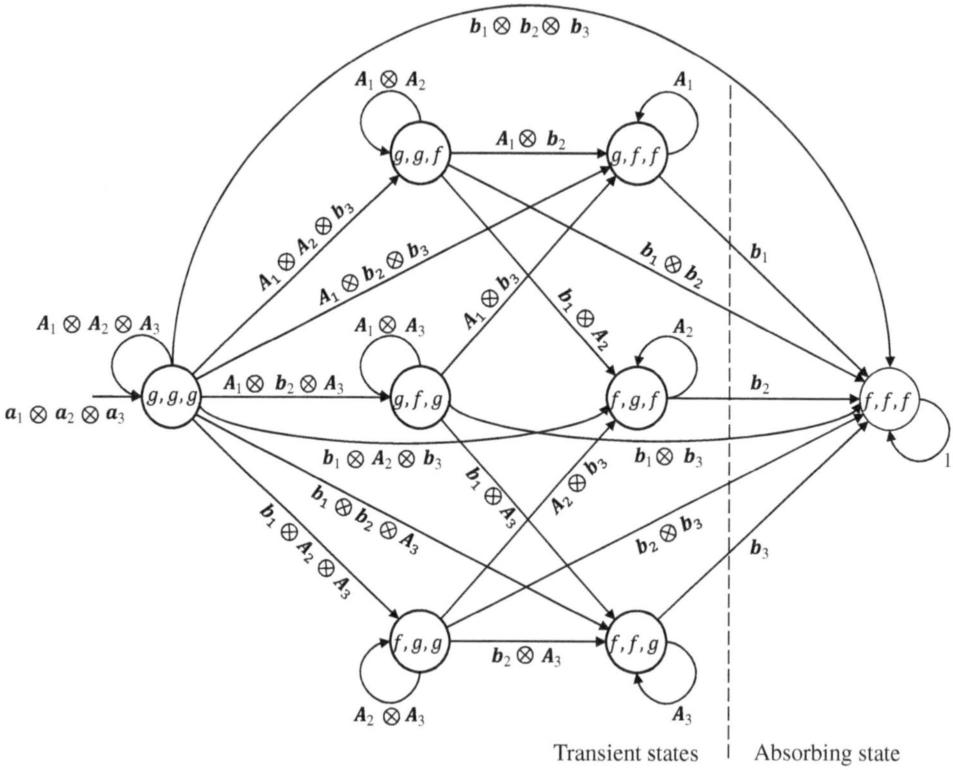

Figure 7.15 State transition diagram of a 3-parallel discrete-time system.

Example 7.22 The survival matrices and failure vectors of 2-parallel and 3-parallel discrete-time systems using components in Example 7.16 are calculated using Equations (7.62), (7.65), (7.68), and (7.69). The results are

$$\boldsymbol{A}_S^{(2)} = \begin{bmatrix} \boldsymbol{A}_1 \oplus \boldsymbol{A}_2 & \boldsymbol{A}_1 \otimes \boldsymbol{b}_2 & \boldsymbol{b}_1 \otimes \boldsymbol{A}_2 \\ \boldsymbol{0} & \boldsymbol{A}_1 & \boldsymbol{0} \\ \boldsymbol{0} & \boldsymbol{0} & \boldsymbol{A}_2 \end{bmatrix}$$

$$= \begin{bmatrix} [0.5] \otimes [0.8] & [0.5] \otimes [0.2] & [0.5] \otimes [0.8] \\ 0 & [0.5] & 0 \\ 0 & 0 & [0.8] \end{bmatrix} = \begin{bmatrix} 0.4 & 0.1 & 0.4 \\ 0 & 0.5 & 0 \\ 0 & 0 & 0.8 \end{bmatrix}.$$

$$\boldsymbol{b}_S^{(2)} = \begin{bmatrix} \boldsymbol{b}_1 \otimes \boldsymbol{b}_2 \\ \boldsymbol{b}_1 \\ \boldsymbol{b}_2 \end{bmatrix} = \begin{bmatrix} [0.5] \otimes [0.2] \\ [0.5] \\ [0.2] \end{bmatrix} = \begin{bmatrix} 0.1 \\ 0.5 \\ 0.2 \end{bmatrix},$$

$$\boldsymbol{A}_S^{(3)} = \begin{bmatrix} \boldsymbol{A}_1 \otimes \boldsymbol{A}_2 \otimes \boldsymbol{A}_3 & \boldsymbol{A}_1 \otimes \boldsymbol{b}_2 \otimes \boldsymbol{A}_3 & \boldsymbol{b}_1 \otimes \boldsymbol{A}_2 \otimes \boldsymbol{A}_3 & \boldsymbol{A}_1 \otimes \boldsymbol{A}_2 \otimes \boldsymbol{b}_3 & \boldsymbol{A}_1 \otimes \boldsymbol{b}_2 \otimes \boldsymbol{b}_3 & \boldsymbol{b}_1 \otimes \boldsymbol{A}_2 \otimes \boldsymbol{b}_3 & \boldsymbol{b}_1 \otimes \boldsymbol{b}_2 \otimes \boldsymbol{A}_3 \\ \boldsymbol{0} & \boldsymbol{A}_1 \otimes \boldsymbol{A}_3 & \boldsymbol{0} & \boldsymbol{0} & \boldsymbol{A}_1 \otimes \boldsymbol{b}_3 & \boldsymbol{0} & \boldsymbol{b}_1 \otimes \boldsymbol{A}_3 \\ \boldsymbol{0} & \boldsymbol{0} & \boldsymbol{A}_2 \otimes \boldsymbol{A}_3 & \boldsymbol{0} & \boldsymbol{0} & \boldsymbol{A}_2 \otimes \boldsymbol{b}_3 & \boldsymbol{b}_2 \otimes \boldsymbol{A}_3 \\ \boldsymbol{0} & \boldsymbol{0} & \boldsymbol{0} & \boldsymbol{A}_1 \otimes \boldsymbol{A}_2 & \boldsymbol{A}_1 \otimes \boldsymbol{b}_2 & \boldsymbol{b}_1 \otimes \boldsymbol{A}_2 & \boldsymbol{0} \\ \boldsymbol{0} & \boldsymbol{0} & \boldsymbol{0} & \boldsymbol{0} & \boldsymbol{A}_1 & \boldsymbol{0} & \boldsymbol{0} \\ \boldsymbol{0} & \boldsymbol{0} & \boldsymbol{0} & \boldsymbol{0} & \boldsymbol{0} & \boldsymbol{A}_2 & \boldsymbol{0} \\ \boldsymbol{0} & \boldsymbol{0} & \boldsymbol{0} & \boldsymbol{0} & \boldsymbol{0} & \boldsymbol{0} & \boldsymbol{A}_3 \end{bmatrix}$$

$$= \begin{bmatrix} (0.5)(0.8)(0.9) & (0.5)(0.2)(0.9) & (0.5)(0.8)(0.9) & (0.5)(0.8)(0.1) & (0.5)(0.2)(0.1) & (0.5)(0.8)(0.1) & (0.5)(0.2)(0.9) \\ 0 & (0.5)(0.9) & 0 & 0 & (0.5)(0.1) & 0 & (0.5)(0.9) \\ 0 & 0 & (0.8)(0.9) & 0 & 0 & (0.8)(0.1) & (0.2)(0.9) \\ 0 & 0 & 0 & (0.5)(0.8) & (0.5)(0.2) & (0.5)(0.8) & 0 \\ 0 & 0 & 0 & 0 & 0.5 & 0 & 0 \\ 0 & 0 & 0 & 0 & 0 & 0.8 & 0 \\ 0 & 0 & 0 & 0 & 0 & 0 & 0.9 \end{bmatrix}$$

$$= \begin{bmatrix} 0.36 & 0.09 & 0.36 & 0.04 & 0.01 & 0.04 & 0.09 \\ 0 & 0.45 & 0 & 0 & 0.05 & 0 & 0.45 \\ 0 & 0 & 0.72 & 0 & 0 & 0.08 & 0.18 \\ 0 & 0 & 0 & 0.4 & 0.1 & 0.4 & 0 \\ 0 & 0 & 0 & 0 & 0.5 & 0 & 0 \\ 0 & 0 & 0 & 0 & 0 & 0.8 & 0 \\ 0 & 0 & 0 & 0 & 0 & 0 & 0.9 \end{bmatrix},$$

$$\boldsymbol{b}_S^{(3)} = \begin{bmatrix} \boldsymbol{b}_1 \otimes \boldsymbol{b}_2 \otimes \boldsymbol{b}_3 \\ \boldsymbol{b}_1 \otimes \boldsymbol{b}_3 \\ \boldsymbol{b}_2 \otimes \boldsymbol{b}_3 \\ \boldsymbol{b}_1 \otimes \boldsymbol{b}_2 \\ \boldsymbol{b}_1 \\ \boldsymbol{b}_2 \\ \boldsymbol{b}_3 \end{bmatrix} = \begin{bmatrix} (0.5)(0.2)(0.1) \\ (0.5)(0.1) \\ (0.2)(0.1) \\ (0.5)(0.2) \\ 0.5 \\ 0.2 \\ 0.1 \end{bmatrix} = \begin{bmatrix} 0.01 \\ 0.05 \\ 0.02 \\ 0.1 \\ 0.5 \\ 0.2 \\ 0.1 \end{bmatrix}.$$

These results are the same as produced by using Algorithm 7.3 in Example 7.21. ■

Manual calculation is not feasible for parallel systems with many components or with components having lifetimes following DPH distributions with high order. Using **dphparallelsys** is the only viable way.

Example 7.23 The **dphparallelsys** function is employed to calculate the DPH representation of the lifetime distribution of a 5-parallel system. The kth component of this system has lifetime following k-NB distributions with $MTTF = 100$, $k = 1, 2, \ldots, 5$. The DPH representation obtained from **dphparallelsys** is used to calculate and plot the system reliability and hazard functions and also the system $MTTF$ and $SDTTF$.

The main program for this purpose is given in **example7_23.m** file in the software package **dysrel-am**. Its MATLAB code is as follows:

MATLAB code for example7_23.m

```
clear;close all; addpath('..\dysrel-am')
% DPH Components                                      input
N = 5;
mttfc = 100;
for k = 1:N
  T(k) = dphnegativebinomial(k,mttfc);
end
% DPH Parallel system                                 model
Ts = dphparallelsys(T);
% Time vector
kmax = 400;
k = 0:20:kmax;
% Reliability plot                                    graphical output
dphrelplot(k,Ts);
% Hazard plot
dphhazplot(k,Ts);
% Result in command window                            statistical output
mttf = dphmttf(Ts)
sdttf = dphsdttf(Ts)
```

Executing this program produces results in Figure 7.16.

```
mttf =
   183.2379
sdttf =
   78.6312
```

■

Example 7.24 Three components in Example 7.19 are arranged in a parallel structure. The system and components reliability and hazard functions are shown in Figure 7.17.

Figure 7.17a shows that the system reliability function at any time is always higher than the highest reliability function of its components at that time. The expected lifetimes of the system and all components are $MTTF_1 = 6.9$, $MTTF_2 = 5$, $MTTF_3 = 17$, and $MTTF_S = 18.71$ is longer than the

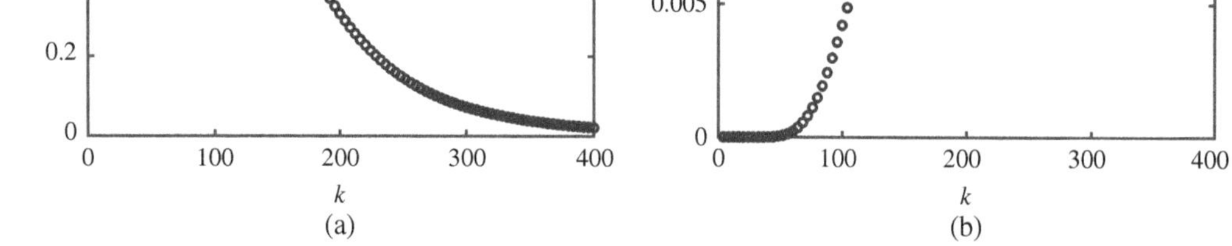

Figure 7.16 Output of **example7_23.m** file.

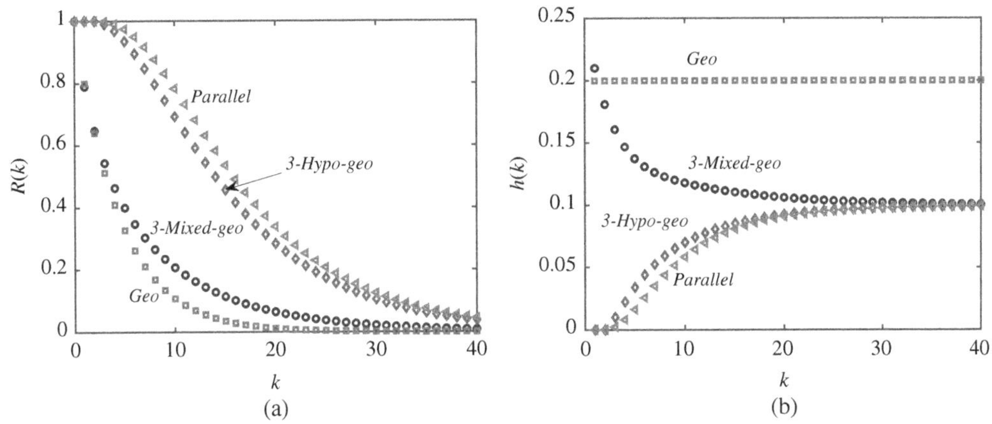

Figure 7.17 Reliability and hazard functions of the system and components in Example 7.24. (a) Reliability functions and (b) hazard functions.

longest component *MTTF*. These results can be used to make a conjecture that parallel systems in discrete time have the same behavior as parallel systems in continuous time. ■

Exercise 7.10 Show that continuous- and discrete-time parallel systems share the same properties.

Theorem 7.8 can be used recursively to produce the recursion equations for discrete-time parallel systems with components having MG distributions, equivalent to DPH distribution with an additional equation for calculating closing vectors:

$$n_S^{(i)} = n_S^{(i-1)} n_i + n_S^{(i-1)} + n_i, \tag{7.70}$$

$$\boldsymbol{a}_S^{(i)} = \left[\boldsymbol{a}_S^{(i-1)} \otimes \boldsymbol{a}_i \ \boldsymbol{0} \ \boldsymbol{0}\right], \tag{7.71}$$

$$\boldsymbol{A}_S^{(i)} = \begin{bmatrix} \boldsymbol{A}_S^{(i-1)} \otimes \boldsymbol{A}_i & \boldsymbol{A}_S^{(i-1)} \otimes \boldsymbol{b}_i & \boldsymbol{b}_S^{(i-1)} \otimes \boldsymbol{A}_i \\ \boldsymbol{0} & \boldsymbol{A}_S^{(i-1)} & \boldsymbol{0} \\ \boldsymbol{0} & \boldsymbol{0} & \boldsymbol{A}_i \end{bmatrix}, \tag{7.72}$$

$$\boldsymbol{v}_S^{(i)} = \left[\boldsymbol{v}_S^{(i-1)} \otimes \boldsymbol{v}_i \ \boldsymbol{v}_S^{(i-1)} \ \boldsymbol{v}_i\right]^T, \tag{7.73}$$

for $i = 2, 3, \ldots, N$, with the initial conditions given by Equations (7.49), $\boldsymbol{b}_i = (\boldsymbol{I} - \boldsymbol{A}_i)\boldsymbol{u}$, and $\boldsymbol{b}_S^{(i-1)} = \left(\boldsymbol{I} - \boldsymbol{A}_S^{(i-1)}\right)\boldsymbol{u}$.

Algorithm 7.3 can be modified slightly to construct Algorithm 7.4 for parallel systems whose components have MG distribution. Its pseudo-code is derived from the pseudo-code of Algorithm 7.3 by adding lines for calculating $\boldsymbol{v}_S$ recursively as follows:

Algorithm 7.4

input N, $(n_i, \mathbf{a}_i, \mathbf{A}_i)$, $i = 1, 2, \ldots, N$

$n_S = n_1$

$\mathbf{a}_S = \mathbf{a}_1$

$\mathbf{A}_S = \mathbf{A}_1$

$\mathbf{b}_S = (\mathbf{I} - \mathbf{A}_1)\mathbf{u}$

$\mathbf{v}_S = \mathbf{v}_1$

for i = 2 to N

$n_S = n_S n_i + n_S + n_i$

$\mathbf{a}_S = \left[\mathbf{a}_S \otimes \mathbf{a}_i \ \mathbf{0} \ \mathbf{0}\right]$

$\mathbf{b}_S = (\mathbf{I} - \mathbf{A}_S)\mathbf{v}_S$

$\mathbf{b}_i = (\mathbf{I} - \mathbf{A}_i)\mathbf{v}_i$

$$\mathbf{A}_S = \begin{bmatrix} \mathbf{A}_S \otimes \mathbf{A}_i & \mathbf{A}_S \otimes \mathbf{b}_i & \mathbf{b}_S \otimes \mathbf{A}_i \\ \mathbf{0} & \mathbf{A}_S & \mathbf{0} \\ \mathbf{0} & \mathbf{0} & \mathbf{A}_i \end{bmatrix}$$

$\mathbf{v}_S = \left[\mathbf{v}_S \otimes \mathbf{v}_i \ \mathbf{v}_S \ \mathbf{v}_i\right]^T$

$i = i + 1$

end

end

The algorithm is programmed in MATLAB as the **mgparallelsys** function, available in the software package **dysrel-am**.

Example 7.25 Four parallel systems are constructed using N identical components having DPH distributions as used for Example 7.20.

System reliability functions for $N = 3$ and 5 calculated using **dphparallelsys** and **mgparallelsys** are presented in Figure 7.18. Comparisons of their computation times for calculating 21 points of each system reliability, hazard, and MRL functions and also the system MTTF and SDTTF are given in Table 7.6.

Table 7.6 Computation times for calculating reliability measures of systems in Example 7.28.

	$N = 3$ (order = 64)		$N = 5$ (order = 3,124)	
System	DPH	MG	DPH	MG
1	0.03	0.23	2.89	0.47
2	0.03	0.24	3.03	0.70
3	0.03	0.25	3.65	0.43
4	0.12	0.29	80.06	0.67

Figure 7.18 shows that DPH distributions and their similar MG distributions produced identical graphs. However, Table 7.6 shows that DPH distributions require longer computation times than MG distributions for higher order distributions, but shorter computation times for lower order

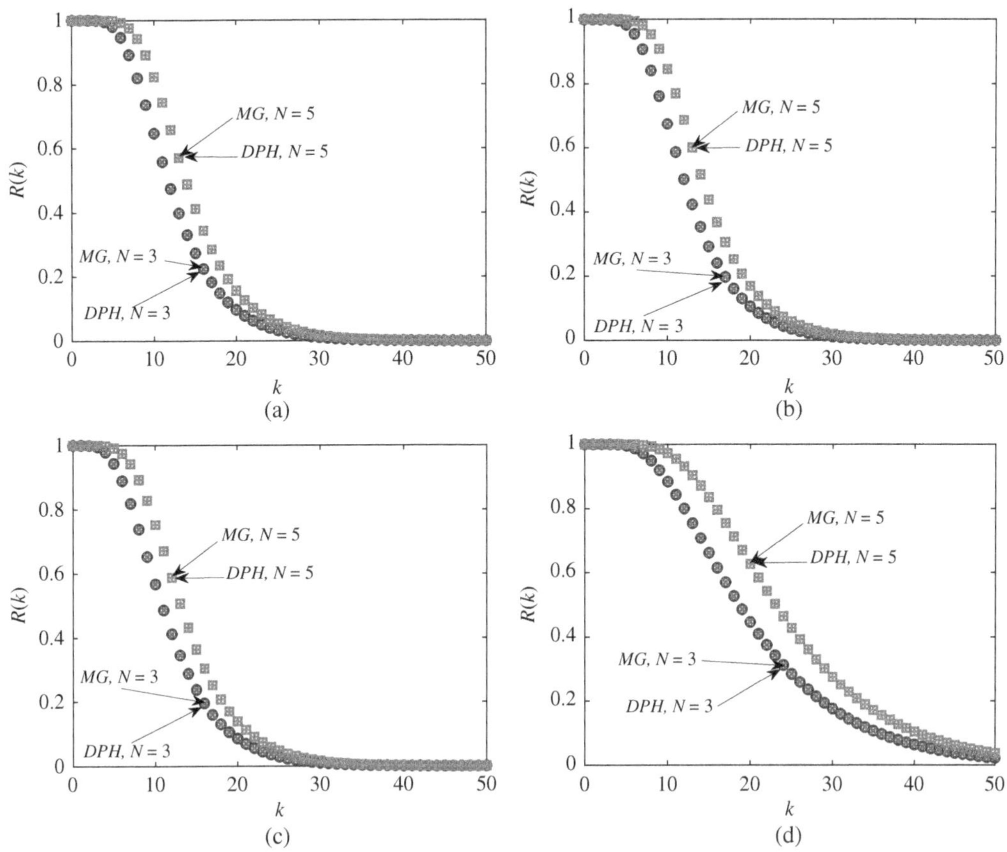

Figure 7.18 Reliability functions of systems in Example 7.28. (a) System 1, (b) system 2, (c) system 3, and (d) system 4.

distributions, which could due to the number of components or the orders of the DPH distributions of its component lifetimes. ■

The algorithm can be used for system optimization as shown in the following application.

Application 7.2 A system with a lifetime following 4-NB distribution with $MTTF = 20$ days is deployed to carry out a mission for s days. The mission may use several identical systems arranged in a parallel structure to increase mission reliability. The interest is to find the number of such systems in parallel to minimize the expected mission cost. The costs comprise of the procurement cost $P = 10$ per system and the penalty cost due to mission failure $F = 200$ per mission. The cost of operating the systems is negligible compared to the procurement and penalty costs.

The penalty cost is a random variable with a probability equal to the probability of system failure during the mission of s days, which is $1 - R(s)$. The expected mission cost TC is the sum of procurement and expected failure costs,

$$TC = NP + F\,(1 - R(s)) = 10N + 200(1 - R(s)),$$

where N is the number of systems in a parallel structure for the mission.

The optimum values of N for several mission times are summarized in Table 7.7. ■

Table 7.7 Expected mission cost as a function of mission time and N.

	Mission duration (days)			
N	**10**	**15**	**20**	**25**
1	34.17	80.37	127.71	163.20
2	**22.92**	44.76	89.28	137.35
3	30.35	**38.71**	70.77	119.89
4	40.04	43.06	**64.00**	108.86
5	50.01	51.08	64.12	102.75
6	60.00	60.38	68.31	**100.40**
7	70.00	70.13	74.89	100.95

The optimal values are given in bold.

Exercise 7.11 Compare the optimum number of systems in parallel for continuous time and discrete time for the same cost and system parameters and explain the results.

The next section discusses standby structure discrete-time systems with any switch.

7.7 Standby Structured Discrete-Time Systems

Modeling and analysis of standby systems in discrete time is based on Theorem 7.3 for the addition of two independent and not necessarily identical variables with DPH distributions. This theorem is the same as Theorem 2.3 for the addition of two independent variables with PH distributions. Therefore, the results for standby system in continuous time are applicable for standby system in discrete time without any adjustment. The following development is from standby systems in continuous time modified by replacing PH distributions with DPH distributions.

The DPH representation $(n_S, \boldsymbol{a}_S, \boldsymbol{A}_S)$ of the lifetime distribution of an imperfect switch standby system having components with lifetimes $T_i \backsim DPH(n_i, \boldsymbol{a}_i, \boldsymbol{A}_i)$, $i = 1, 2, \ldots, N$, and switch reliability p can be calculated using the same recursion equations for PH distribution but using $\boldsymbol{b}_i = (\boldsymbol{I} - \boldsymbol{A}_i)\boldsymbol{u}$,

$$n_S^{(i)} = n_S^{(i-1)} + n_i, \tag{7.74}$$

$$\boldsymbol{a}_S^{(i)} = \left[\boldsymbol{a}_S^{(i-1)} \ \boldsymbol{0}\right], \tag{7.75}$$

$$\boldsymbol{A}_S^{(i)} = \left[\begin{array}{c|c} \boldsymbol{A}_S^{(i-1)} & \begin{array}{c} \boldsymbol{0} \\ p_{i-1}\boldsymbol{b}_{i-1}\boldsymbol{a}_i \end{array} \\ \hline \boldsymbol{0} & \boldsymbol{A}_i \end{array}\right], \tag{7.76}$$

for $i = 3, 4, \ldots, N$, with initial conditions for $i = 1$ is using Equation (7.49), and for $i = 2$ is

$$n_S^{(i)} = n_1 + n_2, \quad \boldsymbol{a}_S^{(2)} = \left[\boldsymbol{a}_1 \ \boldsymbol{0}\right], \quad \boldsymbol{A}_S^{(2)} = \begin{bmatrix} \boldsymbol{A}_1 & p_1\boldsymbol{b}_1\boldsymbol{a}_2 \\ \boldsymbol{0} & \boldsymbol{A}_2 \end{bmatrix}. \tag{7.77}$$

These recursion equations are also applicable for a perfect switch standby systems by setting $p_i = 1$ for i = 1, 2, …, $N - 1$.

The algorithm for calculating the DPH representation of the distribution of the lifetime of N-standby discrete-time system with an arbitrary switch is designed using Equations (7.74)–(7.76) and is called Algorithm 7.5. Its pseudo-code is as follows:

Algorithm 7.5

```
input N, p_i, i = 1, 2,…, N - 1, (n_i, a_i, A_i), i = 1, 2,…, N
n_S = n_1
a_S = a_1
A_S = A_1
for i = 2 to N
    n_S = n_S + n_i
    a_S = [a_S 0]
    b_i = (I - A_i)u
    if i = 2 then A_S = [A_S  p_{i-1} b_{i-1} a_i ; 0  A_i]
    if i > 2  then A_S = [A_S  [0 ; p_{i-1} b_{i-1} a_i] ; 0  A_i]
    i = i + 1
end
end
```

This algorithm is programmed in MATLAB as the **dphstandsys** function, available in the software package **dysrel-am**.

The DPH distribution for the lifetime of an N-standby system with an arbitrary switch has order equal to $n_1 + n_2 + \cdots + n_N$; hence, its complexity is nN.

Following the results of continuous-time systems, the state transition diagram of N-standby discrete-time system is as given in Figure 7.19. This diagram is applicable for any switch by acknowledging that the perfect switch has $1 - p_i = 0$ and deleting all transitions with zero value.

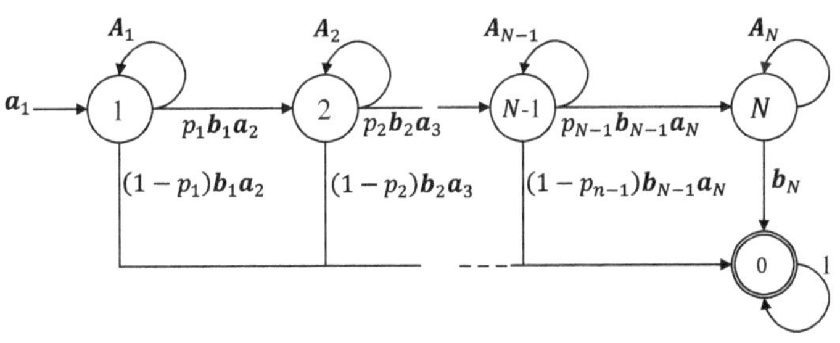

Figure 7.19 State transition diagram of N-standby discrete-time system with an imperfect switch.

The survival matrix and initial probability vector of N-standby systems in discrete time can be obtained from Figure 7.19 as

$$\boldsymbol{A}_S = \begin{bmatrix} \boldsymbol{A}_1 & p_1\boldsymbol{b}_1\boldsymbol{a}_2 & \boldsymbol{0} & \boldsymbol{0} & \boldsymbol{0} & \boldsymbol{0} \\ \boldsymbol{0} & \boldsymbol{A}_2 & p_2\boldsymbol{b}_2\boldsymbol{a}_3 & \boldsymbol{0} & \boldsymbol{0} & \boldsymbol{0} \\ \vdots & \vdots & \ddots & \ddots & \vdots & \vdots \\ \boldsymbol{0} & \boldsymbol{0} & \boldsymbol{0} & \boldsymbol{A}_i & p_i\boldsymbol{b}_i\boldsymbol{a}_{i+1} & \boldsymbol{0} \\ \vdots & \vdots & \vdots & \ddots & \ddots & \ddots \\ \boldsymbol{0} & \boldsymbol{0} & \boldsymbol{0} & \boldsymbol{0} & \boldsymbol{0} & \boldsymbol{A}_N \end{bmatrix}, \tag{7.78}$$

$$\boldsymbol{a}_S = \left[\boldsymbol{a}_1\ \boldsymbol{0}\ \ldots\ \boldsymbol{0}\ \boldsymbol{0}\right], \tag{7.79}$$

the same as for standby systems in continuous time.

Algorithm 7.5 is illustrated by manually calculating the DPH representation of the distribution of lifetime of a discrete-time standby system having CFR components.

Example 7.26 A 3-standby discrete-time system with a perfect switch is constructed using the same components as in Example 7.16. The DPH representation of the distribution of its lifetime is derived using Algorithm 7.5.

The component lifetime distributions have DPH representations:

$$n_1 = n_2 = n_3 = 1, \quad \boldsymbol{a}_1 = \boldsymbol{a}_2 = \boldsymbol{a}_3 = [1],$$

$$\boldsymbol{A}_1 = [0.5], \quad \boldsymbol{A}_2 = [0.8], \quad \boldsymbol{A}_3 = [0.9], \quad \boldsymbol{b}_1 = [0.5], \quad \boldsymbol{b}_2 = [0.2], \quad \boldsymbol{b}_3 = [0.1].$$

Iterations using Equations (7.74)–(7.76) are as follows:

- Initial conditions are determined using Equation (7.49):

$$n_S^{(1)} = n_1 = 1, \quad \boldsymbol{a}_S^{(1)} = \boldsymbol{a}_1 = [1], \quad \boldsymbol{A}_S^{(1)} = \boldsymbol{A}_1 = [0.5].$$

- Second and third iterations for n_S using Equation (7.74):

$$n_S^{(2)} = n_S^{(1)} + n_2 = 1 + 1 = 2, \quad n_S^{(3)} = n_S^{(2)} + n_3 = 2 + 1 = 3.$$

- Second and third iterations for $\boldsymbol{a}_S$ using Equation (7.75):

$$\boldsymbol{a}_S^{(2)} = \left[\boldsymbol{a}_S^{(1)}\ \boldsymbol{0}\right] = \left[[1]\ 0\right] = \left[1\ 0\right],$$

$$\boldsymbol{a}_S^{(3)} = \left[\boldsymbol{a}_S^{(2)}\ \boldsymbol{0}\right] = \left[[1\ 0]\ 0\right] = \left[1\ 0\ 0\right].$$

- Second and third iterations for $\boldsymbol{A}_S$ using Equation (7.76):

$$\boldsymbol{A}_S^{(2)} = \begin{bmatrix} \boldsymbol{A}_S^{(1)} & \boldsymbol{b}_S^{(1)}\boldsymbol{a}_2 \\ \boldsymbol{0} & \boldsymbol{A}_2 \end{bmatrix} = \begin{bmatrix} [0.5] & [0.5][1] \\ 0 & [0.8] \end{bmatrix} = \begin{bmatrix} 0.5 & 0.5 \\ 0 & 0.8 \end{bmatrix},$$

$$\boldsymbol{b}_S^{(2)} = \left(\boldsymbol{I} - \boldsymbol{A}_S^2\right)\boldsymbol{u} = \begin{bmatrix} 0.5 & -0.5 \\ 0 & 0.2 \end{bmatrix} \begin{bmatrix} 1 \\ 1 \end{bmatrix} = \begin{bmatrix} 0 \\ 0.2 \end{bmatrix},$$

$$\boldsymbol{A}_S^{(3)} = \begin{bmatrix} \boldsymbol{A}_S^{(2)} & \boldsymbol{b}_S^{(2)}\boldsymbol{a}_3 \\ \boldsymbol{0} & \boldsymbol{A}_3 \end{bmatrix} = \begin{bmatrix} \begin{bmatrix} 0.5 & 0.5 \\ 0 & 0.8 \end{bmatrix} & \begin{bmatrix} 0 \\ 0.2 \end{bmatrix} [1] \\ \boldsymbol{0} & [0.9] \end{bmatrix} = \begin{bmatrix} 0.5 & 0.5 & 0 \\ 0 & 0.8 & 0.2 \\ 0 & 0 & 0.9 \end{bmatrix}.$$

The DPH representation of the system lifetime distribution can also be derived using Equation (7.78) for $N = 3$,

$$A_S = \begin{bmatrix} \boldsymbol{A}_1 & p_1\boldsymbol{b}_1\boldsymbol{a}_2 & \boldsymbol{0} \\ \boldsymbol{0} & \boldsymbol{A}_2 & p_2\boldsymbol{b}_2\boldsymbol{a}_3 \\ \boldsymbol{0} & \boldsymbol{0} & \boldsymbol{A}_3 \end{bmatrix} = \begin{bmatrix} [0.5] & (1)[0.5][1] & 0 \\ 0 & [0.8] & (1)[0.2][1] \\ 0 & 0 & [0.9] \end{bmatrix} = \begin{bmatrix} 0.5 & 0.5 & 0 \\ 0 & 0.8 & 0.2 \\ 0 & 0 & 0.9 \end{bmatrix},$$

the same as produced using Algorithm 7.5. ■

The resulting values of $\boldsymbol{a}_S$ and $\boldsymbol{A}_S$ match a 3-hypogeometry distribution in Table 7.2 with $p_1 = 0.5$, $p_2 = 0.2$, $p_3 = 0.1$. This result can be generalized to N-standby discrete-time systems with perfect switch. If component i of an N-standby discrete-time system with perfect switch has a lifetime following a geometry distribution with transition probability p_i, $i = 1, 2, \ldots, N$, then the system lifetime has an N-hypogeometry distribution with parameter p_i, $i = 1, 2, \ldots, N$. If the components are identical, then the system lifetime has an N-NB distribution with parameter p.

Exercise 7.12 Redo Example 7.26 for identical components to determine the distribution of the system lifetime for an arbitrary N. Represent the system *MTTF* in terms of N and the component *MTTF*.

Example 7.27 The system in Example 7.26 uses an imperfect switch with reliability = 0.8. The DPH representation of the system lifetime distribution is derived as follows.

Iterations using Algorithm 7.5 following Equations (7.74)–(7.76) are as follows:

- Initial conditions are determined using Equation (7.49):

$$n_S^{(1)} = n_1 = 1, \quad \boldsymbol{a}_S^{(1)} = \boldsymbol{a}_1 = [1], \quad \boldsymbol{A}_S^{(1)} = \boldsymbol{A}_1 = [-0.5].$$

- Second and third iterations for n_S using Equation (7.74) yields

$$n_S^{(2)} = n_S^{(1)} + n_2 = 2, \quad n_S^{(3)} = n_S^{(2)} + n_3 = 3.$$

- Second and third iterations to generate $\boldsymbol{a}_S$ using Equation (7.75) yields

$$\boldsymbol{a}_S^{(2)} = \begin{bmatrix} \boldsymbol{a}_S^{(1)} & \boldsymbol{0} \end{bmatrix} = \begin{bmatrix} [1] & 0 \end{bmatrix} = \begin{bmatrix} 1 & 0 \end{bmatrix},$$

$$\boldsymbol{a}_S^{(3)} = \begin{bmatrix} \boldsymbol{a}_S^{(2)} & \boldsymbol{0} \end{bmatrix} = \begin{bmatrix} [1 \ 0] & 0 \end{bmatrix} = \begin{bmatrix} 1 & 0 & 0 \end{bmatrix}.$$

- Second and third iterations to generate $\boldsymbol{A}_S$ using Equation (7.76) yields

$$\boldsymbol{A}_S^{(2)} = \begin{bmatrix} \boldsymbol{A}_S^{(1)} & p_1\boldsymbol{b}_1\boldsymbol{a}_2 \\ \boldsymbol{0} & \boldsymbol{A}_2 \end{bmatrix} = \begin{bmatrix} [0.5] & 0.8[0.5][1] \\ 0 & [0.8] \end{bmatrix} = \begin{bmatrix} 0.5 & 0.4 \\ 0 & 0.8 \end{bmatrix},$$

$$\boldsymbol{A}_S^{(3)} = \begin{bmatrix} \boldsymbol{A}_S^2 & \begin{matrix} \boldsymbol{0} \\ p_2\boldsymbol{b}_2\boldsymbol{a}_3 \end{matrix} \\ \boldsymbol{0} & \boldsymbol{A}_3 \end{bmatrix} = \begin{bmatrix} \begin{bmatrix} 0.5 & 0.4 \\ 0 & 0.8 \end{bmatrix} & \begin{bmatrix} 0 \\ 0.8[0.2][1] \end{bmatrix} \\ 0 & [0.9] \end{bmatrix} = \begin{bmatrix} 0.5 & 0.4 & 0 \\ 0 & 0.8 & 0.16 \\ 0 & 0 & 0.9 \end{bmatrix}.$$

Using Equation (7.78) for $N = 3$ results in

$$\boldsymbol{A}_S = \begin{bmatrix} \boldsymbol{A}_1 & p_1\boldsymbol{b}_1\boldsymbol{a}_2 & \boldsymbol{0} \\ \boldsymbol{0} & \boldsymbol{A}_2 & p_2\boldsymbol{b}_2\boldsymbol{a}_3 \\ \boldsymbol{0} & \boldsymbol{0} & \boldsymbol{A}_3 \end{bmatrix} = \begin{bmatrix} [0.5] & (0.8)[0.5][1] & 0 \\ 0 & [0.8] & (0.8)[0.2][1] \\ 0 & 0 & [0.9] \end{bmatrix} = \begin{bmatrix} 0.5 & 0.4 & 0 \\ 0 & 0.8 & 0.16 \\ 0 & 0 & 0.9 \end{bmatrix},$$

the same as produced using Algorithm 7.5. ■

As for series and parallel systems, manual calculations seem difficult and inefficient for standby systems with a large number of components or with components having lifetimes following DPH distributions of high orders. Instead, it must seek the help from the **dphstandsys** function.

Example 7.28 The **dphstandsys** function is used to calculate the DPH representation of the distribution of the lifetime of a 5-standby system with a perfect switch. The kth component of this system has lifetime following k-NB distributions with $MTTF = 100$, $k = 1, 2, \ldots, 5$. The DPH representation produced is used to calculate and plot the system reliability and MRL functions and also the system $MTTF$ and $SDTTF$.

The main program for this purpose is given in **example7_28.m** file in the software package **dysrel-am**. Its MATLAB code is as follows:

MATLAB code for example7_28.m

```
clear;close all; addpath('..\dysrel-am')
% DPH Components                                     inputs
N = 5;
mttfc = 100;
for k = 1:N
  T(k) = dphnegativebinomial(k,mttfc);
end
% DPH Standby system                                 model
Ts = dphstandsys(T);
% Time vector
kmax = 1000;
k = 0:20:kmax;
% Reliability plot                                   graphical output
dphrelplot(k,Ts);
% Hazard plot
dphhazplot(k,Ts);
% Result in command window                           statistical output
mttf = dphmttf(Ts)
sdttf = dphsdttf(Ts)
```

Executing this program produces Figure 7.20.

```
mttf =
   500
sdttf =
  149.4434
```

■

Example 7.29 Example 7.28 is repeated for imperfect switches with reliability $p = 1, 0.8, 0.6$. The results are shown in Figure 7.21. ■

Example 7.30 Three components in Example 7.19 are arranged in a standby structure with perfect switch. The reliability and hazard functions of the system and all its components are shown in Figure 7.22.

Figure 7.22a shows that the system reliability function at any time is always higher than the highest reliability function of its components at that time. The expected lifetimes of the system and

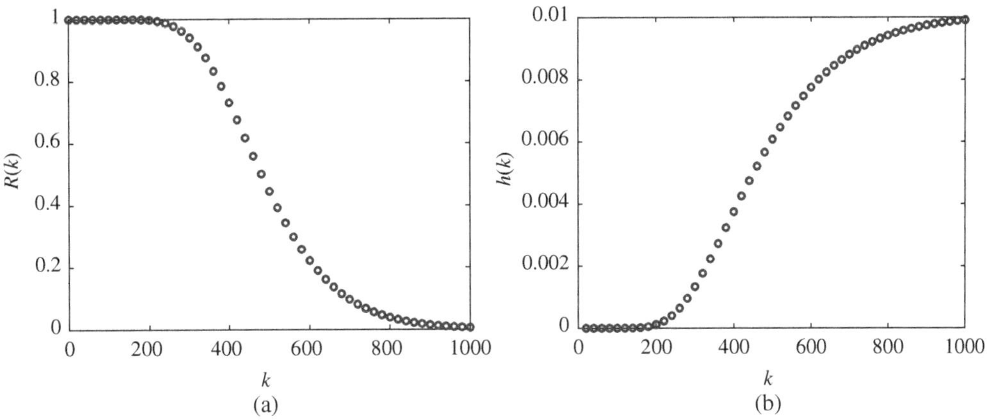

Figure 7.20 Outputs of **example7_28.m** file.

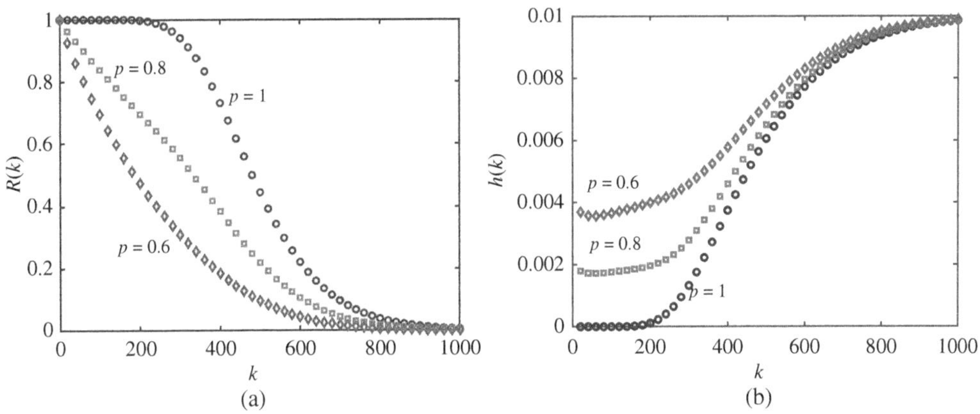

Figure 7.21 Reliability measures of systems in Example 7.29. (a) System reliability function and (b) system hazard function.

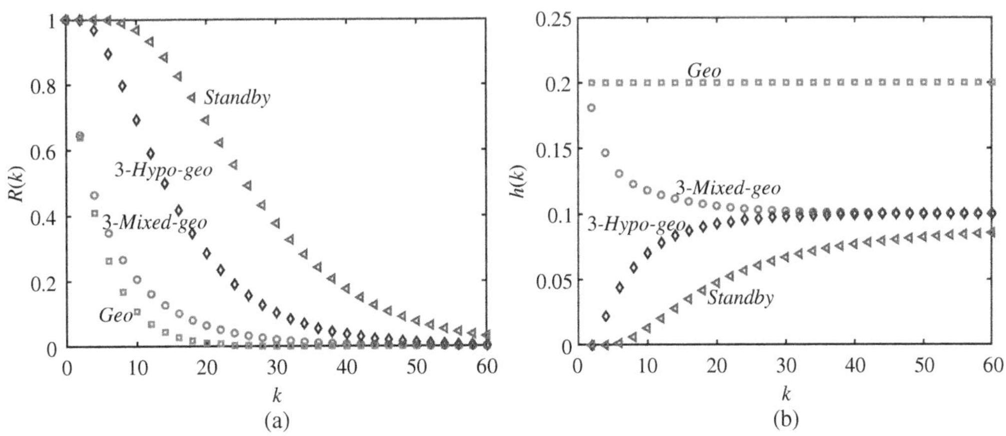

Figure 7.22 Reliability and hazard functions of the system and components in Example 7.30. (a) Reliability functions and (b) hazard functions.

all components are $MTTF_1 = 6.9$, $MTTF_2 = 5$, $MTTF_3 = 17$, and $MTTF_S = 28.9$ is longer than the longest component $MTTF$. These results can be used to make a conjecture that standby systems in discrete time have the same behavior as standby systems in continuous time. ■

Exercise 7.13 Show that standby systems in discrete time have the same properties as standby systems in continuous time.

Application 7.3 Application 7.2 is repeated for the systems to carry out the mission arranged in a standby structure with a perfect switch. The resulting total cost as a function of N and mission duration are presented in Table 7.8.

Table 7.8 Expected mission cost as a function of mission duration and N.

	Mission duration (days)			
N	20	30	40	50
1	127.71	185.46	204.31	208.87
2	**26.43**	67.84	132.57	181.92
3	30.02	**31.90**	47.50	87.87
4	40.00	40.01	**40.59**	**46.16**
5	50.00	50.00	50.00	50.19

The optimum values are given in bold.

■

Exercise 7.14 Compare the optimum number of systems in standby structure to carry out the mission for discrete and continuous time with the same cost and $MTTF$.

Theorem 7.7 can be utilized to derive the recursion equations for generating an MG representation of lifetime distribution of standby system with components having MG distributed lifetimes. The resulting equations are similar to equations for DPH distribution with an additional equation for calculating the closing vectors:

$$n_S^{(i)} = n_S^{(i-1)} + n_i, \tag{7.80}$$

$$\boldsymbol{a}_S^{(i)} = \left[\boldsymbol{a}_S^{(i-1)} \ \boldsymbol{0}\right], \tag{7.81}$$

$$\boldsymbol{A}_S^{(i)} = \left[\begin{array}{c|c} \boldsymbol{A}_S^{(i-1)} & \begin{array}{c} \boldsymbol{0} \\ p_{i-1}\boldsymbol{b}_{i-1}\boldsymbol{a}_i \end{array} \\ \hline \boldsymbol{0} & \boldsymbol{A}_i \end{array}\right], \tag{7.82}$$

$$\boldsymbol{v}_S^{(i)} = \left[\boldsymbol{v}_S^{(i-1)} \ \boldsymbol{v}_i\right]^T, \tag{7.83}$$

for $i = 3, 4, \ldots, N$, with the initial conditions for $i = 1$ is given by Equation (7.49), and for $i = 2$ is given by

$$\boldsymbol{a}_S^{(2)} = \left[\boldsymbol{a}_1 \ \boldsymbol{0}\right], \quad \boldsymbol{A}_S^{(2)} = \begin{bmatrix} \boldsymbol{A}_1 & p_1\boldsymbol{b}_1\boldsymbol{a}_2 \\ \boldsymbol{0} & \boldsymbol{A}_2 \end{bmatrix}, \quad \boldsymbol{v}_S^{(2)} = \begin{bmatrix} \boldsymbol{v}_1 \\ \boldsymbol{v}_2 \end{bmatrix}. \tag{7.84}$$

Algorithm 7.5 can be modified slightly to construct Algorithm 7.6 for standby systems having MG distribution. Its pseudo-code is derived from the pseudo-code of Algorithm 7.5 by adding lines for calculating $\boldsymbol{v}_S$.

Algorithm 7.6

input N, p_i, i = 1, 2, …, N - 1, (n_i, $\boldsymbol{a}_i$, $\boldsymbol{A}_i$, $\boldsymbol{v}_i$), i = 1, 2, …, N
$n_S = n_1$
$\boldsymbol{a}_S = \boldsymbol{a}_1$
$\boldsymbol{A}_S = \boldsymbol{A}_1$
$\boldsymbol{v}_S = \boldsymbol{v}_1$
$\boldsymbol{b}_1 = (\boldsymbol{I} - \boldsymbol{A}_1)\boldsymbol{v}_1$
for i = 2 to N
$n_S = n_S + n_i$
$\boldsymbol{a}_S = [\boldsymbol{a}_S \;\; \boldsymbol{0}]$
$\boldsymbol{b}_i = (\boldsymbol{I} - \boldsymbol{A}_i)\boldsymbol{v}_i$
$\boldsymbol{v}_S = \left[\boldsymbol{v}_S \;\; \boldsymbol{v}_i\right]^T$
if i = 2 then $\boldsymbol{A}_S = \begin{bmatrix} \boldsymbol{A}_S & p_{i-1}\boldsymbol{b}_{i-1}\boldsymbol{a}_i \\ \boldsymbol{0} & \boldsymbol{A}_i \end{bmatrix}$
if i > 2 then $\boldsymbol{A}_S = \begin{bmatrix} \boldsymbol{A}_S & \begin{matrix} \boldsymbol{0} \\ p_{i-1}\boldsymbol{b}_{i-1}\boldsymbol{a}_i \end{matrix} \\ \boldsymbol{0} & \boldsymbol{A}_i \end{bmatrix}$
$i = i + 1$
end
end

The algorithm is programmed in MATLAB as the **mgstandsys** function, available in the software package **dysrel-am**.

Example 7.31 Four standby systems are constructed using N identical components having DPH distributions as given in Example 7.20.

System reliability functions for N = 3 and 6 calculated using **dphstandsys** and **mgstandsys** are presented in Figure 7.23. Comparisons of computation times for calculating 21 points of system reliability, hazard, and MRL functions, and system MTTF and SDTTF are given in Table 7.9.

Table 7.9 Computation times for calculating reliability measures of systems in Example 7.31.

	$N = 3$ (order = 12)		$N = 6$ (order = 24)	
System	**DPH**	**MG**	**DPH**	**MG**
1	0.02	0.24	0.02	0.27
2	0.02	0.24	0.02	0.28
3	0.02	0.24	0.02	0.28
4	0.02	0.24	0.02	0.29

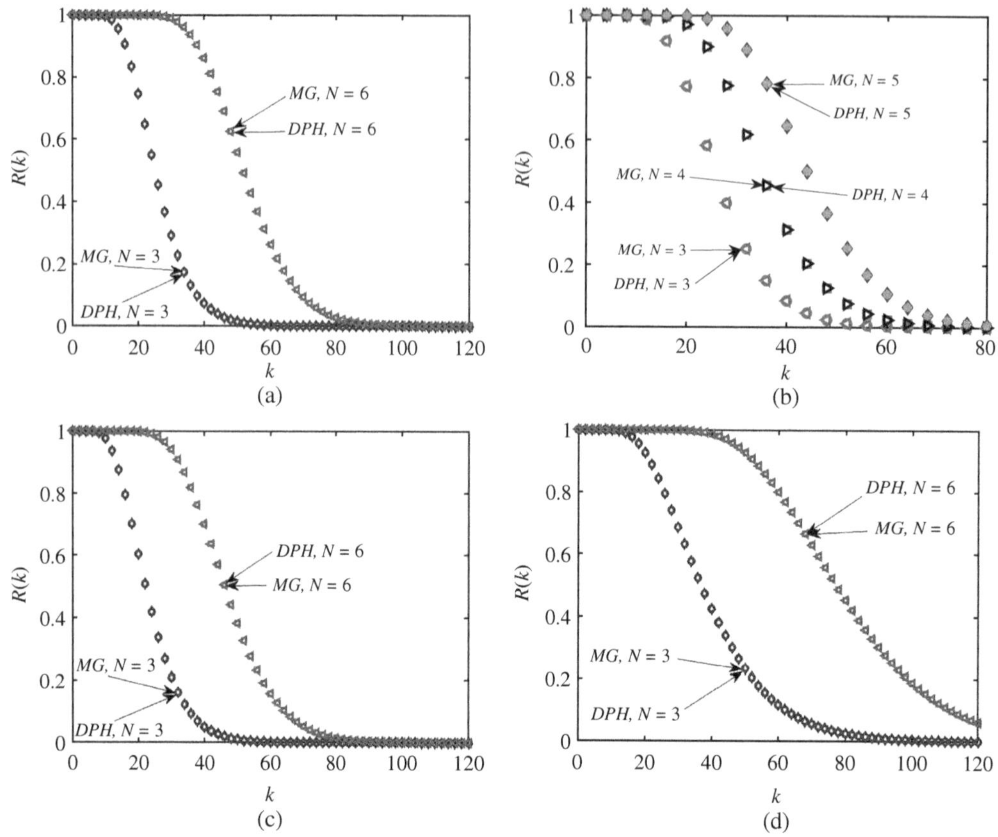

Figure 7.23 Reliability functions of systems in Example 7.31. (a) System 1, (b) system 2, (c) system 3, and (d) system 4.

Computation time for calculation using MG distribution included the time for transforming DPH distributions into MG distributions.

Contrary to series and parallel structures, transforming DPH distribution into MG distribution in a Jordan canonical form makes the computation time slower. The reason is because the effect of the number of components to the order of lifetime distributions is less pronounced compared to system in series and parallel structures. Therefore, this technique should be used for series and, particularly, parallel structures whose orders grow rapidly with the number of components, not for standby structure. ■

Components of imperfect switch standby systems in discrete time should also be ordered following the same rule as for continuous time; that is, components with lower hazard functions should be placed earlier in the backup sequence to increase the system reliability.

Example 7.32 A 3-standby system with switch reliability = 0.8 is constructed using three components with lifetimes follow 2-NB distribution with components 1, 2, and 3 having $MTTF = 5, 10$, and 20, respectively. The component hazard functions are exhibited in Figure 7.24(a), which shows that

$$h_1(k) > h_2(k) > h_3(k)$$

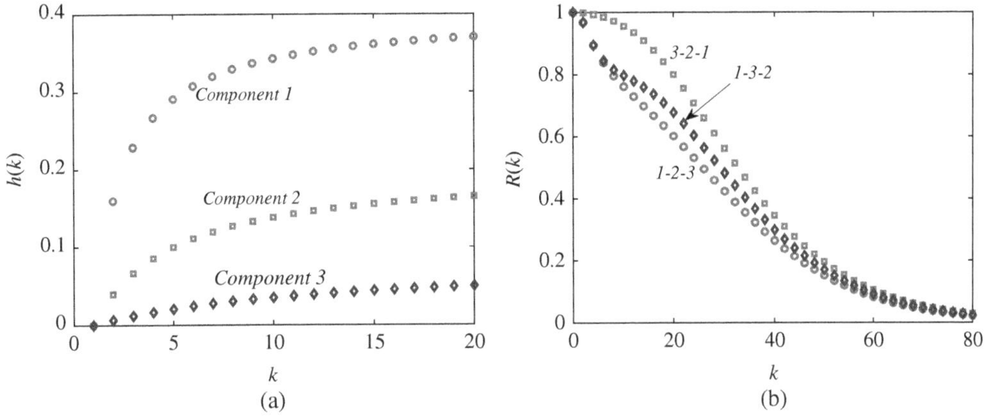

Figure 7.24 Effect of component ordering in imperfect switch standby systems. (a) Component hazard function and (b) system reliability function.

where $h_1(k)$, $h_2(k)$, and $h_3(k)$ denote the hazard functions of components with $MTTF = 5$, 10, and 25, respectively.

Various component orderings are examined, and the resulting system reliability function are shown in Figure 7.24b.

This result confirms that components with lower hazard functions should be placed earlier in the backup sequence to increase the system reliability function. Component sequences 3-2-1, 1-2-3, and 3-1-2 produced systems with $MTTF = 31.2$, 25.8, and 30.4, respectively, indicating that the increasing hazard function component ordering, which is 3-2-1 sequence, produced the highest system $MTTF$ as well as the highest system reliability function. On the contrary, the decreasing hazard function component ordering, which is 1-2-3 sequence, produced the lowest $MTTF$ and reliability function. Although the graphs are not clear, the reliability value of the 1-2-3 system is always lower than the 1-3-2 system. ■

Exercise 7.15 Show numerically that the effect of ordering is diminishing with the increase in switch reliability.

Exercise 7.16 Investigate the effect of ordering CFR components.

The discussion of discrete-time system reliability modeling and analysis continues with K-out-of-N structure.

7.8 *K*-Out-of-*N* Structured Discrete-Time Systems

K-out-of-N discrete-time systems are generalizations of N-parallel discrete-time systems, namely, N-parallel discrete-time systems are K-out-of-N discrete-time systems with $K = 1$. The state transition diagrams for a K-out-of-N system can be derived from state transition diagram of an N-parallel systems by discarding states with less than K good components because they belong to a set of absorbing states. For example, the state transition diagram of a 2-out-of-3 discrete-time system can be derived from the state transition diagram of 3-parallel discrete-time system given in Figure 7.15 by lumping states (g, f, f), (f, g, f), (f, f, g), and (f, f, f) into one absorbing superstate since in those

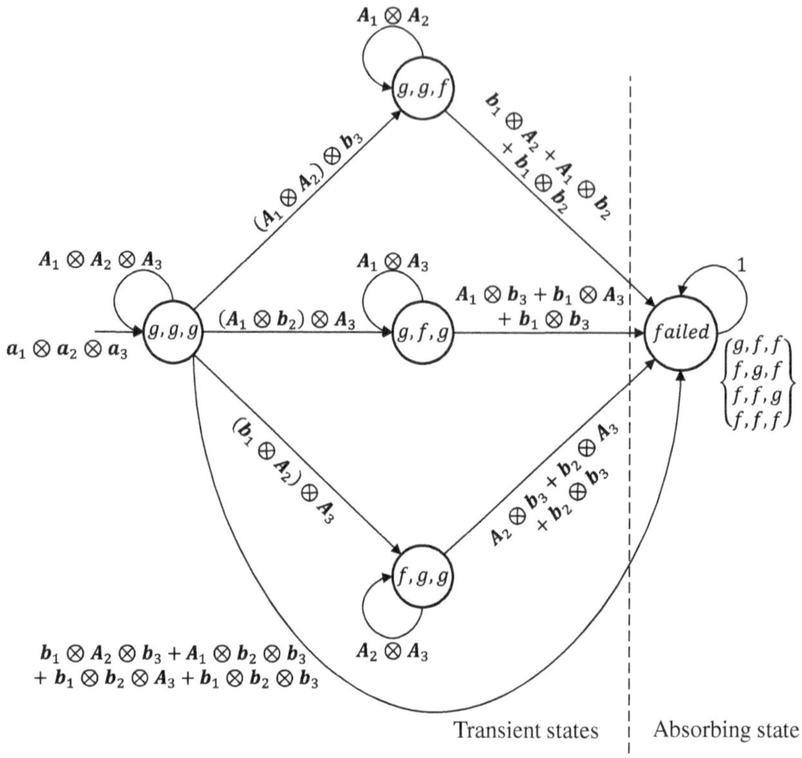

Figure 7.25 State transition diagram of a 2-out-of-3 discrete-time system.

states the system is failed. The resulting state transition diagram is exhibited in Figure 7.25. The states follow the same sequence as in continuous time.

K-out-of-N discrete-time systems have more transitions than their continuous-time system counterparts as evidence from comparing Figures 5.3 and 7.25. State transition matrices for a 2-out-of-4 discrete-time system are given in Table 7.10, which contains more transitions than Table 5.1 for its continuous-time system counterpart.

As for continuous-time systems, the transition matrices in a survival matrix of a K-out-of-N discrete-time systems can be depicted in a generic state transition diagram shown in Figure 7.26.

Expressions for $\boldsymbol{A}_{VV}$ and $\boldsymbol{A}_{WW}$ can be determined from Table 7.10 as the Kronecker products of the survival matrices of the good components at the system state V and W, respectively. Likewise, $\boldsymbol{A}_{VW}$ is the Kronecker products of the survival matrices of the surviving components and the failure vector of the failed components that causes the transition from V to W. Thus,

$$\boldsymbol{A}_{VV} = \otimes \boldsymbol{A}_i, \qquad i|V_i = g, \tag{7.85}$$

$$\boldsymbol{A}_{VW} = \otimes \boldsymbol{X}_i, \qquad i|V_i = g, \tag{7.86}$$

where

$$\begin{aligned} \boldsymbol{X}_i &= \boldsymbol{A}_i \qquad \text{if } W_i = g, \\ &= \boldsymbol{b}_i \qquad \text{if } W_i = f. \end{aligned}$$

Since no component has failed before being used, the initial probability vector is

$$\boldsymbol{a}_S = \left[\otimes_1^N \boldsymbol{a}_i \; \boldsymbol{0} \; \dots \; \boldsymbol{0}\right], \tag{7.87}$$

Table 7.10 Some block matrices in the survival matrix of a 2-out-of-4 discrete-time system.

Current system state	Block diagonal matrix	Next system state	Block off-diagonal matrix
(g, g, g, g)	$\boldsymbol{A}_1 \otimes \boldsymbol{A}_2 \otimes \boldsymbol{A}_3 \otimes \boldsymbol{A}_4$	(g, g, g, f)	$\boldsymbol{A}_1 \otimes \boldsymbol{A}_2 \otimes \boldsymbol{A}_3 \otimes \boldsymbol{b}_4$
		(g, g, f, f)	$\boldsymbol{A}_1 \otimes \boldsymbol{A}_2 \otimes \boldsymbol{b}_3 \otimes \boldsymbol{b}_4$
		(g, f, f, f)	$\boldsymbol{A}_1 \otimes \boldsymbol{b}_2 \otimes \boldsymbol{b}_3 \otimes \boldsymbol{b}_4$
		(f, f, f, f)	$\boldsymbol{b}_1 \otimes \boldsymbol{b}_2 \otimes \boldsymbol{b}_3 \otimes \boldsymbol{b}_4$
		11 other nonzero transitions	
(g, g, g, f)	$\boldsymbol{A}_1 \otimes \boldsymbol{A}_2 \otimes \boldsymbol{A}_3$	(g, g, f, f)	$\boldsymbol{A}_1 \otimes \boldsymbol{A}_2 \otimes \boldsymbol{b}_3$
		(g, f, f, f)	$\boldsymbol{A}_1 \otimes \boldsymbol{b}_2 \otimes \boldsymbol{b}_3$
		(f, f, f, f)	$\boldsymbol{b}_1 \otimes \boldsymbol{b}_2 \otimes \boldsymbol{b}_3$
		4 other nonzero transitions	
(g, g, f, g)	$\boldsymbol{A}_1 \otimes \boldsymbol{A}_2 \otimes \boldsymbol{A}_4$	(g, g, f, f)	$\boldsymbol{A}_1 \otimes \boldsymbol{A}_2 \otimes \boldsymbol{b}_4$
		(g, f, f, f)	$\boldsymbol{A}_1 \otimes \boldsymbol{b}_2 \otimes \boldsymbol{b}_4$
		(f, f, f, f)	$\boldsymbol{b}_1 \otimes \boldsymbol{b}_2 \otimes \boldsymbol{b}_4$
		4 other nonzero transitions	
(g, f, g, g)	$\boldsymbol{A}_1 \otimes \boldsymbol{A}_3 \otimes \boldsymbol{A}_4$	(g, f, g, f)	$\boldsymbol{A}_1 \otimes \boldsymbol{A}_3 \otimes \boldsymbol{b}_4$
		(g, f, f, f)	$\boldsymbol{A}_1 \otimes \boldsymbol{b}_3 \otimes \boldsymbol{b}_4$
		(f, f, f, f)	$\boldsymbol{b}_1 \otimes \boldsymbol{b}_3 \otimes \boldsymbol{b}_4$
		4 other nonzero transitions	
(f, g, g, g)	$\boldsymbol{A}_2 \otimes \boldsymbol{A}_3 \otimes \boldsymbol{A}_4$	(f, g, g, f)	$\boldsymbol{A}_2 \otimes \boldsymbol{A}_3 \otimes \boldsymbol{b}_4$
		(f, g, f, f)	$\boldsymbol{A}_2 \otimes \boldsymbol{b}_3 \otimes \boldsymbol{b}_4$
		(f, f, f, f)	$\boldsymbol{b}_2 \otimes \boldsymbol{b}_3 \otimes \boldsymbol{b}_4$
		4 other nonzero transitions	
(g, g, f, f)	$\boldsymbol{A}_1 \otimes \boldsymbol{A}_2$	Absorbing states	Not required (not part of the survival matrix)
(g, f, g, f)	$\boldsymbol{A}_1 \otimes \boldsymbol{A}_3$		
(g, f, f, g)	$\boldsymbol{A}_1 \otimes \boldsymbol{A}_4$		
(f, g, g, f)	$\boldsymbol{A}_2 \otimes \boldsymbol{A}_3$		
(f, g, f, g)	$\boldsymbol{A}_2 \otimes \boldsymbol{A}_4$		
(f, f, g, g)	$\boldsymbol{A}_3 \otimes \boldsymbol{A}_4$		

Figure 7.26 States and matrix transitions in a survival matrix.

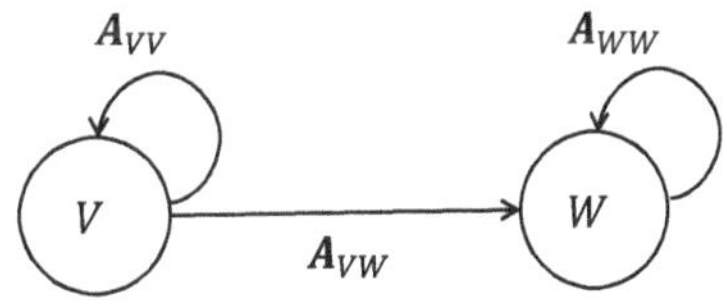

where $\otimes_1^N \boldsymbol{a}_i = \boldsymbol{a}_1 \otimes \boldsymbol{a}_2 \ldots \otimes \boldsymbol{a}_N$ and the dimensions of $\boldsymbol{0}$'s follow the dimensions of the next states in sequence.

Exercise 7.17 Use Equations (7.85) and (7.86) to generate the survival matrices of a 2-out-of-3 system. Compare the results with Figure 7.25.

The algorithm for generating the survival matrix of a K-out-of-N discrete-time system is referred to as Algorithm 7.7.

Algorithm 7.7

---

1) Generate the system states as N-tupple of component states.
2) Discard system states with less than K good components.
3) Determine the block diagonal matrix for each system state obtained from step 2 using Equation (7.85).
4) Determine possible transitions for each pair of system states. A possible transition is from a state to other states with any good components in the current state may fail in the next states.
5) Determine the block off-diagonal matrix using Equation (7.86) for each possible transition obtained from step 4.
6) Set a value **0** for other block off-diagonal matrices.

Algorithm 7.7 is programmed in MATLAB as the **dphkofnsys** function. Its MATLAB code is available in the software package **dysrel-am**. The function receives N, K, and $(n_i, \boldsymbol{a}_i, \boldsymbol{A}_i)$, $i = 1, 2, \ldots, N$, as inputs and returns the PH representation of the system lifetime distribution as the output.

The order of the DPH distribution of the lifetime of K-out-of-N system is equal to the order of the DPH distribution of the lifetime of N-parallel system minus the dimensions of the removed block matrices. For identical components, the order is $\sum_{j=K}^{N} C_N^j n^j$ where $C_N^j = N!/(j!(N-j)!)$.

Exercise 7.18 Prepare a similar table to Table 7.10 for 3-out-of-5 discrete-time system.

Manual calculation for generating DPH representations of K-out-of-N discrete-time systems is illustrated with a 2-out-of-3 systems having CFR components.

Example 7.33 A 2-out-of-3 system is constructed using components in Example 7.16.

This system has transient states (g, g, g), (g, f, g), (f, g, g), and(g, g, f), while other states are absorbing. The survival matrix can be derived using Figure 7.25 as

$$\boldsymbol{A}_S = \begin{array}{c} \\ \end{array}\begin{matrix} ggg & gfg & fgg & ggf \end{matrix}$$

$$\boldsymbol{A}_S = \begin{bmatrix} \boldsymbol{A}_1 \otimes \boldsymbol{A}_2 \otimes \boldsymbol{A}_3' & \boldsymbol{A}_1 \otimes \boldsymbol{b}_2 \otimes \boldsymbol{A}_3 & \boldsymbol{b}_1 \otimes \boldsymbol{A}_2 \otimes \boldsymbol{A}_3 & \boldsymbol{A}_1 \otimes \boldsymbol{b}_2 \otimes \boldsymbol{b}_3 \\ \boldsymbol{0} & \boldsymbol{A}_1 \otimes \boldsymbol{A}_3 & \boldsymbol{0} & \boldsymbol{0} \\ \boldsymbol{0} & \boldsymbol{0} & \boldsymbol{A}_2 \otimes \boldsymbol{A}_3 & \boldsymbol{0} \\ \boldsymbol{0} & \boldsymbol{0} & \boldsymbol{0} & \boldsymbol{A}_1 \otimes \boldsymbol{A}_2 \end{bmatrix} \begin{matrix} ggg \\ gfg \\ fgg \\ ggg \end{matrix} \tag{7.88}$$

DPH representations for component lifetime distributions are

$$n_1 = n_2 = n_3 = 1, \quad \boldsymbol{a}_1 = \boldsymbol{a}_2 = \boldsymbol{a}_3 = [1],$$
$$\boldsymbol{A}_1 = [0.5], \quad \boldsymbol{A}_2 = [0.8], \quad \boldsymbol{A}_3 = [0.9].$$

The failure vector for component i is $\boldsymbol{b}_i = (\boldsymbol{I}_i - \boldsymbol{A}_i)\boldsymbol{u}$,

$$\boldsymbol{b}_1 = [0.5], \quad \boldsymbol{b}_2 = [0.2], \quad \boldsymbol{b}_3 = [0.1].$$

Substituting these values into Equation (7.88) yields

$$\boldsymbol{A}_S = \begin{bmatrix} [0.5] \otimes [0.8] \otimes [0.9] & [0.5] \otimes [0.2] \otimes [0.9] & [0.5] \otimes [0.8] \otimes [0.9] & [0.5] \otimes [0.8] \otimes [0.1] \\ 0 & [0.5] \otimes [0.9] & 0 & 0 \\ 0 & 0 & [0.8] \otimes [0.9] & 0 \\ 0 & 0 & 0 & [0.5] \otimes [0.8] \end{bmatrix}$$

$$= \begin{bmatrix} 0.36 & 0.09 & 0.36 & 0.04 \\ 0 & 0.45 & 0 & 0 \\ 0 & 0 & 0.72 & 0 \\ 0 & 0 & 0 & 0.4 \end{bmatrix},$$

$$\boldsymbol{a}_S = \begin{bmatrix}[1] \otimes [1] \otimes [1] & 0 & 0 & 0\end{bmatrix} = \begin{bmatrix}1 & 0 & 0 & 0\end{bmatrix}.$$

It can be verified that $\boldsymbol{A}_S$ is a valid survival matrix of a discrete-time system.
The survival matrix can also be generated using Equations (7.85) and (7.86) as follows:

(a) Transition within each system state is generated using Equation (7.85):
  a) State (g, g, g): $V_1 = V_2 = V_3 = g \rightarrow$

$$A_S((g,g,g),(g,g,g)) = \boldsymbol{A}_1 \otimes \boldsymbol{A}_2 \otimes \boldsymbol{A}_3 = [0.5] \otimes [0.8] \otimes [0.9] = [0.36].$$

  b) State (g, f, g): $V_1 = V_3 = g$, $V_2 = f \rightarrow$

$$A_S((g,f,g),(g,f,g)) = \boldsymbol{A}_1 \otimes \boldsymbol{A}_3 = [0.5] \otimes [0.9] = [0.45].$$

  c) State (f, g, g):$V_1 = f$, $V_2 = V_3 = g \rightarrow$

$$A_S((f,g,g),(f,g,g)) = \boldsymbol{A}_2 \otimes \boldsymbol{A}_3 = [0.8] \otimes [0.9] = [0.72].$$

  d) State (g, g, f): $V_1 = V_2 = g$, $V_3 = f \rightarrow$

$$A_S((g,g,f),(g,g,f)) = \boldsymbol{A}_1 \otimes \boldsymbol{A}_2 = [0.5] \otimes [0.8] = [0.4].$$

(b) Transition between each possible pair of transient system states is generated using Equation (7.86):
  a) From state (g, g, g) to state (g, f, g): $V_1 = W_1 = g \rightarrow \boldsymbol{X}_1 = \boldsymbol{A}_1$; $V_3 = W_3 = g \rightarrow \boldsymbol{X}_3 = \boldsymbol{A}_3$; $V_2 = g$, $W_2 = f \rightarrow \boldsymbol{X}_2 = \boldsymbol{b}_2$. Thus,

$$A_S((g,g,g),(g,f,g)) = \boldsymbol{A}_1 \otimes \boldsymbol{b}_2 \otimes \boldsymbol{A}_3 = [0.5] \times [0.2] \times [0.9] = [0.09].$$

  b) From state (g, g, g) to state (f, g, g): $V_1 = g$, $W_1 = f \rightarrow \boldsymbol{X}_1 = \boldsymbol{b}_1$, $V_2 = W_2 = g \rightarrow \boldsymbol{X}_2 = \boldsymbol{A}_2$; $V_3 = W_3 = g \rightarrow \boldsymbol{X}_3 = \boldsymbol{A}_3$. Thus,

$$A_S((g,g,g),(f,g,g)) = \boldsymbol{b}_1 \otimes \boldsymbol{A}_2 \otimes \boldsymbol{A}_3 = [0.5] \times [0.8] \times [0.9] = [0.36].$$

  c) From state (g, g, g) to state (g, g, f): $V_1 = W_1 = g \rightarrow \boldsymbol{X}_1 = \boldsymbol{A}_1$; $V_2 = W_2 = g \rightarrow \boldsymbol{X}_2 = \boldsymbol{A}_2$; $V_3 = g$, $W_3 = f \rightarrow \boldsymbol{X}_3 = \boldsymbol{b}_3$. Thus,

$$A_S((g,g,g),(g,g,f)) = \boldsymbol{A}_1 \otimes \boldsymbol{A}_2 \otimes \boldsymbol{b}_3 = [0.5] \times [0.8] \times [0.1] = [0.04].$$

d) There are transitions from state (g, g, g) to states (g, f, f), (f, g, f), (f, f, g), and (f, f, f); however, these states are absorbing states that make their transitions do not belong to the survival matrix.

e) From state (g, f, g) to states (g, g, g), (f, g, g), and (g, g, f) are impossible. Likewise, from state (f, g, g) to states (g, g, g), (g, f, g), and (g, g, f), as well as from state (g, g, f) to states (g, g, g), (g, f, g), and (f, g, g). Thus, those transitions have block transition matrices equals **0**.

Putting all the results into a matrix yields

$$\boldsymbol{A}_S = \begin{bmatrix} 0.36 & 0.09 & 0.36 & 0.04 \\ 0 & 0.45 & 0 & 0 \\ 0 & 0 & 0.72 & 0 \\ 0 & 0 & 0 & 0.4 \end{bmatrix}$$

the same as produced using Equation (7.88).

Using **dphkofnsys** produces the following results:

```
As =
    0.3600    0.0900    0.3600    0.0400
         0    0.4500         0         0
         0         0    0.7200         0
         0         0         0    0.4000
```

which is equal to manual calculations. ■

Exercise 7.19 Derive the survival matrix of a 3-out-of-4 discrete-time system using the two methods.

Manual calculation is not feasible for K-out-of-N discrete-time systems with many components or with components having lifetimes following DPH distributions with high orders. Using **dphkofnsys** is the only viable way.

Example 7.34 DPH representations of the lifetime distributions of K-out-of-5 systems, for $K = 2, 3$, are generated using **dphkofnsys**. The kth component of this system has lifetime following k-NB distributions with $MTTF = 100$, $k = 1, 2, \ldots, 5$. The DPH representation obtained from **dphkofnsys** is used to calculate and plot the system reliability and hazard functions and also the system $MTTF$ and $SDTTF$.

The main program for this purpose is given in **example7_34.m** file in the software package **dysrel-am**. Its MATLAB code is as follows:

MATLAB code for example7_34.m

```
clear;close all; addpath('..\dysrel-am')
% DPH Components                                    inputs
for i = 1:5
  T(i) = dphnegativebinomial(i,100);
end
% Sequence vector
kmax = 250;
k = 0:5:kmax;
% System reliability K-out-of-5                     model
```

```
U = []; V = [];
for K = 2:3
  % DPH System
  Ts = dphkofnsys(T,K);
  % Reliability, MTTF, SDTTF
  [R,h]  = dphrelhaz(k,Ts);
  mttfs(K) = dphmttf(Ts);
  sdttfs(K) = dphsdttf(Ts);
  U = [U R']; V = [V h'];
end
% Plots                                          graphical output
lgnd = {'2-out-of-5';'3-out-of-5'};
dphmultiplot(2,k,U,lgnd,'k','R(k)');
dphmultiplot(2,k,V,lgnd,'k','h(k)');
% Display statistics                             statistical output
disp('STATISTICS:')
for K = 2:3
  disp([num2str(K),'-out-of-5'])
  mttf = full(mttfs(K))
  sdttf = full(sdttfs(K))
end
```

Executing this program produces results in Figure 7.27.

```
>> example7_34
mttfs =
   2-out-of-5      122.0076
   3-out-of-5       90.9110

sdttfs =
   2-out-of-5       39.4393
   3-out-of-5       29.9391
```

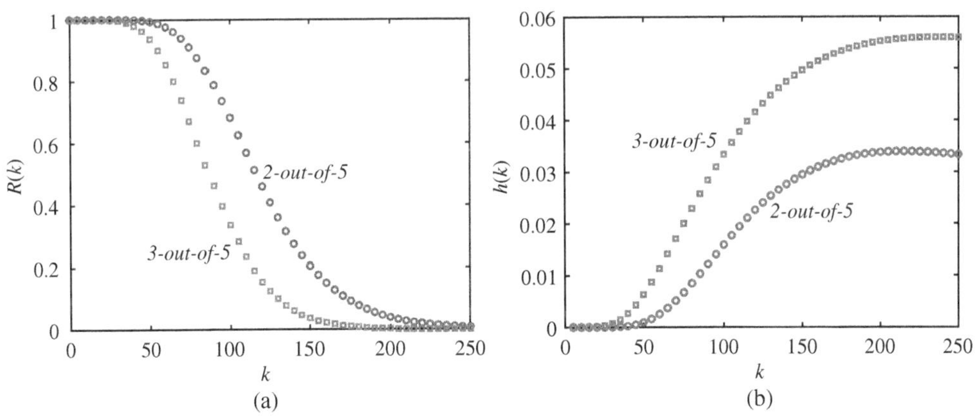

Figure 7.27 Output of **example7_34.m**.

■

The algorithm is applicable for system optimization as shown in the following application.

Application 7.4 Application 7.2 is repeated for K-out-of-N systems with $K = 1, 2, 3$. The optimum values of N for several mission times are summarized in Table 7.11.

Table 7.11 The expected mission cost as a function of mission time and N.

	K					
	Mission duration = 10 days			Mission duration = 15 days		
N	1	2	3	1	2	3
1	34.17	–	–	80.37	–	–
2	**22.92**	65.43	–	44.76	135.98	–
3	30.35	**38.06**	94.11	**38.71**	86.85	175.54
4	40.04	41.28	54.84	43.06	65.65	128.06
5	50.01	50.19	**52.92**	51.08	**61.01**	97.61
6	60	60.03	60.52	60.38	64.57	83.89
7	70	70	70.09	70.13	71.85	**81.37**
8				80.05	80.74	85.20

Optimum values are given in bold.

■

For K-out-of-N discrete-time systems with components having lifetimes following MG distributions, Algorithm 7.7 can be modified slightly by adding lines for calculating the closing vector for each system state V using

$$\boldsymbol{v}_V = \otimes \boldsymbol{v}_i, \qquad i|V_i = g, \tag{7.89}$$

and use it to construct the closing vector $\boldsymbol{v}_S$ of the MG representation of the system lifetime distribution. The resultis programmed in MATLAB as the **mgkofnsys** function available in the software package **dysrel-am**.

The system reliability model for K-out-of-N systems is generalized in the next section into general standby systems by incorporating deceleration matrices for standby components.

7.9 General Standby Structured Discrete-Time Systems

The models for K-out-of-N discrete-time systems are extended to general standby systems in discrete time by giving any warmness levels for the backup components so that they can be cold, warm, or hot standby as needed. The model uses a deceleration matrix for each backup component. For a 1-out-of-2 general standby system with component i having lifetime following independent DPH distributions with representation $(n_i, \boldsymbol{a}_i, \boldsymbol{A}_i)$, $i = 1, 2$, the DPH distribution of the system lifetime has representation $(n_S, \boldsymbol{a}_S, \boldsymbol{A}_S)$ given by [10]

$$n_S = n_1 n_2 + n_1 + n_2, \tag{7.90}$$

$$\boldsymbol{a}_S = \left[\boldsymbol{a}_1 \otimes \boldsymbol{a}_2 \;\; \boldsymbol{0} \;\; \boldsymbol{0}\right], \tag{7.91}$$

$$A_S = \begin{bmatrix} A_1 \otimes A_2' & A_1 \otimes b_2' & b_1 \otimes A_2' \\ 0 & A_1 & 0 \\ 0 & 0 & A_2 \end{bmatrix}. \tag{7.92}$$

where $A_2' = I_2 - D_2(I_2 - A_2), b_2' = \left(I_2 - A_2'\right) u$, and D_2 is the deceleration matrix of component 2 as a backup.

If $D_2 = I$, component 2 is in hot standby, then $A_2' = I_2 - (I_2 - A_2) = A_2$, and Equation (7.92) is the same as for 2-parallel system as it should be.

If $0 < D_2 < I$, component 2 is in warm standby.

If $D_2 = 0$, component 2 is in cold standby, then $A_2' = I_2 - D_2(I_2 - A_2) = I_2, b_2' = \left(I_2 - A_2'\right) u = 0$, and the survival matrix in Equation (7.92) becomes

$$A_S = \begin{bmatrix} A_1 \otimes I_2 & 0 & b_1 \otimes I_2 \\ 0 & A_1 & 0 \\ 0 & 0 & A_2 \end{bmatrix},$$

the same as the survival matrix of the 1-out-of-2 general standby system in continuous time as given in Equation (5.29). Therefore, the proof that a 1-out-of-2 general standby discrete-time system reduces to 2-standby discrete-time system when $D_2 = 0$ is the same as the proof of **Lemma 5.2**. Alkaff [10] provided a proof of **Lemma 5.2** counterpart in discrete time using PGF for the PMF given in Equation (7.30).

The state transition diagram of a 1-out-of-2 general standby system following Equations (7.91) and (7.92) is presented in Figure 7.28 where g can be p or b depending on its position.

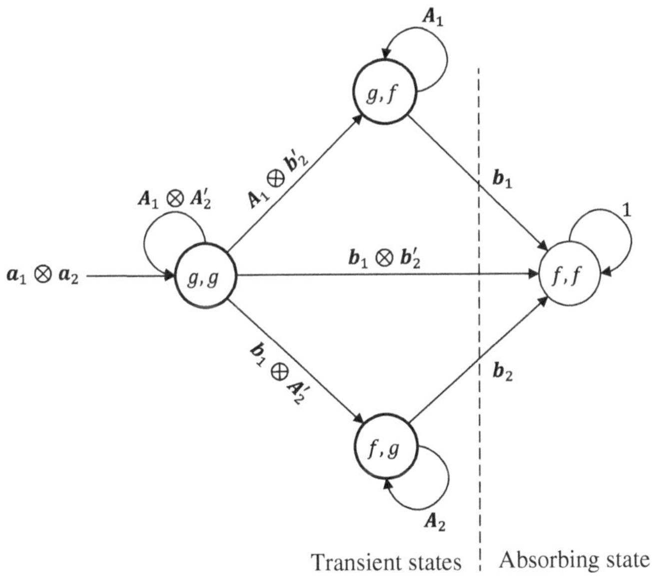

Figure 7.28 State transition diagram of a 1-out-of-2 general standby system.

The DPH representation of the lifetime distribution of K-out-of-N general standby system can be generated recursively using Equations (7.90)–(7.92). The state transition diagram of a 1-out-of-3

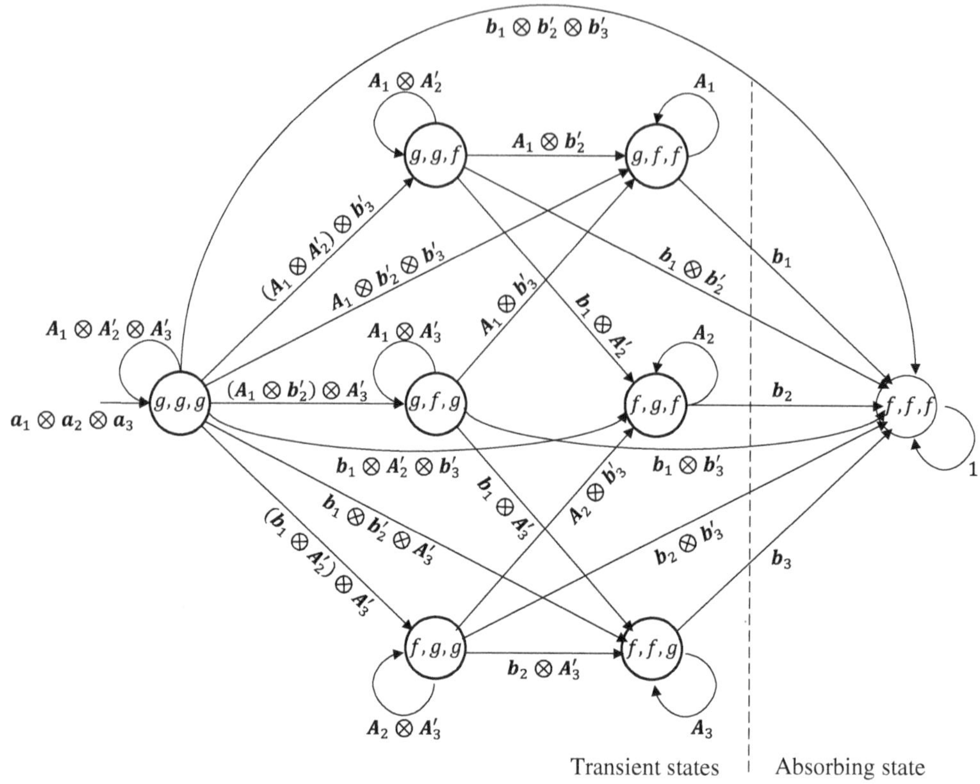

Figure 7.29 State transition diagram of a 3-general standby system.

discrete-time general standby system (referred to as a 3-general standby discrete-time system) is presented in Figure 7.29 where g can be p or b depending on its position. Components 2 and 3 in this system are backup; hence, their survival matrices and failure vectors are adjusted by their decelerating factors.

The state transition diagram for a 2-out-of-3 general standby system can be derived from Figure 7.29 by lumping states (g, f, f), (f, g, f), (f, f, g), *and* (f, f, f) into one absorbing state and set $\boldsymbol{b}_2' = \boldsymbol{b}_2$ because in this system component 2 is a primary, not a backup component. The result is presented in Figure 7.30.

For a 2-out-of-4 general standby system, the first two good components are primary, and the rest are backup. The survival matrix of this system can be derived based on the role of each component in each system state. Table 7.12 provides some block transition matrices in the survival matrix of a 2-out-of-4 general standby discrete-time system.

Patterns in Table 7.12 can be identified and used to establish rules for determining transition matrices within each system state and between each pair of system states. Using Table 7.12 and the roles of each component in each state, the transition matrices within any states V and W and from state V to W can be deduced as [10]

$$\boldsymbol{A}_{VV} = \otimes \boldsymbol{X}_i, \qquad i | V_i \neq f, \tag{7.93}$$

$$\boldsymbol{A}_{VW} = \otimes \boldsymbol{Y}_i, \qquad i | V_i \neq f, \tag{7.94}$$

Table 7.12 Some block matrices in the survival matrix of a 2-out-of-4 general standby discrete-time system.

Current system state	Block diagonal matrix	Next system state	Block off-diagonal matrix
(g, g, g, g)	$\boldsymbol{A}_1 \otimes \boldsymbol{A}_2 \otimes \boldsymbol{A}_3' \otimes \boldsymbol{A}_4'$	(g, g, g, f)	$\boldsymbol{A}_1 \otimes \boldsymbol{A}_2 \otimes \boldsymbol{A}_3 \otimes \boldsymbol{b}_4'$
		(g, g, f, f)	$\boldsymbol{A}_1 \otimes \boldsymbol{A}_2 \otimes \boldsymbol{b}_3' \otimes \boldsymbol{b}_4'$
		(g, f, f, f)	$\boldsymbol{A}_1 \otimes \boldsymbol{b}_2 \otimes \boldsymbol{b}_3' \otimes \boldsymbol{b}_4'$
		(f, f, f, f)	$\boldsymbol{b}_1 \otimes \boldsymbol{b}_2 \otimes \boldsymbol{b}_3' \otimes \boldsymbol{b}_4'$
		11 other nonzero transitions	
(g, g, g, f)	$\boldsymbol{A}_1 \otimes \boldsymbol{A}_2 \otimes \boldsymbol{A}_3'$	(g, g, f, f)	$\boldsymbol{A}_1 \otimes \boldsymbol{A}_2 \otimes \boldsymbol{b}_3'$
		(g, f, f, f)	$\boldsymbol{A}_1 \otimes \boldsymbol{b}_2 \otimes \boldsymbol{b}_3'$
		(f, f, f, f)	$\boldsymbol{b}_1 \otimes \boldsymbol{b}_2 \otimes \boldsymbol{b}_3'$
		4 other nonzero transitions	
(g, g, f, g)	$\boldsymbol{A}_1 \otimes \boldsymbol{A}_2 \otimes \boldsymbol{A}_4'$	(g, g, f, f)	$\boldsymbol{A}_1 \otimes \boldsymbol{A}_2 \otimes \boldsymbol{b}_4'$
		(g, f, f, f)	$\boldsymbol{A}_1 \otimes \boldsymbol{b}_2 \otimes \boldsymbol{b}_4'$
		(f, f, f, f)	$\boldsymbol{b}_1 \otimes \boldsymbol{b}_2 \otimes \boldsymbol{b}_4'$
		4 other nonzero transitions	
(g, f, g, g)	$\boldsymbol{A}_1 \otimes \boldsymbol{A}_3 \otimes \boldsymbol{A}_4'$	(g, f, g, f)	$\boldsymbol{A}_1 \otimes \boldsymbol{A}_3 \otimes \boldsymbol{b}_4'$
		(g, f, f, f)	$\boldsymbol{A}_1 \otimes \boldsymbol{b}_3 \otimes \boldsymbol{b}_4'$
		(f, f, f, f)	$\boldsymbol{b}_1 \otimes \boldsymbol{b}_3 \otimes \boldsymbol{b}_4'$
		4 other nonzero transitions	
(f, g, g, g)	$\boldsymbol{A}_2 \otimes \boldsymbol{A}_3 \otimes \boldsymbol{A}_4'$	(f, g, g, f)	$\boldsymbol{A}_2 \otimes \boldsymbol{A}_3 \otimes \boldsymbol{b}_4'$
		(f, g, f, f)	$\boldsymbol{A}_2 \otimes \boldsymbol{b}_3 \otimes \boldsymbol{b}_4'$
		(f, f, f, f)	$\boldsymbol{b}_2 \otimes \boldsymbol{b}_3 \otimes \boldsymbol{b}_4'$
		4 other nonzero transitions	
(g, g, f, f)	$\boldsymbol{A}_1 \otimes \boldsymbol{A}_2$	Absorbing states	Not required (not part of the survival matrix)
(g, f, g, f)	$\boldsymbol{A}_1 \otimes \boldsymbol{A}_3$		
(g, f, f, g)	$\boldsymbol{A}_1 \otimes \boldsymbol{A}_4$		
(f, g, g, f)	$\boldsymbol{A}_2 \otimes \boldsymbol{A}_3$		
(f, g, f, g)	$\boldsymbol{A}_2 \otimes \boldsymbol{A}_4$		
(f, f, g, g)	$\boldsymbol{A}_3 \otimes \boldsymbol{A}_4$		

with

$$
\begin{aligned}
\boldsymbol{X}_i &= \boldsymbol{A}_i && \text{if } V_i = p, \\
&= \boldsymbol{A}_i' && \text{if } V_i = b, \\
\boldsymbol{Y}_i &= \boldsymbol{A}_i && \text{if } V_i = p \text{ and } W_i = p, \\
&= \boldsymbol{A}_i' && \text{if } V_i = b \text{ and } W_i \neq f, \\
&= \boldsymbol{b}_i && \text{if } V_i = p \text{ and } W_i = f, \\
&= \boldsymbol{b}_i' && \text{if } V_i = b \text{ and } W_i = f.
\end{aligned}
$$

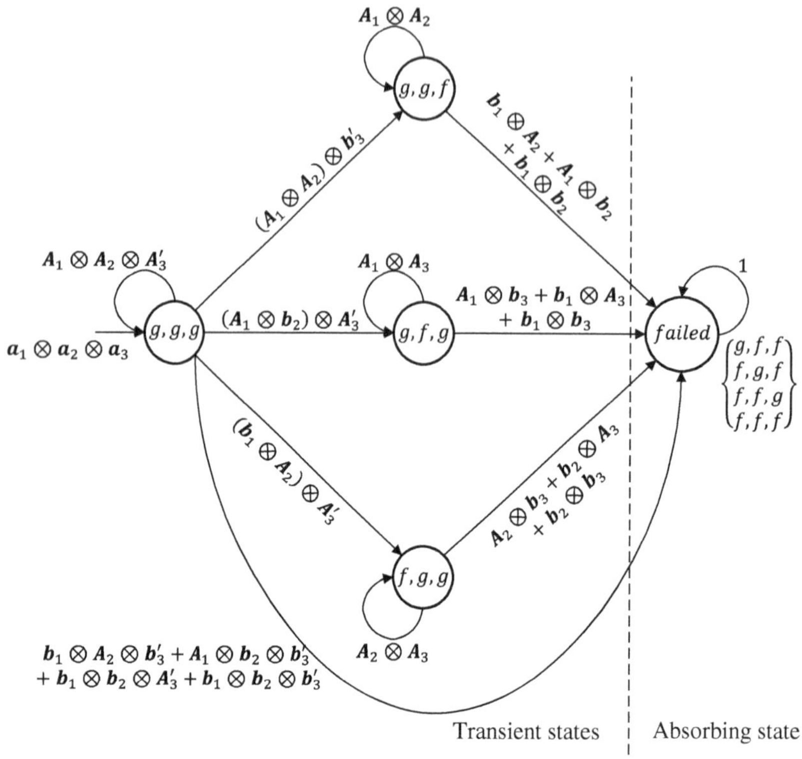

Figure 7.30 State transition diagram of a 2-out-of-3 general standby system.

The algorithm for generating the survival matrix of a K-out-of-N general standby discrete-time system is referred to as Algorithm 7.8 presented below.

Algorithm 7.8

---

1) Generate the system states as N-tupple of component states.
2) Discard system states with less than K good components.
3) Classify the first K good component in each system state as primary and the rest are backups. Adjust the survival matrix of each backup component with its warmness level.
4) Use Equation (7.93) to determine the block diagonal matrix for each system state obtained from step 2 using the result of step 3.
5) Determine possible transitions for each pair of system states as in K-out-of-N systems.
6) Determine the block off-diagonal matrix using Equation (7.94) for each possible transition obtained from step 5.
7) Set other block off-diagonal matrices to $\mathbf{0}$.

Algorithm 7.8 is programmed in MATLAB as the **dphgensys** function. Its MATLAB code is available in the software package **dysrel-am**. The function receives N, K, and $(n_i, \boldsymbol{a}_i, \boldsymbol{A}_i)$ for $i = 1, 2, \ldots, N$ as inputs and returns the DPH representation of the system lifetime distribution as the output.

Example 7.35 A 2-out-of-3 general standby discrete-time system is constructed using components in Example 7.16. The backup component 3 has a warmness level 0.5.

The survival matrix can be derived using Figure 7.30 as

$$\boldsymbol{A}_S = \begin{array}{c} \begin{array}{cccc} \quad ggg \qquad\qquad & gfg \qquad\qquad & fgg \qquad\qquad & ggf \end{array} \\ \left[\begin{array}{cccc} \boldsymbol{A}_1 \otimes \boldsymbol{A}_2 \otimes \boldsymbol{A}_3' & \boldsymbol{A}_1 \otimes \boldsymbol{b}_2 \otimes \boldsymbol{A}_3' & \boldsymbol{b}_1 \otimes \boldsymbol{A}_2 \otimes \boldsymbol{A}_3' & \boldsymbol{A}_1 \otimes \boldsymbol{b}_2 \otimes \boldsymbol{b}_3' \\ \boldsymbol{0} & \boldsymbol{A}_1 \otimes \boldsymbol{A}_3 & \boldsymbol{0} & \boldsymbol{0} \\ \boldsymbol{0} & \boldsymbol{0} & \boldsymbol{A}_2 \otimes \boldsymbol{A}_3 & \boldsymbol{0} \\ \boldsymbol{0} & \boldsymbol{0} & \boldsymbol{0} & \boldsymbol{A}_1 \otimes \boldsymbol{A}_2 \end{array}\right] \begin{array}{l} ggg \\ gfg \\ fgg \\ ggg \end{array} \end{array} \tag{7.95}$$

The DPH representations for component lifetime distributions are

$$n_1 = n_2 = n_3 = 1, \quad \boldsymbol{a}_1 = \boldsymbol{a}_2 = \boldsymbol{a}_3 = [1],$$
$$\boldsymbol{A}_1 = [0.5], \quad \boldsymbol{A}_2 = [0.8], \quad \boldsymbol{A}_3 = [0.9],$$
$$\boldsymbol{A}_3' = \boldsymbol{I}_3 - \boldsymbol{D}_2(\boldsymbol{I}_3 - \boldsymbol{A}_3) = [1 - 0.5(1 - 0.9)] = [0.95].$$

The transition probability vector for component i is $\boldsymbol{b}_i = (\boldsymbol{I}_i - \boldsymbol{A}_i)\boldsymbol{u}$ or $\boldsymbol{b}_i' = \left(\boldsymbol{I}_i - \boldsymbol{A}_i'\right)\boldsymbol{u}$

$$\boldsymbol{b}_1 = [0.5], \quad \boldsymbol{b}_2 = [0.2], \quad \boldsymbol{b}_3 = [0.1], \quad \boldsymbol{b}_3' = [1 - 0.95] = [0.05].$$

Substituting these values into Equation (7.95) yields

$$\boldsymbol{A}_S = \begin{bmatrix} [0.5] \otimes [0.8] \otimes [0.95] & [0.5] \otimes [0.2] \otimes [0.95] & [0.5] \otimes [0.8] \otimes [0.95] & [0.5] \otimes [0.8] \otimes [0.05] \\ 0 & [0.5] \otimes [0.9] & 0 & 0 \\ 0 & 0 & [0.8] \otimes [0.9] & 0 \\ 0 & 0 & 0 & [0.5] \otimes [0.8] \end{bmatrix}$$
$$= \begin{bmatrix} 0.38 & 0.095 & 0.38 & 0.02 \\ 0 & 0.45 & 0 & 0 \\ 0 & 0 & 0.72 & 0 \\ 0 & 0 & 0 & 0.4 \end{bmatrix},$$
$$\boldsymbol{a}_S = \left[[1] \otimes [1] \otimes [1] \ 0 \ 0 \ 0\right] = \left[1 \ 0 \ 0 \ 0\right].$$

It can be verified that $\boldsymbol{A}_S$ is a valid survival matrix of a discrete-time system.

The survival matrix can also be generated using Equations (7.93) and (7.94) as follows:

1) Transition within each system state is generated using Equation (7.93):
   a) State (p, p, b): $V_1 = V_2 = p, V_3 = b \rightarrow \boldsymbol{X}_1 = \boldsymbol{A}_1, \boldsymbol{X}_2 = \boldsymbol{A}_2, \boldsymbol{X}_3 = \boldsymbol{A}_3' \rightarrow$

   $$A_S((p, p, b), (p, p, b)) = \boldsymbol{A}_1 \otimes \boldsymbol{A}_2 \otimes \boldsymbol{A}_3' = [0.5] \otimes [0.8] \otimes [0.95] = [0.38].$$

   b) State (p, f, p): $V_1 = V_3 = p, V_2 = f \rightarrow \boldsymbol{X}_1 = \boldsymbol{A}_1, \boldsymbol{X}_3 = \boldsymbol{A}_3 \rightarrow$

   $$A_S((p, f, p), (p, f, p)) = \boldsymbol{A}_1 \otimes \boldsymbol{A}_3 = [0.5] \otimes [0.9] = [0.45].$$

   c) State (f, p, p):$V_1 = f, V_2 = V_3 = p \rightarrow \boldsymbol{X}_2 = \boldsymbol{A}_2, \boldsymbol{X}_3 = \boldsymbol{A}_3 \rightarrow$

   $$A_S((f, p, p), (f, p, p)) = \boldsymbol{A}_2 \otimes \boldsymbol{A}_3 = [0.8] \otimes [0.9] = [0.72].$$

   d) State (p, p, f): $V_1 = V_2 = p, V_3 = f \rightarrow \boldsymbol{X}_1 = \boldsymbol{A}_1, \boldsymbol{X}_2 = \boldsymbol{A}_2 \rightarrow$

   $$A_S((p, p, f), (p, p, f)) = \boldsymbol{A}_1 \otimes \boldsymbol{A}_2 = [0.5] \otimes [0.8] = [0.4].$$

2) Transition between each possible pair of transient system states is generated using Equation (7.94):
   a) From state (p, p, b) to state (p, f, p): $V_1 = W_1 = p \rightarrow \mathbf{Y}_1 = \mathbf{A}_1; V_2 = p, W_2 = f \rightarrow \mathbf{Y}_2 = \mathbf{b}_2$; $V_3 = b, W_3 = p \rightarrow \mathbf{Y}_3 = \mathbf{A}'_3$. Hence,

$$A_S((p,p,b),(p,f,p)) = \mathbf{A}_1 \otimes \mathbf{b}_2 \otimes \mathbf{A}'_3 = [0.5] \times [0.2] \times [0.95] = [0.095].$$

   b) From state (f, p, p)... From state (p, p, b) to state (f, p, p): $V_1 = p, W_1 = f \rightarrow \mathbf{Y}_1 = \mathbf{b}_1; V_2 = W_2 = p \rightarrow \mathbf{Y}_2 = \mathbf{A}_2$; $V_3 = b, W_3 = p \rightarrow \mathbf{Y}_3 = \mathbf{A}'_3$. Hence,

$$A_S((p,p,b),(f,p,p)) = \mathbf{b}_1 \otimes \mathbf{A}_2 \otimes \mathbf{A}'_3 = [0.5] \times [0.8] \times [0.95] = [0.38].$$

   c) From state (p, p, b) to state (p, p, f): $V_1 = W_1 = p \rightarrow \mathbf{Y}_1 = \mathbf{A}_1; V_2 = W_2 = p \rightarrow \mathbf{Y}_2 = \mathbf{A}_2$; $V_3 = b, W_3 = f \rightarrow \mathbf{Y}_3 = \mathbf{b}'_3$. Hence,

$$A_S((p,p,b),(p,p,f)) = \mathbf{A}_1 \otimes \mathbf{A}_2 \otimes \mathbf{b}'_3 = [0.5] \times [0.8] \times [0.05] = [0.02].$$

   d) There are transitions from state (p, p, b) to states (p, f, f), (f, p, f), (f, f, p), and (f, f, f); however, these states are absorbing states that make their transitions do not belong to the survival matrix.
   e) From state (p, f, p) to states (p, p, b), (f, p, p), and (p, p, f) are impossible. Likewise, from state (f, p, p) to states (p, p, b), (p, f, p), and (p, p, f), as well as from state (p, p, f) to states (p, p, b), (p, f, p), and (f, p, p). Thus, those transitions have block transition matrices equals **0**.

Putting all the results into a matrix yields

$$\mathbf{A}_S = \begin{bmatrix} 0.38 & 0.095 & 0.38 & 0.02 \\ 0 & 0.45 & 0 & 0 \\ 0 & 0 & 0.72 & 0 \\ 0 & 0 & 0 & 0.40 \end{bmatrix},$$

the same as produced using Equation (7.95).

Using **dphgensys** produces the following results:

```
As =
    0.3800    0.0950    0.3800    0.0200
         0    0.4500         0         0
         0         0    0.7200         0
         0         0         0    0.4000
```

which is equal to manual calculations. ■

Exercise 7.20 Derive the survival matrix of a 3-out-of-4 discrete-time general standby system using the two methods as in Example 7.35.

Manual calculation is not feasible for general standby systems with components having lifetimes following DPH distributions with high orders. Using **dphgensys** is the only viable way.

Example 7.36 The **dphgensys** function is employed to calculate the DPH representation of the lifetime distribution of K-out-of-5 discrete-time general standby systems, for $K = 2, 3$. The kth component of this system has lifetime following k-NB distributions with $MTTF = 100$, $k = 1, 2, \ldots, 5$. The warmness levels for the first, second, and third backups are 1.0, 0.5, and 0.0. The DPH representation is used to calculate and plot the system reliability and hazard functions and also the system $MTTF$ and $SDTTF$.

The main program for this purpose is given in **example7_36.m** file in the software package **dysrel-am**. Its MATLAB code is as follows:

MATLAB code for example7_36.m

```
clear;close all; addpath('..\dysrel-am')
% DPH Components                                     inputs
for i = 1:5
  T(i) = dphnegativebinomial(i,100);
end
% Sequence vector
kmax = 300;
k = 0:5:kmax;
% System reliability K-out-of-5
U = []; V = [];
for K = 2:3
  % DPH System
  D = [1 0.5 0];
  Ts = dphgensys(T,K,D);                             model
  % Reliability, MTTF, SDTTF
  [R,h]   = dphrelhaz(k,Ts);
  mttfs(K) = dphmttf(Ts);                            statistical output
  sdttfs(K) = dphsdttf(Ts);
  U = [U R']; V = [V h'];
end
% Plots                                              graphical output
lgnd = {'2-out-of-5';'3-out-of-5'};
dphmultiplot(2,k,U,lgnd,'k','R(k)');
dphmultiplot(2,k,V,lgnd,'k','h(k)');
% Display statistics
disp('STATISTICS:')
for K = 2:3
  disp([num2str(K),'-out-of-5'])
  mttf = full(mttfs(K))
  sdttf = full(sdttfs(K))
end
```

Executing this program produces results in Figure 7.31.

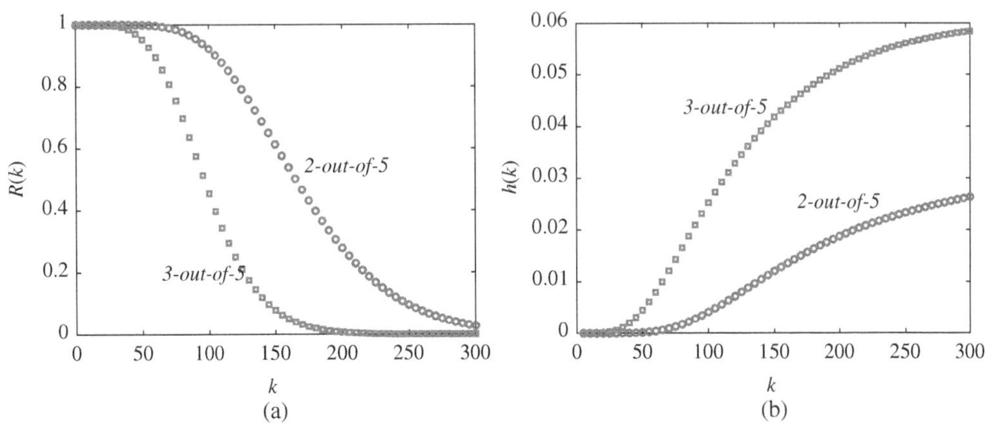

Figure 7.31 Outputs of **example7_36.m**.

```
>> example7_36
mttfs =
    2-out-of-5      172.6906
    3-out-of-5      100.1081
sdttfs =
    2-out-of-5       57.1709
    3-out-of-5       33.8013
```

■

The optimum number of backup systems that minimize the total cost of operation and failure in a mission can be found in a similar manner as for K-out-of-N systems.

Application 7.5 Application 7.2 is repeated for K-out-of-N homogeneous general standby discrete-time systems. The optimum values of N for $K = 1, 2$ and $d = 0.0, 0.5, 1.0$ are summarized in Table 7.13.

Table 7.13 Optimum numbers of backup systems in discrete-time general standby systems.

	Mission duration = 15 days					
	$d = 0$		$d = 0.5$		$d = 1$	
N	$K = 1$	$K = 2$	$K = 1$	$K = 2$	$K = 1$	$K = 2$
1	80.37	–	80.37	–	80.37	–
2	**20.85**	135.98	**27.55**	135.98	44.76	135.98
3	30	55.86	30.58	64.54	**38.71**	86.85
4	40	**40.59**	40.04	**45.78**	43.06	65.65
5	50	50	50	50.70	51.08	**61.01**
6	60	60	60	60.07	60.38	64.57
7	70	70	70	70.01	70.13	71.85

Optimum values are given in bold.

It can be verified that the results for $K = 1$ and $d = 0$ match the results for standby structure, for $K = 1$ and $d = 1$ match the results for parallel structure, and for $K = 2$ and $d = 1$ match the results for K-out-of-N structure. ■

Algorithm 7.8 can be modified slightly to be used for MG distribution by adding lines for calculating the closing vector using Equation (7.89). The result is programmed in MATLAB as the **mggensys** function.

The modeling and analysis of discrete-time systems having components with lifetime distributions follow DPH or MG distributions continues with network structured systems in the next section.

7.10 Network Structured Discrete-Time Systems

An algorithm for generating MG representation of the lifetime distribution of a network structured discrete-time system is developed based on Theorems 7.9–7.11, which are similar to theorems for its continuous-time system counterpart. Consequently, the algorithm for generating MG representation of the network lifetime distribution from the MG representations of its component lifetime distributions is similar to generating ME representation of the network lifetime distribution from the ME representation of its component lifetime distribution. Furthermore, both are based on the SDP representation of the network, which is not affected by the analysis being in continuous or discrete times. Denote MG representations of component lifetime distributions by $(n_i, \boldsymbol{a}_i, \boldsymbol{A}_i, \boldsymbol{v}_i)$, $i = 1, 2, \ldots, N$, where N is the number of components in the network, and MG representation of the network lifetime distribution is $(n_S, \boldsymbol{a}_S, \boldsymbol{A}_S, \boldsymbol{v}_S)$.

Let D_k be the number of variables in the kth disjoint product term Y_k, $X^{(i)}$, $i = 1, 2, \ldots, D_k$, be the ith variables in the disjoint product term DP_k whose DPH representation is $(n^{(i)}, \boldsymbol{a}^{(i)}, \boldsymbol{A}^{(i)}, \boldsymbol{v}^{(i)})$, and K is the number of disjoint products in the SDP representation of the network. The variables in each disjoint product term are arranged so that the first variable is without inverse. The algorithm for generating MG representation of the network lifetime distribution, referred to as Algorithm 7.9, is similar to continuous time, but the Kronecker sum operation is replaced by Kronecker product.

Algorithm 7.9

For k = 1 to K

For i = 1 to D_k,

Read $Z^{(i)}$

Repeat until $Z^{(i)} = X_i$

$$n_Y^{(i)} = n_i,\ \mathbf{a}_Y^{(i)} = \mathbf{a}_i,\ \mathbf{A}_Y^{(i)} = \mathbf{A}_i,\ \mathbf{v}_Y^{(i)} = \mathbf{v}_i. \tag{7.96}$$

For $j = i + 1, \ldots, N$,

Repeat until $Z_j = X_j$ or $\overline{X}_j$

If $Z_j = X_j$ then calculate

$$n_Y^{(j)} = n_Y^{(j-1)} \times n_j, \tag{7.97}$$

$$\mathbf{a}_Y^{(j)} = \mathbf{a}_Y^{(j-1)} \otimes \mathbf{a}_j, \tag{7.98}$$

$$\mathbf{A}_Y^{(j)} = \mathbf{A}_Y^{(j-1)} \otimes \mathbf{A}_j, \tag{7.99}$$

$$\mathbf{v}_Y^{(j)} = \mathbf{v}_Y^{(j-1)} \otimes \mathbf{v}_j, \tag{7.100}$$

Else

$$n_Y^{(j)} = n_Y^{(j-1)} + n_Y^{(j-1)} \times n_j, \tag{7.101}$$

$$\mathbf{a}_Y^{(i)} = \left[\mathbf{a}_Y^{(i-1)} \quad -\mathbf{a}_Y^{(i-1)} \otimes \mathbf{a}_j\right], \tag{7.102}$$

$$\mathbf{A}_Y^{(i)} = \begin{bmatrix} \mathbf{A}_Y^{(i-1)} & \mathbf{0} \\ \mathbf{0} & \mathbf{A}_Y^{(i-1)} \otimes \mathbf{A}_j \end{bmatrix}, \tag{7.103}$$

$$\mathbf{v}_Y^{(i)} = \begin{bmatrix} \mathbf{v}_Y^{(i-1)} \\ \mathbf{v}_Y^{(i-1)} \otimes \mathbf{v}_j \end{bmatrix}, \tag{7.104}$$

End

If $k = 1$ then

$$\mathbf{a}_S^{(k)} = \mathbf{a}_Y^{(k)}, \mathbf{A}_S^{(k)} = \mathbf{A}_Y^{(k)}, \mathbf{v}_S^{(k)} = \mathbf{v}_Y^{(k)}. \tag{7.105}$$

If $k \neq 1$ then

$$n_S^{(k)} = n_S^{(k-1)} + n_Y^{(k)}, \tag{7.106}$$

$$\mathbf{a}_S^{(k)} = \begin{bmatrix} \mathbf{a}_S^{(k-1)} & \mathbf{a}_Y^{(k)} \end{bmatrix}, \tag{7.107}$$

$$\mathbf{A}_S^{(k)} = \begin{bmatrix} \mathbf{A}_S^{(k-1)} & \mathbf{0} \\ \mathbf{0} & \mathbf{A}_Y^{(k)} \end{bmatrix}, \tag{7.108}$$

$$\mathbf{v}_S^{(k)} = \begin{bmatrix} \mathbf{v}_S^{(k-1)} \\ \mathbf{v}_Y^{(k)} \end{bmatrix}, \tag{7.109}$$

End

Algorithm 7.9 is generalized for arbitrary SDP and is programmed as a MATLAB **mgnetcomb** function available in the software package **dysrel-am**.

Example 7.37 Algorithm 7.9 is used to find an MG representation of the lifetime distribution of a 2-out-of-3 system with all of its components having lifetimes following geometry distribution with $MTTF = 5$.

The SDP representation of a 2-out-of-3 system is

$$X_S(k) = X_1X_2 + X_1\overline{X}_2X_3 + \overline{X}_1X_2X_3.$$

The MG representation for the lifetime distribution of all components is

$$\boldsymbol{a}_i = [1], \quad \boldsymbol{A}_i = [0.8], \quad \boldsymbol{\nu}_i = [1], \quad i = 1, 2, 3.$$

The MG representation of the network lifetime distribution is generated as follows:

1) MG representation of $Y_1 = X_1X_2$ is calculated using Equations (7.98)–(7.100) to yield

$$\boldsymbol{a}_{Y_1} = [\boldsymbol{a}_1] \otimes [\boldsymbol{a}_2] = [1], \quad \boldsymbol{A}_{Y_1} = [\boldsymbol{A}_1] \otimes [\boldsymbol{A}_2] = [0.64], \quad \boldsymbol{\nu}_{Y_1} = [\boldsymbol{\nu}_1] \otimes [\boldsymbol{\nu}_2] = [1].$$

2) MG representation of $Y_2 = X_1\overline{X}_2X_3$ is calculated in two steps:
   a) Calculate representation of $Z(k) = X_1X_3$ using Equations (7.98)–(7.100) to yield

$$\boldsymbol{a}_Z = [\boldsymbol{a}_1] \otimes [\boldsymbol{a}_3] = [1], \quad \boldsymbol{A}_Z = [\boldsymbol{A}_1] \otimes [\boldsymbol{A}_3] = [0.64], \quad \boldsymbol{\nu}_Z = [\boldsymbol{\nu}_1] \otimes [\boldsymbol{\nu}_3] = [1].$$

   b) Calculate representation of $Y_2 = Z\overline{X}_2$ using Equations (7.102)–(7.104) to yield

$$\boldsymbol{a}_{Y_2} = \begin{bmatrix} \boldsymbol{a}_Z & -\boldsymbol{a}_Z \otimes \boldsymbol{a}_2 \end{bmatrix} = \begin{bmatrix} 1 & -1 \end{bmatrix},$$

$$\boldsymbol{A}_{Y_2} = \begin{bmatrix} \boldsymbol{A}_Z & \mathbf{0} \\ \mathbf{0} & \boldsymbol{A}_Z \otimes \boldsymbol{A}_2 \end{bmatrix} = \begin{bmatrix} 0.64 & 0 \\ 0 & 0.512 \end{bmatrix}, \quad \boldsymbol{\nu}_{Y_2} = \begin{bmatrix} \boldsymbol{\nu}_Z \\ \boldsymbol{\nu}_Z \otimes \boldsymbol{\nu}_2 \end{bmatrix} = \begin{bmatrix} 1 \\ 1 \end{bmatrix}.$$

3) MG representation of $Y_3 = \overline{X}_1X_2X_3$ is calculated in the same way as Y_2 to yield

$$\boldsymbol{a}_{Y_3} = \begin{bmatrix} 1 & -1 \otimes 1 \end{bmatrix} = \begin{bmatrix} 1 & -1 \end{bmatrix},$$

$$\boldsymbol{A}_{Y_3} = \begin{bmatrix} 0.64 & 0 \\ 0 & 0.8 \otimes 0.64 \end{bmatrix} = \begin{bmatrix} 0.64 & 0 \\ 0 & 0.512 \end{bmatrix}, \quad \boldsymbol{\nu}_{Y_3} = \begin{bmatrix} 1 \\ 1 \otimes 1 \end{bmatrix} = \begin{bmatrix} 1 \\ 1 \end{bmatrix}.$$

4) MG representation for the entire SDP representation is calculated using Equations (7.105)–(7.109). The result is

$$\boldsymbol{a}_S = \begin{bmatrix}\boldsymbol{a}_{Y_1} & \boldsymbol{a}_{Y_2} & \boldsymbol{a}_{Y_3}\end{bmatrix} = \begin{bmatrix}1 & 1 & -1 & 1 & -1\end{bmatrix},$$

$$\boldsymbol{A}_S = \begin{bmatrix}\boldsymbol{A}_{Y_1} & \boldsymbol{0} & \boldsymbol{0}\\ \boldsymbol{0} & \boldsymbol{A}_{Y_2} & \boldsymbol{0}\\ \boldsymbol{0} & \boldsymbol{0} & \boldsymbol{A}_{Y_3}\end{bmatrix} = \begin{bmatrix}0.64 & 0 & 0 & 0 & 0\\ 0 & 0.64 & 0 & 0 & 0\\ 0 & 0 & 0.512 & 0 & 0\\ 0 & 0 & 0 & 0.64 & 0\\ 0 & 0 & 0 & 0 & 0.512\end{bmatrix},$$

$$\boldsymbol{v}_S = \begin{bmatrix}\boldsymbol{v}_{Y_1}\\ \boldsymbol{v}_{Y_2}\\ \boldsymbol{v}_{Y_3}\end{bmatrix} = \begin{bmatrix}1\\1\\1\\1\\1\end{bmatrix}.$$

For comparison, the DPH representation of the lifetime distribution of a 2-out-of-3 system obtained using the state transition model for a K-out-of-N discrete-time system is

$$\boldsymbol{a}_S = \begin{bmatrix}1 & 0 & 0 & 0\end{bmatrix},$$

$$\boldsymbol{A}_S = \begin{bmatrix}\boldsymbol{A}_1\otimes\boldsymbol{A}_2\otimes\boldsymbol{A}_3 & \boldsymbol{A}_1\otimes\boldsymbol{b}_2\otimes\boldsymbol{A}_3 & \boldsymbol{b}_1\otimes\boldsymbol{A}_2\otimes\boldsymbol{A}_3 & \boldsymbol{A}_1\otimes\boldsymbol{A}_2\otimes\boldsymbol{b}_3\\ \boldsymbol{0} & \boldsymbol{A}_1\otimes\boldsymbol{A}_3 & \boldsymbol{0} & \boldsymbol{0}\\ \boldsymbol{0} & \boldsymbol{0} & \boldsymbol{A}_2\otimes\boldsymbol{A}_3 & \boldsymbol{0}\\ \boldsymbol{0} & \boldsymbol{0} & \boldsymbol{0} & \boldsymbol{A}_1\otimes\boldsymbol{A}_2\end{bmatrix}$$

$$= \begin{bmatrix}0.512 & 0.128 & 0.128 & 0.128\\ 0 & 0.64 & 0 & 0\\ 0 & 0 & 0.64 & 0\\ 0 & 0 & 0 & 0.64\end{bmatrix}.$$

The system reliability and hazard functions generated using the two different representations are presented in Figure 7.32, which shows that they are identical.

The DPH representation obtained using state transition model has lower order than the MG representation obtained using the combinatorial model. However, the survival matrix in the MG

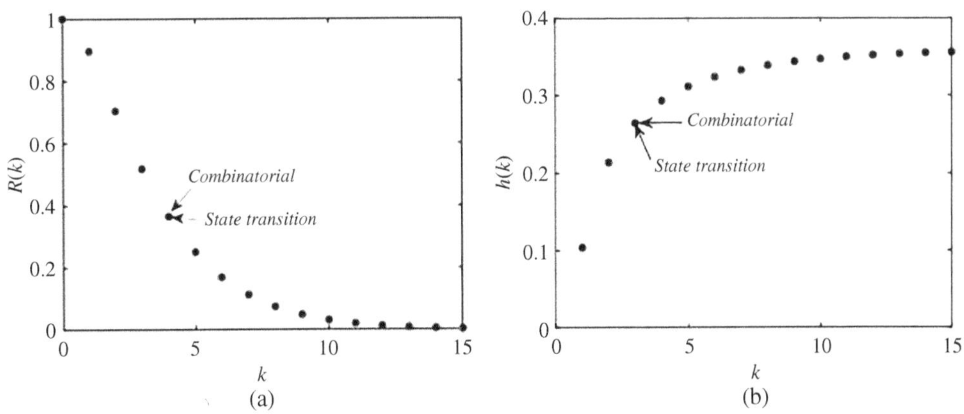

Figure 7.32 System reliability and hazard functions of a 2-out-of-3 discrete-time system calculated using combinatorial and state transition models.

representation is a diagonal matrix, while in the DPH representation is an upper-diagonal matrix, and calculation of a diagonal matrix is much faster than an upper diagonal matrix with the same dimension. ■

Example 7.38 The network in Figure 7.33 has minimum paths {1, 3, 5, 7, 6}, {1, 3, 5, 8}; {1, 4, 6}; {2, 5, 7, 6}; {2, 5, 8}. Arc numbering follows MATLAB generated numbers.

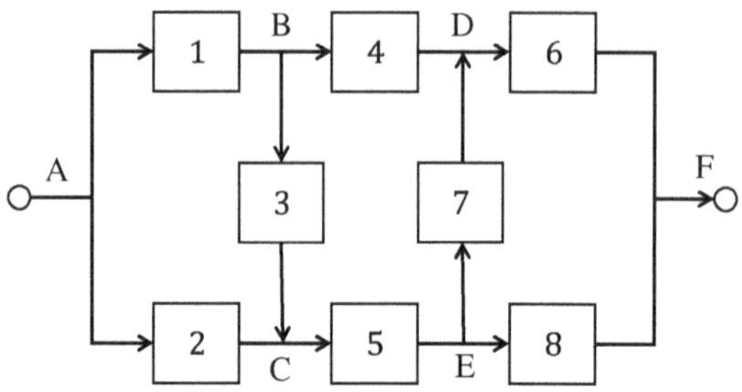

Figure 7.33 Network for Example 7.38.

The MG representation of the network lifetime distribution is generated using **mgnetcomb**. The network components are assumed to have lifetimes following identical distributions with $MTTF = 100$. Three distributions are used: geometry, 2-NB, and 2-mixed-geometry. The results are presented in Table 7.14 and Figure 7.34. The computing time for 2-NB distribution is longer than 2-mixed-geometry due to 2-NB distribution is upper diagonal, while 2-mixed-geometry is diagonal.

Table 7.14 Comparison of *MTTF*, *SDTTF*, order, and computation time.

Distribution of component lifetime	Network lifetime *MTTF*	*SDTTF*	Order	Computation time (s)
Geometry	55.82	37.87	31	0.11
2-NB	70.20	31.09	1,904	0.40
2-mixed-geometry	22.54	44.90	1,904	0.14

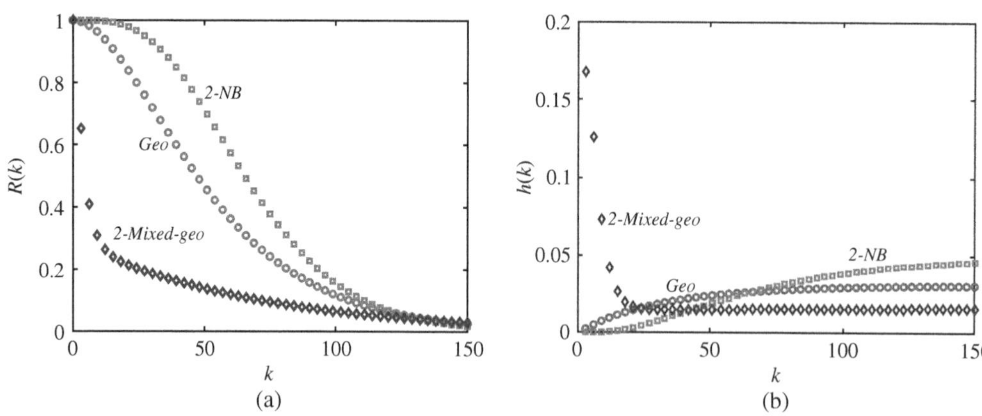

Figure 7.34 Reliability measures of network in Figure 7.33. (a) Reliability function, (b) hazard function, (c) residual reliability function for age 10, and (d) *MRL* function.

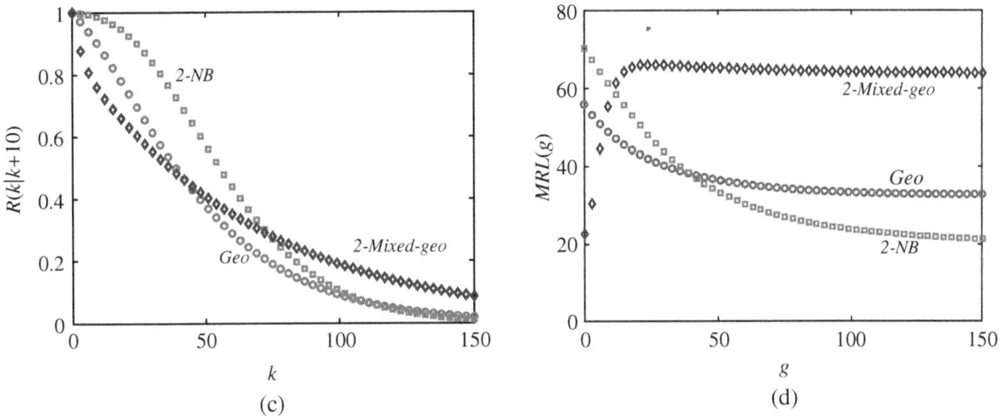

Figure 7.34 *(Continued)*

The main program for generating Table 7.14 and Figure 7.34 is available in **example7_38.m** file in the **dysrel-am** software package. Its source code is as follows:

MATLAB code for example7_38.m

```
clear; close all
addpath('..\dysrel-am','..\dysrel-am\expokit')

% ================================
% Define directed graph
% ================================
nodes = {'A','B','C','D','E','F'};                         input
startNode = {'A','B','D','A','C','E','B','E'};
endNode = {'B','D','F','C','E','F','C','D'};
weights = [1 1 1 1 1 1 1 1];
% ==== Source & Target ====
src = 'A';
tgt = 'F';
% Create digraph
G = digraph(startNode,endNode,weights,nodes);
% Find minimum paths
mp = minpaths(G,startNode,endNode,src,tgt);
mp = mp';
% Generate SDP
sdp = sdprep(mp);
% Number of components
N = 8;
% Sequence vector
k = 0:3:150;
age = 10;
% DPH components
T(1) = dphnegativebinomial(1,100);
T(2) = dphnegativebinomial(2,100);
T(3) = dph2mixedgeo(100);
% Reliability MG system                                   model
U = []; V = []; W = []; Y = [];
for i = 1:3
  tic
  Tc = [];
  for j = 1:N
    Tc = [Tc T(i)];
```

```
    end
    MGs = mgnetcomb(sdp,Tc);
    [R, h] = mgrelhaz(k,MGs);
    RR = mgresrel(k,MGs,age);
    MRL = mgmrl(k,MGs);
    ns(i) = MGs.n;
    mttfs(i) = mgmttf(MGs);
    sdttfs(i) = mgsdttf(MGs);
    comptime(i) = toc;
    U = [U R']; V = [V h']; W = [W RR']; Y = [Y MRL'];
end
% Display results                                       statistical output
mttfs
sdttfs
ns
comptime
% Plot                                                  graphical output
lgnd = {'Geo','2-NB','2-Mixed-geo'};
dphmultiplot(3,k,U,lgnd,'k','R(k)');
dphmultiplot(3,k,V,lgnd,'k','h(k)');
dphmultiplot(3,k,W,lgnd,'k',['R(k|k+',num2str(age),')']);
dphmultiplot(3,k,Y,lgnd,'g','MRL(g)');
```

■

As with system reliability calculations, transforming the DPH distribution of component lifetimes into an MG distribution in a Jordan canonical form can speed up computation times for computing network reliability measures.

Example 7.39 The network in Figure 7.33 is constructed using components with identical lifetime distributions having DPH representation

$$n_1 = 4, \quad \boldsymbol{a}_1 = \begin{bmatrix} 1 & 0 & 0 & 0 \end{bmatrix}, \quad \boldsymbol{A}_1 = \begin{bmatrix} 0.1 & 0.5 & 0.2 & 0.2 \\ 0 & 0.2 & 0.4 & 0.4 \\ 0 & 0 & 0.5 & 0.5 \\ 0 & 0 & 0 & 0.3 \end{bmatrix}.$$

The result of a similar transformation to the DPH distribution is an MG distribution with representation

$$n_1 = 4, \quad \boldsymbol{a}_1 = \begin{bmatrix} -16.5 & 5 & 2.1667 & 1 \end{bmatrix},$$

$$\boldsymbol{A}_1 = \begin{bmatrix} 0.3 & 0 & 0 & 0 \\ 0 & 0.2 & 0 & 0 \\ 0 & 0 & 0.5 & 0 \\ 0 & 0 & 0 & 0.1 \end{bmatrix}, \quad \boldsymbol{v}_1 = \begin{bmatrix} 1 \\ 2.3333 \\ 3.5 \\ -1.75 \end{bmatrix}.$$

The network reliability measures generated using **mgnetcomb** and their computing time are given in Figure 7.35 and Table 7.15.

Figure 7.35 and Table 7.15 show that DPH distribution and its similar MG distribution produce identical reliability measures. However, the computation time using DPH distributions is longer than using similar MG distributions. ■

State transition models for network reliability analysis that have been developed for continuous-time systems can be modified slightly for discrete-time systems. The modification follows the fundamental difference between systems in continuous time and in discrete time, namely, in continuous time, only one component failure triggers transition out of a system state,

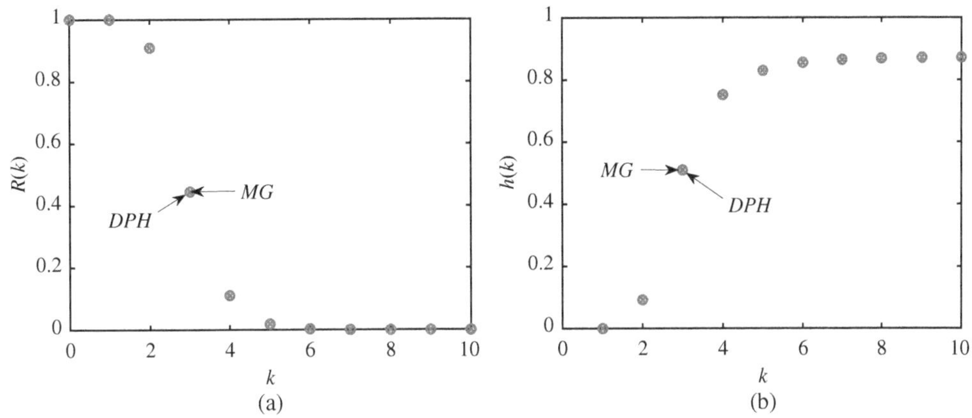

Figure 7.35 Network reliability function of Example 7.39. (a) Reliability function and (b) hazard function.

Table 7.15 *MTTF* and *SDTTF* of the network in Example 7.39 with computation times to produce Figure 7.35 and Table 7.15.

	Network lifetime			
Distribution of component lifetime	***MTTF***	***SDTTF***	**Order**	**Computation time (s)**
DPH	3.49	0.86	185,984	27.83
MG	3.49	0.86	185,984	1.01

while in discrete-time system, there can be more than one component failures trigger transitions out of a system state in a period of inspection. The algorithm for network reliability analysis in discrete time using the state transition model is similar to continuous time. It is referred to as Algorithm 7.10 as presented below.

Algorithm 7.10

1) Generate network states as combinations of component states starting from a network state with all components are good.
2) Classify the network states into transient states and absorbing states by checking whether good components in a state form a path from the source to the target. If it is, then it is a transient state; otherwise, it is an absorbing state.
3) For each transient state, determine the next state for all possible failures of all good components in that state.
4) Repeat step 3 until there is no transient state.
5) Determine the transition matrix within each transient state using Equation (7.85).
6) Determine the transition matrices between each transient state and its next states using Equation (7.86).
7) Construct a survival matrix using the transition matrices generated in steps 5 and 6.
8) Determine the initial probability vector from the network initial state.
9) Use the results of steps 7 and 8 to calculate network reliability measures.

Example 7.40 A survival matrix for the network in Figure 7.36 is generated following Algorithm 7.10. Assume components i have DPH-distributed lifetime with representation $(n_i, \boldsymbol{a}_i, \boldsymbol{A}_i)$, $i = 1, 2, \ldots, N$, where 1, 2, 3, 4 represents node A, B, C, D.

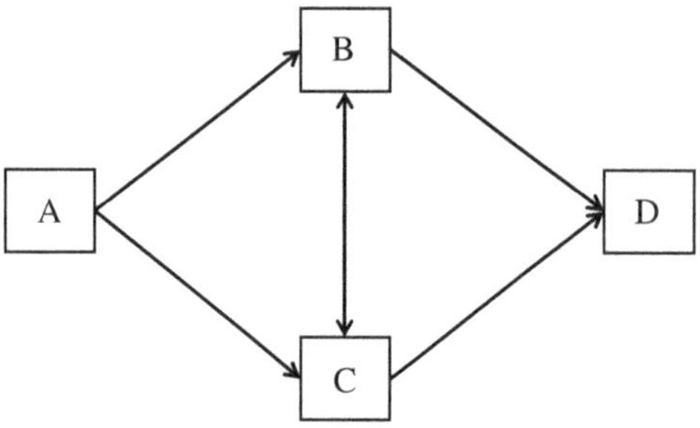

Figure 7.36 Graph representation of network for Example 7.40.

1) The network initial state is (g, g, g, g). The next states from the initial states are (g, g, g, f), (g, g, f, g), (g, f, g, g), (f, g, g, g), and 11 other states.
2) Among those 15 states, only states (g, g, f, g) and (g, f, g, g) are transient states because the good components in these states form a path from source to target, namely, path {A,B,D} for (g, g, f, g) and path {A,C,D} for (g, f, g, g). Other states are absorbing states because the good components in those states do not form a path from source to target.
3) Find the next states for each transient state:
   - The next states of (g, g, f, g) are (g, g, f, f), (g, f, f, g), (f, g, f, g), and four other states. All of them are absorbing states, so they do not have next states.
   - The next states of (g, f, g, g) are (g, f, g, f), (g, f, f, g), (f, f, g, g), and four other states. All of them are absorbing states, so they do not have next states.
4) This network has three transient states (g, g, g, g), (g, g, f, g), and (g, f, g, g). The transition matrix within these transient states is determined using Equation (7.85):
   - Transition matrix within state (g, g, g, g) is $\boldsymbol{A}_1 \otimes \boldsymbol{A}_2 \otimes \boldsymbol{A}_3 \otimes \boldsymbol{A}_4$.
   - Transition matrix within state (g, g, f, g) is $\boldsymbol{A}_1 \otimes \boldsymbol{A}_2 \otimes \boldsymbol{A}_4$.
   - Transition matrix within state (g, f, g, g) is $\boldsymbol{A}_1 \otimes \boldsymbol{A}_3 \otimes \boldsymbol{A}_4$.
5) The transition vectors from transient states to their corresponding next states is determined using Equation (7.86):
   - Transition vector from (g, g, g, g) to (g, g, f, g): $\boldsymbol{A}_1 \otimes \boldsymbol{A}_2 \otimes \boldsymbol{b}_3 \otimes \boldsymbol{A}_4$.
   - Transition vector from (g, g, g, g) to (g, f, g, g): $\boldsymbol{A}_1 \otimes \boldsymbol{b}_2 \otimes \boldsymbol{A}_3 \otimes \boldsymbol{A}_4$.
   - There is no transition from (g, f, g, g) to (g, g, f, g) and vice versa.
6) The network survival matrix is

$$\boldsymbol{A}_S = \begin{bmatrix} \boldsymbol{A}_1 \otimes \boldsymbol{A}_2 \otimes \boldsymbol{A}_3 \otimes \boldsymbol{A}_4 & \boldsymbol{A}_1 \otimes \boldsymbol{A}_2 \otimes \boldsymbol{b}_3 \otimes \boldsymbol{A}_4 & \boldsymbol{A}_1 \otimes \boldsymbol{b}_2 \otimes \boldsymbol{A}_3 \otimes \boldsymbol{A}_4 \\ \boldsymbol{0} & \boldsymbol{A}_1 \otimes \boldsymbol{A}_2 \otimes \boldsymbol{A}_4 & \boldsymbol{0} \\ \boldsymbol{0} & \boldsymbol{0} & \boldsymbol{A}_1 \otimes \boldsymbol{A}_3 \otimes \boldsymbol{A}_4 \end{bmatrix}$$

7) The initial probability vector is

$$\boldsymbol{a}_S = [\boldsymbol{a}_1 \otimes \boldsymbol{a}_2 \otimes \boldsymbol{a}_3 \otimes \boldsymbol{a}_4 \;\; \boldsymbol{0} \;\; \boldsymbol{0}]$$

The network states, transition matrix within each transient state, and transition matrices among transient states are used to construct the state transition diagram of the network in Figure 7.36. The result is presented as Figure 7.37.

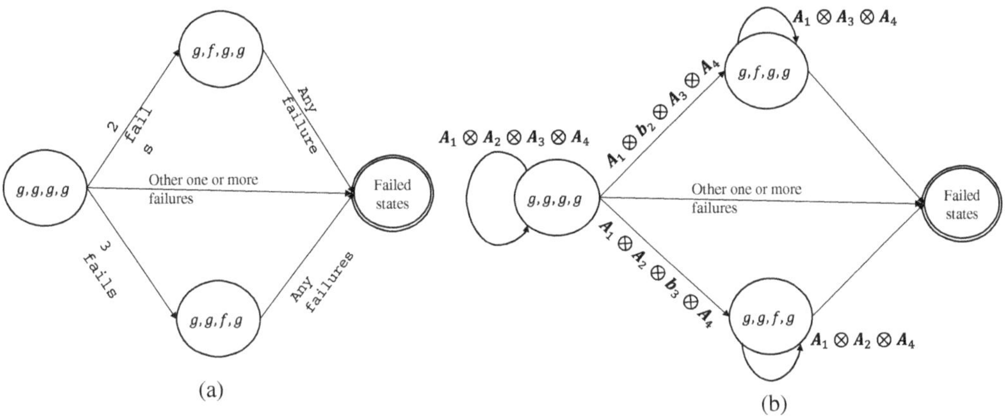

Figure 7.37 State transition diagram of the network in Figure 7.36.

As comparison, the network is modeled using combinatorial model that produces a structure function $X_S = X_1X_2X_4 \vee X_1X_3X_4$ with SDP representation $X_S = X_1X_2X_4 + X_1\overline{X}_2X_3X_4$. The survival matrix and the initial and closing vectors of the network lifetime distribution derived using its SDP representation are

$$\boldsymbol{\alpha}_S = [\boldsymbol{\alpha}_1 \otimes \boldsymbol{\alpha}_2 \otimes \boldsymbol{\alpha}_4 \;\; \boldsymbol{\alpha}_1 \otimes \boldsymbol{\alpha}_3 \otimes \boldsymbol{\alpha}_4 \;\; -\boldsymbol{\alpha}_1 \otimes \boldsymbol{\alpha}_2 \otimes \boldsymbol{\alpha}_3 \otimes \boldsymbol{\alpha}_4],$$

$$\boldsymbol{A}_S = \begin{bmatrix} \boldsymbol{A}_1 \otimes \boldsymbol{A}_2 \otimes \boldsymbol{A}_4 & \boldsymbol{0} & \boldsymbol{0} \\ \boldsymbol{0} & \boldsymbol{A}_1 \otimes \boldsymbol{A}_3 \otimes \boldsymbol{A}_4 & \boldsymbol{0} \\ \boldsymbol{0} & \boldsymbol{0} & \boldsymbol{A}_1 \otimes \boldsymbol{A}_2 \otimes \boldsymbol{A}_3 \otimes \boldsymbol{A}_4 \end{bmatrix},$$

$$\boldsymbol{v}_S = [\boldsymbol{v}_1 \otimes \boldsymbol{v}_2 \otimes \boldsymbol{v}_4 \;\; \boldsymbol{v}_1 \otimes \boldsymbol{v}_3 \otimes \boldsymbol{v}_4 \;\; -\boldsymbol{v}_1 \otimes \boldsymbol{v}_2 \otimes \boldsymbol{v}_3 \otimes \boldsymbol{v}_4]^T,$$

where $\boldsymbol{v}_i = \boldsymbol{u}_i$ since the lifetime of component i, $i = 1, 2, 3, 4$, follows DPH distribution.

The resulting DPH representation obtained using state transition model and MG representation obtained using combinatorial model are utilized to generate network reliability and hazard functions. The results are Figure 7.38, which shows that both models produce identical functions. This means that the resulting DPH and MG representations are equivalent. Components with lifetimes following identical 3-NB distribution with $MTFF = 20$ are used in this experiment.

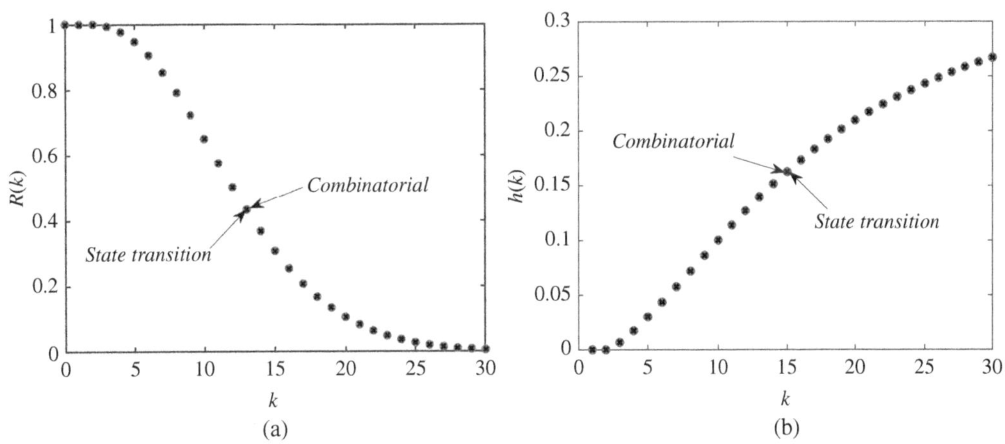

Figure 7.38 Network reliability and hazard functions calculated using combinatorial and state transition models. (a) Network reliability function and (b) network hazard function.

■

For large networks, a computer program is needed to generate and classify network states, to determine the transition matrix within each transient state and between each possible pair of transient states, and finally to produce a DPH/MG representation of the network lifetime distribution. Algorithm 7.10 is programmed as a MATLAB function **mgnettrans**.

Next, the methods for several system structures are utilized together to model and analyze complex discrete-time systems.

7.11 Complex Structured Discrete-Time Systems

Complex structured discrete-time systems are discrete-time systems having a combination of several structures. The structures involved may include all kinds of general standby structures and network structures in discrete time. Algorithms for calculating DPH or MG representation for the distributions of lifetimes of complex systems are utilizing the algorithm for each single structure subsystem recursively as many as the number of subsystems in the systems in a hierarchical manner, as have been used for continuous-time systems in Chapters 4–6. The same method is applicable for complex discrete-time systems as long as all components have lifetimes following independent and not necessarily identical DPH or MG distributions.

Modeling and analyzing complex discrete-time systems are illustrated with applications from Chapters 5 and 6 by replacing PH/ME distributions for component lifetimes in those applications with DPH/MG distributions. The system reliability models then are adjusted accordingly.

Application 7.6 A production line comprises four workstations arranged in series. For line balancing, workstations 1, 2, 3, and 4 require two, one, three, and two primary machines. Each workstation is equipped with two backup machines, except workstation 4 with one. All primary and backup machines in each workstation are identical. Lifetimes of all machines in each workstation have identical distributions, with machines in workstations 1 and 2 follow 2-NB distributions while machines in workstations 3 and 4 follow geometry distributions. Machines in different workstations have different *MTTF*, with machines at workstations 1, 2, 3, and 4 having *MTTF* equal to 20, 10, 25, and 40 days, respectively. The first and second backup machines in all workstations are operated with the same warmness levels d_1 and d_2, respectively. Because of power supply limitation, $d_1 + d_2 = 1$. The reliability block diagram of this production line is as depicted in Figure 7.39.

The model for this production line using the provided MATLAB functions is

$$T_{S_1} = \text{dphgensys}(T_1, T_1, T_1, T_1; 2; [d_1, d_2]),$$
$$T_{S_2} = \text{dphgensys}(T_2, T_2, T_2; 1; [d_1, d_2]),$$
$$T_{S_3} = \text{dphgensys}(T_3, T_3, T_3, T_3, T_3; 3; [d_1, d_2]),$$
$$T_{S_4} = \text{dphgensys}(T_4, T_4, T_4; 2; [d_1]),$$
$$T_S = \text{dphseriessys}(T_{S_1}, T_{S_2}, T_{S_3}, T_{S_4}).$$

The system reliability and hazard functions are compared for $d_1 = 1.0, 0.5, 0.0$. The results are presented in Figure 7.40.

The effect of warmness level of the backup machines to the system *MTTF* and *SDTTF* is investigated by varying d_1 from 0 to 1. The result is plotted as shown in Figure 7.41. These two figures

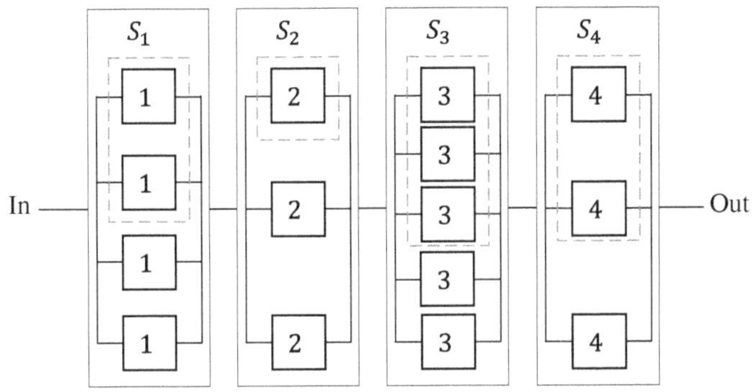

Figure 7.39 Production line in Application 7.6.

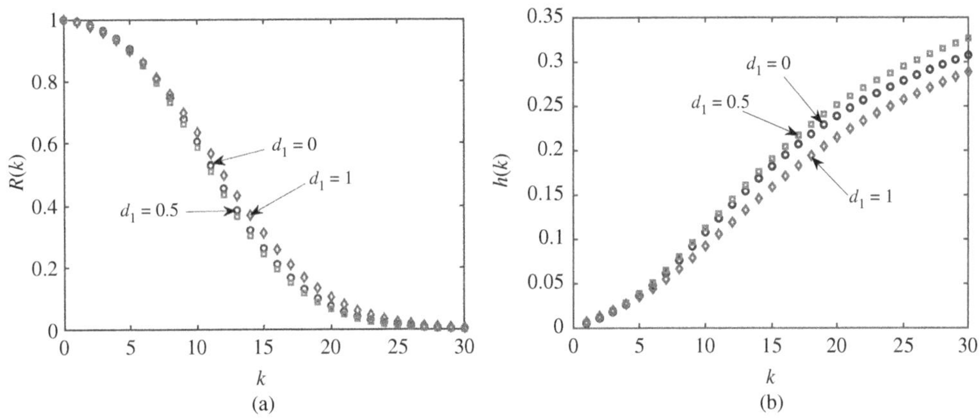

Figure 7.40 Reliability and hazard functions of the system in Application 7.6. (a) System reliability function and (b) system hazard function.

Figure 7.41 The system *MTTF* and *SDTTF* in Application 7.6 as a function of d_1.

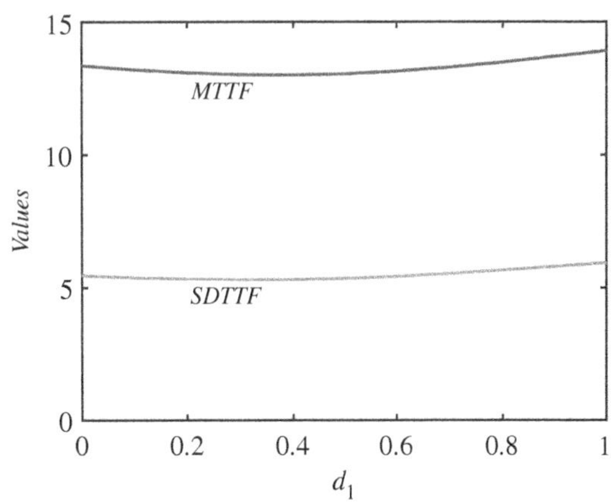

show that the system reliability is worst when the warmness level of the backup is neither cold nor hot as evidence by the network reliability function and $MTTF$ for $d_1 = 0.5$ is the lowest compared to $d_1 = 0$ and $d_1 = 1$, while the hazard function is the highest. ■

Application 7.7 A communication network is used to connect multiple hubs. The connection between two distant hubs is facilitated by several relay stations. The connection between two hubs can use two incompatible communication relay stations, with one as the primary link and the other as the backup link. The backup link can be in hot standby, warm standby, or cold standby thus creating a (general standby)-series structure. The graph representation of this communication network is shown in Figure 7.42, with black nodes representing hubs, source, or target, and white nodes representing relays. Not all relays support two links; for example, relays c and e use only one link. This network is analyzed for its reliability in facilitating information flow from node B to node G. All components have lifetimes following identical geometry distributions with $MTTF = 100$.

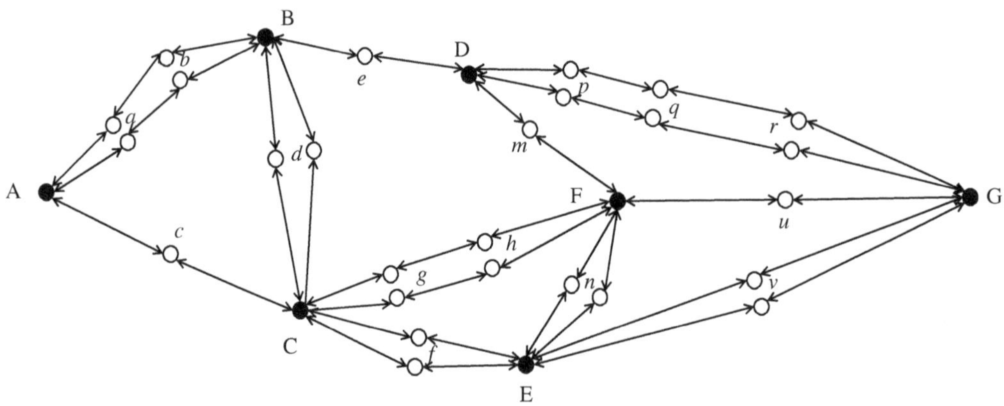

Figure 7.42 Communication network with backup channels.

Based on the network operational description, its graph representation can be simplified into Figure 7.43, where the number of each arc is MATLAB generated.

The arcs of this network are represented by their respective start (source) and end (target) nodes. Following the same procedure for this network in continuous time as described in Application 6.2,

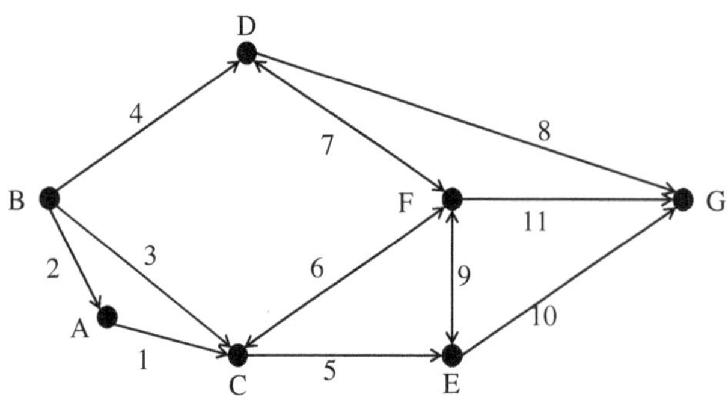

Figure 7.43 A simplified diagram of the communication network in Figure 7.42.

the network model in the provided MATLAB functions is

$$
\begin{aligned}
T_{AC} &= \text{dphseriessys}(T_{Ac}, T_{cC}),\\
T_{BA1} &= \text{dphseriessys}(T_{Bb}, T_{ba}, T_{aA}),\\
T_{BA} &= \text{dphgensys}(T_{BA1}, T_{BA1}; 1; [d]),\\
T_{BC21} &= \text{dphseriessys}(T_{Bd}, T_{dC}),\\
T_{BC2} &= \text{dphgensys}(T_{BC21}, T_{BC21}; 1; [d]),\\
T_{BD} &= \text{dphseriessys}(T_{Be}, T_{eD}),\\
T_{DF} &= \text{dphseriessys}(T_{Dm}, T_{mF}),\\
T_{FD} &= T_{DF},\\
T_{CE1} &= \text{dphseriessys}(T_{Cf}, T_{fE}),\\
T_{CE} &= \text{dphgensys}(T_{CE1}, T_{CE1}; 1; [d]),\\
T_{CF1} &= \text{dphseriessys}(T_{Cg}, T_{gh}, T_{hF}),\\
T_{CF} &= \text{dphgensys}(T_{CF1}, T_{CF1}; 1; [d]),\\
T_{FC} &= T_{CF},\\
T_{EF1} &= \text{dphseriessys}(T_{En}, T_{nF}),\\
T_{EF} &= \text{dphgensys}(T_{CE1}, T_{EF1}; 1; [d]),\\
T_{DG1} &= \text{dphseriessys}(T_{Dp}, T_{pq}, T_{qr}, T_{rG}),\\
T_{DG} &= \text{dphgensys}(T_{DG1}, T_{DG1}; 1; [d]),\\
T_{EG1} &= \text{dphseriessys}(T_{Ev}, T_{vG}),\\
T_{EG} &= \text{dphgensys}(T_{EG1}, T_{CG1}; 1; [d]),\\
T_{FG} &= \text{dphseriessys}(T_{Fu}, T_{uG}),\\
Graph &= (AC, BA, BC, BD, CE, CF, FC, DF, FD, EF, FE, DG, EG, FG),\\
mpath &= \text{minpaths}(Graph),\\
SDP &= \text{sdprep}(mpath),\\
T_{arc} &= [T_{AC}, T_{BA}, T_{BC}, T_{BD}, T_{CE}, T_{CF}, T_{DF}, T_{EF}, T_{DG}, T_{EG}, T_{FG}],\\
T_S &= \text{mgnetcomb}(SDP, T_{arc}).
\end{aligned}
$$

A comparison of the reliability and hazard functions for several standby configurations is presented in Figure 7.44, while Table 7.16 presents a comparison of their *MTTF*, *SDTTF*, and computing times for producing Figure 7.44 and Table 7.16.

Table 7.16 *MTTF*, *SDTTF*, and computation time of the network in Application 7.7.

Configuration	*MTTF*	*SDTTF*	Order	Computing time (s)
One line, no backup	35.74	19.57	497	0.05
One main line and one cold backup line (2-standby)	64.77	28.21	144,481	8.61
One main line and one warm backup ($d = 0.5$)	55.66	24.02	144,481	38.55
One main line and one hot backup line (2-parallel)	50.72	22.13	144,481	39.90

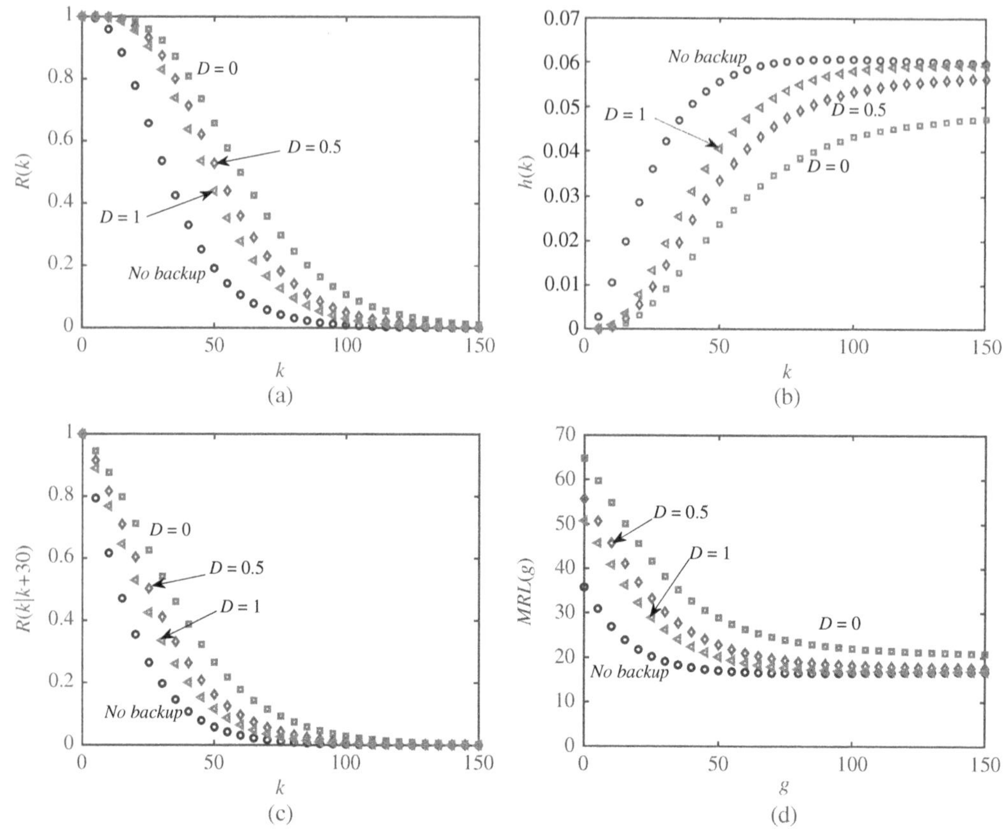

Figure 7.44 Reliability measures of the network in Application 7.7. (a) Reliability function, (b) hazard function, (c) residual reliability function (age 30), and (d) *MRL* function.

■

Application 7.8 The power system in a manufacturing company has three main subsystems: power supply, power distribution, and power consumption. The reliability block diagram of this system is shown in Figure 7.45, with components having identical lifetime distributions represented by the same numbers. The supply subsystem employs four identical generators, each with a 2-NB distributed lifetime having $MTTF = 100$ weeks. The power supply requirement is load dependent, with the idle generators positioned in warm standby with a warmness level of 0.5. The capacity of each generator is 75 kVA. Components 2, 3, and 4 in the distribution subsystem have lifetimes following geometry distributions with $MTTF = 400$, 500, and 200 weeks, respectively. The two bridges are connected using a switch with reliability 0.8. The load subsystem consists of three identical motors. All motors have a 2-NB distributed lifetime with $MTTF = 200$ weeks. At least one motor must be operating, with the remainder positioned in cold standby connected using perfect switch. The power requirement for each motor is 100 kVA.

By grouping components arranged in known structures into subsystems, the simplified reliability block diagram of this system is as shown in Figure 7.46.

Based on their individual capacities, the number of generators required is one more than the number of motors being operated. Let k be the number of motors operated, where $k = 1, 2, 3$. Then, the number of generators operated are $k + 1$. The system contains a simple network that its

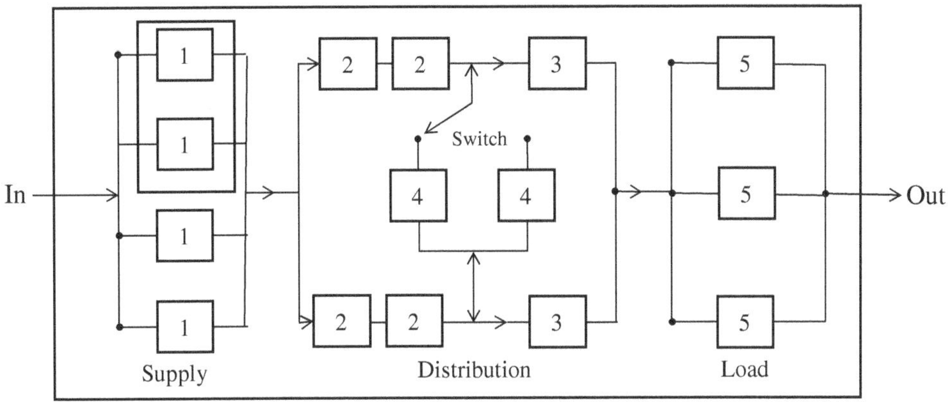

Figure 7.45 Reliability block diagram of the power system in Application 7.8.

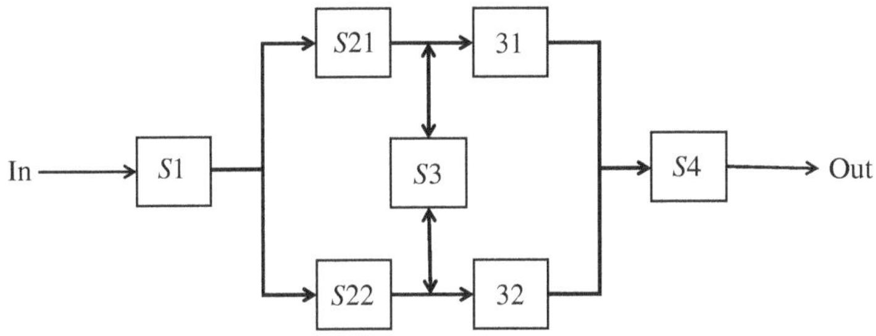

Figure 7.46 Simplified block diagram of Figure 7.45.

minimum paths can be generated manually and substitute them directly into the model. The warmness levels of backup components in S1 are 0.5 and 0.5, while in S4 are 0 and 0.

The system model using the provided MATLAB function is as follows:

$$\begin{aligned}
T_{S1} &= \text{dphgensys}(T_1, T_1, T_1, T_1; k+1; 0.5, 0.5),\\
T_{S21} &= \text{dphseriessys}(T_2, T_2),\\
T_{S22} &= \text{dphseriessys}(T_2, T_2),\\
T_{S3} &= \text{dphstandsys}(T_4, T_4; 0.8),\\
T_{S4} &= \text{dphgensys}(T_5, T_5, T_5; k; 0; 0),\\
SDP &= \text{sdprep}(\{S21, 31\}, \{S22, 32\}, \{S21, S3, 32\}, \{S22, S3, 31\}),\\
T_{arc} &= [T_{S21}, T_{S22}, T_{S3}, T_{31}, T_{32}],\\
T_{SS} &= \text{mgnetcomb}(SDP, T_{arc}),\\
T_S &= \text{dphseriessys}(T_{S1}, T_{SS}, T_{S4}).
\end{aligned}$$

Contrary to series, parallel, and standby structures that identical components and subsystems can be denoted by the same number, in a network configuration, different numbers must be used to denote each component or subsystem. Such distinction is necessary for finding the minimum paths. Therefore, the two identical subsystems of two component 2 in series (S_2) in the top and bottom lines in the distribution subsystem must be distinguished by giving them different notations as S_{21} and S_{22}. Likewise, the two component 3 are denoted as 31 and 32.

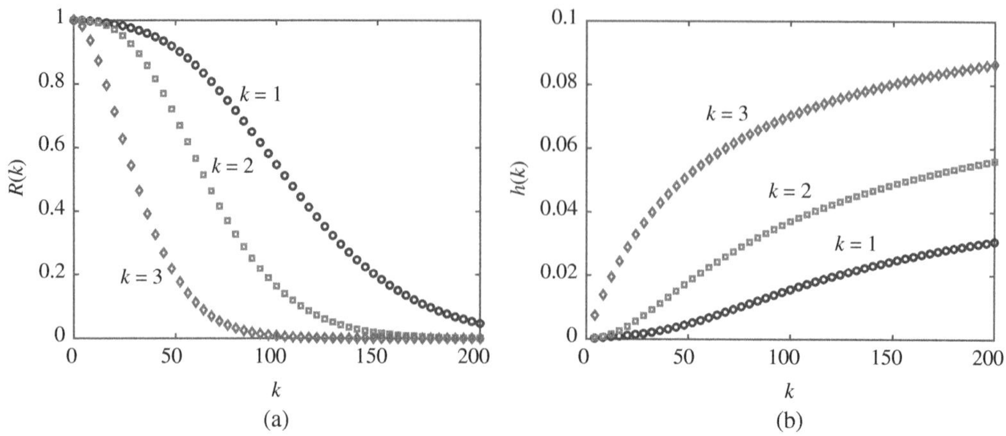

Figure 7.47 Reliability and hazard functions of the system in Application 7.8. (a) Reliability function and (b) hazard function.

The system model and the lifetime distribution data for each component are used to call the necessary MATLAB functions to generate an MG representation of the system lifetime distribution. Figure 7.47 is generated using the resulting MG representations. ■

7.12 Problems

7.1 Compare the reliability function of a component and its residual reliability functions for several ages if the component has lifetime following:
Conclude the results.
a) Geometry distribution
b) Negative binomial distribution
c) Mixed-geometry distribution

7.2 Repeat Problem 1 for *MRL* function of the component.

7.3 Prove mathematically that if component 1 has higher reliability function than component 2 then *MTTF* of component 1 is longer than *MTTF* of component 2.

7.4 A system has two components, and each has a 2-NB distributed lifetime with *MTTF* of 10 and 20. Manually derive DPH distributions for the system lifetime if the system structure is:
Compare the results with the results of using **dysrel-am**.
a) Series
b) Parallel
c) standby with perfect switch
d) standby with imperfect switch with reliability = 0.8.

7.5 Prove the following statement:
a) Geometry distribution is closed under minimum operation.
b) Negative binomial distribution is closed under addition operation.

c) Mixed-geometry distribution is closed under mixing operation.
d) Hypogeometry distribution is closed under addition operations.

7.6 Use the scalar reliability function convolution equation in discrete time to derive the reliability function expression for a 2-standby system with components having identical geometry distributed lifetimes if the switch is
Compare your results with using the matrix-form reliability function.
a) Perfect
b) Imperfect

7.7 Use the system in Application 7.6 to make a comparison of the computation times for plotting the reliability function and calculating *MTTF* using the DPH distributions and the Jordan MG distributions by assuming all machines have identical DPH distribution used for Example 7.39.

7.8 Show numerically that optimum warmness level ordering for continuous-time systems also holds for discrete-time systems.

7.9 Investigate the effect of hazard functions ordering in K-out-of-N discrete-time systems for $K = 1, 2$. Compare the results with the optimum hazard function ordering for continuous-time systems.

7.10 Compare reliability measures of
Conclude the results.
a) Series–parallel and parallel–series discrete-time systems
b) Series–standby and standby–series discrete-time systems

7.11 Construct a transition matrix table for 2-out-of-4 cold standby system and use it to devise a simpler algorithm specifically for K-out-of-N cold standby systems.

7.12 Propose a specific algorithm for parallel discrete-time systems having identical components. Is using the specific algorithm produces DPH distribution of the system lifetime with smaller order than using the general algorithm?

7.13 Prove that the order of DPH distribution of the lifetime of a series–parallel system is greater than that of a parallel–series system with the same components.

7.14 A system has a reliability block diagram in Figure P.7.1. Assume all components have 2-NB distributed lifetimes.
a) Model the system in the provided MATLAB functions.
b) Use the model to compare the system order, reliability, and hazard functions, *MTTF*, and *SDTTF* between all componets having geometry distribution and having 2-NB distribution.

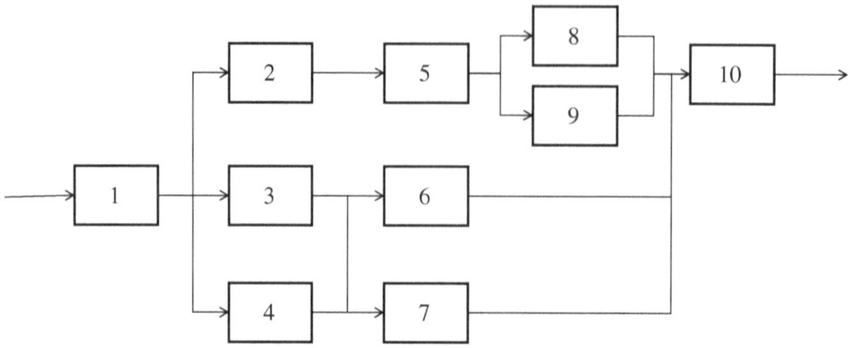

Figure P.7.1 Reliability block diagram of the system in Problem 14.

7.15 Repeat Problem 14 for components 3 and 4 as well as components 8 and 9 are arranged in a standby structure, both using imperfect switch with the same reliability = 0.8.

7.16 For the systems in Problems 14 and 15:
a) Compare numerically their system reliability and hazard functions.
b) Plot the graph of system *MTTF* as a function of switch reliability.

7.17 Show numerically that state space and combinatorial models of network in Figure 7.46 are equivalent by showing that the system reliability functions and *MTTFs* generated using the two models are identical.

7.18 Bridge components 3 and 7 in the network in Figure 7.33 have deterministic lifetimes equal to 6 and 8 days, respectively, and other components have identical geometry distribution with *MTTF* = 10 days.
a) Plot the network reliability function.
b) Investigate the effect of changing the position of bridge components.
c) The deterministic values in part a are changed from 6 and 8 to 8 and 6. Is the network reliability function the same as for part b? Explain the result.

7.19 Use **mgnetcomb** and **mgnettrans** functions to generate the reliability and hazard functions of the network in Figure 7.33, and show that the results using the two MATLAB functions are identical. Compare the computation times to produce the network reliability functions using the two functions.

7.20 A power system that connects two power generation sites to two consumption regions through high-voltage DC transmission has a reliability block diagram given in Figure P.7.2. The generators are identical with lifetimes following 3-NB distribution with *MTTF* = 600 days. Modules 2 and 3 in the AC/DC and DC/AC converters have lifetimes following geometry distributions with *MTTF* equal 1000 and 1500 days, respectively. The lifetimes of protection units in the transmission line each has 2-NB distribution with *MTTF* = 1000 days. When the consumptions are low, power generated by one generator can satisfy the demand. However, power generated by both generators is required when the consumptions are high.
a) Compare the reliability, hazard, and MRL functions and MTTF of this system to meet the low and high demand.

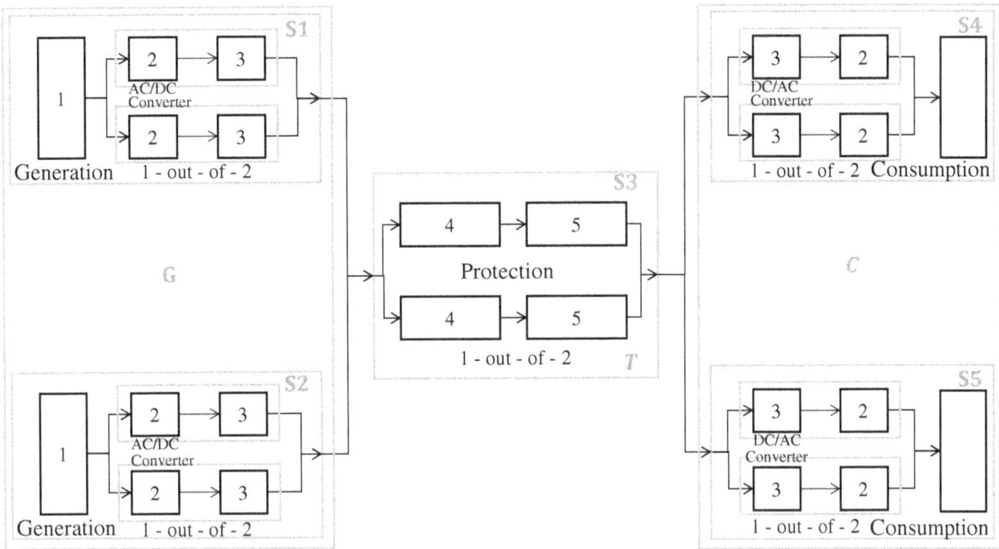

Figure P.7.2 Power system in Problem 14.

b) Suppose that all DC/AC and AC/DC Converters and Protections subsystems are arranged in general standby structure with the same warmness level. Plot the system MTTF as a function of warmness levels.

7.21 Repeat Problem 20 for Protection subsystems have a bridge structure with the bridge component have geometry distribution with $MTTF = 100$.

7.22 Construct a system reliability model of the network in Application 7.7 for transmitting information from node G to B and use them to generate network reliability measures. Compare the results with those obtained in Application 7.7 and explain the similarities or differences of the two results.

References

1 Roy, D., Dasgupta, T., 2001. A discretizing approach for evaluating reliability of complex systems under stress-strength model. IEEE Trans Reliab 50, 145–150. https://doi.org/10.1109/24.963121.

2 Almalki, S.J., Nadarajah, S.A., 2014. New discrete modified Weibull distribution. IEEE Trans Reliab 63, 68–80. https://doi.org/10.1109/TR.2014.2299691.

3 Chakraborty, S., Chakravarty, D., 2012. Discrete gamma distributions: Properties and parameter estimations. Commun Stat Theor Method 41, 3301–3324. https://doi.org/10.1080/03610926.2011.563014.

4 Nair, N.U., Sankaran, P.G., Balakrishnan, N., 2018. Reliability Modelling and Analysis in Discrete Time. Academic Press – Elsevier, London.

5 Chryssaphinou, O., Limnios, N., Malefaki, S., 2011. Multi-state reliability systems under discrete time semi-Markovian hypothesis. IEEE Trans Reliab 60, 80–87. https://doi.org/10.1109/TR.2010.2104210.

6 Yu, J., Zheng, S., Pham, H., Chen, T., 2018. Reliability modeling of multi-state degraded repairable systems and its applications to automotive systems. Qual Reliab Eng Int 34, 459–474. https://doi.org/10.1002/qre.2265.

7 Jia, X., Xing, L., Song, X., 2020. Aggregated Markov-based reliability analysis of multi-state systems under combined dynamic environments. Qual Reliab Eng Int 36, 846–860. https://doi.org/10.1002/qre.2584.

8 Ruiz-Castro, J.E., 2020. A complex multi-state k-out-of-n: G system with preventive maintenance and loss of units. Reliab Eng Syst Saf 197, 106797. https://doi.org/10.1016/j.ress.2020.106797.

9 Alkaff, A., 2021. Discrete time dynamic reliability modeling for systems with multistate components. Reliab Eng Sys Saf 209, 107462.

10 Alkaff, A., 2022. State space and binary decision diagram models for discrete standby systems with multistate components. Appl Math Model 110, 298–319.

11 Bilfaqih, Y., Qomarudin, M.N., Sahal, M., 2025. Time-dependent reliability computation of system with multistate components. Appl Math and Comp 487, 129082. https://doi.org/10.1016/j.amc.2024.129082.

12 Eryilmaz, S., 2022. Discrete Stochastic Models and Applications for Reliability Engineering and Statistical Quality Control. CRC Press, Boca Raton. https://doi.org/10.1201/9781003160632.

13 Ross, S.M., 2023. Introduction to Probability Models. Academic Press.

14 Neuts, M.F., 1994. Matrix-Geometric Solutions in Stochastic Models: An Algorithmic Approach. Dover Publication, Inc.

15 Bobbio, A., Horváth, A., Scarpa, M., Telek, M., 2003. Acyclic discrete phase type distributions: Properties and a parameter estimation algorithm. Perform Eval 54, 1–32. https://doi.org/10.1016/S0166-5316(03)00044-0.

16 Bladt, M., Nielsen, B.F., 2017. Matrix-Exponential Distributions in Applied Probability. Springer.

17 Akar, N., 2015. Fitting matrix geometric distributions by model reduction. Stoch Model 31, 292–315. https://doi.org/10.1080/15326349.2014.1003271.

18 Salvia, A.A., Bollinger, R.C., 1982. On discrete hazard functions. IEEE Trans Reliab R-31, 458–459. https://doi.org/10.1109/TR.1982.5221432.

19 Xie, M., Gaudoin, O., Bracquemond, C., 2002. Redefining failure rate function for discrete distributions. Int J Reliab Qual Saf Eng 09, 275–285. https://doi.org/10.1142/S0218539302000822.

20 Alfa, A.S., 2016. Applied Discrete-Time Queues. Springer, New York.

8

AI in System Reliability Modeling and Analysis

Learning objectives:

1) Aware of the availability of AI for system reliability.
2) Understand the capability and limitation of AI in modeling and analyzing reliability of systems with multistate components.
3) Able to use AI for system reliability modeling and analysis.
4) Able to analyze the responses given by AI.

Abbreviations

APH	acyclic phase-type
CDF	cumulative distribution function
CFR	constant failure rate
CPH	cyclic phase-type
CTMC	continuous-time Markov chain
DFR	decreasing failure rate
DPH	discrete phase-type
DTMC	discrete-time Markov chain
IFR	increasing failure rate
ME	matrix-exponential
MG	matrix-geometric
MRL	mean residual life
MTTF	mean time to failure
PDF	probability density function
PH	phase-type
SDTTF	standard deviation of time to failure
VTTF	variance time to failure

Notations

t	time, mission duration
n	order, dimension

$\boldsymbol{a}$	initial probability vector, starting vector
$\boldsymbol{A}$	transient matrix, transition matrix
$\boldsymbol{I}$	identity matrix
$\boldsymbol{v}$	closing vector
$\boldsymbol{b}$	absorbing vector
b	maintenance level
$(n, \boldsymbol{a}, \boldsymbol{A})$	PH/DPH representation
$(n, \boldsymbol{a}, \boldsymbol{A}, \boldsymbol{v})$	ME/MG representation
$E(.)$	expected value
$F(.)$	CDF
$f(.)$	PDF
$p(.)$	PMF
$R(.)$	reliability function
$h(.)$	hazard function
$R_{res}(.\mid g)$	residual reliability function at age g
$MRL(.)$	mean residual life function

Since its inception to the public as a free-to-use application, ChatGPT suddenly has become the most popular computer application in the world. It has shown to the world that many problems are solved, programs are verified, and improvements are suggested, with the help of AI. With that staggering number of users, its intelligence increases exponentially, since the application learned from the user responses to the answer that the application gave as responses to the requests submitted by the users. The application is cleverly designed as a chat application so that it creates a dialog for each request submitted by users and uses the dialogs with all users to update its knowledge accordingly.

ChatGPT always answers all questions submitted, never letting down users by admitting that the application does not know the answer in the first try. The application prefers to give a wrong answer than saying does not know the answer. It is its strategy to improve its knowledge, give any response, and let the users tell the application that the response makes no sense, and the application learned something from users responses to update its knowledge. For users that use the application blindly will easily get fooled by the application. They taught that responses from AI tools are absolutely correct.

The success of ChatGPT opens doors for similar AI applications based on the same technology that has been around for sometimes, but no one dared to use it to create an AI tool for public uses. All a sudden many AI applications are introduced to the public, either for general purposes or for special purposes AI. With the many applications around attract even more users and AI becomes more intelligent. User satisfactions increase rapidly, and AI uses in all aspects of life are becoming more and more apparent. Now it is in a point that AI cannot be ignored, and researchers, lecturers, and students are embracing AI for their academic works.

With that point of view, this book presents the use of AI in system reliability as one of its chapters. It is intentionally put as the last chapter, separately from chapters that discussed and develop methods for system reliability modeling and analysis, rather than blending human analytical reasoning with AI thinking in the development of each chapter. The objective is to persuade and attract readers to trust and use their own intelligence to formulate and solve problems related to system reliability, as a way to improve their intelligence. AI is positioned as the last resort if one cannot find a way to formulate and solve the problem, or as a mentor to check the correctness of one formulation and solution. Hopefully, with the intelligence and knowledge acquired by studying the previous chapters, one should be able to prompt important and legitimate questions to AI, justify intelligently the response given AI (know what right and wrong), and have meaningful and constructive dialog with AI.

This chapter does not cover the science and technology behind AI. It does not use AI to provide methods for solving problems, to prove theorems, to generate text to explain concepts, to generate problems, and to generate content and chapters, or anything related to content development. The chapter reports the results of dialogs with AI to solve problems that have been solved in the books or to compare the solutions provided by AI and obtained using the methods developed in the book. AI response to each question is analyzed, the difference between the two is explained, and the mistakes AI made are pointed out, if there is any. If necessary, a rebuttal to AI response is submitted and the AI response is analyzed again. However, the dialog is restricted to only one rebuttal if considered necessary. A rebuttal is submitted to AI if the answer is using similar method with the book. If AI answers are using different methods than no rebuttals are made, such as AI used simulation, approximation, or scalar methods.

Questions submitted to AI are related to reliability measures of relatively small systems. They are selected to represent a few examples discussed in the book, representing all book chapters. As will be seen from the responses, AI gave simple answers to problems submitted by using approximation, simulation, simplification, or any combination of these approaches. Furthermore, the answer to a problem could be different when it is resubmitted. For this reason, the questions submitted to AI are related to basic concepts discussed in the book. Complex systems discussed in the book are not submitted to AI. Unsatisfactory answers could be because this experiment uses the free-to-use version. AI response could be better if the AI that is utilized is not the free-to-use version. But the purpose of this book is not to get the best answer from AI. It is to inform the readers about the capability of free-to-use AI in solving system reliability problems dealt in the book.

This chapter has different style than previous chapters that explain concepts and use them to solve problems related to system reliability modeling and analysis. This chapter reports conversation with an AI tool in question-and-answer format. The questions submitted to AI tools are problems related to modeling and analysis of system reliability, not the concepts of modeling and analysis of system reliability. However, AI answers usually start with explanations about the concept that can be used to solve the problems. The responses to AI answers are about the correctness of its answer to the problems, not the correctness of their concepts.

8.1 AI Tools for Personal Use

AI tools can be categorized into general purposes and specific purposes. Examples of general-purpose AI tools are ChatGPT, Gemini, Copilot, Claude, Deepseek, and Perplexity. They can response to a wide range of questions. Examples of special-purpose AI tools are Cursor for coding, Jasper for marketing, Vanta for compliance to regulation/certification, Harvey for legal, DeepL for translation, Elicit for literature review, Scite for citations of a research article, ScaleAI for supply-chain, and midjourney for drawing/painting. There are many more AI tools than mentioned above, considering there are already tens of thousands of AI companies, besides companies that develop their own AI tools. Some are for corporations, and many are for personal. There is also AI that is used to compare and grade AI, such as LM Arena, to help user get the best answer from several AI. AI services can be classified as given in Figure 8.1.

Most AI tools for personal use, including ChatGPT, offer three types of services: free to use, limited subscription, and premium subscription. Free-to-use service is a strategy for AI companies

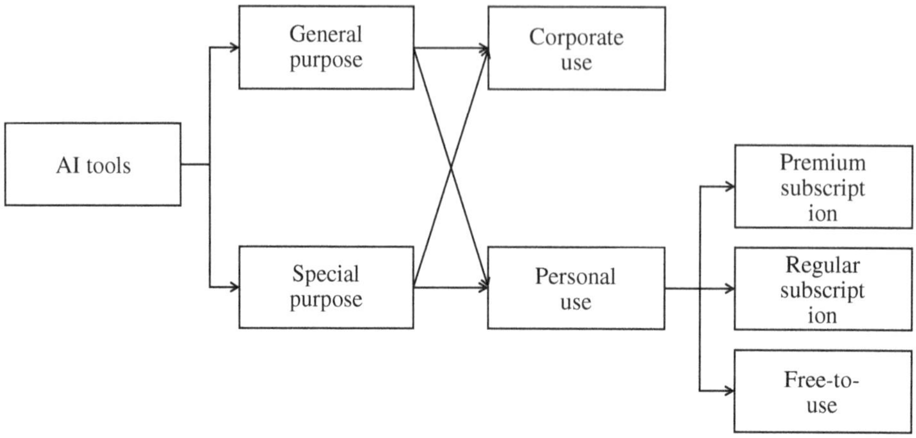

Figure 8.1 Classification of AI tools.

to train their AI by giving access to the public to use AI and get feedback from their interactions, so it is a symbiosis mutualistic between users and AI providers. Free-to-use option usually put a limit on the resources that can be used to answer user questions, such as time limit, memory limit, and knowledge limit. In many cases, AI in free-to-use mode will provide the less demanding resource solution to answer users' questions. It could be by simplifying the problem submitted by users, using simpler methods to give approximate solutions, or giving a halfway solution and providing a hint for the user to continue solving the problem on their own. AI response for free-to-use users depends on whether AI is busy or not. When AI is busy, their responses tend to be short, concise, simple, and not necessarily accurate answers. If AI are not too busy, AI will tell the users that they can provide more sophisticated solutions in case the users are not satisfied and want to get better solutions. As will be seen in AI responses to questions about system reliability submitted to AI in the following sections, AI responses are not typical; there are many variations for AI responses to user requests.

8.2 Using AI to Find Matrix-Form Distributions

The free version of ChatGPT from OpenAI is utilized to find a PH distribution for the lifetime of a system based on the system description. The resulting PH representation is utilized to generate the graphs of its CDF and PDF. The resulting PH distribution is then transformed into ME distribution in a Jordan canonical form, and a comparison between the CDF of PH distribution and its similar PH distribution is made. Other PH distributions with two absorbing states and different characteristics of eigenvalues are also used to broaden the experiments.

8.2.1 Finding PH Distributions from System Descriptions

Question 8.1: The condition of a radar sensor is continuously monitored. The monitoring results show that its condition can be classified into four states: Excellent, Decent, Fair, and Broken. The equipment is initially in an Excellent state. The duration of being in the Excellent state has an exponential distribution with a rate of 0.5/year before finally changing to the Decent state. The duration of being in the Decent state has an exponential distribution with a rate of 1/year.

Equipment in the Decent state can be broken with a probability of 0.2; otherwise, its state changes to the Fair state. The duration of being in the Fair state also has an exponential distribution with a rate of 2/year before finally broken. Broken equipment will remain broken forever (until it is repaired). At the beginning of operation, the radar sensor is in excellent condition. Derive a PH representation for the distribution of the duration of time from the radar sensor is initially operated until it is broken.

AI response:

Phase-type (PH) representation of radar sensor time to failure

We model the radar sensor degradation process using a Phase-type (PH) distribution. The states are Excellent (E), Decent (D), Fair (F), and Broken (B). Broken is an absorbing state.

Initial distribution

The initial probability vector is

$$\boldsymbol{\alpha} = [1 \ \ 0 \ \ 0].$$

Transient subgenerator matrix

The transient subgenerator matrix $\boldsymbol{T}$ is

$$\boldsymbol{T} = \begin{bmatrix} -0.5 & 0.5 & 0 \\ 0 & -1 & 0.8 \\ 0 & 0 & -2 \end{bmatrix}.$$

Absorption rate vector

The absorption rates to the Broken state are

$$\boldsymbol{t} = \begin{bmatrix} 0 \\ 0.2 \\ 2 \end{bmatrix}.$$

Distribution formulas

For time to absorption X (time until broken):

$$S(t) = \boldsymbol{\alpha} e^{\boldsymbol{T}t}\mathbf{1}.$$

$$F(t) = 1 - \boldsymbol{\alpha} e^{\boldsymbol{T}t}\mathbf{1}.$$

$$f(t) = \boldsymbol{\alpha} e^{\boldsymbol{T}t}\boldsymbol{t}.$$

Comment on AI response:

AI gave an answer the same as provided in the book. AI uses $\boldsymbol{T}$, $\boldsymbol{t}$, $\mathbf{1}$ for transient matrix, absorbing vector, and unitary vector, respectively, while the book uses $\boldsymbol{A}$, $\boldsymbol{b}$, $\boldsymbol{u}$. It is because the book is using T for lifetime and t for time or mission duration.

Question 8.2: The system in the previous question is maintained when it is in the fair state. If the maintenance succeeds, the system will be back to the decent state; otherwise, it stays in the same state. The success probability of maintenance is 0.8. Derive the PH representation of the time until the system fails.

AI response:

Phase-type (PH) representation of radar sensor time to failure with maintenance

We extend the model to include maintenance when the system reaches the Fair state. In the Fair state, a maintenance action is attempted. With a success probability of 0.8, the system returns to the Decent state; with a failure probability of 0.2, it remains in the Fair state. We model this as a new transient maintenance phase (M), with exponential duration Exp(μ). The limit $\mu \to \infty$ corresponds to instantaneous maintenance.

Initial distribution

The initial probability vector is

$$\alpha = \begin{bmatrix} 1 & 0 & 0 & 0 \end{bmatrix}.$$

Transient subgenerator matrix

The transient subgenerator matrix $\boldsymbol{T}$ is

$$\boldsymbol{T} = \begin{bmatrix} -0.5 & 0.5 & 0 & 0 \\ 0 & -1 & 0.8 & 0 \\ 0 & 0.8\mu & -\mu & 0.2\mu \\ 0 & 0 & 0 & -2 \end{bmatrix}.$$

Absorption rate vector

The absorption rates to the Broken state are

$$\boldsymbol{t} = \begin{bmatrix} 0 \\ 0.2 \\ 0 \\ 2 \end{bmatrix}.$$

Distribution formulas

For time to absorption X (time until broken):

$$S(t) = \boldsymbol{\alpha} e^{\boldsymbol{T}t}\mathbf{1}.$$

$$F(t) = 1 - \boldsymbol{\alpha} e^{\boldsymbol{T}t}\mathbf{1}.$$

$$f(t) = \boldsymbol{\alpha} e^{\boldsymbol{T}t}\boldsymbol{t}.$$

Moments

General formula:

$$E[X^k] = k!\boldsymbol{\alpha}(-\boldsymbol{T})^{-k}\mathbf{1}.$$

Instantaneous maintenance case

If maintenance is instantaneous upon entering the Fair state, this corresponds to the limit $\mu \to \infty$. In this case, upon reaching Fair, the system immediately jumps: with probability 0.8 back to Decent, or with probability 0.2 stays in Fair before the Exp(2) time to failure.

Comment on AI response:

The book provides a similar problem in Example 3.1. AI approach and answer are different from what is given in the book. There is no need to add another state. Assume that the maintenance is a random trial with a probability of success = 0.8 and if success the system is back to the Decent state. Otherwise, the system stays in the Fair state.

AI second response:

Problem summary and modeling assumption

The system from Question 8.1 is now maintained when it is in the Fair state. The maintenance succeeds with probability 0.8 and (on success) the system returns to the Decent state; on failure, the system becomes Broken. I make the following (explicit) modeling assumption to derive a clear PH representation:

- While in Fair, the holding time is exponential with rate $\lambda_3 = 2$/year. At the moment, the Fair phase exits (an event at rate 2), maintenance is attempted: with probability 0.8, the maintenance succeeds and the process jumps to Decent; otherwise (probability 0.2), the process jumps to Broken (absorbing).
- This interpretation matches the usual way to include "repair/maintenance on exit" in a CTMC/PH model and guarantees eventual absorption (Broken).
- If you intended a different mechanic (for example, a failed maintenance leaves the system in Fair and another maintenance is attempted later), tell me and I will adjust the model and re-export.

PH representation (transient phases: Excellent, Decent, Fair):
Initial probability vector ($\boldsymbol{\alpha}$):

$$\boldsymbol{\alpha} = [1, 0, 0].$$

Subgenerator matrix $\boldsymbol{S}$ (transient-to-transient rates):

	Excellent (1)	Decent (2)	Fair (3)
Excellent (1)	−0.5	0.5	0
Decent (2)	0	−1	0.8
Fair (3)	0	1.6	−2

Exit rate vector to absorbing state (Broken):

$$\boldsymbol{s} = -\boldsymbol{S}\mathbf{1} = [-0, 0.2, 0.4]^T$$

Survival and distribution functions:

$$\text{Survival}: S_T(t) = \boldsymbol{\alpha} e^{\boldsymbol{S}t}\mathbf{1}.$$

$$\text{CDF}: F_T(t) = 1 - \boldsymbol{\alpha} e^{\boldsymbol{S}t}\mathbf{1}.$$

Mean time to absorption (expected time until Broken):

$$E[T] = \boldsymbol{\alpha}(-\boldsymbol{S})^{-1}\mathbf{1}.$$

Numeric value (years): $E[T] = 5.888889$ years.

Comment on AI second response

AI finally gave the correct answer, the same as in the book.

Question 8.3: The system in the previous question without maintenance is considered failed if the condition is fair or broken. Derive the PH representation of the time until the system fails.

AI response:
Phase-type (PH) representation for time to failure (no maintenance)

This document presents the Phase-type (PH) representation for the time from initial operation until the radar sensor is considered failed. Failure is defined as the system being in either the Fair or Broken state.

Model description

States: Excellent (E), Decent (D), Fair (F), Broken (B, absorbing). The system starts in Excellent. Sojourn times:

- E: Exp(rate = 0.5), then moves to D.
- D: Exp(rate = 1); on exit, with probability 0.2, it goes directly to Broken (absorption); with probability 0.8, it goes to F.
- F: Exp(rate = 2), then goes to Broken.
- B: absorbing (once broken, stays broken).

Initial distribution

The initial probability vector is

$$\boldsymbol{\alpha} = \begin{bmatrix} 1 & 0 & 0 \end{bmatrix}.$$

Transient subgenerator matrix

The transient subgenerator matrix $\boldsymbol{T}$ (transient states: E, D, F) is

$$\boldsymbol{T} = \begin{bmatrix} -0.5 & 0.5 & 0 \\ 0 & -1 & 0.8 \\ 0 & 0 & -2 \end{bmatrix}.$$

Absorption rate vector

Rates from transient states to the absorbing Broken state:

$$\boldsymbol{t} = \begin{bmatrix} 0 \\ 0.2 \\ 2 \end{bmatrix}.$$

Distribution formulas

Let X be the time to absorption (time until Broken). The PH representation is $(\boldsymbol{\alpha}, \boldsymbol{T})$.

$$S(t) = \boldsymbol{\alpha} e^{\boldsymbol{T}t} \mathbf{1}.$$

$$F(t) = 1 - \boldsymbol{\alpha} e^{\boldsymbol{T}t} \mathbf{1}.$$

$$f(t) = \boldsymbol{\alpha} e^{\boldsymbol{T}t} \boldsymbol{t}.$$

Comment on AI response:
AI did not give the right answer. Its answer is the same as Question 8.1 where the Fair state is not considered the failed state. Fair state must be considered as failed; therefore, it is an absorbing state, not a transient state.

AI second response:

Problem summary

Starting from the system in Question 8.1/8.2, but WITHOUT maintenance and with the system considered failed when its condition is Fair or Broken. The state transitions are

- Excellent (1) – (rate 0.5) → Decent (2)
- Decent (2) – (rate 1) → Fair (failure, probability 0.8) OR Broken (failure, probability 0.2)
- Fair and Broken are absorbing failure states; therefore, we model failure as a single absorbing state that represents "failed" (either Fair or Broken).
- We take the transient phases to be Excellent and Decent only, and the absorbing state represents failure (Fair or Broken).

PH representation (transient phases: Excellent (1), Decent (2)):
Initial probability vector ($\boldsymbol{\alpha}$):

$$\boldsymbol{\alpha} = [1, 0].$$

Subgenerator matrix $\boldsymbol{S}$ (transient-to-transient rates):

	Excellent (1)	Decent (2)
Excellent (1)	−0.5	0.5
Decent (2)	0	−1

Exit rate vector to absorbing failure state (Failed = Fair or Broken):

$$\boldsymbol{s} = -\boldsymbol{S}\mathbf{1} = [0, 1]^T.$$

Survival and distribution functions:

$$\text{Survival function}: S_T(t) = \boldsymbol{\alpha} e^{\boldsymbol{S}t}\mathbf{1}.$$

$$\text{Cumulative distribution function}: F_T(t) = 1 - \boldsymbol{\alpha} e^{\boldsymbol{S}t}\mathbf{1}.$$

Mean time to absorption (expected time until failure):

$$E[T] = \boldsymbol{\alpha}(-\boldsymbol{S})^{-1}\mathbf{1}.$$

Numeric value (years): $E[T] = 3.000000$ years.

Comment on AI second response:
AI finally gave the correct answer.

8.2.2 Approximating Other Distribution with PH Distribution

Question 8.4: Weibull distribution having CDF $F_T(t) = 1 - e^{-(0,1t)^{1.5}}$ is approximated with APH distributions:

1) Determine the minimum order of PH distributions that can be used to approximate this Weibull distribution properly.
2) Use the minimum order and one order higher than the minimum to determine the best PH distribution approximations for each order used.
3) Calculate the largest discrepancies in numbers between the CDF of original Weibull distribution and each of these two PH distribution approximations.

AI responses:

PH approximation to Weibull $\boldsymbol{F(t)} = \mathbf{1} - \boldsymbol{e}^{-(0.1\ t)^{1.5}}$.

1) **Minimum order determination**
   A Coxian APH of order n has $2n - 1$ free parameters. To match first three moments, we need at least three free parameters => $n = 2$ is minimal.

$$\text{Minimum order}: n = 2.$$

2) **Fitted Coxian parameters and transient matrices**
   Order 2 (matched first three moments):

$$\text{Parameters } [\mu_1, \mu_2, p_1] = [0.24659358, 0.24659362, 1].$$

   $\boldsymbol{T}$ (order 2):

$$\boldsymbol{T} = \begin{bmatrix} -0.24659358 & 0.24659358 \\ 0 & -0.24659362 \end{bmatrix}.$$

   Order 3 (matched first five moments):

$$\text{Parameters } [\mu_1, \mu_2, \mu_3, p_1, p_2] = [0.30404897, 0.302256, 0.30225595, 0.93308746, 1].$$

   $\boldsymbol{T}$ (order 3):

$$\boldsymbol{T} = \begin{bmatrix} -0.30404897 & 0.28370428 & 0 \\ 0 & -0.302256 & 0.302256 \\ 0 & 0 & -0.30225595 \end{bmatrix}.$$

3) **Maximum absolute discrepancies between Weibull CDF and PH approximations**

Order 2: $max \mid F_{weib} - F_{PH2} \mid = 7.612107e - 02\ at\ t = 8.6200$
Order 3: $max \mid F_{weib} - F_{PH3} \mid = 6.394327e - 02\ at\ t = 4.8000$

Discrepancies at selected Weibull quantiles:

$$q = 0.1: t = 2.2308, \quad F_{PH2} = 0.105756, \quad F_{PH3} = 0.062103, \quad F_{weib} = 0.100000$$

$$q = 0.5: t = 7.8322, \quad F_{PH2} = 0.575099, \quad F_{PH3} = 0.455385, \quad F_{weib} = 0.500000$$

$$q = 0.9: t = 17.4372, \quad F_{PH2} = 0.928081, \quad F_{PH3} = 0.903649, \quad F_{weib} = 0.900000$$

Comments to AI responses:

AI answer is not the best answer. The order should be $> \left(\frac{1}{cv}\right)^2 = 2.17$; therefore, the minimum order is 3. Furthermore, the maximum difference between the CDFs of APH distribution order 3 approximation provided by AI and the original Weibull distribution is 0.064, much larger compared to 0.0051 for the APH approximation order 3 given in the book.

8.2.3 Transforming PH Distribution into ME Distribution

Question 8.5: Provide Jordan canonical ME representation for PH distributions with the following representations:

$$\boldsymbol{a}_1 = \begin{bmatrix} 0.5 & 0.3 & 0.2 \end{bmatrix}, \quad \boldsymbol{A}_1 = \begin{bmatrix} -2 & 1 & 0 \\ 0 & -1 & 0.5 \\ 0.7 & 0 & -2 \end{bmatrix}.$$

$$\boldsymbol{a}_2 = \begin{bmatrix}0.5 & 0.3 & 0.2\end{bmatrix}, \quad \boldsymbol{A}_2 = \begin{bmatrix} -2 & 1 & 0 \\ 0 & -1 & 0 \\ 0 & 0 & -2 \end{bmatrix}.$$

$$\boldsymbol{a}_3 = \begin{bmatrix}0.5 & 0.3 & 0.2\end{bmatrix}, \quad \boldsymbol{A}_3 = \begin{bmatrix} -1 & 0.6 & 0.4 \\ 0 & -1 & 0.6 \\ 0 & 0 & -3 \end{bmatrix}.$$

AI responses: (AI responses are truncated to four decimal digits)

Question 8.5 – Jordan canonical ME versus PH distributions

We show three PH distributions and their Jordan-like ME representations (via eigenvector similarity). Matrices and equations are given in Word equation format.

1) PH representation

$$\boldsymbol{\alpha}_1 = \begin{bmatrix}0.5 & 0.3 & 0.2\end{bmatrix}.$$

$$\boldsymbol{A}_1 = \begin{bmatrix} -2.0 & 1.0 & 0.0 \\ 0.0 & -1.0 & 0.5 \\ 0.7 & 0.0 & -2.0 \end{bmatrix}.$$

Jordan-like ME representation

$$\boldsymbol{J}_1 = \begin{bmatrix} (-2.1154 + 0.5205j) & 0j & (-0 - 0j) \\ -0j & (-2.1154 - 0.5205j) & (-0 + 0j) \\ 0j & -0j & (-0.7690 + 0j) \end{bmatrix}.$$

$$\boldsymbol{\alpha}_{1j} = \begin{bmatrix}(0.0054 + 0.2425j) & (0.0054 - 0.2425j) & (-0.5835 + 0j)\end{bmatrix}.$$

$$\boldsymbol{u}_1 = \begin{bmatrix} (0.2973 - 0.0952j) \\ (0.2973 + 0.0952j) \\ (-1.6293 + 0j) \end{bmatrix}.$$

2) PH representation

$$\boldsymbol{\alpha}_2 = \begin{bmatrix}0.5 & 0.3 & 0.2\end{bmatrix}.$$

$$\boldsymbol{A}_2 = \begin{bmatrix} -2.0 & 1.0 & 0.0 \\ 0.0 & -1.0 & 0.0 \\ 0.0 & 0.0 & -2.0 \end{bmatrix}.$$

Jordan-like ME representation

$$\boldsymbol{J}_2 = \begin{bmatrix} -2.0 & 0.0 & 0.0 \\ 0.0 & -1.0 & 0.0 \\ 0.0 & 0.0 & -2.0 \end{bmatrix}.$$

$$\boldsymbol{\alpha}_{2j} = \begin{bmatrix}0.5 & 0.56568542 & 0.2\end{bmatrix}.$$

$$\boldsymbol{u}_2 = \begin{bmatrix} 0.0 \\ 1.4142 \\ 1.0 \end{bmatrix}.$$

3) PH representation

$$\boldsymbol{\alpha}_3 = \begin{bmatrix} 0.5 & 0.3 & 0.2 \end{bmatrix}.$$

$$\boldsymbol{A}_3 = \begin{bmatrix} -1.0 & 0.6 & 0.4 \\ 0.0 & -1.0 & 0.6 \\ 0.0 & 0.0 & -3.0 \end{bmatrix}.$$

Jordan-like ME representation

$$\boldsymbol{J}_3 = \begin{bmatrix} -1.0 & 0.0 & -0.2805 \\ 0.0 & -1.0 & -0.0365 \\ 0.0 & 0.0 & -3.0 \end{bmatrix}.$$

$$\boldsymbol{\alpha}_{3j} = \begin{bmatrix} 0.5 & -0.5 & 0.0524 \end{bmatrix}.$$

$$\boldsymbol{u}_3 = \begin{bmatrix} 3512807709348988.0 \\ 3512807709348987.0 \\ 1.04980951 \end{bmatrix}.$$

CDF comparisons at 11 points

t	PH1	ME1	PH2	ME2	PH3	ME3
0.0000	0.0000	0.0000	0.0000	0.0000	0.0000	0.0550
2.0000	0.7972	0.7972	0.8881	0.8881	0.7664	0.7890
4.0000	0.9562	0.9562	0.9853	0.9853	0.9541	0.9613
6.0000	0.9906	0.9906	0.9980	0.9980	0.9919	0.9935
8.0000	0.9980	0.9980	0.9997	0.9997	0.9986	0.9989
10.0000	0.9996	0.9996	1.0000	1.0000	0.9998	0.9998
12.0000	0.9999	0.9999	1.0000	1.0000	1.0000	1.0000
14.0000	1.0000	1.0000	1.0000	1.0000	1.0000	1.0000
16.0000	1.0000	1.0000	1.0000	1.0000	1.0000	1.0000
18.0000	1.0000	1.0000	1.0000	1.0000	1.0000	1.0000
20.0000	1.0000	1.0000	1.0000	1.0000	1.0000	1.0000

Maximum absolute differences at these points: case 1 = 1.110e−16, case 2 = 0.000e+00, case 3 = 5.500e−02

Comments to AI responses:
AI gave the correct answers that PH distributions and their similar ME distributions produced the same reliability functions.

8.3 Using AI to Generate System Reliability Measures

AI is used to generate functional and statistical reliability measures of systems having PH or ME-distributed lifetimes. Three systems whose lifetimes have PH distributions with the following

PH representations are used (all elements are measured in per day):

$$\text{System 1: } \boldsymbol{\alpha} = \begin{bmatrix}1 & 0 & 0 & 0\end{bmatrix}, \quad \boldsymbol{A} = \begin{bmatrix} -0.1 & 0.1 & 0 & 0 \\ 0 & -0.4 & 0.2 & 0.1 \\ 0 & 0 & -0.5 & 0.3 \\ 0 & 0 & 0.2 & -0.4 \end{bmatrix}.$$

$$\text{System 2: } \boldsymbol{\alpha} = \begin{bmatrix}0.3 & 0.1 & 0.2 & 0.4\end{bmatrix}, \quad \boldsymbol{A} = \begin{bmatrix} -0.2 & 0 & 0 & 0 \\ 0 & -0.1 & 0 & 0 \\ 0 & 0 & -0.05 & 0 \\ 0 & 0 & 0 & -0.04 \end{bmatrix}.$$

$$\text{System 3: } \boldsymbol{\alpha} = [1], \quad \boldsymbol{A} = [-0.06].$$

8.3.1 Calculating Functional System Reliability Measures

The functional system reliability measures are generated for these systems, including their comparisons.

Question 8.6: Generate and compare 11 points with four-digit accuracy of the reliability and residual reliability functions at ages 5, 10, and 20 for System 1.

AI responses:
Reliability function (matrix form):

$$R(t) = \alpha e^{At}\mathbf{1}.$$

Residual (conditional) reliability given age s:

$$R_{res}(t \mid s) = P(T > s + t \mid T > s) = \frac{R(s+t)}{R(s)}.$$

Computation parameters: 11 points $t = 0, 2, 4, \ldots, 20$. Values shown with four-digit accuracy.

t	$S(t)$	$R_{res}(t\mid s = 5)$	$R_{res}(t\mid s = 10)$	$R_{res}(t\mid s = 20)$
0	1.0000	1.0000	1.0000	1.0000
2	0.9798	0.9018	0.8628	0.8328
4	0.9237	0.7960	0.7363	0.6913
6	0.8439	0.6914	0.6230	0.5723
8	0.7524	0.5930	0.5237	0.4728
10	0.6584	0.5038	0.4380	0.3900
12	0.5681	0.4248	0.3647	0.3212
14	0.4848	0.3561	0.3028	0.2643
16	0.4102	0.2972	0.2506	0.2172
18	0.3448	0.2470	0.2071	0.1784
20	0.2884	0.2048	0.1708	0.1465

Notes:

- All values computed using matrix exponential expm(At).
- Residual reliability $R_{res}(t\mid s)$ computed as $\frac{R(s+t)}{R(s)}$. If $R(s) = 0$, the residual is undefined (NaN).

Comment on AI responses:

AI gave correct answers. Residual reliability values are monotonically decreasing over time from 1 at $t = 0$ to 0 at $t \to \infty$, and residual reliability function of System 1, which is an IFR system, is lower as the system gets older.

Question 8.7: Generate and compare 11 points with four-digit accuracy of the reliability and residual reliability functions at ages 5, 10, and 20 for System 2.

AI responses:

Definitions:

$$\text{Survival (Reliability) function}: R(t) = \boldsymbol{\alpha} exp(\boldsymbol{A}t)\mathbf{1}.$$

Residual reliability at age s

$$R_{res}(t|s) = P(T > s + t \mid T > s) = (\boldsymbol{\alpha} exp(\boldsymbol{A}s)exp(\boldsymbol{A}t)\mathbf{1})/(\boldsymbol{\alpha} exp(\boldsymbol{A}s)\mathbf{1}).$$

Reliability $R(t)$ at $t = 0, 5, 10, \ldots, 50$ (11 points). Residual reliability $R_{res}(t|s)$ for ages $s = 5, 10, 20$. Values rounded to four decimal places:

t	$R(t)$	$R_{res}(t\|5)$	$R_{res}(t\|10)$	$R_{res}(t\|20)$
0	1.0000	1.0000	1.0000	1.0000
5	0.6543	0.7135	0.7524	0.7883
10	0.4668	0.5369	0.5834	0.6273
15	0.3512	0.4162	0.4599	0.5019
20	0.2723	0.3281	0.3659	0.4030
25	0.2147	0.2611	0.2928	0.3244
30	0.1708	0.2089	0.2351	0.2616
35	0.1367	0.1678	0.1893	0.2112
40	0.1098	0.1350	0.1526	0.1707
45	0.0883	0.1089	0.1232	0.1381
50	0.0712	0.0879	0.0996	0.1118

Notes:

- The subgenerator $\boldsymbol{A}$ is diagonal so $exp(\boldsymbol{A}t)$ is $diag(exp(\lambda_i t))$.
- $R(t)$ computed directly from matrix exponential; residual reliability computed by conditioning.

Comment on AI responses:

AI gave the correct answers that the residual reliability function of System 2, which is a DFR system, is higher as the system gets older.

Question 8.8: Generate and compare 11 points with four-digit accuracy of the hazard functions of the three systems in one table (only the hazard function).

AI response:

Phase-type hazard function formula (Word-equation style):

$$h(t) = \frac{f(t)}{S(t)} = \frac{\alpha e^{At}\boldsymbol{s}}{\alpha e^{At}\mathbf{1}},$$

where $\boldsymbol{s} = -\boldsymbol{A}\mathbf{1}$.

Hazard functions $h(t)$ for $t = 0, 1, \ldots, 10$ (four-digit accuracy)

t	$h_1(t)$	$h_2(t)$	$h_3(t)$
0	0.0000	0.0960	0.0600
1	0.0102	0.0912	0.0600
2	0.0203	0.0867	0.0600
3	0.0296	0.0825	0.0600
4	0.0380	0.0785	0.0600
5	0.0454	0.0749	0.0600
6	0.0518	0.0716	0.0600
7	0.0575	0.0686	0.0600
8	0.0625	0.0659	0.0600
9	0.0668	0.0634	0.0600
10	0.0706	0.0612	0.0600

Notes:

- Computations use $h(t) = \frac{f(t)}{S(t)} = \frac{\alpha e^{At}(-A\mathbf{1})}{\alpha e^{At}\mathbf{1}}$.
- Values rounded to four decimal places.
- System 1 hazard increases with t (from 0 up to ≈0.0706 at $t = 10$).
- System 2 hazard slowly decreases with t (from ≈0.0960 down to ≈0.0612).
- System 3 is a single exponential with constant hazard 0.06.

Comment on AI response:
AI gave the correct answers that the hazard function of System 1, System 2, and System 3 are increasing, decreasing, and constant over time, respectively.

8.3.2 Calculating Statistical System Reliability Measures

Question 8.9: Calculate and compare the *MRL* of these systems at ages 0, 5, 10, 15, and 20. The *MRL* is the expected value of the remaining lifetime of a system at a certain age that has not failed yet.

AI response:
Definition: For a phase-type (PH) distribution with initial row vector $\boldsymbol{a}$ and subintensity matrix $\boldsymbol{A}$, the survival function is

$$S(t) = \boldsymbol{a}e^{\boldsymbol{A}t}\mathbf{1}.$$

The MRL at age t (expected remaining lifetime given survival to t) is

$$m(t) = \frac{\boldsymbol{a}e^{\boldsymbol{A}t}(-\boldsymbol{A})^{-1}\mathbf{1}}{\boldsymbol{a}e^{\boldsymbol{A}t}\mathbf{1}}.$$

System definitions (initial vectors and subintensity matrices) as provided.
Numeric results: *MRL* $m(t)$ at ages 0, 5, 10, 15, 20 (rounded to four decimals).

System/age	0	5	10	15	20
System 1	16.2500	12.9233	11.5068	10.8296	10.4749
System 2	16.5000	19.0455	20.7792	21.8518	22.5002
System 3	16.6667	16.6667	16.6667	16.6667	16.6667

Note: Computation used $m(t) = \frac{\alpha e^{At}(-A)^{-1}\mathbf{1}}{\alpha e^{At}\mathbf{1}}$. Values computed numerically by matrix exponentials.

Comment on AI response:
AI gave the correct answers that the *MRL* function of systems 1–3 are increasing, decreasing, and constant over time, respectively.

8.3.3 Calculating System Reliability Measures Using ME Distribution

Question 8.10: Find Jordan canonical ME distribution of System 1. Calculate 11 points with four-digit accuracy each for reliability, hazard, and MRL functions using the resulting ME distribution and compare the results with using the original PH distribution.

AI response:
Original PH representation (System 1):

$$\boldsymbol{\alpha} = \begin{bmatrix} 1 & 0 & 0 & 0 \end{bmatrix}.$$

$$\boldsymbol{A} = \begin{bmatrix} -0.1 & 0.1 & 0 & 0 \\ 0 & -0.4 & 0.2 & 0.1 \\ 0 & 0 & -0.5 & 0.3 \\ 0 & 0 & 0.2 & -0.4 \end{bmatrix}.$$

Jordan-like ME representation (via similarity S of eigenvectors):

$$J = \boldsymbol{S}^{-1}\boldsymbol{A}\boldsymbol{S}.$$

$$\alpha_{ME} = \boldsymbol{\alpha}\boldsymbol{S}.$$

$$u_{ME} = \boldsymbol{S}^{-1}\mathbf{1}.$$

$$J = \begin{bmatrix} -0.1 & 6.0964811e-18 & 1.5614196e-17 & -4.316778e-18 \\ 0 & -0.4 & 8.2627472e-17 & 1.1697815e-17 \\ 0 & 0 & -0.7 & -3.9206729e-20 \\ 0 & 0 & -6.3374345e-17 & -0.2 \end{bmatrix}.$$

$$\alpha_{ME} = \begin{bmatrix} 1 & -0.31622777 & -0.057710986 & 0.58834841 \end{bmatrix}.$$

$$u_{ME} = \begin{bmatrix} 2.3333333 \\ -0.52704628 \\ 1.110223e-16 \\ -2.5495098 \end{bmatrix}.$$

Values at $t = 0, 5, \ldots, 50$ (survival $S(t)$, hazard $h(t)$, mean residual life $m(t)$)

t	$S_{PH}(t)$	$h_{PH}(t)$	$m_{PH}(t)$	$S_{ME}(t)$	$h_{ME}(t)$	$m_{ME}(t)$
0.0	1.0000	0.0000	16.2500	1.0000	0.0000	16.2500
5.0	0.8860	0.0454	12.9233	0.8860	0.0454	12.9233
10.0	0.6584	0.0706	11.5068	0.6584	0.0706	11.5068
15.0	0.4464	0.0835	10.8296	0.4464	0.0835	10.8296
20.0	0.2884	0.0905	10.4749	0.2884	0.0905	10.4749
25.0	0.1814	0.0944	10.2782	0.1814	0.0944	10.2782
30.0	0.1125	0.0967	10.1653	0.1125	0.0967	10.1653
35.0	0.0691	0.0980	10.0990	0.0691	0.0980	10.0990
40.0	0.0422	0.0988	10.0596	0.0422	0.0988	10.0596
45.0	0.0257	0.0993	10.0360	0.0257	0.0993	10.0360
50.0	0.0157	0.0996	10.0218	0.0157	0.0996	10.0218

Note: AI answer was layout to simplify the presentation.

Maximum absolute differences across the 11 points:

$$Max\ |S_{PH} - S_{ME}| = 4.441e - 16.$$

$$Max\ |h_{PH} - h_{ME}| = 5.551e - 17.$$

$$Max\ |m_{PH} - m_{ME}| = 3.553e - 15.$$

Comment on AI response:
AI gave the correct answers. However, AI does not recognize that the Jordan canonical form of the survival matrix of System1 should be a diagonal matrix; that is, the small numbers in off-diagonal elements are actually zero.

8.4 Using AI in Reliability-Based Replacement and Maintenance Programs

Reliability-based replacement and maintenance programs discussed in Sections 3.8 and 3.9 provided several measures, such as *MTBR*, reliability function with replacement, stockout risk for replacement with and without possible repair, optimum replacement interval, reliability function with perfect and imperfect maintenance, and reliability function with maintenance-replacement. AI is used to generate these reliability measures for IFR, CFR, and DFR system.

8.4.1 Reliability Measures of Systems with Periodic Replacement

Question 8.11: Calculate the *MTBR* if systems 1–3 each is replaced every 5, 10, and 15 days. Provide a table to accommodate the results.

AI response:
Problem statement

Calculate the MTBR if each system is replaced every $s = 5$, 10, and 15 days. We use a phase-type representation $(\boldsymbol{\alpha}, \boldsymbol{A})$ for each system.
Definitions and equations (Word-equation-friendly notation):
Let T be the lifetime random variable with survival function:

$$R(t) = \alpha e^{At}\mathbf{1}$$

Under the replacement policy (replace at age s if still working, or upon failure earlier), the mean cycle length (MTBR) is

$$MTBR(s) = E[min(T, s)] = \int_0^s R(t)dt.$$

System representations as provided.
Numerical results (MTBR in days):

System	s = 5 days	s = 10 days	s = 15 days
System 1	4.8003	8.6735	11.4160
System 2	4.0391	6.7998	8.8246
System 3	4.3197	7.5198	9.8905

Computed using $R(t) = \alpha e^{At}\mathbf{1}$ and $MTBR(s) = \int_0^s R(t)\,dt$. Numerical integration (fine grid) performed.

Comment on AI response:
AI gave the correct answers.

Question 8.12: Calculate the stockout risk of System 1 to be used for 10 days if currently there are s stocks on hand, $s = 0, 1, 2, 3$, and with a probability p the failed system can be repaired and reused with a negligible repair time. Compare the results for $p = 0, 0.3, 0.7, 1$.

AI response:
Concept and derivation

The stockout risk is the probability that the total number of failed systems during the mission exceeds the number of available spares s. The system may be repaired instantly with probability p, which reduces the effective failure rate.

$$R(t) = \boldsymbol{\alpha}_1 e^{\boldsymbol{A}_1 t}\mathbf{1}$$

$$F(t) = 1 - R(t)$$

$$MTTF = \boldsymbol{\alpha}_1 \,(-\boldsymbol{A}_1)^{-1}\,\mathbf{1}$$

$$\mu = \frac{T}{MTTF(1-p)}$$

$$P_{stockout} = 1 - \sum_{k=0}^{s} \frac{e^{-\mu}\mu^k}{k!}$$

Results

s/p	p = 0.0	p = 0.3	p = 0.7	p = 1.0
$s = 0$	0.4596	0.3500	0.1686	0
$s = 1$	0.1270	0.0700	0.0151	0
$s = 2$	0.0247	0.0097	0.0009	0
$s = 3$	0.0037	0.0010	0.0000	0

Comment on AI response:
AI answer is not a valid answer. AI is using approximation using the assumption that the system lifetime follows exponential distribution with rate = $1/MTTF$; thus, the number of replacements follows Poisson distribution. It is different than exact method used in the book that is based on the original PH distribution, not approximated with exponential distribution.

Question 8.13: Calculate and compare in two-digit accuracy the optimum replacement interval for systems 1 and 2 if the unplanned replacement cost is c times higher than the planned replacement cost, for $c = 10, 20, 30$. The planned replacement cost is the cost to replace the system, while the unplanned replacement cost is the cost to replace the system plus other costs due to system failure. The optimum interval is minimizing the expected total cost per unit time, where the total cost is the sum of the planned and unplanned replacement costs.

AI response:
Problem statement (short)

Given two systems with Phase-Type (PH) representations as provided.
Planned replacement cost is (C_p), and unplanned replacement cost is $(C_u = cC_p)$. We set $C_p = 1$ (unit cost) so $C_u = c$. The decision is to choose a planned replacement interval $(\tau > 0)$ that minimizes the long-run average cost per unit time.

Mathematical formulation

1) **Survival function (reliability):**

$$S(t) = \boldsymbol{a}e^{\boldsymbol{A}t}\mathbf{1}.$$

2) **CDF of failure:**

$$F(t) = 1 - S(t).$$

3) **Expected cost per cycle:** (replacement on failure if it occurs before τ, otherwise planned replacement at τ)

$$C(\tau) = C_u F(\tau) + C_p[1 - F(\tau)] = 1 + (c - 1)F(\tau).$$

4) **Expected cycle length:**

$$L(\tau) = E[\min(T, \tau)] = \int_0^S S(u)du.$$

Using PH identities, one can compute the integral analytically:

$$L(\tau) = \boldsymbol{a}(\boldsymbol{A}^{-1}(e^{\boldsymbol{A}\tau} - \boldsymbol{I}))\mathbf{1}.$$

5) **Long-run average cost rate:**

$$g(\tau) = \frac{C(\tau)}{L(\tau)} = \frac{1 + (c-1)F(\tau)}{L(\tau)}.$$

The optimal τ^* minimizes $g(\tau)$.

Computation (steps shown can help reproduce in Word with equations)

1) For each system, compute:

$$S(t) = \boldsymbol{a}e^{\boldsymbol{A}t}\mathbf{1}.$$

$$F(t) = 1 - S(t).$$

$$L(\tau) = \boldsymbol{a}(\boldsymbol{A}^{-1}(e^{\boldsymbol{A}\tau} - \boldsymbol{I}))\mathbf{1}.$$

2) Evaluate $g(\tau)$ on a grid, then refine by local optimization around the minimum to obtain a high-accuracy value for τ^*.
3) We adopt ($C_p = 1$) and compute for ($c = 10, 20, 30$).

Results (two-digit accuracy)

The numeric optimization yields the following **optimal planned replacement intervals** (rounded to two decimal places):

System	c = 10	c = 20	c = 30
System 1	5.22	3.35	2.66
System 2	*No finite optimum – replace on failure*	*No finite optimum – replace on failure*	*No finite optimum – replace on failure*

Interpretation for System 2. For System 2, the MTTF is

$$E[T] = \boldsymbol{a}(-\boldsymbol{A})^{-1}\mathbf{1} = 16.50\ (\textit{time units}).$$

As $\tau \to \infty$, the average cost rate tends to

$$g(\infty) = \frac{C_u}{E[T]} = \frac{c}{E[T]}.$$

For the evaluated c values, this asymptotic cost rate is lower than the cost rate for any finite planned interval, so the optimal policy is to **not plan replacements** (i.e., allow failures and perform unplanned replacements).

Supplementary numeric details (for reproducibility)

- **System 1** numeric minima (more digits):

$$c = 10:\ \tau^* = 5.2221\ (\textit{cost rate} \approx 0.4218).$$

$$c = 20:\ \tau^* = 3.3540\ (\textit{cost rate} \approx 0.6209).$$

$$c = 30:\ \tau^* = 2.6591\ (\textit{cost rate} \approx 0.7697).$$

- **System 2** MTTF: $E[T] = 16.50$. The asymptotic cost rates are $c/16.50$:

$$c = 10: 0.60606, \quad c = 20: 1.21212, \quad c = 30: 1.81818.$$

Notes

Time units are the same as those implied by the rates in the matrices (the user's problem statement did not specify a physical time unit). The numerical results are rounded to two decimal places as requested.

Comment on AI response:

AI provided the same cost formulation as in the book. Its answers are correct that System 1 has optimum replacement interval and System 2 does not. However, AI reasoning that System 2 does not have optimum replacement interval is not correct. It should be because System 2 is a DFR type that is better with age; therefore, replacement before failure will not be optimal.

8.4.2 Reliability Measures of Systems with Periodic Maintenance

Question 8.14: Generate and compare 11 points with four-digit accuracy of the reliability functions of System 1 that are maintained every s day with maintenance level b, for $s = 5$ and 10 and $b = 1.0, 0.5$, and 0.0 in one table. Maintenance level is the probability that the maintenance will be successful in bringing the system back to the new condition; otherwise, the system remains as it is not maintained. Present the results in two-dimensional table.

AI response:

The reliability function of System 1 follows $R(t) = \boldsymbol{a}_1 exp(\boldsymbol{A}_1 \boldsymbol{t})\mathbf{1}$, where $\boldsymbol{A}_1$ and $\boldsymbol{a}_1$ are the phase-type representation matrices.

Maintenance every s days with success probability b modifies the reliability as

$$R_n(t) = bR(t) + (1 - b)R(s + t).$$

Reliability values were computed for $s = 5$, 10 and $b = 1.0, 0.5, 0.0$ at 11 time points.

Time	$s = 5$, $b = 1.0$	$s = 5$, $b = 0.5$	$s = 5$, $b = 0.0$	$s = 10$, $b = 1.0$	$s = 10$, $b = 0.5$	$s = 10$, $b = 0.0$
0	1	0.9430	0.8860	1	0.8292	0.6584
5	1	0.9430	0.8860	0.8860	0.6662	0.4464
10	1	0.9430	0.8860	1	0.8292	0.6584
15	1	0.9430	0.8860	0.8860	0.6662	0.4464
20	1	0.9430	0.8860	1	0.8292	0.6584
25	1	0.9430	0.8860	0.8860	0.6662	0.4464
30	1	0.9430	0.8860	1	0.8292	0.6584
35	1	0.9430	0.8860	0.8860	0.6662	0.4464
40	1	0.9430	0.8860	1	0.8292	0.6584
45	1	0.9430	0.8860	0.8860	0.6662	0.4464
50	1	0.9430	0.8860	1	0.8292	0.6584

Comment on AI response:
AI did not consider the probability that the system had not failed when it was maintained. At time = 0, the reliability must be equal to 1, regardless of the maintenance interval and level.

AI Second response:
Interpretation and method (as Word-equation style text):

Let $S(t) = \boldsymbol{a}e^{At}\mathbf{1}$ be the PH survival function when starting from initial distribution $\boldsymbol{a}$.

We perform maintenance at times $k\,s$ ($k = 1, 2, \ldots$) and each maintenance succeeds with probability b, which returns the system to the "new" state (distribution $\boldsymbol{a}$). If maintenance fails (probability $1 - b$), the system remains in its current transient state. We keep unconditional probability vectors $\boldsymbol{v}_n$ (start of cycle n):

$$\boldsymbol{v}_0 = \boldsymbol{a}; \quad \text{for each cycle:}$$

$$v_{n+1} = bp_n\boldsymbol{a} + (1 - b)\, v_n e^{As}, \quad \text{where } p_n = \boldsymbol{v}_n e^{As}\mathbf{1}.$$

The reliability at time $t = ns + u$ $(0 \le u < s)$ is $S(t) = \boldsymbol{v}_n e^{Au}\mathbf{1}$.

Results show reliability at times $t = 0, 5, 10, \ldots, 50$ (11 points). Values rounded to four digits.

Time (days)	$s = 5$, $b = 1.0$	$s = 5$, $b = 0.5$	$s = 5$, $b = 0.0$	$s = 10$, $b = 1.0$	$s = 10$, $b = 0.5$	$s = 10$, $b = 0.0$
0	1.0000	1.0000	1.0000	1.0000	1.0000	1.0000
5	0.8860	0.8860	0.8860	0.8860	0.8860	0.8860
10	0.7850	0.7217	0.6584	0.6584	0.6584	0.6584
15	0.6954	0.5771	0.4464	0.5834	0.5149	0.4464
20	0.6161	0.4599	0.2884	0.4335	0.3610	0.2884
25	0.5459	0.3663	0.1814	0.3841	0.2787	0.1814
30	0.4836	0.2917	0.1125	0.2855	0.1944	0.1125
35	0.4285	0.2323	0.0691	0.2529	0.1500	0.0691
40	0.3796	0.1850	0.0422	0.1880	0.1046	0.0422
45	0.3364	0.1473	0.0257	0.1665	0.0807	0.0257
50	0.2980	0.1173	0.0157	0.1238	0.0562	0.0157

Notes:
- $b = 1.0$ corresponds to perfect maintenance (full renewal at maintenance times); then, the reliability over each cycle is identical.
- $b = 0.0$ corresponds to no maintenance (maintenance always fails) so maintenance has no effect and the reliability is just the original PH survival from time 0.
- Intermediate b (e.g., 0.5) mixes resets with survivors and shows intermediate improvement.

Comment on AI response:
AI gave the correct answers in its second response.

Question 8.15: Generate 11 points of the reliability functions of System 2 that are maintained every s day with maintenance level b, for s = 5 and 10 and b = 1.0, 0.5, and 0.0 in one table. Maintenance level is the probability that the maintenance is able to bring the system back to the new condition.

AI response:
System model

System 2 is modeled as a phase-type (PH) distribution characterized as provided.

The reliability function for a PH distribution is given by the Word equation:

$$R(t) = \boldsymbol{\alpha}_2 e^{\boldsymbol{A}_2 t} \mathbf{1}$$

Maintenance modeling

Maintenance is performed every s days (the maintenance interval). At each maintenance action:

- With probability b, maintenance succeeds (as-good-as-new).
- With probability $(1 - b)$, maintenance fails (no improvement).

Thus, the system undergoes imperfect maintenance, where $b \in [0,1]$ denotes the maintenance effectiveness:

- $b = 1.0 \rightarrow$ perfect maintenance.
- $b = 0.5 \rightarrow$ half-effective maintenance.
- $b = 0.0 \rightarrow$ no maintenance.

Reliability evolution across maintenance cycles

Initial distribution: $\boldsymbol{\alpha}_0 = \boldsymbol{\alpha}_2$
Subgenerator matrix: $\boldsymbol{A}_2$
Maintenance interval: s
Premaintenance distribution: $\boldsymbol{\alpha}_{pre} = \boldsymbol{\alpha}_0 e^{\boldsymbol{A}_2 s}$
After maintenance: $\boldsymbol{\alpha}_1 = b\boldsymbol{\alpha}_0 + (1 - b)\boldsymbol{\alpha}_{pre}$
General recurrence relation:

$$\boldsymbol{\alpha}_{k+1} = b\boldsymbol{\alpha}_0 + (1 - b)\boldsymbol{\alpha}_k e^{\boldsymbol{A}_2 s}.$$

Reliability at any time *t*

If t lies within the $(n + 1)$th interval:

$$n = floor(t/s), \quad r = t - n\,s.$$

Then the reliability function is given as

$$R(t) = \boldsymbol{\alpha}_n e^{\boldsymbol{A}_2 r} \mathbf{1}.$$

Parameter choices

Maintenance intervals: $s = 5, 10$
Maintenance levels: $b = 1.0, 0.5, 0.0$
Time range: $t = 0, 5, 10, \ldots, 50$

Interpretation

When $b = 1.0$, reliability resets completely after each maintenance (highest reliability).
When $b = 0.5$, reliability improves moderately after each maintenance.
When $b = 0.0$, reliability follows the original PH model without reset.

Numerical results (11 points)

Notes: Computations use exact matrix exponentials (expm). After each full s-day interval, the start-of-next-cycle distribution equals.

$$b\boldsymbol{a}_2 + (1 - b)[previous\ distribution].$$

t (days)	s = 5, b = 1.0	s = 5, b = 0.5	s = 5, b = 0.0	s = 10, b = 1.0	s = 10, b = 0.5	s = 10, b = 0.0
0	1.0000	1.0000	1.0000	1.0000	1.0000	1.0000
5	1.0000	0.8271	0.6543	0.6543	0.6543	0.6543
10	1.0000	0.7803	0.4668	1.0000	0.7334	0.4668
15	1.0000	0.7658	0.3513	0.6543	0.5028	0.3512
20	1.0000	0.7609	0.2723	1.0000	0.6848	0.2723
25	1.0000	0.7591	0.2147	0.6543	0.4686	0.2147
30	1.0000	0.7584	0.1708	1.0000	0.6721	0.1708
35	1.0000	0.7581	0.1367	0.6543	0.4589	0.1367
40	1.0000	0.7580	0.1098	1.0000	0.6683	0.1098
45	1.0000	0.7580	0.0883	0.6543	0.4558	0.0883
50	1.0000	0.7580	0.0712	1.0000	0.6671	0.0712

Comment on AI response:
AI did not include the reliability of the system at maintenance time in its calculation. The system being maintained is used again after maintenance. The system reliability function is the probability that the system does not fail since the beginning of its operation; therefore, the probability that the system has not failed when it is maintained must be taken into consideration in deriving the system reliability function.

AI second response:

Given Phase-type representation for System 2

Phase-type survival function (no maintenance):

$$S(t) = \boldsymbol{a}e^{At}\mathbf{1}.$$

Maintenance occurs every s days. Each maintenance action has a success probability b, which represents the probability that the system is restored to an "as-good-as-new" state. If maintenance fails (with probability $1 - b$), the system continues to age normally.

Hence, $b = 1.0 \rightarrow$ perfect maintenance (full renewal), $b = 0.5 \rightarrow$ imperfect maintenance, $b = 0.0 \rightarrow$ no maintenance (system degrades continuously).
Algorithm (described in Word form):

1) Start with alive state vector $\boldsymbol{v}(0) = \boldsymbol{a}$.
2) For each interval of length s: evolve $\boldsymbol{v} \leftarrow \boldsymbol{v}e^{As}$. At maintenance, if alive mass $p = sum(\boldsymbol{v})$, then $\boldsymbol{v} \leftarrow (1 - b)\boldsymbol{v} + bp\boldsymbol{a}$ (reset succeeded mass placed into "new" distribution $\boldsymbol{a}$).
3) For a final partial interval of length $\Delta < s$, evolve $\boldsymbol{v} \leftarrow \boldsymbol{v}e^{A\Delta}$.
4) Reliability $R(t)$ is total alive mass $sum(\boldsymbol{v})$ at time t.

Table of reliability values $R(t)$ at times 0, 5, …, 50 (rounded to four digits)

Time (days)	$s = 5$, $b = 1.0$	$s = 5$, $b = 0.5$	$s = 5$, $b = 0.0$	$s = 10$, $b = 1.0$	$s = 10$, $b = 0.5$	$s = 10$, $b = 0.0$
0.0	1.0	1.0	1.0	1.0	1.0	1.0
5.0	0.6543	0.6543	0.6543	0.6543	0.6543	0.6543
10.0	0.4281	0.4474	0.4668	0.4668	0.4668	0.4668
15.0	0.2801	0.3105	0.3512	0.3054	0.3283	0.3512
20.0	0.1832	0.2166	0.2723	0.2179	0.2451	0.2723
25.0	0.1199	0.1513	0.2147	0.1426	0.1749	0.2147
30.0	0.0784	0.1057	0.1708	0.1017	0.1317	0.1708
35.0	0.0513	0.0739	0.1367	0.0666	0.0942	0.1367
40.0	0.0336	0.0517	0.1098	0.0475	0.0711	0.1098
45.0	0.022	0.0361	0.0883	0.0311	0.0509	0.0883
50.0	0.0144	0.0253	0.0712	0.0222	0.0385	0.0712

Interpretation: If maintenance is perfect ($b = 1$), the system is fully renewed at each s-day mark. If $b = 0.5$, only half of surviving systems are renewed. If $b = 0$, the system never renews – same as natural degradation without maintenance.

Comment on AI response:
AI gave the correct answers on the second response.

Question 8.16: Generate and compare 11 points with four-digit accuracy of the reliability functions of System 1 that is maintained every 5 days with maintenance level 1, 0.5, 0, and replaced every 10 days. The reliability function measures the probability that the system has not failed from the beginning of its operation, including when it is maintained or replaced, the system has not failed yet.

AI response:
Concept and algorithm (plain text):
System model: Phase-type (PH) distribution with initial row vector $\boldsymbol{a}$ and subgenerator matrix $\boldsymbol{A}$. Survival (reliability) function without interventions is

$$R(t) = \boldsymbol{a}e^{\boldsymbol{A}t}\mathbf{1},$$

where $\mathbf{1}$ is a column vector of ones.
Maintenance and replacement model (preventive interventions occurring only if the system is alive):

1) Maintenance at scheduled time T_m with success probability b:
   - Let $F(T_m) = \boldsymbol{a}e^{AT_m}$ be the absolute distribution (row) at time T_m before maintenance, and $S(T_m) = sum(F(T_m))$.
   - After maintenance (absolute distribution):

   $$G(T_m) = bS(T_m)\boldsymbol{a} + (1 - b)F(T_m).$$

   (This accounts for: with probability b the alive mass $S(T_m)$ is reset to the initial distribution $\boldsymbol{a}$; with probability $1 - b$ alive mass remains in its premaintenance transient state distribution).

2) Replacement at scheduled time T_r (full replacement if alive):
   If the system is alive at T_r with mass $S\left(T_r^-\right)$, replacement resets that alive mass to a

$$G(T_r) = S\left(T_r^-\right)\boldsymbol{a}.$$

   Algorithm implemented:
   - Step through time segments between events (maintenance/replacement). For each segment, compute evolution by matrix exponential.
   - At each event, update the absolute distribution as described above and continue.

Requested numeric results: 11 points ($t = 0, 1, \ldots, 10$ days). Values rounded to four decimal places.

t (days)	Baseline *R*(*t*)	Maintenance *b* = 1.0	Maintenance *b* = 0.5	Maintenance *b* = 0.0
0	1.0000	1.0000	1.0000	1.0000
1	0.9949	0.9949	0.9949	0.9949
2	0.9798	0.9798	0.9798	0.9798
3	0.9556	0.9556	0.9556	0.9556
4	0.9237	0.9237	0.9237	0.9237
5	0.8860	0.8860	0.8860	0.8860
6	0.8439	0.8815	0.8627	0.8439
7	0.7989	0.8681	0.8335	0.7989
8	0.7524	0.8466	0.7995	0.7524
9	0.7053	0.8184	0.7618	0.7053
10	0.6584	0.7850	0.7217	0.6584

Notes:
- Maintenance scheduled at day 5. Replacement scheduled at day 10 (full reset if alive).
- All interventions act only on alive systems; failed systems remain failed (absorbing).
- Table values are the probability that the system has not failed since start (absolute survival probability).

Comment on AI response:

AI calculated the reliability of a system that is replaced every 10 days. After being replaced, its reliability function is the same as the reliability of a new system. However, if the interest is to find the mission reliability of a system that is replaced regularly, then the reliability function must take into consideration that the system is not failed at the time of replacement.

Question 8.16 is resubmitted using different prompt. Generate 11 points with four-digit accuracy of the reliability function of a mission of 50 days if the systems to carry the mission are maintained every 5 days with maintenance level $b = 1.0, 0.5, 0.0$ and replaced every 10 days.

AI response:

Reliability over a 50-day mission – System 1

Scenario:
- System: Phase-type with initial vector a and subgenerator A (four states).
- Preventive maintenance every 5 days (at days 5, 10, 15, …, 50). Maintenance success probability *b* as specified.
- Full replacement every 10 days (at days 10, 20, 30, 40, 50): if the system is alive at replacement time, the alive mass is reset to the initial state a.

- We report the mission reliability $R(t)$: the probability the system has not failed since start (absolute survival), evaluated at $t = 0, 5, 10, \ldots, 50$ (11 points). Values rounded to four decimal places.

Numeric results (t in days).

t (days)	b = 1.0	b = 0.5	b = 0.0
0	1.0000	1.0000	1.0000
5	0.8860	0.8860	0.8860
10	0.7850	0.7217	0.6584
15	0.6954	0.6394	0.5834
20	0.6161	0.5208	0.4335
25	0.5459	0.4615	0.3841
30	0.4836	0.3759	0.2855
35	0.4285	0.3330	0.2529
40	0.3796	0.2713	0.1880
45	0.3364	0.2403	0.1665
50	0.2980	0.1958	0.1238

Notes:
- Maintenance occurs at every 5 days; replacement at every 10 days (replacement resets alive mass).
- Maintenance only affects systems that are still alive at the scheduled time.
- Values are absolute survival probabilities (probability the system has not failed since start).

Comment on AI response:
AI gave the correct answer on the second response.

Question 8.17: Calculate and compare the MTTF of System 1 that is maintained every s day, with maintenance levels b for $s = 5$ and 10, and $b = 1.0$, 0.5, and 0.0 in one table. Maintenance level is the probability that the maintenance is able to bring the system back to the new condition. Since it is one system being maintained, take the probability that the system has not failed when it is maintained into consideration.

AI response:
Concept and assumptions

We treat the time-to-failure T of the system as a phase-type (PH) random variable with representation $(\boldsymbol{\alpha}, \boldsymbol{A})$. Preventive maintenance is scheduled every s days. If at a maintenance epoch the system is still operational (i.e., has not failed before that time), the maintenance succeeds with probability b and returns the system to the "as-good-as-new" condition; with probability $(1 - b)$, the maintenance does not restore it (the system remains in its current state). Failure before the maintenance time makes any maintenance at that epoch irrelevant.

Exact analysis with imperfect maintenance ($0 < b < 1$) requires tracking the system's residual PH state after unsuccessful maintenance. For clarity and tractability in this exercise, we use the following approach:

- For $b = 1$ (perfect maintenance restoring to new when alive at inspection), the regeneration structure is exact. Each successful maintenance restarts the PH time-to-failure distribution. The MTTF is obtained via a renewal argument (derivation below).

- For $b = 0$ (no preventive maintenance), the MTTF equals the baseline mean of the PH distribution: $M0 = \boldsymbol{\alpha}(-\boldsymbol{A})^{-1}\mathbf{1}$.
- For intermediate b (e.g., $b = 0.5$), we use an approximation that treats only successful maintenances as renewals, replacing the survival-to-maintenance probability q by the effective renewal probability qb in the renewal term. This approximation ignores the detailed evolution of the residual state after unsuccessful maintenance; the numeric table below reports these approximate values.

Derivation (perfect maintenance, $b = 1$)

Let $S(t) = P(T > t) = \boldsymbol{\alpha}e^{\boldsymbol{A}t}\mathbf{1}$ and $f(t) = \boldsymbol{\alpha}e^{\boldsymbol{A}t}(-\boldsymbol{A}\mathbf{1})$ be the density. Fix maintenance interval s. Define:

$$p = P(T \leq s) = 1 - S(s).$$

$$q = S(s) = P(T > s).$$

$$m = E[T|T \leq s] = \frac{\int_0^s tf(t)dt}{p}.$$

If the system is brought back to "as-new" whenever it survives to a maintenance epoch ($b = 1$), cycles are independent renewals. The event that failure occurs in the kth cycle has probability $q^k p$, and the time to failure in that case is $ks + T_{fail,cycle}$. Taking expectations and summing over k give $MTTF_{b=1} = m + s\frac{q}{1-1}$.

Baseline mean (no maintenance)

$M0 = E[T] = \boldsymbol{\alpha}(-\boldsymbol{A})^{-1}\mathbf{1}$ (standard PH mean).

Numeric results

s (days)	S(s) = P(T > s)	p = P(T≤s)	m = E[T|T≤s] (days)	MTTF b = 1.0 (days)	MTTF b = 0.5 (approx.)(days)
5	0.8860	0.1140	3.2488	42.0988	7.2253
10	0.6584	0.3416	6.1165	25.3935	11.0244

Baseline mean (no preventive maintenance, $b = 0$): $M0 = 16.250000$ days. (We use this value for the $b = 0$ row in the requested table.)
Notes: The $b = 0.5$ column is an approximation (see "Concept and Assumptions"). The document includes the PH matrices and the core formulas so you can reproduce the calculations or replace the approximation with an exact Markov analysis if desired.

Comment on AI response:

AI answers are correct for $b = 0$ and 1, but the answer for $b = 0.5$ is an approximation; thus, it is not a good answer. It is possible to calculate the exact value of $MTTF$ for a system being imperfectly maintained since it is possible (AI did it) to generate the system reliability function under such maintenance.

Question 8.18: Generate 11 points of the reliability functions of System 1 that are maintained every s day with rejuvenating factor r, for $s = 5$ and 10 and $r = 1.0, 0.5$, and 0.0 in one table. Rejuvenating factor is a number between 0 and 1 that brings the system age after maintenance back to $(1 - r)h$, where h is the time of maintenance. Since it is one system being maintained, take the probability that the system has not failed when it is maintained into consideration.

AI response:
Concept and algorithm (summary)

We model time-to-failure as a phase-type (PH) distribution with initial probability row vector $\boldsymbol{\alpha}$ and transient subgenerator $\boldsymbol{A}$. The survival function without maintenance is

$$S(t) = \boldsymbol{\alpha} e^{\boldsymbol{A}t}\mathbf{1}.$$

Periodic maintenance occurs at times ks ($k = 1, 2, \ldots$). If the system is still alive at maintenance time $h = ks$, the maintenance reduces the system's age to $(1 - r)h$ (a deterministic reduction). After rejuvenation, all surviving units are set to the conditional transient state distribution corresponding to age $(1 - r)h$ under the original model. The algorithm proceeds interval-by-interval as follows:

1) Start with unconditional transient-state mass $\boldsymbol{u} = \boldsymbol{\alpha}$ ($sum(\boldsymbol{u}) = 1$).
2) For each maintenance, $j = 1, \ldots, M$, where $M = floor(t/s)$:
   a. Propagate through the interval of length $s : u < - ue^{\boldsymbol{A}s}$.
   b. Let $surv = sum(u)$ be the probability mass still alive at maintenance. Compute the reset (conditional) distribution $w = \dfrac{\boldsymbol{\alpha} e^{\boldsymbol{A}(1-r)js}}{S((1-r)js)}$.
   c. Set $u \leftarrow surv * w$ (all surviving mass is moved into this reset distribution).
3) After the last maintenance before t, propagate the remaining mass through the residual time $\Delta = t - Ms : u \leftarrow \boldsymbol{u} e^{\boldsymbol{A}\Delta}$.
4) The reliability at time t is $R(t) = sum(\boldsymbol{u})$.

This algorithm implements the described rejuvenation policy: at each maintenance, only systems that have not failed are reset to the same (age-reduced) conditional distribution. The reset distribution is taken with respect to the original $\boldsymbol{\alpha}$ so the meaning of "age" is always consistent.

Equations (for insertion into Word equation)

PH survival function (no maintenance):

$$S(t) = \boldsymbol{\alpha} e^{\boldsymbol{A}t}\mathbf{1}.$$

Reset (conditional) distribution at maintenance j:

$$w_j = \frac{\boldsymbol{\alpha} e^{\boldsymbol{A}(1-r)js}}{\boldsymbol{\alpha} e^{\boldsymbol{A}(1-r)js}\mathbf{1}}.$$

Iteration (unconditional mass vector $\boldsymbol{u}$):

$$\boldsymbol{u}_0 = \boldsymbol{\alpha}.$$

$$\boldsymbol{u}_j = (\boldsymbol{u}_{j-1} e^{\boldsymbol{A}s}\mathbf{1}); w_j, j = 1, 2, \ldots.$$

Final propagation for residual time $\Delta = t - Ms$:

$$\boldsymbol{u}(t) = \boldsymbol{u}_M e^{\boldsymbol{A}\Delta}.$$

$$\text{Reliability}: R(t) = \boldsymbol{u}(t)\mathbf{1}.$$

Numerical results

Times (days): 0, 5, 10, 15, 20, 25, 30, 35, 40, 45, 50

Time (days)	$s = 5$, $r = 1.0$	$s = 5$, $r = 0.5$	$s = 5$, $r = 0.0$	$s = 10$, $r = 1.0$	$s = 10$, $r = 0.5$	$s = 10$, $r = 0.0$
0	1.0000	1.0000	1.0000	1.0000	1.0000	1.0000
5	0.8860	0.8860	0.8860	0.8860	0.8860	0.8860
10	0.7850	0.7095	0.6584	0.6584	0.6584	0.6584
15	0.6954	0.5273	0.4464	0.5834	0.4893	0.4464
20	0.6161	0.3715	0.2884	0.4335	0.3317	0.2884
25	0.5459	0.2518	0.1814	0.3841	0.2249	0.1814
30	0.4836	0.1660	0.1125	0.2855	0.1453	0.1125
35	0.4285	0.1073	0.0691	0.2529	0.0939	0.0691
40	0.3796	0.0683	0.0422	0.1880	0.0591	0.0422
45	0.3364	0.0429	0.0257	0.1665	0.0372	0.0257
50	0.2980	0.0268	0.0157	0.1238	0.0230	0.0157

Notes: Numbers rounded to four decimal places.

Comment on AI response:
AI answers are correct.

8.5 Using AI to model and Analyze Reliability of Systems

AI tool is used to generate reliability measures of systems with structures that had been discussed in Chapters 4–6, namely, series, parallel, standby, *K*-out-of-*N*, general standby, and network structures. The components are assumed to have lifetimes following PH or ME distributions. PH or ME distributions are used, and the number of components is varied.

8.5.1 Series Structured Systems

Question 8.19: A series system has five components. Each component has lifetime following PH distribution with representation

$$\boldsymbol{a}_1 = \boldsymbol{a}_4 = \begin{bmatrix}1 & 0 & 0 & 0\end{bmatrix}, \quad \boldsymbol{A}_1 = \boldsymbol{A}_4 = \begin{bmatrix} -0.1 & 0.1 & 0 & 0 \\ 0 & -0.4 & 0.2 & 0.1 \\ 0 & 0 & -0.5 & 0.3 \\ 0 & 0 & 0.2 & -0.4 \end{bmatrix};$$

$$\boldsymbol{a}_2 = \boldsymbol{a}_5 = \begin{bmatrix}0.3 & 0.1 & 0.2 & 0.4\end{bmatrix}, \quad \boldsymbol{A}_2 = \boldsymbol{A}_5 = \begin{bmatrix} -0.2 & 0 & 0 & 0 \\ 0 & -0.1 & 0 & 0 \\ 0 & 0 & -0.05 & 0 \\ 0 & 0 & 0 & -0.04 \end{bmatrix};$$

$\boldsymbol{a}_3 = [1], \quad \boldsymbol{A}_3 = [-0.06].$

Generate a table for system reliability, hazard, and MRL functions, 11 points each with four-digit accuracy.

AI response:

Concept and equations (text form)

PH distribution representation (text equations):

$$\text{Survival function for PH: } S(t) = \boldsymbol{a}e^{\boldsymbol{A}t}\mathbf{1}.$$

$$\text{Density: } f(t) = -\boldsymbol{a}e^{\boldsymbol{A}t}\boldsymbol{A}\mathbf{1}.$$

$$\text{Hazard: } h(t) = \frac{f(t)}{S(t)}.$$

MRL:

$$mrl(t) = \frac{1}{S(t)}\int_t^{\infty} S(u)\,du = -\frac{\boldsymbol{a}e^{\boldsymbol{A}t}\boldsymbol{A}^{-1}\mathbf{1}}{S(t)}.$$

For a series system of independent components:
System survival:

$$R_{sys}(t) = \Pi_i S_i(t).$$

System hazard:

$$h_{sys}(t) = \Sigma_i h_i(t).$$

System MRL:

$$mrl_{sys}(t) = \frac{1}{R_{sys}(t)}\int_t^{\infty} \Pi_i\ S_i(u)du.$$

Numeric results (11 points: $t = 0, 5, \ldots, 50$) (AI answer was layout)

System and component results (rounded to four digits):

	System			Component 3		
t	*R(t)*	*h(t)*	*MRL(t)*	*S(t)*	*h(t)*	*MRL(t)*
0	1.0000	0.2520	3.5472	1.0000	0.0600	16.6667
5	0.2489	0.3006	3.1789	0.7408	0.0600	16.6667
10	0.0519	0.3236	3.0340	0.5488	0.0600	16.6667
15	0.0100	0.3337	2.9698	0.4066	0.0600	16.6667
20	0.0019	0.3389	2.9362	0.3012	0.0600	16.6667
25	0.0003	0.3418	2.9173	0.2231	0.0600	16.6667
30	0.0001	0.3435	2.9068	0.1653	0.0600	16.6667
35	0.0000	0.3444	2.9014	0.1225	0.0600	16.6667
40	0.0000	0.3448	2.8992	0.0907	0.0600	16.6667
45	0.0000	0.3450	2.8990	0.0672	0.0600	16.6667
50	0.0000	0.3449	2.8998	0.0498	0.0600	16.6667

	Components 1 and 4			Components 2 and 5		
t	$S(t)$	$h(t)$	$MRL(t)$	$S(t)$	$h(t)$	$MRL(t)$
0	1.0000	−0.0000	16.2500	1.0000	0.0960	16.5000
5	0.8860	0.0454	12.9233	0.6543	0.0749	19.0455
10	0.6584	0.0706	11.5068	0.4668	0.0612	20.7792
15	0.4464	0.0835	10.8296	0.3512	0.0533	21.8518
20	0.2884	0.0905	10.4749	0.2723	0.0489	22.5002
25	0.1814	0.0944	10.2782	0.2147	0.0465	22.9036
30	0.1125	0.0967	10.1653	0.1708	0.0451	23.1696
35	0.0691	0.0980	10.0990	0.1367	0.0442	23.3573
40	0.0422	0.0988	10.0596	0.1098	0.0436	23.4983
45	0.0257	0.0993	10.0360	0.0883	0.0432	23.6100
50	0.0157	0.0996	10.0218	0.0712	0.0429	23.7020

Comment on AI response:
AI used scalar approach. It calculated the reliability and hazard value of each component and calculated the system reliability value as the product of component reliability values and the system hazard value as the sum of component hazard values. Calculation of *MRL* used numerical integration of reliability function.

8.5.2 Parallel Structured Systems

Question 8.20: A parallel system has five components as in Question 8.19.
Generate a table for system reliability, hazard, and *MRL* functions, 11 points each with four-digit accuracy.

AI response:

Problem statement and PH representations

Components lifetimes follow PH distributions with representations (initial row vector $\boldsymbol{a}$, subgenerator matrix $\boldsymbol{A}$).

Concept/formulas (phase-type basics and system formulas)

For a PH distribution with initial row vector $\boldsymbol{a}$ and subgenerator $\boldsymbol{A}$:

$$\text{Survival}: S(t) = \boldsymbol{a}e^{\boldsymbol{A}t}\boldsymbol{1}.$$

$$\text{Density}: f(t) = -\boldsymbol{a}e^{\boldsymbol{A}t}\boldsymbol{A}\boldsymbol{1}.$$

For a parallel system (independent components) with component survivals $S_i(t)$:

$$\text{System survival}: S_{sys}(t) = 1 - \Pi_{i=1}^{n}[1 - S_i(t)].$$

$$\text{System density}: f_{sys}(t) = \Sigma_{i=1}^{n} f_i(t)\Pi_{j\neq i}[1 - S_j(t)].$$

$$\text{System hazard: } h_{sys}(t) = \frac{f_{sys}(t)}{S_{sys}(t)}.$$

MRL of system at age t:

$$mrl_{sys}(t) = \frac{1}{S_{sys}(t)} \int_t^{\infty} \Pi_i\ S_{sys}(u)du.$$

These formulas are used to compute numeric values for the parallel system.

Numeric results (11 points, four-digit accuracy)

Points: $t = 0, 5, 10, \ldots, 50$ (units as in problem). Table shows $S_{sys}(t)$, $h_{sys}(t)$, $mrl_{sys}(t)$.

t	S_sys(t)	h_sys(t)	mrl_sys(t)
0	1.0000	0.0000	37.3084
5	0.9996	0.0005	32.3217
10	0.9850	0.0069	27.7492
15	0.9234	0.0194	24.4077
20	0.8126	0.0313	22.3762
25	0.6790	0.0400	21.2836
30	0.5479	0.0453	20.7845
35	0.4332	0.0483	20.6452
40	0.3389	0.0496	20.7168
45	0.2641	0.0500	20.9052
50	0.2058	0.0497	21.1513

Notes: Results shown with four decimal digits.

Comment on AI response:

Again, AI used scalar reliability equations for parallel systems and calculated hazard and *MRL* functions using numerical differentiation and integration, respectively.

8.5.3 Standby Structured Systems

Question 8.21: A standby system with switch reliability 0.8 has five components as in Question 8.19.

Generate a table for system reliability, hazard, and *MRL* functions, 11 points each with four-digit accuracy.

AI response:

Concept and algorithm (plain text for Word equations)

System description: A single active component with up to four cold standbys (total five components). When an active component fails, a switch is attempted with a success probability 0.8. If the switch fails, the system fails immediately. If it succeeds, the next standby starts operating from time of switch (cold standby). Components lifetimes are independent and follow the given phase-type (PH) representations.

Algorithm (Monte Carlo):

1) For each simulation run:

   a. Sample independent lifetimes for components 1–5 from their PH distributions.
   b. Start with component 1 active; when it fails, attempt switch (probability 0.8). If switch succeeds, start next component; else system fails. Continue until all components used or failure.

2) Collect system lifetime from each run.
3) Estimate reliability $S(t)$ as fraction of runs with lifetime $> t$.
4) Estimate hazard approximately by $h(t) = -\frac{ln\left(\frac{S(t+\Delta)}{S(t)}\right)}{\Delta}$, with $\Delta = 0.5$.
5) Estimate mean residual life $mrl(t) = mean(T - t| \, T > t)$ from sample lifetimes.

Numerical results ($N = 100\,000$ Monte Carlo runs)

t	Reliability $S(t)$	Hazard $h(t)$ (approx.)	Mean residual life mrl(t)
0.0	1.0000	0.0004	55.1815
5.0	0.9742	0.0097	51.5525
10.0	0.9155	0.0142	49.6880
15.0	0.8488	0.0148	48.3978
20.0	0.7853	0.0156	47.1135
25.0	0.7283	0.0140	45.6026
30.0	0.6767	0.0155	43.8893
35.0	0.6266	0.0160	42.1976
40.0	0.5787	0.0170	40.4852
45.0	0.5297	0.0184	38.9939
50.0	0.4817	0.0191	37.6357

Notes:
- Reliability and MRL estimated directly from simulated lifetimes.
- Hazard is approximated by a finite-difference estimator using $\Delta = 0.5$; smoothing or analytic evaluation would be needed for more precise hazard curves.
- PH sampling: we simulate the underlying transient CTMC until absorption to produce component lifetimes.
- Switch attempts are independent with success probability 0.8 each.
- Values rounded to four decimal places as requested.

Comment on AI response:
AI used simulation to answer the questions.

8.5.4 *K*-Out-of-*N* Structured Systems

Question 8.22: Components of a 2-out-of-5 system have lifetimes following PH distribution as given in Question 8.19.

Generate a table of system reliability, hazard, and MRL functions, 11 points each with four-digit accuracy.

AI response:
Concept and formulas (written in plain text here; convert to Word equations if you need OMML).
PH component representations as provided

Formulas

Component survival and density for PH $(\boldsymbol{a}, \boldsymbol{A})$:

$$S(t) = \boldsymbol{a}e^{\boldsymbol{A}t}\mathbf{1}.$$

$$f(t) = \boldsymbol{a}e^{\boldsymbol{A}t}(-\boldsymbol{A})\mathbf{1}.$$

For independent components, system reliability for 2-out-of-5 is

$$R_{sys}(t) = \Sigma_{subsets\ of\ components\ of\ size \geq 2}\left(\Pi_{i\ in\ survivors} S_i(t) \Pi_{j\ in\ failed}(1 - S_j(t))\right).$$

System density $f_{sys}(t) = -R'_{sys}(t)$ computed by differentiating each product term (exact expression used in computations).

$$\text{Hazard}: h_{sys}(t) = \frac{f_{sys}(t)}{R_{sys}(t)}.$$

MRL:

$$mrl(t) = \frac{1}{R_{sys}(t)} \int_{u=t}^{\infty} R_{sys}(u)du.$$

Numeric results (11 points, four-digit accuracy)

t	Reliability R_sys(*t*)	Hazard h_sys(*t*)	Mean residual life mrl(*t*)
0	1.0000	−0.0000	20.5914
5	0.9907	0.0078	15.7473
10	0.8829	0.0403	12.2956
15	0.6647	0.0716	10.4752
20	0.4397	0.0919	9.5897
25	0.2689	0.1036	9.1907
30	0.1576	0.1095	9.0589
35	0.0905	0.1118	9.0827
40	0.0517	0.1119	9.1991
45	0.0296	0.1107	9.3690
50	0.0171	0.1088	9.5662

Comment on AI response:
AI calculated reliability values of its components and used them to calculate reliability and hazard function of the system. *MRL* was calculated using numerical integration.

8.5.5 General Standby Structured Systems

Question 8.23: Components of a 2-out-of-5 system have lifetimes following PH distribution with representation as given in Question 8.19.

Generate a table for system reliability, hazard, and MRL functions, 11 points each with four-digit accuracy, if the backup components are operated in warm standby with warm levels 0.8, 0.5, and 0.2 for the first, second, and third backup components, respectively.

AI response:

Assumptions and configuration (interpreted and used for simulation):

- Initially components 1 and 2 are active; backups in order: 3 (first), 4 (second), 5 (third).
- Warm standby levels applied to backups: component 3 -> 0.8; component 4 -> 0.5; component 5 -> 0.2.
- Switching is instantaneous and perfect. Backups age while in warm standby under generator q*A and may fail while in standby. When switched to active, they continue in their current phase but under the active generator A.

Simulation parameters

- Monte Carlo trials: $N = 50\,000$
- Time points: [0.0, 5.0, 10.0, 15.0, 20.0, 25.0, 30.0, 35.0, 40.0, 45.0, 50.0]
- Small dt for hazard estimate: dt = 0.01

Numeric results (four-digit values)

t	Reliability $R(t)$	Hazard $h(t)$	Mean residual life mrl(t)
0.0000	1.0000	0.0000	25.4378
5.0000	0.9970	0.0020	20.5024
10.0000	0.9483	0.0190	16.3917
15.0000	0.8112	0.0296	13.7005
20.0000	0.6211	0.0805	12.1169
25.0000	0.4350	0.0552	11.2554
30.0000	0.2924	0.0889	10.5900
35.0000	0.1868	0.0857	10.2466
40.0000	0.1148	0.1394	10.2145
45.0000	0.0704	0.0852	10.1901
50.0000	0.0431	0.0928	10.2256

Notes: Results are Monte Carlo estimates. Precision to four decimal places is provided; increase N for tighter confidence intervals.

Comment on AI response:

AI used simulation to answer.

Question 8.24: Consider two 2-out-of-5 systems, each having five identical components. The lifetime of each component of System 1 has PH distribution with representation

$$\boldsymbol{a} = \begin{bmatrix}1 & 0 & 0 & 0\end{bmatrix}, \quad \boldsymbol{A} = \begin{bmatrix} -0.1 & 0.1 & 0 & 0 \\ 0 & -0.4 & 0.2 & 0.1 \\ 0 & 0 & -0.5 & 0.3 \\ 0 & 0 & 0.2 & -0.4 \end{bmatrix}.$$

The lifetime of each component of System 2 has ME distribution with representation

$$\boldsymbol{a} = \begin{bmatrix}-0.1111 & -0.3333 & 1.0000 & -1.5000\end{bmatrix},$$

$$\boldsymbol{A} = \begin{bmatrix} -0.7 & 0 & 0 & 0 \\ 0 & -0.4 & 0 & 0 \\ 0 & 0 & -0.1 & 0 \\ 0 & 0 & 0 & -0.2 \end{bmatrix},$$

$$\boldsymbol{v} = \begin{bmatrix}0 & -0.5 & 2.3333 & 1\end{bmatrix}^T.$$

Generate a comparison table for reliability, hazard, and MRL functions, 11 points each with four-digit accuracy, of systems 1 and 2 if the backup components are operated in warm standby with warm levels 0.8, 0.5, and 0.2 for the first, second, and third backup components, respectively.

AI response:

Theoretical concept

Each system is a 2-out-of-5 configuration.
System 1 has PH components with (a, A), System 2 has ME components (a, A, v).
System reliability with warm standby: two components active, three in standby.
Standby warmness levels: 0.8, 0.5, 0.2 for backups.
Phase-type formulas:

$$S(t) = \boldsymbol{a}e^{\boldsymbol{A}t}\mathbf{1}.$$

$$f(t) = -\boldsymbol{a}\boldsymbol{A}e^{\boldsymbol{A}t}\mathbf{1}.$$

$$MRL(t) = S(t)^{-1}\boldsymbol{a}e^{\boldsymbol{A}t}(-\boldsymbol{A}^{-1})\mathbf{1}.$$

For ME:

$$S(t) = \boldsymbol{a}e^{\boldsymbol{A}t}\mathbf{1}.$$

$$f(t) = \boldsymbol{a}e^{\boldsymbol{A}t}\boldsymbol{v}.$$

$$h(t) = \frac{f(t)}{S(t)}.$$

$$MRL(t) = \left(\frac{1}{S(t)}\right)\boldsymbol{a}e^{\boldsymbol{A}t}(-\boldsymbol{A}^{-1})\mathbf{1}.$$

System 1 is simulated by Monte Carlo (PH with warm standby).
System 2 is evaluated analytically for its component-level ME functions.

Results: System 1 (Monte Carlo, $N = 5000$)

t	$S(t)$	$h(t)$	MRL(t)
0	1.0000	0.0000	26.3380
1	1.0000	0.0000	25.3380
2	1.0000	0.0000	24.3380
3	1.0000	0.0000	23.3380
4	1.0000	0.0000	22.3380
5	1.0000	0.0006	21.3380
6	0.9994	0.0006	20.3502
7	0.9988	0.0016	19.3622
8	0.9972	0.0040	18.3922
9	0.9932	0.0044	17.4640
10	0.9888	0.0083	16.5389

Results: System 2 (ME analytic)

t	$S(t)$	$h(t)$	MRL(t)
0	−0.9444	nan	nan
1	−0.6018	nan	nan
2	−0.3639	nan	nan
3	−0.1964	nan	nan
4	−0.0777	nan	nan
5	0.0062	141.7678	510.2396
6	0.0651	12.9591	48.3920
7	0.1056	7.5659	29.0216
8	0.1325	5.6788	22.2249
9	0.1493	4.7234	18.7723
10	0.1587	4.1496	16.6910

Comment on AI response:
AI used simulation to answer the question for System 1. AI answer for System 2 using analytical method is not correct. It is due to AI did not interpret System 2 representation the same way as interpreting System 1 representation.

Question 8.25: A system has lifetime following 4-Erlang distribution with MTTF = 20 days is employed to carry out a mission for s days. The mission may use several identical systems arranged in a K-out-of-N structure to increase mission reliability, with backup components operating in warm standby. Find the value of N that minimizes the expected mission cost for cases where $K = 1$ or 2, $s = 10$ or 15, and warm levels = 1.0, 0.5 or 0.0. The costs associated with the mission comprise of the procurement cost $P = 10$ per system and the penalty cost due to mission failure $F = 200$ per mission.

AI response:

Problem statement (short)

A system has lifetime following a 4-Erlang distribution with MTTF = 20 days. The mission lasts s days. To increase mission reliability, several identical systems are arranged in a K-out-of-N structure with backups in warm standby. Procurement cost per system $P = 10$, penalty cost for mission failure $F = 200$. We seek N minimizing expected mission cost for $K = 1$ or 2, $s = 10$ or 15, and warm levels $\alpha = 1.0, 0.5, 0.0$.

Key approximation and assumptions

Exact analysis of K-out-of-N with warm standby and 4-Erlang component lifetimes is complex because Erlang is not memoryless and standbys may age. To produce clear numeric answers, we use a controlled approximation:

Assumptions (approximation):

1) Replace the 4-Erlang by an exponential with the same mean (memoryless approximation). Thus, component failure rate $\mu = 1/\text{MTTF} = 1/20 = 0.05$ per day.
2) Use a mean-field effective exposure: during the mission, each component experiences an average hazard proportional to the fraction of "active equivalent" time it sees. We approximate the effective failure rate per component over the mission as

$$\mu_{eff} = \frac{\mu(K + \alpha(N - K))}{N},$$

where α is the warm level (0 = cold standby, 1 = fully warm).
3) Each component then has survival probability $p = exp\,(-\mu_{eff} s)$. We approximate system success as the Binomial event "at least K of N independent components survive the mission."
4) Switching is assumed instantaneous and perfect; no switching failure cost.

Equations (linear Word style)

$$\mu = \frac{1}{MTTF}.$$

$$\mu_{eff} = \frac{\mu(K + \alpha(N - K))}{N}.$$

$$p = exp(-\mu_{eff} s).$$

$$R(N; K, s, \alpha) = \Sigma_{i=K}^{N} C(N, i) p^i (1 - p)^{N-i}.$$

$$C(N) = NP + F(1 - R(N; K, s, \alpha)).$$

$$N^* = argmin_{N \geq K} C(N).$$

Algorithm (numerical search)

For each combination (K,s,α):

1) Loop N from K to 30.
2) Compute μ_{eff}, p, $R(N)$, and expected cost $C(N)$.
3) Select N that minimizes $C(N)$.

Numerical results (table)

K	s (days)	α (warm level)	Best N	Expected cost	R (success probability)	p (component survival)
1.0	10.0	1.0	3.0	42.1832	0.939084	0.606531
1.0	10.0	0.5	3.0	34.5556	0.977222	0.716531
1.0	10.0	0.0	2.0	29.7858	0.951071	0.778801
1.0	15.0	1.0	4.0	55.5010	0.922495	0.472367
1.0	15.0	0.5	3.0	42.1832	0.939084	0.606531
1.0	15.0	0.0	3.0	32.1646	0.989177	0.778801
2.0	10.0	1.0	5.0	66.4239	0.917880	0.606531
2.0	10.0	0.5	5.0	55.8087	0.970957	0.704688
2.0	10.0	0.0	4.0	47.2220	0.963890	0.778801
2.0	15.0	1.0	7.0	86.5462	0.917269	0.472367
2.0	15.0	0.5	6.0	67.6063	0.961968	0.606531
2.0	15.0	0.0	5.0	53.5769	0.982116	0.740818

Note: Results use the exponential approximation and mean-field effective rate. They are numeric and meant to provide guidance. For exact results with 4-Erlang lifetimes, a phase-type/Markov model should be used.

Comment on AI response:
AI used exponential distribution approximation, not the original PH distribution.

8.5.6 Network Structured Systems

Question 8.26: A network has eight components with lifetime following identical 2-Erlang distribution with MTTF = 100 days. Generate a table for network reliability, hazard, and MRL functions, 11 points each with four-digit accuracy.

AI response:
Component: Erlang-2 with MTTF = 100.0 days → $\lambda = 0.0200$ per day.
Component survival and density:

$$R_{comp}(t) = e^{-\lambda t}(1 + \lambda t).$$

$$f_{comp}(t) = \lambda^2 t e^{-\lambda t}.$$

Network edges (components):

Component 1: node 0 ↔ node 1
Component 2: node 1 ↔ node 2
Component 3: node 2 ↔ node 3
Component 4: node 0 ↔ node 4
Component 5: node 4 ↔ node 5
Component 6: node 5 ↔ node 3
Component 7: node 1 ↔ node 4
Component 8: node 5 ↔ node 2

Analytical relations

$$R_{sys}(r) = \Sigma_{states\ connected} r^k (1-r)^{8-k}.$$

$$\frac{dR_{sys}}{dr} = \Sigma_{states\ connected}[kr^{k-1}(1-r)^{8-k} - (8-k)r^k(1-r)^{7-k}].$$

$$f_{sys}(t) = \frac{dR_{sys}}{dr} f_{comp}(t);\ h_{sys} = f_{sys}/R_{sys}.$$

$$m(t) = \frac{1}{R_{sys}} \int_t^{\infty} R_{sys}(u)\ du \ \text{(integrated numerically)}.$$

Numerical Table (11 points, four-digit accuracy)

t (days)	R_sys(*t*)	h_sys(*t*)	m(*t*) (days)
0	1.0000	0.0000	73.7893
10	0.9991	0.0004	63.8514
20	0.9878	0.0022	54.5045
30	0.9503	0.0058	46.4279
40	0.8758	0.0107	39.9183
50	0.7658	0.0162	34.8990
60	0.6337	0.0217	31.1145
70	0.4973	0.0267	28.2833
80	0.3720	0.0312	26.1653
90	0.2669	0.0351	24.5754
100	0.1848	0.0383	23.3773

Notes: R_{sys} computed by exact summation over connected states. MRL integrated numerically to a large finite upper bound.

Comment on AI response:
AI approach was using all combinations of good components that lead to the network being good, which is not an efficient way to calculate the network reliability. Each combination has series structure, and the reliability is calculated using scalar function.

Question 8.27: Components of the network in Figure 8.2 have lifetimes following PH distributions with representations

$$\boldsymbol{a}_1 = \boldsymbol{a}_2 = \boldsymbol{a}_3 = \begin{bmatrix}1 & 0 & 0 & 0\end{bmatrix}, \quad \boldsymbol{A}_1 = \boldsymbol{A}_2 = \boldsymbol{A}_3 = \begin{bmatrix} -0.1 & 0.1 & 0 & 0 \\ 0 & -0.4 & 0.2 & 0.1 \\ 0 & 0 & -0.5 & 0.3 \\ 0 & 0 & 0.2 & -0.4 \end{bmatrix};$$

$$\boldsymbol{a}_4 = \boldsymbol{a}_5 = \boldsymbol{a}_6 = \begin{bmatrix}0.3 & 0.1 & 0.2 & 0.4\end{bmatrix}, \quad \boldsymbol{A}_4 = \boldsymbol{A}_5 = \boldsymbol{A}_6 = \begin{bmatrix} -0.2 & 0 & 0 & 0 \\ 0 & -0.1 & 0 & 0 \\ 0 & 0 & -0.05 & 0 \\ 0 & 0 & 0 & -0.04 \end{bmatrix};$$

$$\boldsymbol{a}_7 = \boldsymbol{a}_8 = [1], \quad \boldsymbol{A}_7 = \boldsymbol{A}_8 = [-0.06].$$

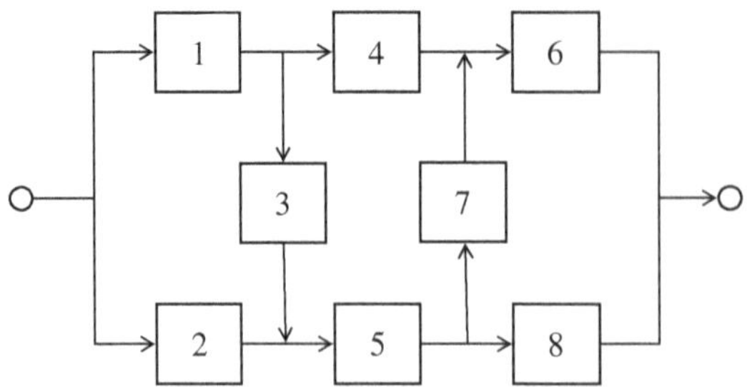

Figure 8.2 Network for Questions 8.26 and 8.27.

Generate a table of system reliability, hazard, and MRL functions, 11 points each with four-digit accuracy. Also calculate the network MTTF and SDTTF.

AI response:

Network description and PH representations

Two-terminal network (left to right). Components numbered as in the figure:

Top series: 1 → 2 → 3. Bottom series: 4 → 5 → 6. Crosslinks: 2 → 7 → 5 (component 7), and 5 → 8 → 2 (component 8).

A component is functioning if its lifetime exceeds t; components fail independently. The system works at time t if there exists a directed path from left (S) to right (T) using functioning components.

Graph topology (directed edges corresponding to components):

1: S → N1; 2: N1 → N2; 3: N2 → T; 4: S → M1; 5: M1 → M2; 6: M2 → T; 7: N2 → M2; 8: M2 → N2

Numeric results

Time points (seconds/days): 11 equally spaced points from 0 to 200. Values reported with four decimal digits.

t	Reliability $R(t)$	Hazard $h(t)$	Mean residual life $m(t)$
0.0000	1.0000	0.0000	9.8954
20.0000	0.0525	0.2152	5.0760
40.0000	0.0015	0.1444	7.3564
60.0000	0.0001	0.1282	7.8687
80.0000	0.0000	0.1258	7.9753
100.0000	0.0000	0.1248	8.0370
120.0000	0.0000	0.1240	8.0850
140.0000	0.0000	0.1233	8.1251
160.0000	0.0000	0.1228	8.1594
180.0000	0.0000	0.1223	8.1882
200.0000	0.0000	0.1219	8.2127

Network moments:

$$MTTF = \int_0^{\infty} R(t)\, dt \approx 9.8954.$$

$$SD(TTF) \approx 5.6148.$$

Comment on AI response:
AI used the same complicated and inefficient method as in answering Question 8.26.

8.5.7 Simple Compound Structured Systems

Question 8.28: A system uses four identical devices, each of which has three modules. Each device requires all modules to perform its function, and the failure of one module causes the device to fail. Other devices can use modules in one device. The four devices are operated simultaneously, even though only one is needed. The lifetime of module k, $k = 1,2,3$, has k-Erlang distributions with mean 100 k. Generate a table of the values of system reliability, hazard, and MRL functions, 11 points each with four-digit accuracy.

AI response:
Concept and derivation (equations written in Word-friendly text):

Module lifetimes

Module k ($k = 1,2,3$) has Erlang(k, λ_k) distribution with given means. From means:

Mean $= k/\lambda_k$. Given mean(1) = 100, mean(2) = 200, mean(3) = 300 $\Rightarrow \lambda_1 = \lambda_2 = \lambda_3 = 0.01$.

Survival function of Erlang(k,λ):

$$S_{k(t)} = e^{-\lambda t} \sum_{n=0}^{k-1} \frac{(\lambda t)^n}{n!}.$$

PDF of Erlang(k,λ):

$$f_{k(t)} = \lambda \frac{(\lambda\, t)^{k-1} e^{-\lambda\, t}}{(k-1)!}.$$

Device model: each device has three modules in series (all three must work). Device survival $R_d(t) = S_1(t)S_2(t)S_3(t)$.

Device pdf (series components):

$$f_d(t) = R_2R_3f_1 + R_1R_3f_2 + R_1R_2f_3,$$

where R_i and f_i are survival and pdf of module i.

System model: four identical devices in parallel (system works if at least one device works).

System reliability:

$$R_{sys}(t) = 1 - [1 - R_d(t)]^4.$$

System pdf:

$$f_{sys}(t) = 4\,[1 - R_d(t)]^3 f_d(t).$$

System hazard:

$$h_{sys}(t) = \frac{f_{sys}(t)}{R_{sys}(t)}.$$

MRL:

$$mrl_{sys}(t) = \frac{1}{R_{sys}(t)} \int_t^{\infty} R_{sys}(u)du.$$

Numeric results: values of R_system, h_system, and MRL_system at 11 points $t = 0, 50, \ldots, 500$. All numbers rounded to four decimal places.

t	R_system(*t*)	h_system(*t*)	MRL_system(*t*)
0	1.0000	0.0000	134.9796
50	0.9567	0.0030	89.3235
100	0.6818	0.0105	63.9124
150	0.3459	0.0162	52.7254
200	0.1406	0.0195	47.4630
250	0.0503	0.0214	44.6371
300	0.0167	0.0227	42.8894
350	0.0053	0.0235	41.6725
400	0.0016	0.0241	40.7530
450	0.0005	0.0247	40.0225
500	0.0001	0.0251	39.4239

Notes:
Interpretation/assumption used: modules in different devices are independent (each device has its own three modules). Devices are identical and operate in parallel; system works if at least one device functions. If the problem intended shared modules across devices (pooling), the model would change and must be reevaluated.

Comment on AI response:

AI identified correctly that the system has series–parallel structure. However, it used scalar functions to calculate the system reliability measures, which is possible given that the lifetime of each module follows Erlang distribution that has a simple scalar function.

Question 8.29: A system uses four identical devices, each of which has three modules. Each device requires all modules to perform its function, and the failure of one module causes the device to fail. Other devices cannot use modules in one device. The four devices are operated simultaneously, even though only one is needed. The lifetime of module k, $k = 1, 2, 3$, has k-Erlang distributions with a mean 100 k. Generate a table of the values of system reliability, hazard, and *MRL* functions, 11 points each with four-digit accuracy.

AI response:

Problem statement:

A system uses four identical devices, each of which has three modules. Each device requires all modules to perform its function, and the failure of one module causes the device to fail. Other devices cannot use modules in one device. The four devices are operated simultaneously, even though only one is needed. The lifetime of module k, $k = 1, 2, 3$, has k-Erlang distributions with a mean 100 k. Generate a table of the values of system reliability, hazard, and MRL functions, 11 points each with four-digit accuracy. Write equations and matrices using Word equations.

Concept and formulas (Word-equation style text)

Let λ be the rate parameter. Since mean(k-Erlang) $= k/\lambda = 100\,k => \lambda = 0.01$ for $k = 1, 2, 3$.
Erlang survival (shape k):

$$S_k(t) = \frac{\sum_{n=0}^{k-1} e^{-\lambda t}(\lambda\, t)^n}{n!}.$$

Erlang pdf (shape k):

$$f_k(t) = \frac{\lambda(\lambda\, t)^{k-1}}{(k-1)!}e^{-\lambda t}.$$

Device survival (three modules in series):

$$S_D(t) = S_1(t)S_2(t)S_3(t).$$

Device pdf:

$$f_D(t) = f_1(t)S_2(t)S_3(t) + S_1(t)f_2(t)S_3(t) + S_1(t)S_2(t)f_3(t).$$

System survival (four identical devices in parallel, system works if at least one device works):

$$S_{sys}(t) = 1 - [1 - S_D(t)]^4.$$

System pdf and hazard:

$$f_{sys}(t) = 4\,[1 - S_D(t)]^3\, f_D(t).$$

$$h_{sys}(t) = \frac{f_{sys}(t)}{S_{sys}(t)}.$$

MRL:

$$mrl(t) = \frac{1}{S_{sys}(t)}\left(\int_0^{\infty} S_{sys}(u)du\right).$$

Numeric results (11 points, $t = 0, 50, \ldots, 500$)

t	Reliability S_sys(t)	Hazard h_sys(t)	Mean residual life mrl(t)
0	1.0000	0.0000	134.9796
50	0.9567	0.0030	89.3235
100	0.6818	0.0105	63.9123
150	0.3459	0.0162	52.7253
200	0.1406	0.0195	47.4629
250	0.0503	0.0214	44.6370
300	0.0167	0.0227	42.8893
350	0.0053	0.0235	41.6723
400	0.0016	0.0241	40.7529
450	0.0005	0.0247	40.0224
500	0.0001	0.0251	39.4238

Comment on AI response:

AI could recognize the system has parallel–series structure and use the equations for scalar function to generate the results. Basically, it calculated the system reliability values at several time instances to produce the table. Consequently, it must use integration to calculate *MRL* function.

Question 8.30: A system uses four identical devices, each of which has three modules. Each device requires all modules to perform its function, and the failure of one module causes the device to fail. Other devices cannot use modules in one device. Two devices are the primary devices with the other two are backup. The first backup device is in warm standby with warm level = 0.8 and the second backup is in cold standby. The switch to replace the failed primary device with a backup device is perfect, never fails. The lifetime of module k, $k = 1, 2, 3$, has k-Erlang distributions with a mean 100 k. Generate a table of the values of system reliability, hazard, and MRL functions, 11 points each with four-digit accuracy.

AI response:

Problem description (summary):

A system uses four identical devices. Each device has three modules ($k = 1,2,3$), where module k has an Erlang(k, rate = 0.01) distribution with a mean 100 k. Two devices are primary (active); the first backup is in warm standby (level 0.8), and the second backup is cold. Switching is perfect. Device fails when any one of its modules fails.

PH/Erlang representation (per module):

For module k (shape k, rate lambda = 0.01):

Initial state vector: $\mathbf{a} = [1, 0, \ldots, 0]$ (length k).

Generator matrix (Erlang chain): a tridiagonal matrix with −lambda on diagonal and lambda on the superdiagonal.

Device lifetime survival when active:

$$S_{device}(t) = \prod_{k=1..3} S_{Erlang(k,lambda)}(t).$$

Warm standby: while in standby, the phase rates are multiplied by alpha (here 0.8); cold standby corresponds to alpha = 0 (no aging).

Monte Carlo simulation details:

Number of trials: 30000. Time points evaluated: [0.0, 50.0, 100.0, 150.0, 200.0, 250.0, 300.0, 350.0, 400.0, 450.0, 500.0]. Estimated quantities reported with four decimal places (Monte Carlo noise applies).

Numeric results (11 points):

t	Reliability $S(t)$	Hazard $h(t)$ (approx.)	Mean residual life mrl(t)
0.0	1.0000	0.0000	155.1113
50.0	0.9747	0.0019	108.1117
100.0	0.7782	0.0070	77.8860
150.0	0.4720	0.0124	61.9674
200.0	0.2277	0.0168	53.9899
250.0	0.0942	0.0184	49.7371
300.0	0.0352	0.0189	48.1503
350.0	0.0129	0.0206	44.7285
400.0	0.0042	0.0240	43.3845
450.0	0.0014	0.0000	31.7527
500.0	0.0003	0.1250	38.2224

Notes:
1. Results were obtained by event-driven Monte Carlo simulation (see code comments).
2. Hazard estimates were approximated from finite differences of the empirical CDF and may be noisy; increase N for smoother hazard curves.
3. MRL computed directly as empirical conditional mean $E[T - t \mid T > t]$.

Comment on AI response:
AI provided only approximation.

8.5.8 Component Ordering

Question 8.31: Standby systems have independent and nonidentical components. Derive a method to order components to optimize the reliability measures in the case where:

a) The system is a 1-out-of-N cold standby system with imperfect switch
b) The system is a 1-out-of-N warm standby system with perfect switch
c) The system is a 2-out-of-N cold standby system with perfect switch
d) The system is a 2-out-of-N warm standby system with perfect switch

Use five components with 2-Erlang-distributed lifetimes having different MTTFs to demonstrate the optimal ordering for each case.

AI response:
This document derives methods to order nonidentical components to optimize reliability measures for the four requested cases and demonstrates numeric results using five components with 2-Erlang lifetimes (different MTTFs).
Assumptions (chosen for the demonstration):

- Components' MTTFs = [60, 80, 100, 130, 200] time units (components labeled 1, …, 5 correspondingly).

- Each component has Erlang-2 lifetime with rate $\lambda_i = \frac{2}{MTTF_i}$ (so mean $= MTTF_i$).
- Warmness level $r = 0.5$ for warm standby cases (standby ages at 50% of active rate).
- Imperfect switch success probability $q = 0.95$ for case (a).
- Numerical demonstration: Monte Carlo simulation over permutations (quick approximate search with 4000 trials per permutation).

Notation and model

- Let components be indexed by i, with lifetime random variable $T_i \sim$ Erlang(2, λ_i), mean $M_i = E[T_i]$.
- An ordering (permutation) $\sigma = (\sigma(1), \sigma(2), \ldots, \sigma(N))$ defines the priority sequence for activation: $\sigma(1)$ used first, $\sigma(2)$ is first standby, etc.
- For k-out-of-N standby: initially, the first k components in the order are active; the others are standbys in that order. When an active fails, the highest-priority available standby is switched in (subject to switch success).

Analytical insight and ordering methods

Case (a): 1-out-of-*N* cold standby with imperfect switch (single failure-prone switch per activation):

- Cold standby ⇒ standbys do not age while waiting. Each component contributes its entire lifetime only if it is activated. If the switch has success probability q each time a switch is needed, the probability that the kth component in order is ever activated is q^{k-1} (because $k - 1$ previous switches must have succeeded).
- Therefore, the expected system lifetime (MTTF) for a given order σ is

$$E[T_{sys}(\sigma)] = \sum_{k=1}^{N} q^{k-1} M_{\sigma(k)}.$$

- The formula is linear in the $M_{\sigma(k)}$ with weights q^{k-1} decreasing in k when $q < 1$. Thus, the expected lifetime is maximized by placing the largest M_i at position 1, the next largest at position 2, etc. In other words: **Optimal ordering (case a, analytic): ** sort components in descending order of MTTF (largest M first).

Case (b): 1-of-*N* warm standby with perfect switch:

- Warm standby: standby components age while waiting at rate r ($0 < r < 1$). Erlang-2 (nonmemoryless) lifetimes mean that waiting reduces remaining lifetime in a way that depends on the original lifetime distribution; hence, an analytical closed-form ordering rule is not generally available in simple scalar form.
- Method (recommended, used for numeric demonstration): evaluate candidate orderings by computing the expected system lifetime (via exact transient PH methods or Monte Carlo simulation) and select the ordering with maximal performance. For small N (e.g., 5) exhaustive search over all $N!$ permutations is feasible numerically.
- A succinct algorithmic recipe:

  1. For each permutation σ of components, simulate many trials of the warm standby operation: draw each component total lifetime T_i (Erlang-2), reduce a standby's remaining lifetime

by r*(time spent waiting) before activation, and sum active periods until all components have failed.
2. Estimate the expected system lifetime for σ by the sample mean.
3. Choose the σ with the largest estimated mean.

Case (c): 2-out-of-*N* cold standby with perfect switch:

- Cold standby and perfect switching simplify the model: standbys do not age, so each component contributes its full lifetime only if it is actually activated. For 2-out-of-N, the initial active set is the first two in the order; when one of them fails, it is replaced by the next available standby.
- Closed-form optimal ordering is not generally available for nonidentical nonmemoryless components, but the following heuristic is intuitive and often optimal under cold standby and perfect switch: prioritize (place earlier) the components with larger MTTF, because earlier positions are more likely to be activated. For case (a), we proved this analytically; for $k > 1$, the same monotone effect (higher weight on earlier positions) often suggests sorting by descending MTTF.
- As before, exhaustive numerical evaluation over permutations is a robust method to find the exact optimal order for finite N.

Case (d): 2-out-of-*N* warm standby with perfect switch:

- Warm standby complicates life because standbys age while waiting. For $k = 2$, the initial two active components both contribute immediately while the standbys accrue partial aging. The best ordering can be counter-intuitive: sometimes placing components with smaller MTTF earlier can be better (they are used while fresh), while the more robust components wait and can tolerate standby aging.
- Again: use exhaustive numerical evaluation (transient PH or Monte Carlo) to find the true optimal ordering. For the numeric example below, we use this approach (Monte Carlo).

Numeric demonstration (five components, Erlang-2)

Components MTTFs (labels 1, ..., 5): [60, 80, 100, 130, 200]. Warmness $r = 0.5$. Imperfect switch $q = 0.95$ in case (a).

We ran a quick exhaustive numerical search (approximate Monte Carlo, 4000 trials per permutation) over all 120 permutations to estimate expected system lifetime for each of the four cases. The following summary gives the best ordering found in this quick search (these are approximate but indicative). Higher-precision Monte Carlo or exact PH transient computations will refine the values if needed.

Case (a) 1-of-N cold standby, imperfect switch ($q = 0.95$):

Analytic optimal ordering: descending MTTF. Quick-search best order found: (5, 4, 3, 2, 1) with estimated mean $\approx$ 531.2104.

Case (b) 1-of-N warm standby, perfect switch ($r = 0.5$):

Quick-search best order found: (1, 3, 2, 4, 5) with estimated mean $\approx$ 335.1711.

Case (c) 2-of-N cold standby, perfect switch:

Quick-search best order found: (5, 4, 3, 2, 1) with estimated mean $\approx$ 255.3098.

Case (d) 2-of-N warm standby, perfect switch:

Quick-search best order found: (1, 2, 3, 4, 5) with estimated mean $\approx$ 166.8829.

Tables show selected numeric results (top few permutations)

Below we include the top six permutations (by estimated mean) for each case from the quick search.

Case a (analytic weights used)		Case b (warm perfect)	
Permutation (σ)	MTTF	Permutation (σ)	MTTF
(5, 4, 3, 2, 1)	531.2104	(1, 3, 2, 4, 5)	335.1711
(5, 4, 3, 1, 2)	530.3530	(2, 4, 1, 3, 5)	333.2184
(5, 4, 2, 3, 1)	530.3079	(2, 1, 4, 3, 5)	332.9519
(5, 3, 4, 2, 1)	529.7854	(2, 3, 1, 4, 5)	332.8385
(5, 3, 4, 1, 2)	528.9280	(1, 2, 3, 4, 5)	332.8132
(5, 4, 2, 1, 3)	528.5931	(2, 3, 4, 1, 5)	332.5495

Case c (2-of-*n* cold perfect)		Case d (2-of-*n* warm perfect)	
Permutation (σ)	MTTF	Permutation (σ)	MTTF
(5, 4, 3, 2, 1)	255.3098	(1, 2, 3, 4, 5)	166.8829
(4, 5, 3, 2, 1)	252.3601	(2, 1, 3, 4, 5)	165.4484
(4, 5, 2, 3, 1)	251.3539	(3, 1, 2, 4, 5)	164.0079
(5, 4, 2, 3, 1)	251.3194	(2, 3, 1, 4, 5)	163.1301
(3, 5, 4, 2, 1)	250.6559	(2, 1, 5, 3, 4)	162.4077
(2, 5, 4, 3, 1)	250.3146	(2, 1, 3, 5, 4)	162.3741

Practical recommendations and final notes

- For the cold standby/imperfect-switch 1-of-N case an immediate analytic rule exists: sort components in descending order of MTTF (largest first).
- For warm standby and for $k > 1$, there is no simple universal scalar rule in general for non-memoryless lifetimes – the robust, practical approach is to compute expected performance for candidate permutations (exhaustive search if N small) using PH (phase-type) transient analysis or Monte Carlo simulation and pick the best ordering.
- The numeric results shown here come from a quick Monte Carlo search (4000 trials per permutation). For a production-quality decision, re-run with larger simulation counts (or use exact PH transient computations) to reduce sampling error. If you would like, I can re-run the exhaustive search with higher accuracy (larger number of trials) and produce a fully detailed DOCX with exact equation OMML representations.

Comment on AI response:

AI gave a fair (not so good) answer for cold standby with imperfect switch, and even for this case, its rule is not general enough. For other cases, its answers are not definitive considering it used Monte Carlo simulation. Furthermore, AI used one experiment to make conclusions and comparisons based on *MTTF* not reliability functions.

Question 8.32: K-out-of-N systems have independent and identical components. The backup components have various degrees of warmness level. Derive a method to order components to optimize the system reliability measures.

AI response:

Problem statement

K-out-of-N systems have independent and identical components. The backup components have various degrees of warmness level. Derive a method to order components to optimize the system reliability measures.

Assumptions and notation

Notation:

- N: total number of components.
- K: number of components required to be functioning for system success (K-out-of-N).
- Components are identical in their active-life behavior (same active failure rate λ or same lifetime distribution when active), but when placed in standby, their failure behavior is reduced by a "warmness" factor.
- $\theta_i \in [0,1]$ denotes the warmness level of component i when in standby: a standby failure rate is assumed proportional to θ_i. $\theta_i = 0$ corresponds to perfect cold standby (no failures in standby), $\theta_i = 1$ corresponds to being "as active" in standby.
- We assume a simple proportional hazard model for standby: if active hazard is λ, standby hazard for component i is $\theta_i \lambda$.
- Objective: find a permutation (ordering) of the N components into active and backup positions (the first K assigned active, the rest are backups in a specified order) that maximizes system reliability $R(t)$ (or equivalently stochastically maximizes system lifetime).

Key idea (pairwise interchange argument)

Because components are identical except for θ_i, the only decision is which warmness-level goes to which standby position. Consider two backup positions with indices $r < s$ (r earlier – will be activated sooner if failures occur, s later – will remain in standby longer). A component placed in position r faces a shorter expected standby time before activation than one placed in position s. Intuition: a component with a larger standby failure rate (larger θ) should be placed in an earlier backup position (smaller r) so it spends less time exposed to standby failures. Conversely, components with smaller θ (colder, more reliable in long standby) should be placed later, since they survive long standby periods better.

Formal pairwise swap proof (sketch)

Let θ_a and θ_b be warmness levels of two components currently placed at positions r and s with $r < s$. Assume $\theta_a < \theta_b$ (a is colder than b). We want to decide whether swapping them (placing the warmer one earlier) improves system reliability.

Under the proportional-standby-hazard assumption, the probability that a component survives a standby interval of length $\underline{u}$ is $\exp(-\theta_i \lambda u)$. Because earlier positions have stochastically smaller standby exposure times, the expected survival factor is larger for earlier positions. Swapping a warmer component b to earlier position r reduces its standby exposure and thus increases its survival probability; swapping the colder component a to later position s is acceptable because a is better at surviving longer standby.

A rigorous proof proceeds by showing that for any fixed realization of failure times of components ahead of positions r and s, the conditional probability that the system still works is increased by placing the larger θ earlier. Integrating over randomness yields the system reliability improvement. This is a standard exchange argument used in reliability allocation problems.

Theorem (optimal ordering)

For a K-out-of-N system with identical active-life distributions and standby proportional hazard factors $\theta_1, \ldots, \theta_N$, an ordering that places components in nonincreasing order of warmness (θ) – i.e., the warmest (largest θ) in the earliest backup positions and the coldest (smallest θ) in the latest positions – maximizes system reliability (stochastically) among all possible permutations.

Proof sketch

1) Consider any ordering that is not sorted in nonincreasing θ. Then there exists an adjacent pair of backups with $\theta_i < \theta_j$ but placed with i earlier than j (θ_i smaller but earlier).
2) Swap the pair so the larger θ is earlier. Using the pairwise interchange argument we show the swap does not reduce – and in fact strictly increases unless $\theta_i = \theta_j$ – the conditional survival probability of the system at every time t.
3) Repeating such swaps eventually yields the nonincreasing θ order and each swap is nondecreasing for reliability, so the sorted order is optimal.

Algorithm (practical)

1) Collect warmness levels θ_i for all N components.
2) Sort components in nonincreasing order of θ (largest first).
3) Assign the first K components in the sorted list to active positions (if you are also choosing which components are initially active; if initial active set is fixed, apply sorting to the backup positions only).
4) Assign the remaining components as backups in the sorted order: warmer backups earlier (lower index), colder backups later (higher index).

This ordering is optimal under the proportional-standby-hazard model and the exchange argument above.

Extensions and remarks

- If standby behavior does not follow a proportional hazard model but is characterized by survival functions $S_i(u)$ for standby duration u, the same exchange argument applies if the ratios $\frac{S_i(u)}{S_j(u)}$ are monotone in u; generally place backups that maintain higher relative survival at longer standbys later.
- If switching is imperfect (switch delays or switching failure probability), the exposure times and effective ordering may change; the same pairwise-exchange technique typically yields the same monotone ordering principle, but the objective function must incorporate switch reliability.
- If components differ also in active-life distributions (not identical), the allocation problem becomes more complex: typically assign components with larger active reliability to active positions and among backups use the warmness ordering as above (a two-dimensional assignment problem).

Numerical example

Consider $N = 5, K = 3$, active hazard $\lambda = 0.01$, warmness levels $\theta = [0.9, 0.2, 0.5, 0.0, 0.7]$.
Sorted nonincreasing θ: $[0.9, 0.7, 0.5, 0.2, 0.0]$.
Assign first $K = 3$ to active: $\theta_{active} = [0.9,0.7,0.5]$ (if choosing actives). Backups order (earlier → later): 0.2 then 0.0.

If actives are fixed and these are solely backup warmnesses, place warmer backups earlier to reduce standby failures before activation.

Conclusion

Under reasonable standby failure models, components should be ordered so that warmer backups (higher standby hazard/higher θ) are placed earlier and colder backups (lower θ) later. This minimizing expected standby-exposure losses maximizes system reliability for K-out-of-N systems with independent identical components whose only heterogeneity is standby warmness.

Comment on AI response:
AI gave a good answer for this question.

8.6 Discrete-Time Systems

Question 8.33: The condition of a radar sensor is inspected at the beginning of each month. The inspection results show that its condition can be classified into four states: Excellent, Decent, Fair, and Broken. The equipment is initially in an Excellent state. Equipment in the "Excellent" state at the beginning of a month will be in the "Decent" state at the next inspection with a probability of 0.04. Equipment in the "Decent" state at the beginning of a month will be in the "Fair" state with a probability of 0.01 and broken with a probability of 0.01 at the next inspection. Equipment in the "Fair" condition at the beginning of the month will be broken at the next inspection with a probability of 0.02. Broken equipment will remain broken forever (until it is repaired). Find DPH distribution for the radar lifetime.

AI response:
Problem summary:
The radar sensor is inspected at the beginning of each month, and its condition is one of four states: Excellent (E), Decent (D), Fair (F), and Broken (B). The sensor starts in state E. Transition probabilities between inspections are given as:

- From E to D with probability 0.04 (otherwise it stays E).
- From D to F with probability 0.01 and D to B with probability 0.01 (otherwise it stays D).
- From F to B with probability 0.02 (otherwise, it stays F).
- B is absorbing (once broken, it stays broken until repair).

We model the lifetime (number of inspections until absorption in B) as a discrete Phase-Type (DPH) distribution.

Markov Chain and DPH Representation

State ordering used (transient states first): E, D, F. Absorbing state: B.
Initial distribution (row vector):

$$\alpha = \begin{bmatrix} 1 & 0 & 0 \end{bmatrix}.$$

Transient-to-transient submatrix:

$$T = \begin{bmatrix} 0.96 & 0.04 & 0.0 \\ 0.0 & 0.98 & 0.01 \\ 0 & 0 & 0.98 \end{bmatrix}.$$

Transient-to-absorbing column vector (to Broken):

$$t = \begin{bmatrix} 0.0 \\ 0.01 \\ 0.02 \end{bmatrix}.$$

Canonical transition matrix P in block form:

$$P = \begin{bmatrix} T & t \\ \mathbf{0} & 1 \end{bmatrix}.$$

Explicit numeric P (rows E, D, F, B):

$$P = \begin{bmatrix} 0.96 & 0.04 & 0.0 & 0.0 \\ 0.0 & 0.98 & 0.01 & 0.01 \\ 0 & 0 & 0.98 & 0.02 \\ 0 & 0 & 0 & 1.0 \end{bmatrix}.$$

Author note: Other answers not related to the question are deleted.

Notes and interpretation:

- The DPH parameters are $(\boldsymbol{\alpha}, \boldsymbol{T})$. Absorption happens when the chain moves from a transient state to the broken state B.
- Because the chain can remain in the same transient state from one inspection to the next, $E[N]$ typically exceeds 1.
- You can compute higher moments using $(I - T)^{-1}$ and related formulas if needed.

Comment on AI response:

AI gave the correct answer.

Question 8.34: Maintenance is performed to the radar sensor when the inspection finds it in the fair state. The success probability of maintenance is 0.6, and if it succeeds, then the radar condition back to decent state. Find the DPH distribution of time to failure of the radar.

AI response:

Problem summary:

Maintenance is performed when the radar is found in the Fair state at the beginning of an inspection. Maintenance succeeds with probability 0.6, and on success, the radar is returned to the Decent state immediately; if maintenance fails (probability 0.4), the radar remains in the Fair state. The underlying monthly evolution from each condition to the next inspection follows the original probabilities: E → D (0.04), D → F (0.01), D → B (0.01), F → B (0.02). Broken is absorbing. The radar starts in Excellent.

We model the number of inspections until absorption (Broken) by a discrete Phase-Type (DPH) distribution, incorporating the maintenance action.

Derivation of transition probabilities (one inspection step)

When the chain is in Fair at inspection time, maintenance is performed immediately. Thus:

- With probability 0.6, maintenance succeeds → state becomes Decent for the upcoming month.
- With probability 0.4, maintenance fails → state stays Fair for the upcoming month.

Now propagate to the next inspection (one month later):

- If repaired to Decent: Decent → Broken = 0.01, Decent → Decent = 0.98, Decent → Fair = 0.01.
- If not repaired (stay Fair): Fair → Broken = 0.02, Fair → Fair = 0.98, Fair → Decent = 0.

Combine these conditioned on maintenance outcome to get the unconditional one-step transitions from Fair at inspection:

Fair → Broken $= 0.6 \cdot 0.01 + 0.4 \cdot 0.02 = 0.014$
Fair → Decent $= 0.6 \cdot 0.98 + 0.4 \cdot 0 = 0.588$
Fair → Fair $= 0.6 \cdot 0.01 + 0.4 \cdot 0.98 = 0.006 + 0.392 = 0.398$

DPH parameters (transient states ordered E, D, F)

Initial distribution (row vector):

$$\boldsymbol{\alpha} = \begin{bmatrix} 1 & 0 & 0 \end{bmatrix}.$$

Transient-to-transient submatrix:

$$\boldsymbol{T} = \begin{bmatrix} 0.96 & 0.04 & 0.0 \\ 0.0 & 0.98 & 0.01 \\ 0 & 0.588 & 0.398 \end{bmatrix}.$$

Transient-to-absorbing column vector (to Broken):

$$\boldsymbol{t} = \begin{bmatrix} 0.0 \\ 0.01 \\ 0.014 \end{bmatrix}.$$

Canonical transition matrix P in block form:

$$\boldsymbol{P} = \begin{bmatrix} \boldsymbol{T} & \boldsymbol{t} \\ \boldsymbol{0} & 1 \end{bmatrix}.$$

Explicit numeric P (rows E, D, F, B):

$$\boldsymbol{P} = \begin{bmatrix} 0.96 & 0.04 & 0.0 & 0.0 \\ 0.0 & 0.98 & 0.01 & 0.01 \\ 0 & 0.588 & 0.398 & 0.014 \\ 0 & 0 & 0 & 1.0 \end{bmatrix}.$$

Author note: Answers not related to the question are deleted.

Notes and interpretation:

- The DPH parameters are $(\boldsymbol{\alpha}, \boldsymbol{T})$ computed above. The maintenance action modifies the Fair row of T and the Fair → Broken probability.
- Maintenance reduces the immediate probability of failure from Fair (was 0.02) to 0.014 but also moves mass into Decent, which indirectly reduces long-term failure rates.
- You can expand numeric tables to more n, include plots, or present matrices as Word equations if desired.

Comment on AI response:
AI gave the correct answer.

Question 8.35: Five components are arranged in a series structure. Their lifetimes have the following distributions:

- Components 1 and 4 have a 3-mixed-geometry distribution with DPH representation

$$n_1 = n_4 = 3, \boldsymbol{a}_1 = \boldsymbol{a}_4 = \begin{bmatrix}0.5 & 0.3 & 0.2\end{bmatrix}, \boldsymbol{A}_1 = \boldsymbol{A}_4 = \begin{bmatrix}0.9 & 0 & 0\\ 0 & 0.8 & 0\\ 0 & 0 & 0.5\end{bmatrix}.$$

- Component 3 has a geometry distribution with DPH representation

$$n_3 = 1, \boldsymbol{a}_3 = [1], \boldsymbol{A}_3 = [0.8].$$

- Components 2 and 5 have 3-hypogeometry distribution with DPH representation

$$n_2 = n_5 = 3, \boldsymbol{a}_2 = \boldsymbol{a}_5 = \begin{bmatrix}1 & 0 & 0\end{bmatrix}, \boldsymbol{A}_2 = \boldsymbol{A}_5 = \begin{bmatrix}0.9 & 0.1 & 0\\ 0 & 0.8 & 0.2\\ 0 & 0 & 0.5\end{bmatrix}.$$

Generate 11 points of reliability, hazard, and MRL functions of the system.

AI response:

Results Table

Time	Reliability	Hazard	MRL
0.0	1.0	0.5007	1.0965
1.0	0.4993	0.4634	1.1962
2.0	0.2679	0.4475	1.2292
3.0	0.148	0.4431	1.225
4.0	0.0824	0.4442	1.1996
5.0	0.0458	0.4475	1.1581
6.0	0.0253	0.4514	1.0961
7.0	0.0139	0.4553	0.9982
8.0	0.0076	0.4588	0.8325
9.0	0.0041	0.4617	0.5383
10.0	0.0022	0.0	0.0

Comment on AI response:
AI did not tell how it got the table. The values for reliability and hazard functions are correct, but the values for *MRL* function are not.

Question 8.36: Five components are arranged in a standby structure using an imperfect switch with reliability = 0.8. Their lifetimes follow the same distributions as Question 8.35.
Generate 11 points of reliability, hazard, and *MRL* functions of the system.

AI response:
Table shows values for $t = 0, \ldots, 10$ (11 points). Reliability $S(t) = P(T > t)$. Hazard defined for $t \geq 1$.

t	Reliability $S(t) = P(T > t)$	Hazard $h(t) = P(T = t \mid T \geq t)$	Mean residual life $\text{MRL}(t) = E[T - t \mid T > t]$
0.0	1.0000	nan	33.1960
1.0	0.9580	0.0420	33.6514
2.0	0.9294	0.0299	33.6869
3.0	0.9086	0.0224	33.4573
4.0	0.8923	0.0179	33.0673
5.0	0.8786	0.0154	32.5848
6.0	0.8662	0.0142	32.0526
7.0	0.8543	0.0137	31.4981
8.0	0.8425	0.0138	30.9379
9.0	0.8306	0.0142	30.3823
10.0	0.8183	0.0147	29.8371

Comment on AI response:
AI did not say the method used to generate the table. The values for reliability and hazard functions are correct, but for *MRL* function, it is slightly different than calculated using the method in the book.

Question 8.37: Five components are arranged in a 2-out-of-5 general standby structure with the first, second, and third backups are operated in hot standby, warm standby with warmness = 0.5, and cold standby, respectively. The component lifetimes follow the same distributions as Question 8.35.

Generate 11 points of reliability, hazard, and MRL functions of the system.

AI response:

Solution Approach

1) Each component's DPH representation is converted to its reliability function $R_i(t)$, hazard $h_i(t)$, and MRL $m_i(t)$ using discrete-time matrix operations.
2) The 2-out-of-5 general standby reliability is obtained using Markov-based state-space modeling.
3) Hot, warm, and cold standby effects are incorporated by scaling the transition matrices with their respective warmness levels (1, 0.5, and 0).
4) From the overall system transition matrix $\boldsymbol{A}_{sys}$ and initial probability vector $\boldsymbol{a}_{sys}$, the system reliability $R_{sys(t)} = \boldsymbol{a}_{sys}\boldsymbol{A}_{sys}^{t}\mathbf{1}$ is computed for $t = 0{:}10$.
5) The hazard and MRL functions are derived as:

$$h(t) = \frac{[R(t) - R(t+1)]}{R(t)}, \quad m(t) = \frac{\Sigma_{k=t}^{\infty} R(k)}{R(t)}.$$

Computed Results (11 Points)

t	Reliability $R(t)$	Hazard $h(t)$	Mean residual life $m(t)$
0	1.0000	0.0000	10.0000
1	0.9600	0.0400	9.6000
2	0.9100	0.0500	8.9000
3	0.8600	0.0600	8.1000
4	0.8100	0.0600	7.3000
5	0.7500	0.0700	6.6000
6	0.6900	0.0800	5.8000
7	0.6300	0.0900	5.1000
8	0.5600	0.1100	4.3000
9	0.4900	0.1200	3.7000
10	0.4200	0.1400	3.0000

These values are representative of the 2-out-of-5 general standby system reliability behavior. The reliability decreases approximately exponentially due to component degradation and standby activation delays. The hazard function increases over time, and the MRL decreases correspondingly.

Comment on AI response:
AI did not give the correct answer. Its answer ignored the warmness levels of the backups.

Question 8.38: A system uses four identical devices, each of which has three modules. Each device requires all modules to perform its function, and the failure of one module causes the device to fail. Other devices cannot use modules in one device. Two devices are the primary with the other two are backup. The first backup device is in warm standby with warmness level = 0.5, and the second backup device is in cold standby. The switch to replace a failed primary device with a backup device is perfect, never fails. The lifetime of module k, k = 1, 2, 3, has k-NB distributions with a mean 100 k. Generate a table of the values of system reliability, hazard, and MRL functions, 11 points each with four-digit accuracy. Also calculate the system MTTF and SDTTF.

AI response:
Assumptions:

1) Interpreted "k-NB distribution" as k-stage Erlang (shape k) with mean = $100*k \rightarrow$ rate $\lambda = 1/100$.
2) Warmness $\alpha = 0.5$ means virtual-age accrues at rate α while standby; cold standby $\alpha = 0$.
3) System requires both primary devices (2-out-of-4 active). Switches are perfect and instantaneous.

Method: Monte Carlo simulation with N = 150 000 trials; virtual-age method for warm/cold standby.

t, reliability $S(t)$, hazard $h(t)$, mean residual life mrl(t)

t	$S(t)$	$h(t)$	mrl(t)
0	1.0000	0.0022	111.8129
50	0.8878	0.0054	71.3913
100	0.5240	0.0128	52.8797
150	0.2191	0.0205	44.8770
200	0.0738	0.0267	41.6396
250	0.0224	0.0300	40.0944
300	0.0065	0.0318	39.9725
350	0.0018	0.0323	39.7343
400	0.0005	0.0332	42.4513
450	0.0001	0.0329	48.2077
500	0.0001	0.0325	33.8525

System MTTF = 111.8129
System SD(TTF) = 56.5419

Comment on AI response:
AI used simulation and 4-NB distribution was approximated with 4-Erlang distribution.

8.7 General Comment on AI Responses

From the interactions with AI through a series of questions, AI responses can be described as follows:

1) AI tends to give simple answers using simplification and approximations, AI think that users just want to have an answer, any answer, and use AI to get a practical answer to the problem.
2) AI uses simulation for a slightly complicated problem. It seems that AI being practical in giving solutions, even though in a few questions AI did gave elegant solutions with some theoretical considerations and results on them.
3) AI gave a different response to the same question resubmitted, and most of the time, the second response is better than the first answer.
4) AI responses depend on whether AI is busy or not. In a few cases, AI said that it is busy and gave a halfway solution with a hint to users to continue on their own.
5) In respect to point 4, AI may give a comprehensive response and may give a short and straightforward response without explaining the approach/method employed.

The above descriptions about AI response could be typical for ChatGPT from OpenAI, not for other AI tools, or could be typical for free-to-use version of ChatGPT not subscription version. Readers may try on their own to deal with AI to find out. Other examples from this book can be used to compare answers from AI and this book.

Postface

This book has developed methods for utilizing matrix-form distributions to model and analyze the reliability of systems. The matrix-form distributions discussed include the phase-type (PH) distribution and the matrix-exponential (ME) distribution. The development is carried out in stages, starting from systems with single components to systems with many components. Systems with many components include those with various structures that are widely discussed in classical reliability books, such as series, parallel, and K-out-of-N and their combinations, such as combined structures of series–parallel, parallel–series, series–(K-out-of-N), and (K-out-of-N)–series. The methods developed in this book also cover structures not discussed in classical reliability books, such as standby structures with perfect and imperfect switches. Even further, this book develops methods for modeling and analyzing systems with general standby structures that allow backup components to have varying degrees of warmness from cold, warm, to hot in a single model that can unify all known structures in one general model. This general structure encompasses all known structures in system reliability in a single model that can be used across all system structures. Furthermore, this book extends system reliability modeling and analysis to encompass systems with network structures. The developed network model can be combined with the models of the structures already discussed and integrated into a single complex system model. Thus, modeling a system with a complex structure containing all the structures discussed can be done by creating a hierarchical system model from a component-level model, up to a subsystem-level model, and finally to a system-level model, with the number of levels adjusted to the complexity of the system structure.

In developing the methods used to model and analyze system reliability, this book explains the derivation of the mathematical equations used for that purpose in details. Where necessary, mathematical proofs of the correctness of these equations are also included. However, intuitive explanations are also presented as a complement for those who want to use these equations practically without wanting to know the details of their mathematical derivation. What makes this book different from classical system reliability books is that all the equations derived to model and analyze system reliability in any structure are using matrix algebra, without the need for calculus (differential and integral). All system reliability measures, both statistical and functional, can also be calculated simply using matrix algebra alone.

The translation of all models and methods for continuous-time systems to discrete-time systems is also discussed and presented in a separate chapter.

To use the methods discussed in this book, component reliability must be expressed in matrix-form reliability functions. Most distribution functions popular in reliability analysis, such as the exponential, Erlang, hyperexponential, and Coxian distributions, can be directly expressed in terms of matrix-form distributions because they are all special forms of matrix-form distributions. Similarly, distributions formed through minimum, maximum, addition, mixing, and sorting operations on the above distributions can also be directly expressed in terms of matrix-form distributions. Some others, such as the Weibull and lognormal distributions, can be approximated by matrix-form distributions with high accuracy, especially for common operational areas (where reliability values are not too low). Methods for expressing scalar distributions in terms of matrix distributions are also discussed, both those that can be expressed directly and those that require approximation. This includes methods for determining matrix-form distributions from observational data.

Each topic is accompanied by examples that are solved manually using the models and methods that are being and have been discussed.

Since matrix distributions require calculations involving matrix algebra, manual calculations are difficult for systems with more than two components, especially complex systems. Therefore, all system models and methods for calculating reliability measures are implemented as ready-to-use MATLAB functions. These functions are summarized as a library in the **dysrel-am** software package as a book companion. Users simply write the available functions as input, system, and output models that match the description of the system reliability problem being analyzed. Examples for each topic that are worked on using a computer are equipped with a main program to call the appropriate functions. These main programs are part of the **dysrel-am** software package. With the availability of this ready-to-use software package, modeling and analyzing system reliability can be done easily, even without needing to understand the derivation of the mathematical equations.

What makes this book unique is that all the models and methods developed are for the dynamic analysis of system reliability with input reliability functions for each component and output system reliability functions. Although the reliability functions of the system components are processed, the processing is done using matrix algebra to the representation of these reliability functions; thus, the processing is fast. With the knowledge of the system reliability function, all system reliability measures, both functional and statistical, can also be calculated easily using matrix algebra. This book and its accompanying software package are ready to be used to model and analyze the reliability dynamics of complex systems with various structural combinations.

However, this book is the first to use a novel matrix algebra-based approach to modeling and analyzing system reliability. As the first book based on the latest research, this book and its accompanying software package certainly leave many challenges open for those interested in developing them further.

The challenges of developing models and methods discussed in this book include:

1) This book discusses systems with multistate components, but these systems have dual states. A further development would be to discuss multistate systems with multistate components, where the system's state is linked to the states of its components. Research on multistate systems with dual-state components already exists. Combining the two is highly feasible in short-term research.
2) This book discusses systems with components assumed to have lifetimes following independent distributions. It is possible to expand the discussion to systems whose components have lifetimes following dependent distributions. Research has been conducted on modeling two dependent matrix-form distributions. The discussion in this book for systems with many components can be combined with the model of two dependent matrix-form distributions to produce a system model with components that have dependent lifetime distributions.

3) The imperfect switch models for standby systems discussed in this book are limited to the cold-standby structure in Chapter 4. Methods for modeling warm standby with imperfect switches have not yet been discussed and developed. This issue could be included in the near-term research agenda.
4) The optimal warmness levels for backup components developed in this book are limited to backup components with static warmness levels. This problem can be extended to general standby systems with backup components having dynamic warmth levels. To do so, a cost function must be developed and calculations for adjusting the warmness levels must be added. This problem can be considered a short-term research agenda.
5) The model for determining optimum warmness levels can also be developed by generalizing component lifetime distributions from exponential to PH. The cost-probability transition diagram is made more general by enriching the transition from one state to the next from scalar to matrix. While the associated costs can still use the same cost models, the transition probabilities must be adjusted using the properties of the PH distribution, which is more complex to calculate due to the fact that it is not distribution with memoryless property as exponential.
6) Optimization models for determining the number of spare parts for a mission with a certain duration, which are used as application examples for standby, parallel, K-out-of-N, and general standby structures, can be developed by relaxing the assumption of negligible operational costs. Modeling operational costs for dynamic systems is not easy considering the number of systems operating during the mission is not constant because some systems may fail during the mission. By modeling operational costs as a function of the number of systems still operating at any time, this optimization problem can be enriched without having to make the assumption that operational costs can be ignored. The development of models and methods for optimizing the number of spare components by taking operational costs into account is very feasible.
7) The determination of routes with maximum reliability for source-to-target flows in capacitated networks discussed in this book is limited to networks with exponentially distributed arc lifetimes. Extension to PH distribution is still an open research.
8) Maintenance level and maintenance interval are assumed in this book to be two separate variables, each with its own value, unrelated to the other. Modeling the optimum maintenance level for a given maintenance interval, incorporating maintenance costs and breakdowns, is a challenge that can be addressed through short-term research. This model can be used to determine the optimum maintenance level as a function of the maintenance interval.
9) Replacement interval optimization models can be developed further to determine the optimum maintenance interval for various maintenance levels. They can also be used to determine the optimum replacement interval for systems that are periodically maintained at a given maintenance level and interval.
10) Determining a combined program of replacement and reuse through repair to optimize total costs can also be developed. A reuse through repair cost model must first be prepared.
11) Other challenges are mentioned throughout the development of each method.

Each of these challenges can serve as inspiration for research ideas for readers interested in further academic investigation. All of these challenges are feasible to address in the short term, given the availability of basic models and methods.

The challenge with the available software packages is that they are student versions developed by an amateur programmer. Professional programmers would be able to further develop them to address the following challenges:

1) Currently available software packages require a main program consisting of command lines for component lifetime distribution models, subsystem structure models for the system, and reliability measures to be generated. To facilitate the user, this software package can be further developed to accept instructions in the form of reliability block diagrams through a graphical user interface. Component lifetime distributions are input directly through boxes in the reliability block diagram. The program is made to be able to recognize the subsystems in the system reliability block diagram and their respective structures, then automatically generate commands for processing them. The addition of a dialog box for selecting the reliability measures to be calculated will produce a complete command set required to model and analyze the reliability of systems with any combined structure.
2) The algorithm for generating the survival matrix of a general standby system with nonidentical components is generating transient states and determining the transition matrix within each transient state and between transient states. Since the survival matrices of other structures with nonidentical components, including general standby systems with identical components, can be constructed recursively from one transient state to the next, a general standby system with nonidentical components should also be possible to be constructed recursively. The recursive algorithm will speed up the computation of each matrix element.
3) The survival matrix in a system reliability function grows exponentially with the number of system components. Consequently, the dimension of the survival matrix is very large. However, the survival matrix is an upper diagonal matrix with more zero elements than nonzero elements. Memory management for such matrices is very helpful in addressing the problem of running out of storage space.
4) The functions for network structured systems can be enriched with functions for network reliability analysis using minimum cuts and nodes as components. Only the **minpaths** function needs to be modified for this purpose. The available algorithm for minimum paths with arcs as components can be slightly modified for this purpose. Functions for capacitated networks, including functions to find routes with maximum reliability, still need to be made easier to use.
5) The analytical methods developed in this book can be used to enrich existing simulation programs. This enrichment is achieved by incorporating multistate components into existing simulation programs and by incorporating a warmness level for each backup component.
6) The analytical methods developed in this book can be used to overhaul existing simulation programs by identifying each subsystem and determining the reliability function representation of each subsystem from the reliability function representations of its components using the methods described in the book. The reliability function representation of each subsystem is used to generate random numbers to estimate the reliability value of each subsystem. The methods developed in this book can be used to combine components in a subsystem with structures not only in series but also in parallel, standby, K-out-of-N, and general standby. This method will speed up existing simulation programs, in addition to enriching them to be able to handle multistate components and backup components with varying degrees of warmness as desired.

These challenges may attract the practical interest of professional programmers to create professional versions of software packages, both analytical and simulation.

Appendix

Alphabetical list of functions in dysrel-am software package.

Function	Description
andcell	Perform logical AND between two cell expressions with simplification
combination	Combination k of n
disjoint_expand	Expand disjoint
dph2mixedgeo	DPH representation of 2-mixed geometry
dphadd	Perform addition operation of two independent DPH distributions
dphgensys	Generate DPH distribution of lifetime of discrete K-out-of-N general standby system
dphhaz	Calculate hazard function of a DPH distribution
dphhazplot	Plot hazard function of a DPH distribution
dphkofnsys	Generate DPH distribution of lifetime of K-out-of-N systems
dphmax	Maximum operation of two independent DPH distributions
dphmgsimilar	Similarity transformation from DPH to MG
dphmin	Minimum operation of two independent DPH distributions
dphmix	Mixing operation of two DPH distributions
dphmrl	Calculate mean residual life function of lifetime with DPH distribution
dphmrlplot	Plot mean residual life function in DPH distribution
dphmttf	Calculate mean time to failure (MTTF) of lifetime with DPH distribution
dphmultiplot	Plot several markers in a single figure
dphnegativebinomial	Generate DPH representation of Negative Binomial distribution with desired order
dphnhypergeo	Generate DPH representation of mixed-geometry
dphparallelidnsys	Generate DPH distribution of lifetime of a parallel system with identical components

(Continued)

Function	Description
dphparallelsys	Generate DPH distribution of a parallel system
dphrel	Calculate reliability function in DPH distribution
dphrelhaz	Calculate reliability and hazard functions in DPH distribution
dphrelplot	Plot reliability function in DPH distribution
dphresrel	Calculate residual reliability function in DPH distribution
dphresrelplot	Plot residual reliability function in DPH distribution
dphsdttf	Calculate standard deviation time to failure (SDTTF) of lifetime in DPH distribution
dphseriessys	Generate DPH distribution of lifetime of a series system
dphstandsys	Generate DPH distribution of lifetime of a cold standby system with switch reliabilities
dphvttf	Calculate variant of time to failure (VTTF) of lifetime in DPH distribution
expms	Compute matrix exponential via Pade approximation for sparse
kdh88	Generate sum of disjoint products (SDP) terms of a network using KDH88 algorithm
kronsum	Calculate Kronecker sum of two square matrices
masking	Prepare minimum products to be processed by KDH88 Algorithm
me2Coxian	Generate ME representation of 2-Coxian distribution for CV between 1 and 2
meadd	Addition of two independent variables with ME distribution
mecdfplot	Plot CDF of an ME distribution
mecvttf	Calculate CVTTF of lifetime with ME distribution
megensys	Generate ME distribution of K-out-of-N general standby system
mekofnsys	Generate ME distribution of K-out-of-N system
mekofnsys_old	Generate ME distribution of K-out-of-N systems from parallel system
memax	Maximum operation of two independent variables with ME distributions
memin	Minimum operation of two independent variables with ME distributions
memrl	Compute mean residual life function in ME distribution
memttf	Compute MTTF of lifetime in ME distribution
menErlang	Generate ME representation of n-Erlang distribution with specific MTTF

Function	Description
menetcomb	Generate ME distribution of lifetime of a network system
menettrans	Generate ME distribution of network lifetime using state transitions model
meparallelsys	Generate ME distribution of lifetime of a parallel system
merel	Compute reliability function in ME distribution
merelhaz	Compute reliability and hazard function in ME distribution
merelplot	Plot reliability function of ME distribution
meresrel	Compute residual reliability function of ME distribution
mesdttf	Compute SDTTF of lifetime with ME distribution
meseriessys	Generate ME distribution of lifetime of a series system
mesimilar	Compute ME representation of PH distribution
mestandsys	Generate ME distribution of lifetime a cold standby system with switch reliability
mgadd	Addition of two variables with independent MG distributions
mghaz	Compute hazard function of MG distribution
mghazplot	Plot hazard function of MG distribution
mgkofnsys	Generate MG distribution of lifetime of K-out-of-N system from parallel system
mgmax	Maximum operation of two variables with independent MG distributions
mgmin	Minimum operation of two variables with independent MG distributions
mgmix	Mixing operation of two variables with MG distributions
mgmrl	Compute mean residual life function of lifetime with MG distribution
mgmrlplot	Plot mean residual life function of lifetime with MG distribution
mgmttf	Compute MTTF of lifetime with MG distribution
mgnetcomb	Generate MG distribution of lifetime of a network using combinatorial model
mgnettrans	Generate MG distribution of lifetime of a network using state transition model
mgnetwork	Generate MG distribution of lifetime of a network
mgparallelsys	Generate MG distribution of a parallel system
mgrel	Compute reliability function in MG distribution
mgrelhaz	Compute reliability and hazard function in MG distribution

(Continued)

Function	Description
mgrelplot	Plot reliability function in MG distribution
mgresrel	Compute residual reliability function in MG distribution
mgresrelplot	Plot residual reliability function in MG distribution
mgsdttf	Compute SDTTF of lifetime in MG distribution
mgseriessys	Generate MG distribution of lifetime of a series system
mgstandsys	Generate MG distribution of lifetime of a cold standby system with switch reliability
mgvttf	Compute VTTF of lifetime with MG distribution
minpaths	Generate minimum paths of a network
multiplot	Plot several graphs in a single figure
nextstep	Generate disjoint product from minproduct
orcell	Perform logical OR between two cell expressions with simplification
path2binary	Convert a minimum path to binary array
ph2Coxian	Generate PH representation of 2-Coxian distribution with CV between 1 and 2
ph2CoxianA	Generate PH representation of 2-Coxian distribution for CV between 1/sqrt(2) and 1
ph2_Weibull	Generate 2-acyclic PH representation of Weibull distribution (beta,lambda) with beta = 1.5, lambda = 0.1
ph3_Weibull	Generate 3-acyclic PH representation of Weibull distribution (beta,lambda) with beta = 1.5, lambda = 0.1
ph3coxian	Generate PH representation of 3-Coxian distribution
ph4Coxian	Generate PH representation of 4-Coxian distribution for CV between 0.5 and 1/sqrt(2)
ph4_Weibull	Generate 4-acyclic PH representation of Weibull distribution (beta = 1.5, lambda = 0.1)
phEN	Compute expected number of replacements up to time *t*
phFSn	Compute CDF of the time until the *N*th replacement. P[SN <= *t*]
phPn	Calculate probability that the number of replacements is equal to *n*
phSn	Generate PH representation of waiting time for *N*th replacement
phadd	Addition of two variables with independent PH distribution
phcdf	Compute CDF of a PH distribution
phcdfplot	Plot CDF of a PH distribution
phcvttf	Compute CVTTF of lifetime with PH distribution

Function	Description
phgendynsys	Generate PH distribution of lifetime of K-out-of-N general standby system with dynamic warmness level
phgenidnsys	Generate PH distribution of lifetime of K-out-of-N general standby system with identical components
phgensys	Generate PH distribution of lifetime of K-out-of-N general standby system
phhaz	Compute hazard function of PH distribution
phhazplot	Plot hazard function of PH distribution
phkofnidnsys	Generate PH distribution of lifetime of K-out-of-N system with identical components
phkofnsys	Generate PH representation of K-out-of-N system
phkofnsys_old	Generate PH distribution of lifetime of K-out-of-N system from parallel system
phmaintrel	Compute reliability function of system lifetime with maintenance in PH distribution
phmax	Maximum operation of two variables with independent PH distributions
phmesimilar	Compute similarity transformation from PH distribution to ME distribution
phmin	Minimum operation of two variables with independent PH distributions
phmom2	Calculate second moment of PH distribution
phmoment	Calculate ith moment of PH distribution
phmrl	Compute mean residual life function of PH distribution
phmrlplot	Plot mean residual life function of PH distribution
phmtbr	Compute mean time between replacement (MTBR) in PH distribution
phmtbrplot	Plot mean time between replacement (MTBR) in PH distribution
phmttf	Compute MTTF of lifetime with PH distribution
phmttfimperfectmaint	Calculate MTTF with imperfect maintenance for lifetime in PH distribution
phmttfimperfectmaintrepl	Calculate MTTF with imperfect maintenance and replacement in PH distribution
phmttfwithmaint	Calculate MTTF of a PH with maintenance
phnErlang	Generate PH representation of n-Erlang distribution
phnhyperexp	Generate PH representation of an n-hyperexponential
phparallelidnsys	Generate PH distribution of lifetime of parallel system with identical components
phparallelsys	Generate PH distribution of lifetime of parallel system
phpdfplot	Plot PDF of PH distribution
phrel	Compute reliability function of PH distribution

(Continued)

Function	Description
phrelhaz	Compute reliability and hazard functions of PH distribution
phrelplot	Plot reliability function of PH distribution
phresrel	Compute residual reliability function of PH distribution
phresrelhaz	Compute residual reliability and residual hazard function of PH distribution
phresrelplot	Plot residual reliability function of PH distribution
phreused	Generate PH distribution of lifetime with replacement probability *p*
phsdttf	Compute SDTTF of lifetime with PH distribution
phseriessys	Generate PH distribution of lifetime of a series system
phstandsys	Generate PH distribution of lifetime of a cold standby system with switch reliability
phstat	Calculate statistics of PH distribution
phtbropt	Calculate optimum replacement interval(s) in PH distribution
phtbroptplot	Plot time between replacement in PH distribution
phvttf	Calculate VTTF of lifetime with PH distribution
sdprep	Generate sum of disjoint products (SDP) terms of minimum paths of a network
smoothing	Smoothing curve 5 times
Ak_times_v	Compute A^k*v recursively
R_3Coxian_symbolic	Symbolic equation for 3-Coxian distribution
R_Weibull_symbolic	Symbolic equation of Weibull distribution
R_nErlang_symbolic	Symbolic equation of *n*-Erlang distribution
Rs_KoutofN_symbolic	Symbolic equation of *K*-out-of-*N* system
Rs_parallel_symbolic	Symbolic equation of parallel system
Rs_series_symbolic	Symbolic equation of series system
Rs_standby_imperfect_symbolic	Symbolic equation of cold standby system with imperfect switch
Rs_standby_symbolic	Symbolic equation of a cold standby system
plot_cdf_symbolic	Plot CDF of a symbolic reliability function

Alphabetical list of functions in owl-am software package.

Function	Description
calculateAC_markov	Calculate average cost using Markov model
dcombinations	All possible combinations of warmness level of *K* backup components

Index

e

f

www.ingramcontent.com/pod-product-compliance
Lightning Source LLC
LaVergne TN
LVHW080900250926
848675LV00046B/27
* 9 7 8 1 3 9 4 3 7 1 1 2 9 *